The Connecticut River Ecological Study (1965–1973) Revisited: Ecology of the Lower Connecticut River 1973–2003

The Connecticut River Ecological Study (1965–1973) Revisited: Ecology of the Lower Connecticut River 1973–2003

Edited by

Paul M. Jacobson
Ecological Assessments, LLC
Post Office Box 376, Old Saybrook Connecticut 06475, USA

Douglas A. Dixon
EPRI (Electric Power Research Institute)
3412 Hillview Avenue, Palo Alto, California 94304, USA

William C. Leggett
Department of Biology, Queen's University, Kingston, Ontario K7L 3N6, Canada

Barton C. Marcy, Jr.
Westinghouse Savannah River Company, Environmental Services
Building 742-A, Post Office Box 616, Aiken, South Carolina 29808, USA

Ronald R. Massengill
2434 Trenton Street, Cambria, California 93428, USA

American Fisheries Society Monograph 9

American Fisheries Society
Bethesda, Maryland, USA
2004

The American Fisheries Society Monograph series is a registered serial. Suggested citation formats follow.

Entire book

Jacobson, P. M., D. A. Dixon, W. C. Leggett, B. C. Marcy, Jr., and R. R. Massengill, editors. 2004. The Connecticut River Ecological Study (1965–1973) revisited: ecology of the lower Connecticut River 1973–2003. American Fisheries Society, Monograph 9, Bethesda, Maryland.

Chapter within the book

Howell, P., and D. Molnar. 2004. Stock assessment of white perch in the lower Connecticut River. Pages 379–390 *in* P. M. Jacobson, D. A. Dixon, W. C. Leggett, B. C. Marcy, Jr., and R. R. Massengill, editors. The Connecticut River Ecological Study (1965–1973) revisited: ecology of the lower Connecticut River 1973–2003. American Fisheries Society, Monograph 9, Bethesda, Maryland.

Front Cover: Aerial view of the Connecticut River looking upstream and northward from the area of Gillette Castle and the Chester-Hadlyme Ferry toward the Haddam Bridge and the Connecticut Yankee Haddam Neck power plant in the far distance. Photo by Robert Perron. Back Cover: Aerial view of the Connecticut River East Haddam Bridge and the barge transporting the decommissioned Connecticut Yankee Haddam Neck nuclear reactor downstream and ultimately toward the Savannah River and final disposal at the low-level radioactive waste disposal site in Barnwell, South Carolina. Photo by Ray Martin.

Printed in the United States of America on acid-free paper.

Library of Congress Control Number 2004113760
ISBN 1-888569-66-2
ISSN 0362-1715

American Fisheries Society website address: www.fisheries.org

American Fisheries Society
5410 Grosvenor Lane, Suite 110
Bethesda, Maryland 20814-2199
USA

Dedication

We dedicate this monograph to Dr. Daniel Merriman (deceased) and Lyle Thorpe, co-directors of the original Study, whose leadership provided the foundation for development and implementation of the 1965–1973 studies. From the start, Dan and Lyle recognized the importance of establishing clearly stated research questions as the basis for scientific investigation. At the same time, they understood and fostered an atmosphere of investigative freedom that allowed for response to emerging areas of inquiry. That the Study accomplished its research objectives can be attributed in great measure to their insight. For that leadership and for their friendship and mentoring, we express our deeply felt gratitude and thanks.

Contents

Preface ... ix
Charles C. Coutant

Original Preface ... xiii
Daniel Merriman and Lyle M. Thorpe

Connecticut River Ecological Study—Revisited: Participant Acknowledgments ... xvii

Peer Reviewer Acknowledgments ... xix

Symbols and Abbreviations ... xxi

Part I: Original Ecological Study

Introduction ... 3
Daniel Merriman and Lyle M. Thorpe

Hydrography ... 27
William A. Boyd

Oxygen Balance in the Condenser-Cooling Water System of the Connecticut Yankee Plant ... 37
Paul M. Jacobson

Benthic Fauna: 1965–1967 versus 1968–1972 ... 43
Ronald R. Massengill

Entrainment of Zooplantkton at the Connecticut Yankee Plant ... 59
Ronald R. Massengill

Fishes of the Lower Connecticut River and the Effects of the Connecticut Yankee Plant ... 65
Barton C. Marcy, Jr.

Planktonic Fish Eggs and Larvae of the Lower Connecticut River and the Effects of the Connecticut Yankee Plant, Including Entrainment ... 125
Barton C. Marcy, Jr.

Early Life History Studies of American Shad in the Lower Connecticut River and the Effects of the Connecticut Yankee Plant ... 155
Barton C. Marcy, Jr.

The American Shad, with Special Reference to Its Migration and Population Dynamics in the Connecticut River ... 181
William C. Leggett

Recurrent Mass Mortalities of the Blueback Herring in the Lower Connecticut River ... 239
Sanford A. Moss, William C. Leggett, and William A. Boyd

Summary and Conclusions 247
Daniel Merriman and Lyle M. Thorpe

Part II: The Connecticut River Ecological Study Revisited

Introduction: Connecticut River Watershed Management—Past, Present, and Future 263
Paul M. Jacobson, Charles G. Fredette, and Nels Barrett

Summary of Water Quality Trends in the Connecticut River, 1968–1998 273
John R. Mullaney

An Overview of the Program to Restore Atlantic Salmon and Other Diadromous Fishes to the Connecticut River with Notes on the Current Status of these Species in the River 287
Stephen Gephard and James McMenemy

Trends in Abundance, Distribution, and Growth of Freshwater Fishes from the Connecticut River in Connecticut (1988–2002) 319
Robert P. Jacobs, William A. Hyatt, Neal T. Hagstrom, Eileen B. O'Donnell, Eric Schluntz, Penelope Howell, and David R. Molnar

Population Estimate and Utilization of the Lower Connecticut River by Shortnose Sturgeon 345
Thomas F. Savoy

Food Habits of Shortnose Sturgeon Collected in the Lower Connecticut River from 2000 to 2002 353
Thomas F. Savoy and Jacqueline Benway

Factors Affecting the Recent Decline of Blueback Herring and American Shad in the Connecticut River 361
Thomas F. Savoy and Victor A. Crecco

Stock Assessment of White Perch in the Lower Connecticut River 379
Penelope Howell and David Molnar

The Impact of Enhancement Initiatives on the Structure and Dynamics of the Connecticut River Population of American Shad 391
William C. Leggett, Thomas F. Savoy, and Christine A. Tomichek

American Shad Early Life History and Recruitment in the Connecticut River: A 40-Year Summary 407
Thomas F. Savoy, Victor A. Crecco, and Barton C. Marcy, Jr.

Effect of Connecticut Yankee Power Plant on Population Dynamics of Asiatic Clams and Their Interactions with Native Bivalves 419
Douglas E. Morgan, Milan Keser, John T. Swenarton, and James F. Foertch

Comparison of Benthic Community Structure in the Lower Connecticut River before, during, and after Operation of the Connecticut Yankee Power Plant 441
Ronald R. Massengill, Joseph M. Vozarik, Douglas E. Morgan, and Milan Keser

Twenty Years of Impingement History: Connecticut Yankee Haddam Neck Nuclear Power Plant 455
Paul M. Jacobson, Christine Tomichek, and Donald J. Danila

Developments in Fishery Science with Special Reference to Power Plant Impact Assessment Methods and the Connecticut River Ecological Study: 1973–Present 475
Saul B. Saila and Ernesto Lorda

Feasibility of Applying a Watershed Model to the Connecticut River for Total Maximum Daily Load Analysis and Watershed Planning 501
Carl W. Chen, Robert A. Goldstein, and Joel Herr

Modeling the Hydraulic Zone of Influence of Connecticut Yankee Nuclear Power Plant's Cooling Water Intake Structure 513
John Richardson and Douglas A. Dixon

Summary and Conclusions 525
Paul M. Jacobson, William C. Leggett, Barton C. Marcy, Jr., and Ronald R. Massengill

Appendix I: Lower Connecticut River Fish Species List 539

Appendix II: Connecticut River Ecological Study Bibliography: Related Papers by the Original (1965–1973) Principle Investigators 543

Preface

CHARLES C. COUTANT
Environmental Sciences Division, Oak Ridge National Laboratory[1]
Oak Ridge, Tennessee 37831-6036, USA

These days, a great many environmental scientists are scratching our heads in bewilderment.

Because of a love for science and discovery, we pursued scientific careers. Because of a love for the natural world around us, we specifically pursued study and training that led us to ecology, fisheries, hydrology, and similar academic fields. And because we saw the damage that humans can do to our environments, we expanded and applied our knowledge to evaluating and rectifying environmental impacts, sometimes to the degree of specializing in subjects like environmental or resource management. We were raised, so to speak, in a "knowledge environment." One read about what was known, made observations, framed hypotheses for what was not known or uncertain, conducted research and more observations to test the hypotheses, encouraged others to do the same, and periodically reviewed the accumulated data to see what was worthy of being considered "knowledge."

To an important degree, we also became environmental activists, warning of perceived risks from planned developments. The warnings often contained a certain amount of hyperbole for sake of getting attention and suggesting what might occur if no action was taken. Despite the initial hyperbole, I think it is fair to say that our concerns were founded on an assumption that, once recognized, the impending risks would be studied, characterized, technically evaluated, minimized, and the remaining effects mitigated. We welcomed landmark environmental legislation in the late 1960s and 1970s as ways to institutionalize those assumptions and set up avenues for regulation of the risks beyond our individual diligence.

Why the bewilderment? Many of us are puzzled by the way things have turned out, at least on some fronts. One of those fronts is the regulation of power plant cooling systems, the underlying topic of this book.

By way of brief background, electricity-generating stations that use fuels (coal, oil, gas, nuclear) to generate heat and steam to drive steam turbines and electricity generators must discharge much of that heat (roughly two-thirds of the total) to the environment after the steam is used and condensed (the other one-third goes to electricity). A common source of "coolness" to condense the steam is cool water pumped from nearby water bodies (rivers, lakes, coastal waters). The water is returned to the water body warmed usually by 10 to 12 degrees Celsius (about 18 to 22 degrees Fahrenheit). In the context of an industry creating pollutants (the initial paradigm for evaluating cooling systems), this was "thermal pollution" (or "calefaction" as Dan Merriman, a principal in this Connecticut River Ecological Study, called it). Somewhat belatedly, it was recognized that the cooling water drew small organisms through the cooling circuit (entrainment) and larger organisms (e.g., fish, crabs) were caught on intake screens (impingement) and that in both instances, some of those affected were injured or killed. Because of the demonstrated or perceived problems with this "open-cycle" cooling system, closed cycle systems consisting of cooling towers were developed and installed at many power stations (cooling towers generally cool recirculating water by evaporation). Cooling ponds are hybrid systems in which an essentially once-through plant design is used with a small, recirculating pond rather than an open-water body. Air cooling (like an automobile radiator) is feasible but expensive.

Concerns over "thermal pollution" were raised initially in Pennsylvania in the mid-1950s, where

[1] Oak Ridge National Laboratory is managed by UT-Battelle, LLC under contract DE-AC05-00OR22725 for the U.S. Department of Energy.

power stations on small rivers heated the water so much that summer temperatures exceeded the tolerance levels of most native aquatic organisms. I was privileged to participate as an undergraduate in one of the early ecological studies of a thermal discharge on the Delaware River (Coutant 1962). Scientific outcries and verbal skirmishes with power company officials led to examination of other instances across the country (e.g., the Ohio River) where too much heat was being added to too small water bodies and river life was suffering. The concerns were elevated when it became known that the federal government was fostering development of commercial nuclear power stations that were not only much larger than any existing fossil-fueled thermal power plants, but discharged a larger proportion of their energy as waste heat. Extrapolating from known study sites, it was feared that larger power stations could raise many rivers to ecologically intolerable temperatures. On the West Coast, the massive thermal discharges to the Columbia River from the Atomic Energy Commission's (AEC) Hanford plutonium production reactors raised concerns over survival of the indigenous coldwater salmon (which I studied intensively there). Similar reactors in South Carolina raised concerns over "hot ponds" on the AEC's Savannah River Reservation. Much of this early history is laid out in testimony for a series of congressional hearings held in 1968 and 1970, after the "thermal pollution" issue had been raised to a national "cause celebre" (Senate Committee on Public Works, Subcommittee on Air and Water Pollution 1968; Joint Committee on Atomic Energy 1970). Concern in the United States soon led to international concern and conferences by the International Atomic Energy Agency (IAEA 1970, 1975).

In a flurry of activity in the early 1970s, cooling-water effects became part of our regulatory institutions. Cooling-water effects analyses became mandated as part of the National Environmental Policy Act's (NEPA) Environmental Impact Statements for nuclear power plants licensed by the Atomic Energy Commission (later the Nuclear Regulatory Commission); the Clean Water Act was amended to include specific stipulations for thermal [316(a)] and intake [316(b)] issues (which then became part of the National Pollutant Discharge Elimination System); and the EPA Water Quality Criteria were rewritten by the National Academy of Sciences to include a major section on heat and temperature and their relationships to power stations (which I authored). After nearly a decade of gestation, international agencies aided by U.S. participants such as myself produced handbooks and manuals on how to evaluate the effects of power plant cooling (UNESCO 1979; IAEA 1980).

In the midst of this history (mid-1960s), with no regulatory strictures yet developed, and facing the prospect of the largest (to that time) commercial nuclear power station in the country that would use cooling water from the lower Connecticut River, Lyle Thorpe (with the state of Connecticut) and Dan Merriman (with Yale University) conceived a grand scientific experiment. It was their premise that rigorous scientific study and monitoring, overseen for relevance and scientific standards by an advisory committee of independent scientists, could produce factual information to judge whether the power station's cooling system was detrimental to the Connecticut River ecosystem. Better yet, such a study could provide useful precedents for guiding other power plant studies elsewhere. Thus was born the Connecticut River Ecological Study, which is the subject of this volume.

Both originators of the study were well-respected professionals with a sense of how good science should be done. Dan Merriman, especially, was well known for his exemplary scientific work on marine and estuarine species. His enthusiasm was infectious and nearly all-consuming (he so enthusiastically led me on a tour of the study sites in the late 1960s that we ran out of gas for the car and had to bum some to get to a gas station!). Consistent with his scientific background, Merriman insisted on scientific credibility and an open process. But the crowning feature of their organizational skills was the ability to select highly qualified and motivated staff. There was a stability of program design and staffing that led to consistent research and monitoring that allowed scientifically rigorous interpretations. And science was the product, both as a monograph published by the American Fisheries Society (1976) and as individual contributions to the scientific literature (which are catalogued in an appendix to this volume). Few, if any, power plant studies could claim such scientific productivity. The original monograph is reprinted as Part I of this volume. Judge for yourself.

But a critical test of any scientific endeavor is whether it weathers the test of time. A notable failing of many ecological studies is that they are not easily replicated as would be a chemistry experiment or physics experiment. The initial work is not easily validated in order to sanctify the results as accepted knowledge. From a strictly scientific perspective, a retrospective analysis of the lower Connecticut River after about 30 years of power station operation is about as good a validation test for the original study conclusions as anyone can do. Part II of this volume is a series of papers that revisits the study components, in some cases reanalyzing the data with new techniques, but most often providing an update on the status of the river based on research and monitoring conducted in the intervening years. Much of the river environment has changed, but much has stayed the same—the effects are not related to the operation of the power station but to other, more pervasive influences.

We have to recognize, however, that the motivation for revisiting the Connecticut River Ecological Study and its evaluation of the aquatic effects of the Connecticut Yankee power plant is not strictly scientific. Environmental protection issues that brewed in the 1970s (and became effectively resolved then with good biological criteria and better engineering designs) have arisen again into the public domain. Although thermal effects issues have been raised in some locations, the most intense activity is related to intake issues of entrainment and impingement. As a result of legal action in the mid 1990s related to enforcement of Clean Water Act Section 316(b) on intake effects, EPA has been undertaking the setting of regulatory standards for water intakes. The current and proposed regulations (described in articles in Part II) presume dire consequences from entrainment and impingement and prescribe what some analysts believe to be draconian measures compared to the actual impacts. Other analysts feel equally strongly that the regulatory controls are needed to protect aquatic life. Sadly, the debate is punctuated by ample evidence that the current promoters of stricter regulations (and some of those arguing against them) are unaware of the large documentation for minimal detrimental effects on aquatic populations and communities of most well-designed, current power station cooling systems.

Thus, the head scratching and bewilderment. How does one inform the seemingly intractable societal arguments when they are based on lack of knowledge of the scientific evidence? For example, in the 1970s and 1980s, the epitome of fisheries population science was brought to bear on these issues. Detailed ecological studies like the Connecticut River Ecological Study (and a number of others) were conducted successfully. Monitoring has continued for more than 25 years in some water bodies to identify any long-term effects (e.g., the Hudson River fish populations). Yet the studies seem to be forgotten or ignored. With our technical, scientific backgrounds, we environmental scientists find this hard to fathom.

This book represents one contribution to a possible solution. It is to reprint the exhaustive Connecticut River study monograph of 1976 and append a thorough reanalysis of information obtained then and in the intervening years. It does so with objective scientific papers on topical areas relevant to the original work, and with summary and analysis articles that provide perspective on the social as well as scientific questions. Much of the credit for reprising the study goes to Doug Dixon of EPRI—the Electric Power Research Institute—and to Paul Jacobson of Ecological Assessments, LLC. It could not have been done without the enthusiastic participation by original researchers Bill Leggett (now of Queen's University), Bart Marcy (now of Westinghouse Savannah River Company), and Ron Massengill (now retired). In the spirit of scientific rigor and openness fostered by Dan Merriman and Lyle Thorpe at the outset in 1965, the retrospective contains many new authors from a variety of organizations who provide their recent information and perspectives.

For one steeped in the scientific tradition, I find fascinating the new information, the reanalyses, the updates, and the candid appraisals of what, on reflection, were strengths of the original study and what were weaknesses or missed opportunities. Historians of environmental science should be delighted. I just hope that present-day regulators will also find the work informative.

References

Coutant, C. C. 1962. Effect of a heated water effluent upon the macro-invertebrate riffle fauna of the Delaware River. Proceedings of the Pennsylvania Academy of Science 36:58–71.

IAEA (International Atomic Energy Agency). 1970. Environmental aspects of nuclear power. Symposium 146/53, Vienna.

IAEA (International Atomic Energy Agency). 1975. Environmental effects of cooling systems at nuclear power plants. STI/PUB/378, Vienna.

IAEA (International Atomic Energy Agency). 1980. Environmental effects of cooling systems. Technical Reports Series No. 202, Vienna.

Joint Committee on Atomic Energy. 1970. Environmental effects of producing electric power. 91st Congress, 2nd Session.

Senate Committee on Public Works, Subcommittee on Air and Water. 1968. Thermal pollution – 1968. 90th Congress, 2nd Session.

UNESCO (United Nations Educational, Scientific and Cultural Organization). 1979. Predicting effects of power plant once-through cooling on aquatic systems. Technical Papers in Hydrology 20, Paris.

Original Preface

This monograph presents a detailed description of the results of an ecological study, 1965–1973, of the effects of the warmwater effluent from the Connecticut Yankee Atomic Power Company's electric generating plant at Haddam Neck on the lower reaches of the Connecticut River, USA; it is accompanied in the Introduction by summary recitals of associated research projects and monitoring programs—microbiology, impingement of fish on the intake screens of the plant, and environmental radiation.[1] The study, which embodies hydrography, the benthos, and the resident and anadromous fishes, is characterized by the fact that it was begun while the plant was still in the early stages of construction and that the field work was only terminated nearly 6 years after the inception of commercial operation. Thus, there were essentially 3 years for the collection of baseline data on the biota under "pre-thermal" conditions for comparison with the situation as it was altered by the large flow of condenser-cooling water from the generating station. For example, under these circumstances, it was possible to determine the migratory habits of tagged anadromous American shad *Alosa sapidissima* past the plant site during their upstream passage to the spawning grounds for three successive springs before the existence of a potential thermal barrier, and to obtain by subsequent comparison definitive evidence that the warmwater effluent does not constitute a significant impediment to the annual run of this valuable species (see Leggett 2004 [1976], this volume).

Some remarks in the preface of a recent book on the Columbia River Estuary (Pruter and Alverson 1972) are so apropos to our studies that we repeat them here:

> Since these studies were completed, there has been a great surge of public interest in environmental quality. This is a healthy interest which in the long run will help to establish realistic guidelines for man's activities as they affect both his own livelihood and the environment. Unfortunately, this interest is sometimes reflected by an expectation that a particular study or group of studies will, at one fell swoop, provide a panacea to our environmental problems. It should be noted, however, that even the best organized and executed studies usually provide only stepping stones of knowledge, the whole of which, if we are lucky, eventually lead to solution of an important problem or problems.

Just as the book on the Columbia deals specifically with the radioecology of the system, so our work on the Connecticut deals essentially with the thermal effects. In a sense, these two series of studies, though dealing with different aspects of bioenvironmental research and conducted on major rivers on opposite sides of the continent, are complementary and a basic principle applies to both—namely that, "People involved with this type of work in other areas will be able to draw general, qualitative lessons from the described studies; however, they will *not* be able to use them as a prescription for their own work" (loc. cit.). In short, the readers are warned against drawing sweeping and indiscriminate conclusions, "particularly as they might be applied to other regions."

It may seem to some that certain aspects of the work reported in this monograph are distinctly peripheral to the main core of the study (i.e., the effects on the aquatic ecosystem of the large volume of heated effluent from a specific nuclear-powered, electric-generating plant of the pressurized-water type at a particular locality). This may be so, depending on one's point of view, but here we have

[1] This monograph does not include a discussion of the general situation with regard to sources of energy in this country. These subjects are dealt with in varying depth by Hubbert (1971); Starr (1971); Weaver and Kristof (1972); Bethlendy (1973); Cook (1973); Gregory (1973); Hammond et al. (1973); Roberts (1973); Seaborg et al. (1973); Weinberg (1973); Auer (1974); Cheney (1974); and Rose (1974a, 1974b). See also *Science* 184 (4134), 19 April 1974; and *Nature* 249 (5459), 21 June 1974.

drawn encouragement and benefit from the complete freedom sanctioned by the supporting Connecticut Yankee Atomic Power Company in determining the scope of our investigations.

In many situations, the distinction between applied and pure science is quite clear. But in research of the type which comprises the main body of this work, the distinction is not simplistic. There is no hard and fast line, and our premise has been that research in "... basic science is the nutrient on which application feeds" and "... that basic knowledge usually does [in the long run] prove relevant to social needs" (Simon 1973). Hence, this study, though marked like most others of its kind by shortcomings, has entertained certain questions, which at first glance might seem relatively immaterial but which in the long run may well become pertinent bits in the highly complex picture of the whole ecosystem. As de Tocqueville wrote in 1835, "Americans judge that the diffusion of knowledge must necessarily be advantageous and the consequences of ignorance fatal" (Cadwalader 1973).

It is not always easy to determine in advance the precise confines of the various areas of an ecological investigation, perhaps especially so when one seeks a clear answer to a pragmatic problem. Superficiality or lack of depth is indefensible. Yet the circumstances of practicability, cost, and time impose certain limitations; these are obvious at the extremes, but within the bounds of feasibility, the exact scope of each particular area can be extremely difficult to define—often increasingly so as the initial results begin to materialize. In this latter respect, the evolution of this study, again like others of its kind, has been revealing in the sense that it lends added credence to the fundamental fact that we know extraordinarily little about the details and interactions among the elements of aquatic ecosystems. As Broeker (1973) has pointed out, on a national scale, "Almost nothing is being spent on research devoted to understanding the ecosystems we seek to save." Even though the word "ecology" was first coined a century ago,[2] this division of biology is still in its minority. Nor is it likely that ecology will ever undergo a period of such spectacular development as the emergence of molecular biology in the past quarter of a century through the combined disciplines of biochemistry, genetics, and microbiology. Its very nature argues against such a course. New technologies and mathematical modeling boost it along the way with increasing momentum, but its complexity in dealing with problems that "... fall between the cracks of 'basic' sciences" reasons against rapid evolvement. Thus it has been said that "... in most of the world, environmental biology has not yet passed the stage of inventory and survey, and is far from ready to grapple with the galloping degradation of the human habitat" (Handler 1970).

The Connecticut River study started as an inventory and survey before the nuclear plant went into operation; for example, while some things were known about the ichthyofauna in a general way, basic knowledge of the elements of the benthos was next to nil. As the study progressed, we were able to document the effects of the thermal effluent on the distribution and seasonal composition of the benthic fauna, and our understanding of the behavior and life histories of the components of the ichthyofauna was greatly increased—to the extent that we are in a far better position to speak with authority about the significant consequences of the man-made thermal plume at Haddam Neck and even to extrapolate to some degree and with circumspection to similar situations that might arise in the general locale or comparable estuarine quarters.

A word of caution needs be added here in the light of the foregoing commentary about the inherent nature of ecology. Because of the complexity of the interrelationships between the various elements in a natural aquatic ecosystem, it is next to impossible to provide a masterwork of the whole. It is not as though the system were stable as with an artist in the delineation of a still life; it is a constantly changing scene from season to season and year to year. Further, as science advances, it characteristically raises more questions to be answered, some large, some small—some capable of reasonable resolution, some quite beyond the bounds of practicality in terms of their immediate elucidation. The matter of entrainment and the mortality of planktonic fish eggs and larvae resulting from their passage through the condenser-cooling system of a steam electric-generating station is a case in point. To ascertain with some measure of exactitude the effects of such man-made mortality on

[2] In 1873, by Ernst Heinrich Haeckel, the first German advocate of organic evolution, in his preface to his "History of Creation."

the adult populations, one needs to know the natural mortality rates (which vary from year to year) of each species over its life span, the competition for food between the different larval types, and a whole host of interelated imponderables—including, for example, other sources of man-made mortality, such as that caused by turbulence resulting from the passage of commercial and pleasure traffic in the areas of spawning and early stages of development. These are problems of such complexity that their detailed solution is next to impossible with reference to the siting and operation of power plants. It is possible to make generalized commentary, such as that by Clark and Brownell (1973), but since each situation is different, even that may be speculative.

The point is that it is all-important for those who judge environmental impact statements to assess accurately the relative importance of the various factors as well as the practical capability of obtaining information in the depth required to make intelligent decision. These are matters for hard-headed, mature judgment—especially since there is ample room for disagreement in areas where the unknowns abound. Time is too precious and experienced personnel too limited to divert attention from the critical issues and thereby decrease the opportunity for objective decision with regard to the relationship between the demand for energy and concern for the environment.

DANIEL MERRIMAN
Director, Connecticut River Ecological Study;
Associate Professor of Biology, Yale University

LYLE M. THORPE
Associate Director, Connecticut River Ecological Study;
Retired Director, Connecticut State Board of Fisheries and Game;
Consultant, Northeast Utilities Service Company

References

Auer, P. L. 1974. An integrated national energy research and development program. Science 184(4134):295–301.
Bethlendy, G. 1973. Environmental aspects of nuclear power stations. Environmental Letters 4(3):151–185.
Broeker, W. S. 1973. Environmental priorities. Science 182(4111):435.
Cadwalader, G. 1973. Freedom of science in the oceans. Science 182(4107):15–20.
Clark, J. R., and W. Brownell. 1973. Electric power plants in the coastal zone: environmental issues. American Littoral Society Special Pubication 7.
Cook, C. S. 1973. Energy: planning for the future. American Scientist 61(1):61–65.
Gregory, D. P. 1973. The hydrogen economy. Scientific American 228(1):13–21.
Hammond, A. L., W. D. Metz, and T. H. Maugh II. 1973. Energy and the future. American Association for the Advancement of Science, Washington, D.C.
Handler, P., editor. 1970. Biology and the future of man. Oxford University Press, New York.
Hubbert, M. K. 1971. The energy resources of the earth. Scientific American 224(3):60–70.
Pruter, A. T., and D. L. Alverson, editors. 1972. The Columbia River Estuary and adjacent ocean waters. University of Washington Press, Seattle.
Roberts, R. 1973. Energy sources and conversion techniques. American Scientist 61(1):66–75.
Rose, D. J. 1974a. Energy policy in the U.S. Scientific American 230(1):20–29.
Rose, D. J. 1974b. Nuclear eclectric power. Science 184(4134):351–359.
Seaborg, G. T., J. L. Bloom, and E. L. Nelson. 1973. Nuclear and other energy. Annals of the New York Academy of Sciences 261:79–88.
Simon, H. A.1973. Relevance—there and here. Science 181(4100):613.
Starr, C. 1971. Energy and power. Scientific American 224(3):36–49.
Weaver, K. F., and E. Kristof. 1972. The search for tomorrow's power. National Geographic 142(5):650–681.
Weinberg, A. M. 1973. Some views of the energy crisis. American Scientist 61(1):59–60.

Connecticut River Ecological Study—Revisited
Participant Acknowledgments

EPRI and the monograph editors acknowledge the services of the following individuals relative to supporting and preparing this revised monograph, including sampling, sample processing, data analysis, preparation of figures, map preparation, authors and coauthors:

Nels Barrett, Natural Resources Conservation Service, Tolland, Connecticut.
Jacqueline Benway, Connecticut Department of Environmental Protection, Old Lyme, Connecticut.
Carl Chen, Systech Engineering. San Ramon, California.
Charles Coutant, Oak Ridge National Laboratory, Oak Ridge, Tennessee.
Victor Crecco, Connecticut Department of Environmental Protection, Old Lyme, Connecticut.
Donald Danila, Millstone Environmental Laboratory, Waterford, Connecticut.
David Dodge, Millstone Environmental Laboratory, Waterford, Connecticut.
James Foertch, Millstone Environmental Laboratory, Waterford, Connecticut.
Charles Fredette, Connecticut Department of Environmental Protection, New Britain, Connecticut.
Richard Gallagher, Millstone Power Station, Waterford, Connecticut.
Stephen Gephard, Connecticut Department of Environmental Protection, Old Lyme, Connecticut.
Robert Goldstein, Electric Power Research Institute, Palo Alto, California.
Kurt Gottschall, Connecticut Department of Environmental Protection, Old Lyme, Connecticut.
Gabe Gries, New Hampshire Department of Fish and Game, Keene, New Hampshire.
Marisa Guillot, Connecticut Department of Environmental Protection, Old Lyme. Connecticut.
Neil Hagstrom, Connecticut Department of Environmental Protection, Hartford, Connecticut.
Joel Herr, Systech Engineering, San Ramon, California.
Penelope Howell, Connecticut Department of Environmental Protection, Old Lyme, Connecticut.
William Hyatt, Connecticut Department of Environmental Protection, Hartford, Connecticut.
Robert Jacobs, Connecticut Department of Environmental Protection, Hartford, Connecticut.
Milan Keser, Millstone Environmental Laboratory, Waterford, Connecticut.
Timothy King, USGS Leetown Science Center, Kearneysville, West Virginia.
Barbara Korzendorfer, USGS, East Hartford, Connecticut.
Deborah Lehman, American Fisheries Society, Bethesda, Maryland.
Ben Letcher, USGS Silvio O. Conte Anadromous Fish Research Center, Turners Falls, Massachusetts.
Ernesto Lorda, Enviro-Stat Consulting Services, Old Lyme, Connecticut.
F. Douglas Martin, Westinghouse Savannah River Company, Aiken, South Carolina.
Ray Martin, Martin Photo and Art, West Dover, Vermont.
William McDermott, Northeast Generation Services, Berlin, Connecticut.
James McMenemy, Vermont Department of Fish and Wildlife, Springfield, Vermont.
David Molnar, Connecticut Department of Environmental Protection, Old Lyme, Connecticut.
Douglas Morgan, Millstone Environmental Laboratory, Waterford, Connecticut.
John Mullaney, U.S. Geological Survey, East Hartford, Connecticut.
Kevin Nebiolo, Connecticut Department of Environmental Protection, Old Lyme, Connecticut.

Eileen O'Donnell, Connecticut Department of Environmental Protection, Hartford, Connecticut.
Deborah Pacileo, Connecticut Department of Environmental Protection, Old Lyme, Connecticut.
Robert Parker, Kleinschmidt Associates, Essex, Connecticut.
Robert Perron, Branford, Connecticut
Ernie Pizzuto, Connecticut Department of Environmental Protection, Hartford, Connecticut.
Donald Pugh, USGS Silvio O. Conte Anadromous Fish Research Center, Turners Falls, Massachusetts.
Joseph Ravita, Connecticut Department of Environmental Protection, Barkhamsted, Connecticut.
John Richardson, Blue Hill Hydraulics, Blue Hill, Maine.
Edward Roseman, Ohio Department of Natural Resources, Sandusky, Ohio.
Steve Roy, United States Forest Service, Green Mountain National Forest, Rutland, Vermont.
Saul Saila, University of Rhode Island. Narragansett, Rhodes Island.
Thomas Savoy, Connecticut Department of Environmental Protection, Old Lyme, Connecticut.
Eric Schluntz, Connecticut Department of Environmental Protection, Hartford, Connecticut.
Tim Sisk, Kleinschmidt Associates, Essex, Connecticut.
Caleb Slater, Massachusetts Division of Fisheries and Wildlife, Westboro, Massachusetts.
John Swenarton, Millstone Environmental Laboratory, Waterford, Connecticut.
Ned Taft, Alden Research Laboratory, Holden, Massachusetts.
Christine Tomichek, Kleinschmidt Associates, Essex, Connecticut.
Joseph Vozarik, Millstone Environmental Laboratory, Waterford, Connecticut.
John Warner, U.S. Fish and Wildlife Service, Concord, New Hampshire.
Kent Zammit, Electric Power Research Institute, Palo Alto, California.

Peer Reviewer Acknowledgments

The monograph editors thank the following individuals who assisted with reviewing and commenting on the draft manuscripts:

Dr. James Bowen, Civil Engineering Department, University of North Carolina, Charlotte.
Dr. James Carscadden, Department of Fisheries and Oceans, White Hills Laboratory, St. Johns, Newfoundland.
Michael Chelminski, PE, Woodlot Alternatives, Inc., Topsham, Maine.
Dr. Carl Chen, Systech Engineering, Inc., San Ramon, California.
Dr. John Clark, Delaware Division of Fish and Game, Little Creek.
Dr. Mark Collins, Marine Resources Institute, South Carolina Department of Natural Resources, Charleston.
Dr. Edward Crossman, Centre for Biodiversity and Conservation Biology, Royal Ontario Museum, Toronto.
William Dey, ASA, Analysis and Communication Inc., Wappingers Falls, New York.
Dr. Ken Frank, Department of Fisheries and Oceans, Bedford Institute of Oceanography, Dartmouth, Nova Scotia.
Beth Goettle, USFWS Silvio Conte National Fish and Wildlife Refuge, Turners Falls, Massachusetts.
Dr. Cyril Hirt, Flow Science, Santa Fe, New Mexico.
Guy Hoffman, Connecticut Department of Environmental Protection, Hartford.
Dr. George Ice, National Center for Atmosphere and Stream Improvement (NCASI), Corvallis, Oregon.
Dr. Francis Juanes, Department of Natural Resources Conservation, University of Massachusetts, Amherst.
Dr. Desmond Kahn, State of Delaware, Natural Resources and Environmental Control Division of Fish and Wildlife, Dover.
Dr. Ben Letcher, USGS Silvio O. Conte Research Center, Turners Falls, Massachusetts.
Dr. Martha Mather, University of Massachusetts, Amherst.
Dr. Rob Neumann, Department of Zoology, Fisheries and Illinois Aquaculture Center, Southern Illinois University, Carbondale.
Vernon Osteen, Westinghouse Savannah River Company, Aiken, South Carolina.
Dr. Mike Paller, Westinghouse Savannah River Company, Aiken, South Carolina.
Dr. Marcel Reichert, Department of Biological Sciences, University of South Carolina, Columbia.
Dr. William Richkus, Versar Inc., Columbia, Maryland.
Janice Rowan, USFWS Connecticut River Coordinator's Office, Sunderland, Massachusetts.
Robert Sadzinski, Maryland Division of Natural Resources, Annapolis.
Dr. Robert Schmidt, Biology Department, Simon's Rock College, Great Barrington, Massachusetts.
Douglas Smith, Department of Biology, University of Massachusetts, Amherst.
Paul Stacey, Connecticut Department of Environmental Protection, Hartford.
Robert Stira, Northeast Generation Services, Berlin, Connecticut.
Steven Summer, South Carolina Electric & Gas Company, Columbia.
Ned Taft, Alden Research Laboratory, Inc., Holden, Massachusetts.
Dr. John Tietjen, Division of Invertebrate Zoology, American Museum of Natural History, New York.
Elaine Trench, U.S. Geological Survey, East Hartford, Connecticut.
Dr. Webb Van Winkle, Van Winkle Environmental Consulting, Boise, Idaho.
Dr. Robert Whitlatch, Department of Marine Sciences, University of Connecticut, Groton.
Dr. Robert Wilce, Department of Biology, University of Massachusetts, Amherst.

Symbols and Abbreviations

The following symbols and abbreviations may be found in this book without definition. Also undefined are standard mathematical and statistical symbols given in most dictionaries.

A	ampere
AC	alternating current
Bq	becquerel
C	coulomb
°C	degrees Celsius
cal	calorie
cd	candela
cm	centimeter
Co.	Company
Corp.	Corporation
cov	covariance
DC	direct current; District of Columbia
D	dextro (as a prefix)
d	day
d	dextrorotatory
df	degrees of freedom
dL	deciliter
E	east
E	expected value
e	base of natural logarithm (2.71828...)
e.g.	(exempli gratia) for example
eq	equivalent
et al.	(et alii) and others
etc.	et cetera
eV	electron volt
F	filial generation; Farad
°F	degrees Fahrenheit
fc	footcandle (0.0929 lx)
ft	foot (30.5 cm)
ft^3/s	cubic feet per second (0.0283 m^3/s)
g	gram
G	giga (10^9, as a prefix)
gal	gallon (3.79 L)
Gy	gray
h	hour
ha	hectare (2.47 acres)
hp	horsepower (746 W)
Hz	hertz
in	inch (2.54 cm)
Inc.	Incorporated
i.e.	(id est) that is
IU	international unit
J	joule
K	Kelvin (degrees above absolute zero)
k	kilo (10^3, as a prefix)
kg	kilogram
km	kilometer
l	levorotatory
L	levo (as a prefix)
L	liter (0.264 gal, 1.06 qt)
lb	pound (0.454 kg, 454g)
lm	lumen
log	logarithm
Ltd.	Limited
M	mega (10^6, as a prefix); molar (as a suffix or by itself)
m	meter (as a suffix or by itself); milli (10^{23}, as a prefix)
mi	mile (1.61 km)
min	minute
mol	mole
N	normal (for chemistry); north (for geography); newton
N	sample size
NS	not significant
n	ploidy; nanno (10^{29}, as a prefix)
o	ortho (as a chemical prefix)
oz	ounce (28.4 g)
P	probability
p	para (as a chemical prefix)
p	pico (10^{212}, as a prefix)
Pa	pascal
pH	negative log of hydrogen ion activity
ppm	parts per million
qt	quart (0.946 L)
R	multiple correlation or regression coefficient
r	simple correlation or regression coefficient
rad	radian
s	second
S	siemens (for electrical conductance); south (for geography)

SD	standard deviation
SE	standard error
s	second
T	tesla
tris	tris(hydroxymethyl)-aminomethane (a buffer)
UK	United Kingdom
U.S.	United States (adjective)
USA	United States of America (noun)
V	volt
V, Var	variance (population)
var	variance (sample)
W	watt (for power); west (for geography)
Wb	weber
yd	yard (0.914 m, 91.4 cm)
α	probability of type I error (false rejection of null hypothesis)
β	probability of type II error (false acceptance of null hypothesis)
Ω	ohm
μ	micro (10^{26}, as a prefix)
′	minute (angular)
″	second (angular)
°	degree (temperature as a prefix, angular as a suffix)
%	per cent (per hundred)
‰	per mille (per thousand)

Part I: Original Ecological Study

American Fisheries Society Monograph 9:3–25, 2004

Introduction

DANIEL MERRIMAN AND LYLE M. THORPE

Connecticut River Ecological Study, Connecticut Yankee Atomic Power Company Box 270, Hartford, Connecticut 06101, USA

"With regard to life on earth, temperature is—next to light—the most potent environmental component" (Kinne 1970a). In this monograph, we are of course concerned with the aquatic element of the earth's environment, and here, "it is generally conceded that 'temperature is the most important single factor governing the occurrence and behavior of life (Gunter 1957)'" (Hedgepeth and Gonor 1969).

It is worth commentary ab initio that in the perspective of the history of man's concern for the well-being of his environment, the popular phrase "thermal pollution" came into wide public prominence in the United States only with the maturation of the nuclear age as applied to the electric power industry. It was almost as though the problem were new rather than one which in reality had its origin in 1765 with the invention of the modern condensing steam engine by James Watt. Except for a few foresighted studies in the 1950s (vide infra), steam-electric power stations that were fossil-fueled had generally failed to arouse broad concern in this country on the score of heat rejection. Thus, for example, in the 1960s, the nine fossil-fueled plants on the lower 140 km of the Connecticut River could at maximum use more than 4.4 million m^3 per day of condenser-cooling water, returning it to its source at an average ΔT (increase above ambient) of more than 10.0°C (Merriman 1971). Yet, apparently there was no great agitation over the potential for detriment to the aquatic environment by these thermal effects—at least, in the case of the Connecticut, over their relationship to the annual run of the valuable anadromous American shad *Alosa sapidissima*. Such a volume could scarcely be considered "relatively small" as Black (1969) put it; the amount is more than twice that utilized since 1969 by the Connecticut Yankee Atomic Company plant at Haddam Neck, some 26 km above the river's mouth at Saybrook light. However, its "effects were not apparent," to the extent that no mention of this heated effluent is made in the earlier studies on the biology of the American shad in the Connecticut River.

It was, in short, the increased magnitude of the condenser-cooling water required by the nuclear-generated power industry and the rapid expansion, both actual and proposed, of the number and size of atomic plants, which finally stirred a public that was rapidly becoming acutely "ecology conscious" to the problems of thermal pollution real and potential. It is a fact that the design of nuclear plants does require some 50% more cooling water than do fossil-fueled plants of equal generating capacity. With a conventional coal- or oil-fueled power station, in general terms, about 40% of the initial thermal energy is converted to electrical energy, and of the 60% waste heat, one-sixth is lost to the atmosphere and the remaining 50% is carried away with the condenser cooling water. With the typical nuclear plant on the other hand, only about 30% of the primary energy (derived from the high-speed fission of uranium atoms) is transformed into electricity and 70% is taken off in the cooling water or thermal effluent. At the present time, the maximum conversion efficiencies are said to be about 42% for fossil-fueled plants as opposed to 33% for both the pressurized- and boiling-water types of nuclear-fueled plants now in operation, with improvements in these efficiencies expected to be minimal in the immediate future (Bell 1971).

With regard to the use of nuclear energy as a means of generating electricity for commercial purposes, it is only the past decade that has witnessed its growth from infancy. Actually, the world's first useable electricity from a nuclear power reactor was generated at the National Reactor Testing Station in Idaho on 20 December 1951 (Seaborg 1972). On 30 March 1953, the Submarine Thermal Reactor, Mark 1, began operation in a mock hull at the same site; this was the prototype of the nuclear power plant, a pressurized-water reactor, used in the submarine U.S. *Nautilus* that could circumnavigate the globe at 20 knots without surfacing and that sailed submerged across the Arctic and past the North Pole in 1958 (Kagan 1966). By 1964, the aircraft carrier *U.S. Enterprise* alone maintained eight pressurized-water reactors; there are now more than 100 power reactors on vessels of the U.S. Navy (Eisenbud 1974). So much for the military. As for civilian commercial installations, by 1963, there were just three in operation in the United States that were of 150 or more megawatt capacity: Dresden, Illinois, 1959, a boiling-water reactor, 200 MW_e; Massachusetts Yankee, Rowe, 1960, the first pressurized-water reactor, 160 MW_e; and Indian Point, New York, 1962, also a pressurized-water reactor (with oil-fired super heat capability), 255 MW_e. By contrast, the Atomic Industrial Forum, Inc. reported that at the end of 1973, the United States had 42 operable nuclear reactors capable of producing electricity on a commercial scale; their combined potential output was more than 25,500 MW_e, constituting 5.6% of the country's electric-generating capacity. The AIF report states that "if this nuclear capability did not exist and had to be supplied instead by fossil-fired generation, it would require some 250 million barrels of oil or 57 million tons of coal annually" (*Scientific American* 230(3):44, 1974). It is estimated that nuclear energy may meet 60% of our electrical energy demand by the year 2000 (Council on Environmental Quality 1973). In New England, the electrical energy produced from nuclear generating units is already more than 20% of the region's total output.

The initial reaction of the public to the use of nuclear energy for the commercial generation of electricity was one of deep apprehension; the realistic fear of contamination or explosion was inescapable in the light of the detonations of 6 and 8 August 1945. To the point of this monograph, however, and as we have already seen, a secondary and widespread reaction in this country was the rapidly rising concern about the potentiality of thermal pollution of the aquatic environment through the large-scale operation of nuclear facilities—a situation that was brought most forcibly to the public attention by Boyle (1969), who wrote in *Sports Illustrated*:

> Thermal pollution from a single nuclear plant can do all sorts of damage to the receiving waters. For instance, thermal pollution decreases the dissolved oxygen content, increases the toxicity of pollutants, makes water turbid (which prevents adequate sunlight penetration), spurs the growth of noxious blue-green algae (the stink of it literally can peel the paint off nearby houses), increases the metabolic rates of fish and other organisms, changes their behavior or interferes with their reproductive cycles, and often kills them outright.

The relationship of temperature to aquatic organisms, particularly fishes, had of course been the subject of scientific observation in nature[1] and of serious investigation by biologists[2] well back into the 19th century. With regard to the effects of thermal effluents from industry on the aquatic biota, among the early papers were those by Cairns (1955, 1956) who cites the use of the term "thermal pollution," Van Vliet (1957), Moore (1958) and Herry (1959) in Warinner and Brehmer (1972), Laberge (1959), Markowski (1959, 1960) dealing with invertebrates and entrainment, Ross (1959), Trembley (1960, 1961, 1965), and Alabaster (1963). The more recent literature is reviewed on pages 6–7.

It is interesting that the discharge of heat by thermal power stations in Britain, which has also been marked by significant increases in the past two decades and where the proportion of nuclear plants in operation has been higher, has not caused anything like the degree of public concern that has been generated in the United States (Clark 1969; Hawes 1970; Langford 1972). Indeed, one might have expected the reverse, since the British people as a whole have always tended to be more environmentally oriented than Americans. Furthermore, their problems are much more acute in the sense that their rivers are much smaller and the population density is much higher. Thus, the total mean annual runoff of all the rivers in En-

[1] Classic examples of temperature-induced catastrophic mortalities are the tilefish (Bigelow and Schroeder 1953) and the periodic effects of the El Niño current on the Peruvian anchovy and associated fauna (Idyll 1973).

[2] The papers by Bullock (1955) and Brett (1956, 1960) provide basic background material.

gland and Wales is a little more than 2,000 m^3, similar in magnitude to the mean discharge alone of the Ohio River at Huntington, West Virginia; and the mean load density in megawatts per unit area in England and Wales was some nine times that in the United States at the end of the last decade (Hawes 1970). Because of this situation, thermal problems in Britain have generally been considered "old hat," and appropriate steps have been taken to insure the necessary cooling-tower systems. In the United States, because of abundance (not alone of water), we have tended to be profligate, and it is only in relatively recent times that we have become generally and widely aware of the necessity for circumspection with regard to our natural resources. Yet, there is a danger in our newly found ecological consciousness and our sometimes hurried desire to make amends for past laxity. Thus, Langford (1972) writes: "In the past the discourse about heat from both conservationists and power companies has been too emotional, considering the small amount of actual recorded ecological damage caused by cooling-water discharges. It is time for a more objective and rational approach."

We have used the word "calefaction" in previous writings (Merriman 1970, 1971, 1972, 1973a, 1973b) in an effort to achieve scientific objectivity by keeping the subject of thermal discharges by power stations in fair perspective. When the expression "thermal pollution" started to achieve popularity in this country, it became a catchword, indeed, essentially a political slogan to some. Thus, all thermal effluents from industrial sources came to be widely considered as *pollutants* in the most serious sense of the word. This in turn led to such extremes as a policy statement by the U.S. Federal Water Quality Administration to the effect that "in no event will heat discharges to Lake Michigan be permitted to exceed 1°F rise over natural to the point of discharge"—a proposal which Langford (1972), with characteristic British understatement, called "particularly unrealistic" and which brings to mind the entry in Thoreau's Journal (11 November 1854) to the effect that "Some circumstantial evidence is very strong, as when you find a trout in the milk." Alternatively, "thermal effects" is a reasonable mode of expression, since some thermal discharges are inoffensive and some may have beneficial aspects. If the calefaction or warming of a body of water is proven to be excessive, it is properly called "thermal pollution," but indiscriminate incrimination is malapropos. Alteration of the environment by man is not always hurtful or unwholesome.

Thus, René Dubos (1973), in his delightful B. Y. Morrison Memorial Lecture on "Humanizing the Earth," elaborates on Barry Commoner's fourth law of ecology, "Nature knows best," and comes to the contrary conclusion "... that nature is incapable, by itself, of fully expressing the diversified potentialities of the Earth." He believes that the interplay between man and nature has often in the past and can in the future take the form of a creative symbiotic relationship—a biological interdependency, if you will, which is mutually beneficial and which facilitates the creation of "... environments which are ecologically stable, economically profitable, esthetically rewarding, and favorable to the continued growth of civilization." Does nature always know best? Mr. Justice Holmes once remarked that "No generalization is worth a damn, not even this one."

Athelstan Spilhaus (1973) has expressed himself forcibly on the general subject of man's expectations and objectives in this country in a manner that is singularly pertinent to the subject of this monograph:

> We are seeing a burgeoning of aspiration and concern of people with regard to the quality of the environment. Such aspirations and concern are excellent.... But there is a real danger that political and public manipulation of them can destroy the very productivity we need to achieve them.
>
> These aspirations deal, on the one hand, with an increasing protective attitude to people, and on the other hand, with the over-concern for an over-clean environment. In both of these worthwhile objectives, there lurks the danger of ignorant overemphasis to the extent of inhibiting national productiveness.
>
> ... [Thus] the naively overstated goal of "zero effluents" prevents our developing productivity—especially the productivity that we need to enhance our environment....
>
> We need to move toward a better public understanding of what I've called an "ecolibrium" position—bal-

> ancing the desired ecology (an harmonious pattern between organisms and their environment) with the necessary economy (the management of affairs with a view to maintaining productiveness)....
>
> Zero effluents? Nonsense! There must be more awareness that in the use of the components of our environment—air, land and water—*there is an acceptable burden of man's wastes of the proper kind that these components can carry, and that this is not zero* (italics ours).

The italicized bit above exactly expresses the purpose of our study—to ascertain as nearly as possible whether or not calefaction by the thermal effluent from the Connecticut Yankee Atomic Power Company plant at Haddam Neck is "an acceptable burden,"[3] which the lower reaches of the Connecticut River can carry.

Review of the Literature

There is an extensive scholarly literature on the relationship between temperature and aquatic life, both with reference to individual organisms and to the major components of the ecosystem; in the past few years, there has also developed an interminable body of writing on the subject of thermal effects—especially those caused by industry—which ranges from quasi-academic to popular and "emotional" (Langford 1972). To attempt to review here the whole sweep would be both impractical and contrary to common sense. Furthermore, each author of the various sections of this work has cited the references that are pertinent to his area of concern, so that the accumulated bibliographies represent a wide coverage of literature. However, it does seem within the bounds of reason to call attention to particular sources of information that are both currently useful and relevant to this study.

Following on from the earlier references cited on page 4, Jones (1964) has a chapter (No. 13) on "Thermal Pollution: the Effect of Heated Effluents" that covers heat tolerance and "acclimation" in fishes from a number of points of view. Naylor (1965) goes into the matter in further detail with marine and estuarine organisms and includes a comprehensive body of reference material. Voluminous bibliographies of the literature on the effects of temperature in general and heated effluents in particular on aquatic life have been prepared by Kennedy and Mihursky (1968), Raney et al. (1973), and Beltz et al. (1974). There is also a bibliography on thermal pollution (Committee on Thermal Pollution 1967), as well as the hearings on the same subject before the United States Senate (1968–1969) and on the environmental effects of producing electric power (1969). The Proceedings of the 2nd Thermal Workshop of the U.S. International Biological Program (*Chesapeake Science* 10(3–4):125–341, 1969) deal with "... advancements in biological understanding in one area of environmental change" and cover a wide range of aquatic temperature surveys and effects. Similarly, the volume edited by Eisenbud and Gleason (1969), which includes a variety of papers presented at a symposium on "Thermal Considerations in the Production of Electric Power" held in Washington, D.C., on 28–30 June 1970, serves to reflect the concerns and general state of knowledge at the end of the last decade. Coutant (1968, 1969, 1970, 1971, 1972a, 1972b), Coutant and Goodyear (1972), Coutant et al. (1972), and Coutant and Pfuderer (1973) have provided extensive information, annotated bibliographies, and invaluable reviews of the literature on the biological aspects of thermal pollution. The matter is also treated in detail by Hawkes (1969) and Krenkel and Parker (1969); the latter reference has particularly useful contributions with relevant bibliographies by Patrick (algae), Hedgepeth and Gonor (marine and estuarine benthos), Alabaster (freshwater fishes), Nakatani (anadromous fishes), and de Sylva (marine fishes). Further representative contributions are those by Clark (1969), Cairns (1971, 1972), Sylvester (1972), and Levin et al. (1972). The symposium on the environmental aspects of nuclear power stations organized by the International Atomic Energy Agency (IAEA 1971) contains considerable information on effluent control and monitoring as well as site selection. Also, the symposium on freshwater biology and electrical power generation in the United Kingdom under the auspices of the Central Electricity Research Laboratories (CERL 1971) includes a wide variety of

[3] "A load, as a measure of quality." Hence "burthen," the carrying capacity of a ship stated as so many tons (Shorter Oxford English Dictionary 1939).

germane material; further, the "Discussion on Freshwater and Estuarine Studies of the Effects of Industry" organized by Sir Frederick S. Russell, F.R.S., and H. C. Gilson contains pertinent papers by Langford and Aston (1972) and Barnett (1972). Last, the Proceedings of the First World Congress on Water Resources (Chow et al. 1973) contains a section on "Thermal Pollution" that includes a broad assortment of 19 papers, and the symposium on "Thermal Ecology" (Gibbons and Sharitz 1974) comprises a gamut of 55 articles.

In a more general way, the September 1970 issue of *Scientific American* on "The Biosphere" is significant, particularly the article by Singer, who discusses the waste products of power generation and states that it would be incautious to assume "... that the heat put into the biosphere as a result of human energy consumption can be neglected because it is so much smaller than the solar input. The atmospheric engine is subtle in its operation and delicate in its adjustments. Extra inputs of energy in particular places can have significant and far-reaching consequences." The September 1971 issue of the *Scientific American* is devoted to "Energy and Power," and here, the papers by Starr and by Hubbert give comprehensive views of the subject that are invaluable for the broad spectrum.

The monumental treatise on "Marine Ecology" by Kinne (1970b) provides extensive general coverage of the biological aspects of temperature in the aquatic environment with ensuing chapters on the effects of temperature on micro-organisms (Oppenheimer), plants (Gessner), invertebrates (Kinne), and fishes (Brett and Garside), and with an expansive bibliography of the whole. Kinne (pp. 323–324), though concerned essentially with marine forms, makes a point that has clear relevance to studies such as those reported in this monograph:

> Studies concerned with responses to temperature of living organisms originally centered almost exclusively on the intact individual. Hence, at the individual level, there exists today an extensive literature on temperature responses of a large variety of marine and brackish-water living organisms, both in the sea and under laboratory conditions. While such research is of great importance, studies at the subindividual and at the supra-individual levels have opened up new avenues for a more complete understanding of temperature effects on living systems. Unfortunately, *there still exists but little information on responses to temperature at the supra-individual level (populations, ecosystems)* [italics ours].

Another general reference is that by F. E. J. Fry (1971), which, while not giving special consideration to the responses of fish to polluted environments, deals in extenso with the effects of environmental factors on the physiology of fishes and also appends an extended list of references.

More to the subject immediately at hand, Langford (1972) compares the thermal effects in Britain and North America. With reference to the United States, he holds to the view "... that 'thermal effluents' have a great deal of expanding to do before they cause drastic ecological changes of the magnitude of those caused by other industries and by the impoundment of rivers." The handbook for the layman, "Nuclear Power Plant Siting" (Meredith 1972), is a thoroughly informed and well-balanced presentation dealing, amongst other matters, with thermal effects and entrainment (pp. 11–16) and coming to the overall conclusion "... that power generation will always have some impact on the environment; there is no escaping the fact. However, power plants can be sited, built, and run to have a minimal impact, and such operations can be accomplished safely, efficiently, and economically." It is of some interest in this connection that the siting of the Connecticut Yankee Atomic Power Company plant at Haddam Neck was determined a long decade ago, well before the public's attention was focused on this new type of energy source. Finally, the publication by Clark and Brownell (1973) dealing with the environmental issues of electric plants in the coastal zone is a signal contribution, and that by Beyers (1974) on the ecological impacts of energy production on rivers and lakes is a general review.

The Connecticut River

The Connecticut River, situated in the northeastern sector of the United States, is the longest river in New England; it exceeds the neighboring and parallel Hudson River in New York by some 160 km. It originates from the several Connecticut Lakes in the upper corner of New Hampshire at the Canadian border just above the 45th latitude. Flowing north–south, it separates Vermont from New Hampshire and bisects Massachusetts and Connecticut (Figure 1). From its source to the sandbar at its wide mouth on Long Island Sound (general salinity 25–31‰), it is 660 km in length, and in this course, it drops some 800 m. With the exception of the Susquehanna, which is more than twice as big, it carries the largest volume of any river on the East Coast of the United States north of Georgia; the monthly average discharge 1965–1967 was 333 m^3/s, as compared with the Hudson (273), Pee Dee, South Carolina (236), Potomac (222), and the Delaware (209 m^3/s) in the same period (UNESCO 1971). The channel of the Connecticut makes the river navigable for a steady commercial traffic (amounting to 3,650,000 tons in 1970) mainly by oil barges and tankers to Hartford 65 km upstream.

The history of the discovery of the Connecticut River is well told by Bacon (1906) and Hard (1947).[4] The English first settled in the Connecticut watershed in 1635 and renamed the river from the Algonquian "Kwinitekwa," hence "Quinnehtukgut," the "Long (Tidal) River." The Indians had traveled the river by birchbark canoe from early times, and the whiteman followed with flatboats and the like. But there were many obstacles in the form of rapids and falls, and the early ferries had no wires to guide them. The first bridge across the river, which connected Walpole, New Hampshire with Bellows Falls, Vermont, was built in 1784. There soon followed dams and canals as aids to navigation.

Thus, in 1795, a wing dam and canal were constructed to bypass South Hadley Falls, Massachusetts. In 1800, at Turners' Falls, Massachusetts, the first dam across the Connecticut River was built and with it a canal and locks. And in 1829, the dam above the Enfield Rapids and the accompanying Windsor Locks canal with its four locks were completed (Sears et al. 1953). It was taken for granted that the Connecticut River waterway would be "the major shipping artery of New England" and the cheapest form of transportation, but Hill (1972) points out that by the middle of the 19th century, "the railroad had changed all that." The first large industrial dam was built at Holyoke, Massachusetts, in 1849, and with it came the incipient portents of modern pollution. There are now 17 dams along the length of the river (McCarry 1972). The first hydroelectric power dam was constructed at Rainbow on the Farmington River some 11 km upstream from its juncture with the Connecticut above Hartford.

Coincident with the turn into the 19th century and the construction of dams on the Connecticut came the demise of the Atlantic salmon *Salmo salar* population in the river. It is said that *S. salar* formerly ran up all suitable rivers as far south on our coast as "... the Housatonic emptying into Long Island Sound; perhaps the Hudson also" (Bigelow and Schroeder 1953). Dymond (1963) writes that "The Connecticut River ... supported a large run until the end of the eighteenth century. In four years, however, following the construction of a dam 16 feet high across the river at Miller's Run in 1798, there were few fish left, and in 12 years there was none." As has been pointed out elsewhere (Merriman 1971), it is largely a matter of conjecture as to precisely how great was the abundance of salmon in the Connecticut prior to 1800; there is little in the way of solid information to go on (Moss 1960), and it is a common failing to enlarge on the glories of the past—or at least to paint them in the most roseate colors (Netboy 1968; Buck 1973). However, since it is generally agreed that the southern limit of the salmon on this side of the Atlantic was latitude

[4] As with the Hudson, the credit for the discovery of the Connecticut belongs to the Dutch. Adrian Block undertook his second voyage from Holland to Manhattan in the summer of 1613 in the *Tiger*. Having taken on a full cargo of peltry, he was ready for the return home when his ship was destroyed by fire. He could have gone back to Holland on a companion ship, but Block and some of his crew stayed on Manhattan over that winter and built a new craft, the *Onrust* or *Restless*, which measured 13.5 m overall, beam 3.5 m, and was 14.5 metric tons burthen. In the spring of 1614, he navigated (and named) Hell Gate, sailed Long Island Sound and put into the mouth of the Connecticut, which he named "De Versche Riviere"—the "Freshwater" River. Giovanni da Verrazano, who had cruised the coast of North America from North Carolina to Newfoundland 90 years earlier and discovered the New York and Narragansett bays and Block Island, apparently missed the Connecticut River. Block sailed up the Connecticut some 110 km to Enfield Rapids, tidehead, returned to the Sound and coasted southern New England, rediscovering Block Island en route to Cape Cod.

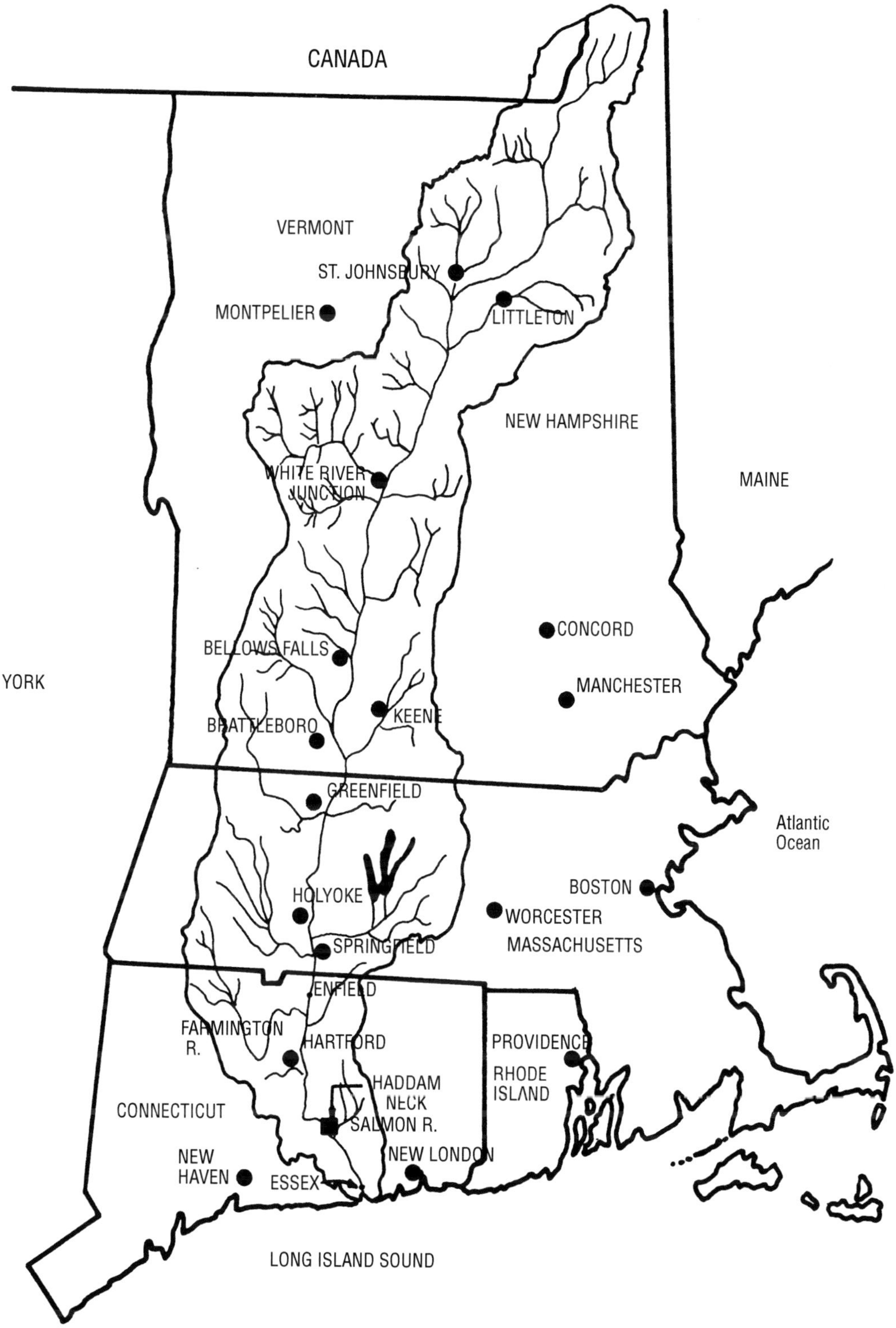

FIGURE 1. The Connecticut River and the area encompassed by its valley.

41°N, which is just south of the mouth of the Connecticut, it seems reasonable to assume that the abundance of this fish at this extreme of its southern range would have been far less than it was north of Cape Cod and especially in Canadian waters. In this light, attempts at restoration of Atlantic salmon to the Connecticut River should not be viewed, at best, with anything more than the prospect of modest success; on the other hand, there is no reason to suppose that the present effluent from the Connecticut Yankee Atomic Power Company plant would constitute a thermal barrier to salmon runs any more than it does to those of the shad (Leggett 2004 [1976], this volume).

The Connecticut River Valley, situated almost exactly half way between the equator and the North Pole, has a basin of about 29,000 km^2 (Figure 1). Its rainfall is generally ample and relatively evenly distributed throughout the year; thus, the average monthly rainfall at Hartford, Connecticut has a range of 7.5–10.0 cm and at White River Junction, Vermont, 5.5–9.5 cm (Sears et al. 1953).[5] Despite this relatively even rainfall, however, the volume of water-flow in the Connecticut River exhibits wide seasonal fluctuations because the winter precipitation is stored on the ground as snow until it melts in March and April with resultant spring runoffs of varying but usually large proportions. Over the past five decades, the average daily rate of flow of the Connecticut as recorded by the U.S. Geological Survey has been about 450 m^3/s; the maximum during this period was some 7,980 m^3/s during the flood on 20 March 1936, and the minimum 27.4 m^3/s on 20 October 1963. In the spring, the average daily flow is of the order of 700 m^3/s, in the late summer 140 m^3/s (Bureau of Outdoor Recreation 1968). It should be noted that these recordings were made above the zone of tidal influence and that at Haddam Neck—the site of the Connecticut Yankee Atomic Power Company plant—the minimum daily average tidal flow provides an additional 425 m^3/s. However, while the site experiences a semidiurnal tidal flow reversal, the last vestiges of the saline waters of Long Island Sound do not usually extend beyond a point approximately 3 km below the plant.

[5] By and large, this situation tends to be advantageous to Connecticut Valley farmers not only because of the steady supply of rain and the smaller likelihood of drought, but also because the soils are less subject to erosion. So it is, for example, that the Valley tobacco crop from mid-Connecticut to northern Massachusetts grown in ground featuring silt to fine sandy loam overlying gravel, silt, and clay from glacial lake and marine sediments, and with a surrounding native vegetation of oak, maple, and white pine, produces $24 million a year in Connecticut alone; it earns more money per acre than any other crop grown in the United States (McCarry 1972).

The Connecticut Yankee Atomic Power Company Plant[6]

The Connecticut Yankee Atomic Power Company is owned by 11 operating utility companies in Maine, Vermont, New Hampshire, Massachusetts, and Connecticut, of which The Connecticut Light and Power Company (25% of the stock), The Hartford Electric Light Company (9.5%), and the Western Massachusetts Electric Company (9.5%) are wholly owned affiliates of Northeast Utilities, Hartford, Connecticut. The output from the Connecticut Yankee plant, located in the township of Haddam, Middlesex County, Connecticut is dispatched as part of the New England Power Pool, in which utilities from all the New England states participate, to points throughout the region as needed. Connecticut Yankee was one of the first large nuclear base load units to go into operation in the United States, and towards the termination of the main body of our study (December 1972), it had produced more power from nuclear fission for commercial use than any other single unit in the world.

The Connecticut Yankee facility was first seriously considered in 1961, and in the spring of 1962, a study of the economic feasibility of a large nuclear-powered generating station in Connecticut was begun. Ground was broken in 1963, and construction was undertaken in accordance with Permit No. CPPR-14 issued on 14 May 1964, by the Atomic Energy Commission. The Con-

[6] Much of the information contained in this section is taken directly from a report prepared by Eugene F. Sturgeon in late 1964, "A River and A Nuclear Power Plant" by Peter H. Judd (private printing, Northeast Utilities Service Company 1970), and from the Environmental Report of the Connecticut Yankee Atomic Power Company, Haddam Neck plant, Operating License Stage, Docket Number 50-213, to the U.S. Atomic Energy Commission, 1972.

necticut River Ecological Study, which was ordered by the Connecticut Water Resources Commission (page 15) and which we have directed throughout, was begun in January 1965. The plant, initially licensed to produce about 490 MW_e of gross electrical power, began commercial operation in January 1968. The design rating of the plant was about 600 MW_e, and on 11 March 1969, it was licensed to operate at this level. Following the first refueling, April–May 1969, it maintained this output, with exceptions for repairs and refueling, with a relatively high degree of consistency (Figure 5 in Boyd 2004 [1976], this volume). The nuclear unit of Connecticut Yankee is a pressurized-water reactor with a "once-through" condenser cooling system. The design and construction of the nuclear steam supply and auxiliary system was the responsibility of the Westinghouse Electric Corporation. Stone and Webster Engineering Corporation undertook the site development, the design of buildings and secondary systems, and all plant construction.

The Connecticut Yankee plant site consists of some 212 ha on the east side of the Connecticut River on Haddam Neck, which is formed by the Salmon River's entry into the Connecticut 2.5 km below the actual facility. The buildings themselves occupy a space on what used to be a small airport of 10.5 ha for private planes. This site is 26 km upstream from Long Island Sound, 38 km as the river flows south and east of Hartford. The demography of the region, present and predicted, is given in the environmental report cited in footnote 7. The great majority of the whole site is of rugged terrain; the Pleistocene glaciation here gave rise to a physiography of erosional surfaces of metamorphic rock outcrops and glacial deposits. Seismic fault systems in the area had their principal activity eons ago, and the nearest earthquakes of marked magnitude in recent times have occurred in the St. Lawrence Valley and off the Grand Banks. Three to 6 m above the river level the hillsides rise steeply, cresting at more than 90 m; they are mainly covered by relatively shallow glacial till, which is underlain by Hebron granitic gneiss. Below the 3–6 m elevation, there is an extensive flood plain; the typical tide-associated mud flats and swamps have soils of interbedded sands and gravels with varying amounts of silts, some of them of a highly compressible organic nature. At the river's edge, the bedrock is generally as much as 9–15 m below the ground surface, a general reflection of the late Pleistocene with lower sea levels and relatively higher drainage gradients. The highlands and slopes of the whole site are dominated by second-growth red and black oak, beech, black birch, red maple, and hemlock. Along the roadways and clearings, the plant species diversity is high, with cherry, poplar, birch, and aspen clumps forming dense thickets along the way. The riparian area along the river and in the vicinity of the plant discharge canal is characterized by the typical alders, willows, cottonwood, rush, sedge, and reed grass, as well as sassafrass, hornbeam, elm, goldenrod, cocklebur, and the switch or panicum grasses. Amongst the aquatic vegetation in the region, the eel grass and duckweed are predominant, with cord and sword grass, cattail, pickerel weed, broad-leaved arrowhead, and water plantain the common emergent species. The more prevalent animal species in the riparian and aquatic areas are the muskrat, water snake, frogs, and such avifauna as heron and the several gulls.

The immediate plant area is filled and graded at the 6.4-m mean sea level, a yard grade of sufficient height to eliminate plant damage from flood situations. On the basis of unpublished data from the U.S. Geological Survey office in Hartford, Connecticut, it appears that the two greatest flows on the Connecticut River in the past century and a half or more occurred just two and a half years apart in the latter parts of March 1936 (a spring flood) and September 1938 (hurricane). The mean sea level height of the 1936 flood at the plant site was 5.9 m, and since that time, a comprehensive program of flood control has been in progress in the Connecticut River watershed. However, taking no credit for an expanding flood plain, and combining the worst conditions ever experienced, namely the 1936 flood with the 1938 coastal hurricane, studies indicate that a site elevation of 6.2-m mean sea level would be reached.

Connecticut Yankee utilizes a four-loop pressurized-water nuclear steam supply system (Figure 2). The fuel used is slightly enriched uranium[7] in the form of small pellets of UO_2, which

[7] The concentration of Uranium-235 in uranium as found in nature is 0.7%; as used at Connecticut Yankee, it is enriched to 3–4%.

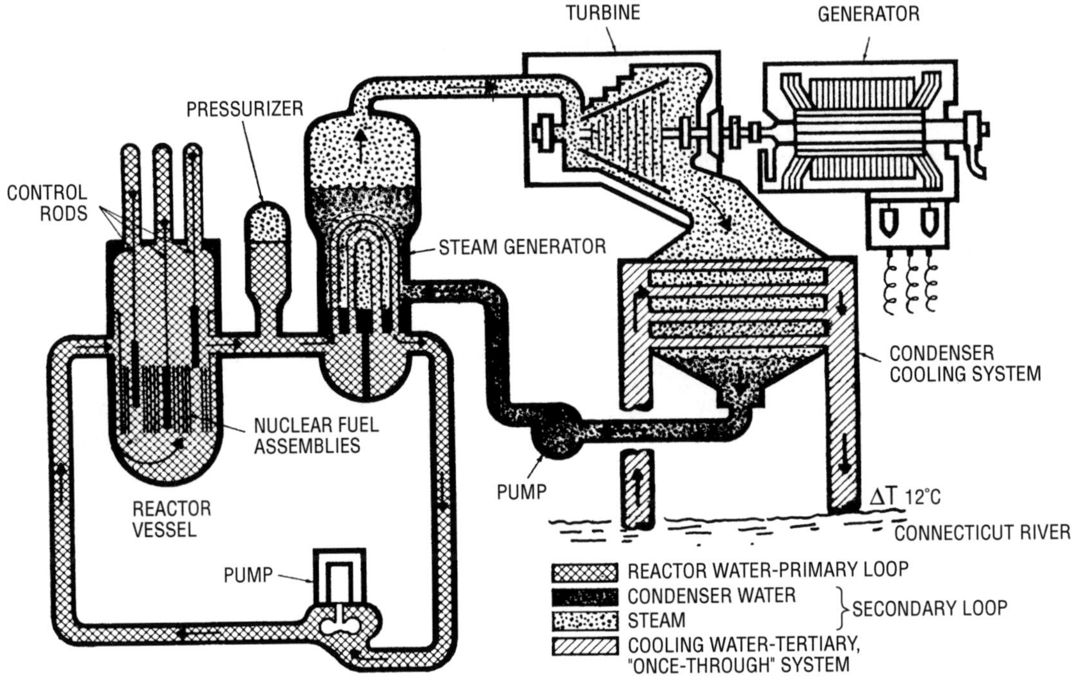

FIGURE 2. Schematic diagram of the Connecticut Yankee pressurized-water nuclear steam supply system with "once-through" cooling.

are enclosed in stainless steel (cladding) tubes. A bundle of these tubes makes up a fuel assembly, and a number of such fuel assemblies, positioned in a grid pattern, form the core of the reactor. This core, the heart of the atomic power plant, which generates heat by the controlled fission[8] of the fuel, is contained in a massive steel tank known as the reactor vessel. The reactor provides heated water at 303°C to four large heat exchangers referred to as steam generators. The name, "pressurized-water reactor" comes from the fact that this water, which is used to cool the reactor and act as a moderator and which circulates in a closed or primary equipment loop, is kept under sufficient pressure (151 kg/cm^2) to keep it from boiling as it picks up heat from the reactor core.[9] This heated water in its closed piping system is returned to the reactor to begin another cycle after it has turned water in the separate closed system into steam. The steam in this second system drives the turbines, which in turn drive the electrical generator; on leaving the turbines, the spent steam is condensed and returned to the heat exchangers to be reconverted to fresh steam. A third, separate flow of water, the one with which we are concerned in this work, is that used to cool the turbine condensers; it comes directly from the Connecticut River and discharges directly back to it as the thermal effluent. The function of this "once-through" system is not only to carry away low-temperature energy from the conventional portion of the plant, but also to dilute the low-level radioactive wastes discharged under the Atomic Energy Commission, Title 10, Code of Federal Regulations. The

[8] Fission occurs when a uranium fuel atom is split by a subatomic particle or neutron; the division in turn releases several neutrons from the nucleus of the atom and these in their turn incite further fission leading to the chain reaction. The rate of fission, and hence the amount of heat produced, is regulated by 45 control rods (80% Ag, 15% In, 5% Cd) which absorb neutrons and which are raised to increase fission and lowered to decrease it. Other controls include the use of boric acid as a liquid neutron absorber and the natural phenomenon of the Doppler effect.

[9] In the boiling water type of nuclear power plant the steam is generated directly in the reactor vessel proper.

river water has no direct connection with the reactor.

The condenser cooling water enters the plant via an intake structure on the east bank of the river. The construction of this structure, with four bays (each 3.7 m wide), required the dredging and removal of approximately 138,000 m^3 of material from the immediate riverbed, for which authorization subject to specific conditions, was granted by the State of Connecticut Water Resources Commission on 21 October 1964 (page 13). The intake structure houses four 352,000 L/min circulating water pumps as well as four 22,700 L/min service pumps for the station. The river water entering the cooling water system at the intake has a velocity of 0.3–0.6 m/s and passes through a traveling screen of 1-cm mesh. It then enters twin single-pass divided water boxes where the velocity may approach 2.4 m/s, takes two 90° turns, and passes through two banks of 26,860 condenser tubes 2.3-cm inside diameter × 15.25 m long, in which the heat exchange takes place. Flow time through the plant is approximately 93 s. The heated water flows out over a weir and down a canal of fairly uniform depth and width (average about 3 × 18 m), which enters the river via a broad mouth (180 m wide), 1.83 km below the plant intake. The average velocity in the canal, varying with tidal stage (range 0.75 m), is 0.3–0.6 m/s dwindling to about 6 cm/s at the discharge mouth; the transit time in the discharge canal thus runs from 50–100 min. The 23.5 m^3/s of cooling water required for condensing the turbine cycle (secondary) system is raised some 12.5°C at full power operation in the course of passage through the plant; Figure 3 shows the rate at which the DT decreases with diminishing power operation. The drop in the temperature of the thermal effluent from the top of the canal to its discharge mouth into the main river is only about 1°C.

The Connecticut River Ecological Study

Before the Connecticut Yankee Atomic Power Company plant could be built at Haddam Neck, it was necessary to obtain various permits from

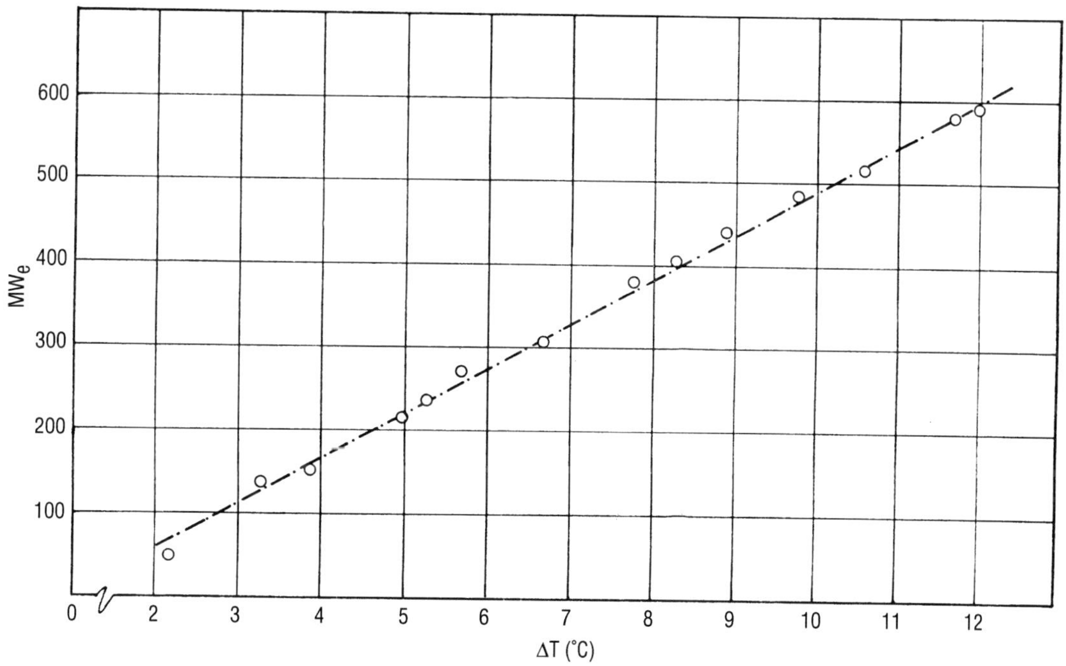

FIGURE 3. The relationship between the temperature increase over ambient (ΔT) and the level of megawatt operation (MW_e). The scatter of data is due to inaccuracies of the temperature sensors. Courtesy of Richard H. Graves, Assistant Plant Superintendent.

federal and state agencies. The first was from the Atomic Energy Commission (AEC), which had to pass on all the specifications for the plant and its safety measures. At that time, however, the AEC did not deal with environmental issues such as the possible effects of the thermal effluent on the aquatic biota. The commission gave preliminary approval on 21 May 1963, and after a public hearing in April, it issued a provisional authorization in May and the required construction permit on 15 June 1964. A second permit from the U.S. Army Corps of Engineers was necessary because there was to be dredging in a navigable waterway; this presented no obstacle, since the dredging was to be done on the opposite shore from the west-side channel at a point where the river is approximately 500 m wide and since the dredged material was to be disposed of on the applicant's property.

The third permit was at the state level. Early in 1964, the Connecticut Yankee Atomic Power Company made application to the State of Connecticut Water Resources Commission for a permit[10] "to construct a water intake gate and screen house and carry out certain dredging operations necessary to obtain water for condenser cooling purposes in the proposed atomic power plant and discharge those waters back into the Connecticut River." A first hearing on this application was held on 20 July 1964; it was inconclusive to the extent that the members of the commission felt they required more factual information before attempting to arrive at a decision on the application. The reason for the commission's hesitancy stemmed from its regulations, which provided that any dredging to be done must have "due regard for the protection of shellfish grounds and fin fish habitats." Thus, though the commission did not at that time have direct jurisdiction over thermal discharge as such, its responsibilities clearly gave it reason for concern with what the effect of the operation of the plant, after the dredging was done, might be on the biota of the river. And in this area, positive data and objective opinion were hard to come by.

In such wise, what had initially seemed to be a relatively insignificant application for a dredging permit resulted in a forum for the discussion of a variety of environmental issues surrounding the siting of the Connecticut Yankee plant at Haddam Neck with particular emphasis on the effect of the thermal effluent. It is of more than passing historical interest and of much credit to the state of Connecticut that these events took place well before the advent of the Water Quality Act (1965), the National Environmental Policy Act (1969), and the requirements for environmental impact studies.

The Water Resources Commission, allowing time for the collection and preparation of pertinent information, called a second public hearing for 17 September 1964, in the Hall of the House in the State Capitol; this meeting, called to order at 1930 hours, was adjourned at 0045 hours. At the first hearing, Dr. Ralph W. Gunwaldsen, Chief Hydraulic Engineer of the Stone and Webster Engineering Corporation, Boston, Massachusetts, had offered some general theoretical predictions on the flow pattern of the thermal discharge from the Connecticut Yankee plant, based primarily on models derived from existing data on the temperature behavior of the effluent from the Middletown, Connecticut, steam-electric power station 15 km upstream. His essential conclusions were that the warm water would mix with the surrounding water without permanent stratification and that the operation of the plant would "... only increase river temperatures moderately above what they would normally be." By the time of the second meeting, he had made further and more extensive investigation—in August, at a time of exceptionally low river flow, enabling him to predict in more detail the pattern of the effluent under extreme conditions. Gunwaldsen's predictions have been generally substantiated by our hydrographic portion of this study (Boyd 2004 [1976], this volume).

The commission took under advisement not only the wide range of testimony at the hearings, but also the advices from numerous outside sources. A dominant theme in this material was the recommendation for a thorough study of the aquatic ecology of the region to accompany the construction and operation of the Connecticut Yankee plant; such counsel came from as far away as the Marine Laboratory at Plymouth, England, from the U.S. Fish and Wildlife Service, from the University of Connecticut's Marine Laboratory, and from other local organizations and individuals. As will appear from what follows, what was "impressive about the hearings in 1964 is the way the checks and bal-

[10] Pursuant to Sections 25-7d and 25-11, 1963 Supplement to the General Statutes.

ances in the system of private decision and public regulation worked to bring about a reasonable result ..." (Judd 1970).

On 21 October 1964, the Connecticut Water Resources Commission gave its authorization to Connecticut Yankee to proceed with the "Proposed Dredging and Screenwell Structure in the Connecticut River," having come to the conclusion that this work was in conformance with the pertinent sections of the 1963 Supplement to the General Statutes and not in violation of any of their provisions or regulations. Moreover, this permit was issued under specific conditions, including the following pertinent matters:

> [A] thorough and comprehensive study of the river conditions and environment undertaken under the auspices and with the support of the applicant shall continue throughout the starting-up and initial operating stages of the plant, and for a period of five (5) years after the plant is placed in full operation, or for such longer period as the Commission may hereafter find necessary. Such study shall continue under an expert and professional staff, qualified to collect relevant data and formulate sound conclusions concerning any changing conditions and their effect. The applicant shall submit to the Commission, for its approval, within ninety (90) days from the date of this permit, a detailed written outline of such study, including therein a list of the persons engaged in same, and a time schedule for the completion of specific sections thereof. The applicant shall supply to the Commission reports concerning the data, findings and conclusions of such continuing studies, at least semi-annually from the date of this permit, and at such more frequent intervals as the Commission may request.

In January 1965, the above-requested outline was submitted to the commission. The introductory material, which bears directly on some remarks contained in our preface, read as follows:

> A number of points need to be made at the outset to explain this report and underline its philosophy.
>
> 1. The breadth and scope of the proposed investigation may appear far wider than that originally envisaged by the Commissioners, Connecticut Yankee, or indeed the interested public. There is no intent to expand the study beyond the limits of knowledge that are needed to solve the basic problem; i.e., to accurately assess the effects of the plant, with particular reference to the warm-water out flow, on the Connecticut River and its environs through a "thorough and comprehensive study of the river conditions and environment." The preliminary outline given on the following pages is considered the minimum to accomplish the purpose and to produce results in which all concerned may take justifiable pride. It should also be pointed out that changes in direction or emphasis or techniques are almost a trademark of such studies, and flexibility to drop unproductive investigation and to explore promising new avenues of inquiry within the broad objective is the surest path to the desired result.
> 2. It is essential that all studies on the Connecticut River that have bearing on the problem at hand be completely coordinated. Neither competitive surveys nor useless and costly duplication of effort are desired. The success of the investigation will be commensurate with the degree of cooperation from all sources: from federal and state agencies, from utility companies, from educational and research institutions, from private sources, from within the staff conducting the various aspects of the study, etc.
> 3. The hallmarks of this investigation must be scientific competence and integrity. Nothing short of the highest standards will suffice in these regards. No reputable scientist would want a part in the investigation unless free to publish the results, whatever they may be, and to defend their interpretation as is appropriate.

With regard to item 3 above, it is worth noting in passing that we encountered a certain skepticism throughout the entire course of the study—namely, that since the Connecticut Yankee Atomic Power Company financed the investigation from its inception to its completion, the conclusions might be partial to the utility's interests. The point of course is that the company was *required* to support a "thorough and comprehensive study of the river conditions and environment" (vide ante). This was in line with the general public attitude, then and now, that the utility companies ought to bear the cost of such environmental impact studies. If this is so, however, the view that company-financed investigations are necessarily company-oriented is at odds with the desired legalities. There are obvious safeguards that can and should be taken, thus assuring impartiality and integrity. Further, the Atomic Energy Commission now requires such environmental-impact studies routinely; these are subject to rigorous review, and our semi-annual progress reports to the state have also undergone close scrutiny at the federal level.

By the spring of 1965, the Connecticut River Ecological Study was underway. A preliminary compilation of relevant published and unpublished information had been made, and a working staff was assembled. The areas of investigation under initiatory scrutiny were hydrology, benthos, planktonic fish eggs and larvae, "resident" ichthyofauna, and the juvenile and adult anadromous species with special emphasis on the American shad because of its high value both as a commercial and a sports fish. The independent Essex Marine Laboratory, although still under construction, was initially and has been throughout the course of the study the base of operations. The first meeting of the Scientific Advisory Committee was held on 23–24 April 1965, at which time its members expressed general satisfaction with the proposed investigation to the extent that the conviction was stated that "... the project is well-planned, is exceedingly well-staffed and will produce the answers to the questions that have been raised."

In accordance with the original directive from the Water Resources Commission, we have submitted 16 semi-annual reports on the progress of the study; totaling some 800 pages of public record, they provide a chronological account of the proceedings of the whole investigation. In the course of the study also, the staff has published some two dozen scientific papers both in the United States and abroad.

It is of some interest to speculate on exactly what led the Water Resources Commission to stipulate that the study should continue "... for a period of five (5) years after the plant is placed in full operation." Whatever the reason—whether 5 years just seemed a sensible round number or whether that period of time was selected because it was general knowledge that most female shad return to the river to spawn for the first time when they are 5 years old—that part of the field work dealing with the adult shad population in the spring and the migration of the juveniles to sea as late as the first week in November was carried into the sixth year after the plant officially began commercial operation. This extension allowed an analysis of the virgin component of the shad population spawned in the spring of 1968 and of the progeny resulting from the 1973 spawning (Marcy et al. 2004 [1976], this volume). Actually, Connecticut Yankee's original time schedule called for the plant to be commercially operative early in 1967, but this objective was retarded by construction delays; thus, though initial criticality was attained on 24 July 1967, the production of electricity for commercial use did not begin until the following January. From the point of view of our study, this was advantageous in that there was the better part of 3 years instead of two in which to conduct our pre-thermal-effluent investigations and collect baseline data for subsequent comparison. Each of these 3 years also happened to differ, one from another, in general meteorological character: 1965 was the last year of a drought period in the northeast United States, 1966 was one of "average" precipitation, and 1967 was unusually wet—a fortuitous combination of events with regard to our opportunities for observation. By coincidence, too, there were large fish kills on the river in the general vicinity of the Connecticut Yankee plant in 1965 and 1966; these provided the opportunity for documentation in the preoperative period (Moss et al. 2004 [1976], this volume).

So it came about that the Connecticut River Ecological Study has had certain distinctive features. It has been long-term and "before-and-after" in character, and as such, has drawn great benefit from the cohesion of its staff and the sustained catholicity of its support. Nor, as the following pages show, has it been narrow in the scope of its inquiry. And in a geographical sense, too, the study has extended its limits far beyond the local aquatic regimen as affected by the thermal discharge. Quite naturally, the main effort centered on an 8-km stretch of the Connecticut River in the vicinity of the plant, but in order to obtain pertinent information, especially on fish populations, the river has been sampled from 140 km upstream down to its mouth; furthermore, the movements of the American shad have been studied in the neighboring waters of Long Island Sound, and shad that have been tagged in the Connecticut River have been caught as far away as Albemarle Sound, North Carolina, in the south, the Bay of Fundy in the north, and off-shore south of Nantucket Island by Russian and Roumanian trawlers.

Finally, one might be led to speculate as to how or in what light this study could be viewed by the chronic skeptic. In the most general way, it has corroborated the broad predictions offered at the public hearing on 17 September 1963 (vide ante) in that the flow pattern of the thermal effluent has not seriously impeded the passage of adult or juvenile shad and that the changes in the local semi-resident fresh and brackish water populations have not been of such magnitude as to be cause for great apprehension or immediate and broad remedial action. From a narrow viewpoint, it might seem that this was a minuscule return on an investment of several million dollars—the cost of this study and associated projects. However, it is more than likely that in the absence of the study and with the impact of the ecology movement, the Connecticut Yankee Company would have been forced into the position of adding a cooling tower; this could only have been done at vastly greater expense (Cairns 1972), which would have been passed along to the consumer, and with attendant environmental stresses and esthetic problems—in short, a monumental and wholly unnecessary outlay. Furthermore, this study demonstrates with crystal clarity the fallacy involved in attempting to make universal regulations for thermal discharges from electric-generating plants without regard for their situation; as Bell (1971) has put it, "... suitable regulations should incorporate a large measure of flexibility because of the importance of local site conditions on effluent dilution and distribution."

The results of the study to be described in the following pages have gone far beyond the "inventory and survey" stage and have added materially to our knowledge of the aquatic faunal regime before and after the Connecticut Yankee plant went into action. Even though no academician would grant that we have provided a full understanding of all the ramifications of the ecosystem, the interaction of its various elements, or latent synergistic effects, we hope it may be fairly said that as environmental biologists we have produced definitive information along the lines ordered by the Water Resources Commission—namely, "to collect relevant data and formulate sound conclusions concerning any changing conditions and their effect." At the very least, the fundaments for future observation, comparison, and further elaboration as needed are now available.

Microbiological Studies

Coincident with the first year of the Connecticut River Study (1965), the Marine Research Laboratory (John S. Rankin, Jr., Director), University of Connecticut, was awarded a contract by the Federal Water Pollution Control Administration,[11] U.S. Department of the Interior. This contract (No. 14-12-177) provided support for developing fundamental data and information on the microbiology and chemistry of the river as related to the effects of the thermal discharge from the Connecticut Yankee Atomic Power Company plant. The resultant study, carried on in close rapport with our work, has been summarized by Buck (1970), Buck and Rankin (1972), and Rankin et al. (1974).

The study period ran from October 1965 through September 1969, thus providing baseline data on the populations of bacteria and phytoplankton and the chemical conditions for 27 months before the plant achieved commercial operation for comparison with a 20-month "thermal" period. The authors stress the fact that their observations were relatively short-term and in particular inadequate with regard to the evaluation of prethermal conditions where the prevailing climatological conditions were of undoubted significance. However, under the given conditions, their "... approach to the problem was sound," and since the sampling procedures and techniques were constant throughout, they are not

[11] Subsequently the "Federal Water Quality Administration."

only comparable for all periods of collection, but also for parallel studies in the future. Within the limitations of the study, Buck and Rankin (1972) conclude that "... it cannot be stated that any drastic changes have occurred at the present time in the overall microbial ecology of the Connecticut River as a direct result of the addition of thermal effluent."

The routine chemical determinations of river water and sediment samples collected on a bi- or tri-weekly schedule included salinity, conductivity, pH, dissolved oxygen, nitrate, ortho- and total phosphate, and ammonia and total nitrogen. The salinity, routinely less than 1‰, was not considered to be a significant ecological parameter in the area of the power plant. The pH ranged from 6.4 to 8.4 in the preoperative period and 6.2–7.9 after 1967; there were no uniform fluctuations nor any correlation with thermal elevation from the plant effluent. The dissolved oxygen values, however, were consistently higher at all stations following the onset of plant operation; this may be attributed to the turbulence of the discharged water passing over a large man-made weir as it leaves the plant and by an increased phytoplankton population (see also Boyd 2004 [1976]). The nitrate values ranged up to 2 ppm (average 1.0) and the total phosphate up to 1.5 ppm (average 1.0) over the entire study period with the lower values in the warmer months and the higher in the winter—a reflection of algal activity on the one hand and increased runoff on the other; the values in the thermal period tended to show a somewhat greater range, but these variations were not ascribed to plant operation. Similarly, the ammonia and total nitrogen values gave no indication of thermal effects.

It was evident from analysis of the water samples that the total number of phytoplankters increased over the whole period of study. The examination of plankton tows and phycoperiphyton (attached algae) taken in the summer of 1968 and winter of 1969 revealed that the largest number of species were diatoms except in August 1968 when there was a "bloom" of green alga *Scenedesmus quadricauda* and the blue-green alga *Microcystis aeroginosa* in a limited area of the mixing zone 0.6 km above the thermal outfall; there were no apparent comparable shifts in the plankton populations 1.6 km upstream or downstream of the outfall. Buck (1970) states that "... this study presents what may be definite effects of the thermal discharge.... This was most apparent for phycoperiphyton, where there was a shift from a diatom-dominated community at station 1" (1.6 km above outfall) "to a blue-green dominated community at station 3" (0.6 km above outfall). Buck concludes that under the existing circumstances, it is reasonable to expect that the measurable thermal effects on phytoplankton populations would not extend beyond the 0.6-km zone directly upstream from the outfall.

Field and laboratory studies were also carried out on the heterotrophic saprophytic bacteria found in the Connecticut River. Counts of the total mesophilic populations over the course of the 4-year investigation showed a random variation from 1,000 to 250,000/mL with no pronounced seasonal effects or significant differences between the plant's pre- and postoperative periods. The thermal effluent appeared to have no effect on the numbers of amylolytic, lipolytic, or proteolytic bacteria, nor on psychrophilic or thermophilic forms; similar observations involved the anaerobes and the pollution indicators (total and fecal coliforms and fecal streptococci). Laboratory studies indicated that in the immediate vicinity of the outfall from the canal, there may be some changes in relative abundance of the dominant genera; thus, the "... thermal effluent may be selecting (stimulating? supporting?) the *Achromobacter-Alcaligenes* and coryneform bacteria while suppressing the *Vibrio-Aeromonas* group and the enteric bacteria." In general, however, the chemostat studies tended to confirm the field observations that no pronounced microbiological imbalance had occurred in the presence of added heat, and such generic shifts as were noted or postulated could not be characterized as either deleterious or beneficial. "It seems likely that the effect of increased temperature alone is minimal on the general bacterial condition of the Connecticut River" (Buck 1970).

While emphasizing with due caution that the studies under this FWQA contract following the initiation of plant operation lasted only 21 months, the summary report states that "...we cannot conclude from the thermal discharge from the CYAP nuclear plant has grossly or significantly altered the general ecology of the Connecticut River as of this date...."

Impingement of Fish at the Intake

Shortly after the Connecticut Yankee plant began operation in the fall of 1967, it became apparent that substantial numbers of fish were being drawn into the intake and impinged on the traveling screens. The clarity of hindsight reveals that this might have been anticipated, but it was not. Other power plants on the river did not take any substantial number of fish, and during the two public hearings before the State Water Resources Commission, environmental concern focused upon thermal effects. Thus, when the plant was in its formative stages, no one questioned the standard intake design that was used.

At this point, it is desirable to pause for definition of terms. "Impingement" is used herein to refer to fish that enter the intake and are too large to pass through the 1-cm mesh of the traveling screens, which protect the condensers. Some authors have used "entrapment" in this same context, although logically this term might apply to any organism drawn into the intake. In this report, the term "entrainment" refers to forms of a size that pass through the screens and condensers (see Marcy 2004 [1976], this volume). It may also be noted in passing that there is little consideration in this study of the fishes that fall in the sizes between the post yolk-sac larvae that are entrained and the smaller of the individuals that are impinged. The fact is that there is no evidence that this size-category (1.5–5.0 cm) is here subject to entrainment in significant numbers.

Later, when this study to measure the effect of the plant's operation on the aquatic biota was being planned, the potential for fish impingement was overlooked and was not included as a facet of the study design. So when the problem became manifest by the presence of substantial numbers of fish in the screen washings, and in the absence of knowledge as to the extent or importance of the mortality, the matter of dealing with the problem fell to company personnel rather than to the fully committed River Study group.

Fish movement in this reach of the river exhibits great variation with the seasons. Of course, the anadromous species come and go in the expected and predictable rhythms of their life cycles, but many of the so-called resident species respond with almost equal regularity to seasonal changes, modified somewhat in phase by unusual temperature or river-flow conditions. Movement to and from winter quarters, spawning and nursery areas, and feeding grounds results in waves of fish being within the influence of the plant intake at certain times, while at other times, relatively few fish are in this area. Because of this complicated pattern of fish movement, coupled with the somewhat erratic operation of the plant during the shakedown period, an understanding of the magnitude of the impingement losses evolved only gradually. As soon as there was an awareness of a probable fish-impingement problem, a system of counting fish from the screens was initiated, state fishery authorities were notified, and various ameliorating possibilities were investigated. By the summer of 1969, it became possible to fix the probable annual impingement losses at between 125,000 and 150,000 fish.

All of the data from the counts of impinged fish indicate that no correlation exists between take of fish and such factors as tidal state or meteorological conditions. It appears, with minor exceptions, that fish are impinged in direct ratio to their presence within the zone of influence of the pump-generated currents (0.3–0.6 m/s at the intake) setting in toward the plant. The exceptions involve common carp *Cyprinus carpio*, spottail shiner *Notropis hudsonius*, suckers, golden shiner *Notemigonus crysoleucas*, and sunfishes (primarily the pumpkinseed *Lepomis gibbosus*), which do not enter the intake in numbers commensurate with their abundance in the river.

White perch *Morone americana*, catfishes, yellow perch *Perca flavescens*, and American eel *Anguilla rostrata* together comprise about 65% of the "resident" fish biomass of this reach of the river, and this group makes up the bulk of the impingement losses. For example, in the12-month period November 1968–October 1969, counts of impinged fish showed that about 12% were yellow perch, 40% were white perch, and 10% were white catfishes *Ameiurus catus*. Great northern pike *Esox lucius* are taken in small absolute numbers (about 230 per year), but there are no data to relate this to the population in the river.

Adult shad are not vulnerable to the plant, apparently because they tend to follow the navigation channel on the opposite side of the river from the intake; 123 were taken in 4 years, and these were almost all spent fish on their way back to the sea, many of which were obviously dead on arrival. Juvenile shad are taken in small numbers, about 200 per year. River herring (alewife *Alosa pseudoharengus* and blueback herring *A. aestivalis*) are seasonally abundant near the plant, and the juveniles particularly are occasionally taken in considerable numbers. In 1971, juvenile menhaden invaded this reach of the river for the only time during the nearly 9 years of the study, and an estimated 600,000 were impinged on the screens. A few rainbow smelt *Osmerus mordax* in the spring and occasional "school" striped bass *Morone saxatilis*, some sturgeon, largemouth bass *Micropterus salmoides*, and chain pickerel *Esox niger* appear in the counts each year.

Observations here provide some interesting sidelight information on the effect of intake velocity on fish impingement. Some fish came into the intake when only the service pumps were operating and intake velocity was about 2.1 cm/s—well below the avoidance capability of very small fish. This suggests that some fish will permit themselves to be carried into an intake simply because they happen into its field of influence and further, that some, at least, will not swim back through the trash racks although they could easily do so. This raises the question as to whether intake velocity alone plays a primary role in impingement losses, a theory now with many adherents.

With the passage of time and the accumulation of knowledge as to the magnitude of the problem, various exclusion devices were tested. Of these, electrode arrays composed of three banks of vertical aluminum tubing placed across the entrance of each of the four intake bays gave significant results. Specifically, this device reduced the loss of fish greater than 7.5 cm (total length) by about 75%. By accident, it was discovered that fish were diverted when the electrodes were not energized. No certain explanation of this result is possible, but it seems probable that it is a combination of factors (i.e., low-frequency sound waves generated by the aluminum tubes vibrating in their holders, visual recognition of an obstacle, and the louver effect of the current changing direction as it passes through the array). Data analysis of a careful test program revealed that energizing the electrodes gave a further reduction of about 20% in the impingement of the catfishes and eels; other species did not seem to be diverted by the charged field. It was speculated that the silt in front of the electrodes was charged and that this factor accounted for the reaction of the bottom-dwelling species.

This summary of the fish impingement problem is based largely upon tests and data analysis of a University of Rhode Island research team headed by Dr. Saul B. Saila. This group has designed and installed a new electric guiding device, which is about to undergo testing by the research team. The new design consists of horizontal electrodes, in the configuration of an equilateral triangle. The apex is opposite the center of the intake structure with the legs of the triangle terminating at the sides of the intake structure. The positive electrode is an aluminum pipe suspended from a floating oil-slick boom; the ground electrode is an aluminum pipe laid on the bottom, parallel to and slightly outboard of the positive electrode. The designers are optimistic that this device will attain its objective of excluding 95% of all fish greater than 5 cm in total length (Hyman et al. 1975).

Environmental Radiation Studies

The monitoring of the environs of the Connecticut Yankee Atomic Power Company plant at Haddam Neck for radioactive wastes has at no time been a part of the assignment of our River Study group. Our sole contribution in this critical sector has been the collection of various elements of the aquatic biota as a convenience to those responsible for the radiation studies. The brief summation that follows is included in this report to provide basic information for those who are concerned with this facet of the plant's environmental impact. By way of introduction, and in the light of the recurrent and widespread argument about the safety of present-day nuclear reactors in general, the reader is referred to the following sources: IAEA (1973), Eisenbud (1974), Hammond (1974), and Lapp (1974).

In order to ensure that radionuclides released to the region by the Connecticut Yankee Nuclear Station remained at acceptable levels, an environmental monitoring program by an independent company was started in January 1964, 4 years before the plant began commercial operation. The objective of this early program was to establish the levels of natural and man-made (fallout from weapons testing) radioactivity, as well as the seasonal variations of natural background radioactivity in the various environmental media before plant operation. The radiological monitoring program has continued to maintain a close and regular surveillance of both the terrestrial and aquatic environs; it will continue to do so for the life of the plant. Personnel from the Essex Marine Laboratory assisted in the routine collections of river water, bottom sediment, fish, and clams until December 1970. Since then, samples have been collected by employees of the Northeast Utilities Service Company. The sampling locations are at the condenser cooling discharge to the river and a few kilometers upstream and downstream of the discharge point. In the terrestrial environment, samples of well water, vegetation, soil, milk, and air are collected at stations as far as 19 km away from the plant. Gamma dose rates are recorded continuously using thermoluminescent dosimeters (TLD) at sampling locations chosen on the basis of prevalent wind directions, local topography, tidal conditions, and population distribution. The samples are analyzed for gross beta and gross gamma radioactivity; a qualitative gamma spectrum is also obtained. In addition, the milk samples are analyzed for iodine-131 and the water samples for tritium. The program is continually modified to reflect the changing state of the art. Thus, an expanded program for specific isotopic identification (e.g., strontium-90, cesium-137, etc.) in the various samples was added in August 1973. The results to date have shown no unacceptable build-up of radioactive material discharged from the plant in the various environmental media.

There have also been exacting studies of the uranium fuel cycle under the auspices of the Office of Radiation Programs, National Environmental Research Center, Cincinnati, Ohio (Kahn et al. 1970, 1971). We have been privileged to see the manuscript of the results of the radiological surveillance study dealing with the Connecticut Yankee Atomic Power Company plant, where detailed measurements and field collections were made in 1970 and 1971 (Kahn et al. 1974).

These authors have estimated the annual radionuclide releases of liquid waste from the plant for 22 isotopes. Of these, ^{131}I was at the level of 6% of the AEC limit for aqueous discharges, ^{133}I at 0.9%, ^{3}H at 0.6%, ^{135}I at 0.07% and the remaining 18 nuclides at 0.003% or lower. The magnitude of these individual radionuclide discharges as indicated has no direct health implications. Water collected at sampling localities beyond the point of release during and after the controlled discharges of test tank waste solution into the Connecticut River contained less than 3% of the annual average limit of permissible discharge levels (Title 10, Code of Federal Regulations). The concentrations of radionuclides, other than ^{3}H in the canal and river water, were generally near or below minimum detectable levels. However, bottom samples collected from the mouth of the canal and the adjacent area that were analyzed by gamma-ray spectrometry did contain some radionuclides[12] attributable to plant operation. These were associated with particulate sediment that is apparently swept down the canal at a rate too high for immediate deposition. These radionuclide concentrations in the sediment are considered to be too low to result in any detectable radiation exposure to humans.

Samples of algae collected from the log boom across the mouth of the discharge canal contained relatively high concentrations of certain radionuclides,[13] and the investigators point out that such samples are useful as radioisotopic indicators. Similarly, plankton from the discharge canal contained a number of radionuclides[14] that were not detected in a control sample taken some 7 km upstream; these were indicative of reactor waste concentration. Also, the aquatic plant *Vallisneria americana* from near the intake structure contained measurable quantities of radionuclides[15] discharged by the station. Analysis of molluscan samples from the Connecticut River Estuary disclosed no radionuclides attributable to the nuclear power station.

The following fishes were collected for analysis from the canal: white catfish, brown bullhead *Ameiurus nebulosus*, and common carp *Cyprinus carpio*. Samples of the catfishes, sunfish spp., yellow perch, and shad were also taken at a number of localities from the Connecticut River Estuary to Rocky Hill, 25 km upstream from the plant. The fish were dissected for radionuclide analysis of muscle, bone, kidney and liver, thyroid (heart and tissue from the vicinity of the anterior kidney), and alimentary tract. Kahn et al. (1974) present a table of the hypothetical radionuclide concentrations in the various organs as calculated

---

12 ^{58}Co, ^{60}Co, ^{134}Cs, ^{137}Cs, ^{40}K, ^{54}Mn, ^{226}Ra, ^{90}Sr, ^{232}Th.
13 ^{14}C, ^{58}Co, ^{60}Co, ^{134}Cs, ^{55}Fe, ^{131}I, ^{54}Mn.
14 ^{58}Co, ^{60}Co, ^{134}Cs, ^{131}I, ^{54}Mn, ^{95}Nb.
15 ^{58}Co, ^{60}Co, ^{131}I, ^{54}Mn, ^{32}P, ^{90}Sr.

from known approximate concentration factors in freshwater fish and in the discharge water, based on the assumption that isotopic equilibrium had been attained in the canal species. From these data, it would appear that ^{32}P and iodine-131, -133, and -135 provide the highest gamma ray dose rates to various critical organs of the fishes; the ^{32}P from bone is estimated to be 3.9% of the recommended limit, the total of the radioisotopes of iodine from the thyroid 1.3% and the remaining 18 radionuclides each less than 0.05%. Actual measurements of radionuclide concentrations in canal-fish muscle included ^{3}H, ^{14}C, and cesium-134 and -137; ^{90}Sr in muscle was inferred from bone analyses. The data indicate that the radiation dose (including that from fallout) from the consumption of canal-caught catfishes and carp is far less than that from eating clams and oysters not exposed to the Connecticut Yankee nuclear plant outfall.

References

Alabaster, J. S. 1963. The effect of heated effluents on fish. International Journal of Air and Water Pollution 7:541–563.

Bacon, E. M. 1906. The Connecticut River. C. P. Putnam's Sons, New York.

Barnett, P. R. O. 1972. Effects of warm water effluents from power stations on marine life. Proceedings of the Royal Society of London B 180(1061):497–509.

Bell, W. H. 1971. Thermal effluents from electrical power generation. Fisheries Research Board of Canada Technical Report 262.

Beltz, J. R., J. E. Johnson, D. L. Cohen, and F. B. Pratt. 1974. An annotated bibliography of the effects of temperature on fish with special reference to the freshwater and anadromous species of New England. University of Massachusetts, Massachusetts Agric. Exp. Station, Res. Bulletin no. 605, Amherst.

Beyers, R. J. 1974. Ecological impacts of energy production on rivers and lakes. Pages 87–117 *in* A. J. Finkel, editor. Energy, the environment, and human health. Publishing Sciences Group, Inc., Acton, Massachusetts.

Bigelow, H. B., and W. C. Schroeder. 1953. Fishes of the Gulf of Maine. Fishery Bulletin of the Fish and Wildlife Service 74(53), Washington, D.C.

Black, D. S. 1969. Keynote address. Pages 3–9 *in* P. A. Krenkel and F. L. Parker, editors. Biological aspects of thermal pollution. Vanderbilt University Press, Nashville, Tennessee.

Boyd, W. A. 2004. Hydrography. Pages 27–36 *in* P. M. Jacobson, D. A. Dixon, W. C. Leggett, B. C. Marcy, Jr., and R. R. Massengill, editors. The Connecticut River Ecological Study (1965–1973) revisited: ecology of the lower Connecticut River 1973–2003. American Fisheries Society, Monograph 9, Bethesda, Maryland. (Originally published in 1976)

Boyle, R. H. 1969. The nukes are in hot water. Sports Illustrated 30(3):24–28.

Brett, J. R. 1956. Some principles in the thermal requirements of fishes. Quarterly Review of Biology 31(2):75–87.

Brett, J. R. 1960. Thermal requirements of fish—three decades of study, 1940–1970. Pages 110–117 *in* C. M. Tarzwell, editor. Biological problems in water pollution. Trans. Second Seminar Biological Problems in Water Pollution, April 20–24, 1959, Public Health Service, U.S. Dept. Health, Education, and Welfare, Robert A. Taft Sanitary Engineering Center, Cincinnati, Ohio.

Buck, J. D. 1970. Connecticut River microbiology summary report. Contract No. 14-12-177, Federal Water Pollution Control Administration, U.S. Department of the Interior. (June 1).

Buck, J. D. , and J. S. Rankin. 1972. Thermal effects on the Connecticut River: bacteriology. Journal of the Water Pollution Control Fedation 44(1):47–64.

Buck, R. A. 1973. 40,000 salmon looking for a home. Yankee 37(4):93–101, 110–114.

Bullock, T. H. 1955. Compensation for temperature in the metabolism and activity of poikilotherms. Biological Reviews 30(30):311–342.

Bureau of Outdoor Recreation. 1968. New England heritage. The Connecticut River national recreation area study. Bureau of Outdoor Recreation, U.S. Department of the Interior, Washington, D.C.

Cairns, J., Jr. 1955. The effects of increased temperatures upon aquatic organisms. Pages 346–353 *in* Proceedings of the 10th Industustrial Waste Conference, Purdue University, West Lafayette, Indiana.

Cairns, J., Jr. 1956. Effects of heat on fish. Industrial Wastes. 1(5):180–183.

Cairns, J., Jr. 1971. Thermal pollution—a cause for concern. Journal of the Water Pollution Control Federation 43(1):55–66.

Cairns, J., Jr. 1972. Coping with heated waste water discharges from steam-electric power plants. BioScience 22(7):411–419, 423.

CERL (Central Electricity Research Laboratories). 1971. Symposium on freshwater biology and electrical power generation. Central Electricity Research Laboratories, Lab. memo. no. RD/L/M312, Leatherhead, Surrey, England.

Cheney, E. S. 1974. U.S. energy resources: limits and

future outlook. American Scientist. 62(1):14–22.

Chow, V. T., et al., editors. 1973. Water for the human environment. Thermal pollution. Proceedings of the First World Congress. Water Resources IV, International Water Resources Association.

Clark, D. 1969. Marine biology and electricity planning. Pages 4–7 *in* Symposium on marine biology. Central Electricity Research Laboratories, Lab. memo. no. RD/L/M269, Leatherhead, Surrey, England.

Clark, J. R. 1969. Thermal pollution and aquatic life. Scientific American 220(3):19–27.

Clark, J. R., and W. Brownell. 1973. Electric power plants in the coastal zone: environmental issues. American Littoral Society Special Pubication 7.

Committee on Thermal Pollution. 1967. Bibliography on thermal pollution. Journal of the Sanitary Engineering Division, Proceedings of the American Society of Civil Engineers 93(SA3):85–113.

Council on Environmental Quality. 1973. Energy and the environment: electric power. U.S. Government Printing. Office, Washington, D.C.

Coutant, C. C. 1968. Thermal pollution—biological effects. A review of the literature of 1967. Journal of the Water Pollution Control Federation 40(6):1047–1052.

Coutant, C. C. 1969. Thermal pollution—biological effects. A review of the literature of 1968. BNWL-2376, Battelle-Northwest, Richmond, Washington.

Coutant, C. C. 1970. Biological aspects of thermal pollution I. Entrainment and discharge canal effects. CRC Critical Reviews in Environmental Control 1(3):341–381.

Coutant, C. C. 1971. Thermal pollution—biological effects. Journal of the Water Pollution Control Federation 43(6):1292–1334.

Coutant, C. C. 1972a. Biological aspects of thermal pollution II. Scientific basis for water temperature standards at power plants. CRC Critical Reviews in Environmental Control 3(1):1–24.

Coutant, C. C. 1972b. Index for annual reviews 1967–1970. Thermal pollution—biological effects. Oak Ridge National Laboratory, ORNL-EIS-71-25, Oak Ridge, Tennessee.

Coutant, C. C., and C. P. Goodyear. 1972. Thermal effects. Journal of the Water Pollution Control Federation 44(6):1250–1294.

Coutant, C. C., E. E. Huber, and H. A. Pfuderer. 1972. Thermal effects on aquatic organisms—annotated bibliography of 1971 literature. Oak Ridge National Laboratory, ORNL-EIS-72-28, Oak Ridge, Tennessee.

Coutant, C. C., and H. A. Pfuderer. 1973. Thermal effects on aquatic organisms—annotated bibliography of 1972 literature. Oak Ridge National Laboratory, ORNL-EIS-73-28, Oak Ridge, Tennessee.

Dubos, R. 1973. Humanizing the earth. Science 179(4075):769–772.

Dymond, J. R. 1963. Atlantic salmon. Pages 460–498 *in* H. B. Bigelow et al., editors. Fishes of the western north Atlantic. Memoir 1, part 3. Sears Foundation for Marine Research, Yale University, New Haven, Connecticut.

Eisenbud, M. 1974. Health hazards from radioactive emissions. Pages 163–179 *in* A. J. Finkel, editor. Energy, the environment, and human health. Publishing Sciences Group, Inc., Acton, Massachusetts.

Eisenbud, M., and G. Gleason, editors. 1969. Electric power and thermal discharges. Gordon and Breach, New York.

Fry, F. E. J. 1971. The effect of environmental factors on the physiology of fish. Pages 1–98 *in* W. S. Hoar and D. J. Randall, editors. Fish physiology. VI. Academic Press, New York.

Gibbons, J. W., and R. R. Sharitz, editors. 1974. Thermal ecology. USAEC Technical Information Center, Oak Ridge, Tennessee.

Gunter, G. 1957. Temperature. Pages 159–184 *in* J. W. Hedgepeth, editor. Treatise on marine ecology and paleoecology. 1. Ecology. Geological Society of America Memoir 67.

Hammond, R. P. 1974. Nuclear power risks. American Scientist 62(2):155–160.

Hard, W. 1947. The Connecticut. Rinehart & Company, Inc., New York.

Hawes, F. B. 1970. Thermal problems: 'old hat' in Britain. Elect. World 173(14):40–42.

Hawkes, H. A. 1969. Ecological changes of applied significance induced by the discharge of heated waters. Pages 15–57 *in* F. L. Parker and P. A. Krenkel, editors. Engineering aspects of thermal pollution. Vanderbilt University Press, Nashville, Tennessee.

Hedgepeth, J. W., and J. J. Gonor. 1969. Aspects of the potential effect of thermal alteration on marine and estuarine benthos. Pages 80–118 *in* P. A. Krenkel and F. L. Parker, editors. Biological aspects of thermal pollution. Vanderbilt University Press, Nashville, Tennessee.

Herry, S. 1959. Pollution of rivers by heated discharge. Bull. Cent. belge Étud. Docum. Eaux 46:226–235.

Hill, E. 1972. The Connecticut River. Wesleyan University Press, Middletown, Connecticut.

Hyman, M. A. M., W. H. Mowbray, and S. B. Saila. 1975. The effects of two electrical barriers on the entrainment of fish at a freshwater nuclear power plant. Pages 205–237 *in* S. B. Saila, editor. Fisheries and energy production: a symposium. Lexington Books, D. C. Heath and Co., Lexington, Massachusetts.

IAEA (International Atomic Energy Agency). 1971. Environmental aspects of nuclear power stations. Proceedings Ser., International Atomic Energy Agency, Symposium 146, STI/PUB/261, Vienna, Austria.

IAEA (International Atomic Energy Agency). 1973. Nuclear power and the environment. International Atomic Energy Agency, STI/PUB/321 (Rev.), Vienna, Austria.

Idyll, C. P. 1973. The anchovy crisis. Scientific American 228(6):22–28.

Jones, J. R. E. 1964. Fish and river pollution. Butterworths, London.

Judd, P. H. 1970. A river and a nuclear power plant. Northeast Utilities Service Co. (private printing.).

Kagan, H. H., editior. 1966. Atlas of United States history. American Heritage Publishing Co., Inc., New York.

Kahn, B., R. L. Blanchard, H. L. Kreiger, H. E. Kolde, D. B. Smith, A. Martin, S. Gold, W. J. Averett, W. L. Brinck, and G. J. Karches. 1970. Radiological surveillance studies at a boiling water nuclear power reactor. U.S. Public Health Service Report, BRH/DER 70-1.

Kahn, B., et al. 1971. Radiological surveillance studies at a pressurized water nuclear power reactor. U.S. Environmental Protection Agency Report RD 71-1.

Kahn, B., R. L. Blanchard, et al. 1974. Radiological surveillance study at the Haddam Neck PWR Nuclear Power Station. U.S. Environmental Protection Agency—520/3-74-007, Cincinnati, Ohio.

Kennedy, V. S., and J. A. Mihursky. 1968. Bibliography on the effects of temperature in the aquatic environment. Pages 471–568 *in* Thermal pollution—1968 (Part 1). Hearings Subcommittee. Air and Water Pollution, Committee of Public Works., U.S. Senate, 90th Congress. U.S. Government Printing. Office, Washington, D.C.

Kinne, O. 1970a. Temperature. General introduction. Pages 321–346 *in* O. Kinne, editor. Marine ecology. I(1). Wiley-Interscience, New York.

Kinne, O. 1970b. Marine ecology. I(1). Wiley-Interscience, New York.

Krenkel, P. A., and F. L. Parker, editors. 1969. Biological aspects of thermal pollution. Vanderbilt University Press, Nashville, Tennessee.

Laberge, R. H. 1959. More thermal discharges. Water & Sewage Works 106:536–540.

Langford, T. E. 1972. A comparative assessment of thermal effects in some British and North American rivers. Pages 319–351 *in* R. T. Oglesby, C. A. Carlson, and J. A. McCann, editors. River ecology and man. Academic Press, New York.

Langford, T. E., and R. J. Aston. 1972. The ecology of some British rivers in relation to warm water discharges from power stations. Proceedings of the Royal Society of London B 180(1061):407–419.

Lapp, R. E. 1974. Nuclear salvation or nuclear folly? New York Times (February 10):12–13, 64–66, 69, 72–73.

Leggett, W. C. 2004. The American shad, with special reference to its migration and population dynamics in the Connecticut River. Pages 181–238 *in* P. M. Jacobson, D. A. Dixon, W. C. Leggett, B. C. Marcy, Jr., and R. R. Massengill, editors. The Connecticut River Ecological Study (1965–1973) revisited: ecology of the lower Connecticut River 1973–2003. American Fisheries Society, Monograph 9, Bethesda, Maryland. (Originally published in 1976)

Levin, A. A., T. J. Birch, R. E. Hillman, and G. E. Raines. 1972. Thermal discharges: ecological effects. Environmental Science & Technology 6(3):224–229.

Marcy, B. C., Jr. 2004. Planktonic fish eggs and larvae of the lower Connecticut River and the effects of the Connecticut Yankee plant, including entrainment. Pages 125–153 *in* P. M. Jacobson, D. A. Dixon, W. C. Leggett, B. C. Marcy, Jr., and R. R. Massengill, editors. The Connecticut River Ecological Study (1965–1973) revisited: ecology of the lower Connecticut River 1973–2003. American Fisheries Society, Monograph 9, Bethesda, Maryland. (Originally published in 1976)

Markowski, S. 1959. The cooling water of power stations: a new factor in the environment of marine and freshwater invertebrates. Journal of Animal Ecology 28(2):243–258.

Markowski, S. 1960. Observations on the response of some benthonic organisms to power station cooling water. Journal of Animal Ecology 29(2):349–357.

McCarry, C. 1972. Yesterday lingers along the Connecticut. National Geographic 142(3):334–369.

Meredith, D. L. 1972. Nuclear power plant siting: a handbook for the layman. Mar Advis. Service Sea Grant, Marine Bulletin 6, University of Rhode Island, Kingston.

Merriman, D. 1970. The calefaction of a river. Scientific American 222(5):42–52.

Merriman, D. 1971. Does industrial calefaction jeopardize the ecosystem of a long tidal river? Pages 507–533 *in* Environmental aspects of nuclear power stations. Proceedings Ser., International Atomic Energy Agency, Symposium 146, STI/PUB/261, Vienna, Austria.

Merriman, D. 1972. Power stations and aquatic ecology. Proceedings of the American Power Conference 34:48–52.

Merriman, D. 1973a. Calefaction of the Connecticut river, U.S.A. Transactions of the New York Academy of Science, series 11, 35(1):59–65.

Merriman, D. 1973b. Some biological aspects of aquatic thermal discharges. Pages 154–164 *in* Proceedings of the First World Congress, Water Resources IV, International Water Resources Association.

Moore, E. W. 1958. Thermal "pollution" of streams. Industrial Engineering and Chemistry 50:1–4, 87A.

Moss, D. D. 1960. A history of the Connecticut river and its fisheries. Connecticut Board of Fish and Game, Hartford.

Moss, S. A., W. C. Leggett, and W. A. Boyd. 2004. Recurrent mass mortalities of the blueback herring in the lower Connecticut River. Pages 239–246 *in* P. M. Jacobson, D. A. Dixon, W. C. Leggett, B. C. Marcy, Jr., and R. R. Massengill, editors. The Connecticut River Ecological Study (1965–1973) revisited: ecology of the lower Connecticut River 1973–2003. American Fisheries Society, Monograph 9, Bethesda, Maryland. (Originally published in 1976)

Naylor, E. 1965. Effects of heated effluents upon marine and estuarine organisms. Advances in Marine Biology 3:63–103.

Netboy, A. 1968. The Atlantic salmon: a vanishing species? Faber and Faber, London.

Raney, E. C., B. W. Menzel, and E. C. Weller. 1973. Heated effluents and effects on aquatic life with emphasis on fishes—a bibliography. Ichthyological Associates Bulletin no. 9, and AEC Technical Information Services Report no. TID-3918, Oak Ridge, Tennessee.

Rankin, J. S., J. D. Buck, and J. W. Foerster. 1974. Thermal effects on the microbiology and chemistry of the Connecticut River—a summary. Pages 350–355 *in* J. W. Gibbons and R. R. Sharitz, editors. Thermal ecology. USAEC Technical Information Center, Oak Ridge, Tennessee.

Ross, F. F. 1959. The operation of thermal power stations in relation to streams. Journal and Proceedings of the Institute of Sewage Purification 1:16–26.

Seaborg, G. T. 1972. Nuclear milestones. W. H. Freeman and Company, San Francisco.

Sears, P. B., et al. 1953. The Connecticut river: your valley, your future. Connecticut River Watershed Council, Inc., Greenfield, Massachusetts.

Spilhaus, A. F. 1973. Engineering, ecology—and use of the sea. National Fisherman Yearbook 54(13):38–43.

Sylvester, J. R. 1972. Possible effects of thermal effluents on fish. Environmental Pollution 3(3):205–215.

Trembley, F. J., editor. 1960. Research project on effects of condenser discharge water on aquatic life. Progress report, 1956–1959. Lehigh University, Inst. Res., Bethlehem, Pennsylvania.

Trembley, F. J., editor. 1961. Research project on effects of condenser discharge water on aquatic life. Progress report., 1960. Lehigh University, Inst. Res., Bethlehem, Pennsylvania.

Trembley, F. J. 1965. Effects of cooling-water discharge from steam-electric power plants on stream biota. Pages 334–346 *in* C. M. Tarzwell, editor. Biological problems in water pollution. 3rd Seminar, August 3–17, 1962. U.S. Department of Health, Education, & Welfare, Public Health Service Publication 999-WP-25.

UNESCO. 1971. Discharge of selected rivers of the world. III. Mean monthly and extreme discharges (1965–1969). Unesco, Paris.

United States Senate. 1968–1969. Thermal pollution—1968 (Parts 1–4). Hearings Subcommittee. Air and Water Pollution, Committee of Public Works, U.S. Senate, 90th Congress. U.S. Government Printing Office, Washington, D.C.

United States Senate. 1969. Environmental effects of producing electric power. Part 1. Hearings Joint Committee. Atom. Energy, 91st Congress. U.S. Government Printing Office, Washington, D.C.

Van Vliet, V. 1957. Effect of heated condenser discharge water upon aquatic life. American Society of Mechanical Engineering 57:1–10.

Warinner, J. E., and M. L. Brehmer. 1972. The effects of thermal effluents on marine organisms. Pages 381–393 *in* R. F. Ford and W. F. Hazen, editors. Readings in aquatic ecology. W. B. Saunders Co., Philadelphia.

American Fisheries Society Monograph 9:27–36, 2004

Hydrography

WILLIAM A. BOYD
Essex Marine Laboratory, Inc., Essex, Connecticut 06426, USA

The hydrographic portion of the Connecticut River Ecological Study was designed to serve two broad purposes: (1) to determine and describe any physical or chemical changes in the Connecticut River resulting from the operation of Connecticut Yankee Atomic Power Company's plant at Haddam Neck, and (2) to provide background data on river conditions for the benefit of the various biological studies that comprise the main part of this report. Both of these objectives were well served by a chain of five monitoring stations that furnished physical and chemical data on water affected and unaffected by the plant. The locations of these stations are shown in Figure 1. In addition, detailed studies were made of the distribution of temperature and dissolved oxygen within the effluent plume.

This report summarizes the results and conclusions of these studies, with appropriate references to the more detailed reports which have been published concerning each.

Operating History of Connecticut Yankee

As shown in Figure 2, Connecticut Yankee began operation in October of 1967 and performed at a nominal rate of 480 MW_e for the remainder of 1967 and 1968, except for periods of shutdown for inspection and repairs. After its first refueling in April and May, 1969, the permissible output was increased to 600 MW_e, a level that the plant succeeded in achieving with considerable regularity. Refueling was accomplished in April 1970 and 1971, and June 1972. It should be noted here that the time scale on all figures in this section of this report runs from the beginning of 1965 to the end of 1972, which is the duration of the major portion of the Connecticut River Ecological Study. This emphasizes the fact that the study had nearly 3 years in which to establish baseline conditions in the Connecticut River prior to the operation of the Connecticut Yankee plant, facilitating comparisons of preoperative and postoperative conditions.

Connecticut River Discharge and Climatological Conditions

In 1967, the Hartford office of the United States Geological Survey developed a method for determining the monthly discharge of the Connecticut and other rivers draining into Long Island Sound. As shown in Figure 3, there was a wide variation in peak mean discharge during the study period. Fortunately, the period preceding plant operation had both a low and a relatively high peak mean discharge, affording the opportunity for comparisons with similar years in the postoperative phase. Minimum discharges fell rather uniformly in the 140–200 m^3/s range. Superimposed on these discharges are tidal flows ranging up to more than 850 m^3/s (Water Resources Commission, state of Connecticut 1964). These tidal flows are difficult to describe with any accuracy due to the lack of sufficient data at the Connecticut Yankee plant site. The tidal influences are large in relation to natural flow during periods of low river discharge, and absent or nearly absent during freshet conditions. They therefore represent a seasonally compensating volume of mixing waters for the plant effluent.

The flow and temperature of the Connecticut River are obviously functions of precipitation and air temperature, respectively. Figures 4 and 5 give values for these climatological factors, obtained from the annual summaries prepared by the National Oceanic and Atmospheric Administration, 1965–1972.

Hydrographic Monitoring System

A five-station monitoring system constituted the main source of hydrographic data for the Connecticut River Ecological Study. Four stations

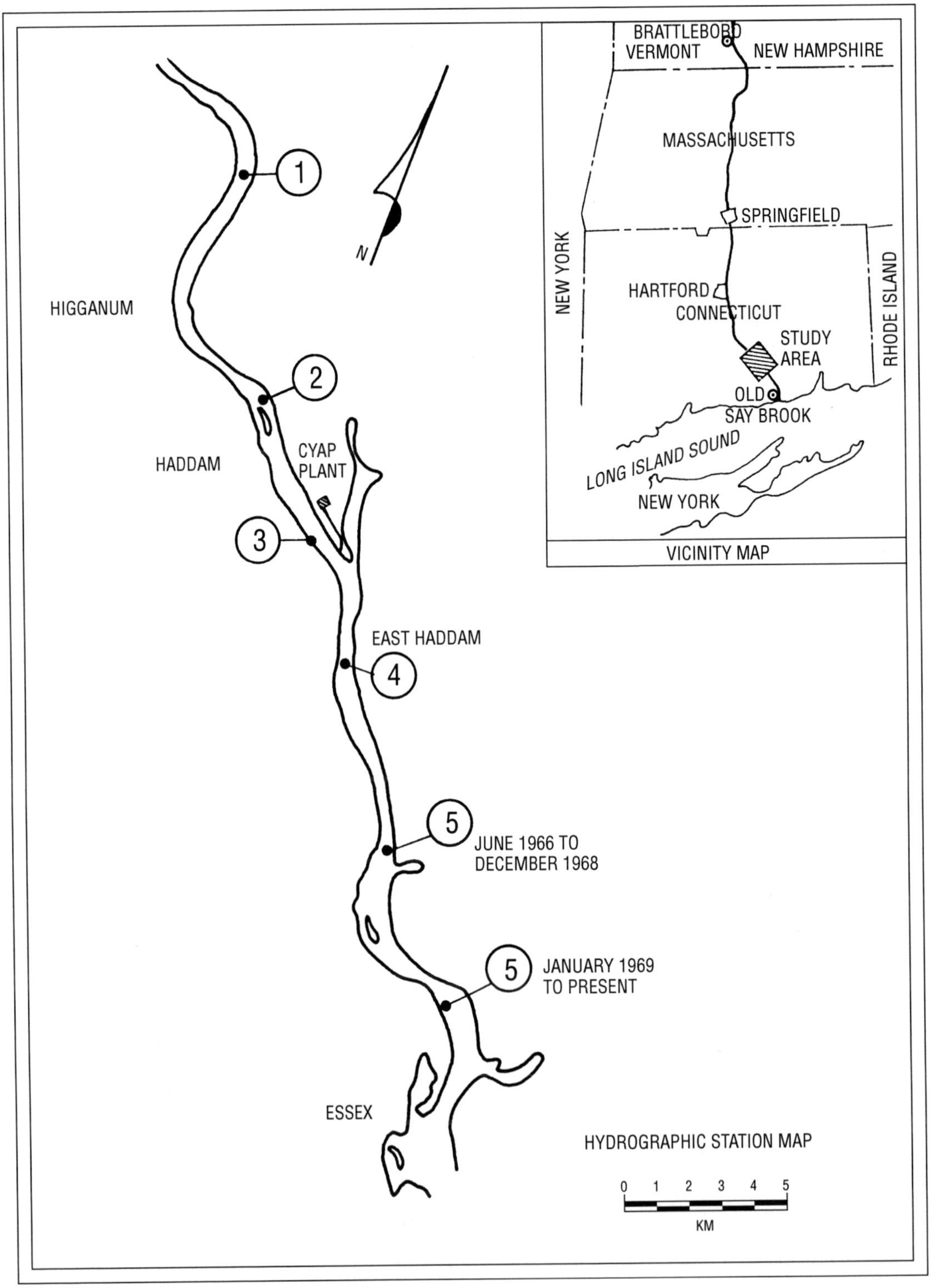

FIGURE 1. Hydrographic station map. CYAP = Connecticut Yankee Atomic Power Company.

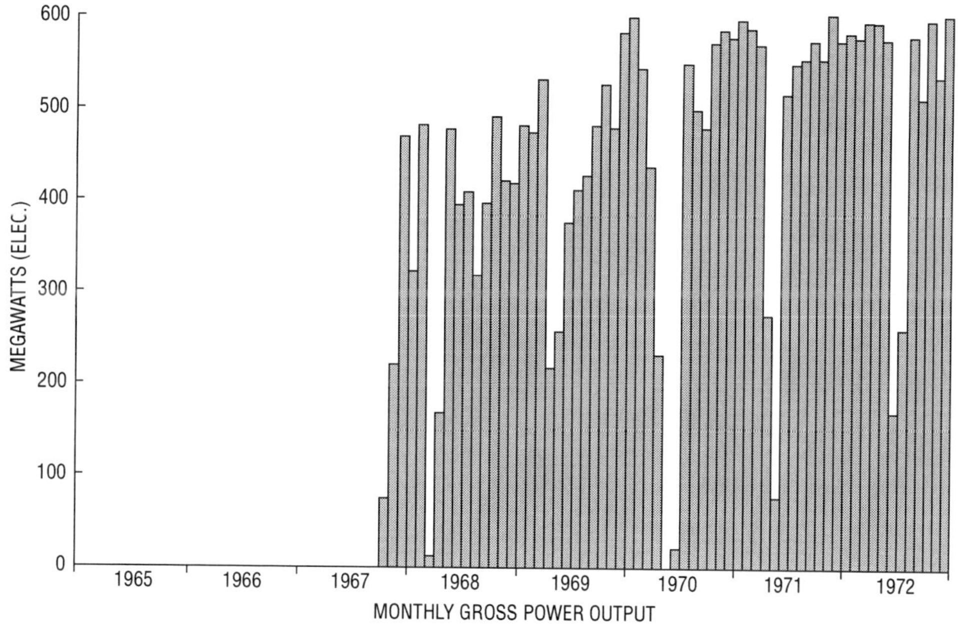

FIGURE 2. Gross power output of Connecticut Yankee. Data from Connecticut Yankee Atomic Power Company.

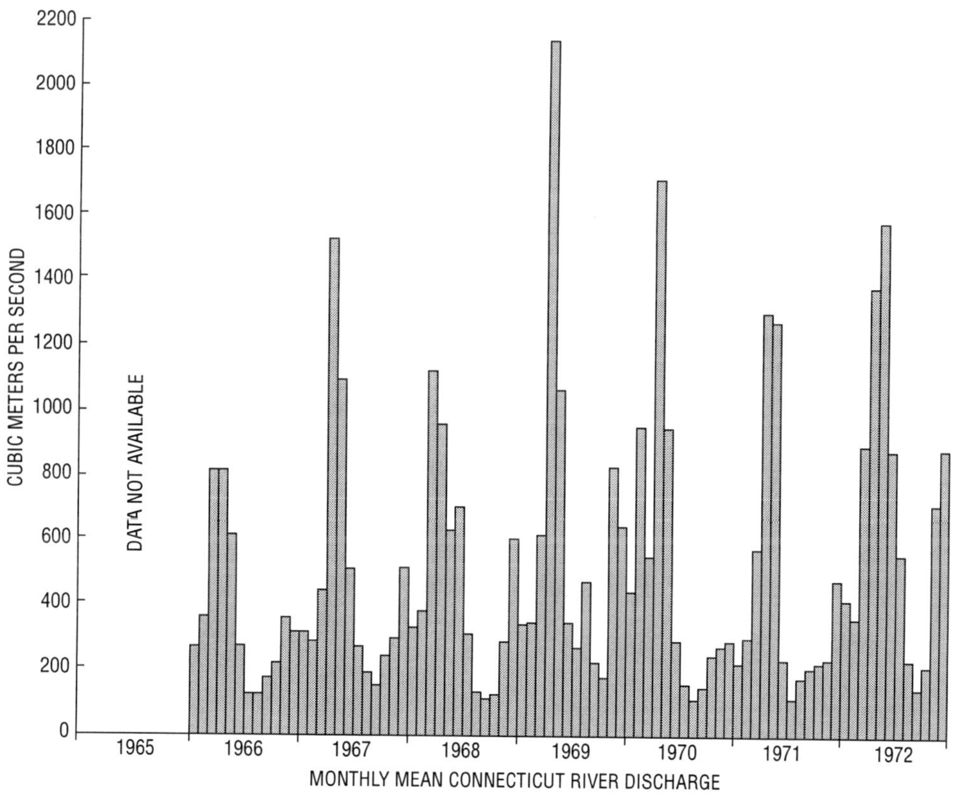

FIGURE 3. Monthly mean discharge of the Connecticut River. Source: U.S. Geological Survey.

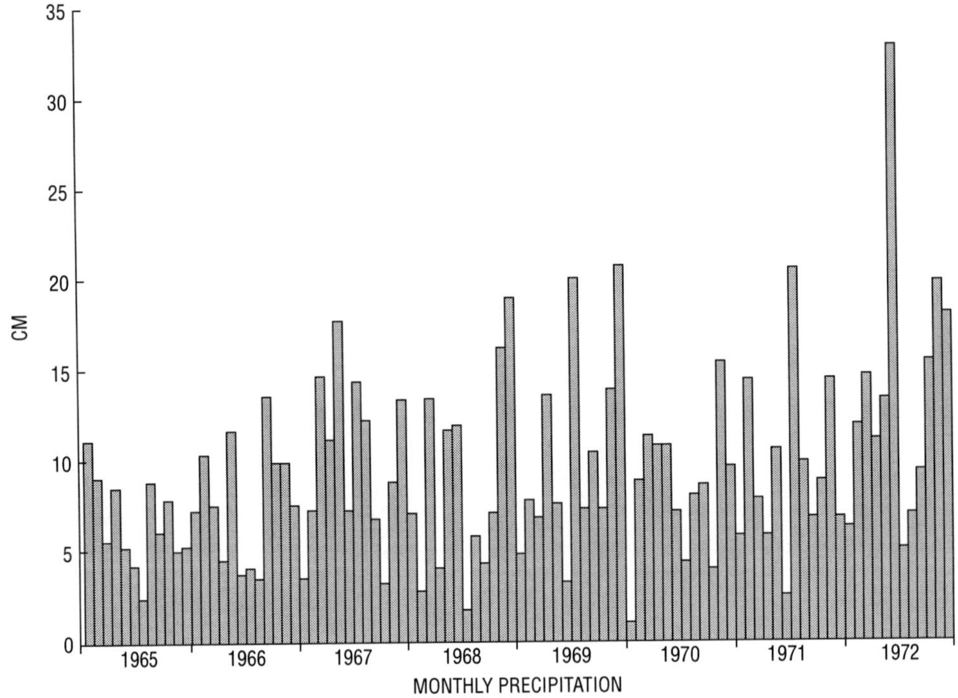

FIGURE 4. Monthly precipitation at Westbrook, Connecticut.

were established in June 1966, and the fifth station (station 1) was established on 1 July 1966. Considerable time was devoted to design and construction of the system. While today there is available a plethora of integrated environmental monitoring systems, in 1965 and 1966 there was none that was deemed suitable for a long-term study. Therefore, the system had to be designed (integrating components from several suppliers), constructed, tested, and finally installed, all without much benefit from prior work. The monitoring system has performed most satisfactorily, and its performance compares favorably with that of more modern systems. At the end of 1970, three stations were decommissioned in accordance with recommendations accepted by the Scientific Advisory Committee. One of these stations was sold to the Swedish Institute of Meteorology and Hydrology for use in a program designed to evaluate hydrographic monitoring systems for use in Swedish waters. The station was operated in friendly competition with a British-made monitoring system of 1970 vintage that cost nearly twice as much as the Essex Marine Laboratory system. In his master's thesis, Holmström (1972) reported the performance of the Essex system as superior to the British system in most respects.

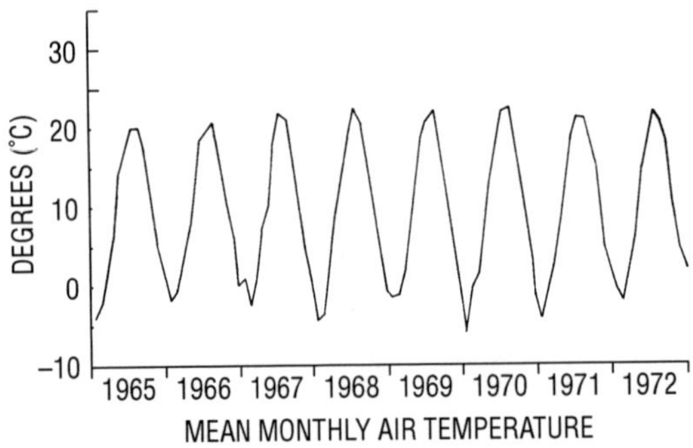

FIGURE 5. Monthly mean air temperature at Westbrook, Connecticut.

Each station in the Connecticut River had the capability of recording six channels of analog data. At stations 1, 2, 4, and 5, records were made of surface and bottom temperature, dissolved oxygen, and conductivity. In addition to these, at station 3, pH was recorded from June 1966 through December 1969. At this time, again with the concurrence of the Scientific Advisory Committee, pH monitoring was terminated, in small part due to the difficulty of maintaining the pH electrodes, but mainly because no significant changes in pH had been observed after the startup of Connecticut Yankee's plant. A detailed description of the system, its components, and its installations can be found in Boyd (1967) and in Holmström (1972).

In brief, each station consisted of a sensor for each parameter, signal conditioning circuitry to yield a 0–10-mV analog output for each channel, and a 6-channel strip chart recorder. Design specifications called for accuracies of ±0.1°C for the temperature measurements, ±1.0 mg/L for the oxygen concentrations, and ±4 mmhos/cm for the conductivity measurements. Field calibrations against portable standards indicated that the system consistently achieved these accuracies. The scan rate at all stations was 12 s per channel, yielding a total of 201,722,400 observations, which filled nearly 4 mi of strip charts.

Table 1 gives the monitoring system's operating record on a yearly basis. Of the 1,814 channel days lost over the 7-year span, 1,051, or 57.9%, were lost due to power failures. These should not necessarily be construed as outages of The Connecticut Light and Power Company, but include such mischances as lightning strikes, severance of the power cord by lawn mowers, and deliberate disconnection by persons desirous of borrowing our power. If the channel days lost due to lack of power are excluded, each year's data acquisition rate exceeds 95%, and the overall data acquisition rate for the period from 1966 through 1972 reaches 97.44%.

Analog values of daily highs, means and lows for each parameter were punched onto tabulating cards and converted to digital reports on an IBM 1130 computer. This method of data reduction has become obsolete even during the short span of 7 years; today's technology would make possible, and most desirable, a monitoring system in which the data would be recorded in digital form on computer-compatible magnetic tape.

Conductivity

Conductivity was monitored for the purpose of determining the spatial limits of the saltwater incursion from Long Island Sound. From the amassed data (1966–1972), it is evident that at periods of low flow, there were large increases in salinity at station 5, and that these occurred with considerable regularity. Such large and regular occurrences were completely absent at station 3, indicating that the normal upstream limit of significant salinity lay between these stations. This conclusion was borne out by a series of 25 surveys, utilizing portable equipment, carried out between August 1966 and November 1968, each of which measured conductivity at three depths at 17 stations located along the lower 34 km of the river. Further confirmation of these salinity distributions can be found in Meade (1966).

Salinity is clearly a variable having little direct relation to the effects of the Connecticut Yankee plant or the maximum up- and downstream extent of its thermal plume. The sole bearing of salinity on the subject of this study is that the proximity of brackish water permits the occasional presence of holeuryhaline species as well as the anadromous fishes.

Dissolved-Oxygen Concentrations

Figure 6 portrays the monthly means of daily high, mean, and low concentrations of dissolved oxygen at station 3 from June 1966 through December 1972. Oxygen sensors for all continuous-monitoring stations were located near the bottom of the river. A more detailed depiction of the dissolved oxygen data was reported by Boyd (1972a), in which the daily high, mean, and low oxygen concentrations were presented as a series of isopleths plotted against time and river location.

TABLE 1. Acquisition of hydrographic data.

	1966	1967	1968	1969	1970	1971	1972	Totals
Available channel days	3,984	5,095	4,985	5,310	5,213	2,591	2,653	29,831
Channel days acquired	3,802	4,858	4,605	4,795	4,962	2,439	2,556	28,017
Percent acquired	95.4	95.3	92.4	90.3	95.2	94.1	96.3	93.9

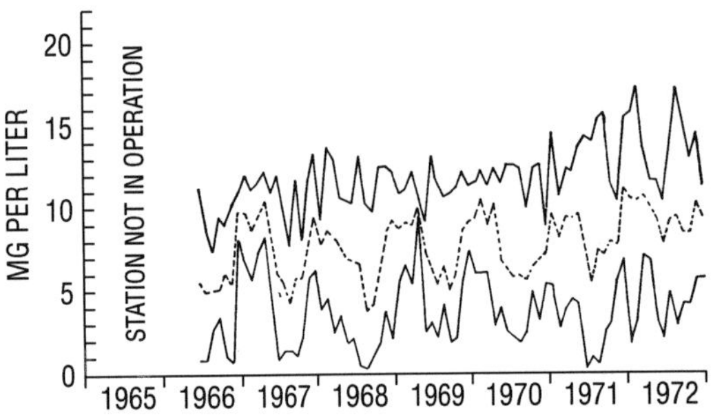

FIGURE 6. Mean monthly high, mean, and low dissolved oxygen at station 3.

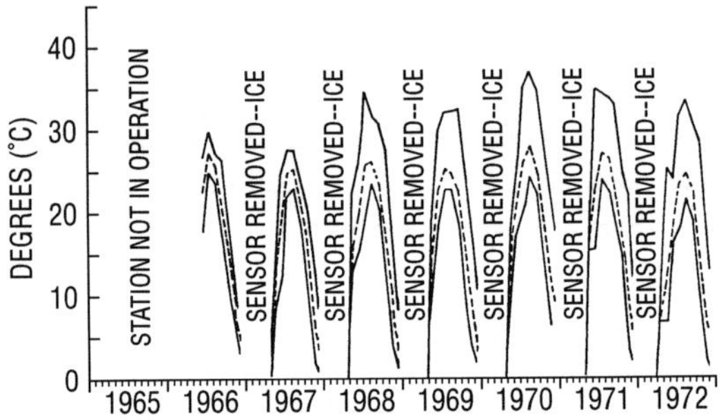

FIGURE 7. Mean monthly high, mean, and low surface temperature at station 3.

The values shown in Figure 6 indicate a small but steady increase in dissolved oxygen during the period of the study. A linear regression of the monthly means from June 1966 through December 1972 yields the following expression:

$$O_2 = 0.03226\, M + 6.56122\ ,$$

where O_2 is oxygen concentration in mg/L, and M is the number of months after June 1966.

In as much as the oxygen electrode for the continuous monitor was located near the bottom of the river, it was seldom affected by the effluent plume, which was shown by temperature data to occupy station 3 only at or near the surface. In order to determine what, if any, differences in oxygen concentration existed within the plume, a short-term experiment was conducted involving the use of an oxygen electrode mounted on a float so as to sample the oxygen concentration of the plume at such times as it was present at station 3, the presence or absence of the plume being evidenced by the surface temperature.

The results of both the continuous monitoring and the short-term studies of oxygen concentrations show no measurable effect attributable to Connecticut Yankee's operation.

River Temperatures

Figures 7 and 8 portray the monthly means of daily high, mean, and low surface temperatures at stations 3 and 5. These stations were chosen to portray the differences between locations, which were, respectively, most and not at all affected by plant operation. Effects of the heated effluent can be seen in the temperature graph of station 3, where the mean temperature was closer to the low value than to the high value. In contrast, the mean temperature at station 5 was nearly intermediate to the high and low. This difference between the relative positions of the two means resulted from the fact that the heated effluent reached the surface temperature sensor at station 3 for brief periods of time shortly after tidal reverses in the flow of the river; due to the short duration of these events, the mean temperature was relatively unaffected.

Temperature distribution for the years 1966 through 1969 in the instrumented reach of the river have been reported in Boyd (1970). The report consists of three isothermal plots for each month: surface temperature, bottom temperature, and the difference between these two. Plots of temperature differences clearly delineated the temperature elevation caused by Connecticut Yankee's operation, both in terms of areal extent and magnitude of temperature elevation. Bottom temperatures in the vicinity of the hydrographic stations appear to be essentially unaffected by

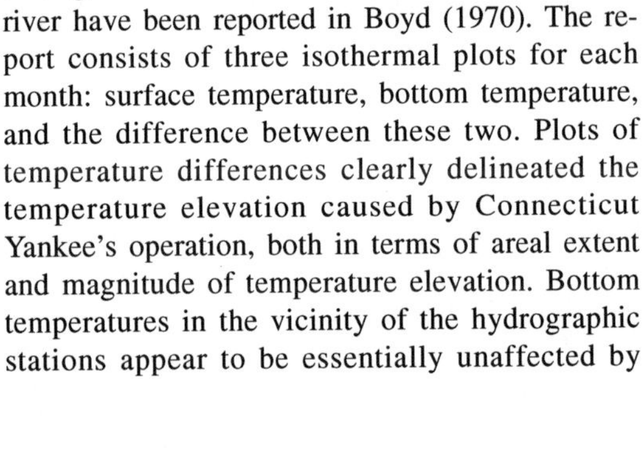

FIGURE 8. Mean monthly high, mean, and low surface temperatures at station 5.

plant operation. This is not to say that bottom temperatures are unaffected in the shallow-water area of the outfall, where observations by the benthic division have noted distinct temperature elevations. Furthermore, under conditions of high flow, the warmwater plume was not detected at any station. However, under normal or low-flow conditions, surface temperatures showed elevations of as much as 8°C at station 3, but these diminished to 1–3°C before reaching stations 2 or 4. It should be emphasized that the warmwater plume was not continually present at station 3, but that under the normal or low-flow conditions, it was usually present at least once each day.

Detailed Surveys of the Effluent Plume

Because of its importance to the biological portions of the study, the temperature distribution within the effluent plume was surveyed in considerable detail, not only by Essex Marine Laboratory, but also by the U.S. Geological Survey (USGS), the Connecticut Department of Environmental Protection (CTDEP), and Travelers Research Company (TRC).

The USGS-supported aerial surveys used advanced infrared scanning systems. The results of these surveys have not been published, but the USGS has permitted inspection of the resulting films by the author, and they support the conclusions reported below.

The CTDEP surveys were conducted from a helicopter using infra-red radiometry. Their data were processed at Essex Marine Laboratory, and they also support our conclusions (R. A. Jones, personal communication).

The TRC surveys were accomplished with thermistors, lowered at selected points on a surveyed grid. These required subsequent adjustment to achieve a near-synoptic presentation of the temperature distribution at a given time (TRC 1968).

A boat-mounted system for the simultaneous, continuous measurement of water temperature at five depths was developed at Essex Marine Laboratory and was used to conduct four sets of surveys of the effluent plume. Each set consisted of four surveys, one at high tide, one at mid-ebb tide, one at low tide, and one at mid-flood tide. These surveys were carried out during October 1969, and August, September, and November 1970. The October 1969 survey was intended primarily as a demonstration of the capability of the system, and to provide sample data with which to work out a data-reduction process. The August 1970 surveys were designed to delineate the plume under conditions of low flows and high temperatures. The September 1970 surveys were timed to coincide with the downstream migration of juvenile shad, and the November 1970 surveys were undertaken to yield information on the plume during low temperature and medium flows.

The design of the measuring system permitted speeds through the water sufficient to yield near-synoptic values for a given state of tide. A survey could usually be accomplished in approximately 1 h; for example, to determine the distribution of the heated plume at high tide, the survey would commence 1.5 h before high tide and would be concluded approximately 1.5 h after high tide.

The results of the surveys have been presented by Boyd (1972b) in the form of 80 isothermal plots representing four surveys, each at four stages of tide, and at five depths. Figures 9 and 10 are examples of these isothermal plots, at high tide on 11 November 1970 and at low tide on 23 September 1970 respectively; Figures 11 and 12 are the corresponding cross-sectional representations of each situation. The portion of Connecticut Yankee's effluent plume that retained a temperature of 2°C or more above ambient river temperature was largely found at depths of less than 1.2 m. On the surface, the effluent plume extended upriver an average distance of some 950 m and downriver some 990 m; however, the ranges of the extrema were large.

The surveys showed no temperature distributions that would constitute a thermal barrier to the passage of migratory or resident fish, assuming their ability to sense the heated plume and to swim under or around it.

References

Boyd, W. A. 1967. Hydrology report. Pages 13–22 *in* D. Merriman, editor. The Connecticut River investigation. Connecticut Water Resources Commission, 4th semiannual progress report, Hartford.

Boyd, W. A. 1970. Water temperatures, 1966–1969. Connecticut River Ecological Study. Essex Marine Laboratory, Report no. 1, Essex, Connecticut.

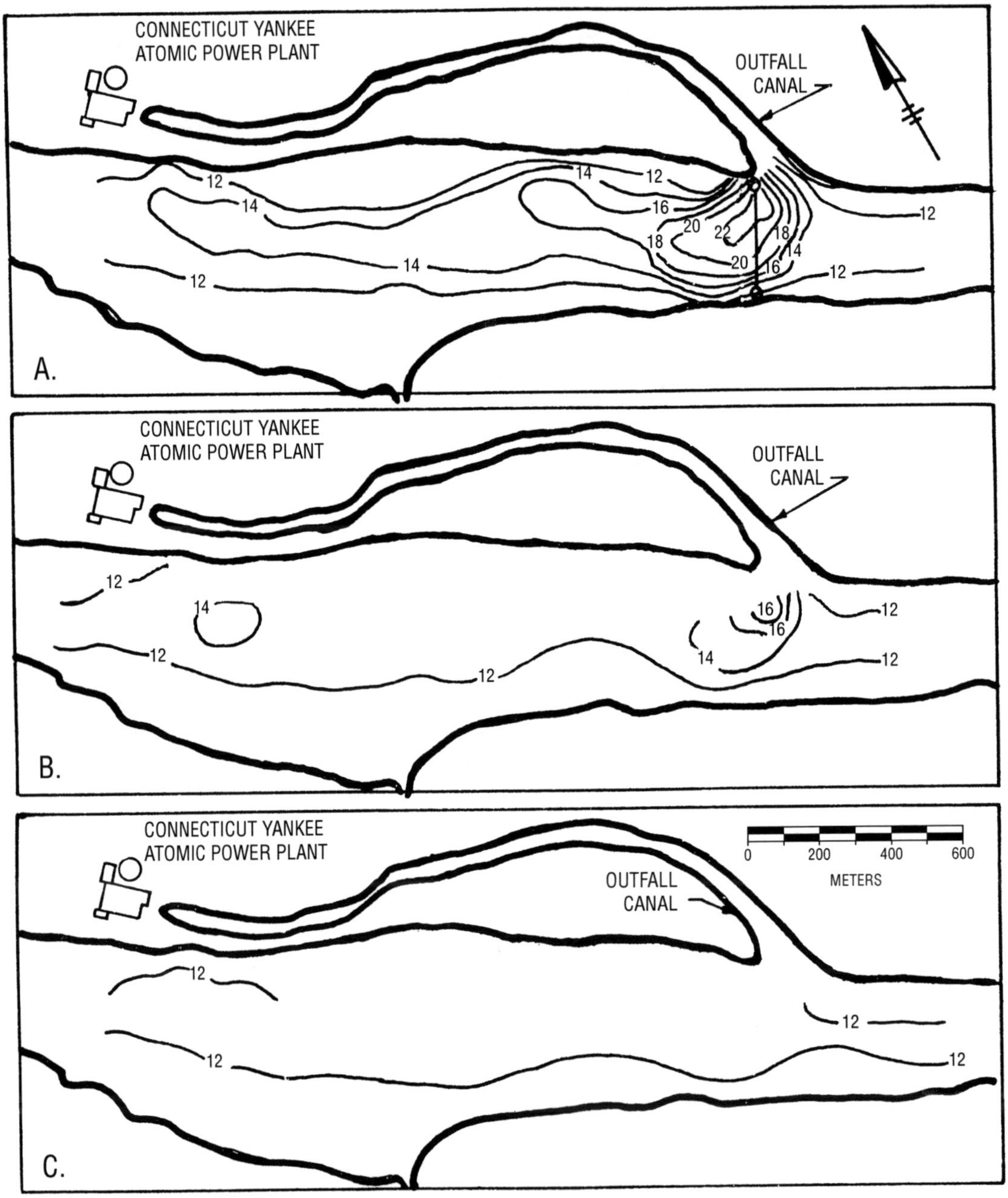

FIGURE 9. Effluent plume temperatures, high tide, 11 November 1970. A., Surface; B., 1.2 m; C., 2.5 m in depth. Isotherms in degrees C, ambient 12°C. Stippled areas less than 1.2 (B.) and 2.5 m (C.) in water depth.

Boyd, W. A. 1972a. Dissolved oxygen concentrations 1966–1970. Connecticut River Ecological Study. Essex Marine Laboratory, Report no. 3, Essex, Connecticut.

Boyd, W. A. 1972b. The extent and temperature of the effluent plume from Connecticut Yankee Atomic Power Company's generating plant at Haddam Neck, Connecticut. Connecticut River Ecological Study. Essex Marine Laboratory, Report no. 4, Essex, Connecticut.

Holmström, H. 1972. Test of two automated water quality monitors under field conditions. Notiser oct

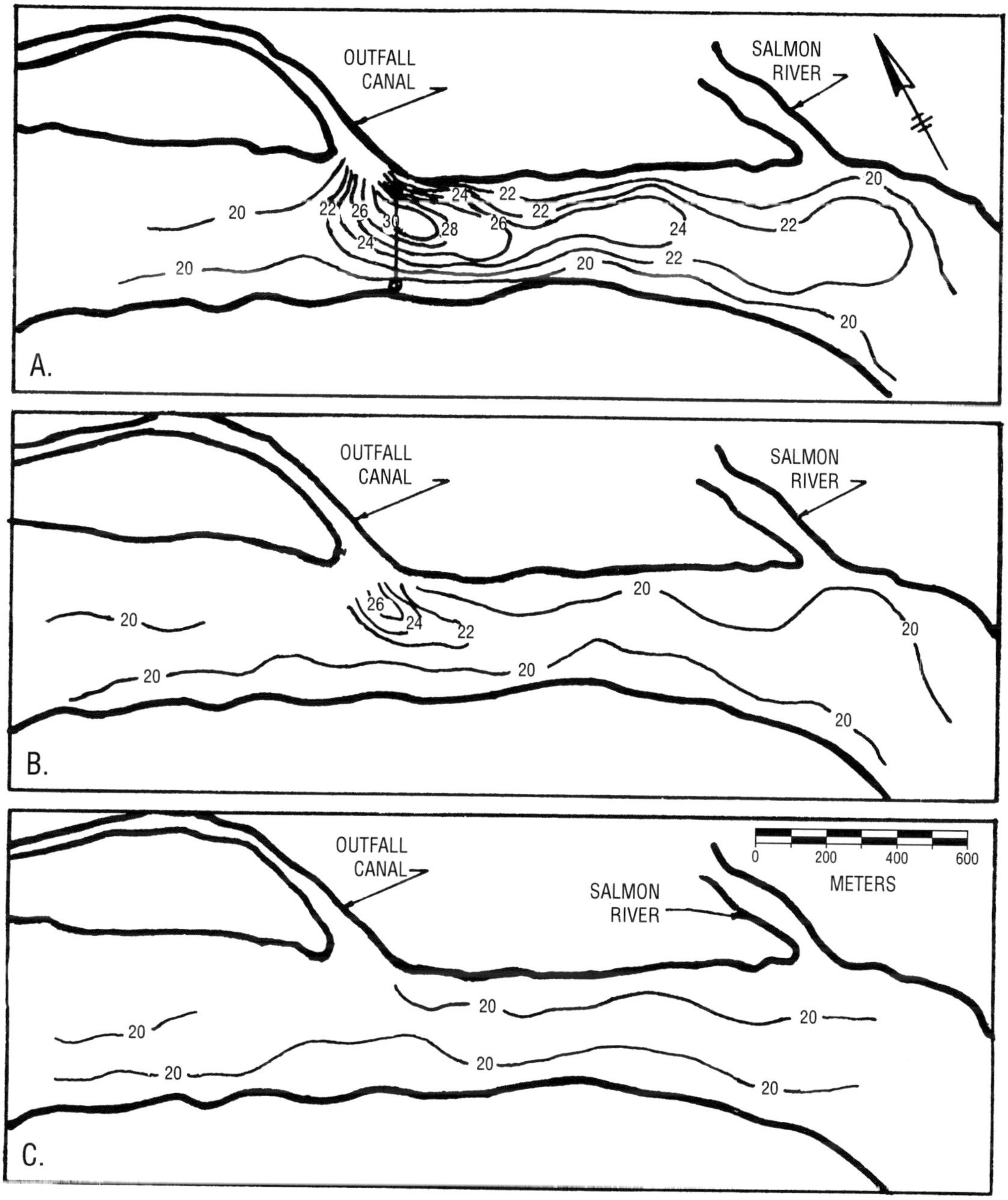

FIGURE 10. Effluent plume temperatures, low tide, 23 September 1970. A., Surface; B., 1.2 m; C., 2.5 m in depth. Isotherms in degrees C, ambient 20°C. Stippled areas less than 1.2 m (B.) and 2.5 m (C.) in water depth.

preliminara rapporter Serie HYDROLOGI. Nr 19 Sveriges Meteorlogiska oct Hydrologiska Institut.

Meade, R. H. 1966. Salinity variations in the Connecticut River. Water Resources Research 2:567–579.

TRC (Travelers Research Co.) 1968. Connecticut River temperature survey in vicinity of Haddam, Connecticut, September 9, 1968. Report to CYAP Co., prepared by G. F. Collins, TRC Service Corp., Hartford, Connecticut.

Water Resources Commission, State of Connecticut. 1964. Minutes of Public Hearing of July 20, 1964.

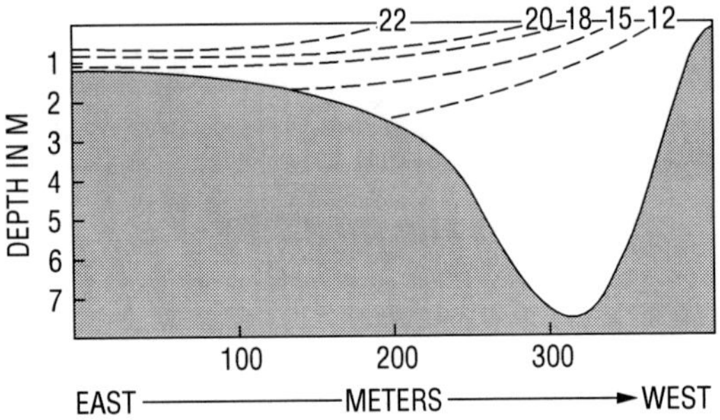

FIGURE 11. Cross-sectional representation (vertical line O——O, Figure 9A) of the temperature distribution, high tide, 11 November 1970.

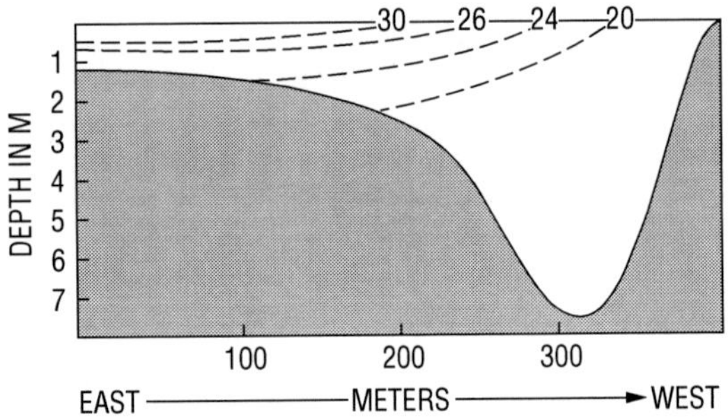

FIGURE 12. Cross-sectional representation (vertical line O——O, Figure 10A) of the temperature distribution, low tide, 23 September 1970.

American Fisheries Society Monograph 9:37–41, 2004

Oxygen Balance in the Condenser-Cooling Water System of the Connecticut Yankee Plant

PAUL M. JACOBSON
Essex Marine Laboratory, Inc., Essex, Connecticut 06426, USA

A study of oxygen balance in the condenser-cooling water system was undertaken to determine the extent and nature of any changes in dissolved oxygen concentrations that might occur in the cooling water during its passage through the plant and down the canal to its discharge back into the river. Oxygen measurements were made at selected locations throughout the system, and estimates were made of community metabolism (primary production and respiration) in the effluent canal and in the Connecticut River at the Connecticut Yankee intake structure.

Methods

Four sampling locations were selected: (A) intake; (B) at the head of the canal immediately above the rip-rap weir; (C) 84 m below the weir; and (D) 1,729 m below the weir near the end of the effluent canal.

The volume of the effluent canal between stations C and D was determined from morphometric measurements made on 20 May 1971. Permanent water level markers were placed simultaneously at stations C and D. Morphometry was determined at eight cross sections between these stations by stretching a rope across the canal and measuring the depth at 1-m intervals. Distances between cross sections were measured with an engineer's odometer. The area of each cross section was calculated as follows: Area = $(n_1/2) + n_2 + \ldots + (n_i y/2)$, where $n_1 \ldots, n_i$ were depths and y was the distance to shore from the last meter mark on the rope. The volume between cross sections was calculated by averaging the areas of each pair and multiplying the average area by the distance between the pair. Total volume was determined by adding the individual volumes (Table 1).

The retention time (min) between stations C and D was calculated by dividing flow (1,500 m^3/min) into total volume. Changes in volume due to tidal fluctuations were determined by multiplying the water level change at station C by the canal's surface area (59,610 m^2) and adding or subtracting the volumetric change. Near the end of the retention time, similar calculations were made based on the water level change at station D. A mean volume change was then found by averaging the two, and the retention time adjusted accordingly.

Three dissolved oxygen measurements were made at each station every 3 h for 24 h once each month from 14 September 1971 to 24 August 1972. A single water sample was taken at each station with a Frautschy bottle and divided into three 300-mL biological oxygen demand bottles for analysis. Dissolved oxygen was measured using the Alsterberg modification of the Winkler Method (American Public Health Association 1965). Beginning near dawn, samples were taken sequentially at stations A, B, and C. Station D was sampled after the mean retention time between stations C and D. Temperatures at stations A and B were obtained from the Connecticut Yankee hourly log sheets; a hand-held thermometer was used at stations C and D.

Rates of community metabolism (photosynthesis and respiration) were calculated following the method of Odum (1956) as modified by Copeland and Dorris (1962). The single curve method was used to estimate metabolism at the intake, while the upstream–downstream double curve method was applied to estimate metabolism in the canal.

Results and Discussion

Daily temperature, dissolved oxygen, and percent saturation for each station and sampling date are presented in Table 2. Ambient intake tempera-

TABLE 1. Morphometry of the effluent canal between stations C and D, 20 May 1971.

Station	Depth (m) Mean	Depth (m) Max.	Width (m)	Distance between stations (m)	Cross-sectional area (m^2)	Volume between stations (m^3)
1(C)	2.3	3.5	36.7		81.6	
				253		18,747
2	2.2	3.4	30.0		66.6	
				249		17,853
3	2.3	3.3	32.0		76.8	
				205		16,995
4	2.7	4.1	31.0		89.0	
				244		20,728
5	1.9	3.2	37.5		80.9	
				246		19,815
6	2.2	3.2	31.4		80.2	
				229		19,396
7	2.1	3.0	35.0		89.2	
				218		22,083
8(D)	1.6	2.2	56.4		113.4	
Total				1,644		135,617

TABLE 2. Mean daily temperature, dissolved oxygen, and percent saturation of the condenser-cooling water from intake to discharge, 14 September 1971 to 24 August 1972. Bracketed number is the mean daily megawatt output.

Station[a]	Temperature (°C) Mean	Temperature (°C) Range	Dissolved oxygen (ppm) Mean	Dissolved oxygen (ppm) Range	Mean saturation (%)	Saturation range (%)
14, 15/9 (592)						
A	21.8	21.2–22.8	4.6	4.3–5.0	51.7	48.3–56.8
B	33.4	32.8–34.4	4.6	4.4–4.9	63.6	60.3–67.8
C	33.4	32.8–34.3	5.3	5.0–5.7	73.4	68.5–78.1
D	33.3	32.8–34.2	5.3	4.9–5.6	73.2	68.0–77.4
28, 29/10 (601)						
A	14.9	14.4–16.1	6.0	5.6–6.3	58.9	55.4–64.0
B	26.9	26.7–27.8	5.9	5.7–6.2	73.3	70.4–77.2
C	26.9	26.7–27.8	6.6	6.3–6.9	81.3	77.8–85.8
D	26.7	26.4–27.5	6.5	6.1–7.0	79.9	75.3–85.4
23, 24/11 (604)						
A	6.8	6.1–7.8	8.5	8.1–9.0	69.7	66.4–73.9
B	19.0	16.1–20.0	8.4	8.1–8.9	88.9	85.0–94.7
C	19.0	16.1–20.0	8.6	8.4–8.9	90.9	88.0–95.2
D	18.5	15.8–19.3	8.5	8.2–8.8	89.9	85.5–93.6
28, 29/12 (606)						
A	2.1	1.7–2.8	12.5	12.2–12.8	91.0	88.4–92.7
B	14.3	13.9–14.4	11.9	11.5–12.4	115.0	110.6–119.2
C	14.6	13.6–15.3	11.4	11.1–11.7	111.3	108.6–115.3
D	14.2	13.2–15.4	11.2	10.8–11.5	108.2	103.8–113.7
13, 14/1 (578)						
A	2.1	1.7–2.8	11.8	11.3–12.1	85.5	81.9–87.7
B	14.1	13.3–14.4	11.2	11.0–11.5	107.7	104.7–110.6
C	14.3	13.3–15.2	10.8	10.5–11.0	104.2	101.4–106.4

TABLE 2. Continued

Station[a]	Temperature (°C) Mean	Temperature (°C) Range	Dissolved oxygen (ppm) Mean	Dissolved oxygen (ppm) Range	Mean saturation (%)	Saturation range (%)
D	13.7	12.8 14.5	10.6	10.4–11.0	101.4	98.1–106.2
24, 25/2 (601)						
A	1.0	0.6–1.1	12.0	11.5–12.6	84.9	81.3–88.7
B	14.0	13.9–14.4	11.6	11.2–12.0	111.5	107.7–115.4
C	14.2	13.4–14.9	11.0	10.6–11.4	106.2	100.0–111.8
D	14.0	12.7–15.6	10.9	10.6–11.4	104.7	100.0–112.0
23, 24/3 (535)						
A	3.0	2.8–3.3	12.4	12.1–12.7	92.2	90.4–94.1
B	15.7	15.6–16.1	11.9	11.6–12.3	118.8	117.0–123.0
C	15.7	15.2–16.1	11.4	11.0–12.0	114.3	110.0–120.0
D	15.7	14.8–16.3	11.1	10.9–11.5	110.5	107.8–113.0
19, 20/4 (600)						
A	5.9	5.6–6.1	11.8	11.7–11.9	94.5	93.6–95.2
B	17.6	17.2–17.8	11.3	11.0–11.4	117.9	113.4–121.0
C	17.4	16.7–18.0	11.1	10.6–11.4	114.9	109.3–120.0
D	16.4	15.4–17.4	10.8	10.1–11.2	109.6	105.5–113.9
25, 26/5 (588)						
A	15.3	14.4–16.1	7.6	7.4–7.8	75.6	73.1–78.0
B	27.2	26.1–28.3	7.5	7.1–7.7	93.2	88.3–95.6
C	27.7	23.5–29.8	7.5	7.3–7.7	93.3	85.0–98.7
D	26.7	23.8–28.6	7.3	6.7–7.7	90.6	84.7–93.9
29, 30/6 (0)[b]						
A	17.2	16.1–17.8	7.4	7.2–7.8	76.9	75.0–80.4
B	17.2	16.1–17.8	7.6	7.3–7.9	78.8	76.3–81.4
C	17.3	16.2–17.8	8.0	7.8–8.2	83.0	80.9–85.3
D	17.5	16.6–18.0	8.0	7.6–8.3	83.0	80.0–84.2
27, 28/7 (581)						
A	23.9	23.3–24.4	8.3	7.2–9.1	96.9	84.7–111.8
B	35.7	35.0–36.1	7.8	6.7–8.4	110.7	95.1–121.4
C	36.4	35.2–37.0	7.4	6.6–8.1	106.0	94.3–115.7
D	36.0	35.1–36.6	7.4	6.8–8.0	105.8	95.8–114.3
24, 25/8 (581)						
A	23.4	22.8–24.4	6.3	5.6–6.8	73.8	64.4–78.2
B	35.4	35.0–36.1	6.2	5.7–6.6	87.6	80.3–92.9
C	35.7	34.9–36.6	6.2	6.0–6.5	88.9	84.5–92.8
D	35.5	34.7–36.6	6.1	5.7–6.4	87.0	81.4–91.4

[a] Station A is plant intake; B is immediately above weir; C is 83.8 m below weir; D is 1,728.8 m below weir (near end of canal).

[b] Power plant shutdown during refueling, intake pumps operating.

tures ranged between 0.6°C in February and 24.4°C in July and August. The maximum observed ΔT through the plant was 13.0°C. Temperatures in the discharge canal reached a maximum of 37.0°C.

Dissolved oxygen concentrations measured at the intake ranged between 4.3 ppm in September and 12.8 ppm in December. Oxygen concentrations were found to decrease throughout the cooling system when intake waters became supersaturated upon the addition of heat. These losses ranged between 0.9 and 1.3 ppm and usually occurred at ambient water temperatures below 6°C and at ambient dissolved oxygen concentrations above 11.8 ppm. Similar losses also occurred on 27 July when the mean ambient temperature was 23.9°C and the mean dissolved oxygen concentration at the intake was 8.3 ppm.

Dissolved oxygen levels remained virtually unchanged after passage through the condenser when percent oxygen saturations in the heated water were below 100%. Increases in dissolved oxygen (0.4 to 0.7 ppm) were noted below the weir (station C) when percent saturations were near or below 83.0 in the heated water.

The influence of condenser passage on dissolved oxygen concentrations has been studied at other power plants with varying results. Adams (1969), studying two tidewater power plants in California, noted supersaturation of oxygen in the discharge waters but found no changes in dissolved oxygen concentrations. Trembley (1960), investigating the Martins Creek plant on the Delaware River reported losses of dissolved oxygen up to 4.5 ppm across the condenser; turbulent mixing at the discharge structure reduced the losses in the effluent canal. Markowski (1959) listed dissolved oxygen levels and percent saturations in discharge and intake waters of six power plants in the United Kingdom; these results showed that intake waters near or slightly below saturation became supersaturated upon the addition of heat. Changes in the actual dissolved oxygen levels ranged from +0.4 to –0.15. At Connecticut Yankee, the gain or loss of oxygen in the system was directly related to supersaturation.

Estimates of community metabolism in the effluent canal and in the Connecticut River at the Connecticut Yankee intake are shown in Table 4. Both production and respiration were usually higher in the canal than in the river. The percent increase across the plant followed a seasonal pattern common to the results of other similar studies, the largest increases occurring at seasonally lower water temperatures (Warinner and Brehmer 1966; Morgan and Stross 1969; Brooks et al. 1970; Hamilton et al. 1970; Smith and Brooks 1971). Respiration, unlike production, was lower in the canal at high summer water temperatures.

The measured dissolved oxygen levels and calculated production/respiration (P/R) ratios indicate that metabolism in the canal has little influence on dissolved oxygen levels in the heated waters discharged to the river. The P/R ratios also show that except for July, there was no net deposition of organics from the canal to the river.

References

Adams, J. R. 1969. Thermal power, aquatic life, and kilowatts on the Pacific Coast. Nuclear News (September):75–79.

American Public Health Association. 1965. Standard

TABLE 4. Estimates of community metabolism at the intake and the heated effluent canal, Connecticut Yankee atomic power plant—14 September 1971 to 24 August 1972.

Date	Production ($gO_2/m^2/d$)			Respiration ($gO_2/m^2/d$)			Prod./resp. ratio		Mean discharge temp. (°C)	Mean ΔT (°C)
	Canal	Intake	% change	Canal	Intake	% change	Canal	Intake		
14–15 Sept.	3.1	4.6	–33	22.5	27.4	–18	0.14	0.17	33.3	11.6
28–29 Oct.	2.4	1.7	+41	19.3	17.9	+8	0.12	0.09	26.7	12.0
23–24 Nov.	3.9	0.9	+333	9.4	6.4	+47	0.41	0.14	18.5	12.2
28–29 Dec.	1.1	1.9	–42	4.6	2.7	+70	0.24	0.70	14.2	12.2
13–14 Jan.	1.9	0.8	+137	4.0	1.4	+186	0.43	0.57	13.7	12.0
24–25 Feb.	1.3	0.4	+225	4.3	3.4	+26	0.30	0.12	14.0	13.0
23–24 Mar.	2.0	1.2	+67	10.6	3.3	+221	0.19	0.36	15.7	12.7
19–20 Apr.	4.9	1.4	+250	6.3	5.1	+23	0.78	0.27	16.4	12.7
25–26 May	3.8	1.1	+245	13.4	4.0	+235	0.28	0.27	26.7	11.9
29–30 June	4.3	2.2	+95	10.9	9.4	+16	0.39	0.23	17.5	0
27–28 July	3.2	3.0	+7	1.6	8.3	–81	2.0	0.36	36.0	11.8
24–25 Aug.	2.4	1.2	+100	12.7	20.2	–37	0.18	0.06	35.5	12.0

methods for the examination of water, sewage, and industrial wastes. American Public Health Association Inc., New York.

Brooks, A. S., G. S. Brush, R. M. Davies, R. A. Smith, and B. K. Sullivan. 1970. Plankton studies. Pages 138–168 *in* L. D. Jensen. Interim report, Site 12, James River Chesterfield Station, Virginia Electric and Power Company. Johns Hopkins University Cooling Water Discharge Project, Department of Geography and Environmental Engineering, Baltimore, Maryland.

Copeland, B. J., and T. C. Dorris. 1962. Photosynthetic productivity in oil refinery effluent holding ponds. Journal of the Water Pollution Control Federation 34:1104–1111.

Hamilton, D. H., Jr., D. A. Flemer, C. W. Keefe, and J. A. Mihursky. 1970. Power plants: effects of chlorination on estuarine primary production. Science 169:197–198.

Markowski, S. 1959. The cooling water of power stations, a new factor in the environment of marine and fresh water invertebrates. Journal of Animal Ecology 28(2):243–258.

Morgan, R. P., and R. G. Stross. 1969. Destruction of phytoplankton in the cooling water supply of a steam electric station. Chesapeake Science 10:165–171.

Odum, H. T. 1956. Primary productivity of flowing water. Limnology and Oceanography 1:102–107.

Smith, R. A., and A. S. Brooks. 1971. Primary production. Pages 32–37 *in* R. W. Koss. Interim report, environmental responses to thermal discharges from Marshall Steam Station, Lake Norman, North Carolina. Johns Hopkins University Cooling Water Discharge Project, Department of Geography and Environmental Engineering, Baltimore, Maryland.

Trembley, F. J., editor. 1960. Research project on effects of condenser discharge water on aquatic life. Lehigh University, Inst. Res., Progress report 1956–1959, Bethlehem, Pennsylvania.

Warinner, J. E., and M. L. Brehmer. 1966. The effects of thermal effluents on marine organisms. International Journal of Air and Water Pollution 10:277–289.

American Fisheries Society Monograph 9:43–58, 2004

Benthic Fauna: 1965–1967 versus 1968–1972

RONALD R. MASSENGILL
Essex Marine Laboratory, Inc., Essex, Connecticut 06426, USA

In order to assess the changes in benthic invertebrate populations in the vicinity of the Connecticut Yankee plant from 1965 to 1972, we made quantitative studies of species composition, frequency of occurrence of species, and abundance and biomass of individual species. Comparisons of the benthic populations (1) in the river before (1965–1967) and after (1968–1972) the plant went into operation, (2) in the discharge canal, and (3) at stations unaffected by the thermal plume, have made it possible to determine the effects of changes in temperature and sediments on abundance and species composition in the various biotopes of the river bottom.

More than 4,700 samples have been collected since the field work began in June 1965 (Table 1). The plant began to operate at low power levels in October 1967, and except for refueling periods and shutdowns for inspection and repairs, it maintained near peak roads of 590 MW_e from January 1968 to December 1972 (see Figure 5 in Boyd 2004 [1976], this volume).

Methods

It was important at the outset to select sampling locations that would provide the data necessary to assess the changes in the biota that might result several years later when the plant was in operation. To this end, we were fortunate in having the predicted configuration of the thermal plume as developed by Stone and Webster (1964). On this basis, we chose station sites that should represent areas of maximum temperature and flow increase, areas of intermediate increase, and control areas outside the influence of the plume; to this pattern, we added supplementary sampling points to insure thorough coverage (Figure 1). Emphasis was placed on sampling the area that was most likely to be affected—the stretch of the river immediately downstream from the discharge at the mouth of the canal and the sandy shallow flat upstream off the east bank. Samples were taken at the more critical stations every 2 weeks from January 1966 through December 1970, except during January and February when ice restricted work to stations within the heated area adjacent to the mouth of the discharge canal. In 1971 and 1972, stations 2 and 6–9 were eliminated (vide infra). In the course of the study, the area in direct contact with the plume (stations 4, 5, and 10) received 39% of the sampling effort; an equivalent effort (33%) was devoted to adjacent and control stations 3, 11, 12, 13, and 16. The canal, which represents a special situation, received 18% of the total sampling effort.

Bottom dwelling specimens were collected with a modified van Veen grab that sampled an area of 0.1 m^2. The depth of penetration varied with sediment composition and compaction of the substratum. Generally the grab sampled to a depth of 10 cm in sand, 5 cm in gravel, and 10 cm in silt. The grab was modified by installing a 0.5-mm screen mesh in the doors, which permitted the grab to approach the bottom without disturbing the substratum. The grab was lowered with an electric winch at a rate of 4 m/min. Upon contact of the grab with the bottom, a spring released the two halves of the bucket. As the grab was raised it closed. Up to 20 L of sediment were brought to the surface from each grab. The water from the sediment was passed through a 0.5-mm sieve, and the animals caught on the sieve were returned to the sample and transported to the laboratory in 20-L containers.

In the laboratory, organisms in each sample were separated from the sediment by washing on a 1.0-mm sieve. They were then preserved in 70% isopropyl alcohol for future quantitative study. When possible, all animals were identified to species, counted, weighed wet, and weighed after drying at 90°C for 24–48 h. Decalcification of mollusks with N HCl preceded drying.

Sediment samples were dried at 90°C, and

TABLE 1. Numbers of benthic samples, by station. Asterisk (*) indicates visual observation on rocky substrate. Dash (–) indicates no sampling.

Station	1965	1966	1967	1968	1969	1970	1971	1972	Total
1	16	12	28	–	–	–	–	–	56
2	14	24	28	12	32	6	–	–	116
3	14	30	29	37	80	32	34	40	296
4	27	66	90	178	80	75	71	84	671
5	22	52	51	50	78	71	72	83	479
6	30	28	28	24	12	12	–	–	134
7	*	*	*	*	*	*	–	–	–
8	30	28	28	24	12	12	–	–	134
9	*	*	*	*	*	*	–	–	–
10	21	54	43	87	78	83	72	84	522
11	22	48	38	44	76	71	71	84	454
12	18	35	33	34	34	36	36	40	266
13	12	34	29	30	34	36	34	40	249
16	15	18	17	22	43	48	60	74	297
17	–	–	–	–	18	42	60	76	196
Canal	–	–	70	160	168	212	112	140	862
Total	241	429	512	702	745	736	622	745	4732

100 g were removed and placed in Tyler sieves (2.0–0.044 mm) for 10 min of mechanical shaking, and the resulting fractions were weighed.

Temperatures were obtained with a temperature-oxygen meter at all stations. Vertical temperature profiles were established from measurements taken at 15-cm intervals. Additional temperature data were provided by the hydrographic section of the study and by the Northeast Utilities Service Company. Water velocity was measured at various depths with a Gurley meter.

Biotopes of the Study Area

Four major bottom types were sampled: gravel, rock, sand, and silt. More than 550,000 organisms involving 72 species were collected from these biotopes from 1965 through 1972 (Table 2), the single most abundant animal living on or in the substratum being the annelid *Limnodrilus hoffmeisteri*.

Gravel (Channel) Biotope

The channel bottom in the Haddam reach is predominantly gravel and constitutes approximately half of the bottom area of the 170 ha in the whole study area. In this reach, the channel is adjacent to the west shore, and its bottom was never directly affected by the thermal plume. The highest flow velocities were located in the channel portion of the river. At 5 cm above the substratum, flow velocities of 1.1 cm/s were measured, while in the shoal and rocky and silty areas, at the same time, the velocities were 0.5 cm/s. Sampling on the channel stations commenced in June in 1965 and continued through 1970 on a quarterly schedule, a total of 134 samples having been collected from each of stations 6 and 8 in this period. Exploratory and scheduled sampling demonstrated that invertebrates were consistently sparse in the channel area, and aside from occasional animals in drift or solitary animals in the substratum, no seasonal patterns or channel invertebrate communities were found. Sampling was, therefore, discontinued after 1970.

Rock Biotope

Rocky areas having a total of 7.3 ha were sampled at Rock Landing, 5.2 km upstream of the discharge point, and Beaver Meadow Creek, 1,150 m above the discharge point on the west shore. These areas are characterized by vertical ledges from the surface to 3 m and boulders and cobble stones near the bottom at 4–5 m. Sand and silt were found between the crevasses. Algae were commonly found on the rock substratum. Unionids (*Ellipitio*, *Anodonta*, and *Lampsilis*) were collected from silt deposits on the walls.

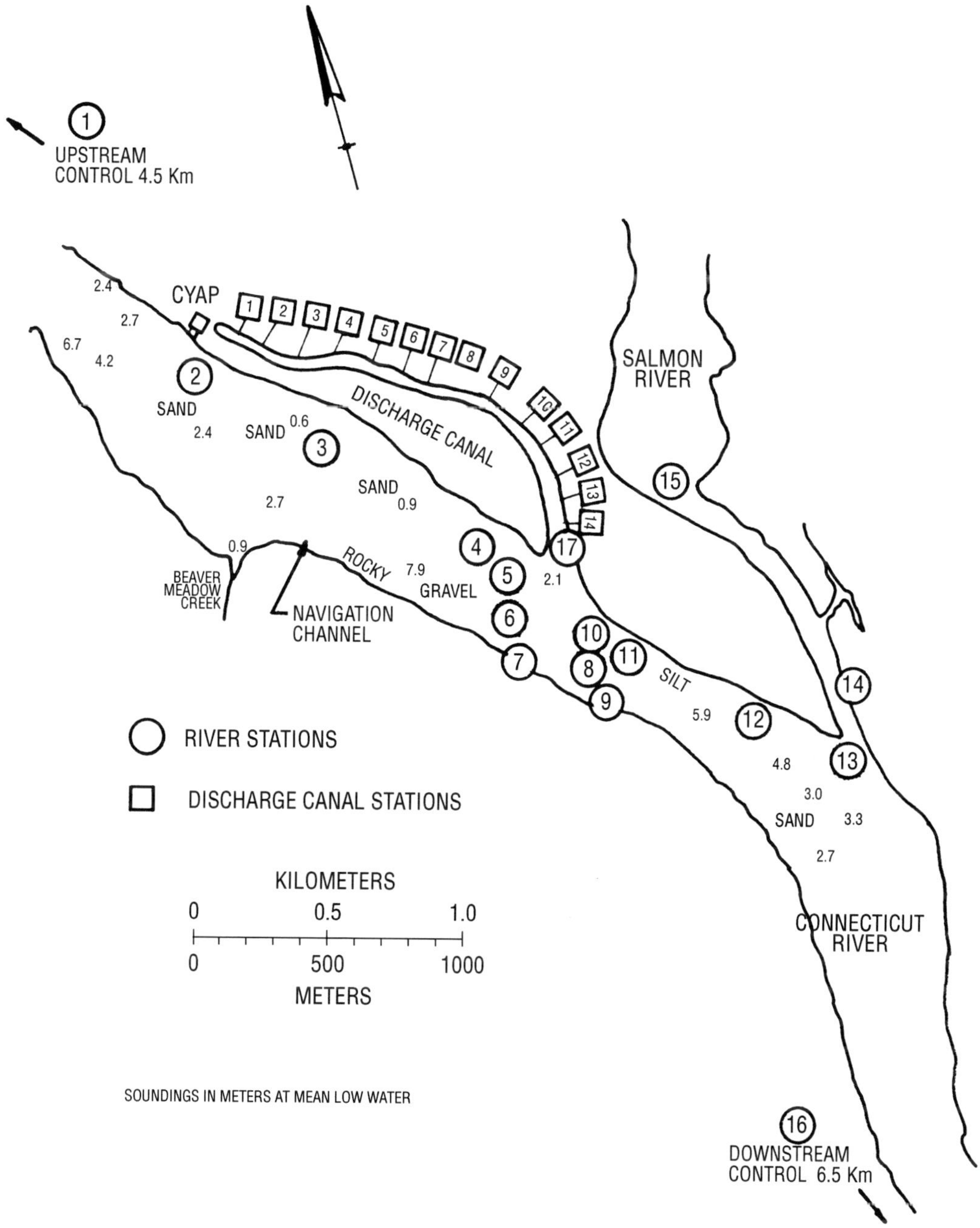

FIGURE 1. Location of river benthic stations 1–17, encircled. CYAP = Connecticut Yankee Atomic Power Company

Tube-building insects were found attached to the rock substratum. No quantitative studies were made of the rocky reaches. Stations 7 and 9, across the river from, and downstream of, the discharge point, were sampled quarterly by SCUBA diving from 1965 through 1970.

Sand Biotope

The sandy habitat stations 3–5 comprised approximately 35% of the study area (60 of the 170 ha). This substratum type is perhaps the most complex of the four biotopes discussed. During the spring, the finer sediments are washed away leav-

TABLE 2. Organisms collected from channel, sand, silt, and rock 1965–1972.

Taxon	Channel	Sand	Silt	Rock
Coelenterata				
Hydrozoa	X	X	X	X
Platyhelminthes				
Turbellaria	X	X	X	X
Nemathelminthes				
Nematoda	X	X	X	X
Bryozoa		X		
Annelida				
Polychaeta				
Scolecolepides viridis	X	X	X	X
Oligochaeta				
Limnodrilus hoffmeisteri	X	X	X	X
Hirudinea				
Helobdella ornata		X	X	X
Helobdella stagnailis		X	X	
Haemopis plumbeus		X	X	
Mollusca				
Gastropoda				
Ferrissia sp.		X		X
Lymnaea sp.		X		X
Physa hetherostropha		X	X	
Gyraulus sp.		X	X	
Pelecypoda				
Pisidium adamsi		X		
Pisidium aequilaterale	X	X		
Pisidium conventus		X		
Pisidium compressum		X		
Pisidium nitium		X		
Pisidium punctiferum		X		
Pisidium sp.		X		
Sphaerium corneum		X		
Sphaerium lacustre		X		
Sphaerium occidentale		X		X
Sphaerium sulcatum		X		X
Sphaerium transversum		X		
Sphaerium spp.		X	X	X
Anodonta sp.	X	X		X
Elliptio complantatus	X	X		X
Lampsilis radiata	X	X		X
Lampsilis cariosa	X	X		X
Arthropoda				
Crustacea				
Ilyocryptus acutifrons		X		
Asselus sp.	X	X		
Cyathura polita		X		
Gammarus fasciatus	X	X	X	X
Arachnoidea		X	X	
Insecta				
Ephemeroptera				
Baetidae		X		X
Odonata				
Anisoptera		X		
Zygoptera		X		

TABLE 2. Continued

Taxon	Channel	Sand	Silt	Rock
Limnocharidae				
Macromia sp. (A)	X	X	X	
Macromia sp. (B)		X	X	X
Gomphus sp.		X	X	
Coenagrionidae		X	X	X
Plecoptera			X	
Hemiptera		X	X	
Megaloptera		X	X	
Neuroptera		X	X	
Trichoptera		X		
Hydropsychidae		X		
Leptoceridae		X		
Ishnura sp.		X	X	X
Oecetis sp.		X	X	X
Lepidoptera		X		
Hymenoptera		X		
Coleoptera		X		
Emilidae				
Berosus sp.	X	X		
Diptera				
Tipulidae		X		
Culicidae		X		
Ceratopogonidae		X		
Psychodidae		X		
Simuliidae		X		
Heleidae		X		
Tendipedidae		X	X	
Paratendipes sp.		X	X	
Harnischia sp.		X	X	X
Psectrotanypus sp.		X	X	X
Cryptochironomus sp.		X	X	X
Glyptotendipes lobiferus	X	X		
Chironomus attenuatus		X	X	
Microtendipes sp.		X	X	
Polypedilum laetum		X	X	
Tanytarsus sp.		X	X	
Glyptotendipes senilis	X	X	X	X
Tendipes plumosus		X	X	X

ing coarse sand and gravel (1.5–2.0-mm particle size). During the summer, 1 to 5 cm of 0.044-mm silt is deposited on the sand substrate. As aquatic plants flourish, additional deposition occurs. In the fall, leaves and detritus collect in depressions on the bottom. Ice during the winter time gouges and plows the sandy bars.

A feature that the sand biotope shared with the other areas was the predominance of *Limnodrilus hoffmeisteri*. However, a particular attribute of this biotope was the occurrence of a variety of Sphaeriidae (fingernail clams). Eleven species were identified along the reach between the Connecticut Yankee Plant intake bays and the discharge point (1.2 km), and at times, the clams made up more than 50% of the numbers and biomass of this area. During the summer dense growths of the flowering plant *Vallisneria americana* were present, providing additional habitats for insects and other invertebrates. Almost every species of insect encountered in the entire study area was found in this biotope. Members of the Sphaeriidae and *Vallisneria* were not found in the sandy reaches of Haddam Island 4.7

km upstream of the discharge point, nor those of Rich-Lord Islands 4.3 km downstream.

Silt Biotope

The sloping bottom along the east bank of the river downstream of the discharge point is typically silt (<0.044 mm). Silt deposits are also found downstream of the Haddam Island and Rich-Lord Island areas. Approximately 18 ha or 11% of the study area was estimated to be comprised of silt, although during low flow months (July to September), from 1 to 3 cm of silt also covered the sandy portions of the bottom.

Silt stations 10, 11, 12, 13, and 16 were sampled from June 1965 through 1972; of the 4,732 samples collected from all stations, 38% (1,788) were taken from this biotope. Again the most abundant organism was *L hoffmeisteri*. Mollusks were rarely collected in the silty reaches, although occasional unionid clams were encountered. Near the upstream end of the silty area, fingernail clams were occasionally taken, but seldom more than one or two specimens in a series of samples. Detritus and silt formed hummocks and mounds and stair-type ridges descending the banks, and the variety of insects collected on the silt stations was associated with the variety of habitats available.

Biotopes Combined

For the 91 months of the study, the annelid *L. hoffmeisteri* accounted for 81% of all the organisms collected, *Pisidium* spp. (fingernail clam) 6%, the insect larvae *Psectrotanypus* sp. and *Cryptochironomus* spp. 6% and 4%, and all other organisms combined 3%.

Results

Benthic Fauna before Power Plant Operation

For 29 months (June 1965–October 1967), 0.1-m^2 grab samples were taken from the four sampling areas to establish a basis for comparison with conditions to be observed after the plant went into operation (October 1967–December 1972).

Sand Flat Biotope

Station 2 was the upstream sampling point of the sandbar that extends from the intake to the mouth of the discharge canal; it was 50 m from the intake at a depth of 3 m. At the onset of sampling, June–August 1965, approximately 1,000–1,500 organisms/m^2 were found at this point. However, construction and dredging had begun on the intake by June 1965, and because of the close proximity of these operations, the 1965–1966 data from station 2 do not represent "control" conditions prior to power plant operation. A characteristic common to all the stations on the sand flat was the presence of a mixed population of *Pisidium* clams. As construction continued, the number of fingernail clams at stations 2 decreased from 617/m^2 in 1965 to 178 and 37 in 1966 and 1967, a reduction from 43% of all the animals collected to 32% and 6% during late 1966 and through 1967 (Table 3). For the 3 years before power plant operation, the order of rank of dominant forms at station 2 by weight was the same as by number: *L hoffmeisteri* 52%, fingernail clams 40%, and insects 8%.

Station 3 was located 800 m downstream of station 2, approximately in the middle of the 60-ha sand-flat area. It was often difficult of access because of its extreme shallowness and the growth of the grass *Vallisneria*, which occupied approximately 18 ha. The highest number of organisms (>1,000/m^2) occurred in July–August 1966. In general, the benthic samples from this area showed similar trends in mean numbers and species composition to those from the other sand stations. *L. hoffmeisteri* increased from 24% to 38% and 55% of the composition, while *Pisidium* decreased from 73% to 59% and 29% (Table 3). The insects remained about the same (3–5%). In 1967, the character of the sediment changed from sand to silt; although dredging occurred only at the intake and discharge areas, silting could be detected along the entire 1.18-km sand flat. In the pre-plant-operation period the *Vallisneria* beds were utilized by insects, and their eggs, pupae, and larvae were abundant on the grass blades (Table 4). The root region of the grass also contained an abundant and diverse population, with fingernail clams (*Sphaerium* spp.) and annelids (*L. hoffmeisteri*) being especially numerous.

Stations 4 and 5 contained similar substrata and showed similarities to each other in mean

TABLE 3. Mean annual numbers of organisms, dry weights, numbers of species, and percent composition of the dominant taxa in the sand biotopes, 1965–1972.

	Pre-operation			Post-operation				
	1965	1966	1967	1968	1969	1970	1971	1972
			Station 2					
Numbers/m^2	1435	556	617	357	0	1066		
Dry weight, g/m^2	0.50	0.19	0.09	0.13	0	0.22		
Number of species	8	10	10	7	0	6		
Percent composition:								
L. hoffmeisteri	55	64	86	62	0	85		
Pisidium spp.	43	32	6	8	0	0		
Insects	2	4	8	28	0	15		
			Station 3					
Numbers/m^2	279	766	297	801	410	609	746	368
Dry weight, g/m^2	0.07	0.58	0.08	0.17	0.06	0.04	0.07	0.04
Number of species	5	9	10	11	13	9	10	12
Percent composition:								
L. hoffmeisteri	24	38	55	61	79	65	58	65
Pisidium spp.	73	59	29	12	2	14	23	6
Insects	3	3	5	27	18	21	19	29
			Station 4					
Numbers/m^2	723	855	338	507	234	260	275	395
Dry weight, g/m^2	0.39	0.31	0.12	0.18	0.05	0.04	0.04	0.07
Number of species	10	15	12	18	12	9	8	11
Percent composition:								
L. hoffmeisteri	37	64	55	76	80	71	70	70
Pisidium spp.	60	32	35	11	7	14	9	3
Insects	3	4	4	13	12	15	21	27
			Station 5					
Numbers/m^2	725	986	529	660	294	265	303	320
Dry weight, g/m^2	0.30	0.45	0.15	0.26	0.08	0.04	0.05	0.05
Number of species	8	10	12	18	13	12	10	9
Percent composition:								
L. hoffmeisteri	35	72	70	74	72	68	71	67
Pisidium spp.	60	24	15	4	4	7	3	4
Insects	5	4	14	22	24	25	26	29

TABLE 4. Epifauna on each of 10 *Vallisneria* blades 30 cm in length. + = present in large numbers.

	Sample no.									
Species	1	2	3	4	5	6	7	8	9	10
Heleidae (pupae)	80	40	60	81	8	6	13	19	40	73
Psectrotanypus sp. (larvae)	16	31	41	9	17	3	26	44	9	4
Glyptotendipes sp. (larvae)				17			1			1
Chironomus sp. (larvae)	4	1	3	1		1	2	1	7	1
Ferrissia sp.	6	1	2			22			4	7
Lymnaea sp.	3	7		1	1		6	8	4	1
Nematoda	+	+	+	+	+	+	+	+	+	
L. hoffmeisteri	1			2		1		1	1	

numbers, species composition, and biomass of organisms 1965–1967. *Vallisneria* growth was sparse on and around these stations. Station 5, 150 m upstream from the discharge point, was affected by the dredging in the canal and had more silt than the preceding sand stations. Both stations had higher numbers of organisms in 1966 than in 1965 or 1967. In 1965 and 1967, also, the percent composition of *L. hoffmeisteri* was lower than the 1966 population. On both stations 4 and 5, the numbers of *Pisidium* decreased. Insect larval composition remained about the same on station 4 from 1965 to 1967 but increased to 14% of the species collected from station 5 in 1967. The mean biomass at station 4 decreased from 0.3909 g/m^2 in 1965 to 0.1170 g/m^2 by 1967. At station 5, the biomass increased in 1966, but was reduced by 50% of the 1965 level in 1967. Also at station 5, as the silt began to increase the number of species increased.

Silt Flat Biotope

Station 10 was located on the downstream end of the sandy area 100 m below the discharge point. This station was consistently siltier than the sand stations. Annelids made up as much as 93% of all the organisms collected. Fewer fingernail clams were collected here than at stations 4 and 5, and a similar trend of a decreasing population occurred each year from 123/m^2 in 1965 to 27/m^2 in 1966 to 2 in 1968. The percent composition of insects increased with the change in substratum from 5% in 1966 to 23% in 1967. Due to the change in insect composition, the biomass increased from 0.1753 g/m^2 in 1965 to 0.3675 g/m^2 in 1967 (Table 5).

Station 11, about 300 m downstream from the discharge canal, was selected as being out of the path of the predicted heated effluent. The entire sloping bank of the east side of the river from the discharge point to the entrance of Salmon River (1,000 m) was unconsolidated silt and detritus. The silt stations differed from the sand stations primarily in the absence of fingernail clams and the higher number of both insects and annelids. At station 11, between 91% and 75% of the animals were *L. hoffmeisteri*, and in 1967, 23% of the 988 animals collected were insect larvae. No changes in sediment composition could be found along the silt quadrant. When scouring occurred in the sand stations during freshets, organisms in the uppermost layers (1–4 cm) of silt were carried away at current velocities of 10–15 cm/s. At these velocities on the silt stations, the organisms were often carried away, but a silt substratum remained. Summer population maxima at station 11 were similar to those in sand stations 4 and 5, except that they were denser. Mean annual numbers were 1,295/m^2 in 1965, 1,247/m^2 in 1966, and 988/m^2 in 1967, and the mean biomass decreased from 0.6342 g/m^2 in 1965 to 0.32 g/m^2 and 0.3105 g/m^2 in 1966 and 1967.

Stations 12, 13, and 16 had a sediment composition that was entirely silt. Samples contained 87–97% *L. hoffmeisteri* and 2–14% insect larvae 1965–1967. The mean number of organisms at station 12 decreased from 1,295/m^2 to 640/m^2, although no change in the sediment was evident. Biomass in the three sampling areas showed a general increase from 1965 through 1967.

Rock and Gravel (Channel) Biotopes

Sampling of these two biotopes was discontinued after 1970, by which time it had been established that the relatively sparse fauna in each was at no time in regular contact with the thermal plume and that neither had showed any substantial change since 1965.

Benthic Fauna during Power Plant Operation

The Connecticut Yankee plant began discharging heated effluent intermittently during October 1967. In this year, a series of stations was established in the discharge canal after construction dredging had been discontinued.

Sand Flat Biotope

Station 2 was directly off the intake structure (Figure 1). As a result of river water being pumped into the plant at the rate of 23.5 m^3/s, the bottom was scoured, and the organisms inhabiting that area declined sharply in abundance. Sampling was continued at the station through 1970; no organisms were collected from the substrate while the plant was operating. During the October start-up, the population fell to zero. Populations re-established themselves only for the duration of rela-

TABLE 5. Mean annual numbers of organisms, dry weights, numbers of species, and percent composition of the dominant taxa in the silt biotopes, 1965–1972.

	Pre-operation			Post-operation				
	1965	1966	1967	1968	1969	1970	1971	1972
			Station 10					
Numbers/m²	823	1355	986	1470	961	688	1137	828
Dry weight, g/m²	0.18	0.32	0.37	0.46	0.24	0.18	0.24	0.23
Number of species	8	14	20	22	14	11	13	10
Percent composition:								
L. hoffmeisteri	81	93	77	87	80	81	68	90
Pisidium spp.	15	2	0.1	1	0	0	0	0
Insects	4	5	23	12	20	19	32	10
			Station 11					
Numbers/m²	1295	1247	988	1506	834	1409	1855	1138
Dry weight, g/m²	0.63	0.32	0.31	0.41	0.20	0.28	0.49	0.36
Number of species	8	13	17	18	14	18	12	10
Percent composition:								
L. hoffmeisteri	88	91	75	74	83	90	73	80
Pisidium spp.	0	0	0	0	0	0	0	0
Insects	8	7	23	26	17	10	27	20
			Station 12					
Numbers/m²	1295	1026	640	2999	1207	1730	2136	1604
Dry weight, g/m²	0.17	0.22	0.19	0.90	0.49	0.34	0.51	0.46
Number of species	7	13	12	16	12	14	14	16
Percent composition:								
L. hoffmeisteri	90	95	90	92	85	86	72	89
Pisidium spp.	0	0	0	0	0	0	0	0
Insects	14	4	10	8	15	14	28	11
			Station 13					
Numbers/m²	595	852	558	2788	1387	1420	1179	1582
Dry weight, g/m²	0.12	0.24	0.32	0.95	0.43	0.27	0.28	0.45
Number of species	8	13	9	16	16	14	12	21
Percent composition:								
L. hoffmeisteri	87	91	88	92	88	88	79	82
Pisidium spp.	0	0	0	0	0	0	0	0
Insects	11	8	10	8	12	12	21	18
			Station 16					
Numbers/m²	258	785	827	2275	1282	828	865	1294
Dry weight, g/m²	0.07	0.38	0.59	0.90	0.49	0.15	0.25	0.39
Number of species	5	4	9	13	16	16	23	24
Percent composition:								
L. hoffmeisteri	94	97	97	92	87	87	69	75
Pisidium spp.	0	0	0	0	0	0	0	0
Insects	2	2	3	8	13	13	30	25

tively short-interval shutdowns. The area thus affected was 0.5 ha ± 500/m².

Station 3 was characterized by the absence of *Vallisneria* beds on the sand flat during the postoperational period, an estimated 18 ha being lost in 1968. After freshets in the spring of 1968, up to 3 cm of silt were deposited on the sand flat, and in general, the silt component tended to be higher each year, except in 1971. Over the 5-year period (1968–1972), the biomass at station 3 showed a sharp drop after the first year and relatively low values thereafter (Table 3). The overall diversity on this station was high, with an average of 11 species a year, and the average number of individuals low, 587/m^2.

Station 4 temperatures increased as much as 4°C during the flood phase of the tide. The mean numbers of organisms showed an overall decrease in the postoperational years, with the lowest average in 1969. Low values occurred during the winter months and spring freshets (April and May), and highest numbers in the summer. Annelids comprised between 70% and 80% of the number of animals, and fingernail clams from 3% to 14%; insect larvae increased from 13% to 27% of the animals collected. During the years 1965–1968, biomass had varied seasonally with winter minima and summer maxima, but from 1969 onwards, the summer peaks failed to appear, and biomass values remained persistently low with no marked seasonal fluctuation. During the postoperational years (1968–1972), the largest biomass was in 1968 (0.1769 g/m^2). For this period (1968–1972), 74% of the biomass was *Limnodrilus*, 1.0% the fingernail clam, and 25% insect larvae.

Station 5, located 150 m upstream of the discharge, typically received water up to 6°C warmer than ambient for up to 45 min during the flood phase of the tide. Between 67% and 74% of the organisms collected were annelids. Yearly biomass values were somewhat greater on station 5 than on 4, except in 1970 and 1972. A notable decrease in biomass and numbers occurred during 1969 (Table 3).

Silt Biotope

The station 10 area had a higher increase in bottom temperature (up to 10°C) than any other river station. The sediment composition of sand became overlain by silt and detritus by 1968; sand was sparse through 1972. Corresponding with this change, the kinds and numbers of species decreased from 1968 to 1972. The mean number of animals decreased from 1,470/m^2 in 1968 to 828/m^2 in 1972. Annelids comprised from 68% to 90% of the animals collected on a yearly basis. The highest yearly proportion of insects collected in the entire course of this study was on this station in 1971 (32%). The clams made up only 1% of the population in 1968 and were absent from 1969 through 1972 (Table 5).

Station 11 was located out of the direct path of the plume; no increase in temperature was recorded at the bottom or in the substratum, which was here composed of silt (<0.044 mm) throughout the 91 months of the study. *L. hoffmeisteri* comprised from 73% to 90% of the animals collected, insects from 10% to 27%; mollusks were rare. The mean number of organisms collected during the postoperational phase remained comparatively high with the exception of 1969 (834). The mean yearly and monthly biomass was higher than at the sand stations, with a mean of 0.4877 g/m^2 dry weight recorded in 1971.

Stations 12, 13, and 16 were also located out of the influence and path of the plume. The greatest number of organisms collected at any of the stations was taken here, and *L. hoffmeisteri* was again the dominant form. In 1971, the insects comprised as much as 30% of the animals collected at station 16, and 28% and 21% respectively on stations 12 and 13. The annual concentrations of organisms from 1968 to 1972 showed no clear trend on stations 10 and 11, but there was a marked increase on stations 12, 13, and 16 (Table 5). Most species tended to be present in the fall and the smallest number in the spring. Ice conditions restricted winter sampling, especially on these stations outside the heated effluent.

These three silt stations were all beyond the direct influence of the thermal plume, and the data from station 12 for 1967–1972 are discussed here as an example of a "control" in the Haddam Neck area. It was located over 1,050 m downstream of the discharge canal on the east side of the river at a depth of 3.1 m (mean low water). The substratum composition is primarily silt and detritus at all seasons. However, during the spring, the freshets have a marked effect when the flow velocities exceed 0.7 m/s, although the bottom is quickly recolonized with organisms from upstream when the freshets subside. Annual variation in temperature at this station was 0°C to 30°C, with a daily range of approximately 1°C. From 1967 through 1972, 213 samples were collected. They contained 29 species and 36,749 organisms. More than 85% of these were *L. hoffmeisteri* (31,236 individuals), and approximately 3,307 or 9% were dipteran larvae. Sampling results were affected by high-flow velocities, which varied in degree from year to year so that the collections each winter and spring were different. On the other hand, from early summer, low flow-rates prevailed and year-to-year differences were relatively slight. During the early spring of 1968, the highest densities of invertebrates (monthly mean, 9,965/m^2) occurred at this station. Annelids comprised 92% of the population that year, and insect larvae were abundant that fall and early winter. In 1969, the maximum population was collected in the spring before freshets (2,018/m^2); then the population decreased to approximately 400/m^2 and increased to a summer average of 1,418/m^2. Similar trends followed in 1970, 1971, and 1972. Seasonal and annual changes in biomass paralleled the changes in numbers. Of the total biomass taken from this station over the 6-year period, 89% was contributed by *L.hoffmeisteri* and 6% by insects (*Psectro-tanypus* sp. and *Chironomus* spp.).

Benthos in the Discharge Canal

The discharge canal leading from the Connecticut Yankee facility to the river (Figure 1) was constructed in order to carry the thermal effluent to a distance downstream that would ensure minimal recirculation of any heated water through the plant under flood-tide conditions. On completion, it measured 1.9 km in length and averaged 3 m in depth and 18 m in width, except where it entered the river via a 180-m mouth. The average velocity in the canal varies with the tide and ranges from 0.3–0.6 m/s, diminishing to 6 cm/s at the mouth. The 23.5 m^3/s of condenser cooling water is raised a maximum of 12.4°C in its passage through the plant, and it loses only about 1°C on its way down the canal.

As might be expected, the most rapid and drastic changes in biota observed in the entire study were those in the discharge canal after the plant went into operation. Not only did this habitat experience the extremes of heat and current velocity, but it was also the area most affected by the sudden changes caused by plant shutdown and start-up.

Prior to plant operation, the 4-ha canal area was a man-made cove of the river excavated through the alluvial sandy silt of the flood plain. The canal underwent the final stages of excavation in July 1966, and that fall and winter was a period of rapid change as the sides of the freshly dug canal were eroded by fluctuating water levels resulting from tidal action. Actually, the canal was in a state of transition from 1967 through 1969, but early in 1970 and through 1972, the sediment and biological characteristics were more stable, similar to those of the river control stations. Approximately 75% of the floor of the canal was silt, although the lower two-thirds of the canal (except the mouth) had heavy detrital deposits at all seasons 1967–1972. Nearly the whole of the submerged sides of the canal was silt. Before the addition of the heated water from the plant, the temperature in the canal with within 2°C of ambient.

Canal, 1967–1969

From 936 samples, a total of 33,107 organisms, including 22 species of recurring organisms, were collected. The annelid *L. hoffmeisteri* accounted for 80% of these. Ranking next in abundance were the larvae of three species of tendipedid flies, viz. Psectrotanypus sp. (6%), Chironomus spp. 5%, and Glyptotendipes lobiferus (3%). The remaining 6% consisted of dragonflies, naids, trichopterans, coleopterans, dipterans, gastropods, and other assorted specimens. The biomass (dry weight) totaled just over 36 g, of which 75% or 27 g was *L. hoffmeisteri*, 6% was Psectrotanypus sp. and 6% Chironomus spp., 5% Glyptotendipes lobiferus, and 8% other organisms.

Sediment variation at each station was pronounced during 1967 through 1969, and erosion of the canal sides and floor was apparent as judged by particle size distribution at each station. Construction dredging was in progress until late 1966, and sampling before January 1967 was exploratory. From January through September 1967, prior to power plant operation, 126 samples were obtained in an effort to show what effect the initial heat and flow had on the canal population. During January, February, and early March 1967, the power plant's circulation pumps were operating intermittently; on these occasions, the ice was flushed out of the canal and sampling was possible. More than 95% of the population collected was *L. hoffmeisteri*, and the remainder were insect larvae. The population increased through the summer and declined in September. During October, water heated to 6°C above ambient passed through the canal, and the population immediately decreased from 3,100/m^2 to 0. Through November and December (28 samples), no organisms were collected.

In January–February of 1968, the density of the population was approximately 300/m^2; it increased in May after a shutdown in March and April 1968. Plant operation was variable throughout 1968, particularly in the summer. The population mean in August was 610/m^2, at which time the canal temperature was 31.5°C. During the winter of 1968–1969, the population was lower than the previous year at 96/m^2, with no organisms being taken in December of 1968 (56 samples).

Canal, 1970–1972

Fourteen taxa of macro-invertebrates representing four phyla were collected from the bottom in this period (Table 6). The water flea *Chydorus* sp. (smaller than 1.0 mm and hence not a macro-invertebrate) was conspicuous in numbers high enough to be considered an important species in the winter bottom populations 1970–1972.

During February of 1970, canal stations 8, 9, and 10 had the highest populations of the 14 stations (Figure 2). Figure 3 illustrates the trends of the main elements of the mid-canal biota from January through November 1970 as characterized by the ubiquitous *Limnodrilus* and by *Psectrotanypus*, *Glyptotendipes*, and *Cryptochironomus*. The power plant was operating at 598 MW$_e$, with an 11°C increase above river temperatures, except for a 67-d scheduled shutdown during April, May, and June. With no flow over the weir, and little or no velocity in the canal, some annelids and unionid mussels (*Lampsilis*) were found within 150 m of the weir at the top of the canal. Also, stations in the canal became generally similar to station 5 in the river during the shutdown period.

In terms of total numbers, the largest population at station 8 for all four seasons occurred on 22 July with a canal water temperature of 37°C (Figure 2). Limnodrilus was the single species found at that temperature. Four days later, at 40°C, its numbers were reduced from 900/m^2 to 30/m^2; it would appear that the maximum temperature these annelids can tolerate is slightly above 37. Insects were sparse during the summer temperatures. However, an increase in the variety of insects occurred in the fall when the temperatures were going down. In November, the canal was 23°C, the plant load was 605 MW$_e$, while the river temperature was 13°C. At that time, the highest population of Psectrotanypus, Cryptochironomus, and Glyptotendipes for the year occurred on canal stations 7, 8, 9, 10, and 11. Beetle larvae and caddis fly larvae were found repeatedly during the fall.

Limited investigation of fly larvae passing through the power plant's condenser indicated that most survived the 93-s trip at fall temperatures of 20–30°C. in the river, maximum populations of predominantly tendipedid larvae also occurred at this season, suggesting that recruitment of populations may result not only from eggs laid in the canal, but also from organisms which had passed through the plant.

There follows a brief account of the major biotic changes in the canal from 1970 to 1972. A shift in dominance from an annelid toward an insect population had already begun in 1967, and it continued until 1971. *Limnodrilus* comprised 46% of the population and 50% of the organisms

TABLE 6. Seasonal occurrence of species in the discharge canal, 1970–1972. a = abundant; s = sparce; o = absent.

Taxon	Winter	Spring	Summer	Fall
L. hoffmeisteri	s	s	a	a
Chydorus sp.	a	s	o	o
Gammarus fasciatus	s	o	o	s
Hydropsyche sp.	o	o	o	s
Oectis sp.	o	o	o	s
Berosus sp.	s	s	o	a
Tendipedidae	s	s	o	a
Psectrotanypus sp.	s	s	o	a
Cryptochironomus fulvus	s	s	o	a
Chironomus plumosa	s	s	o	s
Glyptotendipes lobiferus	s	s	o	a
Ceratopogonidae	s	o	o	s
Ferrissia sp.	s	s	o	s
Lampsilis cariosa	o	s	o	o

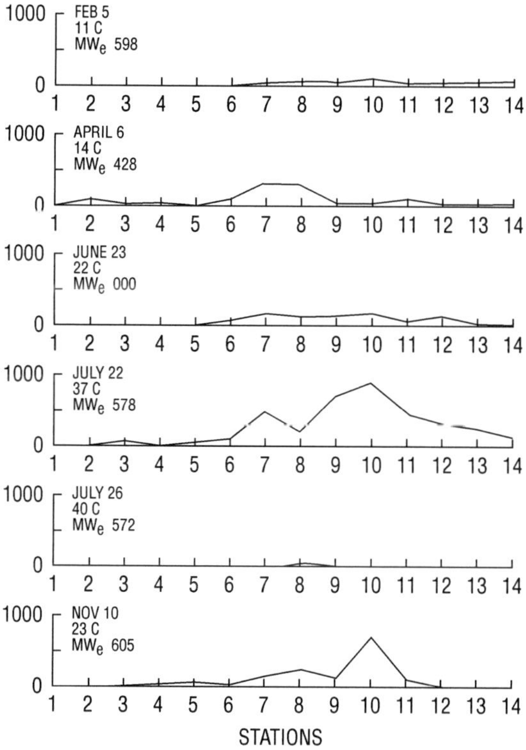

FIGURE 2. Total benthos (numbers/m^2) in the discharge canal by months, 1970. The canal stations, spaced equidistantly, are numbered from near the top to bottom (mouth); see Figure 1, canal stations in squares.

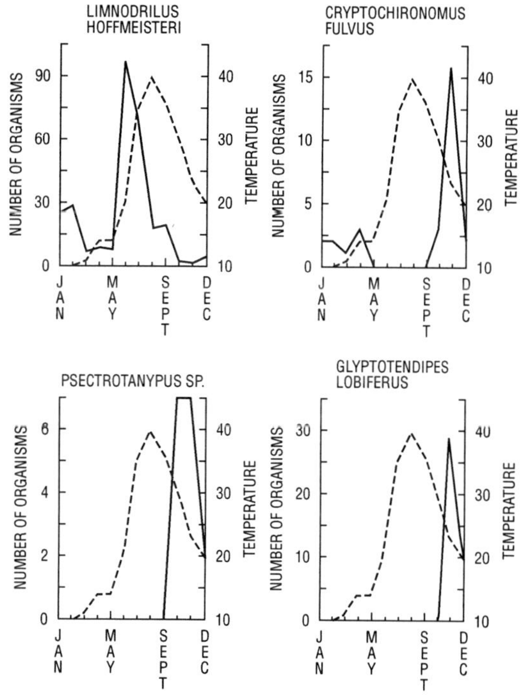

FIGURE 3. Abundance of four principal species by months at canal station 8, 1970, in numbers per 0.1 m^2. Solid lines represent organisms and broken lines water temperature (°C).

were insect larvae. During the scheduled shutdown for refueling in April 1971, the total population burgeoned, *Limnodrilus* numbers increasing from 28/m^2 to 618/m^2, larvae of Psectrotanypus from less than one to more than 839/m^2, and *Cryptochironomus* from 16/m^2 to 35/m^2. However, the May population decreased from 1,532/m^2 (during plant shutdown), to 291 in June (the plant at 516 MW_e), to 0/m^2 in July (549 MW_e, 34°C). The 1972 scheduled shutdown occurred in June and July, and the population decreased in the summer instead of the spring, from 788/m^2 (0 MW_e) to 39/m^2 and 6.4/m^2 in August and September.

The mean number of animals in the canal during the 9 months before plant operation was significantly higher than in each of the 5 years that followed. There were little differences in species composition and diversity between the preoperative and the postoperative months of the study. From 1970 through 1972, there was a total of 22 species collected from the 464 samples. The most common animal was *Limnodrilus*, which occurred in 63% of the samples over the 3 years. It was followed by *Cryptochironomus* spp. (37%) and *Psectrotanypus* sp. (28%); larvae of the midge *Glyptotendipes* sp. occurred in 7% of the samples, and the remaining 18 species in less than 3%.

Discussion

The area adjacent to the intake (station 2) was almost completely depopulated. Heated water did not reach the bottom at this point. Hundreds of cubic meters of sediment were removed during construction of the power plant, leaving only coarse sand and gravel. In the Haddam reach of the river, this type of sediment supports very few organisms. The velocity of the water moving toward the intake has been measured at 0.3–0.6 m/s, and its scouring action is clearly visible. Algal growth at depths of 4 m or more was markedly reduced, which helps to explain the absence of epiphytic feeders.

The extreme sparcity of benthic animals near

the plant intake contrasts with the situation at a coastal power plant in the United Kingdom that has been studied extensively by Markowski (1959, 1960). Over a 5-year period of plant operation, denser animal populations were found at the intake station (19 species) than in the discharge area (26 species). Markowski attributed this to lower velocities (0.36 m/s) at the intake than at the discharge (0.89 m/s) and less competition from algae, which were very abundant at the outfall. A difference in sampling method probably accounts for much of the difference between Markowski's findings and ours: he used stable surface test panels rather than direct sampling of the sediments. Markowski did not know whether settlement on the panels took place during plant operation or during the brief daily shutdown.

Stations 5, 4, and 3, respectively 150 m, 300 m, and 1,000 m upstream of the discharge point, received progressively less heat from the plant. This was an area of sandy flats. For purposes of comparison, similar environments were sought upstream as far as Bodkin Rock (15.5 km) and downstream to Eustasia Island (12.6 km). However, because most reaches undergo rapid sedimentary changes and due to the unusual abun- dance of *Vallisneria* in the vicinity of the power plant, no other parts of the river precisely similar in species composition could be found. On station 3, between pre- and postoperation periods, there was a slight increase in numerical abundance of benthic animals, due to *L. hoffmeisteri.* On stations 4 and 5, abundance decreased, and on all three stations, the operation of the power plant was accompanied by a decrease in biomass (Table 7). On all three stations, the most abundant species, *L. hoffmeisteri,* increased its dominance, while insects gained in percentage of the total population; these shifts were at the expense of *Pisidium* spp., the second most abundant taxon, which consistently declined. For the entire area represented by these stations (21.7 hectares), *L. hoffmeisteri* increased by 23% and *Pisidium* decreased by 77%, while the total benthic biomass decreased from an estimated 5,755 g to 1,547 g, a loss of 73%.

The beds of *Vallisneria* that formerly characterized this area were separated by meandering channels several meters wide that guided and funneled the flow of water. While the beds collected silt, the channels were swept clear and were sandy. When the power plant went into operation, the *Vallisneria* beds were destroyed, evidently in direct consequence of the heated effluent. The silt was now more evenly distributed, blanketing the entire area, thus favoring *L. hoffmeisteri* and discouraging the fingernail clams, which prefer a sandy habitat. Changes in sediment were especially pronounced at station 5, 150 m upstream from the discharge point. Here, in addition to the removal of the grass, direct effects of the discharge were evident. During the first 24 months of plant operation, an area 25–50 m wide along the east shore and extending 200 m upstream and 200 m downstream from the discharge point received large volumes of silt and detritus from the scouring of the canal. Hummocks formed, and the character of the bottom became more complex. The increase in *L. hoffmeisteri* and the corresponding decline in *Pisidium* spp. are attributed to the increase in silt content. The increase in number of species resulted from the greater complexity of the substratum. Evidently, the hummock habitat favored the insects, for *Psectrotanypus* and *Chironomus* both increased, and in the postoperational period, other insects from a variety of orders comprised about 11% of the animals collected.

Station 10, 100 m downstream from the discharge point, was covered by water 10°C above ambient during all phases of the tide and represents the maximally affected area of the river. The *Vallisneria* beds had not extended this far south. Here, the sediment was siltier than at stations 3, 4, and 5 both before and during plant operation. Annelids were correspondingly abundant and fingernail clams scarce. The specimens of *Pisidium* spp. found here were often collected in plankton samples of the bottom water along with moving detritus: apparently, they had drifted downstream from the sandy flats north of the discharge. Although the temperature change here was maximal, the sediments of this grass-free area were altered less than those at the preceding stations. The decreases in animal abundance and biomass and the changes in the percentage composition of the dominant taxa were commensurately moderate. This indicates that the principal effects of the power plant on the benthos were mediated primarily by changes in the sediment composition rather than by direct thermal stress.

The dominance of *L. hoffmeisteri* was greater in the vicinity of the discharge point than along the flats upstream, amounting to 86% before and 81% after the power plant went into operation. It is possible that peculiarities in the life cycle of this species contributed to the constancy of the benthos at station 10 in the face of environmental changes. Aston (1972) found higher numbers of sexually mature *L. hoffmeisteri* downstream than upstream of a power plant on the River Trent (England) at temperatures of 25°C. Under controlled conditions, its rate of egg production was a function of temperature up to 25°C (the limit of his observations), while a related tubificid worm maintained a fairly constant rate over the range 10–25°C. Aston was led to believe that *L. hoffmeisteri* responded more to elevated temperatures than to other factors associated with the power plant effluent. However, the applicability of his findings to the present work is limited. At station 10, the water temperature while the plant is running is 20–25°C for brief periods in the spring and the fall, is more than 25°C for about 120 d of the year, and even reaches 40°C; there are indications that temperatures of 30°C are supraoptimal for *L. hoffmeisteri*. Kennedy's (1966) observations on the life cycle of this species appear to have greater significance for the Connecticut Yankee population. He showed that breeding may take place throughout the year or be strictly seasonal or show seasonal peaks in activity; the young take from 6 months to 2 years to attain maturity. According to Kennedy, it can adapt its breeding to take advantage of local conditions, thereby augmenting its reproductive potential. This capability of the animal, comprising more than four-fifths of the population, might in part account for the stability of the benthic community in the maximally affected portion of the river.

Before the plant began to operate, station 10 was similar to stations 11, 12, and 13 in depth, sediment, and species composition. The region of the river represented by these three stations, beyond the influence of the plume, provides a basis for comparison with the maximally affected area. Although the temperature regime was not altered in this region, benthic abundance and biomass, while remaining nearly constant at station 11, increased substantially at stations 12 and 13. This is believed to have been caused by drift of animals from the variable habitat upstream, especially from the hummocks near the discharge point. In 1966, each of these three stations supported 13 species. By 1972, the number at station 11 had dropped to 10 species; farther downstream, there was a progressive increase to 16 species at station 12 and 21 at station 13.

The population of the discharge canal itself could not be surveyed adequately during construction and in 1967–1968 because conditions were changing rapidly. From 1970 to 1972, the substratum became more stable and uniform, but its biota was subject to rapid changes due to (1) sudden fluctuations (12°C) in temperature as the plant started up or shut down, and (2) the extraordinary seasonal impact of predatory fishes. Both the annelids and insects fared best during periods when the plant was not running. Following these annual maxima during plant shutdowns, in most years the annual minima occurred in late summer, at temperatures of 35–40°C, after resumption of plant operation. Trembley (1960, 1961) reported similar seasonal changes in the benthos affected by the Martin Creek Steam Electric Station in Pennsylvania, where river temperatures varied from 0°C to 30°C; during the summer, macro-invertebrates decreased in the area heated 10°C above ambient, and the bottom was repopulated during the colder months. In the Connecticut Yankee discharge canal, the insect population increased from summer to fall, but a potentially rich winter benthos was checked by an unusually dense concentration of bottom-feeding fishes (brown bullhead *Ictalurus nebulosus* and white catfish *Ameiurus catus* [also known as *I. catus*]) that were attracted to the canal during November and remained through March (Marcy 1972 and 2004 [1976], this volume).

Acknowledgments

I have received help from many sources in the course of this study, particularly in the matter of identification: E. L. Bousefield, National Museum of Canada, Ottawa (amphipods); R. O. Brinkhurst, Department of Zoology, University of Toronto (annelids); W. Brungs, National Environmental Research Center, U.S. Environmental Protection Agency, Cincinnati (algae); D. R. Davis and O. S. Flint, Jr., Department of Entomology, U.S. National Museum, Washington (insects); J. D. George, Department of Zoology, British Museum (Natural History), London (annelids); H. B. Herrington, Westbrook, Ontario (mollusks); R. I. Johnson, Museum of Comparative Zoology, Harvard University (mollusks); K. V. Krombein, Department of Entomology, U.S. National Museum, Washington (insects); E. L. Mills, Institute of Oceanography, Dalhousie University, Halifax (amphipods); P. A. Morris, Peabody Museum, Yale

University (mollusks); M. H. Pettibone, Division of Worms, U.S. National Museum, Washington (annelids); E. H. Reed, Springfield Science Museum, Massachusetts (mollusks); P. J. Spangler, Department of Entomology, U.S. National Museum, Washington (insects); J. E. Sublette, Eastern New Mexico University, Portales, New Mexico (insects); and W. W. Worth, Department of Entomology, U.S. National Museum, Washington (insects).

I owe a special debt of gratitude to Drs. Melbourne R. Carriker and David M. Pratt (p. xi), Daniel Merriman, and Lyle M. Thorpe for their painstaking guidance throughout, and I here record my appreciation of the help received from the entire staff of the Essex Marine Laboratory, Inc.

References

Aston, R. J. 1972. Field and experimental studies on the effects of a power station effluent on Tubificidae (Annelida-Oligochaeta). Proceedings of the Royal Society of London B 180:407–419.

Boyd, W. A. 2004. Hydrography. Pages 27–36 *in* P. M. Jacobson, D. A. Dixon, W. C. Leggett, B. C. Marcy, Jr., and R. R. Massengill, editors. The Connecticut River Ecological Study (1965–1973) revisited: ecology of the lower Connecticut River 1973–2003. American Fisheries Society, Monograph 9, Bethesda, Maryland. (Originally published in 1976)

Kennedy, C. R. 1966. The life history of *Limnodrilus hoffmeisteri* Clap. (Oligochaeta; Tubificidae) and its adaptive significance. Oikos 17:158–168.

Marcy, B. C., Jr. 1972. Pages 20–30 *in* D. Merriman, editor. The Connecticut River investigation. Connecticut Department of Environmental Protection, 15th semi-annual progress report, Hartford.

Marcy, B. C., Jr. 2004. Fishes of the lower Connecticut River and the effects of the Connecticut Yankee plant. Pages 65–124 *in* P. M. Jacobson, D. A. Dixon, W. C. Leggett, B. C. Marcy, Jr., and R. R. Massengill, editors. The Connecticut River Ecological Study (1965–1973) revisited: ecology of the lower Connecticut River 1973–2003. American Fisheries Society, Monograph 9, Bethesda, Maryland. (Originally published in 1976)

Markowski, S. 1959. The cooling water of power stations: a new factor in the environment of marine and freshwater invertebrates. Journal of Animal Ecology 28(2):243–258.

Markowski, S. 1960. Observations on the response of some benthonic organisms to power station cooling water. Journal of Animal Ecology 29(2):349–357.

Stone & Webster Engineering Co. 1964. Design memorandum. Investigation of thermal effects of Connecticut Yankee Atomic Power Plant on Connecticut River for Connecticut Yankee Atomic Power Company. Boston, Massachusetts.

Trembley, F. J., editor. 1960. Research project on effects of condenser discharge water on aquatic life. Lehigh University, Inst. Res., Progress report 1956–1959, Bethlehem, Pennsylvania.

Trembley, F. J., editor. 1961. Research project on effects of condenser discharge water on aquatic life. Lehigh University, Inst. Res., Progress report 1960, Bethlehem, Pennsylvania.

American Fisheries Society Monograph 9:59–64, 2004

Entrainment of Zooplankton at the Connecticut Yankee Plant

RONALD R. MASSENGILL
Essex Marine Laboratory, Inc., Essex, Connecticut 06426, USA.

One of the more obvious ways in which a power plant can be expected to influence the environment is in its effects upon zooplankton entrained in the condenser cooling-water system. For example, Raymont (1964) and Raymont and Currie (1964) noted the consequences of such entrainment on the species composition of zooplankton. Much of the work on this subject has been reviewed by Naylor (1965) and Coutant (1970). The present report describes mechanical and thermal factors that may affect survival of zooplankton (apart from fish eggs and larvae) during passage through the cooling water condenser and heated water outfall of the Connecticut Yankee plant. Estimates are made of the proportion of the river zooplankton entrained by the plant.

Methods

The Connecticut Yankee plant withdraws approximately 25.5 m^3/s of water from the river and raises its temperature to a maximum of 12.5°C above ambient before returning it to the river. This diversion amounts to as much as 6% of the river flow during low-flow conditions. The passage of water through the plant and down the discharge canal is described by Merriman and Thorpe (2004 [1976], this volume).

Zooplankton samples were collected at specific depths both by day and night (vide infra) with nets of three mesh sizes: No. 2 (0.37 mm), No. 10 (0.16 mm), and No. 20 (0.08 mm). Because of the high concentration of seston and the low concentration of zooplankton, only the No. 2 nets were used for comparative purposes. Flows through these nets were metered. The finer mesh nets clogged with detritus within seconds after being lowered into the river. After 10–30 s, these nets were removed and the contents were examined quantitatively. Periodically, 20-L bucket samples were taken as a control on organisms missed or damaged by the nets. A 1-L aliquot from each bucket sample was condensed with a Sedwick Rafter funnel. From a 5-mL concentrate, 1 mL was examined at 200×. Both preserved and live samples were examined.

In the field, fresh plankton samples were examined with a dissecting microscope at 70×. Counts of living and dead specimens were recorded from 25-mL aliquots. Only those specimens that remained immobile after being prodded with a needle were considered dead. All samples were preserved in 2% neutral formalin.

The weight of seston (total particulate matter) retained by No. 2, No. 10, and No. 20 nets was determined by drying at 90°C for 24 to 48 h and weighing on a precision balance to the nearest milligram.

The portion of zooplankton in the river that was entrained through the plant from 1 June to 25 July 1972, was estimated from four series of samples taken by day and two 24-h series, a total of 198 samples collected along a transect of the river and at each of the four intake bays. Each series was collected in the river 50 m upstream of the intake on the ebb tide or downstream at slack tide (see Figure 2 in Marcy 2004 [1976], this volume). Each transect was composed of 11 stations, approximately 50 m apart. Net collections were made at 1, 3, and 5 m below the surface, or 0.5 m off the bottom in shallow portions of the river. Sampling at each bay in the intake was made at 1, 3, and 5 m from the surface simultaneously with sampling in the river. Animals were entrained at different concentrations in each bay depending on the phase of the tide. During flood tide, more animals were collected in the downstream bay; during ebb tide, the upstream bay had higher numbers.

The percentages of zooplankton entrained

were calculated using Marcy's (1973) formula for entrained fish larvae:

$$\% \text{ entrained} = 100\ N_i F_i / N_r F_r\ ,$$

where N_i equals mean number of zooplankton per m^3 in the intake, F_i equals rate of flow into intake bays (constant 25 m^3/s), N_r equals mean number of zooplankton per m^3 in the river transect, and F_r equals mean flow rate (m^3/s) in the river.

Survival and mortality at various temperatures in the discharge canal were determined from samples collected in 1970, 1971, and 1972.

Estimates of mortality caused by mechanical damage during passage through the plant were made on 27 June and 13 July 1972, when the pumps were running but no heat or sodium hypochlorite was being added to the intake water. Plankton collections were taken from the discharge side of the condenser with a vacuum pump. At this site, only 4 L could be collected within 10 min. Additional samples were obtained with nets from three discharge stations in the canal.

In an attempt to estimate mortality due to chlorination, we took zooplankton samples on 13–14 July 1972 from (a) the pump house in front of the condenser, (b) the vacuum pump on the outlet side of the condenser, and (c) the discharge canal tunnel. Sodium hypochlorite (13% by weight of available chlorine) was injected as a regular matter of plant operation at a rate of 19 L/min for 5–10 min into one bay at a time, usually daily at 1300 hours, in warm weather, and several times a week during cooler weather. This chlorination method usually gave the following approximate concentrations of residual chlorine: 2 ppm before the traveling screens, 2 ppm before the condenser, 1 ppm at the discharge canal. The resulting uncertainties in concentrations of zooplankton and of chlorine were, unfortunately, not understood at the time and would require a much longer series of samples to yield a reliable estimate of biocide-induced mortality. For these reasons, the effects of chlorine on the zooplankton were not assessed in this study.

Results

The zooplankton population was dominated by entomostracan crustaceans (Table 1) and included fish larvae, larvae, and pupae of *Psectrotanypus* sp., *Cryptochironomus* spp., *Chironomus* spp., and *Glyptotendipes* spp., as well as Diptera, rotifers, and protozoans.

TABLE 1. Dominant zooplankton.

Cladocera
Leptodora kindtii
Sida crystallina
Daphnia pulex
D. ambigua
D. dubia
D. galeata mendotae
D. parvula
D. catawaba
Ceriodaphnia reticulata
Simocephalus vetulus
Bosmina longirostris
Illyocryptus sordidus
I. acutifrons
Leydigia quadrangularis
Alona affinis
Calanoida
Osphranticum labronectum
Diaptomus pallidus
D. birgei
D. pygmaeus
Cyclopoida
Cyclops vernalis
C. bicuspidatus thomasi
Eucyclops speratus
Macrocyclops albidus
Mesocyclops edax

Vertical and Horizontal Distribution of the Zooplankton in the River

Deep water in the river is located in the channel near the opposite shore from the power plant. During the day, on the average, 28% of the zooplanktonic organisms were found at 1 m, 33% at 3 m, and 39% at 5 m. The largest number of organisms collected during the day was 149/m^3 and occurred in the bottom water of the channel (Table 2). At night, the distribution was reversed and 56% were collected at 1 m, 31% at 3 m, and 13% at 5 m. The largest number of organisms collected in the evening was 580/m^3 in the channel, near the surface, approximately 100 m from the west shore. Samples from shore stations 1 and 11 (see Figure 3 in Marcy 2004 [1976]) taken either by day or night generally contained lower numbers of zooplankton than the channel stations.

TABLE 2. Vertical and horizontal distribution of zooplankton passing by the power plant, 1 June–25 July 1972; numbers/m^3.

Date	Hour and tide	Depth (m)	River stations										
			11	10	9	8	7	6	5	4	3	2	1
1 June	1111–	1	0	11.5	33.6	6.0	13.4	24.0	22.0	17.0	17.0	15.0	1.3
	1342	3		21.3	19.4	13.0	23.7	25.8	21.6	32.0			
	Ebb	5		75.0	33.0	60.3							
27 June	0933–	1	64.8	30.5	11.9	4.1	12.3	0	5.6	0	11.8	20.7	12.7
	1300	3		61.6	39.9	16.3	24.0	0	0				
	Ebb	5		148.9	19.6	37.0	41.9	55.6					
27 June	2055–	1	83.9	53.0	140.6	579.8	101.9	71.0	154.0	94.7	48.0	53.4	94.7
	2356	3		112.0	60.6	79.5	146.0	187.0	124.0	146.3			
	Ebb	5		0	2.0	85.4	98.0	180.0					
13 July	0946–	1	0	0	0	5.9	4.9	12.6	10.9	7.8	8.0	8.5	3.2
	1343	3		0	0	3.9	21.5	6.7	10.2	34.3			
	Slack	5		0	5.9								
13–14 July	2126–	1	0	0	4.2	14.6	3.4	5.2	1.8	4.6	23.2	94.4	7.5
	0100	3		3.8	8.1	7.0	14.0	7.4	6.5	6.2	2.6		
	Ebb	5		9.0	2.5	7.0	5.0						
25 July	1400–	1	0	7.5	5.8	4.9	0.4	7.3	6.4	3.3	2.6	0.5	1.3
	1715	3		4.3	14.4	6.1	3.7	6.6	2.0	2.4	5.0		
	Ebb	5		6.2	16.5	4.0	12.0	14.7					

Seasonal Distribution

Monthly means of abundance of zooplankton as sampled in the river from March to December varied from less than 10 to more than 1000 animals per m^3. Cladocerans, principally *Daphnia* spp., were by far the dominant group throughout the year; the cyclopoid copepods were numerically important during the spring and summer (Table 3).

Percentage Entrained

Table 4 shows the distribution of zooplankton in the intake bays. The percentage of zooplankton entrained during the 6 dates sampled for this purpose ranged from 1.2% to 14.9%. The mean number of zooplankton entrained was 4.2%. The maximum percentage, 14.9%, occurred on slack tide. On the other five dates, the tide was ebbing.

Mechanical Effects

Estimates of zooplankton mortality due to mechanical damage during passage through the condenser cooling-water system were less than 1%. With the samples collected during no addition of heat to cooling water and pumps running, nearly all of the insect pupae and a predaceous cladoceran, *Leptodora kindtii*, (less than 1% of the population entrained) were killed by the time they reached the end of the discharge canal. On 27 June and 13 July 1972 when river temperatures were 21.5°C and 23.0°C, more than 85% of the population were cladocerans (*Daphnia* spp.) and approximately 15% were cyclopoid copepods. No mortality occurred in these two groups. Furthermore, egg cases of egg-bearing cladocerans collected in the samples showed no apparent physical damage from their passage through the plant.

On 8 June 1972, 12 seston samples were collected at the intake and 23 in the river at regular plankton stations. Only about 2% (dry weight) of the total particulate matter retained by the nets was carried at the surface of the river, 5% at 3 m, and 93% near the bottom. Similar percentages were observed in samples from the intake. This evidence of a dense layer of detritus near the bottom was confirmed by direct observation (SCUBA). The average seston concentration in the river water was 0.2306 g/m^3 and in the plant cooling water was 0.2338 g/m^3. The organic material carried by the river may be an important

TABLE 3. Abundances of total zooplankton in the river, and abundances and percentages of cladocerans, calanoid copepods, and cyclopoid copepods, 1970–1972.[a]

Date	Total no./m^3	Cladocerans No./m^3	Cladocerans % of total	Calanoid No./m^3	Calanoid % of total	Cyclopoid No./m^3	Cyclopoid % of total	River temp. (°C)
1970								
30 June	3,319	3,253	98	30	+	36	1	21.0
2 July	3,652	2,309	63	986	27	357	10	22.0
6 July	1,030	751	73	213	21	66	6	23.2
8 July	4,163	2,575	62	1,089	26	437	11	23.2
29 July	79	6	8	73	92			29.4
1971								
19 May	479	282	59	22	5	155	32	13.7
2 June	334	169	51	15	5	149	45	16.4
10 June	1,778	1,276	72	161	9	340	19	22.7
16 June	3,094	2,629	83					25.0
24 June	61,671	40,914	66	4,171	7	16,576	27	26.0
29 June	9,510	3,564	37	33	+	5,913	62	26.4
6 July	345	220	64			125	36	27.7
15 July	1,062	446	42	112	11	499	47	27.3
22 July	32	32	100					25.6
28 July	850	398	47			451	53	27.2
3 Aug.	413	83	20	37	9	293	71	27.4
13 Aug.	259	188	73	15	6	55	21	26.8
17 Aug.	4,056	3,994	98	55	1	16	+	26.5
24 Aug.	256	132	51	67	26	57	22	24.6
30 Aug.	5,043	3,175	63			1,867	37	21.2
10 Sept.	87	87	100					23.6
15 Sept.	668	98	15	427	64	142	21	21.5
28 Sept.	40	40	100					18.2
5 Oct.	96	55	57			41	43	18.7
15 Oct.	0							14.6
15 Nov.	11			3	27	8	73	6.0
15 Dec.	6					6	100	2.1
1972								
12 Mar.	80	49	61	1	1	30	38	1.5
10 Apr.	60	58	97	1	2	1	1	5.8
10 May	90	70	78			20	22	7.6
1 June	572	361	63	13	2	175	31	18.7
27 June	5,226	4,445	85	8	+	103	2	22.5
13 July	931	501	54					26.0
14 July	213	35	16					23.2
25 July	277	211	76	8	3	34	12	25.0
15 Aug.	26	6	23			18	69	22.6
15 Sept.	0							20.7
25 Oct.	699	41	6	42	59	242	35	11.6
28 Nov.	45					44	98	4.1
19 Dec.	323	162	50			161	50	0.4

[a] + = less than 1%. When percentages do not total 100%, it is due to the presence of other organisms.

TABLE 4. Vertical and horizontal distribution of zooplankton entering the power plant, 1 June–25 July 1972, and estimated percentages entrained from the river; numbers/m^3.

Date	MW_e	Temp. (°C) river (canal)	River flow (m^3/s)	Depth (m)	Intake stations 1	2	3	4	% entrained from % river	Confidence limits (95%)
1 June	577	19.9	743	1	0.6	2.4	19.4	11.5		
		(29.5)		3	1.4	5.5	24.0	10.7	1.2	±2.5
				5	1.4	4.8	32.0	10.8		
27 June	0	20.5	1,053	1	9.2	18.6	26.7	49.5		
		(20.8)		3	32.9	47.0	30.0	44.1	3.6	±2.5
				5	6.6	27.0	87.0	117.6		
27 June	0	21.5	1,076	1	47.7	39.0	175.0	61.0		
		(21.0)		3	79.2	78.0	117.0	62.0	1.7	±8.8
				5	111.0	65.5	109.0	86.0		
13 July	0	23.0	704	1	19.3	18.0	4.4	2.7		
		(23.2)		3	27.9	225.0	7.1	4.7	14.9	±4.7
				5	31.7	20.0	1.8	2.9		
13–14 July	0	23.5	779	1	9.1	15.0	6.8	6.0		
				3	40.2	11.0	8.0	5.5	1.7	±2.5
				5	30.7	8.8	14.0	2.9		
25 July	577	25.0	561	1	1.5	1.8	1.4	11.2		
		(37.0)		3	2.7	2.3	2.2	6.8	2.2	±3.3
				5	1.0	2.0	1.6	0.9		

factor in damage of certain organisms being pumped through the plant.

Heat Effects

When the temperature of the discharge water ranged between 31.0°C and 39.4°C, no cladocerans or copepods survived passage down the discharge canal to the river in the years 1970, 1971, and 1972. Insect larvae survived temperatures up to 33°C but were not collected alive or dead from river or canal samples at higher temperatures. Nearly 100% mortality of cladocerans and copepods was observed every week on periodic inspections during the summer-time (June through September) period of 120 d (31.2°C to 40.0°C).

Discussion

Previously published reports of entrainment of zooplankton have not specified the proportion of the total population thus affected (Markowski 1959; Barnett and Hardy 1969; Churchill 1969; Heinle 1969; Young 1969; and Normandeau—no date).

The number of zooplankton entrained at the intake of the Connecticut Yankee plant averaged about 4% of those passing by in the river. Although the criterion for survival through the cooling system was mobility of organisms at the moment of examination, we do not know if some mobile individuals later succumbed, nor if some immobile individuals recovered from the entrainment. Mortality of zooplankton passing through the condenser at temperatures above 31°C was 100%. These temperatures prevailed for about 120 d a year. During cooler periods, with the exception of dipteran pupae and *Leptodora kindtii,* most of the zooplankton survived. Mortalities were attributed almost wholly to heat. Mechanical damage resulting in death of delicate forms (which might well have been killed by heat alone) may have been due entirely to the high velocities of flow (244 cm/s) through the condenser tubes, although their bombardment and entanglement by detritus may also have been a factor.

References

Barnett, P. R. O., and B. L. S. Hardy. 1969. The effect of temperature on the benthos near the Hunterston Generating Station, Scotland. Chesapeake Science 10:255–256.

Churchill, M. A. 1969. Effects of heated discharges: the TVA experience. Nuclear News (September):80.

Coutant, C. C. 1970. Biological aspects of thermal pollution. I. Entrainment and discharge canal effects. CRC Critical Reviews in Environmental Control 1(3):341–381.

Heinle, D. R. 1969. Temperature and zooplankton. Chesapeake Science 10:186–209.

Marcy, B. C., Jr. 1973. Vulnerability and survival of young Connecticut River fish entrained at a nuclear power plant. Journal of the Fisheries Research Board of Canada 30(8):1195–1203.

Marcy, B. C., Jr. 2004. Planktonic fish eggs and larvae of the lower Connecticut River and the effects of the Connecticut Yankee plant, including entrainment. Pages 125–153 *in* P. M. Jacobson, D. A. Dixon, W. C. Leggett, B. C. Marcy, Jr., and R. R. Massengill, editors. The Connecticut River Ecological Study (1965–1973) revisited: ecology of the lower Connecticut River 1973–2003. American Fisheries Society, Monograph 9, Bethesda, Maryland. (Originally published in 1976)

Markowski, S. 1959. The cooling water of power stations. A new factor in the environment of marine and freshwater invertebrates. Journal of Animal Ecology 28:243–258.

Merriman, D., and L. M. Thorpe. 2004. Introduction. Pages 3–25 *in* P. M. Jacobson, D. A. Dixon, W. C. Leggett, B. C. Marcy, Jr., and R. R. Massengill, editors. The Connecticut River Ecological Study (1965–1973) revisited: ecology of the lower Connecticut River 1973–2003. American Fisheries Society, Monograph 9, Bethesda, Maryland. (Originally published in 1976)

Naylor, E. 1965. Effects of heated effluents upon marine and estuarine organisms. Pages 63–103 *in* Sir F. S. Russell, editor. Advances in marine biology, Volume 3. Academic Press, London.

Normandeau, D. A. (No date). The effects of thermal releases on the ecology of the Merrimack River. Report to the Public Service Company of New Hampshire, mimeo. Normandeau Associates, Bedford, New Hampshire.

Raymond, J. E. G. 1964. The marine fauna of Southampton waters. Pages 118–120 *in* F. J. Monkhouse, editor. A survey of Southampton and its region. British Association for the Advancement of Science, Southampton, UK.

Raymond, J. E. G., and H. Currie. 1964. The production of zooplankton in Southampton water. Int. Rev. Ges. Hydrobiol. Hydrogr. 49:185–232.

Young, W. D. 1969. Milliken Station studies. Pages 315–336 *in* R. T. Oglesby and D. J. Allee, editors. Ecology of Cayuga Lake and the proposed Bell station. Cornell University, Water Resources and Marine Science Center, Publication No. 27, Ithaca, New York.

American Fisheries Society Monograph 9:65–124, 2004

Fishes of the Lower Connecticut River and the Effects of the Connecticut Yankee Plant

BARTON C. MARCY, JR.
Ecological Sciences Division, NUS Corporation
1910 Cochran Road, Pittsburgh, Pennsylvania 15220, USA

From 1965 through 1972, a comprehensive survey was conducted of the fishes in the lower Connecticut River between Higganum (km 29.8) and Essex (km 11.3). For a period of 3 years before the Connecticut Yankee Atomic Power Company (CYAP) plant began to operate, collections were repeatedly made in the area to determine the status of the fish populations living there. These assumed-to-be-normal data could then be compared with similar data collected after plant operation had commenced in 1968 in order to measure any changes that had taken place in species composition, abundance, growth, movement, distribution, and catchability. Prior to this study, no data had ever been gathered on the resident fish species inhabiting the lower Connecticut River, nor was there any information available on the size or importance of the sport fishery in the Haddam area. This study describes changes in the population dynamics and the sport fishery of fishes in the vicinity of the plant, some of which result from the wintering of fish in the heated discharge.

Fish Populations of the Lower Connecticut River

Methods

Three collecting stations (6, 6A, and 4A) were located within the area of increased water temperature as predicted by Stone and Webster Engineering Co. (1964). Stations 6B and 6C were established in 1968 to sample the discharge canal after its construction. The limits of this distribution pattern of stations were defined by two control stations, one (9) located 5.7 km upstream and the other (2) located 12.8 km below the discharge canal. A third control station (4C) was located in the Salmon River cove and was used to indicate whether warm water passing by the mouth of the Salmon River would cause any change in the movement of fishes up or down the Salmon River. The locations, depths, years sampled, and gear used at each station are indicated in Table 1 and the location of each station is shown in Figure 1.

Initially, in 1965, 12 additional stations were sampled outside the model area, between Essex and Enfield Dam. These stations were eliminated in 1966 when it was found that their species composition did not differ significantly from control stations 9 and 2. In 1966, a separate program was initiated to sample juvenile shad at 16 stations from Essex (km 11.3) to Northampton, Massachusetts (km 138).

The gear used in the repetitive sampling during the survey years consisted of a bag seine to sample the shore areas and a bottom otter-door trawl to sample the channel areas. Stations 9, 6, 6A, and 2 were sampled by bag seine with a corresponding trawl in order to cover both the shore and channel areas. Station 4C was sampled only by bag seine and station 4A by trawl. The nylon bag seine was 14.2 m long and 2.4 m deep and was made of 1-cm knotless bar mesh. Its bag was 2.4 m in length and was buoyed, and the seine was hauled by lines 30.5 m long. The 4.9-m semi-balloon mid-water trawl was towed with 25 m lines at 1–2 knots between the same two points at each station on each sampling day. The bag seine and the trawl were fished in the same fashion and at the same location on each sampling date so that the use of the gear would reflect a "standard unit of effort"; for example, if seasonal differences occurred from year to year at a particular station, the catch would reflect them, whether they resulted from differences in year-class strength or other causes.

TABLE 1. Location of bag-seine and trawl stations in the lower Connecticut River between km 11.3 and 29.8, 1965–1972.

River km	Station[c]	Town	Location	Average depth (m)	Gear[b] used
29.8	9	Higganum	Higganum Creek—mouth (west bank)	4.6	1, 2
24.9	6	East Haddam	Flats north of CYAP[a] discharge point (east bank)	0.9	1
24.1	6B	East Haddam	CYAP canal—log boom to 600 m up canal	1.5	2
24.1	6C	East Haddam	CYAP canal—log boom to mouth (150 m)	1.5	2
24.1	6A	Haddam	Opposite CYAP discharge point	7.6	1, 2
24.9	4C	East Haddam	Salmon River—Salmon Cove (east bank)	1.2	1
22.5	4A	East Haddam	Mouth of Salmon River—cans "47" and "45" (east bank)	3.0	2
11.3	2	Essex	Brockway Island (east bank)	3.0	1, 2

[a] CYAP = Connecticut Yankee Atomic Power Company.

[b] 1 = Bag seine; 2 = Trawl.

[c] All stations were sampled twice a month from April through November, during daylight hours. Samples were also taken at 6B and 6C during the winter months. Sampling was increased to a weekly schedule in 1970 at stations 6, 6A, 6B, 6C, and 4A, to monitor more adequately the rapid changes in population density that often occur within the area affected by the heated discharge.

The record made of each collection noted the fish species, temperature, dissolved oxygen, and tidal phase. Counts of all fishes usually were made in the field, but when large numbers were collected, they were returned to the laboratory for sorting and counting. Specimens to be saved were fixed in 10% neutral formalin, later rinsed in water, and preserved in 40% isopropyl alcohol. In 1966 and 1967, total wet weights were taken of each species collected at each station for biomass estimates. Length, weight, stomach contents, and scale and spine samples were taken from white perch *Morone americana*, brown bullhead *Ameiurus nebulosus*, white catfish *A. catus*, and spottail shiner *Notropis hudsonius* periodically throughout the survey years.

Results and Discussion

During the survey years (1965–1972), 229,933 fish representing 44 species and 21 families were collected in 1,398 bag-seine and trawl hauls between Higganum (km 29.8) and Essex (km 11.3) (Table 2). Six additional species have been reported in the commercial and sport fisheries, making a total of 50 fish species known to inhabit this area of the lower Connecticut River. A listing of these species and the families to which they belong, with description of their principal use of the lower Connecticut River, is provided in Appendix II.

Ten of the species collected at station 2 at Essex (km 11.3) in the lower river were of marine origin and did not move far enough into freshwater to come under the influence of the heated discharge. These species are indicated in Table 2. Except for the fishes indicated as marine, all species listed in Table 2 have been collected in the vicinity of the plant at some time during the investigation. Species that were taken in the general vicinity of the plant, but not in the heated discharge, were the Atlantic menhaden *Brevoortia tyrannus*, banded killifish *Fundulus diaphanus*, redfin pickerel *Esox americanus*, fallfish *Semotilus corporalis*, chain pickerel *E. niger*, rock bass *Ambloplites rupestris*, creek chub *Semotilus corporalis*, and threespine stickleback *Gasterosteus aculeatus*.

Nine of the 44 species made up 95% of the total number of the young (young of the year and subadults—fish not sexually mature) and adult fishes caught: blueback herring *Alosa aestivalis* (34%), Atlantic menhaden (18%), white perch (14.7%), spottail shiner (12.9%), brown bullhead (7.1%), white catfish (4%), alewife *A. pseudoharengus* (2.1%), golden shiner *Notemigonus crysoleucas* (1.3%), and American shad *A. sapidissima* (1.3%) (Table 2). The abundance of the American shad is incorrectly represented by the above percentage because adults of this species are rarely caught by the methods employed in this study.

Five of the nine most abundant species were permanent residents; the others were either anadromous or summer migrants that were mostly

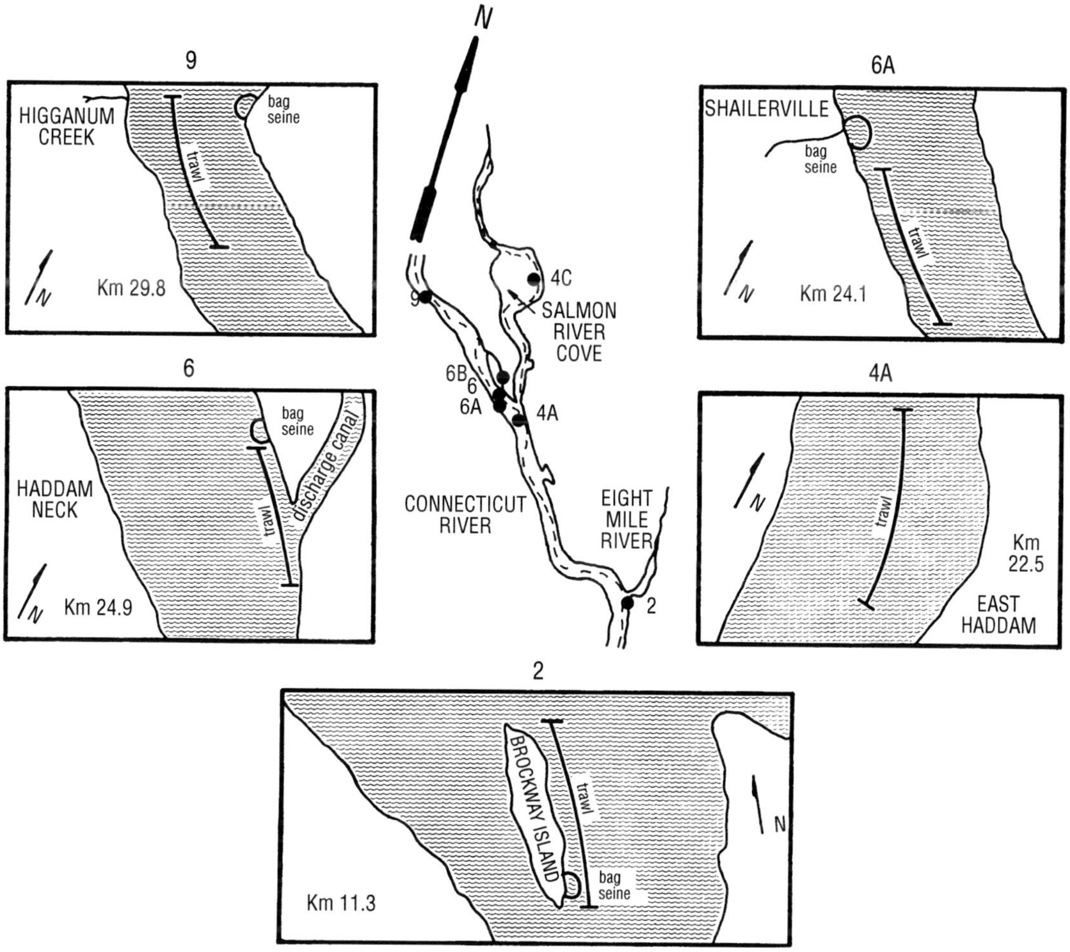

FIGURE 1. Location of control fish sampling stations 9, 2, and 4C and stations within the area of increased temperature, 6, 6A, and 4A. Trawl and bag-seine sampling locations are indicated for each station.

represented by young-of-the-year fish. The white perch was the dominant resident species, followed by the spottail shiner, brown bullhead, white catfish, and golden shiner; together they comprised 40% of the total catch.

In the samples, there were 43,039 adult fishes represented by 43 species (Table 2). The five most abundant were all permanent adult residents of the river: spottail shiner (32%), brown bullhead (23%), white catfish (12.4%), white perch (10.5%), and golden shiner (5.4%).

The young of the year totaled 186,894 and represented 35 species (Table 2). Anadromous species and marine species that utilized the river as a nursery area made up the majority of the production of young of the year in the river (67%). The blueback herring was by far the most abundant of these species (41.2%), followed by the menhaden (22.1%), alewife (2.5%), and American shad (1.5%). Other anadromous or marine species with young stages that utilized the lower Connecticut River included the mummichog *Fundulus heteroclitus*, Atlantic silverside *Menidia menidia*, bay anchovy *Anchoa mitchilli*, bluefish *Pomatomus saltatrix*, rainbow smelt *Osmerus mordax*, hogchoker *Trinectes maculatus*, Atlantic tomcod *Microtgadus tomcod*, striped killifish *F. majalis*, crevalle jack *Caranx hippos*, fourspine stickleback *Apeletes quadracus*, and Atlantic herring *Clupea harengus*. These species, however, accounted for only 1% of the total catch of young in the river. Resident fish

TABLE 2. Summary of the relative abundance of young and adult fishes collected between Higganum (km 29.8) and Essex (km 11.3) in the lower Connecticut River, 1965–1972.

Species	Young		Adults		Young and adults	
	Number	Percent	Number	Percent	Number	Percent
Blueback herring	76,927	41	1,209	3	78,136	34
Atlantic menhaden	41,341	22	215	1	41,556	18
White perch	29,203	16	4,543	11	33,746	15
Spottail shiner	15,882	9	13,725	32	29,607	13
Brown bullhead	6,198	3	10,065	23	16,263	7
White catfish	3,858	2	5,352	12	9,210	4
Alewife	4,680	3	227	1	4,907	2
Golden shiner	644	0.3	2,310	5	2,954	1
American shad	2,852	2	40	0.1	2,892	1
Banded killifish	1,537	1	495	1	2,032	1
Bluegill *Lepomis macrochirus*	1,218	0.7	587	1	1,805	1
Mummichog	569	0.3	909	2	1,478	1
Yellow perch *Perca flavescens*	165	0.1	872	2	1,037	1
Johnny darter *Etheostoma nnigrum*	256	0.1	730	1.7	986	0.4
American eel *Anguilla rostrata*	16	–[b]	761	1.8	777	0.3
Atlantic silverside[a]	560	0.3	89	0.2	649	0.3
Bay anchovy[a]	324	0.2	24	0.1	348	0.2
Common carp *Cyprinus carpio*	6	–	308	0.7	314	0.1
Rainbow smelt	132	0.1	131	0.3	263	0.1
White sucker *Catostomus commersonii*	109	0.1	122	0.3	231	0.1
Bluefish	228	0.1	1	–	229	0.1
Hogchoker	105	0.1	39	0.1	144	0.1
Common shiner *Luxilus cornutus*	7	–	86	0.2	93	–
Northern pike *Esox lucius*	18	–	53	0.1	71	–
Redfin pickerel	16	–	42	0.1	58	–
Black crappie *Pomoxis nigromaculatus*	6	–	22	0.1	28	–
Atlantic tomcod[a]	13	–	9	–	22	–
Largemouth bass *Micropterus salmoides*	7	–	12	–	19	–
Fallfish			18	–	18	–
Channel catfish *Ictalurus punctatus*	5	–	13	–	18	–
Crevalle jack[a]	5	–	4	–	9	–
Striped bass *Morone saxatilis*			7	–	7	–
Striped killifish[a]	5	–			5	–
Chain pickerel			4	–	4	–
Fourspine stickleback[a]	1	–	3	–	4	–
Northern pipefish[a] *Syngnathus fuscus*			4	–	4	–
Weakfish[a] *Cynoscion regalis*			3	–	3	–
Rock bass			2	–	2	–
Creek chub			1	–	1	–
Threespine stickleback			1	–	1	–
Atlantic herring[a]	1	–			1	–
Sea lamprey *Petromyzon marinus*			1	–	1	–
Totals	186,894	100	43,039	100	229,933	100

[a] Marine species found outside the influence of the plant toward the mouth of the river.

[b] Dash denotes less than 0.05%.

species made up 32% of the total production of young of the year in the river. The white perch showed the highest production (15.6%), followed by the spottail shiner (8.5%), brown bullhead (3.3%), and the white catfish (2.1%). The other 16 species accounted for only 2.5% of the total catch of young resident fish species.

Total weights were taken of the adults and young of each species collected during 1966 and 1967 at all stations. During these 2 years, 160,793 fish representing 25 species weighed 1,153 kg. Twelve species made up 98% of the total biomass of fish collected in the lower Connecticut River (Table 3). The white perch was the dominant species by weight and, with the brown bullhead and common carp, made up the more than half of the measured biomass of fish collected in the lower Connecticut River.

TABLE 3. Percentage of total weight for 12 of 25 fish species collected at all sampling stations in the lower Connecticut River during 1966 and 1967.

Species	Percent total weight
White perch	26.7
Brown bullhead	18.3
Common carp	12.6
Alewife and blueback herring	10.1
White catfish	9.2
American eel	5.7
Spottail shiner	5.1
Yellow perch	3.6
White sucker	3.6
Golden shiner	1.5
Sunfishes	1.3
Northern pike	0.4

The abundance and composition of fish species at each station are given in Tables 4 and 5. The abundance of each species varied considerably among stations, and specific fluctuations and occurrences at each station can be found in Table 5. The following is a discussion of more pertinent comparisons among stations.

The largest number of fish collected per net haul (253) and the highest diversity of species (33) was found at control station 2 (km 11.3) in the lower estuary (Tables 4 and 5). As mentioned previously, 10 marine species, not collected farther upriver, were collected here. Young-of-the-year Atlantic menhaden dominated the catch. This species and the rainbow smelt were dominant at this station as compared with other stations. Station 4C in the Salmon River cove had the second highest number of species collected (28) and the largest number of dominant species (14). (Dominance refers to those species that exhibited the highest abundance at a single station when compared to all stations.) Fallfish, chain pickerel, and rock bass were found only at this station. Station 9 (km 29.8), located 4.9 km upstream from the CYAP plant, had the third largest number of fish per net haul (207) and ranked third in number of species. This station, however, had only five dominant species, notably the river herrings and northern pike.

TABLE 4. Abundance of all fish species at each station and sampling year between km 29.8 and 11.3, lower Connecticut River, 1965–1972.

Station	1965	1966	1967	1968	1969	1970	1971	1972	All years
9	564(17)[a]	226(33)	243(26)	226(27)	25(18)	34(32)	407(30)	19(29)	207(212)
6A	100(17)	70(31)	325(26)	103(27)	131(19)	11(48)	85(49)	10(43)	87(260)
6	294(21)	352(32)	187(26)	187(28)	398(29)	103(49)	344(48)	31(44)	224(277)
6B	NO SAMPLE			238(11)	51(18)	33(39)	35(37)	69(31)	100(136)
6C	NO SAMPLE						105(35)	95(31)	100(66)
4A	543(14)	271(17)	110(14)	88(14)	17(15)	29(34)	58(35)	8(31)	120(174)
4C	392(15)	91(21)	78(12)	66(8)	53(4)	58(9)	36(12)	27(8)	120(89)
2	25(5)	264(30)	110(26)	285(20)	465(15)	54(30)	461(30)	269(28)	253(184)
All stations	31,079	35,335	24,990	24,212	22,819	11,125	54,172	15,425	219,157
Effort	89	164	130	135	118	241	276	245	1,398
Catch per haul	349	215	192	179	193	46	196	63	156

[a]Numbers represent numbers caught per bag seine and trawl haul, and in parentheses, the number of collections. Trawl and bag-seine data have been pooled where both were taken concurrently.

TABLE 5. Abundance and composition of fish species (number per net haul) collected at each station between km 29.8 and 11.3 of lower Connecticut River, 1965–1972. Numbers in parentheses identify the station with the greatest abundance.

Species	Station 9 km 29.8	Station 6A km 24.1	Station 6 km 24.9	Station 4A km 22.5	Station 4C km 24.9	Station 2 km 11.3
Blueback herring and alewife	(80.15)	50.61	77.75	10.43	79.35	79.56
Atlantic menhaden[a]	71.78	9.75	1.84	0.93	1.80	(124.40)
White perch	11.07	4.26	41.73	(52.73)	4.52	32.60
Spottail shiner	19.42	9.21	(56.57)	17.46	10.72	4.21
Brown bullhead	3.32	0.76	(22.80)	5.41	2.52	0.31
White catfish	6.11	5.07	3.69	(12.31)	0.17	1.74
Golden shiner	1.16	0.42	(4.95)	0.29	2.02	0.11
American shad[a]	1.00	0.82	(4.70)	4.13	0.94	0.90
Banded killifish	2.32	0.22	2.20		(8.38)	0.05
Bluegill	1.86	0.60	2.45	0.03	(3.97)	0.04
Mummichog	0.50	0.32	0.81		(4.69)	1.81
Yellow perch	0.55	0.04	(2.00)	0.65	0.72	0.18
Johnny darter	0.53	0.25	1.02	0.01	(4.89)	0.02
American eel	0.31	0.78	0.34	(1.53)	0.16	0.59
Atlantic silverside						(2.35)
Bay anchovy						(1.89)
Common carp	0.03	0.01	0.26	0.07	(0.29)	0.01
Rainbow smelt	0.08	0.15	0.19	0.10	0.03	(0.07)
White sucker	0.19	0.03	0.14	0.09	(0.37)	
Bluefish						(1.22)
Hogchoker	0.02	0.05	0.15	(0.16)	0.09	0.11
Common shiner	0.02		(0.09)			0.03
Northern pike	(0.15)	0.02	0.06	0.01	0.03	
Redfin pickerel	(0.02)	+[b]	0.12		(0.02)	
Black crappie	0.02	+			(0.11)	0.01
Atlantic tomcod						(0.12)
Largemouth bass	+		0.02		(0.07)	
Fallfish					(0.08)	
Striped killifish						(0.03)
Crevalle jack						(0.03)
Striped bass		+	0.01	+		(0.01)
Chain pickerel					(0.04)	
Fourspine stickleback						(0.02)
Northern pipefish						+
Weakfish						(0.02)
Rock bass					(0.02)	
Creek chub	+					
Threespine stickleback			+			
Atlantic herring						+
Number of species	27	25	27	21	28	33
Number of dominant species	5	0	7	4	14	12
Number of net hauls	222	260	277	174	89	184

[a] Young of the year only.

[b] + indicates present, but in minimal numbers.

Of the three stations located within the influence of the heated discharge, other than those in the discharge canal itself, station 6A near the bank opposite the discharge canal had the lowest recorded catch per net haul (87). Of the 25 species collected at this station, none dominated. Station

6, located on the flats just north of the discharge canal, had the second largest number of fish caught (224 per haul) including seven dominants and 27 species. Many of the important resident fish species were represented in the catch at this station (i.e., brown bullhead, spottail shiner, yellow perch, and golden shiner). Two important resident species, the white catfish and white perch, were most abundant at station 4A located at the confluence of the Connecticut and Salmon rivers. Here, the catch averaged 120 fish per haul, the number of dominant species was four, and the number of species was 21, the lowest species diversity of all stations.

Young fishes belonging to 33 species were collected between km 29.8 and 11.3 during the survey. The seasonal occurrence and abundance of each species can be found in Table 6. Most species first became available in the collections in June, the young of northern pike and spottail shiner being the only two species collected in May. The greatest variety of species was col-

TABLE 6. Seasonal occurrence and abundance (number per net haul) of young fishes collected at stations 9, 6, 6A, 4A, 4C, and 2 between km 11.3 and 29.8 of lower Connecticut River, 1965–1972. Numbers in parentheses indicate the month of peak abundance.

Species	May	June	July	August	Sept.	Oct.	Nov.	Dec.
Blueback herring and alewife		1.38	115.86	(182.09)	115.15	29.58	12.04	
Atlantic menhaden			5.38	116.57	(136.36)	4.48	4.94	
White perch		0.30	25.32	(66.05)	51.02	15.46	12.39	2.90
Spottail shiner	+[a]	0.42	27.57	(28.10)	14.53	6.16	8.45	7.75
Brown bullhead		0.02	9.01	(11.42)	2.07	1.62	0.27	0.02
White catfish		0.02	4.00	(10.32)	4.66	2.14	0.45	
Golden shiner		0.04	0.30	1.28	(2.18)	0.45	0.06	0.02
American shad			1.81	4.03	(7.43)	3.58	1.26	
Banded killifish		+	0.43	(4.65)	3.06	0.97	0.26	
Bluegill		0.02	0.29	(3.33)	2.28	0.99	0.05	
Mummichog		0.02	0.51	(1.80)	0.20	0.60	0.12	0.07
Yellow perch		0.08	(0.17)	0.14	0.11	0.07		0.02
Johnny darter		0.05	0.19	(0.66)	0.30	0.25	0.09	0.07
American eel (elvers)		+	(0.07)	+	0.01			
Atlantic silverside			0.46	0.41	0.48	(0.98)		
Bay anchovy			0.06	0.09	(2.27)	0.09		
Common carp			+					(0.08)
Rainbow smelt			0.31	0.04	0.13	(0.35)	0.01	
White sucker		(0.14)	0.03	+			+	0.02
Bluefish			0.01	(0.92)	0.40			
Hogchoker			+	0.08	0.22	0.24	0.03	
Common shiner		(0.03)			0.02			
Northern pike	+	0.03	+	0.02	(0.06)	0.00		
Redfin pickerel		0.03	0.02	(0.04)				
Black crappie					(+)			
Atlantic tomcod				0.01	0.01	(0.06)		
Largemouth bass					(+)			
Crevalle jack				0.01	(0.02)			
Fourspine stickleback				(+)				
Atlantic herring					(0.01)			
Number of species	2	19	26	28	29	22	18	9
Number of species at maximum abundance	0	2	2	15	8	4	0	1
Number of net hauls	147	154	180	189	124	151	114	40

[a] + indicates present, but in minimal numbers.

lected in September and more species reached maximum abundance in August than in other months.

The seasonal occurrence and abundance of adult fish species collected during the survey years are shown in Table 7. The maximum number of species (31) occurred in July and more species (13) reached maximum abundance in this month than in any other. The number of fishes recorded was lowest in December, when most summer mi-

TABLE 7. Seasonal occurrence and relative abundance (number per net haul) of adult fishes collected at stations 9, 6, 6A, 4A, 4C, and 2, between km 11.3 and 29.8 of lower Connecticut River, 1965–1972. Numbers in parentheses indicate month of peak abundance.

Species	March	April	May	June	July	August	Sept.	Oct.	Nov.	Dec.
Spottail shiner	6.71	3.03	5.96	4.76	6.31	(26.55)	10.30	7.20	14.04	6.98
Brown bullhead	(29.86)	1.40	3.73	5.75	8.20	3.88	3.03	1.19	0.25	0.12
White catfish	1.14	0.25	0.80	2.20	(4.98)	2.55	4.20	0.53	0.21	
White perch	0.14	0.81	4.16	(4.27)	4.09	3.21	1.21	1.93	1.49	0.05
Golden shiner	0.43	0.48	0.40	0.61	(2.83)	1.55	1.21	1.32	0.61	0.17
Blueback herring		0.02	1.22	2.47	(2.57)	0.02	0.03	0.55		
Mummichog		0.03	0.95	0.54	0.46	(1.38)	0.29	0.25	0.32	
Yellow perch	0.86	0.50	1.05	0.98	(1.09)	0.48	0.35	0.30	0.23	0.35
American eel		0.04	0.11	0.59	(1.34)	0.84	1.12	0.32	0.04	0.02
Johnny darter	0.28	0.28	0.45	0.27	(1.61)	0.80	0.26	0.10	0.10	0.27
Bluegill	0.14	0.01	0.15	0.22	(0.95)	0.56	0.31	0.16	0.03	
Banded killifish		0.13	0.69	(1.14)	0.69	0.24	0.06	0.03	0.02	0.05
Common carp	(2.71)	0.01	0.05	0.11	0.12	0.12	0.20	0.02	0.02	0.07
Alewife		0.10	0.54	(0.64)	0.12		+[a]	0.02	0.01	
Atlantic menhaden								(1.39)	0.04	
Rainbow smelt	1.00	(1.09)	0.05		+	0.02	0.02	0.01	+	
White sucker	(2.43)	0.02	+	0.14	0.17	0.05	0.09	0.01	0.07	0.10
Atlantic silverside					+	0.03	0.13	(0.22)	0.08	
Common shiner				(0.24)	0.32	0.14				
Northern pike			0.01	0.06	0.03	(0.07)	0.05	0.04	+	
Redfin pickerel			+	0.05	0.04	0.08	0.06			(0.13)
Hogchoker					0.04	(0.05)	0.02	0.01		
Bay anchovy					0.01	0.02		(0.11)		
Black crappie	(1.00)		0.02		+	0.01	0.01			
Fallfish			0.05		(0.06)					
Largemouth bass	(0.14)			+	0.01	+	+			
Atlantic tomcod							+	+	(0.06)	
Striped bass				(0.02)	0.01		+			
Chain pickerel	(0.57)									
Northern pipefish					(0.02)					
Crevalle jack					(+)					
Weakfish							(0.02)			
Fourspine stickleback							(0.02)			
Rock bass			(0.01)							
Threespine stickleback			(+)							
Creek chub					(+)					
Number of species	16	18	24	22	31	25	28	24	21	11
Number of species at maximum abundance	6	1	2	5	13	4	2	3	1	1
Number of net hauls	7	92	147	154	180	189	124	151	114	40

[a] + indicates present, but in minimal numbers.

grants and anadromous fishes had left the river and most resident species had moved into the coves and tributaries to overwinter.

The abundance of all fish species combined for each sampling year is shown in Table 4. The greatest number per net haul was caught in 1965 and the lowest in 1970. The volume of water discharge from the Connecticut River into Long Island Sound each year may be responsible for some of the variation in catch observed over the survey years. The gear used would be more efficient during years of lower flow because it can be set more precisely and because the fish would occupy a smaller area and be more available. The flow during 1965 was much lower than in 1970.

The occurrence and abundance of the young of the year of each species collected during each year are shown in Table 8. There was a greater number of species (27) and more reached maximum abundance in 1966 than in any other year. At the other extreme stands 1972, when only 12 species were collected and none reached maximum abundance. Again, the difference in the flow rates of the river between 1966 and 1972 may have influenced the efficiency with which the fishing gear could be operated. Table 8 is useful in determining the abundance of each species

TABLE 8. Abundance (number per net haul) and occurrence of young-of-the-year fishes collected at stations 9, 6, 6A, 4A, 4C, and 2 between km 11.3 and 29.8 of lower Connecticut River for each sampling year. Numbers in parentheses indicate year of peak abundance.

Species	1965	1966	1967	1968	1969	1970	1971	1972
Blueback herring and alewife	38.61	51.96	102.75	55.61	(168.67)	15.56	28.31	3.50
Atlantic menhaden	15.75	0.32	4.42	55.43	10.77	4.19	(114.13)	39.07
White perch	(92.13)	50.55	6.93	20.08	7.54	4.80	29.79	2.44
Spottail shiner	(27.10)	25.34	19.25	10.91	11.05	1.80	13.36	
Brown bullhead	1.36	(11.49)	0.10	1.00	0.09	0.06	11.27	
White catfish		1.78	(17.00)	2.52	5.41	0.27	0.05	0.27
Golden shiner	0.96	(1.17)	0.61	0.06	0.02	0.76	0.07	
American shad	0.28	0.36	0.73	0.58	1.94	3.04	(6.83)	1.20
Banded killifish	0.16	0.81	(1.94)	0.41	0.13	0.08		
Bluegill	0.68	(2.07)	1.09	0.40	0.05	0.24	1.20	0.06
Mummichog	0.07	(1.20)	0.78	0.48	0.24	0.01	0.39	0.64
Yellow perch	(0.19)	0.04	0.11	0.04	0.04	0.10	0.06	0.05
Johnny darter	0.14	(0.69)	0.38	0.23	0.18	0.02	0.01	0.01
American eels (elvers)		0.01			0.02		(0.04)	
Atlantic silverside	0.01	(2.41)	0.06	0.02		0.02		
Bay anchovy		0.13	0.36	0.27		(1.06)		0.04
Common carp						(0.01)		
Rainbow smelt		(0.82)	0.01	0.09				
White sucker	(0.25)	0.01		0.02	0.02			
Bluefish		0.01		0.01		0.02	(1.05)	
Hogchoker	0.08	0.14	0.08	(0.17)			0.11	
Common shiner			(0.03)					
Northern pike		(0.05)			0.01	0.01	0.01	
Redfin pickerel		(0.09)	0.01					
Atlantic tomcod		(0.04)				0.02		
Largemouth bass			(0.02)					
Crevalle jack		(0.02)						
No. of species	19	27	23	22	19	22	19	12
No. of species at maximum abundance	4	14	3	1	2	2	4	0
No. net hauls	89	164	130	124	100	202	204	183

through the years, since it reflects the annual production or year-class strength of the fishes found in the lower Connecticut River during the survey years. Table 9 provides an overview of the occurrence and abundance of individual adult species from 1965 through 1972. The number of adult fishes collected remained relatively constant throughout the sampling period (Table 9). The year in which the most species reached maximum abundance was 1967, and the lowest was 1968.

Life History Studies on Major Resident Fishes

During the course of this investigation, life history data were obtained on the four most abundant species of fishes residing in the lower Connecticut River: white perch, spottail shiner, brown bullhead, and white catfish. These four species represented 88% of the total resident fish species collected in the lower river, and the catfishes and white perch constituted 69% of the sport

TABLE 9. Abundance (number per net haul) and occurrence of adult fishes collected at stations 9, 6, 6A, 4A, 4C, and 2 between km 11.3 and 29.8 of lower Connecticut River for each sampling year. Numbers in parentheses indicate year of peak abundance.

Species	1965	1966	1967	1968	1969	1970	1971	1972
Spottail shiner	33.30	37.97	30.81	15.98	14.49	10.50	(39.31)	3.90
Brown bullhead	3.42	7.12	(7.91)	3.99	5.42	0.65	1.02	1.22
White catfish	(5.36)	4.94	3.81	2.44	2.87	0.66	0.49	0.31
White perch	2.94	(6.96)	5.57	5.05	0.88	0.67	1.57	0.55
Golden shiner	0.40	1.06	1.12	0.66	0.67	1.39	1.17	(1.84)
Blueback herring	0.02	0.34	(4.97)	0.55	0.25	0.52	0.76	0.38
Mummichog		0.54	0.41	0.42	0.52	(1.47)	0.32	0.03
Yellow Perch	0.40	(1.51)	1.50	1.08	0.25	0.15	0.27	0.35
American eel	2.14	(1.86)	0.96	0.64	0.28	0.07	0.09	0.04
Johnny darter	1.35	0.49	(2.30)	0.16	0.34	0.31	0.15	0.09
Bluegill	0.28	0.59	0.42	0.22	(0.73)	0.14	0.18	0.26
Banded killifish	0.16	0.81	(1.94)	0.41	0.13	0.08		
Common carp	(0.13)	0.09	0.12	0.12	0.07	0.06	0.05	0.11
Alewife	0.01	0.09	(1.00)	0.19		0.10		0.04
Atlantic menhaden			0.06	0.03		(0.99)		
Rainbow smelt	0.04	0.24	0.14	(0.50)				
White sucker	(0.26)	0.13	0.17	0.04	0.06		0.01	
Atlantic silverside		0.06	0.01	0.08		(0.16)	0.05	
Common shiner						0.01	(0.12)	
Northern pike	0.03	0.02		0.01	0.03	0.01	(0.05)	0.03
Redfin pickerel	0.03	(0.11)	0.01					
Hogchoker		(0.09)	0.04		0.03			0.02
Bay anchovy			(0.12)				0.03	
Black crappie								(0.03)
Largemouth bass			(0.02)					
Atlantic tomcod				(0.05)	0.01			
Striped bass								(0.01)
Crevalle jack					(0.01)			
Weakfish					(0.01)			
Fourspine stickleback							(0.01)	
Threespine stickleback							(0.01)	
No. of species	19	22	24	22	21	20	21	19
No. of species at maximum abundance	3	5	7	2	5	3	5	3
No. net hauls	89	164	130	124	100	202	204	183

fishery catch in the vicinity of the plant (1965–1969) and 76% of the catch in the discharge canal (1972–1973).

Prior to this study, the only life history information ever published for fishes of the lower Connecticut River has concerned the shad, alewife, and blueback herring. Life-history information on the white perch and spottail shiner follows; the brown bullhead and white catfish are treated on pages 79–109.

White Perch

The dominant resident fish in the lower Connecticut River is the white perch. It comprised 14.7% of the total catch and ranked first in total weight of both young and adults. It ranked third in the trawl catches in the discharge canal, second in the sport fishery catch in the discharge canal and comprised 26% of the sport fishery catch of 25 species taken in the vicinity of the Connecticut Yankee plant. Its eggs and larvae ranked second among the 16 species collected in the ichthyoplankton. The white perch was found in all sections of the lower 74 km of the river, and it does not exhibit a definite spawning run from brackish to fresh water. Spawning occurred in May and June, 72% of it in the latter month and 99% above river kilometer (rkm) 41.8. Young white perch were found in the CYAP discharge canal in temperatures up to 40°C, and all sizes were available in the canal during all months except in January and February. Apparently, it did not overwinter there.

Stomach contents were analyzed from 759 young-of-the-year white perch collected in biweekly bag-seine and trawl hauls at six sampling stations between Higganum (km 29.8) and Essex (km 11.3) from June through December, 1966–1969. Lengths of 12,457 and weights of 553 young white perch were taken. Standard length was measured to the nearest millimeter and weight to the nearest gram. Frequency of occurrence of food organisms in all stomachs containing food were analyzed, and the fish were grouped into three size-classes (0–40, 41–60, and 61–100 mm) in order to reveal changes in feeding with increase in size.

Copepods, insects (primarily chironomids), and cladocerans were the food organisms most frequently consumed, but oligochaetes, amphipods, ostracods, polychaetes, isopods, and fish were also taken in significant quantities (Table 10).

Young white perch fed less on plankton and more on benthic organisms as their size increased (Table 11). Occurrence of cladocerans declined from 24.2% to 9.7%, copepods from 40.7% to 20.4%, and amphipods from 21.2% to 4.6%. The occurrence of insects in the stomachs increased from 0.4% to 31.7%, oligochaetes from 6.0% to 10.7%, and polychaetes from 2.6% to 20.4%. White perch in the 0.40-mm class fed primarily on plankton (i.e., copepods [40.7%], cladocerans [24.2%], and amphipods [21.2%]). At 41–60 mm, the principal food organisms were chironomids (31.7%), copepods (26.6%), and cladocerans (16.5%). In fish of the 61–100-mm class, copepods and insects occurred in 20.4% and 23.6% of the stomachs, respectively. At this size feeding preferences shifted to benthic organisms. For example, oligochaetes and polychaetes made up 31.1% of the diet, as compared with 9–10% for fish in the 0–60-mm class. White perch that were 41–100 mm had a more diversified diet and consumed such foods as isopods, mysids, arachnids, fish eggs, and plant tissue. Fish were consumed with about equal frequency by all size-classes. One 36-mm specimen contained a white perch larva 13.5 mm long, the largest encountered in any stomach.

Webster (1942) reported that the most important food organisms consumed by young white perch in a Connecticut lake were cladoceran and chironomid larvae, and Richards (1960) and Taub (1966) found that copepods, amphipods, and cladocerans were the major items consumed by young white perch in lakes in New York and Massachusetts, respectively. Richards noted that plankton consumption decreased, while the consumption of benthic organisms increased as the fish increased in size.

The mathematical relationships of body length to weight and to size of scale are given in Table 12. The formula to convert standard length to total length is also given.

Young-of-the-year white perch were somewhat larger, the nearer the river mouth they were collected (Figure 2). Mansueti (1961), Miller (1963), and Wallace (1971) have also noted an upstream-to-downstream increase in size of young white perch in other Atlantic Coast rivers. The greatest increase in growth of young white perch occurred during August and September (Figure 3). This species grew faster and attained a larger size in each successive year from 1966 through 1969. Mansueti (1961) suggested that the growth

TABLE 10. Food of young white perch in the lower Connecticut River, 1965–1969, in percent frequency of occurrence.

Item	Year 1966	1967	1968	1969
Number of stomachs	161	198	200	200
Number empty	6	9	4	16
Tardigrada			0.5	
Oligochaeta	33.5	19.0	9.7	
Polychaeta (*Scolecolepides viridis*)	7.7	2.6	13.3	0.5
Cladocera (Chydoridae; Daphnidae; Macrothricidae; Bosminidae)	63.2	62.4	59.7	6.0
Ostracoda	17.4		7.7	20.7
Copepoda	87.1	65.6	84.2	17.9
Insecta (Trichoptera: Hydropsychidae; Diptera: Chironomidae pupae and larvae)	67.1	70.4	55.1	56.5
Nematoda			0.5	
Isopoda (*Idotea* spp.; *Cyathura polita*)	3.2	0.5	2.0	
Mysidacea (*Neomysis americana*)	3.9	0.5		
Amphipoda (Gammaridae)	17.4	12.7	19.4	9.2
Decapoda (*Crangon* spp.)	0.6			
Arachnoida (*Acarina* spp.)		0.5		
Invertebrate eggs			70.4	3.3
Fish (*Notropis hudsonius; Anguilla rostrata; Morone americana* larvae)	3.9	3.2	1.5	
Fish eggs			1.0	
Plant matter		0.5	8.7	46.2
Sediment	71.0		17.3	14.1

TABLE 11. Food of young white perch in the lower Connecticut River according to size. Percentages refer to stomachs that contained food.

Item	Total length (mm) 0–40	41–60	61–100
Number of stomachs	245	231	99
Number empty	14	13	6
Oligochaeta	6.5%	6.0%	10.7%
Polychaeta	2.6	4.1	20.4
Cladocera	24.2	16.5	9.7
Copepoda	40.7	26.6	20.4
Insecta	0.4	31.7	23.6
Isopoda		0.5	1.1
Mysidacea		0.5	1.0
Amphipoda	21.2	4.6	6.5
Arachnida		0.5	
Fish	2.6	1.2	2.2
Fish eggs		0.5	
Plant matter		0.9	1.1
Sediment	1.8	6.4	3.2

of young white perch was negatively correlated with the number of individuals (i.e., large populations were associated with a small amount of growth during the first year). He suggested that competition for available food may be a primary growth-limiting factor. The same trend has been noted in the present study. The catch of young-of-the-year white perch per net haul was 87.1 in 1965, 48.5 in 1966, 6.9 in 1967, 20.0 in 1968, and 7.5 in 1969, thus indicating a continuous reduction in population size at the same time the rate of growth was steadily increasing.

Data on the age and growth of the white perch in the lower Connecticut River have been published by Marcy and Richards (1974). The following is a brief summary of the results of this study. These data were obtained from 1,706 white perch collected in a 63-km section (Wilson to Essex) of the lower Connecticut River between 1966 and 1969. The formation of an annulus on the scales began at least as early as the end of March and was completed by the third week in June. Most fish showed a completed annulus be-

TABLE 12. Growth relationships in young-of-the-year white perch in the lower Connecticut River.

Growth relationships	Equation	Correlation coefficient
Body–scale[a]	SL = 20.04 + 37.00 R	(r = 0.91)
Length conversion[b]	TL = 1.57 + 1.2 SL	
Length–weight[c]	log W = 4.6977 + 3.0212 log SL	(r = 0.97)

[a] Body length–scale size relationship for 717 young-of-the-year white perch collected between 1966 and 1969. SL is standard length in millimeters and R is scale radius in millimeters. The presence or absence of scales was determined in 32 individuals between 14 and 22 mm SL. Those between 14 and 16 mm did not possess scales. Scales first appeared on fish at 17 mm and at that time averaged 0.20 mm in diameter. Fish between 18 and 22 mm SL possessed three to six well-defined circuli, which agrees well with the intercept (20.0 mm) determined by a regression analysis. Scales were taken from the side of the body under the posterior two-thirds of the dorsal fin above the lateral line (Mansueti 1960). Scales were measured along the axis from the focus to the right anterior scale margin.

[b] Conversion of standard length (SL) to total length (TL) for 171 young white perch between 28 and 106 mm.

[c] Length–weight relationships for 551 young white perch, where W is weight in grams and SL is standard length in millimeters.

tween late April and mid-May. The mathematical relationship between standard body length (L, cm) and scale radius (R, mm) for 930 fish of age I and older (70 mm and longer) was $L = 4.44 + 3.39\ R$ (r = 0.88). The radius of a typical scale was proportional to an increase in standard length in the older age-groups. More than half of the recorded growth (53%) occurred in the first 2 years of life. The conversion from total length (TL) to standard length (SL) is given by the equation SL = −1.31 + 0.83 TL. The growth rate was faster and the lifespan shorter (maximum age 8 years) in the Connecticut River and other waters of Connecticut than in 10 other populations inhabiting waters from Lake Ontario to North Carolina (Table 13). Growth in the upper Connecticut River (km 119–176) was similar to that in the lower river for the first 2 years of life, but thereafter, none of the 10 populations compared grew as fast as did the white perch in the lower Connecticut River. The maximum empirical length at age VII, found in this study, was 285 mm SL, and this was also higher than any given for the 10 other populations.

There was a reduction in the average calculated length in all age-classes in each succeeding

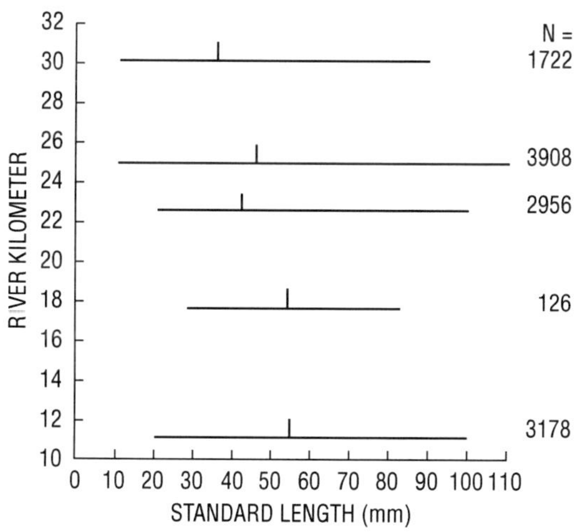

FIGURE 2. Mean standard length (vertical lines) and range in length (horizontal lines) of young-of-the-year white perch collected at various distances from the mouth of the Connecticut River, 1966–1969.

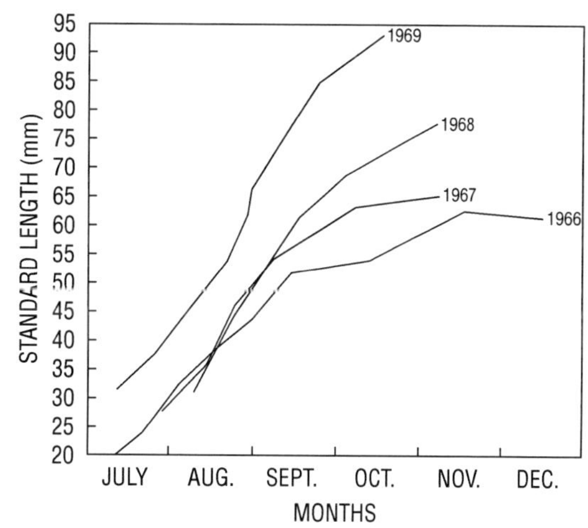

FIGURE 3. Growth in length of young-of-the-year perch in the lower Connecticut River: 1966 (n = 8,236), 1967 (n = 881), 1968 (n = 2,469), and 1969 (n = 871).

TABLE 13. Mean calculated standard lengths (mm) at annulus formation for white perch in various waters of eastern North America.

Annulus formation	Location and authority[a]										
	A	B	C	D	E	F	G	H	I	J	K
1st	64	74	81	71	66	73	68	73	61	64	56
2nd	105	127	146	148	107	117	110	113	99	97	87
3rd	136	171	177	186	137	138	130	135	124	120	120
4th	156	193	204	211	160	153	144	151	144	142	146
5th	174	209		230	184	167	154	164	159	159	168
6th	187	222		255	202	181	162	181	172	176	188
7th	202	233		282		193	170	198	186	193	204
8th	219	251				204	174	209	199	209	218
9th	232	264				248		221	214	220	
10th		232	273				257		241	222	229

[a] A = Lake Ontario—Sheri and Power (1969).
B = Quabbin Reservoir, Massachusetts—Taub (1966).
C = Connecticut River (km 119–176)—Roger Reed (personal communication).
D = Connecticut River (km 11–74)—Present study.
E = State of Connecticut—Whitworth and Sauter (1972).
F = Delaware River Estuary—Miller (1963).
G = Delaware River near Artificial Island—Wallace (1971).
H = Patuxent Estuary, Maryland—Mansueti (1961).
I = James River, Virginia—St. Pierre and Davis (1972).
J = York River, Virginia—St. Pierre and Davis (1972).
K = Roanoke River, North Carolina—Conover (1958).

year from 1959 to 1965 (Table 14). This trend is the opposite of what Lee's phenomenon would indicate, and it suggests that stunting has occurred. White perch may characteristically form stunted populations. That such a phenomenon may exist calls for further research on the effec-

TABLE 14. Calculated lengths of white perch in the lower Connecticut River.[a]

Year-class	Number	Calculated standard length (mm) at each annulus						
		1	2	3	4	5	6	7
1965	110	69.8						
1964	80	68.1	138.2					
1963	71	70.5	142.1	173.9				
1962	60	72.6	156.4	190.1	206.1			
1961	27	83.6	169.6	201.2	216.5	225.8		
1960	7	74.2	172.9	207.2	228.9	242.7	251.8	
1959	1	96.8	199.5	225.1	250.7	265.0	273.5	282.1
Weighted mean		71.2	148.4	185.9	211.2	230.3	254.5	282.1
Increment		71.2	77.2	37.5	25.3	19.1	24.2	27.6
% of growth to the seventh year		25.2	27.4	13.3	9.0	6.7	8.6	9.8
Number		356	246	166	95	35	8	1

[a] No significant difference between the sexes was observed when the lengths and weights of 989 fish of age I and older were compared. With data for sexes combined, the length–weight relationship for white perch in the lower Connecticut River was $\log W = -1.425 + 2.891 \log L$ ($r = 0.96$), where W is weight in grams and L is standard length in centimeters.

tive utilization and management of the white perch—for example, the possibility of opening the entire lower Connecticut River to commercial fishing for this species.

Spottail Shiner

This species was the second most abundant species of resident fish in the lower Connecticut River and made up 12.9% of the total catch. It ranked fourth among the 25 species collected in the CYAP discharge canal. It was present during all months in the canal and was most abundant at station 6 on the flats just north of the canal. It was collected in the canal in water temperatures as high as 39.2°C.

The stomach contents were examined from 123 spottail shiners collected during 1968 at four stations (6, 6A, 6B, and 4A) within the area warmed by the heated discharge. The major items found were aquatic insects (1.0%), diptera (9.6%), fish eggs (1.9%), plant tissue (1.9%), sediment (1.9%), and digested matter (84%). Spottail shiners collected in the discharge canal (6B) contained 30% more aquatic insects (station 6B = 33.3%, 6 = 2.1%, 6A = 3.6%, 4A = 0%) than at any of the three stations outside the canal, indicating that this species apparently utilized the initial increase in food organisms occurring in the discharge canal in 1968, the first year after plant operation (Massengill, personal communication).

Spottail shiners were the major food of the adult brown bullheads that overwintered in the discharge canal; for example, they comprised 96% of the total dry weight found in bullhead stomachs in March 1972 (Massengill 1973). In the densely populated canal, the normal bullhead diet of invertebrates shifted to the spottail shiner during the winter months. Trawl hauls made during February and March at five stations along the canal at 152-m intervals from its mouth along 762 m of its length showed that spottail shiners were distributed in a pattern similar to that of the brown bullhead, that is, more abundant near the mouth of the canal (Table 15).

The length–weight relationship for 126 spottail shiners (42–107 mm) collected at the four stations in the Haddam area during 1968 is described by the regression $\log W = -3.0460 + 2.0552 \log L$ ($r = 0.71$), where W is weight in grams and L is standard length in millimeters.

Movement of Tagged Fishes in the Vicinity of the Connecticut Yankee Plant

Six species of fishes were tagged in the vicinity of the CYAP plant in order to obtain some measure of the mobility of the fish populations in the Haddam area. Because of the abundance and importance of the catfishes in this part of the river, our tagging program concentrated on the brown bullhead and white catfish, and the catfish population was used as an indicator of changes taking place after the commencement of plant operation. These two species made up 36% of the total adult fish population in the lower Connecticut River. Brown bullheads constituted 67% and white catfish 19% of the 25 species that were collected in the discharge canal after the plant began to operate, and they ranked second and third in the sport fishery of the canal. In addition, the movement of tagged yellow perch, white perch, northern pike, and channel catfish are briefly considered.

Tagging Methods

A fluorescent orange "T-bar" spaghetti tag (70 mm long) with a return message "return—Box 367, Essex, Ct.—reward" was implanted with a tagging gun just below the dorsal fin. The tag and tagging gun have been described by Dell (1968). Total lengths were taken of all fish and the weight of many. A one-dollar reward, a thank-you letter, and information pertaining to the original capture of the fish was forwarded to each fisherman whenever a tag was returned. Colored posters that

TABLE 15. Distribution of spottail shiners based on trawl hauls made during February and March 1972 at 152-m intervals in the CYAP discharge canal.

	Distance from mouth of canal (m)				
	0–152	152–305	305–457	457–610	610–762
No. per trawl haul	5.3	4.1	2.5	2.0	1.4

requested tag-return information were placed in all fishing areas on the lower river.

Tag Retention

Thirty white catfish and brown bullheads were tagged and held in the laboratory for 2 months to observe the healing of the wound caused by tagging and also how well the tags were retained. A shallow wound was first observed about 2 weeks after tagging and this began to heal in about a month. Shallow, round wounds were observed in both species of catfishes in the field up to a year after tagging, but such wounds were not seen in the other four species. This was probably the result of the abrasive contact with vegetation that resulted from the bottom-dwelling nature of these catfishes. Nevertheless, there was no evidence to suggest either a significant tag loss or fish mortality, and all fish recaptured after a year were in good condition with no visible wounds. Bullheads were recaptured up to 3 years and white catfish up to 4 years after tagging. The length frequency of recaptured fish indicated no differential mortality associated with the size of the fish tagged. Stobo (1972) applied the same size tag to yellow perch and found that retention was good and that mortality was insignificant. In the present study, yellow perch were recaptured in good condition up to 2 years after tagging.

Analysis of Tagging Results

Of those fishes tagged each year, the percentage recaptured ranged from 3.8% to 10.7%. The lowest return was associated with the first year's tagging and undoubtedly resulted from lack of publicity and the fact that no reward was offered. Between 1969 and 1972, the percentage of tags returned was relatively constant (7.3% to 10.7%). Of 9,817 fish tagged, 918 were recaptured, an average return of 9.4% (Table 16). White catfish showed the highest percentage return (10.5%), bullheads were second at 9.6%, followed by northern pike, channel catfish, yellow perch, and white perch. The low percentage of white perch recaptured was probably due to the few specimens tagged in relation to the extremely large population of white perch in the lower Connecticut River. More fish were returned through the sport fishery than by our own recaptures (495 vs. 423). It is interesting to note that fewer bullheads were caught by sport fishermen than by trawl, while fewer white catfish were caught by trawl. Perhaps bullheads spend more time near the bottom and are thus more vulnerable to the trawl.

That brown bullheads and white catfish were recaptured as long as 3 years and 4 years after tagging, respectively (Table 17), does not seem to be remarkable, since La Faunce (1965) recaptured white catfish up to 10 years and brown bullheads 8 years after tagging. In the Connecticut

TABLE 16. Number of fish tagged and recaptured in the lower Connecticut River, CYAP discharge canal, and Salmon River, 1968–1973.[a]

	Number tagged			Number recaptured	
Species	Conn. R., CYAP canal	Salmon R.	Percent recaptured	Sport fishing	Trawling
Brown bullhead	5,871	636	9.6	297	325
White catfish	2,181	199	10.5	171	81
Channel catfish	11		9.1	1	0
White perch	232	339	2.1	7	5
Yellow perch	211	126	8.9	18	12
Northern pike	11		9.1	1	0
Totals	8,517[b]	1,300[b]	9.4	495	423

[a] All tagging took place from March through May except during November 1969. Six different species were tagged in 1968 and 1969, but only bullheads and white catfish were tagged in 1970 and 1971 and only bullheads, white catfish, and northern pike in 1972. Fish were collected for tagging in Higganum Creek (km 29.8), the CYAP discharge canal (km 24.1), the mouth of the Salmon River (km 22.5), and the narrows located 2.4 km up the Salmon River.

[b] Of the 9,817 fish tagged, 7,525 were in the CYAP discharge canal.

TABLE 17. Numbers recaptured in each year for the six species of fishes tagged in the lower Connecticut River and Salmon River between 1968 and 1973.

Species	Number tagged	Tagging year	Recapture year 1968 N	1969 N	1970 N	1971 N	1972 N	1973 N	Total recaptured in all years N	%
Brown bullhead	174	1968	1	2	1	0	1	0	5	2.9
	3,484	1969		177	99	64	22	0	362	10.4
	539	1970			30	17	8	0	55	10.2
	1,146	1971				67	37	0	104	9.1
	1,164	1972					94	2	96	8.2
White catfish	137	1968	9	4	0	0	0	0	13	9.5
	1,163	1969		36	82	40	9	2	169	14.5
	967	1970			31	24	3	0	58	6.0
	87	1971				9	2	0	11	12.6
	26	1972					0	0	0	0.0
Channel catfish	0	1968								
	11	1969		0	1	0	0	0	1	9.1
White perch	506	1968	5	1	0	0	0	0	6	1.2
	65	1969		3	3	0	0	0	6	9.2
Yellow perch	206	1968	8	5	1	0	0	0	14	6.8
	131	1969		13	2	1	0	0	16	12.2
Northern pike	2	1968	0	1	0	0	0	0	1	50.0
	7	1969	0	0	0	0	0	0	0	0.0
	0	1970								
	0	1971								
	2	1972	0	0	0	0	0	0	0	0.0

River, the white catfish showed a higher percent return in subsequent years (Table 17), indicating that this species may not be vulnerable to recapture for some time after tagging or as age increases. McCammon and Seeley (1961) also found that the number of returns of white catfish increased successively for 4 years after tagging.

In studying the population dynamics of the major fish species possibly affected by the CYAP discharge, particularly the white catfish and brown bullhead, it became necessary to determine how homogeneous the populations were within the Haddam area or how much mixing occurred between fish inhabiting the overwintering areas (e.g., the CYAP canal, the Salmon River, and Chapman Pond) (Table 18). Of the fishes tagged in the CYAP canal, 35.1% returned to the canal and 15% were recaptured in the Salmon River. Of those tagged in the Salmon River, 68% remained in that river and 5.5% were recaptured in the discharge canal. Fish migrated from both the CYAP canal and the Salmon River to Chapman Pond with about equal frequency. Of the fish tagged in the canal, 48% were recaptured in the Connecticut River, while only 25% of the fish tagged in the Salmon River were recaptured in the Connecticut. The indications are that a large portion of the population in the Salmon River remains stationary. Within this population, however, white catfish are evidently a more migratory species than are brown bullheads, since 47.1% of the white catfish and 18.1% of the bullheads tagged in the Salmon River were recaptured in the Connecticut. A greater percentage of the white catfish was also recaptured in Chapman Pond. McCammon and Seeley (1961) have noted that the bullheads they tagged in the arms of a lake tended to remain there. Yellow perch also migrated from area to area in the lower Connecticut; 20% of those tagged in the canal were later recaptured in the Salmon River, and 40% tagged in the Salmon River were recaptured in the canal. Considerable mixing of the species within all of the tagging areas is indicated.

Movements of Catfishes

Returns from the four major species tagged in the Haddam area, including the Salmon River, indicate that they are far from stationary and that they could traverse the entire 85.3 km of the lower

TABLE 18. Number and percentage of recaptures in each area for the five species of fishes tagged in the CYAP canal and the Salmon River, 1968–1973.[a]

Species	Area tagged	Canal Number	Canal %	Salmon River Number	Salmon River %	Chapman Pond Number	Chapman Pond %	Connecticut River Number	Connecticut River %
Brown bullhead	CYAP canal	216	43.0	79	15.8	5	1.0	202	40.2
	Salmon River	6	5.0	90	75.0	2	1.7	22	18.3
White catfish	CYAP canal	36	17.8	25	12.3	8	3.9	134	66.0
	Salmon River	1	2.8	17	47.3	1	2.8	17	47.2
White perch	CYAP canal	0		2	66.7	0		1	33.3
	Salmon River	0		2	66.7	0		1	33.3
Yellow perch	CYAP canal	1	10.0	2	20.0	0		7	70.0
	Salmon River	2	40.0	2	40.0	0		1	20.0
Northern pike	CYAP canal	0		0		0		1	100.0
	Salmon River	0		0		0		0	
All species	CYAP canal	253	35.2	108	15.0	13	1.8	345	48.0
	Salmon River	9	5.5	111	67.7	3	1.8	41	25.0

[a]The Oneida Lake trap nets and mid-water trawls that were used to collect fish for tagging were assumed to be nonselective for fish 12 cm or more. Fish were tagged as they became available, regardless of species and without selection for size, but only fish in good condition were chosen.

Connecticut River from Old Saybrook to Enfield Dam after tagging (Table 19). The mean distance traveled from the Haddam area for all five species was 20.7 km upstream and 5.6 km downstream, but twice as many fish moved in a downstream direction. The white catfish moved the farthest, that is 23 km downstream and 61.2 km upstream, an average of 15.4 km from the tagging site. In a California river, Pelgen and McCammon (1955) found that tagged white catfish moved an average of 16.3 km upstream and 14 km downstream. In our study, the yellow perch moved 54.7 km upstream and 23.3 km downstream and averaged 13.5 km from the tagging site. The brown bullhead moved the least of all, averaging only 3.6 km from the tagging site. McCammon and Seeley (1961) similarly reported a mean movement of tagged brown bullheads of 2.7 km with a maximum of 26 km. One channel catfish was recaptured 54.7 km and a northern pike 29 km upstream from the Haddam area. Fish tagged in the lower Salmon River made rather rapid early spring movements both into the Connecticut River and up the Salmon River.

The swift rates of movement of some of the tagged fish were unexpected, although Pelgen and McCammon (1955) reported that a white catfish moved 40 km in 8 d and Pelgen (1954) reported a movement of 88 km in 60 d. In the present study, one white catfish moved 66 km upstream in 74 d during spring freshet, a yellow perch moved 60 km in 59 days, and a brown bullhead 18 km in 51 d.

White catfish and brown bullheads, tagged in the CYAP discharge canal during the winter months, moved considerable distances in the

TABLE 19. Summary of mean and maximum distances traveled by the four major species of fishes of the Haddam Neck tagging area, 1968–1973.

Species	Upstream Number of fish	Upstream Mean km traveled	Downstream Number of fish	Downstream Mean km traveled	Combined Number of fish	Combined Mean km traveled	Maximum km traveled Up-stream	Maximum km traveled Down-stream
Brown bullhead	44	6.1	156	3.4	296	3.6	45.0	23.3
White catfish	75	28.5	93	8.5	196	15.4	61.2	23.3
White perch	5	16.6			11	8.8	28.9	0.8
Yellow perch	5	36.5	18	10.1	28	13.5	54.7	23.3
Totals	129	20.7	267	5.6	531	8.6	61.2	23.3

lower Connecticut River after leaving the canal in the spring as a result of plant shutdown or high water temperatures. White catfish exhibited a seasonal pattern of migration (Figure 4). Since sport fishing begins throughout the lower river at about the same time each year, fishing effort was not considered to be significantly greater in any one part. After leaving the canal, white catfish moved downstream toward the river mouth in April and May, perhaps to feed. From May through August, movement was more upstream toward Enfield Dam, perhaps as a result of spawning. During the fall, fish returned closer to the Haddam area, and as winter approached, they began to be recaptured in the CYAP canal, Salmon River, and Chapman Pond, thus indicating a movement back into the Haddam area and out of the main river during the winter months. In contrast, McCammon and Seeley (1961) found no evidence of seasonal migration in the white catfish from Clear Lake, California. Brown bullheads moved downstream after leaving the canal and stayed relatively close to the Haddam area through May (Figure 5). They made more extensive movements from June through August and were again found mostly in the Haddam area by fall. In early winter, they moved into the CYAP canal, Salmon River, and Chapman Pond.

Population Changes in the Vicinity of the Plant

Before the CYAP plant began to operate, baseline data on the population of fishes in its vicinity were obtained by means of repeated, systematically executed collections made at several critically located stations. During this period, all the stations served as controls. These data could then

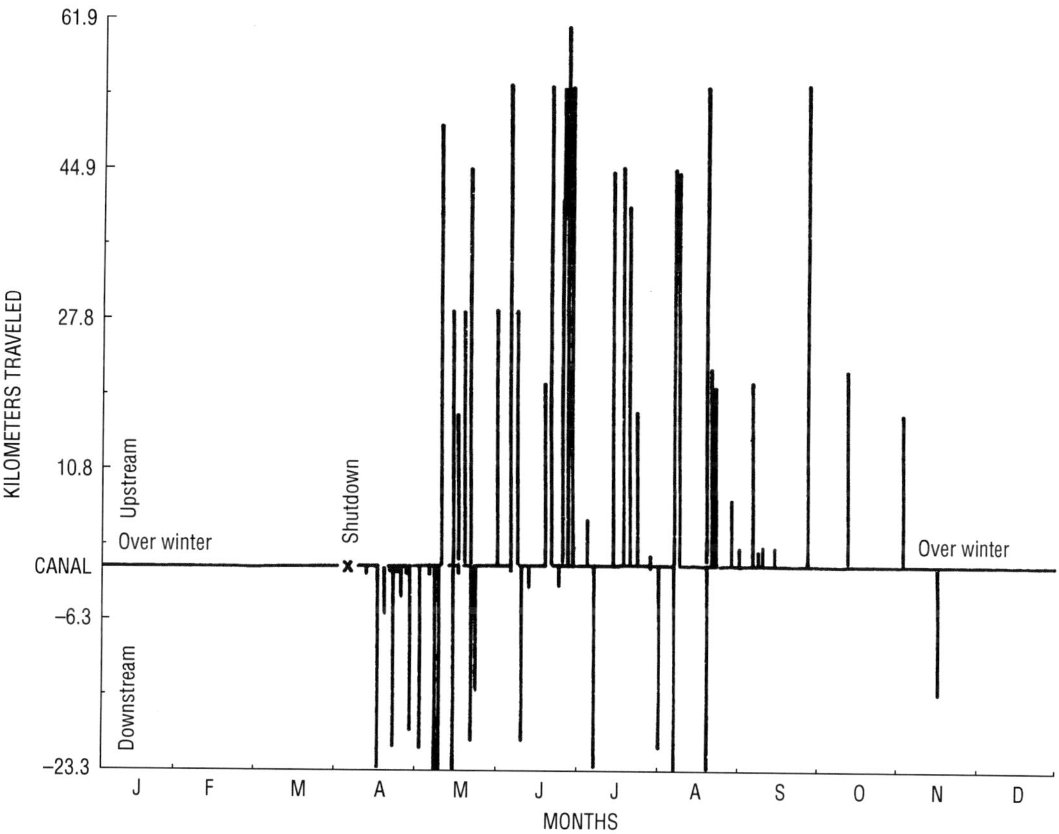

FIGURE 4. Seasonal movement of white catfish tagged in the CYAP discharge canal (horizontal line) and recaptured (vertical lines) at various distances away in the lower Connecticut River, 1968–1972.

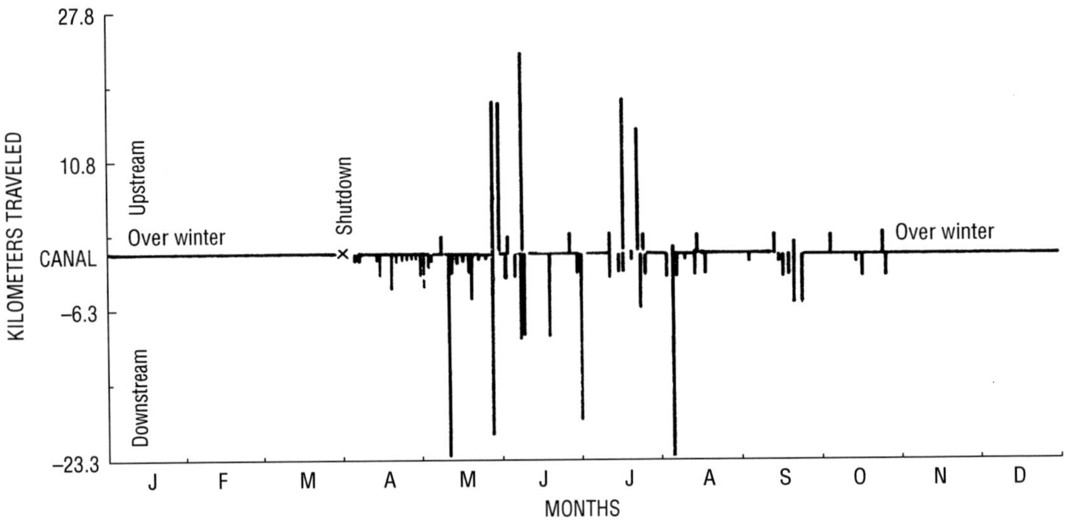

FIGURE 5. Seasonal movement of brown bullheads tagged in the CYAP discharge canal (horizontal line) and recaptured (vertical lines) at various distances in the lower Connecticut River, 1968–1972.

be compared with data collected in the same way and at the same stations after the plant commenced operation. Stations 6, 6A, and 4A then began to receive water at increased temperatures from the heated discharge, while control stations 9, 4C, and 2 did not. The basic reasoning was that any significant changes in the catch per unit effort (catch per net haul)—that is, changes in the species composition, relative abundance, and distribution of the fishes—at the affected (heated) stations could be assumed to have been caused by the heated discharge of the CYAP plant. Data from stations 6B and 6C, which were located within the discharge canal, could not be used in the foregoing statistical evaluations of before-vs-after effects because there was no discharge canal prior to 1968.

The monthly occurrence and abundance of all species of fishes, young and adults combined, for the years before and after plant operation are illustrated in Figures 6 and 7. These show the relationships between water temperatures and catch per net haul at each station and provide a visual means of comparing the changes between years as well as stations.

Control Stations

The catch data were fairly uniform throughout the before-operation years (1965–1967) at control station 9, with similar summer peaks (Figure 6). The greatest abundance occurred in 1968 (the first year after plant operation) followed by irregular fluctuations and an obvious decline in numbers collected during postoperation years. It is possible that many fish moved from this area above the plant into the heated zone, particularly into the discharge canal, and that this caused the obvious decrease in abundance after the plant commenced operation. Station 4C showed peaks of summer abundance before operation with lower, more scattered peaks afterwards. The observed decline in numbers both at station 4C and, more obviously, at station 4A may have been the result of fish not moving into the Salmon River in as large numbers because of the passage of heated water by its mouth. Many species (e.g., white catfish and brown bullhead) move into the Salmon River from the Connecticut River in early fall in order to overwinter. Since many fish overwintered or remained in the discharge canal after plant operation, the increased temperatures of the discharge undoubtedly drew fish from these areas, subsequently reducing their numbers.

There appeared to be little change or perhaps even an increase in abundance at station 2 near Essex after the plant began its operation. The addition of heated water appears not to have affected the fish at this station.

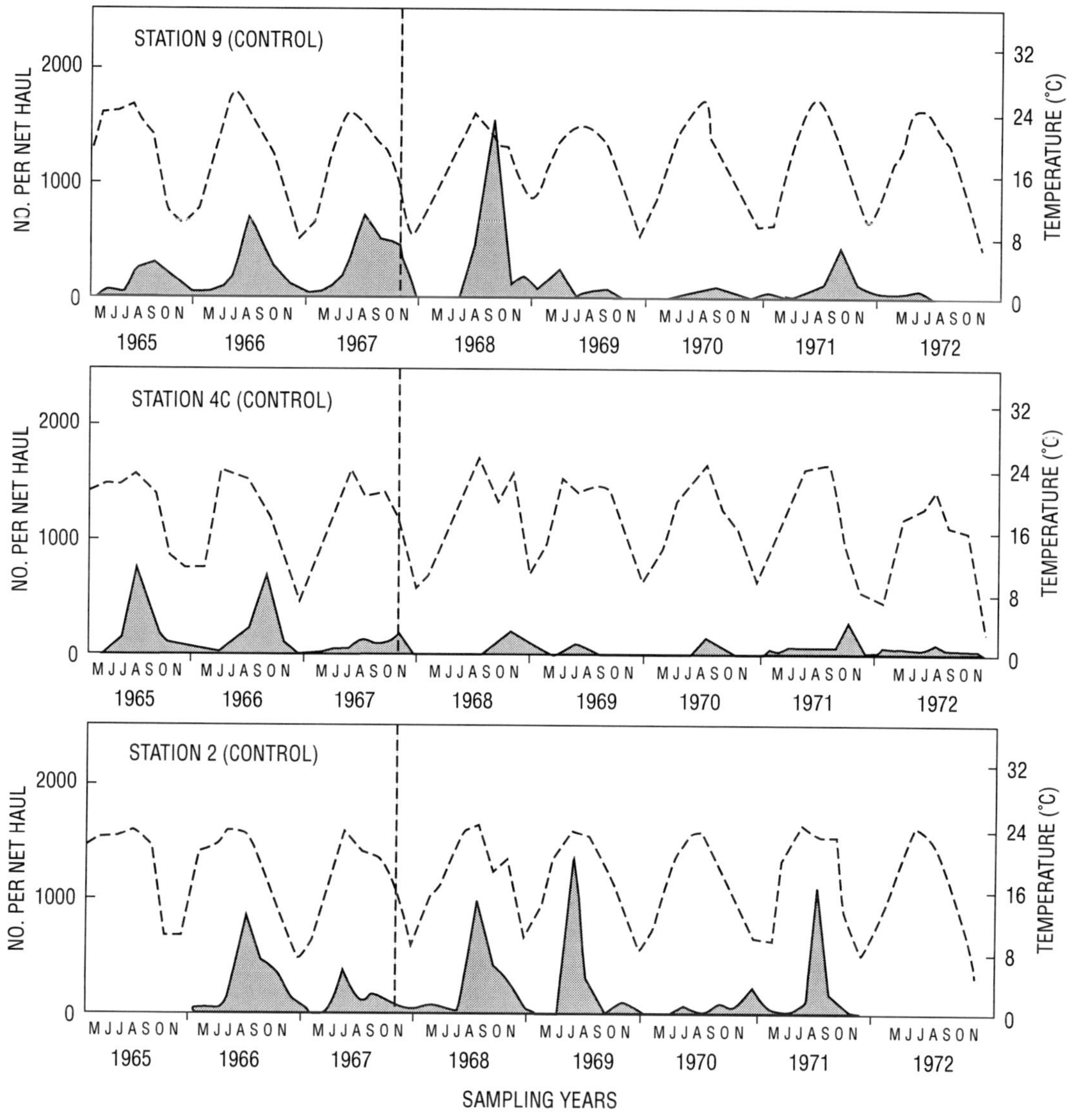

FIGURE 6. Monthly occurrence and relative abundance of all fishes collected at control stations 9, 4C, and 2 from 1965 through 1972. Solid line represents number of fish per bag-seine and trawl haul. Dotted line represents mean monthly river temperature. Vertical dashed line denotes the start of plant operation.

Heated Stations

Station 6 showed uniform peaks of abundance before plant operation and erratic peaks thereafter (Figure 7). The erratic pattern was probably caused by the movement of fish to and from the area as they practiced approach/avoidance behavior in relation to the increased temperatures. The greatest abundance of fish at station 6B, which is within the discharge canal, occurred in October and November 1968. At this time, large numbers of fish for the first time moved into the discharge canal to overwinter. Subsequent years showed rather uniform peaks, indicating a stabilization of abundance. Station 6A showed a slight post-operation decline, while station 4A at the mouth of the Salmon River showed the most obvious decline of all. It may be that fish were attracted away from this station toward or into the discharge canal with greater frequency than at other heated stations. Heated water extends over this station for longer periods than at any other.

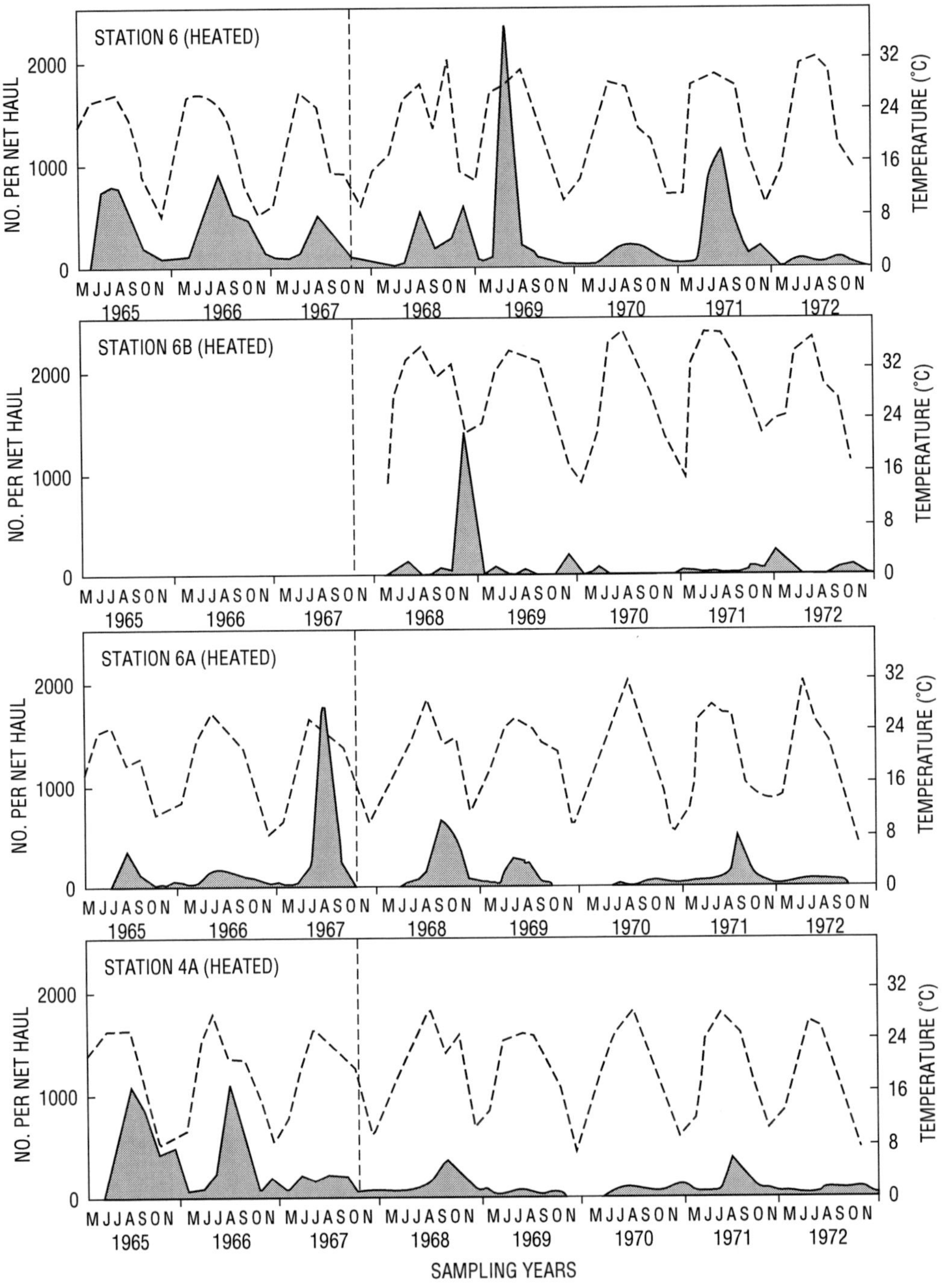

FIGURE 7. Monthly occurrence and relative abundance of all fishes collected at heated stations 6, 6B, 6A, and 4A from 1965 through 1972. Solid line represents number of fish per bag-seine and trawl haul. Dotted line represents mean monthly water temperature. Vertical dashed line denotes the start of plant operation.

Comparisons of the changes in relative abundance of young fishes with those of adults can be seen in Figures 8 and 9. After the plant commenced operation, the adult fishes showed greater declines at most stations than did the young and the young of certain species tolerated higher temperatures in

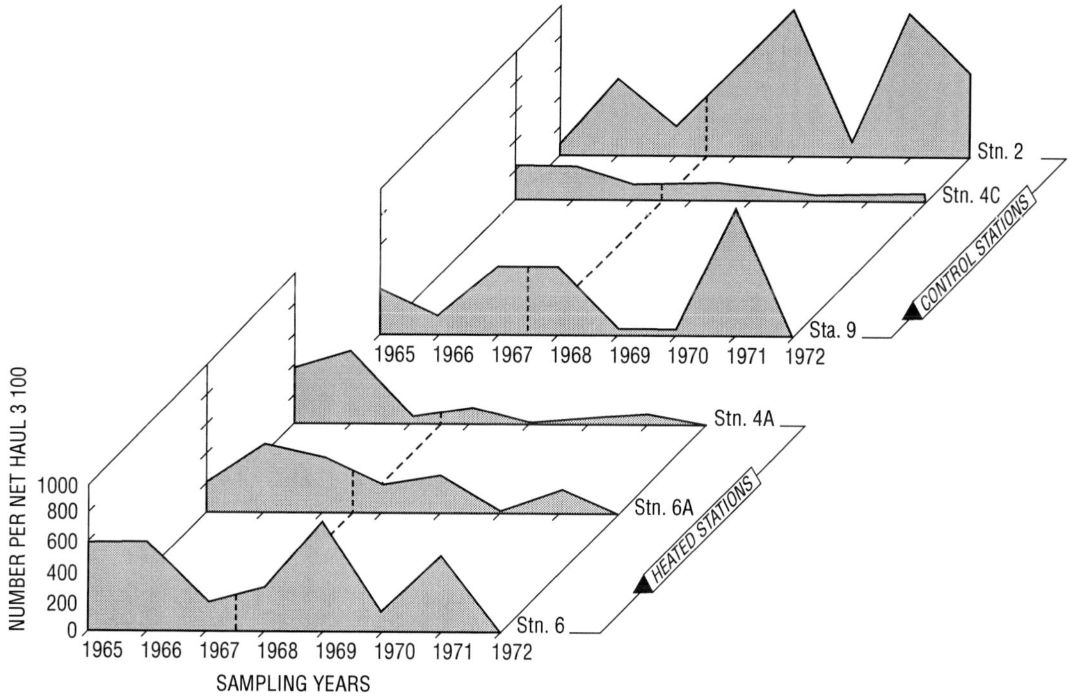

FIGURE 8. Comparison of the relative abundance of the young of the year of all fishes collected each year at each station. Dashed line denotes the start of plant operation.

the discharge canal than did the adults. Adults seem to be less tolerant of thermal stress. It should be remembered that Figure 8 also reflects relative year-class strengths, which differed from year to year.

The number of species collected per net haul,

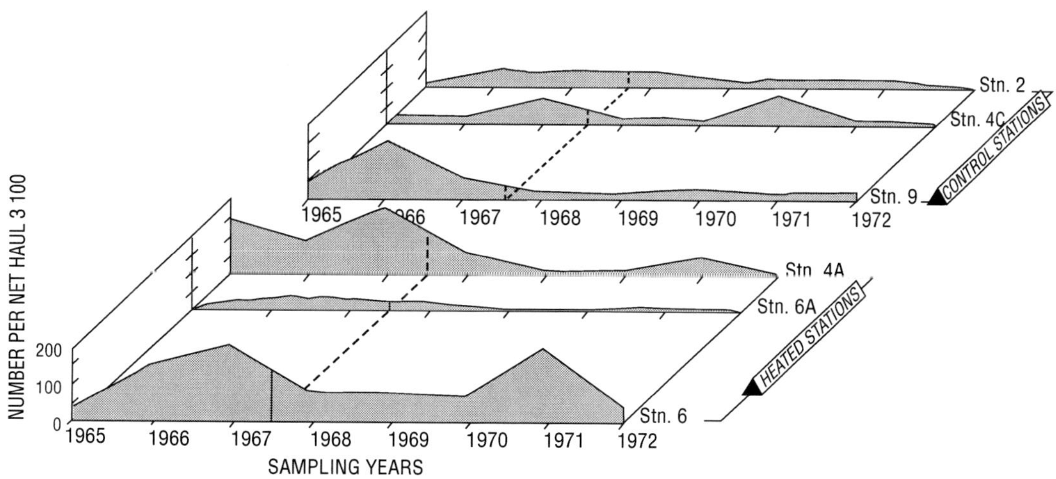

FIGURE 9. Comparison of the relative abundance of the adults of all fish species collected each year at each station. Dashed line denotes the start of plant operation.

that is the diversity, declined at both the control and heated stations after plant operation (Figure 10). Before the plant started to operate, the number of species was higher at the heated stations than at the control stations, and it remained higher through 1969 but then dropped below the number found at the control stations. As the environment stabilized after the addition of heat to the river, fewer species were found in the peripheral warmed areas, but the number of species increased in the discharge canal.

The percentage decrease in the catch per net haul for 21 important fish species before vs. after plant operation was greater at the heated stations than at the control stations (Table 20); in fact, there was an 11% greater decrease in the catch at the heated, that is, 18% as compared with 6.6% at the control stations. Several of these species showed significantly larger decreases at the heated stations than at the controls, namely white and yellow perch, American eel, white catfish, brown bullhead, and sunfishes. These species made up the vast majority of the fish that were attracted to, and then resided in, the discharge canal. Much of the decrease observed at the heated stations in the vicinity of the discharge canal was most likely caused by the movement of these fishes from the peripheral heated areas to the preferred higher temperatures within the canal. Species abundant at control stations 9 and 4C before the beginning of plant operation were probably also attracted to the discharge canal, and this may explain the decreases that occurred at those stations. For example, the decreases in brown bullhead and white catfish at control station 9 were 41% and 62.5%, respectively, and both species moved into the canal in large numbers after plant operation. Many individuals of these two species probably were attracted to the canal from the peripheral heated stations and from upriver (including station 9) because the downriver control stations 4C and 2 showed only a 5.7% and 6.4% decrease in bullheads and a 0.8% and 0% decrease for white catfish, respectively. Similarly, the catch of white perch decreased by 62.5% at control station 9, and this species moved into the canal in late summer in large numbers.

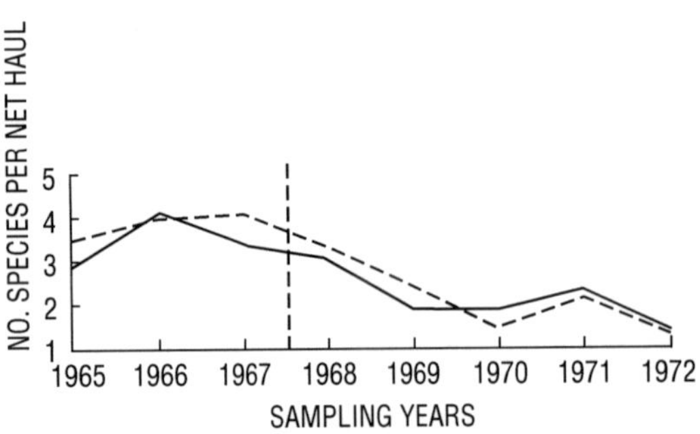

FIGURE 10. Number of species collected per net haul at the control (solid line) and heated (dashed line) stations for all sampling years. Vertical dashed line denotes the start of plant operation.

The catch of spottail shiner decreased by 34% at both the heated and control stations (Table 20). This species moved into the canal and was utilized as food by brown bullheads during the winter months (Massengill 1973). The species that increased in numbers at both the heated and control stations were young menhaden and American shad. The menhaden increased by 56% after plant operation at the control stations, but only by 18% at the heated stations, and no menhaden was ever collected in the discharge canal. Young menhaden increased the most (49%) at station 6A (the least-heated station) and decreased by 14% at the most heated station (4A) and by 2.2% at station 6. All this indicates an avoidance by the fish of water of higher temperatures.

Of the heated stations, station 4A (at the mouth of the Salmon River) showed the greatest percentage decrease in the catch per net haul after the plant commenced operation (Table 20). This station receives heated water over a longer period than do stations 6 or 6A. Station 6A receives the least heated water, and it showed the smallest percentage decrease. Station 6 declined by 15.4%. Of the control stations, station 2 showed the only increase in catch after plant operation (+8.2%), while the catch at station 4C declined the most (−16%) (Table 20). Control station 4C in the Salmon River and heated station 4A located at its mouth declined the most of all stations. As mentioned previously, the plume of heated water that passed by the mouth of the Salmon River may have prevented some fish from entering that tributary and fish may have been drawn from these areas into the warmer waters.

To determine if there were any statistical differences in the abundance of young or adult fishes in the vicinity of the discharge point after the plant operation began, factorial analysis-of-variance tests were applied to the data: 8 (years) × 46 (species) × 2 (age—adults and young) × 6 (stations).

TABLE 20. Differences between the combined mean catch per net haul before (1965–1967) and after (1968–1972) plant operation for the more important fish species collected at the heated and control stations. Numbers represent percentage increase or decrease after plant operation.[a]

Species	Heated stations 6	6A	4A	Combined	Control stations 9	4C	2	Combined
Menhaden[b]	−2.2	+49.0	−14.1	+17.7	+32.4	+1.7	+79.5	+56.0
American shad[b]	+37.1	0.0[c]	+30.8	+25.1	+6.6	+5.6	+10.2	+7.4
Alewife	−37.6	−16.2	−8.9	−23.8	−15.1	−32.6	−22.3	−24.4
Blueback herring	+11.2	−36.4	−25.2	−17.2	−23.9	−47.1	+1.5	−23.6
American smelt	−0.9	−3.8	−2.9	−2.5	−2.9	0.0	−14.1	−6.2
Northern pike	−1.9	+0.9	−0.9	−0.7	+3.9	+0.9	0.0	+1.6
White sucker	−1.1	−1.9	−8.3	−5.3	−8.2	−0.9	0.0	−3.2
Common carp	+1.9	−0.9	−3.8	−1.3	+0.9	−1.9	0.0	−0.3
Golden shiner	−3.9	−2.7	−3.7	−3.5	−1.7	+3.7	−1.0	+0.3
Spottail shiner	−28.0	−27.4	−46.9	−34.1	−42.0	−35.0	−14.8	−34.3
American eel	−13.7	−19.3	−36.3	−24.6	−12.0	+1.9	−10.1	−7.1
White catfish	−23.2	−43.4	−57.2	−45.3	−62.5	−0.8	0.0	−36.2
Brown bullhead	−33.0	−3.4	−32.6	−28.1	−41.0	−5.7	−6.4	−23.2
Mummichog	+6.6	−11.3	0.0	−1.5	+12.9	+50.0	−28.9	+16.8
Banded killifish	−50.0	−13.1	0.0	−28.1	−21.2	−53.2	−2.9	−35.2
Yellow perch	−23.5	0.0	−18.2	−15.7	+2.7	+4.3	−3.8	+1.2
Johnny darter	−30.2	−4.5	0.0	−13.9	−12.0	−25.3	0.0	−15.8
Sunfishes	−13.8	−0.8	−0.6	−5.0	+2.9	−4.6	−0.2	−0.8
White perch	−31.8	−20.0	−64.9	−45.2	−14.8	−30.9	−20.0	−12.7
Striped bass	+0.9	+0.9	+0.9	+0.9	0.0	0.0	0.0	0.0
Hogchoker	−4.7	−2.9	−1.9	−3.2	−0.8	+3.9	0.0	+1.0
Percentage lost or gained	−15.4	−12.7	−25.4	−18.0	−11.6	−16.0	+8.2	−6.6

[a] Percentage difference for species and stations was calculated by comparing the mean catch per net haul before plant operation with the mean catch after operation. The increase or decrease was then converted to a percentage difference. A square-root transformation was applied to the catch data to correct for the non-normal distribution of the percentage data.

[b] Young of the year only.

[c] 0.0 indicates no difference detected.

The analysis of variance of the catch per net haul for the adults and young of all species collected at all stations during all years showed significant differences for every interaction, and all except one were significant at the 0.01 level (Table 21). There was a significant difference between adults and young, across years, among species, among stations, and in the interaction of each with the other.

The analysis of variance of the catch per net haul for adults and young treated separately also showed significant differences for each factor and all its interactions (Tables 22 and 23). Results from the analysis-of-variance tests on the catch of young and adult fish at both control and heated stations across years and among species are summarized in Table 24. The catch of adult fish was significantly different both among species and across all sampling years at each individual station. The catch of young of the year was significantly different among species at all stations. There was, however, no significant difference in the catch of young across the years at the control stations or at heated station 6A, the one that was least affected by increased water temperatures. The F value for station 4A (the most heated station) was also higher than at station 6, the catch at these two stations being significantly lower. These results again suggest that the young were more temperature-tolerant than the older fish and therefore less affected by the changes caused by the heated water from the plant.

Using orthogonal contrasts, the mean catch per net haul of the important species of resident fishes made during the years before the plant began to operate was compared with the mean catch

TABLE 21. Analysis-of-variance table for adult and young fishes collected at all stations during all years with catch per net haul (transformed by square root of catch per net haul + 1).[a] Asterisks denote significance at $P < 0.05$ (*) or $P < 0.01$ (**).

Source of variation	Degrees of freedom	Sum of squares	Mean square	F
Age (adults and young)	1	54.0221	54.0221	116.54**
Years (1965–1972)	7	48.9185	6.9883	15.08**
Species	45	1,660.2645	36.8948	79.59**
Stations	5	33.9933	6.7987	14.67**
Age vs. years	7	18.3442	2.6206	5.65**
Age vs. species	45	740.2889	16.4509	35.45**
Age vs. stations	5	10.5762	2.1152	4.56**
Years vs. species	315	714.0590	2.2669	4.89**
Years vs. stations	35	32.2672	0.9219	1.99**
Species vs. stations	225	692.3943	3.0773	6.64**
Age vs. years vs. species	315	554.0959	1.7590	3.79**
Age vs. years vs. stations	35	25.7489	0.7357	1.59*
Years vs. species vs. stations	1,575	841.6137	0.5344	1.15**
Age vs. species vs. stations	225	289.8557	1.2882	2.78**
Error	1,575	730.0813	0.4635	
Total	4,415	6,466.5237		

[a] The four-way interaction term was used as the experimental error because the mean catch per net haul for each year was used in the analysis. The error mean square was used to calculate F values. The original data of catch-per-net-haul were transformed by taking the square root of the catch per net haul + 1 (Snedecor and Cochran 1972). The purpose of this mild transformation was to stabilize the variance, as the original data when plotted tended toward dependence (i.e. variances were proportional to the means). Transformed data when plotted showed means and variances as independent and more normally distributed. Roessler (1965) showed that the individuals caught per trawl sample followed a negative binomial distribution.

for subsequent years at the heated and the control stations (Table 25). After the plant began to operate, there was a significant reduction in the number of white perch, brown bullheads, white catfish, and spottail shiners at the heated stations. There was also a significant reduction in the number of white catfish and spottail shiners at the control stations. Only the white perch and brown bullheads were significantly reduced in number at the heated stations and not at the control stations after plant operation. Although the other species seem to have been reduced in numbers, none was significantly so.

Orthogonal comparisons were also made, using the transformed data for mean catch per haul for combined adult and young fishes of all spe-

TABLE 22. Analysis-of-variance table for adult fishes collected at all stations during all years with catch per net haul [transformed by $(x + 1)^{1/2}$]. Asterisks denote significance at $P < 0.01$ (**).

Source of variation	Degrees of freedom	Sum of squares	Mean square	F
Years (1965–1972)	7	16.9594	2.4228	21.70**
Species	45	408.7608	9.0836	81.38**
Stations	5	16.2414	3.2483	29.19**
Years vs. species	315	116.4701	0.3697	3.32**
Years vs. stations	35	8.4405	0.2412	2.16**
Species vs. stations	225	199.8678	0.8883	7.96**
Error	1,575	175.8109	0.1116	
Total	2,207	942.5509		

TABLE 23. Analysis-of-variance table for young-of-the-year fishes collected at all stations during all years with catch per net haul [transformed by $(x + 1)^{1/2}$]. Asterisks denote significance at $P < 0.05$ (*) or $P < 0.01$ (**).

Source of variation	Degrees of freedom	Sum of squares	Mean square	F
Years (1965–1972)	7	50.3033	7.1862	8.11*
Species	45	1991.7927	44.2621	49.94**
Stations	5	28.3281	5.6656	6.39**
Years vs. species	315	1151.6838	3.6561	4.13**
Years vs. stations	35	49.5756	1.4164	1.60*
Species vs. stations	225	782.3821	3.4773	3.92**
Error	1,575	1395.8841	0.8863	
Total	2,207	5449.9497		

cies at the heated and control stations before and after plant operation. The mean catch per net haul for all fish decreased significantly after plant operation ($t = 47.25$, df = 1,575, $P < 0.01$) when control and heated stations were combined for comparison. There was a significant reduction in the mean catch per net haul at the heated stations ($t = 5.48$, df = 1,575, $P < 0.01$), but not at the control stations where no significant difference was found between before-and-after plant operation years ($t = 1.84$, df = 1,575, $P > 0.05$).

We may conclude that the operation of the CYAP plant significantly reduced or displaced approximately 18% of the fish population originally found at the stations in the vicinity of the discharge point, an area of approximately 154 ha or 380 acres, that now received additional heat. The catch at the control stations declined by 6.6% and the difference attributable to heat was a decline of about 11% (Table 20).

Attraction of Fishes to the Heated Discharge

Many authors have described seasonal concentrations of fish around heated discharges especially during the cooler months, and Coutant (1970) and Langford (1972) have presented comprehensive reviews of their studies. Few field studies, however, have dealt with the effects of long-term exposures of fish, especially warmwater species, to elevated temperatures in heated discharge canals.

Since the Connecticut Yankee plant began

TABLE 23. Results of analysis-of-variance tests (F values) performed on the transformed catch-per-net-haul data for adults and young at each station for all years and all species. Asterisks denote significance at $P < 0.05$ (*) or $P < 0.01$ (**); unmarked values represent nonsignificant differences.

Stations	Adults: Across[a] years	Adults: Among[b] species	Young: Across years	Young: Among species
Controls				
9	2.89**	13.64**	1.74	5.29**
4C	2.03*	6.88**	1.98	7.44**
2	3.06**	8.72**	1.23	7.87**
Heated				
6	4.27**	22.25**	2.25*	9.43**
6A	5.21**	13.04**	1.65	9.37**
4A	5.74**	15.45**	2.41*	6.26**

[a] df = 7 and 1,575.
[b] df = 45 and 1,575.

TABLE 24. Comparison of the mean catch per net haul of the important species of resident fishes before and after plant operation at the heated and control stations, using orthogonal contrasts. Asterisks denote significance at $P < 0.05$ (*) or $P < 0.01$ (**); other comparisons were not significantly different.

Species	Computed t values: Heated stations[a]	Computed t values: Control stations[a]
White perch	9.51**	1.87
Brown bullhead	3.39**	1.61
White catfish	5.47**	2.10*
American eel	1.71	0.05
Spottail shiner	7.18**	9.02**
Common carp	0.05	0.05
Golden shiner	0.23	0.93
Yellow perch	1.02	0.05

[a] df = 1,575.

TABLE 25. Summary of the abundance of adults and young (young of the year and subadults) of fishes inhabiting the CYAP discharge canal, 1968–1972. Dashes (–) denote less than 0.005%.

Adults			Young			Adults and young combined		
Species	Number collected	Percentage total catch	Species	Number collected	Percentage total catch	Species	Number collected	Percentage total catch
Brown bullhead	16,290	59.77	Brown bullhead	9,356	86.57	Brown bullhead	25,646	67.32
White catfish	7,154	26.25	White perch	698	6.45	White catfish	7,305	19.17
Spottail shiner	1,037	3.80	Blueback herring	260	2.40	White perch	1,733	4.54
White perch	1,035	3.79	White catfish	151	1.39	Spottail shiner	1,176	3.08
Golden shiner	920	3.37	Spottail shiner	139	1.28	Golden shiner	922	2.42
Common carp	433	1.58	American shad	118	1.09	Common carp	442	1.16
Yellow perch	81	0.29	Alewife	23	0.21	Blueback herring	338	0.88
Blueback herring	78	0.28	Hogchoker	21	0.19	American shad	119	0.31
Pumpkinseed	75	0.27	Pumpkinseed	19	0.17	Pumpkinseed	94	0.24
Common shiner	37	0.13	Common carp	9	0.08	Yellow perch	82	0.21
Channel catfish	27	0.09	Channel catfish	5	0.04	Common shiner	39	0.10
Northern pike	24	0.08	Black crappie	5	0.04	Channel catfish	32	0.10
Mummichog	22	0.08	Common shiner	2	0.01	Alewife	32	0.08
White sucker	17	0.06	Golden shiner	2	0.01	Hogchoker	24	0.06
American eel	15	0.05	Yellow perch	1	–	Northern pike	24	0.06
Alewife	9	0.03	Bluegill	1	–	Mummichog	23	0.06
Redbreast sunfish	8	0.02	Mummichog	1	–	White sucker	17	0.04
Black crappie	5	0.01	Atlantic menhaden	1	–	American eel	15	0.03
Largemouth bass	4	0.01				Black crappie	10	0.02
Hogchoker	3	0.01				Redbreast sunfish	8	0.02
Johnny darter	3	0.01				Largemouth bass	4	0.01
Rainbow smelt	1	–				Johnny darter	3	–
Striped bass	1	–				Bluegill	2	–
American shad	1	–				Rainbow smelt	1	–
Sea lamprey	1	–				Sea lamprey	1	–
Totals	27,281			10,812			38,093	

commercial operation in 1967, large numbers of fish have gathered in its discharge canal, especially during the cooler months. Of particular concern, has been the possibility of fish kills occurring in the canal and the possible disruption of the winter phase in the life cycle of those species of fishes that now inhabit the canal during the cold season.

Sampling Methods

The CYAP discharge canal was sampled, generally on a weekly schedule, throughout each year from June 1968 through December 1972. Weekly sampling was considered minimal in order to define the rapid fluctuations in species composition that often result from the oscillating temperatures and rates of flow typically found in heated discharge canals. Special studies of 24-h duration were carried out at four stations within the canal. Many fishes were tagged, and matched samples with areas outside the canal were obtained to monitor such changes more accurately. A 4.9-m semi-balloon midwater trawl was used exclusively, and it was towed in a downstream direction to sample the bottom. Two stations were sampled regularly, one (6B) from a point 600 m up the canal down to the log boom (located near the mouth of the discharge canal) and one (6C) from the boom to a point approximately 150 m south of the discharge canal (see Figure 18). Temperature and dissolved oxygen were recorded for each collection.

Species Composition and Seasonal Occurrence

During the survey years (1968–1972), 38,093 fish representing 25 species were collected in 441 trawl hauls made in the discharge canal. Brown bullheads dominated the catch (67.3%) and together with white catfish made up 86.5% of the total catch. Six of the 25 species comprised 97.7% of the total catch (Table 25). About one-third of the fishes caught were young of the year, and of 18 species and 10,812 specimens that were collected, brown bullheads dominated the catch (86.6%) and, with white perch and blueback herring, constituted 95.5% of the total catch.

Large numbers of fish were first observed in the canal in the early fall of 1968, and trawl catches much greater than any previous ones were recorded. For example, the 1,466 fish per 5-min trawl haul taken in November (Table 26) represented 75% more fish than had been caught at any other of the five trawl stations sampled in the vicinity of the plant during the preoperative years. The average number of fish per trawl haul in 1968 was double that recorded in any other year. After the initial surge of fish into the canal in late 1968, the number per trawl haul in 1969 dropped to about half that recorded in 1968, and the catch per trawl haul dropped further in 1970 and 1971. It increased slightly in 1972.

The period between 1968 and 1969 was a time of transition and readjustment for the fish population in the Haddam area as well as in the discharge canal. Large numbers of fish entered the canal, but only a few species were represented. Although the total number of fish was highest in 1968, the number of species was the lowest ever recorded. The fish population became more stable after 1969 as the total number of fish decreased and more species must have filled a greater number of niches in the canal. In 1971, the fewest fish were caught, but the number of species was the highest of any year. The number of species caught was 10 in 1968, 18 in 1969 and 1970, 24 in 1971, and 19 in 1972. It is most interesting that the invertebrates in the canal showed similar periods of readjustment and stabilization (Massengill 2004 [1976], this volume).

Two distinct seasonal peaks occurred in the fish population in the canal, the higher one in March and April and another, smaller peak in November and December (Table 26). The number of fishes remained relatively constant from May through October. Brown bullheads and white catfish were most abundant in March and April, golden shiners in June, white perch and spottail shiners in August, and common carp in December (Table 27). These six species made up 97.7% of the total catch. The occurrence of all species during each month over the sampling years is shown in Table 28. Golden shiners, white catfish, brown bullheads, common carp, and spottail shiners were the only species found in the canal during all months. More species were collected in June than in any other month.

TABLE 26. Relative abundance (number of fish caught per net haul) of all species of fishes caught in the CYAP discharge canal for each month and year, 1968–1972. Blank spaces denote periods without sampling.

Months	Sampling years 1968	1969	1970	1971	1972	All years combined
January			65		35	47
February			461	52	22	74
March			202	80	183	146
April		241	146	89	150	156
May		18	10	26	132	57
June	66	99	109	27	75	64
July	170	1	0	44	35	46
August	7	79	0	136	21	59
September	72	8	31	64	37	42
October	30	13	3	81	37	43
November	1,466	222	80	58	59	99
December	400		24	165	166	85
Mean	239	115	81	69	100	90

Temperature Tolerances

The range and type of response to the changes in water temperature in the canal varies with the season and species as well as from individual to individual, depending on its physiological condition, age, and previous acclimation to water temperatures. Generally, each species left the heated discharge in late spring and early summer (except during those years of early spring plant shutdown) as the rising canal temperature approached its upper temperature tolerance limit. The range of temperatures at which each species was collected in the canal and its apparent upper limit of tolerance are given in Table 28. Five species (pumpkinseed, white perch, golden shiner, white catfish, and brown bullhead) have been collected in the discharge canal in water up to 40°C. The majority of the fishes left the canal when the water reached about 35°C and returned immediately after as little as a 1-degree drop in temperature (34°C). The same critical temperature and response was noted in

TABLE 27. Monthly abundance (number of fish per net haul) of the six most abundant fish species inhabiting the CYAP discharge canal, 1968–1972. Parentheses denote month of greatest abundance.

Species	Months											
	J	F	M	A	M	J	J	A	S	O	N	D
Brown bullhead	19	53	(128)	116	31	18	8	15	2	12	63	60
White catfish	22	19	16	(34)	16	21	16	13	4	3	21	17
White perch	1	0	1	1	1	2	12	(25)	18	18	2	1
Spottail shiner	2	1	1	2	2	6	1	(8)	7	2	6	2
Golden shiner	1	1	1	1	1	(13)	7	7	2	1	1	1
Common carp	2	1	1	1	1	1	1	1	1	1	1	(5)

the most thermally tolerant species in the Wabash River by Gammon (1971). A quantitative, ecological study made in South Biscayne and Card Sound, Florida, in the vicinity of the Turkey Point power plant, showed that half of the total number of species were excluded from the discharge canal by temperatures higher than 32–33°C (Bader et al. 1971; Roessler 1971). Generally, the smaller or younger fish remain in the CYAP canal at higher temperatures than do adults of the same species. The tendency for young fish to remain in warmer waters than older, larger individuals has also been observed by Fry (1937), Doudoroff (1938), Proffitt (1969), and Gammon (1971).

There has been no evidence to suggest that the species of fishes that inhabit the CYAP discharge canal during the warmer season suffer any mortality because of increased water temperature. The fishes either avoid the canal or leave it as the

TABLE 28. Minimum, maximum, and mean temperatures for fish species (ranked from most to least tolerant) and their seasonal occurrence in the CYAP discharge canal, 1968–1972.

	Temperature (°C)			Occurrence												
				Winter months					Warmer months							
Species	Minimum	Maximum	Mean	N	D	J	F	M	A	M	J	J	A	S	O
Pumpkinseed	11.9	40.0	28.4	X	X	X	X	X		X	X	X	X	X	X
White perch	5.7	40.0	27.5	X	X			X	X	X	X	X	X	X	X
Golden shiner	6.7	40.0	24.0	X	X	X	X	X	X	X	X	X	X	X	X
White catfish	4.5	40.0	20.2	X	X	X	X	X	X	X	X	X	X	X	X
Brown bullhead	5.1	40.0	19.6	X	X	X	X	X	X	X	X	X	X	X	X
Redbreast sunfish	10.1	39.2	27.0		X		X						X	X	
Common carp	5.1	39.2	21.6	X	X	X	X	X	X	X	X	X	X	X	X
Spottail shiner	5.1	39.2	20.1	X	X	X	X	X	X	X	X	X	X	X	X
Mummichog	14.7	37.0	22.8							X	X	X			
Bluegill		35.0	35.0									X			
American eel	11.9	35.0	20.5			X		X	X	X	X				
Hogchoker	32.0	32.5	32.3								X			X	X
Blueback herring	6.7	32.5	20.7						X	X		X			
Alewife	5.7	31.0	15.8						X	X		X			
Johnny darter	20.1	28.8	24.5												
Largemouth bass	14.8	28.7	21.3	X	X					X					
Channel catfish	5.1	28.5	15.2	X	X	X	X	X	X		X	X		X	
Black crappie	16.5	27.3	21.3	X						X	X				
Northern pike	12.9	22.0	16.0		X	X	X	X	X		X				
American shad		22.0	22.0								X				
Common shiner		20.0	20.0								X				
White sucker	10.0	19.9	14.4	X	X	X	X	X	X						
Rainbow smelt		5.7	5.7						X						

rising temperature approaches their upper limits of tolerance.

Effects on Fishes of Overwintering in the Discharge Canal

Of the many fishes that were attracted to the warm waters of the discharge canal, 15 species overwintered there during the cooler months, that is, November through March. During the winter months, 24,934 fishes were collected in 261 trawl hauls in the canal (Table 29). Four species (brown bullhead, white catfish, spottail shiner, and common carp) made up 99% of the total fish overwintering in the canal. The majority of the fishes were adults, but the young of five species were represented, 79% of them being brown bullheads. An obvious question was whether the fishes that had become acclimated to the warm temperatures in the canal during the winter months would be physiologically incapable of leaving the canal for other cooler portions of the river. Of even greater concern was the possible disruption of the winter phase in the life cycle of these fishes, especially the brown bullheads and white catfish, which represented 95% of the total fish population inhabiting the canal during the winter months. We therefore concentrated our studies on the overwintering catfish populations with the intention of also using our findings as an indicator of possible change in other fishes inhabiting the canal.

Seasonal Abundance and Responses to Temperature of Catfishes

Large numbers of brown bullheads and white catfish were first observed in the CYAP discharge canal in the early fall of 1968, the first fall after the plant started to operate. The highest recorded trawl catches occurred at this time, as compared with subsequent years, that is, 1,331 bullheads and 119 white catfish in one 5-min trawl haul. Brown bullheads made up 67% and white catfish 19% of the total catch of 26 species (Table 25). The total catch of adult bullheads was 16,290 (an average of 37/net haul) and white catfish 7,164 (16/net haul). During the winter months (November–March) bullheads made up 68% (11,173) and white catfish 28% (4,562) or 95% of the total adult catch of 15 species (Table 29).

Normally most brown bullheads and white catfish overwinter between November and March in the coves and attached ponds of the Connecti-

TABLE 29. Summary of the abundance of adults and young of the year of fishes inhabiting the CYAP discharge canal during the winter months, November through March, 1968–1972. Dashes (–) denote less than 0.005%.

Adults			Young			Adults and young combined		
Species	Number collected	Percentage total catch	Species	Number collected	Percentage total catch	Species	Number collected	Percentage total catch
Brown bullhead	11,173	67.59	Brown bullhead	8,150	97.26	Brown bullhead	19,323	77.49
White catfish	4,562	27.54	Spottail shiner	116	1.39	White catfish	4,625	18.54
Common carp	319	1.92	White catfish	63	0.76	Spottail shiner	406	1.62
Spottail shiner	290	1.75	White perch	42	0.50	Common carp	327	1.31
Golden shiner	90	0.54	Common carp	8	0.09	Golden shiner	90	0.36
Yellow perch	39	0.23				White perch	59	0.24
White perch	17	0.10				Yellow perch	39	0.16
Northern pike	17	0.10				Northern pike	17	0.06
Pumpkinseed	15	0.09				Pumpkinseed	15	0.06
Channel catfish	14	0.08				Channel catfish	14	0.05
White sucker	13	0.07				White sucker	13	0.05
American eel	2	0.01				American eel	2	–
Redbreast sunfish	2	0.01				Redbreast sunfish	2	–
Largemouth bass	1	–				Largemouth bass	1	–
Sea lamprey	1	–				Sea lamprey	1	–
Totals	16,555			8,379			24,934	

cut River (see pages xx–xx). Here, they either remain inactive on the bottom with a decreased muscle-tissue metabolism (Oros 1968) or burrow in sediment as described by Loeb (1964). Beginning in 1968, however, these species moved into the canal to overwinter each year between mid-October and early November. Both species consistently entered the canal when the water temperature there dropped to about 20°C (range, 19–23°C). Figures 11 through 15 show the seasonal abundance of the two species in the canal for each sampling year in relation to the plant's power output and discharge and the ambient water temperature. The number of catfishes collected by trawl decreased considerably in the Connecticut River in the vicinity of the plant whenever the ambient river temperatures were below 10°C. Trawl collections and the sport fishery have shown that large numbers of both species move from the main river into the Salmon River at this time.

Catfishes of both species reached their greatest abundance in the canal between mid-February and early April. With the exception of periods of plant shutdown for refueling purposes in the spring, both species generally remained in the canal for about 8 months, or through June, when discharge temperatures rose above 35°C. The catfishes that inhabited the canal during the winter therefore could be considered to belong to a "closed" population until early spring when ambient river temperatures rose above 4°C. At this time, both species began to move back into the river. With a few exceptions, neither species was found in the canal again until the following October.

The maximum temperature tolerance for white catfish and brown bullheads, as indicated by their presence in the canal, was 35.2°C and 33.6°C, respectively. White catfish remained in the discharge canal longer and evidently were able to tolerate higher temperatures than brown bullheads. Generally, white catfish would move into the canal at 34°C, but would avoid the canal or move out of it at temperatures higher than 35°C. Brown bullheads were found in the canal at 32°C but were not found there above 33.6°C. A difference of about only a single degree seemed to regulate the movement of these fishes in and out of the canal when temperatures approached the upper limit of tolerance for each species. The upper temperature limits of the white catfish and brown bullhead have been fairly well documented in the literature, at least for certain acclimation temperatures. The following are a few pertinent examples. Trembley (1960) observed brown bullheads swimming in the heated discharge at 37–38°C and reported that the fish entered 40°C water for food. He found that the LT_{50} for brown bullheads was 37.2°C. An upper lethal temperature of between 33°C and 35°C has been reported by Brett (1944), Bardach and Bjocklund (1957), and Carlander (1969) for bullheads acclimated to temperatures between 20°C and 25°C. These temperatures are similar to the ambient river temperatures recorded in the present study at the time when 33.6°C was reached in the discharge canal. Bullheads have survived temperatures of 33.9°C and

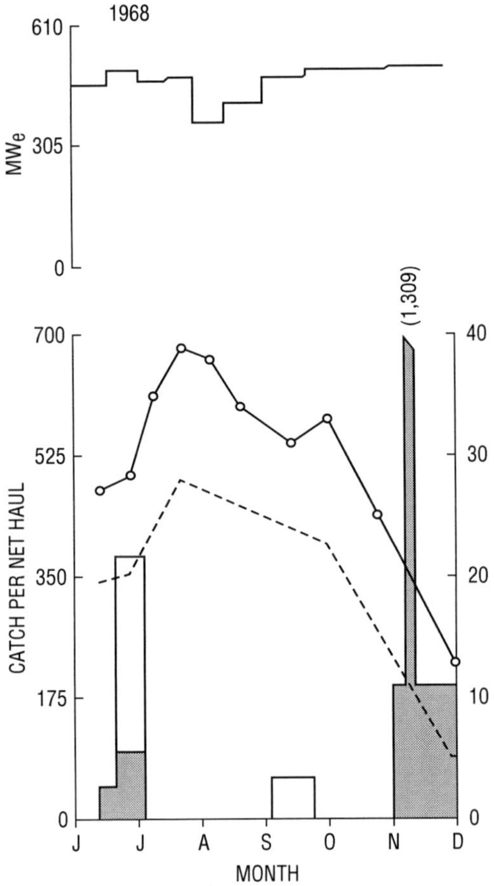

FIGURE 11. Seasonal abundance of brown bullheads (solid bars) and white catfish (open bars) collected per trawl haul in the CYAP discharge canal during 1968. Dotted lines represent ambient river temperatures and solid lines represent discharge-canal temperatures. Circles on the solid lines represent dates of sampling. The seasonal power output (MW_e) of the plant is shown in the upper graph of each figure.

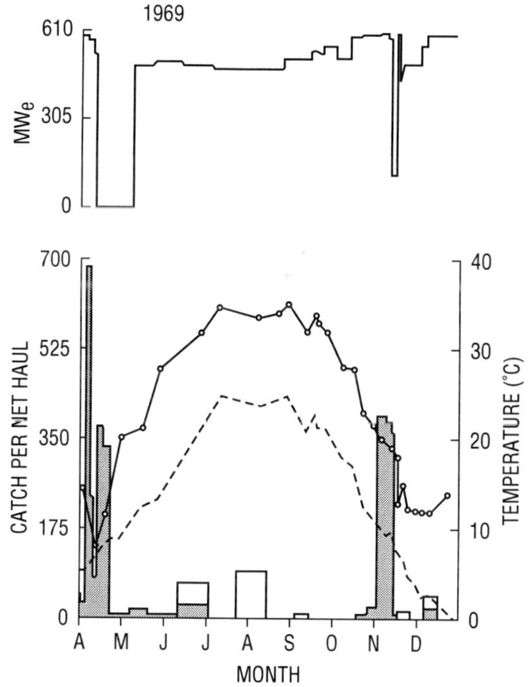

FIGURE 12. Seasonal abundance of brown bullheads and white catfish in the CYAP discharge canal during 1969. See Figure 11 for conventions.

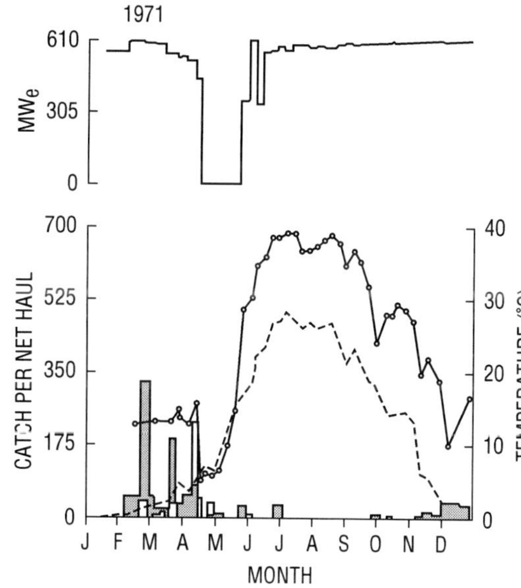

FIGURE 14. Seasonal abundance of brown bullheads and white catfish in the CYAP discharge canal during 1971. See Figure 11 for conventions.

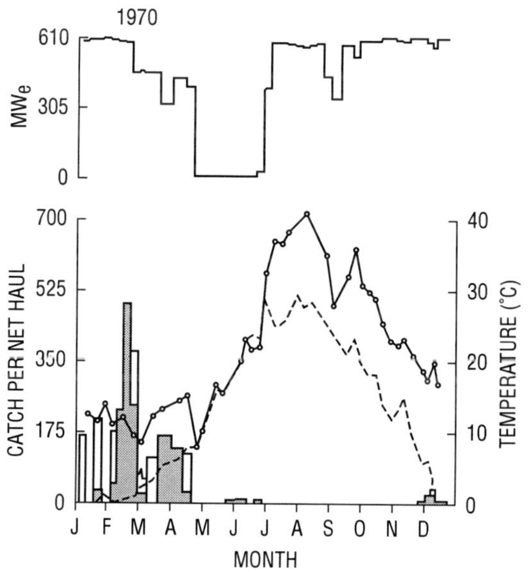

FIGURE 13. Seasonal abundance of brown bullheads and white catfish in the CYAP discharge canal during 1970. See Figure 11 for conventions.

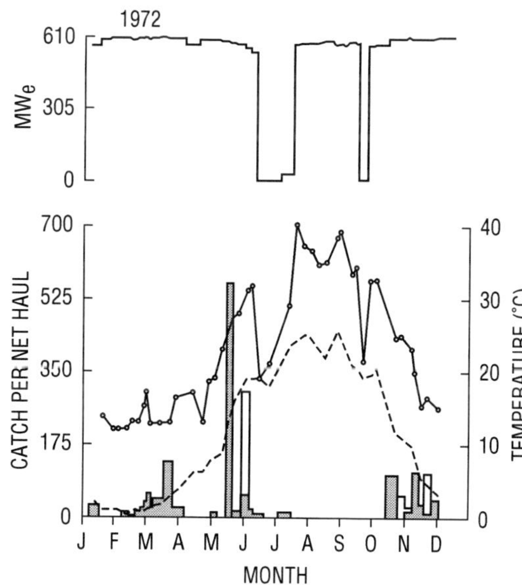

FIGURE 15. Seasonal abundance of brown bullheads and white catfish in the CYAP discharge canal during 1972. See Figure 11 for conventions.

white catfish exhibited an LT_{50} of 36.1°C when acclimated to 11.1°C, according to Cairns (1956). White catfish acclimated at 20°C had an LT_{50} (12 h) of 31°C for smaller and 29.2°C for larger fish (Kendall and Schwartz 1968).

Power plant shutdowns had a striking effect on the abundance of catfishes in the canal. Each year, except 1968, a major plant shutdown in the spring or early summer has caused both species to leave the canal almost immediately (Figures 11–15). After each shutdown, the two species were observed at the surface exhibiting lethargic behavior—as if they had been stunned, probably cold shock—while being carried out of the canal with the 0.3–0.9 m/s current. Figure 13 illustrates an example of a reduction in catch that resulted from a stepwise drop in power that occurred between early March and the end of April 1970. The stepwise drop in power was consistently matched by a drop in the catch per trawl haul, until no fish were collected in the canal when the power level reached 0 and canal temperatures dropped to ambient. The movement of tagged fish as a result of power plant shutdown is described on pages 82–83. No evidence was found to suggest that either species of catfish suffered any mortality as a result of these sudden drops in temperature. On several occasions during mid-winter, more than 50 specimens of each species were collected in the canal (13–15°C) and placed in a live-car, which was then towed out into the river and anchored in water of 2.3°C. Movements of the body and operculum of these fishes became much reduced, and their coloration took on a "washed out" appearance, but no deaths occurred within a 12-h period. Fish in the live-car that were shifted back and forth between the canal and the river several times within a short period also showed no serious ill effects. No data on the effects of cold shock on the other species is available. Nevertheless, deaths resulting from the inability of fish to acclimate rapidly to lowering temperatures due to sudden power plant load reduction have been reported (Edsall and Yocom 1972; Silverman 1971; and the Smith-sonian Institution 1972).

Distribution of Catfishes within the CYAP Discharge Canal

On 16 February and 30 March 1971, 24-h studies were made of the brown bullheads and white catfish in the discharge canal in order to determine their distribution, diurnal movements, feeding habits, sex ratio, gonadal development, weight compared with length, and possible effects of the tidal cycle. Trawl hauls were made at five stations (marked at night by reflectors) along the canal at 152-m intervals some 762 m up the canal from its mouth. Trawls were taken every 4 h during each 24-h period. Numbers, lengths, and weights were taken of all catfishes, and their stomachs were removed and processed. The ovaries of most of the females were weighed.

A description of the distribution of catfishes, as indicated by the two 24-h sampling periods, follows:

(1) White catfish utilized the upper section of the canal where the flow rates were faster, while brown bullheads were found in the lower, broader section, particularly near the mouth where flows were much reduced (Table 30).

(2) Both species were most readily caught in the early morning hours between 0300 and 0700 hours and least abundant around 2300 hours on both dates.

(3) White catfish were most readily caught on ebb tide and bullheads on flood tide (Table 30). A difference in the swimming ability of

TABLE 30. Number of white catfish and brown bullheads per net haul collected at five stations along the CYAP discharge canal at 152-m intervals; collections were made every 4 h during two 24-h periods on 16 February and 30 March 1971. The comparison of the catches of both species (number per net haul) on each tide is based on 24-h and regular weekly discharge-canal sampling (1968–1972).

	Distance from mouth of canal (m)					Tidal stage	
						Flood	Ebb
	0–152	152–305	305–457	457–610	610–762	(n = 110)	(n = 130)
White catfish	0.4	1.3	2.2	3.4	9.0	10.7	16.3
Brown bullhead	60.9	23.5	8.7	5.7	4.3	48.6	35.8

the two species or the sampling methods used could account for some or all of these differences.

(4) There was no evidence of differential distribution by sex along the canal. The ratio of males to females was 1.37 for bullheads (n = 762) and 1.16 for white catfish (n = 1,021) throughout.

(5) There was little evidence of segregation along the canal according to size, although the majority of the young-of-the-year brown bullheads were found only near the extreme lower end of the canal.

Number of Catfishes in the CYAP Discharge Canal

A mark–recapture tagging program was initiated in the early spring of 1969 to estimate the size of the populations of brown bullheads and white catfish inhabiting the canal discharge, as well as to provide information on changes in length and weight and on movement and distribution after plant shutdown or when upper temperatures became critical. Estimates of the populations of the two species were made in each subsequent year, generally in March, when ambient river temperatures were below 4°C and trawl collections showed that the largest number of fish were in the heated discharge; using the same method, Behmer (1969) was able to estimate the size of a population of river carpsuckers in a section of the Des Moines River with reasonable success.

Ricker (1958) has pointed out five conditions necessary for the estimation of populations from tag-and-recapture data, the Schnabel method, most of which we were able to meet reasonably well: (1) Marked fish must suffer the same natural mortality as the unmarked. Since our entire mark–recapture procedure took place within a few days, natural mortality was not considered to be a factor. Moreover, no mortalities of marked fish were ever observed. (2) Marked and unmarked fish must be equally vulnerable to recapture. Fish were collected by trawl and tagged at the lower end of the discharge canal, then put into large tubs and distributed throughout the lower section of the canal by boat prior to the next trawl haul. It was assumed the marked and unmarked fish were sufficiently mixed. (3) Fish should not loose their marks. We observed only eight specimens that had lost their tags during all our years of tagging. The fact that fish with tags were recaptured as long as 4 years after tagging attests to how well the tags were retained. (4) All marks should be recognized and reported. Since all of the recaptured fish were collected by our own trawl, reporting was not a problem for the canal population estimates. Subsequently, however, some fish that were at large for a considerable period after leaving the canal showed a dense algal growth on the bright orange tag that made it difficult to see. In addition, although most tags were reported by sport fishermen as a result of the reward and publicity, some of them were not because some fishermen treated them as collector's items. (5) There must be negligible recruitment during the sampling. The canal population was considered to be a "closed" one during most of the winter months when ambient river temperatures were below 4°C. Fish evidently were present only in the well-defined area of the heated plume, for trawls just outside of the plume produced negative results, and extensive trawling in the vicinity of the plant during the winter failed to produce many fish. At water temperatures above 4°C, however, the catfishes tended to mix with the river population, while other species moved into the canal.

In this study, 5,158 brown bullheads and 1,335 white catfish caught in the canal were tagged, and the estimates were based on a return of 4.1% of the bullheads and 1.1% of the white catfish. The estimated average number of catfishes that overwintered in the canal (mean of 4 years) was 28,500 (with 95% confidence limits of 17,600–49,000). There were about twice as many brown bullheads as white catfish, the average population consisting of 18,700 and 9,800 fish, respectively. The yearly population estimates of both species ranged between 17,000 and 39,700 (95% confidence limits: 10,800–74,000). It is estimated that, in the 4 years under consideration, a total of 74,800 bullheads and 38,900 white catfish spent their winters in the canal.

A reversal in the number of the two species occurred in the 1969–1970 and 1970–1971 winter seasons (Table 31). Although the total population remained about the same, the brown bullhead became the dominant species in 1970–1971 in about the same ratio that the white catfish had dominated during the previous winter. The similarity of the total estimates for both years suggests that the carrying capacity of the canal may have remained at about 40,000 individuals and

TABLE 31. Yearly population estimates of brown bullheads and white catfish overwintering in the CYAP discharge canal.

	Time interval (tagging and recapture)	Brown bullhead No. collected	No. tagged	No. recaptured	Population estimate	95% confidence interval
1968–1969	3–11 April 1969	1,110	1,110	24	14,334	9,632–21,330
1969–1970	4–6 March 1970	553	553	12	7,700	4,406–13,470
1970–1971	2 March–23 April 1971	2,397	1,165	54	31,921	24,000–40,864
1971–1972	1–29 March 1972	2,098	1,168	76	20,913	13,563–41,117
Totals		5,158	3,996	166	74,868	

	Time interval (tagging and recapture)	White catfish No. col- lected	Popula- No. tagged	95% recap- tured	Popula- tion estimate	95% confidence interval	Combined species Popula- tion estimate	95% confidence interval
1968–1969	3–11 April 1969	255	255	4	2,737	1,178–4,294	17,071	10,810–25,624
1969–1970	4–6 March 1970	964	964	9	32,059	16,866–60,935	39,759	21,272–74,405
1970–1971	2 March–23 April 1971	151	90	2	4,117	1,019–13,548	36,038	25,019–54,412
1971–1972	1–29 March 1972	34	26	0			20,913	13,563–41,117
Totals		1,404	1,335	15	38,913		113,781	70,664–195,558

the ratio between the species may have depended upon which one entered the canal first in considerable numbers. Such chance would result from the concatenation of the timing of fish movements and water temperatures and flow.

Percentage of the Catfishes in the Haddam Area Affected by Plant Operation

During the winter of 1968–1969, it was possible to calculate the percentage of all of the white catfish and brown bullheads in the vicinity of the plant that overwintered in the discharge canal. These percentages give an indication of what portion of the Haddam area populations of the two species are subject to overwintering emaciation and changes in ovarian development. The total catch of both species was estimated from data provided by the 1969 creel survey. Since 52.8% of the tagged catfishes obtained from the sport fishery were taken by fishermen in the Haddam area and 47.2% were recaptured by fishermen in other areas of the river, the number of fish tagged in the canal (1,365) was adjusted to represent the number of tagged fish that were actually vulnerable to recapture by fishermen in the Haddam area (1,365 × 52.8%). This figure (721) was multiplied by the 1969 estimated catch of all sport fish in the Haddam area (4,641) (see pages 104–105), and the product was divided by the number of catfish tags returned by Haddam area fishermen (28) to obtain an estimate of the size of the catfish population inhabiting the Haddam area. The population estimate calculated for the discharge canal between 3 and 11 April 1969 (17,071) was divided by the estimate in the vicinity of the plant to obtain the percentages. The fishes in the canal were included as part of the total estimate of fishes in the Haddam area because after being tagged in the canal in March of 1969, most fish left to mix with the Haddam area population after the plant shutdown on 11 April. Very few fish were caught in the canal after this shutdown.

The estimated percentage of the total population of white catfish and brown bullheads in the 386-ha area in the vicinity of the plant that was subject to overwintering stress in the CYAP canal in 1968–1969 was 14.3% with 95% confidence limits of 9.9–20.7%. These percentages were calculated by dividing the estimated number of catfishes overwintering in the canal during 1968–1969 (17,071) by the estimated size of the catfish population in the Haddam area during 1969 (119,500, 95%-confidence limits: 82,600–172,600).

The plant's intake was estimated to have entrapped and subsequently killed about 5% of the total population of the two species of catfishes in the Haddam area in 1969. This percentage was calculated by dividing the recorded number of both species taken on the intake screens in 1969 (5,900) by the estimated population of both species in the Haddam area in 1969 (119,500). The exploitation of the catfish population in the Haddam area by sport fishermen in 1969 was estimated to be 4% (4,650 divided by 119,500). Intake mortality and sport fishing thus accounted for mortality of approximately 9% of the total catfish population with an additional 14% of the population subject to overwintering stress in the discharge canal.

Loss in Weight and Coefficient of Condition of Catfishes in the CYAP Discharge Canal

The coefficient of condition (see footnote, Table 32) and growth of fish that inhabit areas warmed by thermal discharges appears to vary among species and location. Bennett (1972) described adult black crappies and bluegills inhabiting a heated discharge pond in South Carolina as being in better condition than those living at ambient temperatures. Nakatani (1969) found that salmonids reared in warm water below Hanford on the Columbia River grew faster than those in the cooler water. Shcherbukha (1971) stated that nase in a heated discharge improved in condition during the period of hibernation as compared with the period of feeding. On the other hand, Cairns (1956) demonstrated a lack of growth in fishes subjected to certain increased water temperatures, even when their food intake was increased. Cragg-Hine (no date) reported that the growth rate of roach and bream inhabiting a discharge canal in England was depressed as a result of overcrowding and competition for food. Few field studies, however, have dealt with the effects of relatively long-term exposures of fish to elevated temperatures in discharge canals during the cooler months.

TABLE 32. Average weight and condition factor (K)[a] for brown bullheads collected under winter conditions, in the vicinity of the CYAP plant, 1965–1972, and from the CYAP canal, 1968–1972.

Length (cm)	Number collected		Average weight (g)		K-factor (TL)	
	Canal	River	Canal	River	Canal	River
11	26	2	18.5	20.0	1.44	1.36
13	55	2	28.6	25.0	1.33	1.20
14	38	2	28.2	29.0	1.03	1.10
15	21	2	39.7	42.5	1.22	1.23
16	57	5	53.1	62.0	1.30	1.49
17	97	2	65.9	60.0	1.34	1.24
18	224	13	69.9	111.5	1.20	1.90
19	316	20	83.4	122.6	1.22	1.75
20	502	46	99.6	125.1	1.25	1.56
21	549	74	117.8	137.8	1.28	1.49
22	559	86	137.5	154.0	1.30	1.46
23	574	109	156.5	183.0	1.29	1.51
24	553	99	177.2	199.8	1.29	1.45
25	515	122	196.8	210.9	1.27	1.36
26	500	133	218.4	243.7	1.25	1.39
27	464	111	242.0	267.9	1.24	1.36
28	341	111	268.9	290.2	1.23	1.33
29	248	74	291.3	329.0	1.21	1.35
30	128	47	318.2	364.4	1.19	1.36
31	53	20	349.9	389.7	1.19	1.32
32	13	12	364.2	444.3	1.14	1.36
33	7	3	502.1	490.0	1.42	1.38
34	4	1	465.0	700.0	1.23	1.74
39	2	2	317.5	637.5	0.53	1.07

[a] Calculated by the formula $K = 100W/TL^3$, where W equals weight in grams and TL, total length in centimeters.

The amount of weight lost or gained by the two species of catfishes as a result of overwintering in the discharge canal was determined by means of recaptures of tagged individuals that had been originally caught in the discharge canal. To determine the loss or gain in weight of fish that had spent one winter in the canal, lengths and weights were taken of brown bullheads and white catfish tagged from November through December, 1969. The canal was then sampled weekly by trawling, and tagged fish were recaptured through 16 April 1970. The percentage of weight lost or gained over the winter was determined by dividing the percent difference between the weight at tagging and at recapture by the number of days each fish was free. This percentage of weight lost or gained per day for all fish was algebraically added, and the sum divided by the total number of fish recaptured. The mean percentage weight lost or gained per day was multiplied by the maximum number of free days recorded for a recaptured fish through the 16 April 1970 plant shutdown. The result was an estimate of the total percentage weight lost or gained during one overwintering. The total weight lost or gained was also estimated from tag returns of individuals of both species that had spent more than one winter in the canal.

Efforts were made to estimate whether fish that had spent one winter in the canal made up any weight loss after leaving the canal in the spring to spend the summer outside the canal. The average weight and condition factor at recapture for fish that had been tagged in the canal in March and recaptured during the following summer were statistically compared with the average weight and condition factors of fish of similar lengths collected in the vicinity of the plant during the same seasonal period before plant operation (1965–1967).

From the beginning of November to the middle of April for the years 1965–1972, the length and weight of 6,944 brown bullheads were taken in the vicinity of the plant and for the years 1968–1972 in the CYAP discharge canal (Tables 32 and 33). The dates encompass the maximum length of time that fish were subjected to winter conditions (ambient river temperature less than 5°C) in the canal during the 5 years after the beginning of plant operation. The average weights and condition factors for fish of similar lengths were compared for both areas by using the paired *t*-test. The brown bullheads inhabiting the heated discharge during the winter had significantly lower weights (t = 2.79, 24 df, P < 0.005) and were in significantly poorer condition (t = 3.17, 34 df, P < 0.005) than fish of similar lengths collected in the cooler water outside the canal.

In December 1971 and March 1972, matched trawl samples were taken from the canal and Chapman Pond in order to compare the condition, winter food, and ovarian development of the populations of brown bullheads and white catfish than had overwintered in the heated discharge area with fish from a location experiencing normal winter temperatures. The control area, a cove of the river called Chapman Pond, is open to tidal influence from the river at two places and measures approximately 1.5 by 0.3 km and about 1.2 m deep. The cove is located 8 km downstream from the power plant, that is, 3.2 km below any detectable heated water. Both species of catfishes have been tagged in the canal and in the Salmon River and subsequently recaptured in Chapman Pond, and this is strong evidence that the catfish population it contains is not an isolated one. The total length, weight, sex, and ovarian weight of each specimen was determined. The relative condition of fishes in the canal and Chapman Pond

TABLE 33. Average weight and condition factor (K) for white catfish collected under winter conditions, in the vicinity of the CYAP plant, 1965–1972 and from the CYAP canal, 1968–1972 (see footnote, Table 32).

Length (cm)	Number collected		Average weight (g)		K-factor (TL)	
	Canal	River	Canal	River	Canal	River
16	4	1	47.5	60.0	1.17	1.45
17	4	1	115.0	12.0	2.33	0.26
18	10	2	152.0	44.0	2.53	0.71
19	14	1	144.3	80.0	2.14	1.16
20	16	3	100.0	83.3	1.25	1.02
21	29	3	116.2	93.3	1.27	1.03
22	69	9	137.7	128.4	1.29	1.22
23	105	13	144.1	149.9	1.18	1.23
24	159	19	154.8	157.6	1.12	1.16
25	266	18	175.3	194.8	1.13	1.24
26	252	25	194.6	239.9	1.11	1.36
27	278	30	220.4	223.5	1.13	1.13
28	267	30	242.9	263.1	1.12	1.20
29	199	33	278.7	283.4	1.15	1.16
30	188	22	305.4	326.2	1.14	1.20
31	144	17	342.9	366.9	1.16	1.23
32	114	15	382.6	408.5	1.18	1.25
33	77	17	416.4	466.8	1.17	1.31
34	74	20	467.9	492.8	1.20	1.25
35	37	11	523.8	515.3	1.23	1.20
36	38	21	580.9	589.9	1.25	1.26
37	22	19	628.9	622.5	1.24	1.23
38	23	13	637.6	713.8	1.17	1.30
39	15	11	742.3	816.4	1.26	1.38
40	8	5	806.3	770.0	1.28	1.20
41	6	11	828.3	739.0	1.21	1.07
42	3	6	850.0	919.7	1.18	1.25
43	1	5	1050.0	1072.5	1.36	1.35
44	5	2	988.0	1072.5	1.18	1.28
46	2	2	1165.0	1192.5	1.41	1.21

was measured by means of the condition factor described by Ricker (1970), which is an index of the "relative robustness" or plumpness of a fish. Condition factors and ovarian weights from fish of similar lengths could then be statistically tested to reveal any significant differences between the fishes collected in the two areas. Stomach contents were analyzed and the two areas compared (Massengill 1973). Water temperature and dissolved oxygen were recorded periodically while collecting the fish.

The average weights and condition factors of 400 brown bullheads between 14 and 31 cm and 93 white catfish between 25 and 37 cm, taken in the matched trawl samples were compared by using the paired *t*-test for fish of the same lengths. Again, the brown bullheads inhabiting the canal had significantly lower weights (t = 4.88, 17 df, P < 0.0005) and condition factors (t = 10.85, 11 df, P < 0.0005) than those from outside the canal, in this case Chapman Pond.

Comparisons were made between catfish that had spent the winter in the canal and ones that had overwintered in two adjacent coldwater areas, the Salmon River and Chapman Pond. The combined average weights and condition factors for fishes in the same length-classes of both species in the canal in April 1969 and March 1972 were compared with those from the Salmon River in April 1969 and Chapman Pond in March 1972. The members of both species that were living in the canal exhibited significantly lower average weights and condition factors compared with those in the two cooler overwintering areas outside the canal. Figure 16 illustrates the differences in the average weight of brown bullheads from the canal and from the two overwintering areas outside the canal.

The percentage of weight lost by both species during one overwintering period was determined by tagging catfish of known weights in the canal in the fall of 1969 and recapturing and weighing some of them through the early spring of 1970. Of 924 brown bullheads tagged, 17 were recaptured, a 1.8% return. The average weight loss from tagging to recapture was 0.21%/d for the 141-d period, that is, a 30% (range 0–60%) weight loss over one winter. Twenty-two white catfish were recaptured of the 759 tagged, a 2.9% return. The average weight loss was 0.14%/d or 20.3% (range, +50% to –37.5%) for the winter season. Brown bullheads that had spent one winter in the discharge canal apparently made up the loss in weight and condition after leaving the canal in the spring and spending the summer outside the canal. The average weight and condition for 25 fish tagged in the canal and recaptured in the Connecticut River through the following summer were not significantly different (P < 0.05) from fish of similar length that had not spent a winter in the canal, that is, tagged fish recaptured in the vicinity of the plant during the summers prior to plant operation, 1965–1967 (n = 40). Too few white catfish were recaptured the following summer to establish a statistical comparison, even though white catfish lost 20% of their weight during a single overwintering period in the canal. It has been assumed that this species also makes up whatever weight is lost after one winter in the canal.

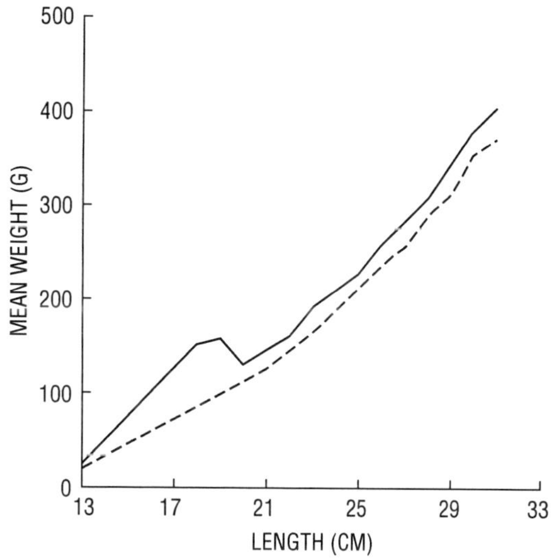

FIGURE 16. Comparison of the average weight of brown bullheads collected in the CYAP discharge canal (n = 1,510) in April, 1969 and March, 1972 (dashed line) and those collected in the Salmon River in April, 1969 and Chapman Pond in March 1972 (n = 627), (solid line).

Twenty brown bullheads that had been subjected to overwintering stress for two winters in the canal (tagged in the canal in March and recaptured in the canal the following March) gained only 175 g or 8.8 g/fish over a period of 351 d, that is, about a year. This yearly increase is much lower than that represented by any of the differences in weight between the age-classes of this species (see page 104). Four white catfish, recaptured after spending two winters in the canal, lost

14 g/fish over a period of 363 d. The effect of this poor growth rate on the total population of catfishes was considered minimal, however, because only 2.5% and 0.5% of the tagged population of brown bullheads and white catfish returned to the canal the next winter.

In summary, possible reasons for the loss in weight and condition of the catfishes that overwinter in the discharge canal are (1) the metabolic rate undoubtedly is higher in the heated canal than in the cove areas at ambient temperatures, since metabolic rate in brown bullheads has been shown to be positively correlated with temperature (Fry 1947); (2) a greater expenditure of energy is necessary for the fish to maintain their position in the continuous current of 0.3 to 0.9 m/s; (3) high population densities in the canal result in overcrowding with increased competition for food; and (4) biased sampling of age structure is a possibility.

Age and Growth Determinations and Weight Loss of Catfishes

The effects of warm water on the life history of these catfishes could not be understood without knowing the age of the fishes involved. Catfishes are scaleless, but Muncey (1959), Stevens (1959), and Schoffman (1961) have been able to use the annual growth marks that are formed within the spines of catfishes to determine their age. In our procedure, the left pectoral spine was cut off as close to its base as possible, and three sections (0.4–0.5 mm) were made of each spine using a jeweler's saw (Vince Kranz, personal communication). The sections were cleared with 5% acetic acid for 4 h, placed on an acetate slide, and covered with mounting medium. Whitworth and Sauter (1972) have provided a photograph of the cross section of the pectoral spine of a bullhead from Connecticut waters; in our specimens, the annuli appeared as dark lines separated by broad clear areas, and they were read on the dorsal side of the spine just anterior to the posteriorly projecting ridge. All cross sections were read independently by three investigators, with an 81% agreement. No age-datum based upon spines about which the readers disagreed was used. These bullheads ranged between 9.5 and 32 cm in length and between 10 and 455 g in weight, the white catfish between 21.7 and 42.6 cm (70 to 1,190 g).

The ages of 122 bullheads and 48 white catfish, collected from the CYAP canal in December 1971 and March 1972, and of 106 bullheads and 26 white catfish, collected from Chapman Pond at the same time, were determined from their pectoral spines. The oldest brown bullheads in Chapman Pond were 7 years and 5 in the canal (Table 34). Half the fish in the canal belonged to age-classes III and IV, while 79% of those in Chapman Pond belonged to IV and V. Males were both longer and heavier than females except at the age of 2 years. Some bullheads were found to be mature at about 15 cm or age II. Fish from Chapman Pond slightly exceeded those collected at the same time from the canal both in length and weight.

Compared with fish from the cooler water outside the canal, brown bullheads lost the equivalent of 1 to 1.5 years of growth in weight as a result of overwintering in the discharge canal. This conclusion follows from a comparison of the mean lengths and weights of similar age-groups from Chapman Pond and the canal (Table 34). At similar lengths, the mean weights for each age-class were lower in the canal than in Chapman Pond. The mean weight of age-class V (249 g) in the canal was comparable to a mean weight between those of age-classes III and IV in Chapman Pond. Combining the age-classes, there was a difference of 88 g between the means calculated for Chapman Pond and the canal. Weight loss was higher in the older or larger fish, probably due to a greater loss of fat deposits.

White catfish with a maximum age of 7 years were caught both in Chapman Pond and the CYAP canal (Table 35). Females exceeded males both in length and weight for all age-classes in both areas. Males were larger among the bullheads. Unlike bullheads, it was not possible to sex white catfish less than 17 cm long. White catfish spending one winter in the canal lost about one half a year's growth in weight compared with that of fish remaining in the cooler water outside the canal. Age-classes IV and V were used to compare the differences in mean weight between the two areas, and they were lower in the canal than in Chapman Pond. A difference of 105 g was found between the combined age-class means. The mean weight of age-class V (367 g) in the canal was comparable to a mean weight between those of age-classes IV and V in Chapman Pond (Table 35).

TABLE 34. Number, mean length and weight, and range of each age-class by sex for brown bullheads collected in the CYAP canal and Chapman Pond, 22 December 1971 and 24 March 1972.

Sex	Age-class	Number of fish	Mean length (cm)	Length range (cm)	Mean weight (g)	Weight range (g)
			CYAP canal			
Female	I	5	19.8	18.1–22.5	98.0	60–150
	II	13	19.5	17.4–21.1	88.5	50–110
	III	12	23.6	18.0–27.2	177.5	100–320
	IV	23	26.5	23.2–29.6	218.5	140–300
	V	14	26.2	23.1–30.5	221.1	140–320
Male	I	4	19.9	18.0–21.6	85.0	60–110
	II	5	20.1	17.2–24.0	104.0	60–170
	III	11	23.0	20.4–31.5	235.0	100–380
	IV	18	27.4	20.0–32.0	254.0	100–360
	V	8	28.0	25.6–30.4	276.3	160–360
	0	9	13.1	09.5–17.5	30.0	10–60
			Chapman Pond			
Female	II	5	23.6	21.5–25.0	166.0	130–200
	III	3	24.0	21.6–25.8	175.0	135–215
	IV	25	24.7	22.0–29.4	229.0	70–390
	V	15	26.5	24.1–30.5	243.0	180–350
	VI	3	25.3	25.0–25.5	226.0	210–250
	VII	1	24.0	24.0	195.0	195
Male	II	1	21.2	21.2	130.0	130
	III	5	26.8	24.8–28.5	257.0	160–300
	IV	30	27.8	24.4–30.5	290.0	170–400
	V	14	29.3	28.0–31.0	337.5	250–455
	VI	2	29.7	29.5–29.8	335.0	330–340
	0	2	13.3	13.0–13.6	32.5	25–40

Seasonal Fluctuations in Weights and Condition of Catfishes

Seasonal fluctuations in the mean weight and condition of 1,378 bullheads collected in the Haddam area of the river between 1965 and 1972 and 5,981 collected in the canal between 1968 and 1972 were compared. In the river, mean weight and condition fluctuated considerably throughout the seasons (Figure 17). They were highest in May and June, undoubtedly reflecting the prespawning condition. Through the summer, they declined and reached their nadir in October and November, when feeding probably decreased. Weight and condition of the fish in the canal continuously declined from January to April, in contrast to the fish outside, and these factors were also significantly lower in the canal fish from October to December.

The average weight and condition factor for each length-class of the bullhead population in the lower Connecticut River was calculated.[1] In general, weight increased with length, while condition decreased. In two other populations of bullheads, Priegel (1966) and Swingle (1965) found that condition increased with length. Bullheads in the lower Connecticut River had a higher aver-

[1] Conversion of standard length (SL) to total length (TL): TL = 1.16 cm SL, n = 16. W. Whitworth (personal communication) reported the conversion from TL to SL for the Thames River (32 km east of the Connecticut River) as SL = −0.40506 + 0.86199 TL, r = 0.96, n = 259. The maximum age reached in the Thames River was IV (4.7–34.2 cm). A bullhead tagged in the Connecticut River in 1969 at 34.0 cm (age III) was recaptured at 28.5 cm in 1973 (age VIII).

The length–weight relationship for brown bullheads collected in the lower Connecticut River between 1965 and 1972 was $\log W = -1.3248 + 2.6217 \log \mathrm{TL}$ (r = 0.89, n = 1,377). Bullheads in the Thames River showed slightly better growth than those in the Connecticut: $\log W = -1.77657 + 2.89118 \log \mathrm{TL}$ (r = 0.94, n = 186) (W. Whitworth, personal communication).

TABLE 35. Total number, mean length and weight, and range of each age class by sex for white catfish collected in the CYAP canal and Chapman Pond, 22 December 1971.

Sex	Age-class	Number of fish	Mean length (cm)	Length range (cm)	Mean weight (g)	Weight range (g)
			CYAP canal			
Female	III	1	26.3	26.3	190.0	190
	IV	4	28.0	25.3–30.0	227.5	160–270
	V	2	37.9	30.3–45.6	735.0	280–1190
	VI	4	33.5	29.5–36.1	415.0	260–520
	VII	2	34.5	34.2–34.7	435.0	420–450
Male	II	1	19.0	19	70.0	70
	III	7	24.6	21.7–26.5	151.4	100–180
	IV	12	27.3	24.0–29.6	210.0	140–270
	V	9	29.7	27.0–36.8	285.6	200–560
	VI	3	31.4	28.5–32.8	336.7	220–410
	VII	3	29.4	23.8–34.0	393.3	310–500
			Chapman Pond			
Female	IV	5	27.9	24.8–30.4	276.0	170–375
	V	4	33.6	29.0–36.5	514.0	335–680
	VI	0				
	VII	1	38.4	38.4	720.0	720
Male	IV	3	26.0	23.0–28.4	233.0	180–300
	V	7	32.0	27.5–35.5	443.0	240–600
	VI	4	36.5	31.1–42.6	685.0	400–1040
	VII	2	40.0	37.9–42.0	885.0	650–1120

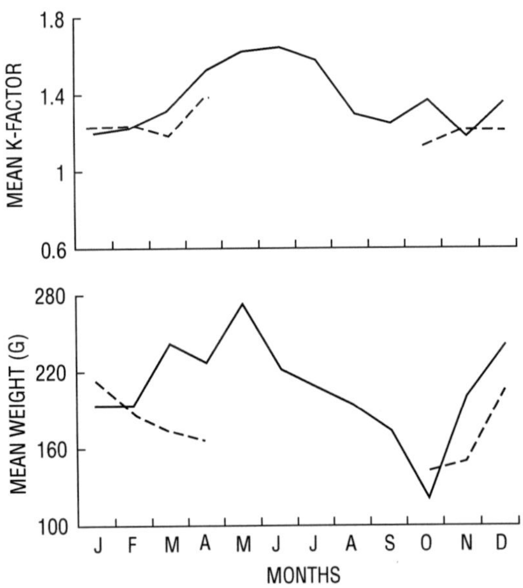

FIGURE 17. Comparison of the seasonal variation in the mean weights and condition factors (K) for 7,359 brown bullheads (17.5–30.4 cm) collected in the CYAP discharge canal (dashed line) and in the vicinity of the plant (solid line), 1965–1972.

age condition factor at a mean length of 22 cm than did those from the Thames River, Connecticut. The mean condition factor at 22 cm for Connecticut River fish was 1.51 (n = 114) and in the Thames River this was 1.30 (n = 186) (W. Whitworth, personal communication).

White catfish reached a peak mean weight and condition in April and May in the river, indicating that spawning probably occurred shortly thereafter. The condition and average weight of this species increased with length.[2] Although the condition

[2] The conversion from standard length to total length for white catfish is given by the expression TL = 1.17 cm SL, n = 78. W. Whitworth (personal communication) reported a conversion factor from total length to standard length for the Thames River, Connecticut as SL = 1.49799 + 0.88666 TL, r = 0.99, n = 644. He also reported a maximum age of 10 years in the Thames.

The length–weight relationship for white catfish collected in the lower Connecticut River between 1965 and 1972 was $\log W = -1.7745 + 2.9056 \log TL$ (r = 0.97, n = 1,096). White catfish in the Thames River exhibited better growth than those in the Connecticut: $\log W = -2.27181 + 3.23850 \log TL$ (r = 0.98, n = 473) (W. Whitworth, personal communication).

factor was slightly higher among the smaller fish, it remained relatively constant (range, 1.12–1.97). The condition factor of 1.24 at 25 cm TL was lower in the Connecticut River than that found in the Thames River (1.55 at 25 cm TL). The white catfish dominates the brown bullhead in the adjacent Quinnebaug and Thames rivers, whereas the opposite is true for the Connecticut.

Accelerated Ovarian Development in Catfishes

Ovaries were taken from brown bullheads and white catfish collected on various dates from November through April in the CYAP canal and the vicinity of the plant. Their relative development was assessed by comparing the ovarian weights of fish of similar lengths inhabiting the heated discharge canal with those from the cooler waters outside the canal. The weight of the ovaries and the size of the ova of brown bullheads, collected in the CYAP canal and Chapman Pond on 22 December 1971 and 24 March 1972, were also compared. After being cleaned of excess tissue, the ovaries were weighed in fresh condition and then stored for several months in Gilson's fluid to loosen the ova from the ovarian tissue (Simpson 1951). Ten ova were randomly selected from each pair of ovaries and measured to the nearest millimeter. As an estimate of fecundity, counts were made of all the ova in both ovaries of three brown bullheads and two white catfish.

The ovaries of female brown bullheads and white catfish develop significantly faster, that is, unseasonably earlier, during the winter in the discharge canal as compared with females overwintering in the cooler waters outside. The weight of the ovaries of 649 brown bullheads with the same length frequencies were compared, about two-fifths of the specimens having been collected in the canal and the rest in the vicinity of the plant during the winter months (1 November to 17 April) of 1965–1972 (Table 36). The ovarian weights of fish inhabiting the canal during the winter were significantly larger ($t = -4.77$, 16 df, $P < 0.0005$) than those from river water outside the canal. The ovarian weights of 527 white catfish were similarly compared and those from fish in the canal were found to be significantly larger ($t = -4.41$, 19 df, $P < 0.0005$) (Table 37).

More detailed data were obtained from female bullheads collected in the canal and

TABLE 36. Comparison of average ovarian weights of brown bullheads collected in the CYAP canal, 1968–1972, and in the vicinity of the plant between 1 November and 17 April 1965–1972.

Length (cm)	Number collected		Mean ovarian weight (g)	
	River	Canal	Canal	River
13	2	1	0.10	0.04
15	1	1	0.20	0.30
16	2	3	0.70	0.37
17	1	1	0.40	0.50
18	5	3	0.76	0.43
20	15	4	1.05	0.58
21	50	5	1.53	0.78
22	76	13	1.51	1.21
23	82	20	1.75	1.23
24	79	20	2.06	1.57
25	63	35	2.73	1.50
26	42	34	2.92	1.75
27	32	15	3.35	2.07
28	6	12	3.69	2.05
29	9	9	3.91	2.54
30	4	4	3.53	2.96

TABLE 37. Comparison of average ovarian weights of white catfish collected in the CYAP canal, 1968–1972, and in the vicinity of the plant between 1 November and 17 April 1965–1972.

Length (cm)	Number collected		Mean ovarian weight (g)	
	River	Canal	Canal	River
20	3	1	2.23	0.47
22	9	3	5.17	1.82
23	19	3	2.21	1.83
24	24	6	2.09	2.22
25	46	3	2.76	1.89
26	44	6	3.05	1.86
27	63	4	4.75	2.87
28	64	7	4.87	3.04
29	54	12	5.80	3.53
30	36	7	5.55	3.81
31	28	2	6.84	3.83
32	22	3	7.95	5.58
33	16	1	10.48	6.60
34	11	2	8.80	6.76
35	5	2	11.37	6.73
36	6	3	9.12	6.75
37	4	2	10.52	7.43
38	2	1	18.15	8.60
41	2	1	35.78	10.90

Chapman Pond in December 1971 and in March 1972. Table 38 shows a comparison of the mean body weight, ovarian weight, and size of ova from the ovaries of bullheads taken from both areas. The body weight of brown bullheads from the canal were consistently lower, but their ovarian weight ($t = -3.07$, 12 df, $P < 0.0005$) and egg size were significantly larger. Ovaries were relatively heavier and the ova relatively larger in the longer or older fish, that is, the difference between the fish from the two areas increased with size.

Since the ovaries of both brown bullheads and white catfish developed significantly earlier in the canal during the winter as compared with fish from the vicinity of the plant, it is possible that they might consequently spawn earlier than normal and their young may therefore be produced out of phase with their normal foods. Several factors indicate, however, that this phenomenon probably has a minimal effect on the entire population of catfishes in the lower Connecticut River. (1) If fish did spawn earlier after leaving the canal, this could not occur before April, at which time sufficient numbers of zooplankton presumably would be available in the river to support the young. (2) In 1972, there was no evidence to suggest that catfishes spawned in the canal earlier than the plant shutdown in June. In all other years, both species generally left the canal in early April due to plant shutdowns. It is probable that the ova development slowed when the fish entered ambient water after the early spring plant shutdowns and that this would cause most fish from the canal to spawn at a time closer to their normal spawning season. (3) The possibility exists that bullheads spawn twice yearly in the Connecticut River, as they do in some other areas (Breder 1935), and this would further minimize the potential effects of earlier maturation. (4) No damage to the ovaries as the result of overwintering was ever observed in the canal fish. Fish that had been tagged in the canal and 1 and 2 years later returned to the canal exhibited ovarian weights and egg sizes similar to fish of comparable lengths from the canal samples taken in December 1971 and March 1972. This suggests that the ovaries function and develop normally after a previous winter's sojourn in the canal. It is not known, however, if deviations from optimum temperature for development would result in a greater number of abnormal fry.

Fish inhabiting other waters warmed by heated discharges during the cooler winter months have shown accelerated gonadal development. Adair and DeMont (1971) noted that certain warmwater species in North Carolina spawned earlier in a discharge canal and heated cove than in a cove with ambient temperatures. Langford (1972) reported that in a lake receiving heated water from a power plant, roach spawned 6 to 8 weeks earlier than they normally do in the heated part of the River Witham, England. White suckers Catostomus commersoni spawned earlier in the discharge canal at the Martin's Creek power plant than elsewhere in the Delaware River (Trembley 1960). Witt (1971) showed that the heated water from a power station affected the gonadal development and accelerated the time of spawning of Missouri largemouth bass.

TABLE 38. Number, average length, weight, ovarian weight, and size of eggs of brown bullheads collected on 22 December 1971 and 24 March 1972 in the CYAP discharge canal and Chapman Pond.

	CYAP canal					Chapman Pond				
Length (cm)	Number	Mean length (cm)	Mean weight (g)	Mean ovarian weight (g)	Mean ovum size (mm)	Number	Mean length (cm)	Mean weight (g)	Mean ovaian weight (g)	Mean ovum size (mm)
17.0–18.9	4	18.1	77.5	0.59	0.62	1	17.1	35.0	0.30	0.46
19.0–20.9	10	19.9	98.0	0.64	0.63	0				
21.0–22.9	3	21.1	121.7	0.90	0.61	4	22.0	125.0	1.35	0.56
23.0–24.9	12	23.8	150.4	1.85	0.71	11	24.0	188.6	1.14	0.60
25.0–26.9	20	26.0	212.8	2.36	0.73	27	25.7	227.6	1.68	0.64
27.0–28.9	13	27.8	258.8	3.03	0.78	9	27.5	269.4	1.73	0.61
29.0–39.0	3	29.6	286.7	4.65	1.00	4	29.6	361.3	3.05	0.64

Fecundity and Spawning of Catfishes

Based on the counts of ova from five ovaries, the estimated fecundity of brown bullheads in the lower Connecticut River was roughly 3,400 and 2,200 for white catfish (Table 39). The reported range in fecundity for brown bullheads lies between 2,000 and 13,800 (Carlander 1969), and for white catfish was between 1,000 (LaRivers 1962) and 4,000 (Carlander 1969). Bullheads spawn from March through September (Carlander 1969) throughout their range and white catfish spawn from late May to early June in Pennsylvania (Fowler 1917). Spawning occurs at temperatures around 21°C (Titcomb 1920; Breder 1935). Both species must spawn in May and June in the lower Connecticut River because young of the year have been collected near the end of June for several years.

TABLE 39. Estimated fecundity of brown bullheads and white catfish from counts of ova in ovaries from specimens collected in the lower Connecticut River.

Species	Length (cm)	Weight (g)	Ovarian weight (g)	Mean ovum size (mm)	Number ova in pairs of ovaries
Brown	24.5	200	1.8	0.66	2,804
bullhead	26.2	270	3.2		2,319
	26.7	270	3.6	0.56	5,085
White	23.8	140	3.3	0.78	1,955
catfish	25.8	200	3.3	0.70	2,501

The Sport Fishery in the Vicinity of the Connecticut Yankee Plant, Including the New Winter–Spring Fishery in Its Heated Discharge

The purpose of this study was to determine the effect of the operation of the Connecticut Yankee plant on the sport fishery in the lower Connecticut and Salmon rivers at Haddam, Connecticut and to assess the newly established winter–spring sport fishery in the plant's heated discharge.

Sport Fishing in the Vicinity of Connecticut Yankee and Plant Effects

The sport fishery in the Haddam area is primarily a shore-based fishery. A few boats fish the upper and lower Salmon River, but there is little fishing activity from boats on the Connecticut River. A creel survey was conducted each year from 1965 through 1969 in the vicinity of Haddam at nine shore locations within an area of 960 acres (389 ha). The three areas covered in the survey ranged from the southern end of Haddam Island (rkm 27.4) to the mouth of the Salmon River (km 22.5) and 4.8 km up the Salmon River to Leesville Dam (Figure 18). The survey was conducted from April through November in 1966, 1968, and 1969, from June through November in 1965, and from July through November in 1967. Survey days were chosen at random, with the restriction that the areas be sampled two out of five weekdays and one of the two weekend days. Each survey day was divided into four 4-h check periods and one 4-h period was chosen at random: 0600 to 1000 hours, 1000 to 1400 hours, 1400 to 1800 hours, or 1800 to 2200 hours. Usually three rounds were made of the three check areas during the randomly selected 4-h period. The information obtained from interviewing fishermen included the number of rods used, the number of hours fished, the number of species of fish caught (as counted by the interviewer himself), and the type of bait used. Frequently, the fish could be weighed and measured.

The total number of fish caught for each survey year was estimated by multiplying the average number of rod-hours within each 4-h period by the total number of available fishing days and by the average catch per rod-hour. Totals were figured separately for weekdays and weekends and summed. The procedures used to estimate the total number of fishermen and fishing effort and other survey information are described in Table 40. During the survey years, 2,590 fishermen were interviewed on 1,016 survey days. The observed catch totaled 6,316 fish in 5,720 rod-hours for an average catch rate of 1.1 fish per rod-hour. The catch rate was highly variable from one day to the next depending on the weather, the skill of the fishermen, and, apparently, the behavior of the fish. The average yearly catch rate, however, remained stable both before and after the operation of the plant, ranging between 1.02 and 1.18 fish/rod-hour. Fishermen caught 6,316 fish in 3,091 man-hours, an average catch per man-hour of 2.04

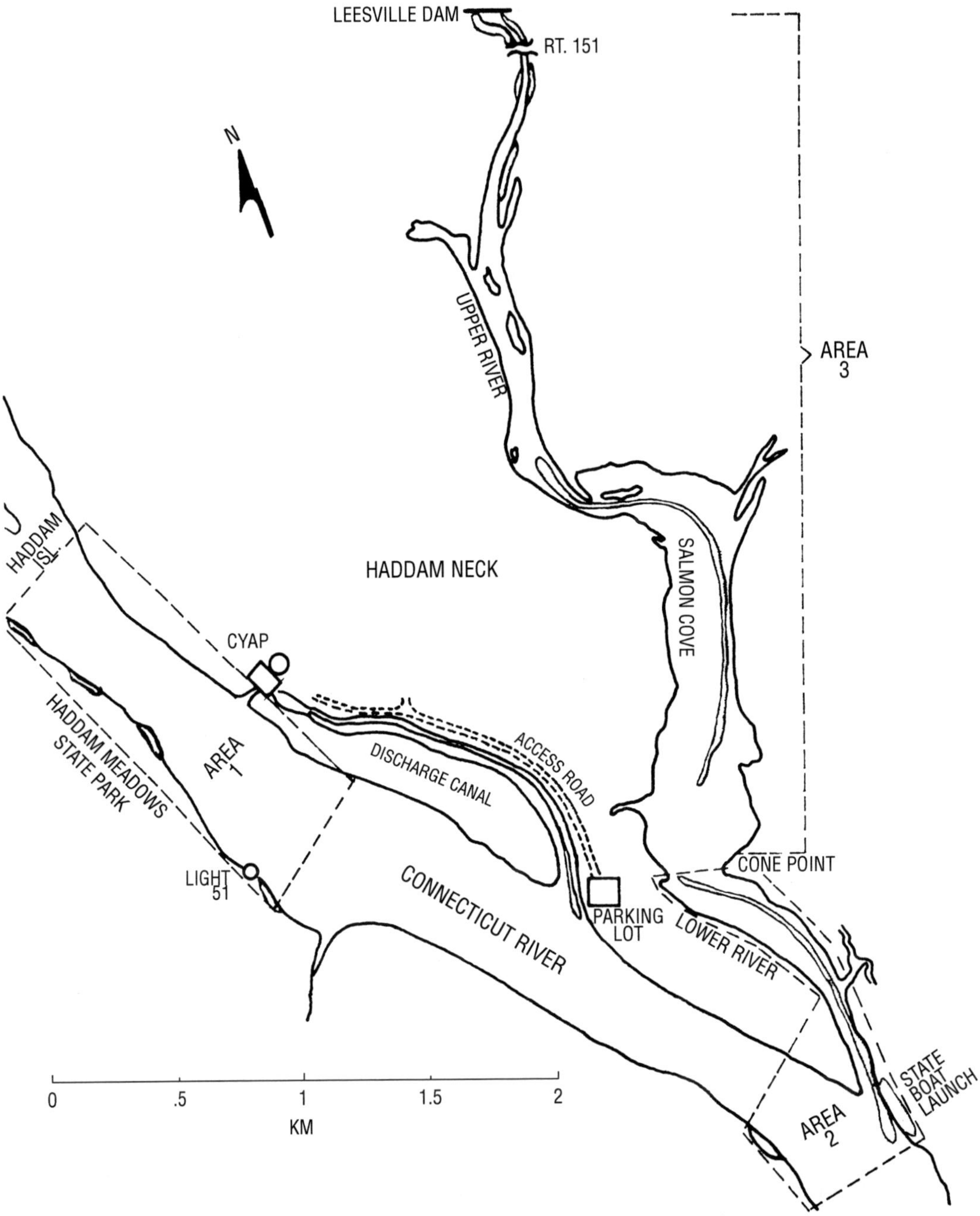

FIGURE 18. Creel survey areas of the Connecticut River at Haddam (1), lower Salmon River (2), upper Salmon River (3), and discharge canal of the Connecticut Yankee Atomic Power Company. Stippled areas indicate areas of concentrated fishing.

fish (range, 1.53–2.47). They used 3,083 rods or 1.19 rods per man, and 61% of them were successful in catching one or more fish.

A few more fishermen were interviewed during the week than on weekends. The largest number of people fished between 1800 and 2200 hours and the smallest number between 600 and 1000 hours. Fishing effort was slightly higher during

TABLE 40. Summary of creel survey results, April through November 1966, 1968, and 1969, June through November 1965, and July through November 1967.

	1965	1966	1967	1968	1969	Total
Number of survey days	153	244	152	225	242	1,016
Number of fishermen interviewed	403	825	281	719	362	2,590
(Estimated)[a] number of fishermen	6,580	10,858	3,817	10,093	5,637	36,985
Number of man hours fished	375	783	433	1,113	387	3,091
Number of rods used	451	1,141	332	773	385	3,082
Number of rod-hours	821	1,650	980	1,601	668	5,720
(Estimated)[b] rod-hours	14,431	21,158	13,888	21,945	19,342	90,764
Number of actual fish caught	895	1,934	996	1,703	788	6,316
(Estimated)[c] number of fish caught	10,607	22,737	14,091	22,393	11,076	80,904
Average number of fish caught/rod-hour	1.09	1.17	1.02	1.06	1.18	1.10
Average number of fish caught/man-hour	2.39	2.47	2.30	1.53	2.04	2.04

[a, b] Calculated by multiplying the average number of fishermen interviewed (or rod-hours) for each of the 4-h time periods for weekdays and for weekends by the number of available days. Totals of the eight estimates have been summed.

[c] Calculated by multiplying the average number of rod-hours in each 4-h time period by the total number of available fishing days and by the average number of fish per rod-hour. Totals were figured separately for weekdays and weekends and summed.

the week. Peak effort was reached between 1000 and 1400 hours during the week and 1800 and 2200 hours on weekends. There was no significant difference in the catch rates between weekdays and weekends at the three survey locations under different weather conditions or at different tides. The catch per rod-hour increased through the spring, reached a peak in May and June, decreased in July, and then increased through September and October until the maximum was attained in November. The peak seasonal catch rate occurred in the fall, but the peak monthly catch varied from year to year (Figure 19).

The estimated number of fishermen was 36,985, and they caught an estimated 80,904 fish in 90,764 rod-hours over the five survey years (Table 40). The average annual estimated catch in the 389 ha covered by the survey was 41.6 fish (9.2 kg) in 46.7 rod-hours per hectare. In an independent set of calculations, Richards (1967) estimated the 1966 catch for the Haddam area at 41.4 fish at 6.3 kg in 45.7 rod-hours per hectare. The remarkable similarity between these data and our data for all 5 years combined provides further evidence of the stability of the sport fishery in the Haddam area.

The only other creel survey in the Connecticut River was conducted by McCraig (1965) in a 28-km area between Northfield and Greenfield, Massachusetts, 140 km upstream from the Haddam area. The estimated 1964–1965 catch for this area was 8,486 fish in 37,251 h or 0.23 fish per hour. The average annual catch was calculated to be 13.6 fish and 3.1 kg in 54.1 rod-hours per hectare. Fishermen in the Haddam area caught more than three times as many fish per hectare and nearly four times as many fish per hour of effort as did those angling upstream between Northfield and Greenfield, Massachusetts.

A comparison of the three survey areas within the Haddam area shows that the river itself received the most fishing pressure (2,113 rod-hours) with the lower Salmon River second (1,778 rod-hours) and the upper Salmon River third (1,500 rod-hours). The highest catch of fish per rod-hour, however, was 1.41 in the lower Salmon River, as compared with 1.23 in the Connecticut River and 0.81 in the upper Salmon River (Table 41). The catch rates by year and survey area are shown in Table 42.

Table 43 shows that the catch of 25 different species was dominated by the brown bullhead (27.0%), white perch (26.0%), and white catfish (16.0%), these three species constituting almost 70% of the total catch. The American eel, pumpkinseed, and yellow perch made up 22.0%. Together, both groups constituted more than 90% of all the fish caught by sport fishermen in the Haddam area.

In all, 25 species were caught in the upper

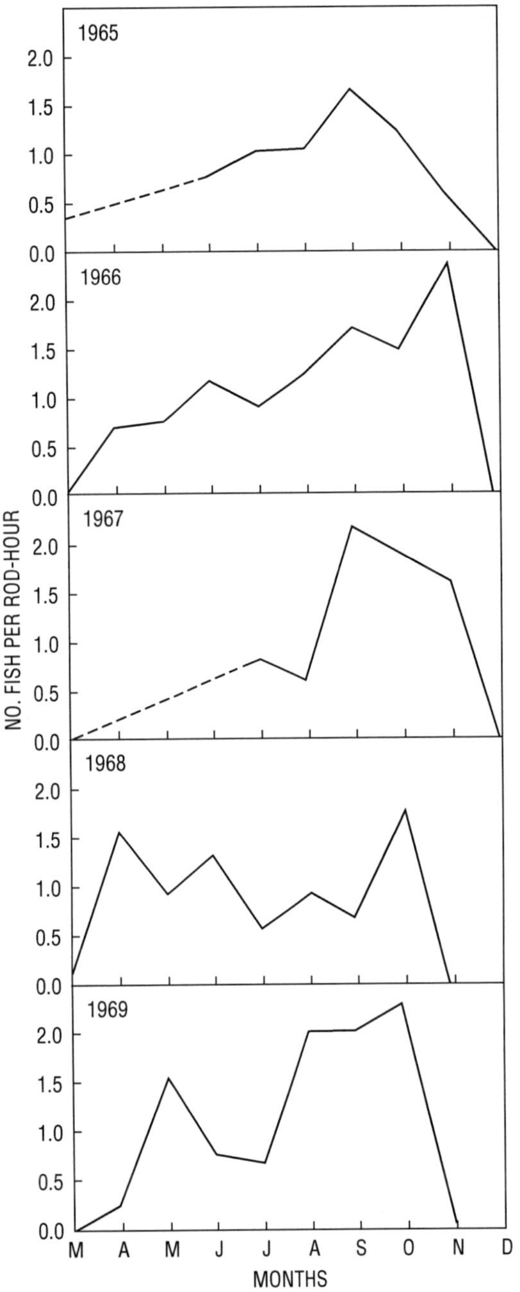

FIGURE 19. Number of fish caught per rod-hour of effort for each month of the creel survey, 1965–1969. Dashed lines indicate no data available.

Salmon River, 14 in the lower Salmon River, and 12 in the Connecticut River (Table 41). In the Connecticut River survey area, 90% of the catch consisted of white perch, brown bullhead, American eel, and white catfish (Table 44). In the lower Salmon River, 86% of the catch consisted of brown bullhead, white catfish, white perch, and American eel (Table 45). In the upper Salmon River, the alewife was most abundant in the catch followed by five other species that were caught with about the same frequency: white perch, brown bullhead, yellow perch, pumpkinseed, and blueback herring (Table 46). The large number of brown trout Salmo trutta caught in 1968 in the upper Salmon River was largely the result of the survey having been conducted on the opening day of trout season, and most of the trout caught were fin-clipped or stocked fish and constituted the basis for a rather short-lived "put and take" fishery in this area. It was estimated that 1,537 brown trout, 728 American shad, 324 largemouth bass, 162 striped bass, 49 rainbow trout Salmo gairdneri, and 24 northern pike were caught over the entire 5-year period (Table 41); these represented only 3.5% of the total catch.

The analysis of variance of the catch per rod-hour for all species combined showed no significant differences among the three different creel-survey areas for the before-operation years (1965–1967) vs. the after-operation years (1968–1969) [F (2,6) = 0.55; $P > 0.05$]. None of the major species (brown bullhead, white catfish, white perch, American eel, and yellow perch), showed any significant differences in catch per rod-hour among the three creel-survey areas for the before- vs. after-operation years (Table 47). We may conclude that the sport fishing in the vicinity of the plant did not change significantly as a result of its operation.

Sport Fishing in the Discharge Canal and Plant Effects

The heated discharge of the Connecticut Yankee atomic power plant is now being utilized by a large number of fishermen during the winter and spring months. Fishermen first gained access to the east bank of the CYAP canal in the fall of 1971 when a road was constructed along its entire length terminating near the discharge point. Previously, the only access had been by boat from the Connecticut River, and this was limited by a boom near the mouth of the canal. Before the construction of the road, very little fishing had occurred in the vicinity of the discharge. For example, from December 1967 through March 1968, in a biweekly creel survey that was conducted in the Haddam area, only three fishermen

TABLE 41. Summary of creel survey data by species, area, and estimated catch, 1965–1969.

Species	Connecticut River (Haddam area) Total no. caught	Percentage total catch	No. fish/ rod-hour	Lower Salmon River Total no. caught	Percentage total catch	No. fish/ rod-hour
Brown bullhead	689	26.5	0.33	863	34.5	0.49
White perch	996	38.3	0.47	503	20.1	0.28
White catfish	345	13.4	0.16	577	23.0	0.32
American eel	298	11.5	0.14	219	8.7	0.12
Yellow perch	162	6.2	0.08	132	5.3	0.07
Alewife	1	0.0	0.00	1	0.0	0.00
Blueback herring	82	3.2	0.04	151	6.0	0.08
Brown trout	8	0.3	0.00	7	0.3	0.00
American shad	0			5	0.2	0.00
Bluegill	0			33	1.3	0.02
Largemouth bass	0			3	0.1	0.00
White sucker	1	0.0	0.00	4	0.2	0.00
Striped bass	9	0.3	0.00	0		
Common carp	6	0.2	0.00	6	0.2	0.00
Rainbow trout	0			0		
Rock bass	0			0		
Black crappie	0			1	0.0	0.00
Sea lamprey	0			0		
Northern pike	2	0.1	0.00	0		
Chicken pickerel	0			0		
Creek chub	0			0		
Redbreast sunfish	0			0		
Fallfish	0			0		
Smallmouth bass	0			0		
Totals	2,599	100.0	1.23	2,505	100.0	1.41

Species	Upper Salmon River Total no. caught	Percentage total catch	No. fish/ rod-hour	All areas combined Total no. caught	Percentage total catch	No. fish rod-hour	Estimated total catch
Brown bullhead	146	12.1	0.10	1,698	26.9	0.315	21,763
White perch	148	12.2	0.10	1,647	26.1	0.305	21,116
White catfish	87	7.2	0.06	1,009	16.0	0.187	12,905
American eel	56	4.6	0.04	575	9.1	0.107	7,362
Yellow perch	144	11.8	0.10	436	6.9	0.081	5,532
Alewife	288	23.8	0.20	290	4.6	0.054	3,722
Blueback herring	124	10.2	0.08	357	5.7	0.066	4,612
Brown trout	101	8.3	0.07	117	1.9	0.022	1,537
American shad	50	4.1	0.03	55	0.9	0.010	728
Bluegill	14	1.2	0.01	47	0.7	0.009	566
Largemouth bass	20	1.7	0.01	23	0.4	0.004	324
White sucker	9	0.7	0.00	14	0.2	0.002	162
Striped bass	4	0.3	0.00	13	0.2	0.002	162
Common carp	2	0.1	0.00	13	0.2	0.002	162
Rainbow trout	4	0.3	0.00	4	0.0	0.001	49
Rock bass	4	0.3	0.00	4	0.0	0.001	49
Black crappie	3	0.2	0.00	4	0.0	0.001	49
Sea lamprey	2	0.2	0.00	2	0.0	0.000	24
Northern pike	2	0.1	0.00	2	0.0	0.000	24
Chicken pickerel	2	0.1	0.00	2	0.0	0.000	24
Creek chub	1	0.1	0.00	1	0.0	0.000	8
Redbreast sunfish	1	0.1	0.00	1	0.0	0.000	8
Fallfish	1	0.1	0.00	1	0.0	0.000	8
Smallmouth bass	1	0.1	0.00	1	0.0	0.000	8
Totals	1,212	100.0	0.81	6,316	100.0	1.10	80,904

TABLE 42. Catch rates (fish per rod-hour) from 1965 to 1969 in the three creel survey areas in the vicinity of the CYAP plant.

	Creel survey areas		
Year	Connecticut River	Lower Salmon River	Upper Salmon River
1965	1.36	1.20	0.78
1966	1.54	1.65	0.69
1967	1.25	1.72	1.09
1968	1.08	1.01	0.94
1969	0.99	1.79	0.75
Mean	1.23	1.41	0.81

were interviewed during the entire 3-month period, and none had caught any fish. In order to assess the newly established winter–spring sport fishery in the CYAP canal, Marcy and Galvin (1973) analyzed creel-survey data collected in 1972. The following discussion is based on these data and a 1973 winter–spring creel survey of the discharge canal.

Little information is presently available regarding the effect of heated water on sport fishing, especially during the cooler months. In Maryland and Texas, respectively, Moore and Frisbie (1972) and Landry and Strawn (1973) recorded greater fishing pressure in a heated discharge between December and April than at any other time. Elser (1965) found that a heated discharge in Maryland yielded substantially more fish per trip at all times except July through September, when the highest water temperatures occurred. In South Carolina, Gibbons et al. (1972) found that significantly more largemouth bass were caught per cast during the winter in the vicinity of a power plant's discharge when the discharge was heated than when it was not.

TABLE 43. Summary of creel survey data from the Connecticut River at Haddam and the upper and lower Salmon River, 1965–1969. Species listed by order of abundance for all years combined.

	1965		1966		1967		1968		1969		All years	
Species	Total no.	% of catch	Total no.	% of catch	Total no.	% of catch	Total no.	% of catch	Total no.	% of catch	Total no.	% of catch
Brown bullhead	182	20.3	696	36.0	205	20.6	413	24.3	202	25.5	1,698	26.9
White perch	194	21.8	305	15.9	441	44.4	475	27.9	232	29.4	1,647	26.1
White catfish	219	24.6	363	18.9	71	7.1	227	13.3	129	16.4	1,009	16.0
American eel	134	15.0	107	5.5	111	11.1	138	8.1	85	10.8	575	9.1
Yellow perch	55	6.1	161	8.3	62	6.2	115	6.9	43	5.5	436	6.9
Pumpkinseed	66	7.4	80	4.1	57	5.7	85	5.0	69	8.8	357	5.7
Alewife	0		132	6.8	0		30	1.8	4	0.5	166	2.6
Blueback herring	19	2.1	17	0.9	1	0.1	87	5.1	0		124	2.0
Brown trout	1	0.1	17	0.9	2	0.2	92	5.4	5	0.6	117	1.9
American shad	11	1.2	36	1.9	0		8	0.5	0		55	0.9
Bluegill	6	0.7	2	0.1	28	2.8	8	0.5	3	0.4	47	0.7
Largemouth bass	2	0.2	1	0.0	13	1.3	4	0.2	3	0.4	23	0.4
White sucker	2	0.2	6	0.3	0		4	0.2	2	0.3	14	0.2
Striped bass	0		3	0.2	0		5	0.3	5	0.6	13	0.2
Common carp	1	0.1	2	0.1	5	0.5	2	0.1	3	0.4	13	0.2
Rainbow trout	0		0		0		2	0.1	2	0.3	4	0.0
Rock bass	0		2	0.1	0		1	0.0	1	0.1	4	0.0
Black crappie	0		0		0		4	0.2	0		4	0.0
Sea lamprey	2	0.2	0		0		0		0		2	0.0
Chain pickerel	0		0		0		2	0.1	0		2	0.0
Northern pike	1	0.1	1	0.0	0		0		0		2	0.0
Creek chub	0		0		0		1	0.0	0		1	0.0
Redbreast sunfish	0		1	0.0	0		0		0		1	0.0
Fallfish	0		1	0.0	0		0		0		1	0.0
Smallmouth bass	0		1	0.0	0		0		0		1	0.0
Totals	895	100.0	1,934	100.0	996	100.0	1,703	100.0	788	100.0	6,316	100.0

TABLE 44. Summary of creel survey data from the Haddam area of the lower Connecticut River (Haddam Meadows State Park to opposite bank Salmon River boat launch), 1965–1969. Species listed by order of abundance for all years combined.

	1965		1966		1967		1968		1969		All years	
Species	Total no.	% of catch	Total no.	% of catch	Total no.	% of catch	Total no.	% of catch	Total no.	% of % of	Total no.	% of catch
White perch	117	29.0	198	30.0	323	50.9	225	39.3	133	39.6	996	38.5
Brown bullhead	92	22.8	184	27.9	132	20.9	164	28.6	117	34.8	689	26.5
White catfish	98	24.3	118	17.9	44	6.9	61	10.7	24	7.1	345	13.4
American eel	53	13.3	66	10.0	77	12.1	76	13.3	26	7.7	298	11.5
Yellow perch	28	6.9	61	9.3	37	5.8	23	4.0	13	3.9	162	6.2
Pumpkinseed	19	3.7	22	3.2	14	2.2	15	2.6	16	4.8	82	3.2
Striped bass	0		0		0		4	0.7	5	1.5	9	0.3
Brown trout	0		1	1.1	0		1	0.2	0		8	0.3
Common carp	0		1	0.2	3	0.5	1	0.2	1	0.3	6	0.2
Northern pike	0		1	0.2	0		1	0.2	0		1	0.1
White sucker	0		0		0		0		1	0.3	1	0.0
Blueback herring	0		0		1	0.2	0		0		1	0.0
Totals	403	100.0	658	100.0	631	100.0	571	100.0	336	100.0	2,599	100.0

A creel survey of the sport fishery was conducted in the discharge canal from January through June in 1972 and 1973. Survey days were chosen at random, with the restriction that the area be sampled two out of five weekdays and one of the two weekend days. Each survey day was divided at random into two 4-h check periods: 0900 to 1300 hours and 1300 to 1700 hours. Hourly checks were made along the canal within each 4-h period. In all, 47 weekdays and 20 weekend days were sampled in 1972 and 52 weekdays and 25 weekend days in 1973. The information obtained from interviewing fishermen included the number of rods used, the number of hours fished, the number of species of fish caught (as counted by the interviewer himself), and the type of bait used. Daily mean temperatures of the discharge water were obtained from plant records.

TABLE 45. Summary of creel survey data from the lower Salmon River (Salmon River boat launch to Cove Point), 1965–1969. Species listed by order of abundance for all years combined.

	1965		1966		1967		1968		1969		All years	
Species	Total no.	% of catch	Total no.	% of catch	Total no.	% of catch	Total no.	% of catch	Total no.	% of catch	Total no.	% of catch
Brown bullhead	65	19.3	432	50.1	71	24.8	230	34.0	65	19.0	863	34.5
White catfish	104	30.9	215	24.9	23	8.0	135	20.0	100	29.2	577	23.0
White perch	64	19.0	84	9.7	115	40.4	177	26.2	63	18.5	503	20.1
American eel	65	19.3	34	3.9	23	8.0	47	7.0	50	14.6	219	8.7
Pumpkinseed	25	7.3	42	4.9	8	2.8	26	3.8	50	14.6	151	6.0
Yellow perch	11	3.3	49	5.7	21	7.3	40	6.0	11	3.2	132	5.3
Bluegill	1	0.3	0		23	8.0	7	1.0	2	0.6	33	1.3
Brown trout	0		1	0.1	0		6	0.9	0		7	0.3
Common carp	1	0.3	1	0.1	2	0.7	1	0.1	1	0.3	6	0.2
American shad	0		5	0.6	0		0		0		5	0.2
White sucker	1	0.3	0		0		3	0.4	0		4	0.2
Largemouth bass	0		0		0		3	0.4	0		3	0.1
Black crappie	0		0		0		1	0.1	0		1	0.0
Blueback herring	0		0		0		1	0.1	0		1	0.0
Totals	337	100.0	863	100.0	286	100.0	677	100.0	342	100.0	2,505	100.0

TABLE 46. Summary of creel survey data from upper Salmon River (Salmon Cove to Leesville Dam), 1965–1969. Species listed by order of abundance for all years combined.

Species	1965 Total no.	1965 % of catch	1966 Total no.	1966 % of catch	1967 Total no.	1967 % of catch	1968 Total no.	1968 % of catch	1969 Total no.	1969 % of catch	All years Total no.	All years % of catch
Alewife	0		132	32.0	0		30	6.6	4	3.7	166	13.7
White perch	13	8.4	23	5.6	3	3.8	73	16.0	36	33.0	148	12.2
Brown bullhead	25	16.2	80	19.4	2	2.5	19	4.2	20	18.3	146	12.1
Yellow perch	17	11.0	51	12.3	4	5.1	52	11.4	19	17.5	143	11.8
Pumpkinseed	26	16.8	16	3.9	35	44.3	44	9.7	3	2.8	124	10.2
Blueback herring	18	11.6	17	4.1	0		86	19.0	0		122	10.1
Brown trout	1	0.6	8	1.9	2	2.5	85	18.7	5	4.6	101	8.3
White catfish	17	11.0	30	7.3	4	5.1	31	6.8	5	4.6	87	7.2
American eel	16	10.3	7	1.8	11	13.9	15	3.3	8	6.4	56	4.6
American shad	11	7.1	31	7.5	0		8	1.8	0		50	4.1
Largemouth bass	2	1.3	1	0.2	13	16.5	1	0.2	3	2.8	20	1.7
Bluegill	5	3.2	2	0.5	5	6.3	1	0.2	1	0.9	14	1.2
White sucker	1	0.6	6	1.5	0		1	0.2	1	0.9	9	0.7
Rainbow trout	0		0		0		2	0.4	2	1.8	4	0.3
Rock bass	0		2	0.5	0		1	0.2	1	0.9	4	0.3
Striped bass	0		3	0.7	0		1	0.2	0		4	0.3
Black crappie	0		0		0		3	0.7	0		3	0.2
Sea lamprey	2	1.3	0		0		0		0		2	0.2
Smallmouth bass	0		1	0.2	0		0		0		1	0.1
Northern pike	1	0.6	1	0.2	0		0		0		2	0.1
Fallfish	0		1	0.2	0		0		0		1	0.1
Redbreast sunfish	0		1	0.2	0		0		0		1	0.1
Creek chub	0		0		0		0		1	0.9	1	0.1
Common carp	0		0		0		0		1	0.9	1	0.1
Chain pickerel	0		0		0		2	0.4	0		1	0.2
Totals	155	100.0	413	100.0	79	100.0	455	100.0	110	100.0	1,212	100.0

TABLE 47. Analysis-of-variance table for catch per rod-hour for all species combined and for some important species. Tests include the three creel-survey areas in the vicinity of the CYAP plant for the before-operation years (1965–1967) vs. the after-operation years (1968–1969). Asterisk denotes significance at $P < 0.05$; unmarked values of F are not significant.

Sources of variation	Degrees of freedom	All species Mean square	All species F	Brown bullhead Mean square	Brown bullhead F	White perch Mean square	White perch F	White catfish Mean square	White catfish F	American eel Mean square	American eel F
Before years (1965–1967) vs. after years (1968–1969)	1	0.0743	1.05	0.0046	0.27	0.0000	0.00	0.0030	0.29	0.0085	0.66
Before years (1965–1967)	2	0.0427	0.61	0.0540	3.16	0.0519	3.42	0.0261	2.46	0.0064	0.49
After years (1968–1969)	1	0.0275	0.39	0.0000	0.00	0.0019	0.12	0.0087	0.82	0.0536	4.12
Areas	2	0.4788	6.78*	0.1464	8.56*	0.1326	8.73*	0.0975	9.18*	0.0014	0.10
Before vs. after years vs. area	2	0.0392	0.55	0.0092	0.54	0.0197	1.30	0.0126	1.19	0.0158	1.21
Error	6	0.0706		0.0171		0.0152		0.0106		0.0130	
Total	14										

The total number of fish caught was estimated by multiplying the average number of rod-hours within each 4-h period by the total number of available fishing days and by the average catch per rod-hour. Separate totals were calculated for weekdays and weekends, and these were summed. The number of fishermen interviewed and the procedure used to estimate their total numbers and other survey information are shown in Table 48.

During the first year of the new fishery (1972), 597 fishermen were interviewed; the majority (303) fished on weekends and the smallest number on Fridays (24). The fishermen interviewed had caught 1,872 fish in 790 h (Table 48). The average number of fish caught per hour of effort ranged from 0.56 in February to 4.11 in May, and averaged 2.37 over the 6-month period. The fishermen used 375 rods, or 1.47 rods per man, and they averaged 2.07 fish per rod-hour. Seventy-two percent (431) of the fishermen were successful in catching one or more fish. An estimated 3,166 fishermen caught an estimated 11,090 fish (Table 49) over the 6-month period in 1972 (Tables 48 and 49).

During the second year of the winter–spring fishery (1973), 1,057 fishermen were interviewed, and about half (539) fished on weekends. The fishermen interviewed had caught 4,383 fish in 3,910 h. The average number of fish caught per hour of

TABLE 48. Summary of creel survey results in the CYAP heated discharge canal, January through June 1972 and 1973.

	January		February		March		April	
	1972	1973	1972	1973	1972	1973	1972	1973
Number of survey days	8	13	12	12	12	13	11	13
Actual number of fishermen interviewed	72	138	24	120	24	183	130	251
(Estimated)[a] number of fishermen	(331)	(486)	(98)	(685)	(143)	(883)	(700)	(800)
Number of successful fishermen	48	79	11	57	15	151	91	236
Number of man hours fished	116	331	52	225	28	390	134	703
Number of rods used	96	166	56	131	35	280	145	378
Number of rod-hours	119	414	56	276	29	476	135	828
Number of actual fish caught	120	243	29	225	26	779	252	1,470
Number of fish caught per rod-hour	1.01	0.59	0.52	0.82	0.90	1.64	1.87	1.77
Number of fish caught per man-hour	1.03	0.73	0.56	1.00	0.93	2.00	1.88	2.09

	May		June		Total		Total
	1972	1973	1972	1973	1972	1973	both years
Number of survey days	12	13	12	13	67	77	144
Actual number of fishermen interviewed	243	274	104	91	597	1,057	1,654
(Estimated)[a] number of fishermen	(1,357)	(510)	(537)	(382)	(3,166)	(3,746)	(6,912)
Number of successful fishermen	200	237	66	60	431	820	1,251
Number of man hours fished	289	1,010	171	251	790	2,910	3,700
Number of rods used	330	347	213	312	875	1,614	2,489
Number of rod-hours	353	1,271	214	369	906	3,634	4,540
Number of actual fish caught	1,187	1,456	258	210	1,872	4,383	6,255
Number of fish caught per rod-hour	3.36	1.15	1.21	0.57	2.07	1.21	1.38
Number of fish caught per man-hour	4.11	1.44	0.99	0.84	2.37	1.51	1.69

[a]Calculated by multiplying the average number of actual fishermen for each of the two 4-h time periods for weekdays and weekends by the number of available fishing days. Totals of the four estimates have been summed.

TABLE 49. Estimated[a] number of fish caught in the CYAP heated discharge canal, January through June 1972 and 1973.

Species	January		February		March		April	
	1972	1973	1972	1973	1972	1973	1972	1973
White catfish	119	334		403	4	1,959	447	2,012
White perch			3		69	28	460	577
Brown bullhead	173	71	40	427	8	1,018	300	1,096
Common carp	327	347	35	81	142	223	95	230
American eel						73	105	147
Striped bass							15	135
Pumpkinseed	6					11	25	172
Northern pike	38	48	16	181		87	10	25
Yellow perch	6	6	5		23	11	35	23
Channel catfish						27	10	180
Black crappie						5		26
Bluegill						8		
Largemouth bass				4		3	5	11
Blueback herring								
Golden shiner								11
Chain pickerel					23	5		
Smallmouth bass					23			
White sucker								
Totals	669	806	99	1,096	292	3,458	1,507	4,645

Species	May		June		Total		Grand total
	1972	1973	1972	1973	1972	1973	
White catfish	2,292	877	497	352	3,359	5,937	9,296
White perch	2,886	2,518	190	315	3,608	3,438	7,046
Brown bullhead	620	334	157	54	1,298	3,000	4,298
Common carp	58	317	39		696	1,198	1,894
American eel	499	236	57		661	456	1,117
Striped bass	275	333	244	19	534	487	1,021
Pumpkinseed	326	334	71	16	428	533	961
Northern pike	4	4		7	68	352	420
Yellow perch	125	40	27	3	221	83	304
Channel catfish	8	23	41		59	230	289
Black crappie	8	58			8	89	97
Bluegill	4		40	24	44	32	76
Largemouth bass		38	2	7	7	63	70
Blueback herring	24	37			24	37	61
Golden shiner	16	3	5		21	14	35
Chain pickerel					23	5	28
Smallmouth bass	4				27		27
White sucker			4		4		4
Totals	7,149	5,152	1,374	797	11,090	15,954	27,044

[a] Calculated by multiplying the average number of rod-hours in each 4-h time period by the total number of available fishing days and by the average number of per rod-hour. Totals were figured separately for weekdays and weekends and summed.

effort ranged from 0.73 in January to 2.09 in April, and averaged 1.51. Fishermen used 1,614 rods, or 1.53 rods per man, and averaged 1.21 fish per rod-hour. Seventy-eight percent (820) were successful in catching one or more fish. An estimated 3,746 fishermen caught an estimated

15,974 fish during the first 6 months of 1973 (Tables 48 and 49).

Twice as many fishermen were interviewed in 1973 as in 1972 (1,057 and 597, respectively). The fishermen interviewed during 1973 expended four times the effort (3,634 rod-hours and 906 rod-hours, respectively) but caught only twice the number of fish as those who fished in 1972 (4,383 and 1,872, respectively). It may be that increased fishing effort (especially for northern pike), competition among the more numerous fishermen, and a decrease in the number of available fish resulted in the decline of the catch per rod- and man-hour to about half that recorded during the first year. Nevertheless, in 1973, the fishermen were 6% more successful in catching one or more fish than they had been the previous year. It was estimated that in 6,912 trips, fishermen caught an estimated 27,044 fish during both years or almost 4 fish for each fishing trip. Landry and Strawn (1973) reported an overall winter–spring catch of 2.5 fish per fisherman in a Texas heated discharge. The most striking difference between the 2 years was the considerable increase in both the estimated number of fishermen and the estimated number of fish caught during the winter months (January through March) (i.e., about 3.6 times more fishermen and 5 times more fish). The fishing success rose slightly from 61.7% to 65.1%, and both the catch per rod- and man-hour increased in 1973 (Table 48).

In 1972, 18 species of fish were caught and 16 in 1973 (Table 49). White catfish, white perch, and brown bullhead made up more than three-quarters of the catch in both years. White perch dominated the catch in 1972, and white catfish dominated in 1973. Other species caught in relatively high numbers were carp, American eel, striped bass, pumpkinseed, and northern pike. White catfish (0.2 to 1.6 kg), brown bullheads, and carp (0.5 to 18.1 kg) dominated the winter catch (January through March). In both years, the greatest number of fish (an estimated 45% of the total) was caught during May. Northern pike and striped bass tended to be preferred species. Most pike ranged between 43 and 84 cm and weighed between 0.9 and 4.0 kg. Six pike were recorded at 6.4 kg, and there was a reliable but unconfirmed report of one weighing 8.2 kg. Many more fishermen fished exclusively for northern pike during 1973 than 1972, generally near the mouth of the canal. These fishermen average 0.52 pike/rod-hour and 0.56/man-hour from January through March, the months when this species was most available. This pike fishery may represent one of the best in the northeast during winter months. It was estimated that 352 pike were caught in 1973, a significant increase over the 68 pike that were estimated to have been caught during 1972. More than 1,000 striped bass were caught during both years. One striped bass weighed 4.5 kg, but 85% were just under legal size (40.6 cm).

Still fishing with shiners, canned sweet corn, potatoes, and worms was the most popular method. A small number of fishermen used artificial lures for game species such as northern pike and striped bass.

Temperatures in the canal ranged from 4.8°C to 30.0°C, and the number of fish caught per rod-hour ranged from 0 to 9.0 (Figures 20 and 21), but no clear-cut relationship between water temperature and catch rate seems evident, except on those days when the power level of the plant was reduced enough to lower the water temperature sharply. On such days, or immediately thereafter, the catch rate markedly declined. Several hundred white catfish, bullheads, and carp were observed swimming on the surface, and they exhibited a stunned or lethargic behavior while being carried out of the canal by the current. During these periods of sudden temperature drop, many carp schooled along the banks of the canal with their backs out of water. Although they showed no apparent reaction to touch, no mortality from the presumed shock of cold stress was noted. It is probably also significant that the catch rate increased sharply in April, when temperatures approached 17°C, and generally dropped when temperatures exceeded 22°C (Figures 20 and 21). Temperatures became warmer earlier in 1973, probably causing the catch rate to drop off earlier than in 1972. There was an increase in the catch rate for a short period after the plant shutdowns on 10 June 1972 and 4 June 1973. These increases were probably due to the return of fish from the river after high temperatures had caused them to leave. After plant start-up in late June, few fish were caught in the canal. Extensive trawls made in early July, when temperatures were above 32°C, yielded no fish.

The creel census that was carried out for 5 years in the 386-ha area around the plant provides a good measure of the success of the sport

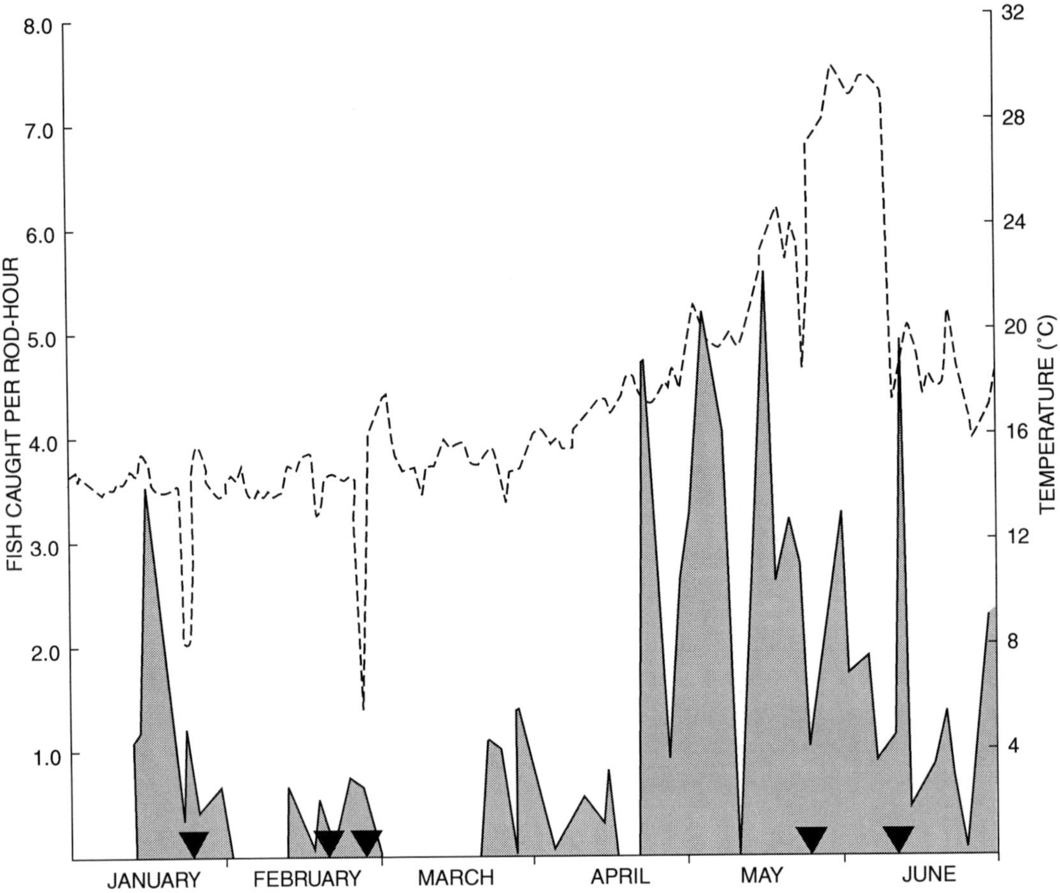

FIGURE 20. Mean daily water temperatures (dotted line) and fish caught per rod-hour of effort (solid line) in the Connecticut Yankee heated discharge canal, January through June 1972. Triangles indicate dates of major drops in power output.

fishery there. An estimated density of 19 fishermen per hectare each year spent an average of 46.7 rod-hours in catching an average of 42 fish weighing a total of 3.6 kg. These figures are vastly different from the average annual estimate of the creel census of the winter–spring fishery in the discharge canal (2.95 ha): 1,171 fishermen per hectare who caught 4,584 fish at 1,743 kg in 3,405 rod-hours. The number of fishermen interviewed per survey day was 2.5 in the vicinity of the plant and 11.5 in the discharge canal. In the only other survey conducted in the Connecticut River, McCraig (1965) reported a winter catch of 872 fish in 1,540 h, or 0.57 fish/h.

Moore and Frisbie (1972) conducted a similar survey in Maryland on the Chalk Point power plant's heated discharge canal of similar design. Their census involved 20,335 angler trips and more than 58,453 fish. They found that 51% of the fishermen caught one or more fish and that they averaged 0.71 fish/h from January through April. They also reported sport fishing to be negligible in the winter and early spring in the vicinity of the Chalk Point plant. Elser (1963) reported a yearly average catch per hour of effort of 1.0 in the vicinity of the Chalk Point plant, and the average in the vicinity of the Connecticut Yankee plant was also 1.0. Unlike the present study, Moore and Frisbie (1972) found no relationship between water temperature and fish caught per hour. They recorded nine species that were caught during the winter–spring months, while 18 were caught in the present study. In their study, white perch dominated the catch. In the present study, white catfish dominated the catch followed by white perch and brown bullheads.

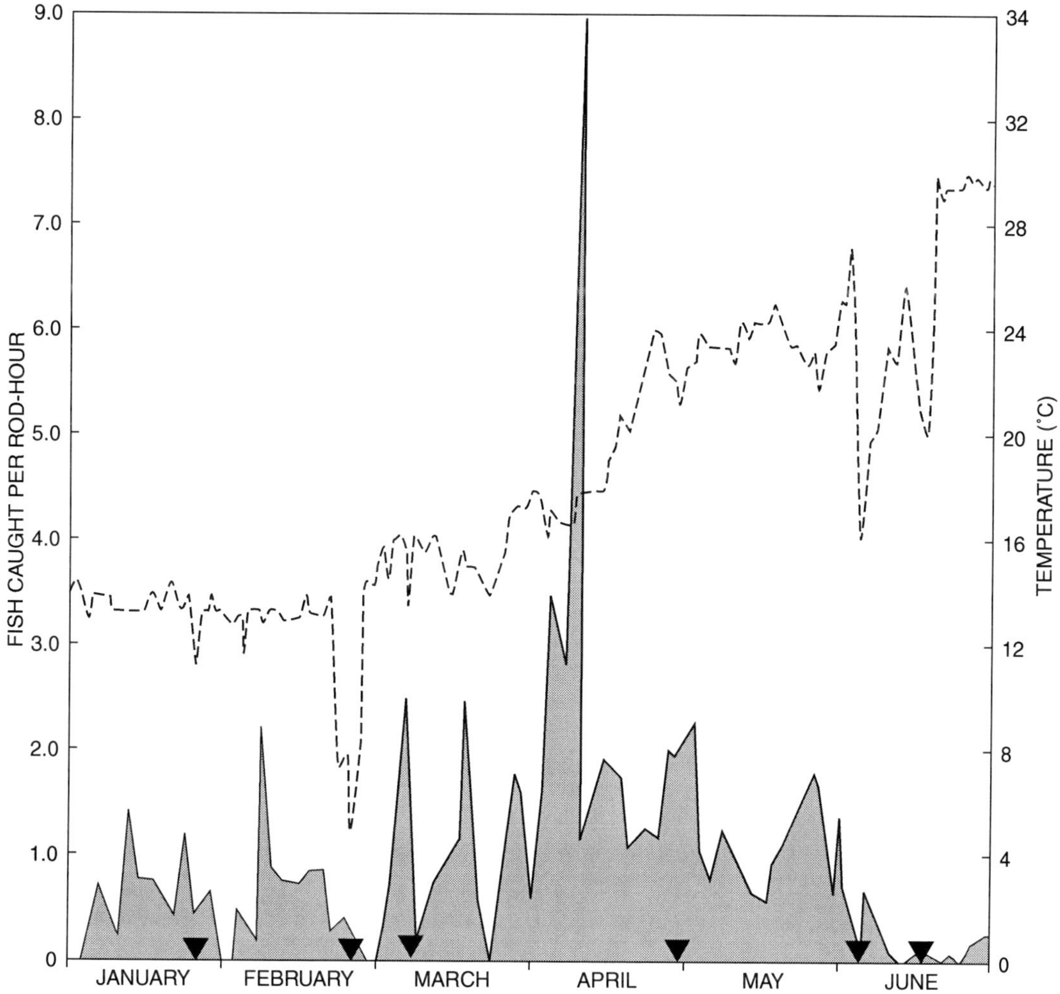

FIGURE 21. Mean daily water temperatures (dotted line) and fish caught per rod-hour of effort (solid line) in the Connecticut Yankee heated discharge canal, January through June 1973. Triangles indicate dates of major drops in power output.

It appears that one of the most successful of all winter–spring fisheries has developed in the discharge canal of the Connecticut Yankee plant. Fishermen who utilize its heated discharge during the cooler months are able to catch twice as many fish-per-unit-effort as those who fish in the Haddam area from April through November. These anglers also caught up to 10 times more fish/h of effort than the fishermen interviewed by McCraig (1965) in his 1964–1965 creel census, performed 141 km upstream, and up to four times more fish/h as compared with his winter estimate. Fishermen along the canal were between 11% and 17% more successful in catching one or more fish than those fishing in the general vicinity of the plant, and they averaged between 1.51 and 2.37 fish/h over the 6-month period. The northern pike provides a striking example of their increasing success. Five times as many pike were caught in 1973 as in the previous year (one pike for every two fishing hours), and this species has become a major attraction for fishermen during the winter months. The fact that anglers fishing in the canal had to expend about half the effort to catch the same number of fish as fishermen at other places in the vicinity of the plant, is a good indication of the unique fishery created after plant operation. Whether this fishery will become depleted or will eventually change the predator–prey balance in fish populations in the whole Haddam area, however, is not presently known.

References

Adair, W. D., and D. J. Demont. 1971. In R. W. Koss, editor. Environmental responses to thermal discharges from Marshall Steam Station, Lake Norman, North Carolina. Cooling water studies, Edison Electric Co., The Johns Hopkins University, Baltimore, Maryland.

Bader, R. A., et al. 1971. Ecological study of South Biscayne Bay and Card Sound, Florida. Rosenstiel School of Marine and Atmospheric Science, University of Miami, Miami.

Bardach, J. E., and R. G. Bjocklund. 1957. The temperature sensitivity of some American freshwater fishes. American Naturalist 91:233–251.

Behmer, D. J. 1969. Schooling of river carpsuckers and a population estimate. Transactions of the American Fisheries Society 98(3):520–523.

Bennett, D. H. 1972. Length-weight relationships and condition factors of fishes from a South Carolina reservoir receiving thermal effluent. Progressive Fish-Culturist 34:85–87.

Breder, C. M., Jr. 1935. The reproductive habits of the common catfish, *Ameiurus nebulosus* (Le-Sueur) with a discussion of their significance in ontogeny and phylogeny. Zoologica 19:143–185.

Brett, J. R. 1944. Some lethal temperature relations of Algonquin Park fishes. University of Toronto Stud. Biological Service 53, Publ. Ontario Fisheries Research Laboratory 64:9–28.

Cairns, J., Jr. 1956. Effects of increased temperature on aquatic organisms. Industrial Wastes 1:150–152.

Carlander, K. D. 1969. Handbook of freshwater fishery biology. The Iowa State University Press, Ames.

Coutant, C. C. 1970. Biological aspects of thermal pollution I. Entrainment and discharge canal effects. CRC Critical Reviews in Environmental Control 1(3):341–381.

Cragg-Hine, D. (no date). Proceedings of the 5th coarse fish conference. University of Liverpool, Liverpool, UK.

Dell, M. B. 1968. A new fish tag and rapid, cartridge-fed applicator. Transactions of the American Fisheries Society 97(1):57–59.

Doudoroff, P. 1938. Reactions of marine fishes to temperature gradients. Biological Bulletin 75:494–509.

Edsall, T. A., and T. G. Yocom. 1972. Review of recent technical information concerning the adverse effects of once-through cooling on Lake Michigan, prepared for the Lake Michigan Enforcement Conference, September 19–21, 1972, Chicago, Ill. Great Lakes Fishery Laboratory, Ann Arbor, Michigan.

Elser, H. J. 1963. Patuxent River creel census. University of Maryland, Natural Resources Institute, Reference No. 63-53.

Elser, H. J. 1965. Effect of a warmed water discharge on angling in the Potomac River, Maryland, 1961–1962. Progressive Fish-Culturist 27:79–86.

Fowler, H. W. 1917. Some notes on the breeding habits of local catfishes. Copeia 42:32–36.

Fry, F. E. J. 1937. The summer migration of cisco, *Leucichthys artedi* (LeSueur), in Lake Nipissing, Ontario. University of Toronto Studies, Biological Series 44.

Fry, F. E. J. 1947. Effects of the environment on animal activity. University of Toronto Studies, Biological Series 55, Publication of Ontario Fisheries Research Laboratory 68.

Gammon, J. R. 1971. The response of fish populations in the Wabash River to heated effluents. Pages 513–523 in Proceedings of the 3rd National Symposium on Radioecology. May 10, 1971; Oak Ridge, Tennessee.

Gibbons, J. W., J. T. Hook and D. L. Forney. 1972. Winter responses of largemouth bass to a heated effluent from a nuclear reactor. Progressive Fish-Culturist 34:88–90.

Kendall, A. W., and F. J. Schwartz. 1968. Lethal temperatures and salinity tolerances of white catfish, *Ictalurus catus,* from the Patuxent River, Maryland. Chesapeake Science 9:103–108.

La Faunce, D. A. 1965. Long-term retention of tags by some freshwater fish. California Fish and Game 51:52–53.

Landry, A. M., Jr. and K. Strawn. 1973. Annual cycle of sportfishing activity at a warmwater discharge into Galveston Bay, Texas. Transactions of the American Fisheries Society 102(3):573–577.

Langford, T. E. 1972. A comparative assessment of thermal effect in some British and North American Rivers. Pages 319–351 *in* River ecology and man. Proceedings of an International Symposium on River Ecology and the Impact of Man. Academic Press, New York.

LaRivers, I. 1962. Fishes and fisheries of Nevada. Nevada State Fish and Game Commission, Reno.

Loeb, H. A. 1964. Submergence of brown bullheads in bottom sediment. New York Fish and Game Journal 11(2):119–124.

Mansueti, R. J. 1961. Movements, reproduction, and mortality of the white perch, *Roccus americanus,* in the Patuxent Estuary, Maryland. Chesapeake Science 2:142–205.

Marcy, B. C., Jr., and R. C. Galvin. 1973. Winter-spring sport fishery in the heated discharge of a nuclear power plant. Journal of Fish Biology 5:541–547.

Marcy, B. C., Jr., and F. P. Richards. 1974. Age and growth of the white perch in the lower Connecticut River. Transactions of the American Fisheries Society 103(1):117–120.

Massengill, R. R. 1973. Change in feeding and body condition of brown bullheads overwintering in the heated effluent of a power plant. Chesapeake Science 14:138–141.

Massengill, R. R. 2004. Entrainment of zooplankton at the Connecticut Yankee plant. Pages 59–64 *in* P. M. Jacobson, D. A. Dixon, W. C. Leggett, B. C. Marcy, Jr., and R. R. Massengill, editors. The Connecticut River Ecological Study (1965–1973) revisited: ecology of the lower Connecticut River 1973–2003. American Fisheries Society, Monograph 9, Bethesda, Maryland. (Originally published in 1976)

McCammon, G. W., and C. M. Seeley. 1961. Survival, mortality, and movements of white catfish and brown bullheads in Clear Lake, California. California Fish and Game 47:237–255.

McCraig, R. S. 1965. Connecticut River harvest and population studies, 1964–1965. Massachusetts Division of Fish and Game, Bureau of Wildlife Research and Management, Project No. F-25-R-1.

Miller, L.W. 1963. Growth, reproduction and food habits of the white perch, *Roccus americanus* (Gmelin) in the Delaware River Estuary. M.S. thesis. University of Delaware, Newark.

Moore, C. J., and C. M. Frisbie. 1972. A winter sport fishing survey in a warm water discharge of a steam electric station on the Patuxent River, Maryland. Chesapeake Science 13:110–115.

Muncey, R. J. 1959. Age and growth of channel catfish from the Des Moines River, Boone County, Iowa, 1955 and 1956. Iowa State Journal of Science 34(2):127–137.

Nakatani, R. E. 1969. Effects of heated discharges on anadromous fishes. Pages 294–353 *in* P. A. Krenkel and F. L. Parker, editors. Biological aspects of thermal pollution. Vanderbilt University Press, Nashville.

Oros, I. 1968. Aspecte metabolice la somnul pitic (*Amiurus nebulosus*). In Sezonul Rice. Studia Univ. Babes-Bolyai, Ser. Biol. (Cluj, Transylvania Univ. Babes Bolyai). 13(2):133–138. (English summary).

Pelgen, D. E. 1954. Progress report on the tagging of white catfish (*Ictalurus catus*) in the Sacramento-San Joaquin Delta. California Fish and Game 40:313–321.

Pelgen, D. E, and G. W. McCammon. 1955. Second progress report on the tagging of white catfish (*Ictalurus catus*) in the Sacramento-San Joaquin Delta. California Fish and Game 41:261–269.

Priegel, G. R. 1966. Age-length and length-weight relationships of bullheads from Little Butte Des Mortes, 1959. Wisconsin Conservation Department, Res. Rep. Fish. 17.

Proffitt, M. A. 1969. Effects of heated discharge upon aquatic resources of White River at Petersburg, Indiana. Water Resources Research Center Report Invest. No. 3.

Richards, S. W. 1967. Resident fish report. In D. Merriman, editor. The Connecticut River investigation. Connecticut Water Resources Commission, 4th semi-annual progress report, Hartford.

Richards, W. J. 1960. The life history, habits and ecology of the white perch, *Roccus americanus* (Gmelin) in Cross Lake, New York. M.S. thesis. New York State University, New York.

Ricker, W. E. 1958. Handbook for computations for biological statistics of fish populations. Bulletin of the Fisheries Research Board of Canada 119.

Ricker, W. E. 1970. Methods for assessment of fish production in fresh waters. IBP handbook, No. 3. Biddles Ltd., Norfolk, UK.

Roessler, M. A. 1971. Environmental changes associated with a Florida power plant. Marine Pollution Bulletin (G.B.) 2.

Schoffman, R. J. 1961. Age and growth of the channel catfish in Reelfoot Lake, Tennessee, for 1953 and 1960. Journal of the Tennessee Academy of Science 36:5–11.

Shcherbukha, A. Y. 1971. A contribution to the biology of *Chondrostoma nasus variabile* (Pisces, Cyprinidae) from the Seversky Donetz Thermal Region. Zoologicheskii Zhurnal 50:713–723 (English Summary).

Silverman, M. J. 1971. Tragedy at Northport. Bulletin of the American Littoral Society 7:15–18.

Simpson, A.C. 1951. The fecundity of the plaice. Fishery Investigations, London Series 2, 17.

Smithsonian Institution. 1972. Report by Center for Short-lived Phenomena. February 7, 1972.

Snedecor, G. W., and W. G. Cochran. 1972. Statistical methods. The Iowa State University Press, Ames.

Stevens, R. E. 1959. The white and channel catfishes of the Santee-Cooper Reservoir and Tailrace Sanctuary. Proceedings of the S.E. Association of Game and Fish Commissioners 13:203–219.

Stobo, W. T. 1972. The effects of dart tags on yellow perch. Transactions of the American Fisheries Society 101(2):365–366.

Stone and Webster Engineering Co. 1964. Design memorandum. Investigation of thermal effects of the Connecticut Yankee power plant on the Connecticut River for Connecticut Yankee Power.

Swingle, H. A. 1965. Length-weight relationships of Alabama fishes. Auburn University Agric. Exp. St. Zool. Ent. Series Fish. 3.

Taub, S. H. 1966. Some aspects of the life history of the white perch, *Roccus americanus* (Gmelin) in Quabbin Reservoir, Massachusetts. M.S. thesis. University of Massachusetts.

Titcomb, J. W. 1920. Some fish-cultural notes. Transactions of the American Fisheries Society 50:200–211.

Trembley, F. J., editor. 1960. Research project on effects of condenser discharge water on aquatic life. Progress Report 1956–1959. Inst. Res., Lehigh University, Bethlehem, Pennsylvania.

Wallace, D. C. 1971. Age, growth, year-class strength, and survival rates of the white perch, *Morone americana* (Gmelin) in the Delaware River in the vicinity of Artificial Island. Chesapeake Science 12:205–218.

Webster, D. A. 1942. Food progression in young white perch, *Morone americana* (Gmelin) from Bantam Lake, Connecticut. Transactions of the American Fisheries Society 72:136–144.

Whitworth, W. R., and S. Sauter. 1972. Age and growth of some common fishes of Connecticut. University of Connecticut, Storrs Agr. Exp. Station, Res. 37, Storrs.

Witt, A., Jr. 1971. The evaluation of environmental alteration by thermal loading and acid pollution in the cooling reservoir of a steam electric station. NTIS, PB-197, Missouri Water Resources Research Center, Columbia, Water Resources Abstract 4. W71-04019.

American Fisheries Society Monograph 9:125–153, 2004

Planktonic Fish Eggs and Larvae of the Lower Connecticut River and the Effects of the Connecticut Yankee Plant, Including Entrainment

BARTON C. MARCY, JR.
Ecological Sciences Division, NUS Corporation
1910 Cochran Road, Pittsburgh, Pennsylvania 15220, USA

Ichythyoplankton

Fish eggs and larvae are generally planktonic or free-floating at various stages of their development and depend upon current or tidal drift for their seasonal dispersion throughout a river system. They are carried passively from spawning grounds to nursery areas and are generally not well adapted for avoiding hazardous environmental conditions such as sudden shifts in water temperature or passage through a power plant. Surprisingly little information is available on the early life history of fishes in the rivers of northeastern United States. To define the spawning and nursery grounds of the more important species of fishes in the Connecticut River relative to the location of the Connecticut Yankee plant, an investigation of the planktonic fish eggs and larvae—the ichthyoplankton—of the lower Connecticut River and its major tributaries and coves was conducted from 1965 through 1969. This study is the first one ever made to provide basic information on the distribution, seasonal occurrence, and relative abundance of the ichthyoplankton of the Connecticut River.

The study area (Figure 1 and Table 1) included the lower Connecticut River from river kilometer (rkm) 11.3 to 85.3, the lower section of the Farmington River (at rkm 77.2), the Salmon River (rkm 27.4), and the Eight Mile River (km 12.9). Two major coves connected to the main river were also investigated: Wethersfield Cove (rkm 69.2) and Chapman Pond (rkm 20.1).

Methods

Surface and bottom plankton samples were collected at each station, and some mid-water and oblique samples were also taken (Table 1, Figure 1). Samples were collected biweekly (twice per month) from March or April through August or September, except in 1965 when they were extended through December. All tows were taken during daylight hours at a speed of 1–2 knots against the current on both tides.

Sampling gear consisted of 0.5-m cone-shaped plankton nets constructed of No. 2 mesh (0.39-mm) nylon. The nets were attached to stainless steel rings with impeller-type flowmeters at their centers. A 1-L plastic container was attached to the cod end of each net. Each net was towed for a 5-min period, during which time the flowmeters indicated that approximately 60 m^3 of water had been strained. Ten paired 5-min tows, taken in June 1968, showed no significant difference between flowmeter readings ($t = 1.94$, df = 18, $P > 0.05$), but the coefficient of variation between different flowmeters was approximately 4.5%. Temperature and dissolved-oxygen content of the water were recorded at each station in all years; secchi disk readings for water transparency were made only during 1965 and 1966.

Samples were fixed in 2% formalin, neutralized with methenamine, and dyed with rose bengal to facilitate processing. In the laboratory, a total count was made of all the eggs and larvae of each species in each sample. The total number of eggs and larvae of each species per tow was used as a measure of relative abundance.

To determine development rates and to help distinguish the larvae of the shad from those of the alewife *Alosa pseudoharengus* and blueback herring *A. aestivalis*, 200 to 300 shad eggs were stripped from ripe fish in 1967, fertilized, and reared in the laboratory to recognizable fry. Samples were taken of known blueback herring

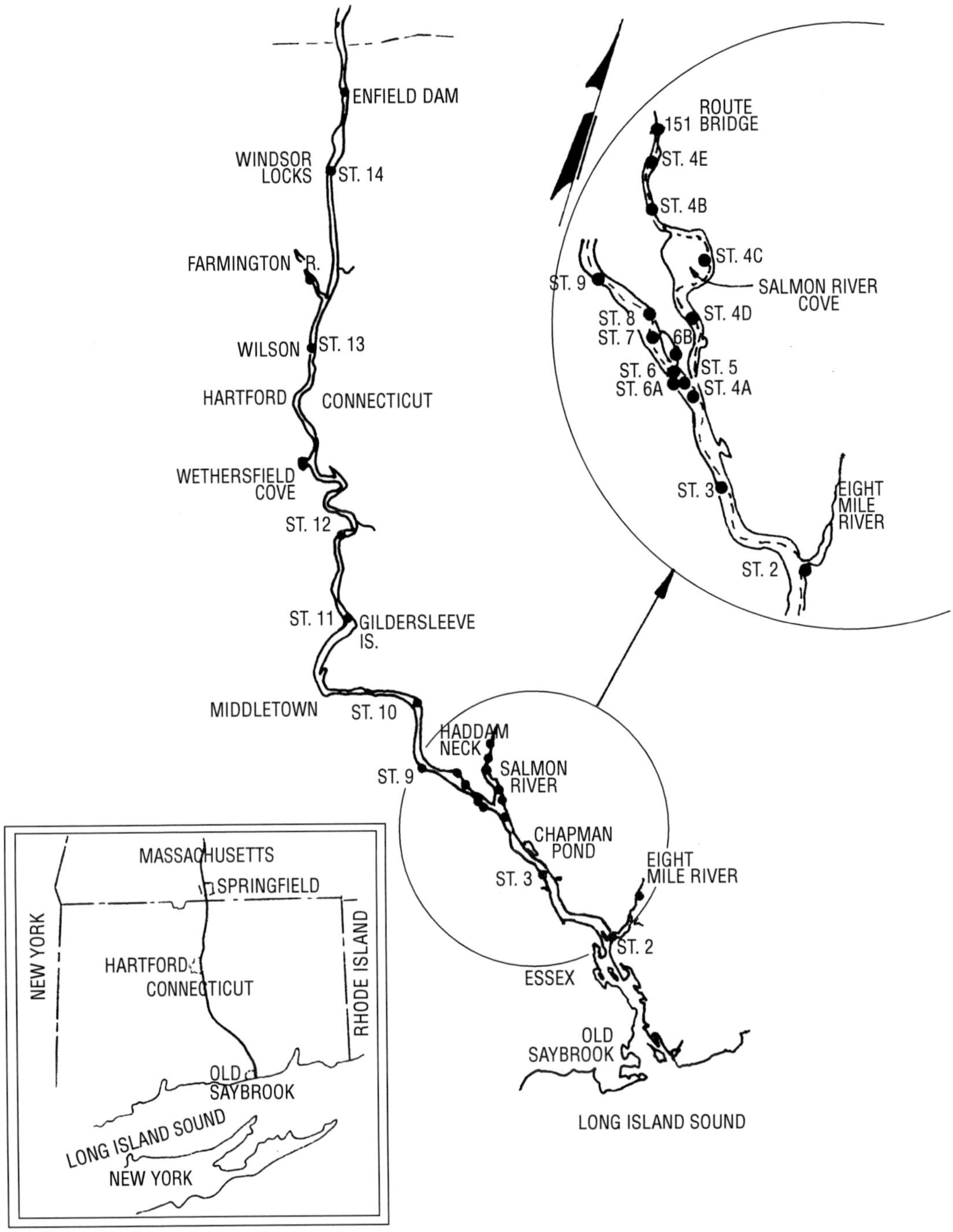

FIGURE 1. Map of the study area in the lower Connecticut River and its tributaries and coves showing stations sampled from 1965 through 1969.

and alewife larvae reared from eggs by Cianci (1969) from Roaring Brook in Glastonbury (rkm 51.5) and Higganum Creek (km 29.8), respectively, both of which are small tributaries of the Connecticut River. The number of myomeres, from the preanal muscle segment forward to the

TABLE 1. Location of plankton sampling stations in the lower Connecticut River, 1965–1969.

River km	Station	Town	Location	Average depth (m)	Year
85.3	15[a]	Suffield	Enfield Pool—0.8 km above dam (west bank)	2.4	1966–1967
78.8	14	Windsor Locks	Townline Marina (west bank)	1.2	1966–1967
71.0	13	Wilson	Above Bissell Bridge (west bank)	3.6	1966 1967
51.5	12	Rocky Hill	Glastonbury bar (west bank)	3.6	1966–1967
41.8	11	Cromwell	Gildersleeve Island—north end (center)	3.0	1966–1967
35.4	10	Middletown	Dart Island (center)	4.3	1966–1967
29.8	9	Higganum	Higganum Creek—mouth (west bank)	4.6	1965–1969
27.4	8	East Haddam	Haddam Island (center)	1.8	1965
25.7	7	East Haddam	CYAP[c] intake to river center (east bank)	2.4	1965–1969
24.9	6	East Haddam	Flats north of CYAP discharge point (east bank)	0.9	1965–1969
24.1	6A	Haddam	Opposite CYAP discharge point (west bank)	7.6	1965
24.1	6B	East Haddam	CYAP discharge canal (east bank)	1.5	1968–1969
23.3	5	East Haddam	Reflector marker "48" (east bank)	0.6	1965
22.5	4A	East Haddam	Mouth of Salmon River—cans "47" to "45" (east bank)	3.0	1965–1969
17.7	3	Chester	Fort Hill—Chester ferry slip (west bank)	2.4	1965–1969
11.3	2	Essex	Brockway Island (east bank)	3.0	1965–1967
77.2	FA[b]	Windsor	Farmington River—Route 5 bridge (west bank)	1.2	1966
69.2	WC	Wethersfield	Wethersfield Cove (center)	3.0	1966
27.4	151	East Haddam	Salmon River—Route 151 bridge (center)	1.2	1966, 67, 68, 69
26.5	4E	East Haddam	Salmon River—Frank Davis Resort (center)	1.5	1965, 67, 68, 69
25.7	4B	East Haddam	Salmon River—narrows at north end of cove	2.4	1965, 67, 68, 69
24.9	4C	East Haddam	Salmon River—Salmon Cove (east bank)	1.2	1955, 67, 68, 69
24.1	4D	East Haddam	Salmon River—Cones Point at south end of cove	3.0	1965, 67, 68, 69
20.1	CHA	East Haddam	Chapman Pond (center)	2.4	1965
12.9	8MR	Lyme	Eight Mile River—mouth at upper Hamburg Cove	1.2	1965–1966

[a] River stations.
[b] Tributary and cove stations.
[c] CYAP (Connecticut Yankee Atomic Power Company).

operculum, was counted from representative samples of the three species and these counts were used to separate shad from river herring larvae. Shad larvae had previously been separated from alewife larvae by Leim (1924) but not from blueback herring.

The major reference sources used in the separation and identification of eggs and larvae included Bayliff (1950); Bigelow and Schroeder (1953); Cianci (1969); Hale (1960); Herman (1963); Hildebrand and Cable (1930); Kuntz (1914); Leim (1924); A. J. Mansueti (1964); R. J. Mansueti (1964); Mansueti and Hardy (1967); McKenzie (1964); Pearson (1941); Scotton et al. (1973); and Whitney (1961).

Results and Discussion

Combined Species

During the investigation, 112,896 eggs and larvae, representing 16 species of fishes, were collected in 690 tows at the 25 sampling stations. The numbers of different eggs and larvae collected are shown in Table 2. Eggs of 11 species and larvae of 16 species were represented in the collections. Of the 11 species, our collections indicate that the alewife, blueback herring, white perch *Morone americana*, and yellow perch *Perca flavescens* made up approximately 98% of all the eggs sampled in the lower Connecticut River and its tributaries and coves. Because of the nature of their spawning habits and early life histories, many species had eggs and larvae that were un-

TABLE 2. Species composition of fish eggs and larvae collected with half-meter nets from 1965 to 1969 in the lower Connecticut River.

Species	Number eggs	Total catch (%)	Number larvae	Total catch (%)
Alewife, blueback herring	24,350	78.75	71,224	86.89
White perch	1,058	3.43	8,526	10.41
Yellow perch	5,028	16.26	1,270	1.55
Rainbow smelt *Osmerus mordax*	221	0.72	494	0.61
Common carp *Cyprinus carpio*	92	0.29	262	0.32
White sucker *Catostomus commersonii*	82	0.27	52	0.06
American shad *Alosa sapidissima*	76	0.24	34	0.04
Common shiner *Luxilus cornutus*, spottail shiner *Notropis hudsonius*, golden shiner *Notemigonus crysoleucas*	14	0.04	31	0.03
Pumpkinseed *Lepomis gibbosus*, bluegill *L. macrochirus*, redbreast sunfish *L. auritus*			77	0.09
Hogchoker *Trinectes maculatus*			3	0.00
Atlantic silverside *Menidia menidia*				
Total	30,921		81,975	

available in the plankton. Based on the fact that young of the year were collected in regular biweekly bag seine and trawl hauls made during this investigation, 13 additional species were presumed to have spawned in the study area: redfin pickerel *Esox americanus*, northern pike *E. lucius*, chain pickerel *E. niger*, creek chub *Semotilus atromaculatus*, fallfish *S. corporalis*, white catfish *Ictalurus catus*, brown bullhead *I. nebulosus*, channel catfish *I. punctatus*, banded killifish *Fundulus diaphanus*, johnny darter *Etheostoma nigrum*, rock bass *Ambloplites rupestris*, black crappie *Pomoxis nigromaculatus*, and largemouth bass *Micropterus salmoides*. Although school-size striped bass *Morone saxatilis* have been collected as far north as Enfield Dam, we found no evidence that this species utilizes the Connecticut River for spawning purposes.

Eggs were present in the river and in the tributaries and coves from April through August (Table 3). The peak of egg production occurred in June in the river and about a month earlier in the tributaries and coves. Generally, spawning began earlier in the tributaries and coves, spread throughout the river during late April and May, became concentrated above Middletown toward Windsor Locks through the peak in June, and then fell off in July. Few eggs were collected during August (Table 4). The mean catch per-unit-effort for all areas sampled during the 5 years was 44.8 eggs/tow from April through August. In the tributaries and coves, egg production, as expressed in mean catch per-unit-effort, was 80.7, and this was more than twice the number found in the river (35.5). Tables 5and 6 present a summary of where the eggs and larvae of each species were collected.

On the basis of catch per tow, 87.5% of the eggs were taken above Middletown (km 35.4). Peak egg catch occurred in the Windsor Locks area (km 78.8). Enfield Dam (km 83.7) is a deterrent for upstream migration as exemplified by the great difference between the total number of eggs per tow collected below the dam at Windsor Locks (station 14) and at station 15 above the dam. Station 14 averaged 929/tow and station 15 averaged less than 4 (Table 4). Station 4A is located at the confluence of the Connecticut and Salmon rivers, and the higher number of eggs per tow at this station, as compared with neighboring river stations, resulted from the addition of spawning products from the Salmon River. Of the tributaries and coves sampled, the Salmon River and Eight Mile River showed the highest egg production.

TABLE 3. Total number of eggs and larvae collected for each month, 1965–1969.

Months sampled	Number eggs	Percent	Number larvae	Percent
April	3,361	10.87	222	0.27
May	6,926	22.40	6,445	7.86
June	20,284	65.60	62,017	75.66
July	350	1.13	12,616	15.39
August	0	0	675	0.82
Total	30,921		81,975	

TABLE 4. Number of fish eggs of all species taken per tow for each station and month from 1965 to 1969.

River km	Total on	April	May	June	July	August	no./tow
85.3	15[a]		1.33	0.25	0.00	0.00	3.62
78.8	14	0.00	28.00	2,069.00	1.50		929.33
74.0	13	1.50	37.33	874.67	10.17	0.50	258.38
51.5	12		2.00	553.25	17.00	3.25	153.07
41.8	11		107.67	174.75	18.50	0.00	78.29
35.4	10	0.00	19.33	269.50	0.00	0.00	54.27
29.8	9	2.20	10.36	30.45	0.05	0.00	9.05
27.4	8		1.50	2.50	0.00	0.00	1.00
25.7	7	2.17	9.33	2.45	0.00	0.00	2.98
24.9	6	0.60	7.18	0.70	0.00	0.00	1.75
24.1	6A		4.75	3.66	0.00	0.00	1.58
24.1	6B		1.50	0.67	0.00	0.00	0.38
23.3	5		2.50	1.00	0.00		1.00
22.5	4A	3.38	14.50	6.95	0.00	0.00	5.31
17.7	3	0.56	3.23	0.42	0.65	0.00	1.01
11.3	2	0.75	2.17	0.17	0.00	0.00	0.65
77.2	FA[b]		33.00	274.50	0.00		127.20
69.2	WC	0.00	0.00	37.67	0.50		19.00
27.4	151	326.29	418.60	169.50	16.29	0.00	203.59
26.5	4E		0.00	68.33	12.00	0.00	21.70
25.7	4B	0.00	1.37	3.33	0.75	0.00	1.17
24.9	4C	0.00	0.22	17.88	0.00	0.00	5.62
24.1	4D		0.00	14.00	0.00		5.60
20.1	CHA		15.33	41.00	0.00		21.33
12.9	8MR	333.33	391.00	24.00	0.00		201.90

[a] River stations.
[b] Tributary and cove stations.

TABLE 5. Summary of the occurrence (+) of fish eggs (E) and larvae (L) at each sampling station within the lower Connecticut River from 1965 to 1969.

	E	L	E	L	E	L	E	L	E	L	E	L	E	L	E	L	E	L	E	L	E	L	E	L	E	L	E	L	E	L
Alewife	+	+	+	+	+	+	+	+	+	+	+	+	+	+	+	+	+	+	+	+	+	+	+	+	+	+	+	+	+	+
Blueback herring		+	+	+	+	+	+	+	+	+		+	+	+	+	+	+	+	+	+	+	+		+	+	+	+	+		+
White perch	+	+	+	+	+		+				+	+	+				+	+		+					+	+	+	+	+	+
American shad	+	+	+		+				+	+			+	+				+		+					+	+	+	+	+	+
Yellow perch					+	+				+				+				+		+		+				+		+		+
Rainbow smelt				+										+						+		+		+		+	+	+		+
Minnows		+		+		+		+	+	+		+		+				+		+						+		+		
Carp		+		+										+														+		
Sunfishes						+								+													+			
White sucker																														+
Atlantic silverside																														+
Station	15		14		13		12		11		10		9		8		7		6		6A		5		4A		3		2	
Km	85.3		78.8		74.0		51.5		41.8		35.4		29.8		27.4		25.7		24.9		24.1		23.3		22.5		17.7		11.3	

TABLE 6. Summary of the occurrence (+) of fish eggs (E) and larvae (L) at each sampling station in tributaries and coves from 1965 to 1969.

Species	E	L	E	L	E	L	E	L	E	L	E	L	E	L	E	L	E	L
Alewife	+	+	+	+	+	+		+	+	+	+	+	+	+	+	+	+	+
Blueback herring																		
White perch	+		+	+	+	+	+	+		+	+	+				+	+	
American shad													+					
Yellow perch				+		+				+		+				+	+	+
Rainbow smelt		+		+	+	+				+		+				+	+	+
Minnows						+	+					+						+
Carp				+		+	+			+		+			+			
Sunfishes		+		+		+				+	+	+						+
White sucker	+	+					+	+										
Station	FA		WC		151		4E		4B		4C		4D		CHA		8MR	
Km	77.2		6.92		27.4		26.5		25.7		24.9		24.1		20.1		12.9	

Unlike egg distribution, the number of larvae generally increased from upriver to downriver stations and was greatest in an area between rkm 29.8 and 22.5 (Higganum-Salmon River) (Table 7). The mean number per tow was 30% higher in this area than in the river as a whole. Larvae be-

TABLE 7. Number of fish larvae of all species taken per tow for each station and month from 1965 to 1969.

River km	Station	April	May	June	July	August	September	Total no./tow
85.3	15[a]		0.66	10.50	2.75	1.00	0.00	4.39
78.8	14	0.00	0.00	302.50	120.00			147.77
74.0	13	0.00	0.66	522.83	32.33	0.33	0.00	158.75
51.5	12		0.00	483.75	95.25	2.75	0.00	155.13
41.8	11		7.66	392.00	97.25	4.50	0.00	141.42
35.4	10	0.00	0.33	77.00	0.00	0.00	0.00	14.09
29.8	9	0.70	28.36	547.00	85.23	7.00	0.00	141.82
27.4	8		83.75	412.50	2.00	0.00	0.00	105.81
25.7	7	1.50	37.75	843.45	321.20	6.77	0.00	255.15
24.9	6	3.80	40.72	708.60	126.45	34.22	0.00	181.41
24.1	6A		57.50	998.30	38.00	0.00	0.00	175.73
24.1	6B		23.00	458.00	235.33	0.75		163.75
23.3	5		355.00	1,842.00	34.50			436.83
22.5	4A	0.62	25.50	424.59	100.30	2.37	0.00	127.13
17.7	3	0.77	47.81	233.26	75.25	4.75	0.00	73.71
11.3	2	0.75	4.00	256.83	20.75	0.66	0.33	63.61
77.2	FA[b]		10.50	9.00	1.00			8.00
69.2	WC	9.00	38.00	424.66	51.00			177.87
27.4	151	22.14	12.70	3.87	2.14	0.00	0.00	9.17
26.5	4E		3.25	8.00	46.00	0.00		8.30
25.7	4B	1.50	2.37	137.83	59.50	9.20	0.00	39.17
24.9	4C	1.25	155.42	191.11	101.50	0.00	0.00	104.00
24.1	4D		0.00	161.00	1.50			65.00
20.1	CHA		103.33	36.50	0.00			63.83
12.9	8MR	0.00	106.00	14.50	3.00			32.09

[a] River stations.

[b] Tributary and cove stations.

came available in April and were more abundant in the Haddam area than in others. In May, the greatest concentration was found below km 29.8, and by June, peak abundance had been reached in all areas with higher percentages still in the downstream section. Generally, larvae were abundant earlier downstream, and their abundance upstream increased as spawning became more concentrated there. Larvae became less abundant through July and August in the upstream areas and were available only at the southern-most station by September. The bulk of the larvae appeared to be concentrated, through drift from the upstream spawning areas, in a nursery area that stretched from south of Middletown to the mouth of the Salmon River.

The concentration of larvae was more than two and a half times greater in the river (136 larvae/tow) than in the tributaries or coves (50 larvae/tow), as larvae were constantly added to the river through drift. Larval abundance peaked and declined earlier in the tributaries than in the river, and larvae apparently left the tributaries about a month earlier. Larvae were available in the upper sections of the Salmon River in April, increased in the lower sections through May, and peaked there in June, becoming less abundant from July through August.

Comparisons of catches between 242 surface and 211 bottom tows, taken one immediately after the other between 1965 and 1969, were made to determine the vertical distributions of eggs and larvae in the lower Connecticut River. These collections were made without closing devices on the nets, and bottom samples were therefore briefly subject to vertical contamination as the net passed through the surface waters. Eggs were roughly four times more abundant near the bottom than near the surface (Table 8). They were distributed almost exclusively near the bottom from Rocky Hill to Windsor Locks in the upper river but were found near the surface in the Haddam area and were more evenly distributed or mixed from the mouth of the Salmon River southward. Larvae were 11 times more abundant near the bottom than near the surface. They were most abundant in the bottom samples from Higganum to Windsor Locks. In the Haddam area, they were more abundant in the surface samples, more evenly distributed or mixed from the mouth of the Salmon River southward. There appeared to be a tendency for the larvae of most species to become more pelagic as they drifted downstream. Of the 13 species examined for vertical distribution, only the eggs of the American shad were evenly distributed between surface and bottom samples. Only the larvae of American shad, yellow perch, and the shiners were found with greater frequency in the surface samples. The remaining eight species occurred predominantly in the bottom samples.

During the sampling years, most of the 74,864 larvae collected were examined for the presence of a yolk sac, and 1,639 larvae were measured to determine size of yolk sac (Table 9). The percentage of larvae with yolk sacs decreased through the reproductive season. The lower percentage of larvae

TABLE 8. Vertical distribution of all eggs and larvae from sequential surface and bottom tows taken in the lower Connecticut River from 1965 to 1969.

River km	Sampling station	Mean number per tow at each station: Eggs Surface	Eggs Bottom	Larvae Surface	Larvae Bottom
78.8	14	1.50	839.00	57.00	101.60
74.0	13	110.58	527.80	55.91	324.80
51.5	12	6.00	249.77	32.25	229.88
29.8	9	9.29	2.31	65.98	132.88
24.9	6	2.23	0.00	131.42	101.16
22.5	4A	3.22	5.00	105.08	94.61
17.7	3	0.61	1.00	48.53	67.42
Total number collected at all stations		2,228	9,611	2,288	23,579
Total number tows at all stations		242	211	242	211
No. per tow at all stations		9.21	45.55	9.45	112.75

TABLE 9. Frequency of larvae with yolk sacs and yolk sac size for each month in the lower Connecticut River.

Sampling month	Percent with yolk sac		Mean yolk sac size (mm)	
April	27.3[a]	(4.8)[b]	0.71[a]	(0.56)[b]
May	39.7	(81.3)	1.26	(0.93)
June	68.0	(15.3)	1.46	(0.99)
July	7.6	(7.4)	0.56	(0.00)
August	1.5	(0.0)	0.51	(0.00)

[a] Main river.
[b] Tributaries and coves

with yolk sacs in the tributaries and coves in April indicates that the larvae were older and that some spawning had occurred earlier there. As shown by the higher percentage of yolk-sac larvae, spawning had reached its peak in the tributaries and coves by May and in the river a month later. The two areas became evenly matched by July, but younger larvae remained in the river into August. After June, larvae in the tributaries and coves had virtually no yolk sacs, while a few larvae in the river still retained them. There was a higher percentage of larvae with yolk sacs upriver above Higganum (63% to 98.4%) than from Higganum southward (1.2% to 36%). Larvae were generally in the post-yolk-sac stage from the Haddam area southward, indicating that this is a major nursery area in which feeding probably begins.

As shown in Table 10, the eggs of the 11 species of fishes were found in water between 5.8°C and 31.0°C. Seventy-eight percent of the eggs occurred within a 7-degree (19–25°C) temperature range, and 15% were found in the lower range of 7–11°C (Table 11). The larvae of the 16 species of fishes were found in water between 7.5°C and 31°C (Table 12). Of these, 95% occurred within an 11-degree (19–29°C) temperature range and 77% were found between 21°C and 27°C (Table 13).

Fish eggs were collected in dissolved oxygen concentrations of between 4 and 13 ppm, the majority between 5 and 11 ppm. Fish larvae were collected in concentrations between 1 and 13 ppm, 84% between 5 and 8 ppm.

Secchi disk readings made during 1965 and 1966 indicated that visibility approached 2 m only in early spring. The average at that time was 1.5 m, and visibility declined to less than 1 m in the summer. These changes reflect the seasonal variation in silt carried by the river and the amount of plankton produced.

River Herrings

The alewife and the blueback herring could not be distinguished from each other at either the egg (0.8–1.9 mm) or larval (2–15 mm) stage. At sizes greater than 15 mm, the two species can generally be separated on the basis of peritoneum color and eye size.

American shad larvae can generally be distinguished from river herring larvae up to 13 mm in length on the basis of differences in the total number of myomeres counted from the preanal muscle segment to the operculum. River herring larvae usually have between 37 and 40 myomeres and shad larvae 41 and up as determined from laboratory rearing experiments (Table 14).

As indicated by the collections of young of

TABLE 10. Size, occurrence, abundance, and temperature range of pelagic fish eggs from the lower Connecticut River.

Number[a] Species	Average measured	size (mm)	Size range (mm)	Dates	Average abundance (no./tow)	Temperature range (°C)
Alewife, blueback herring	1,032	1.16	0.80–1.94	April–August	35.30	7.0–31.0
White perch	147	0.90	0.77–1.30	May–June	1.53	8.9–27.0
Yellow perch	35	3.53	3.10–3.90	April–May	0.04	7.8–9.6
Rainbow smelt	20	1.33	1.22–1.40	April–May	0.31	5.8–10.5
Common carp	23	1.70	1.55–1.86	May–July	0.13	22.0–27.0
American shad	57	2.83	1.10–3.80	May–June	0.13	7.5–22.7
White sucker	4	3.03	2.00–4.00	May–June	0.12	12.2–24.0
Minnows (*Notropis* spp.)	12	2.32	1.84–2.80	May–July	0.02	27.0

[a] Ten eggs were measured of each species from each sample on the major axis to the nearest tenth of a millimeter.

TABLE 11. Distribution of fish eggs relative to water temperature (°C) in the lower Connecticut River (1965–1969).

Species	Temperature (°C)						
	5.0–6.9	7.0–8.9	9.0–10.9	11.0–12.9	13.0–14.9	15.0–16.9	17.0–18.9
Alewife, blueback herring		2,739	1,078	59	279	231	336
American shad		2	16	7	5	4	
White perch		2	10	253	3	7	1
Yellow perch		6	14				
Rainbow smelt	109	100	12				
Minnows (*Notropis* spp.)							
Common carp							
White sucker				1			1
Totals	109	2,849	1,130	320	287	242	338
% total	0.42	11.05	4.38	1.24	1.11	0.94	1.31

Species	Temperature (°C)						
	19.0–20.9	21.0–22.9	23.0–24.9	25.0–26.9	27.0–28.9	29.0–30.9	31.0–32.9
Alewife, blueback herring	797	13,121	5,401	187	1	1	1
American shad	6	22					
White perch	593	49	1	133			
Yellow perch							
Rainbow smelt							
Minnows (*Notropis* spp.)					12		
Common carp		19	1		72		
White sucker			80				
Totals	1,396	13,211	5,483	320	85	1	1
% total	5.42	51.26	21.28	1.24	0.33	0.01	0.01

the year and adults, the blueback herring is by far the dominant species, making up more than 80% of the catch of the two species in the lower Connecticut River. Eggs and larvae of alewife and blueback herring were present in all areas sampled, and they dominated the samples in all years, comprising 78.7% of the total egg catch and 86.9% of the total larval catch (Table 2). River

TABLE 12. Size, occurrence, abundance, and temperature range of fish larvae from the lower Connecticut River.

Species	Number[a] measured	Average size (mm)	Size range (mm)	Dates	Average abundance (no./tow)	Temperature range (°C)
Alewife, blueback herring	2,793	6.88	1.80–21.60	April–August	104.2217	9.0–31.0
White perch	1,239	3.88	1.44–10.40	May–August	12.3536	10.0–28.2
Yellow perch	138	6.69	3.19–12.30	April July	1.7565	7.5 25.3
Rainbow smelt	233	7.03	5.47–10.30	April–May	0.7304	8.2–18.2
Common carp	101	5.22	2.75–7.20	May–August	0.3942	13.8–26.0
Sunfishes (*Lepomis* spp.)	65	5.71	3.40–9.20	June–August	0.1115	21.5–29.0
Hogchoker	3	5.30	4.70–5.90	July	0.0043	26.0–27.0
Atlantic silverside	2	6.80	6.80	July	0.0028	27.0
White sucker	19	12.30	7.20–15.10	May–June	0.0753	14.8–24.0
American shad	33	10.66	4.92–14.60	May–August	0.0449	9.7–30.0
Minnows (*Notropis* and *Notemigonus* spp.)	27	5.36	2.40–8.50	May–June	0.0434	15.0–23.0

[a]Ten larvae of each species were measured (total length) from each sample.

TABLE 13. Distribution of fish larvae relative to water temperature (°C) in the lower Connecticut River (1965–1969).

Species	Temperature (°C)						
	5.0–6.9	7.0–8.9	9.0–10.9	11.0–12.9	13.0–14.9	15.0–16.9	17.0–18.9
Alewife, blueback herring			30	119	511	405	206
American shad			1	2			
White perch			2		27	135	26
Yellow perch		177	142	806	7	1	
Rainbow smelt		4	286	121	6		2
Minnows (*Notropis* and *Notemigonus* spp.)							73
Common carp					1	2	
Sunfishes (*Lepomis* spp.)							
White sucker					1		8
Totals	0	181	461	1,048	553	550	245
% total		0.26	0.66	1.49	0.79	0.78	0.35
Species	Temperature (°C)						
	19.0–20.9	21.0–22.9	23.0–24.9	25.0–26.9	27.0–28.9	29.0–30.9	31.0–32.9
Alewife, blueback herring	6,403	15,351	8,234	25,523	3,228	1,711	704
American shad	2	15		3	3	2	
White perch	922	1,788	1,569	1,346	59		
Yellow perch	1	1	7	10			
Rainbow smelt							
Minnows (*Notropis* and *Notemigonus* spp.)		4	13	2			
Common carp	10	46	193	7			
Sunfishes (*Lepomis* spp.)		10	17	17	24	6	
White sucker			43				
Totals	7,342	17,224	10,065	26,906	3,314	1,719	704
% total	10.44	24.50	14.31	38.47	4.71	2.44	1.00

herring eggs were 23 times more numerous per tow and larvae 8 times more numerous per tow than the white perch, which ranked second in abundance (Tables 10 and 12).

Bigelow and Schroeder (1953) state that the alewife spawns at temperatures between 12.8°C and 15.6°C. Cianci (1969) reported that the temperatures for optimum spawning of the blueback

TABLE 14. Comparison of larval myomere counts (pre-anal to operculum) in shad, alewife, and blueback herring reared from eggs collected in the Connecticut River in 1967.

Species and collection location	Number of myomeres											Total
	37	38	39	40	41	42	43	44	45	46	47	
Number of shad[a] Wilson, Connecticut				1	4	5	10	19	14	9	1	63
Number of alewives[b] Higganum, Connecticut	6	84	508	164	26							788
Number of blueback herring[b] Glastonbury, Connecticut	1	4	1	1								7

[a] Measurements made from laboratory-reared specimens from present study.
[b] Measurements made from laboratory-reared specimens provided by Cianci (1969).

herring in the Connecticut River ranged between 21°C and 24°C, and Loesch (1968) observed that spawning stopped at 27°C in Connecticut waters. Cooper (1961) reported that upstream migrations of alewives in Rhode Island ceased at 21°C. In the present study, eggs of blueback and alewife were collected in water between 7°C and 31°C (Table 10). In comparing the distribution of eggs with water temperature (Table 11), two general peaks occurred, one between 7°C and 10.9°C (16%) and one between 21°C and 24.9°C (76%). The majority of alewives probably spawn between 7°C and 10.9°C and bluebacks between 21°C and 24.9°C.

Blueback herring generally utilized the river and mouths of tributaries for spawning purposes, while the alewife utilized the small streams and upper portions of tributaries and coves. We have seined ripe female blueback herring in the study area as early as May and have observed spawning bluebacks in the lower Salmon River in early September. This indicates an extended spawning season. Based on egg collections, the herrings spawned from April through August, most of them in June (Table 15). Generally, egg production occurred later at upstream stations than downstream ones (Table 16). Spawning began in April in the lower river (below Higganum) as well as in the upper Salmon River. By May, reproduction had peaked in the Salmon River and was distributed throughout the Connecticut River with more spawning above Middletown (km 35.4). In June, spawning reached its peak in the river and the tributaries and coves, with 99.5% of the egg production occurring above Higganum. At this time, spawning was concentrated in an area between Wilson and Windsor Locks. By July, almost all the egg production occurred above Middletown, and in August even farther upriver above Rocky Hill.

TABLE 15. Monthly occurrence of eggs and larvae of alewife and blueback herring in the lower Connecticut River.

Month	Eggs		Larvae	
	Number	Percent	Number	Percent
April	1,136	4.7	6	<0.1
May	3,541	14.5	4,146	5.8
June	19,321	79.3	54,573	76.6
July	337	1.4	11,832	16.6
August	15	0.1	667	0.9
Total	24,350	100.0	71,224	99.9

Enfield Dam (km 83.7) exerts a significant effect on river herring spawning. The total number of eggs per tow above the dam was 0.08, as compared with 871.78 at Windsor Locks (Table 16). The majority of spawning occurred above Middletown with the greatest concentration in the Windsor Locks area. The Farmington River and Salmon River had the highest egg production with the Eight Mile River second.

Larvae were most abundant in June (76.6%), disappearing after August in all sampling years (Tables 15 and 17). In May, the majority of larvae were found below Higganum (km 29.8). Larvae became more abundant in the upper areas in June, but the majority were still found below Higga-num. The majority remained below km 25.7 from July through August. The major concentration of herring larvae occurred in a 4-km area near the Connecticut Yankee Atomic Power Company (CYAP) plant, between Haddam Island and the mouth of the Salmon River. River herring larvae were about 3.5 times as prevalent in the river (120.8 larvae/tow) as in the tributaries and coves (35.3 larvae/tow). River herring larvae were collected in water between temperatures of 9°C and 31°C (Table 12), most of them between 25°C and 26.9°C (Table 13). Edsall (1970) and Marcy (1971, 1973) noted a minimal survival of river herring larvae above 28°C.

River herring eggs were 5 times and larvae 1.3 times more abundant near the bottom than near the surface. Eggs were found almost exclusively near the bottom between Windsor Locks and Rocky Hill, but became more abundant near the surface between Higganum and East Haddam, and from the mouth of the Salmon River southward, they were evenly distributed in the water column. Larvae occurred near the bottom from stations 14 to 9, but became more abundant near the surface between stations 6 and 4A and evenly mixed at station 3. The river herring larvae nearer the bottom were younger than those near the surface, since the mean yolk sac size in sequential surface and bottom samples was 0.7 mm and 2.3 mm, respectively. Larvae that were located near the surface had smaller yolk sacs below Higganum (km 29.8) than above, suggesting they become more pelagic as they drift downstream toward the river mouth.

Yolk-sacs were found among river herring larvae between 14.1°C and 26°C, and the mean yolk sac size (0.4 mm) occurred at 19.1°C. Cianci (1969) showed that in the Connecticut River, yolk sac absorption occurred in alewives in 72 h at

TABLE 16. Number of eggs of alewife and blueback herring taken per tow for each station and month from 1965 to 1969.

River km	Station	April	May	June	July	August	Total no./tow
85.3	15[a]		0.00	0.25	0.00	0.00	0.08
78.8	14	0.00	18.30	1951.50	3.00		871.78
74.0	13	0.00	27.70	1032.60	10.20	0.67	252.81
51.5	12		1.00	548.70	17.00	3.30	151.80
41.8	11		38.30	167.00	18.30	0.00	61.10
35.4	10	0.00	19.00	269.50	0.00	0.00	54.20
29.8	9	1.70	9.10	25.60	0.05	0.00	7.70
27.4	8		1.00	2.50	0.00	0.00	0.82
25.7	7	2.20	8.60	2.20	0.00	0.00	2.75
24.9	6	0.60	7.10	0.70	0.00	0.00	1.73
24.1	6A		4.00	3.70	0.00	0.00	1.42
24.1	6B		3.00	0.67	0.00	0.00	0.61
23.3	5		2.50	0.50	0.00		1.00
22.5	4A	3.00	13.30	6.70	0.00	0.00	4.92
17.7	3	0.56	2.90	0.37	0.65		0.92
11.3	2	0.25	2.00	0.17	0.00	0.00	0.58
77.2	FA[b]		30.50	274.50	0.00		122.60
69.2	WC	0.00	0.00	37.00	0.50		14.00
27.4	151	153.30	218.40	165.00	16.30	0.00	120.30
26.5	4E		0.00	0.00	0.00	0.00	0.00
25.7	4B	0.00	1.38	3.30	0.75	0.00	1.20
24.9	4C	0.00	0.22	17.80	0.00	0.00	5.60
24.1	4D		0.00	11.50	0.00		4.60
20.1	CHA		9.00	1.00	0.00		4.80
12.9	8MR	0.00	56.30	23.00	0.00		19.50

[a] River stations.
[b] Tributary and cove stations.

11.7°C and that blueback herring larvae absorbed their yolk sacs in 72 h at 23.6°C. As determined from measurements taken in the present study, most alewives and blueback herring larvae absorbed the yolk sacs when they reach 9.8 mm in length. Most larvae were in the post-yolk-sac stage by the time they had drifted downstream and begun active feeding in the nursery areas south of km 29.8. The percentage of larvae with yolk sacs below km 29.8 ranged between 0.8% and 34.1%, while north of this area the percentage ranged between 66.8% and 99.5%.

River herring eggs ranged in size between 0.8 and 1.9 mm, averaging 1.2 mm (Table 10). Mean egg size decreased through the season as temperature increased. The mean size was 1.23 mm in April, 1.17 mm in May, 1.09 mm in June, and 0.88 mm in July. Mean egg size decreased from upriver to downriver, being 1.27 mm above km 29 and 1.15 mm below. The size of river herring larvae ranged between 1.8 and 21.6 mm total length (TL), averaging 6.8 mm (Table 12). Larvae were larger toward the river mouth, especially in July and August. Their mean size was also larger in the river than in the tributaries or coves. Loesch (1968) and Kissil (1968) have shown that juvenile bluebacks remain in the Connecticut River about a month longer than juvenile alewives before going to sea. Juveniles of both species (aged as I+) have been found in the lower river in early spring, and this indicates that many probably spend their first winter close to the mouth of the river (Marcy 1969b).

White Perch

This fish was the third most abundant species; it comprised 3.4% of the eggs and 10.4% of the larvae collected in the lower Connecticut River (Table 2). The occurrence of white perch eggs and larvae is shown in Tables 5 and 6.

TABLE 17. Number of larvae of alewife and blueback herring taken per tow for each station and month from 1965 to 1969.

River km	Station	April	May	June	July	August	Total no./tow
85.3	15[a]		0.00	2.75	0.00	0.00	0.85
78.3	14	0.00	0.00	298.80	113.00		145.33
74.0	13	0.00	0.00	490.00	32.30	0.33	149.30
51.5	12		0.00	431.30	95.30	2.80	141.30
41.8	11		1.30	312.00	89.50	4.50	115.60
35.4	10	0.00	0.00	68.00	0.00	0.00	12.40
29.8	9	0.00	24.70	522.00	82.00	6.90	134.90
27.4	8		69.00	374.50	2.00	0.00	93.50
25.7	7	0.00	31.30	681.30	318.60	6.10	217.98
24.9	6	0.00	36.10	694.80	121.90	34.10	176.30
24.1	6A		33.30	924.70	37.70	0.00	158.90
24.1	6B		21.00	456.00	235.00	0.50	162.92
23.3	5		353.50	916.50	34.50		434.83
22.5	4A	0.00	18.90	360.80	86.20	2.20	107.60
17.7	3	0.00	33.90	171.20	64.10	4.80	55.86
11.3	2	0.00	0.50	219.50	19.50	0.67	53.80
77.2	FA[b]		0.50	8.00	0.00		3.40
69.2	WC	0.00	0.50	400.00	44.00		161.10
27.4	151	0.00	0.80	2.75	4.10	0.16	1.50
26.5	4E		3.00	7.30	43.00	0.00	7.70
25.7	4B	0.75	1.10	118.00	52.30	9.20	33.70
24.9	4C	0.25	23.10	171.80	93.50	0.00	66.90
24.1	4D		0.00	161.00	0.67		65.00
20.1	CHA		79.00	14.00	0.00		44.20
12.9	8MR	0.00	9.30	12.50	2.50		5.30

[a] River stations.

[b] Tributary and cove stations.

Spawning was confined to May and June (Table 10) with 72.5% taking place in June. During both months, about 99% of the spawning occurred above Cromwell (km 41.8), most of it in the vicinity of Windsor Locks (km 78.8). No eggs were collected above Enfield Dam, however. Spawning began about a month later in the tributaries and coves than in the river. Of the tributaries and coves, the upper Salmon River showed the highest egg production, and no eggs were collected in Chapman Pond. The tributaries and coves showed slightly fewer eggs (1.22 eggs/tow) than did the river (1.61 eggs/tow).

White perch larvae were present from May through August (Table 12), 83.4% of them in June. Larvae were distributed in the lower river sections of the river in May, and by June, they were most prevalent in the Haddam area. In July, no larvae were collected above km 51.5, and they were most abundant below the Salmon River. Among tributaries and coves, the highest larval abundance occurred in Chapman Pond and Wethersfield Cove. Peak abundance was reached earlier in Chapman Pond (i.e., in May), followed by peaks in Wethersfield Cove and the lower Salmon River in June. By July, larvae were no longer available in the tributaries and coves except for the lower Salmon River. No larvae were collected in Farmington or Eight Mile rivers.

White perch eggs were 3 times and larvae 2.5 times more abundant in the bottom sequential samples than on the surface. Eggs were more abundant near the bottom from km 78.8 to km 23.3, becoming equally distributed from km 17.7 southward. Larvae were collected with greatest frequency in the bottom samples at all stations.

Spawning occurred between 8.9°C and 27°C (Table 10), and eggs were most abundant between 11°C and 12.9°C, 74% between 19°C and 27°C (Table 11). Larvae were collected between 10°C and 28.2°C (Table 12), and 89% were collected between 21°C and 27°C. None of the white perch larvae (2.0–3.9 mm) survived passage through the Connecticut Yankee plant above a temperature of 28°C (Marcy 1971, 1973).

White perch eggs ranged between 0.8 and 1.3 mm, averaging 0.9 mm (Table 10). There was little evidence of decrease in egg size with either time or location, egg diameters averaging 0.89 mm in May and 0.86 in June. Larvae ranged between 1.4 and 10.4 mm, averaging 3.9 mm (Table 12). They were larger toward the river mouth, and their mean size was larger in the river than in the tributaries or coves. Mean yolk sac size decreased from 2.17 mm in May to 1.38 mm in June and 0.37 mm in July. In the river, 37% of the white perch larvae had yolk sacs in May, 41% in June, and 9% in July. These data provide further evidence for later spawning in tributaries and coves. Larvae exhibited yolk sacs between 4.1°C and 26°C, and the mean yolk sac size (0.89 mm) occurred at 20.5°C. The largest larva with a yolk sac was 4.38 mm long when collected. Most white perch larvae were in the post-yolk-sac stage by the time they had drifted downstream and begun to feed actively in the nursery areas south of Higganum (km 29.8).

Yellow Perch

This species made up 16.3% of the total egg catch and 1.5% of the larval catch (Table 2). The occurrence of the eggs and larvae of this species is shown in Tables 5 and 6. Spawning occurred over a short period from April through May (Table 10), 60% in May. Eggs are laid in masses or in long ribbons over vegetation (A. J. Mansueti 1964) and were not available to plankton nets in many areas. In April, most spawning occurred between Wilson (km 74) and Windsor Locks (km 78.8) with some activity at Higganum, the mouth of the Salmon River, and Brockway Island. By May, spawning was almost totally confined above Enfield Dam (km 85.3). The only evidence of spawning in the tributaries and coves was in Eight Mile River (km 12.9), in which a higher egg production was measured than in the river.

Larvae were abundant from April through July (Table 12), 97.3% of them in May. Larvae were distributed well upriver in May, but by June, they were found only below Higganum (km 29.8). Throughout the season, larvae were most abundant at Cromwell and in the Haddam area. Major larval production was recorded in the tributaries and coves, reaching a maximum of 27.6/tow in the lower Salmon River and 25.5/tow in the Eight Mile River, as compared with a catch of only 0.64/tow in the river. Yellow perch eggs were six times more abundant in bottom sequential samples than surface ones and were found almost entirely near the surface downstream from Higganum (km 29.8).

Spawning took place within a very narrow temperature range of 7.8°C and 9.6°C, the peak occurring at 9.4°C (Tables 10 and 11). Larvae were found between 7.5°C and 25.3°C, the peak occurring between 11°C and 12.9°C (Tables 12 and 13).

The largest larva with a yolk sac was 7.5 mm long. Unlike most of the species collected, yellow perch larvae still retained their yolk sacs (mean, 1.04 mm) between Higganum (km 29.8) and Brockway Island (km 11.3). There are perhaps two reasons for this: (1) the majority of yellow perch were produced in the Salmon and Eight Mile rivers, and thus, younger larvae were contributed to the lower sections of the Connecticut River through drift; and (2) larvae hatched at such low temperatures probably retain their yolk sacs for an extended period. Breder and Rosen (1966) and A. J. Mansueti (1964) have stated that the yellow perch has the longest incubation time (27 d) of any common estuarine fish. It may well be that larval yellow perch do not begin to feed until they reach the extreme lower sections of the river.

Eggs ranged in size between 3.1 and 3.9 mm, averaging 3.5 mm (Table 10), and mean egg size remained the same in the river during April and May (3.47 mm). Larvae ranged in size from 3.2 and 12.3 mm, averaging 6.7 mm (Table 12). Little difference in mean length was noted between upstream and downstream areas either in the river or the tributaries or coves.

Rainbow Smelt

This species was fifth in abundance; its eggs and larvae comprised 0.72% and 0.61% of the total collected (Table 12). The occurrence of smelt eggs and larvae is shown in Tables 5 and 6.

Spawning occurred over a short period between April and May (Table 10), 96% in April. Smelt spawned primarily in the tributaries and coves, especially at the northern end of the Salmon River in April and in the Eight Mile River in May. In the main river, eggs were collected only at km 74 and during the month of April.

Larvae were present from April through May (Table 12), 60% of them in May. They were most abundant in the Haddam area in April and became more abundant toward the river mouth during the following month. Larval abundance peaked in the upper sections of all the tributaries and coves in May and in the lower sections of the river in June. Larvae were most abundant in Wethersfield Cove and the Salmon River. Smelt larvae remained near the bottom throughout most of the river.

Spawning occurred between temperatures of 5.8°C and 10.5°C (Table 10), most of it between 5.8°C and 8.9°C (Table 11). Larvae were found between 8.2°C and 18.2°C (Table 12), 68% of them between 9°C and 10.9°C (Table 13). In the Connecticut River, 75.5% of the smelt larvae had yolk sacs (about 0.7 mm in size) in April, and 44.3% had them in May (about 0.6 mm). Younger larvae were observed in the tributaries and coves, as 64.2% had yolk sacs in May. The number of larvae with yolk sacs decreased from 100% at rkm 74 to 27% at rkm 11.3.

Smelt egg size ranged between 1.2 and 1.4 mm, averaging 1.3 mm (Table 10). Egg size decreased from a mean of 1.39 mm in April to 1.29 mm in May. Smelt larvae ranged in size between 5.5 mm and 10.3 mm, averaging 7 mm (Table 12).

Common Carp

Eggs and larvae of the common carp amounted to 0.29% of all the eggs and 0.32% of the larvae (Table 2). Spawning occurred between May and July, 78.3% in June. Apparently, little spawning takes places in the Connecticut River, as most eggs occurred in Chapman Pond in May and in the upper Salmon River in June. More than twice as many eggs were collected in the upper Salmon River as in all other areas. Larvae were taken from May through August (Table 12), but 93% of them occurred in June. No larvae were found below km 17.7; the majority remained above Middletown (km 35.4), and the greatest number were found above Enfield Dam in June and in the Wilson area (km 74) or above in July. Common carp larvae were twice as abundant in bottom samples (4.0/tow) than in surface ones (2.3/tow).

Common carp spawned at temperatures between 22°C and 27°C (Table 10), the greatest number at 27°C (Table 11). Larvae were collected between 13.8°C and 26°C (Table 12), most of them between 23°C and 24.9°C (Table 13). Larvae were collected at lower temperatures than eggs, and this may have been the result of an earlier spawning, since carp are known to spawn more than once in some areas (Swee and McCrimmon 1966). Mean yolk sac size decreased from 0.94 mm in June to 0.30 mm in July, and larvae exhibited yolk sacs between 13.8 and 24.9 C. The largest larva to retain its yolk sac was 6.7 mm long. Below km 29.8, most larval common carp were in the post-yolk-sac stage. Eggs ranged from 1.5 to 1.9 mm, averaging 1.7 mm (Table 10), and larvae from 2.8 to 7.2 mm, averaging 5.2 mm (Table 12).

White Sucker

Eggs and larvae of the white sucker made up 0.27% and 0.06% of the total catch, respectively (Table 2). Spawning occurred between May and June (Table 10), almost all of it in June and in tributaries and coves. The major spawning area was in the upper Salmon River. Spawning occurred in May in the Farmington River and in June in the Salmon River. A limited amount took place at km 17.7 in the Connecticut River itself. Larvae were collected in May and June (Table 12), 82.7% of them in June. A few larvae occurred at Wilson on the Connecticut River, but the majority remained in the tributaries and coves. Spawning occurred between 12.2°C and 24°C (Table 10), 97.5% at 23.4°C. Larvae were collected between 14.8°C and 24°C (Table 12), most of them at 23.8°C (Table 13). Egg sizes ranged between 2 and 4 mm, averaging 3.0 mm (Table 10) and larvae from 7.2 to 15.1 mm, averaging 12.3 mm (Table 12).

Minnows

Three species, common shiner, spottail shiner, and golden shiner, are grouped under *Notropis* spp. because time and technique did not permit accurate separation of them. All three species spawn in the lower Connecticut River, as indicated by the presence of older fish.

These minnows made up 0.04% and 0.03% of the total eggs and larvae collected (Table 2). Spawning occurred in May and July (Table 10), nearly 86% in the latter month. The only eggs collected in the Connecticut River itself were at km 17.7 in May. The major spawning took place in July in the upper Salmon River. Larvae were collected in May and June (Table 12), 61.3% in May. Larvae occurred only below Higganum (km

29.8) in May but became more abundant upstream in June and were taken most frequently at Windsor Locks (km 78.8). Minnow larvae were most abundant in the Eight Mile and Salmon rivers. They were more or less evenly mixed in the water column at all stations in the Connecticut River and occurred at temperatures between 15°C and 23°C (Table 12), peaking at 21.3°C (Table 13).

Sunfishes

The sunfishes, pumpkinseed, bluegill, and redbreast sunfish, have been grouped together as Lepomis spp. because of the difficulty in separating their early stages and because hybrid adults have been observed in the Connecticut River. All three species have produced hybrids when experimentally crossed with one another (Childers and Bennett 1961; Heckman 1969).

No sunfish eggs were collected during the survey, undoubtedly because of the nest-building habits of these species and the adhesive properties of their eggs. Sunfish larvae made up 0.09% of the total catch (Table 2); their occurrence is shown in Tables 5 and 6. These species spawn mostly during June and July, but spawning may extend into August (Whitworth et al. 1968). Larvae were found from June through August (Table 12), 79.2% occurring in July. In July, larvae were found only above km 74 and below km 17.7. By August, however, a few were collected at km 29.8, and the majority above Enfield Dam. Larvae were collected about a month earlier in the tributaries and coves where they were much more abundant than in the Connecticut River itself. They were present in all tributaries and coves except Chapman Pond, the majority in Wethersfield Cove and the lower Salmon River. The larvae were collected between 21.5°C and 29°C (Table 13). Larval size ranged from 3.4 to 9.2 mm, averaging 5.7 mm (Table 12), and the average size was larger in the tributaries and coves than in the river. Sunfish larvae lost their yolk sacs at about 5.5 mm.

Hogchoker

Three larvae of the hogchoker (4.7–5.9 mm) were taken at Brockway Island (km 11.3) in 1966. They were collected on a flood tide in July, in water temperatures of 26°C to 27°C (Table 12). Dovel et al. (1969) showed that the hogchoker spawns in salinities greater than 9% and that following hatching, during July and August, larvae move upstream and congregate in a low salinity nursery area close to the salt–freshwater interface where they remain during the winter. We have found adults of this species in freshwater as far upstream as 25.7 km in the Connecticut River.

Atlantic Silverside

Two larvae (mean, 6.8 mm) of the Atlantic silverside were collected at Brockway Island on a flood tide at a temperature of 27°C (Table 12). Kuntz and Radcliffe (1918) stated that this species spawns in June and July, that the eggs are 1.1 to 1.2 mm in diameter, and that the yolk sac is absorbed before hatching when the larvae are between 3.9 and 5 mm long.

Summary and Conclusions

From 1965 through 1969, an investigation was conducted to determine the species composition, distribution, abundance, and development of the planktonic fish eggs and larvae, as well as the location of the spawning and nursery areas of fishes in the lower Connecticut River and its tributaries and coves. Eggs and larvae (112,896) representing 16 species of fish were collected in 690 biweekly plankton tows at 16 sampling stations on the river and at 9 stations in 5 tributaries or coves between km 11.3 and 85.3 in the lower Connecticut River. Spawning occurred from April through August and peaked in June in the river and in May in the tributaries and coves. Egg production in the tributaries and coves was higher than that found in the river, the Salmon River being the most productive. In the river, 87% of the egg production occurred above Middletown (km 35.4), the greatest quantity in the Windsor Locks area (km 78.8). Larval abundance in the river was more than twice that found in the tributaries or coves; as a result of drift from freshwater upstream spawning areas, the majority of the larvae became concentrated in a low salinity nursery area between Higganum (km 29.8) and Essex (km 11.3). Eggs were 4 times and larvae 11 times more abundant near the bottom than near the surface in the upper river, but the larvae of most species became more pelagic as they drifted downstream.

Planktonic fish eggs and larvae collected between km 11.3 and 85.3 in the lower Connecticut River represent fish species that are almost entirely freshwater spawners. The alewife, blueback herring, white perch, and yellow perch contributed 98% of the total egg production. The eggs of these species hatch upstream in the Connecticut River and its tributaries and coves, and the larvae are carried downstream by freshwater runoff in a planktonic stage. These larvae possess yolk-sac food reserves and probably do not depend on the nutritional capacity of their spawning grounds; active feeding probably does not begin for most of these fish until they move downstream toward waters of low salinity. In the lower Connecticut River, the nursery ground for most species lies in an area between Higganum (km 29.8) and Essex (km 11.3). In this area, young of the year make up 81% of the total catch of fish collected (see pages 64–66). This area of freshwater–brackish water interface (0 to 15‰) has been described by Dovel (1971) and Massmann (1963a) as the critical zone of the estuary, a zone in which high turnover rates of food organisms support large numbers of maturing fishes, especially during the spring and summer months. Dovel (1971) has presented an indepth discussion and schematic overview of the importance of the common estuarine nursery area, and he suggested that the downstream movement of newly hatched larvae of freshwater spawners is not only beneficial to them, but necessary for their survival.

No biological data are presently available regarding the occurrence, distribution, and abundance of any young fishes from km 11.3 downstream to the mouth of the Connecticut River. At km 11.3, the young of 10 marine species were collected by seine and trawl from 1965 through 1972: bay anchovy *Anchoa mitchilli*, striped killifish *Fundulus majalis*, crevalle jack *Caranx hippos*, northern pipefish *Syngnathus fuscus*, weakfish *Cynoscion regalis*, Atlantic herring *Clupea harengus*, Atlantic silverside, bluefish *Pomatomus saltatrix*, Atlantic tomcod *Microgadus tomcod*, and fourspine stickleback *Apeletes quadracus*. None of these species was ever collected farther north. The area from km 11.3 to the river mouth appears to be an important nursery area for many marine species.

Power Plant Effects, Including Entrainment

Of the total egg production of 16 species of fishes in the lower Connecticut River (km 11.3 to 85.3), 87% occurred above Middletown (km 35.4). Of the egg production in the lower Connecticut River and its tributaries and coves, 77% occurred above the Connecticut Yankee plant (km 25.7). The number of eggs collected near the Connecticut Yankee intake represented only 0.2% of the egg production in the lower Connecticut River and its tributaries and coves. For this reason, the potential effects of entrainment on fish eggs at the intake of the plant was considered minimal.

In contrast, the majority of the fish larvae were found to concentrate, as a result of drift from freshwater upstream spawning areas, in a low salinity nursery area between Higganum (km 29.8) and Essex (km 11.3). Of the fish larvae in the lower Connecticut River and its tributaries and coves, 48% were collected between Higganum and the mouth of the Salmon River. Excluding the tributaries and coves, 61% of all larvae collected in the lower Connecticut River were found near the plant site, that is, between Higganum and the mouth of the Salmon River. Since a large portion of the larval population was potentially subject to possible detrimental effects by passage through the condenser cooling-water system of the Connecticut Yankee plant, studies on the effects of entrainment of young fishes were conducted between 1970 and 1972. The following section is based on these studies, which have been reported in more detail by Marcy (1971, 1973).

The fate of young fish passing through the condenser cooling-water systems of power plants has been a relatively recent concern in environmental impact evaluations. Coutant (1970, 1971) and Marcy (1971, 1973, 1975) have provided the most comprehensive reviews of entrainment studies. This section describes a study of mechanical, chemical, and thermal factors that influence the survival of small fish during their passage through the condenser and discharge-canal system of the Connecticut Yankee plant. It also includes estimates of the percentage of nonscreenable, living fish larvae passing by the plant that become entrained.

Methods

At peak load, the plant pumps approximately 25 m^3/s of water heated about 12.5°C above ambient water temperature. The plant uses approximately 6% of the average available river flow at the intake point during extreme low-flow conditions.

The river water that enters the cooling-water system of the Connecticut Yankee station passes through (1) an intake structure at a velocity of 0.3–0.6 m/s, (2) a 1-cm mesh traveling-screen, (3) a variable axial-flow water pump of 352,000 L/min capacity, (4) twin, single-pass, divided water boxes in which the velocity may approach 2.4 m/s, (5) two right-angle turns, and (6) two banks of 26,860 condenser tubes (2.3 cm ID × 236 cm long) in which the exchange of heat takes place. Total elapsed time during passage through the plant was calculated to be 93 s. The heated water passes out of the building and over a weir and then continues down a canal 18.3 m wide and 3 m deep at an average velocity of 0.3–0.6 m/s. The intake and discharge velocities were synonymous, fluctuating congruently with tidal stage. The water is returned to the river 1.83 km (1.14 mi) downstream from the plant. The drop in temperature between the upper and lower ends of the canal is approximately one degree C. An entrained organism is exposed to the elevated temperatures in the canal for 50–100 min.

For the collection of the small fishes, 0.5-m, No. 2 mesh (0.39-mm), cone-shaped nylon plankton nets with No. 313 TSK flowmeters at their centers were used. A 1-L plastic container was attached to the cod end of each net. A specially constructed observation device was used for maintaining water at ambient and at elevated temperatures during the counting of living and dead specimens immediately after their collection from the river or discharge canal. This consisted of a portable, wooden water-table (64.5 × 36.5 × 10.0 cm deep) provided with a diaphragm water pump of 17 L/min capacity that drew water through a 2.5-cm-diameter hose to the table. Water depth in the table was maintained at 5.0 cm. Two 19-cm-diameter blackened preparation dishes were placed side by side in the water table to hold net-captured fish specimens for counting (at night by red light). Only those specimens that had begun to turn opaque were considered dead. All samples were preserved in 2% neutral formalin with rose bengal dye added to facilitate separation and counting in the laboratory.

As the river passes the plant, a proportion of the young fish small enough to pass through the traveling screens is carried into the intake. The percentage of young fish that were entrained was estimated in the following manner. Between 1 June and 25 July 1972, seven day-and-night series of samples were taken on a grid cross section across the width of the river (502 m) and at each of the four intake bays of the plant. Each sampling of a cross section of the river was conducted 50 m upstream or downstream from the intake on ebb or slack tide, respectively (Figures 2 and 3). Cross sections of the river were divided into nine stations, approximately 50 m apart and two additional stations were located in the shallows near the east shore. At each of the nine stations, stationary plankton nets were operated for five minutes 1 m below the surface, 3 m below the surface, and 1 m above the bottom. Similarly, 5-min stationary net collections were made at 1 m and 3 m below the surface and 1 m above the bottom in front of each of the power plant's intake bays at the same time as the corresponding river collections were being made. These bays are each 3.7 m wide, and the water enters them at the rate of 0.3 to 0.6 m/s.

With the aid of the flowmeter, the number of larvae and early juvenile fish per cubic meter of water sampled could be calculated for each sample. An estimate of the percentage of living fish (those surviving natural or net-caused mortality) collected each sampling day was obtained by using the portable water table to count the number of living and dead fish in the samples collected from the middle of the river. The percentage of living fish in the combined samples from the three depths was determined. This percentage was multiplied by the total number of fish collected in the river and the intake cross sections in order to estimate the number of living fish collected in each cross section on a particular sampling day. In all, 4,480 larvae and early juveniles were collected near the plant's intake, and 3,664 were alive when captured. The mortality due to combined netting and natural causes was 18%. It is our belief that the use of No. 2 mesh kept net mortality to a minimum. The total number of cubic meters of water recorded on all the net flowmeters used in each cross section was divided into the total number of living fish collected by all nets to obtain the average number of living fish per cubic meter in each cross section.

The percentages of living small fish that were entrained, under conditions of unidirectional net tidal flow, were calculated by the formula,

$$\text{Percent} = 100\, N_i F_i / N_r F ,$$

where N_i = average number of living fish per cubic meter in the intake cross section, F_i = flow rate into

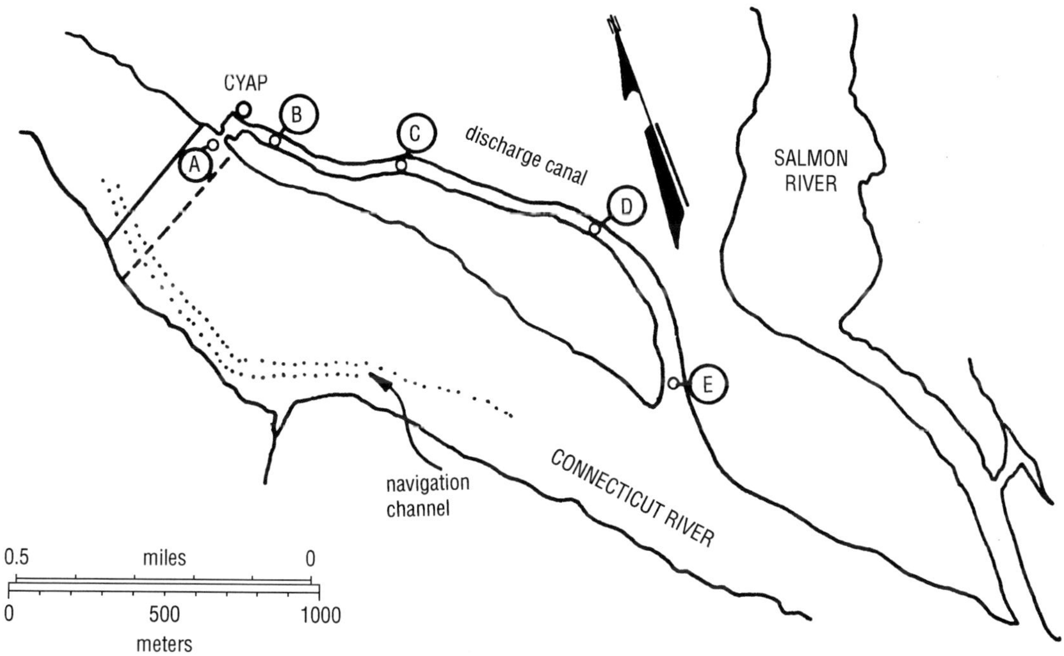

FIGURE 2. Location of sampling stations A–E at the Connecticut Yankee Atomic Power Company (CYAP). The solid line across the river indicates the upstream (ebb tide) cross section sampled and the dashed line the downstream (flood tide) cross section.

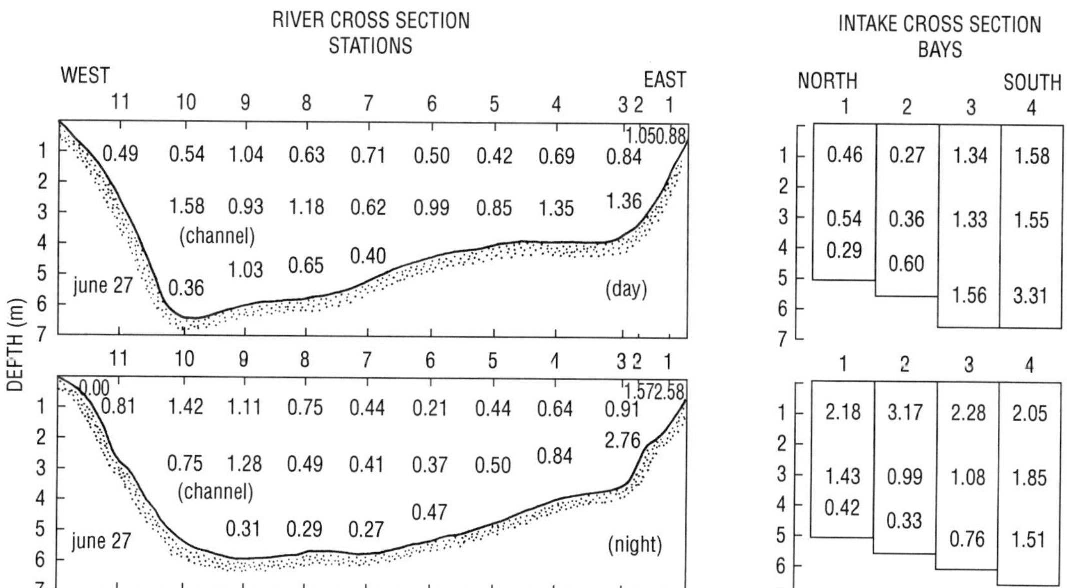

FIGURE 3. Intake and river cross section showing the number of fish larvae collected per cubic meter of water sampled, during day and night, at each station and depth, 27 June 1972.

intake in m^3/s, N_r = average number of living fish per cubic meter in the river cross section, and F_r = average net tidal flow rate (m^3/s) in the river near the plant during the time period sampled. The intake flow rates (F_i) were obtained from plant records, and the net tidal flow rates (F_r) for the river were obtained from the U.S. Geological Survey, Hartford, Connecticut. The latter were determined at the Canel station (8 km upstream from the plant's intake) and are not considered to be significantly different from those at the plant site (J. Stackpole personal communication). The methods used to calculate these flow rates have been described by Lai et al. (1971). River net tidal flow rate records (F_r) were available for each 15 min while the plankton nets were taking samples. All 15-min records were totaled for each sampling period and were divided by the number of records to obtain an average river net rate of tidal flow during each sampling period. The net tidal flow rates were unidirectional, that is, downstream during the entire experiment. The mean June flow rate was high in 1972 (892 m^3/s), as compared with other study years (518 in 1967, 714 in 1968, 357 in 1969, 300 in 1970, and 244 m^3/s in 1971). When flow rates are less than 425 m^3/s, there is an upstream tidal flow past the plant (F. Ruggles personal communication). Under these condition, an organism that drifts with currents will be swept alternately upstream and downstream past the plant's intake as tidal flows change from ebb to flood tide. These organisms, therefore, may be vulnerable to entrainment several times before they pass downstream farther than the distance of one tidal excursion from the intake. Percentage calculations using the above formula, therefore, relate only to the situation of a single tidal pass in which the larval fishes were assumed to be carried by the plant only once, thus representing minimum estimates.

The survival of fishes in the discharge canal in relation to heat was determined from plankton net collections made from 30 June to 29 July 1970 and on 2 and 24 June 1971, at five different places: the plant's intake (A), below the outfall weir (B), and along the discharge canal at 457 m (C), 1,200 m (D), and 1,800 m (E) (Figure 2). Triplicate 5-min stationary samples at the surface, middle, and bottom depths were taken successively at the five stations. The data from each of the three collections per station were combined for comparisons.

The percentage survival at each discharge canal station was calculated by dividing the number of living fish per cubic meter counted at that station by the number of living fish per cubic meter counted at the intake for each sampling day. Survival was based on the number of living fish because sampling for dead fish in the canal often gave inconsistent results. For example, fewer dead fish were collected as the sampling approached the lower end of the canal, indicating that their bodies were settling out. Such inconsistency is probably due to the variable nature of the distribution of the fish in the water column further enhanced by the turbulent mixing and changes in the specific gravity of the water in the canal due to temperature changes.

The mortality caused by mechanical damage brought about by passage through the cooling system was estimated on 27 June and 13 July 1972 at a time when all intake pumps were running at capacity but no heat or hypochlorite was being added. Five-minute net collections were made at all canal stations immediately after the intake had been sampled. The estimated percentage of living fish at the intake was multiplied by the total number of fish collected in each of the four intake cross sections and divided by the total number of cubic meters of water sampled to obtain the number of living fish per cubic meter sampled at the intake.

The mortality caused by the injection of sodium hypochlorite into the cooling water was calculated from data collected on 13 and 14 July 1972 when all intake pumps were running and no heat was being added. Sampling procedures and calculations were the same as those described for the evaluation of mechanical damage. The biocide, sodium hypochlorite, containing approximately 13.5% by weight of available chlorine, is injected at a rate of 19 L/min for 10 min into one intake bay at a time in front of each traveling screen, generally once per day. This "split stream" method yields maximum concentrations of residual chlorine of 2 ppm before reaching the condenser, 1 ppm at the condenser outlet, and less than 0.1 ppm at a point 45 m below the outfall weir (Anonymous 1972).

Results

The species entrained were alewives, blueback herring, white perch, common carp, white catfish, American eel *Anguilla rostrata*, spottail shiner, johnny darter, and American shad. In 1971 and 1972, 1,068 and 10,271 young fish, both larvae and juveniles, were collected. Of the 2,681 larvae and juveniles that were sorted, identified, and measured during 1970, 97.5% were in the post-yolk-sac larval stage (2.6–15.0 mm TL), and 2.5% were early juveniles (15.1–40.0 mm TL) (Table 18). Alewives and blueback herring made up 97.6% of the total catch and white perch 1.3%. The six other species were represented by quantities of less than 1%. The percentages of larvae and early juveniles that survived to various points in the discharge canal are shown in Table 19; the species involved were alewives, blueback herring, and common carp. No living white perch larvae were observed after passage through the plant (Stations B–E). The only survivors taken in the canal at a canal temperature (37°C) were the two small American eels; it is not known, however, whether these fish had come in through the intake or had entered the canal from the river.

TABLE 18. Percentage of the total catch represented by the various species, and total lengths of the fish, 30 June–29 July 1970.

Date	Number collected	Percentage of total catch	Mean length (mm)	Range
		Blueback herring and alewife		
June 30	239	91.6	7.8	4.7–14.0
July 2	134	92.4	6.5	4.0–11.5
6	833	99.4	5.9	3.6–11.3
8	1277	97.9	9.3	4.0–40.0
22	115	100.0	10.1	5.5–28.5
29	18	100.0	8.4	5.0–13.5
		White perch		
June 30	21	8.0	3.1	2.0–03.9
July 2	9	6.2	3.5	3.2–03.7
6	5	0.6	3.0	2.8–03.3
		Carp		
June 30	1	0.4	7.5	7.5
July 2	1	0.7	5.5	5.5
8	10	0.7	3.0	2.1–05.0
		White catfish		
July 2	1	0.7	10.6	10.6
8	2	0.1	17.2	13.0–21.3
		American eel		
July 8	2	0.1	100.7	100.5–100.8
		Spottail shiner		
July 8	12	0.9	24.5	19.2–30.0
		Johnny darter		
July 8	1	0.1	28.5	28.5
		American shad		
July 8	1	0.1	16.0	16.0

Figure 3 illustrates the distribution of young fish in two of the river cross sections, in this case taken during the day and night of 27 June 1972. As in all river cross sections, the fish tended to be somewhat more abundant in the shallower areas near the east shore. The fish also tended to be more abundant near the bottom during the day and between mid-water and the surface at night. The proportion of nonscreenable young, living fish that were entrained on the seven sampled days ranged between 1.7% and 5.8% (Table 20). The mean percentage, calculated in the sample computation, was approximately 4% (95% confidence limits: 3.3–4.5%). The maximum percentage entrained (5.8%) occurred when downstream river flow was at a minimum, on a brief slack tide.

Mechanical Effects on Entrained Fish

Estimates of fish mortality resulting from mechanical damage during passage through the condenser cooling-water system ranged between 72% and 87%, averaging approximately 80% (Table 21). Mortality percentages were consistent (71–72%) on all dates except the 27–28 June night sample when the percentage reached 87%, probably as a result of a greater proportion of larger fish (20–40 mm TL) in the sample, as compared to a majority of less than 15 mm in other samples. Increased mechanical injury with increased size during condenser passage has been reported by Markowski (1962), Oglesby and Allee (1969), and Marcy (1971). Different species and life-history stages also show different susceptabilities to mechanical damage, and the high percentages noted in the present study may have been influenced by the fact that 97.5% of the fish entrained were in post-yolk-sac stages and that fragile clupeids made up 97.6% of all the fish entrained. In addition, a high detritus load passing through the condensers with the organisms may have increased mortality by causing abrasion.

The percentage of the total mortality caused by mechanical damage has not been adequately assessed at most power plants. If studies reveal high percentages of mechanical damage relative to thermal or chemical effects, future alternative

TABLE 19. Percentage of larvae and early juveniles surviving at the various stations in the discharge canal, and environmental conditions, 30 June–29 July 1970, and 2 and 24 June 1971.

Station[a]	Time (hours) and power level (MW_e)	Temp (°C)	Number living/m³	Number dead/m³	Percent survival/m³
30 June 1970					
A	0900–1200	22.2	1.74	0.91	65.5[b]
B	300	28.2	0.60	2.54	34.5
C		28.2	0.49	0.52	28.2
E		28.2	0.13	0.24	07.5
2 June 1971					
A	1000–1535	16.0	0.06	0.00	100.0
B	610	29.0	0.04	0.48	66.6
C		29.0	0.02	0.42	33.3
D		28.8	0.00	0.33	00.0
E		28.6	0.00	0.29	00.0
2 July 1970					
A	1200–1600	22.0	0.54	0.30	64.8
B	588	33.5	0.09	1.64	16.6
C		33.5	0.10	1.22	18.5
D		33.0	0.00	0.96	00.0
E		33.0	0.00	0.82	00.0
24 June 1971					
A	1120–1535	26.0	3.92	2.23	63.7
B	579	35.0	0.00	3.08	00.0
C		35.0	0.00	2.45	00.0
D		34.8	0.00	3.23	00.0
E		34.2	0.00	0.89	00.0
6 July 1970					
A	1000–1500	23.9	1.89	0.90	67.7
B	587	35.5	0.00	1.12	00.0
C		35.2	0.00	0.97	00.0
D		35.0	0.00	1.76	00.0
E		35.4	0.00	0.55	00.0
8 July 1970					
A	1430–1800	29.0	0.92	0.14	86.8
B	584	38.5	0.00	0.96	00.0
C		38.0	0.00	0.78	00.0
D		38.0	0.00	0.92	00.0
E		37.8	0.00	0.26	00.0
8 July 1970					
A	2100–0130	25.0	0.48	0.49	49.5
B	584	36.0	0.00	1.26	00.0
C		36.0	0.00	0.74	00.0
D		36.0	0.00	1.12	00.0
E		36.0	0.00	0.43	00.0
9 July 1970					
A	0230–0630	25.5	1.41	0.86	62.1
B	581	37.0	0.03	2.19	02.6
C		37.0	0.00	0.97	00.0
D		36.5	0.00	2.06	00.0
E		36.5	0.00	1.19	00.0

TABLE 19. Continued

Station[a]	Time (hours) and power level (MW_e)	Temp (°C)	Number living/m^3	Number dead/m^3	Percent survival/m^3
22 July 1970					
A	1300–1600	26.1	0.32	0.89	28.8
B	578	38.2	0.00	0.30	00.0
E		38.0	0.00	0.00	00.0
29 July 1970					
A	0900–1100	30.0	0.04	0.00	100.0
B	569	41.0	0.00	0.33	00.0

[a] Station A is plant intake; B is outfall weir; C is 457 m below weir; D is 1200 m below weir; E is 1,800 m below weir (end of canal).

[b] Intake (control) percent that survived netting and natural mortality factors. Percent survival at stations B through E was calculated by dividing the number of living larvae per cubic meter at each canal station by the number of living larvae per cubic meter at the intake (control) station.

strategies include the raising of condenser ΔTs while lowering the intake water volume (i.e., decreasing the number of entrained organisms vulnerable to mortality). Among recent power plant studies, the following have shown mechanical damage to be the highest single cause of entrainment mortality: Flemer et al. (1971a) (striped bass eggs); Edsall and Yocum (1972) (menhaden larvae); Tarzwell (1972) (menhaden larvae); Lauer et al. (1973) (river herring, hogchocker, white perch, striped bass); Schubel (1973) (blueback herring, striped bass, white perch, American shad); Austin et al. (1973) (menhaden, weakfish, tautog, cunner); Carpenter et al. (1974) (zooplankton); Grosse Ile Laboratory (1972) (zooplankton); McNaught (1972) (zooplankton); Industrial Bio-Test Laboratory, Inc. (1972) (zooplankton); Beck and Miller (1974) (zooplankton); Flemer et al. (1971b) (phytoplankton); and Marcy (1975).

Fish, in particular, seem most vulnerable to mechanical stress. Flemer et al. (1971a) found extremely high mortality (99.7%) of striped bass eggs passing through the Vienna, Maryland generating station and that a high percentage of the eggs had disintegrated. The death of 164.5 million menhaden larvae during one day at the Brayton Point Plant (Rhode Island) was attributed to mechanical damage (Edsall and Yocum 1972). Austin et al. (1973) reported 27–57% maceration (mechanical) mortality to four species of juvenile marine fish after plant passage at Northport, New York. Tarzwell (1972) reported that the only identifiable parts of larval menhaden (25–38 mm) after passage through a fossil-fuel plant were their eyeballs. Lauer et al. (1973) reported the lowest (7–39%) mechanical damage mortality of entrained ichthyoplankton of all studies reviewed. Schubel (1973) stated that mechanical damage may have a much greater impact on development and hatching success than does temperature (i.e., ΔTs up to 10°C were not detrimental to egg development or hatching success of blueback herring, alewife, American shad, white perch, and striped bass).

Two recent studies have shown much higher mortality of zooplankton from the mechanical component than previously reported in earlier power plant studies. For example, Carpenter et al. (1974) reported that 70% of the copepods entering the cooling water system of the Millstone Point Plant at Niantic, Connecticut were killed by mechanical or hydraulic stresses of passage. Beck and Miller (1974) cited unpublished data, which demonstrated that mechanical damage caused 60% mortality of phytoplankton and zooplankton, with the primary contributed being pumping effects. Prior to the pumps, the average mortality was 10%; after pumps, 50%; after condenser passage, 60%.

Morgan et al. (1973) have measured the mortality caused by the shear fields (variation in water velocity with respect to time, expressed as units of force per unit area, that is dynes/cm^2) that are created by water movement over the surface of fish eggs and larvae. These experiments were performed on white perch and striped bass. The data

TABLE 20. Percentages of young fish entrained and associated environmental conditions, 1 June–25 July 1972.[a]

Date	Time (hours)	Tide	Temp (°C)	Intake: No. samples	No. collected	m³ sampled	No./m³	% living	No. living	N_i^b	F_i^c
1 June	1111–1342	Ebb	19.9	12	247	751.8	0.33	85.4	211	0.28	25.00
8 June	1053–1310	Ebb	19.0	12	1,650	734.6	2.25	83.9	1,384	1.88	25.00
27 June	0933–1300	Ebb	20.5	12	464	432.8	1.07	73.9	343	0.79	25.00
27 June	2055–2356	Ebb	21.5	12	713	486.2	1.47	100.0	713	1.47	25.00
13 July	0946–1343	Slack	26.0	12	298	477.3	0.62	50.0	149	0.31	25.00
13–14 July	2126–0100	Ebb	25.5	12	357	592.3	0.60	100.0	357	0.60	25.00
25 July	1400–1715	Ebb	28.0	12	24	364.8	0.06	60.0	14	0.04	25.00
Total				84	3,753	3,839.8			3,171		

Date	Time (hours)	Tide	Temp (°C)	River: No. samples	No. collected	m³ sampled	No./m³	% living	No. living	N_r^d	F_r^e	% living entrained	Confidence limits (95%)
1 June	1111–1342	Ebb	19.9	21	804	1,222.4	0.66	85.4	686	0.56	743.49	1.68	0.91–2.45
8 June	1053–1310	Ebb	19.0	23	2,782	1,512.8	1.84	83.9	2,334	1.54	741.15	4.11	2.79–5.43
27 June	0933–1300	Ebb	20.5	23	960	1,141.5	0.84	73.9	709	0.62	1,052.92	3.02	2.34–3.70
27 June	2055–2356	Ebb	21.5	24	1,069	1,375.6	0.78	100.0	1,069	0.78	1,076.43	4.38	2.78–5.98
13 July	0946–1343	Slack	26.0	24	243	636.0	0.38	50.0	122	0.19	704.12	5.79	3.99–7.60
13–14 July	2126–0100	Ebb	25.5	23	562	997.3	0.56	100.0	562	0.56	778.82	3.44	2.20–4.67
25 July	1400–1715	Ebb	28.0	24	98	1,572.5	0.06	60.0	59	0.04	561.22	4.45	1.90–7.00
Total				162	6,518	8,458.1			5,541			3.91[a]	3.34–4.48

[a] Sample percent computation for all seven dates combined: $100 N_i^b F_i^c / N_r^d F_r^e = 100(0.8258)(25.00)/[(0.6551) \cdot (805.11)] = 3.91\%$.

[b] Total number of living larvae divided by the total number of cubic meters of water sampled in the intake cross section.

[c] Flow rate into intake (m³/s).

[d] Total number of living larvae divided by the total number of cubic meters of water sampled in the river cross section.

[e] Average net tidal flow rate (m³/s) in river near the power plant from 93 15-min records prepared during all sampling time periods.

TABLE 21. Percentages of larval and early juvenile mortality in the entire discharge canal due to mechanical, chemical, and thermal damage, and environmental conditions during June and July 1970–1972.

Date	Time (h)	Power level (MW_e)	River temp	Intake No. samples	Intake No. collected	Intake No. living	Intake m^3 sampled	Intake % living	Intake No. living/m^3
			Mechanical (no heat)						
27 June 1972	1130–1430	0	20.5	12	464	343	432.75	73.9	0.7929
27–28 June 1972	1040–1300	0	21.5	12	713	713	486.22	100.0	1.4664
13 July 1972	1500–1840	0	26.0	12	298	149	477.26	50.0	0.3122
14 July 1972	0210–0320	0	25.5	12	357	357	592.29	100.0	0.6027
			Biocide (no heat)						
13 July 1972	1700–1830	0	26.0	12	298	238	477.26	80.0	0.4995
14 July 1972	0340–0510	0	25.5	12	357	357	552.29	100.0	0.6027
			Heat (28–29°C)						
30 June 1970	0900–1200	300	22.2	3	32	21	12.10	65.6	1.7355
2 June 1971	1000–1530	610	16.0	3	3	3	53.92	100.0	0.0556
			Heat (33.5°C)						
2 July 1970	1200–1600	588	22.0	4	14	9	19.96	64.3	0.4509
			Heat (35.0°C)						
6 July 1970	1000–1500	587	23.9	6	258	175	92.17	67.8	1.8986
24 June 1971	1120–1535	579	26.0	3	270	172	43.90	63.7	3.9179

Date	Time (h)	Power level (MW[a])	Canal temp	Discharge canal No. samples	Discharge canal No. collected	Discharge canal No. living	Discharge canal m^3 sampled	Discharge canal No. living/m^3	Discharge canal % mortality
			Mechanical (no heat)						
27 June 1972	1130–1430	0	20.5	4	53	42	192.19	0.2185	72.4[a]
27–28 June 1972	1040–1300	0	21.5	4	34	31	160.50	0.1931	86.8
13 July 1972	1500–1840	0	26.0	4	67	29	318.46	0.0910	70.9
14 July 1972	0210–0320	0	25.5	4	76	40	255.21	0.1776	70.5
			Biocide (no heat)						
13 July 1972	1700–1830	0	26.0	4	127	50	418.49	0.1194	76.1
14 July 1972	0340–0510	0	25.5	4	100	42	361.88	0.1160	80.8
			Heat (28–29°C)						
30 June 1970	0900–1200	300	28.2	9	311	74	164.69	0.4493	74.1
2 June 1971	1000–1530	610	29.0	12	147	5	371.04	0.0134	75.9
			Heat (33.5°C)						
2 July 1970	1200–1600	588	33.5	15	257	12	219.17	0.0547	87.9
			Heat (35.0°C)						
6 July 1970	1000–1500	587	35.0	15	413	0	379.12	0.0000	100.0
24 June 1971	1120–1535	579	35.0	12	648	0	245.37	0.0000	100.0

[a] Total percentage mortality in the discharge canal as calculated by dividing the total number of living larvae per cubic meter in the discharge canal (Stations B–E) by the total number of living larvae per cubic meter at the intake (A).

obtained were used to predict the shear or mechanical stress created by the water passing through the Chalk Point fossil-fuel plant. Results indicated conservative estimates of mortality between 20% and 50% for both species (R. P. Morgan and R. E. Ulanowicz, personal communi- cations). Morgan, in a paper presented at the Northeast Division of the American Fisheries Society meeting, February 1974, calculated the shear or mechanical force of water passing through the Calvert Cliffs nuclear plant and estimated that shear stress alone would be enough to cause 50–100% mortalities of entrained eggs and larvae.

Biocidal Effects on Entrained Fish

The range of mortality percentages during periods of hypochlorite injection were not measurably different from those calculated for mechan- ical damage (Table 21). The chlorination procedure at Connecticut Yankee is similar to that discussed by Draley (1972), in which the addition of chlorine before the water reaches the condensers and the splitting of the coolant stream into parts (chlorination of one bay at a time) allows for its destructive removal by reaction with slower (organic) chlorine-demand constituents when remixing occurs. Any small mortalities that might have been caused by injection of sodium hypochlorite in the present study probably went undetected because the samples taken involved only 25% of the total treated water volume at places where residual chlorine was recorded at less than 0.1 ppm.

Thermal Effects on Entrained Fish

When heat was added to the cooling system and the discharge temperature reached 28–29°C, no fish survived passage to the end of the canal on 2 June 1971, while 7.5% survived at the same temperatures on 30 June 1970 (Table 19). The reason no fish survived to reach the end of the canal on 2 June 1971 was the increased heat shock caused by a greater difference between ambient and canal temperatures (ΔT) and a lower acclimation (ambient) temperature.

The survival rate of fish immediately after passing through the plant on 30 June 1970 (Station B) was 34.5%, and this percentage continued to drop to 7.5% at the end of the discharge canal. The survival rate on 2 July 1970 at Station B dropped to 16.6% as the power level and ΔT doubled. When the plant was at full power and canal temperatures were above 28°C, no living fish were taken at the end of the canal. At canal temperatures of 29°C to 33.5°C, 100% mortality occurred during passage through the plant. It was estimated that approximately 20% of the total mortality that occurred in the canal was caused by heat shock and prolonged exposure to temperatures elevated above 28°C (Tables 19 and 21).

Of the 81,975 fish larvae representing 16 species that were collected in 690 tows in the Connecticut River between km 11.3 and 85.3, 95% were present during an 11-degree (19–29°C) temperature spread (see page 122). If the plant operated continuously at full power at a ΔT of 12.5°C, it would then follow that discharge canal temperatures would remain above the upper natural temperatures encountered by the young of most species during at least 95% of their seasonal occurrence. During normal operation, however, the power level often fluctuates, causing a corresponding change in the ΔT. Discharge canal temperatures were recorded daily during June and July of 1968 and 1969 when 95% of the larvae were abundant in the intake area (Marcy 1971, 1973). Discharge canal temperatures remained above 28°C during 44 and 56 d (June and July) in 1968 and 1969, respectively.

Discussion and Conclusions

This study shows that all fish larvae and early juveniles that passed through the plant were dead by the time they reached the end of the discharge canal. Sampling during June and July, when 95% of the nonscreenable fish were abundant near the intake, showed that approximately 80% of the total mortality in the canal was caused by mechanical damage and 20% was attributed to heat shock and prolonged exposure to temperature elevated above 28°C. There was no measurable mortality due to injection of sodium hypochlorite as a biocide. The number of nonscreenable young living fish entrained at the intake averaged about 4% (range, 1.7% to 5.8%) of all those passing by the plant under conditions of unidirectional net tidal flow.

Little information was available in the past in the literature on the entrainment of young fish at power stations. Both Kerr (1953) and Markowski (1962) showed no harmful effects on fish passed through condensers. In both studies, however, discharge temperatures were no greater than 28°C and total exposure time was less than 4 min. Carlson and McCann (1969) estimated that 4% of the eggs and 12.2% of the larvae of striped bass were subject to withdrawal from a river cross section near the intake of a pump storage plant during a single tidal pass. It does not seem possible to screen fish larvae effectively. Kerr (1953), for example, found that impingement of striped bass eggs and larvae on fine mesh screening for even a short time was fatal.

Any estimate of the effect of mortality from entrainment on the total fish population must take into account the natural mortalities during specific life stages of each individual species, since entrainment mortality adds to the rate of natural mortality. Natural mortality is high in prolific species. For example, Leggett (1969) estimated a 0.001% survival rate from egg to adult stages for the American shad in the Connecticut River, and Kissil (1974) reported a 0.0014% freshwater survival from eggs to emigrating juveniles for an anadromous alewife population in Connecticut. Until natural mortality information is available on the life stages of the various species encountered in this study and more is known about the carrying capacity of the river system, no inferences can be made as to the impact that the 4% entrainment has on the ultimate population of adults. These larvae drifting by the Connecticut Yankee plant represent about 61% of all larvae collected in the lower 74 km of the Connecticut River (see page 130). The significance of the entrainment impact is related to the loss of 4% of that percentage (61%) of the total river larval population available in the vicinity of the plant.

References

Anonymous. 1972. Nonradioactive chemicals and elements. Connecticut Yankee Atomic Power Company, Haddam Neck Plant, environmental report to A.E.C., operating license stage. Docket no. 50-513. Section 5.4.4:76–77.

Austin, H. M., J. Dickinson, and C. Hickey. 1973. An ecological study of ichthyoplankton at the Northport power station, Long Island, N.Y. N.Y. Ocean Science Laboratory, Montauk, New York, prepared for Long Island Lighting Co.

Bayliff, W., Jr. 1950. The life history of the silverside, *Menidia menidia* (Linnaeus). Chesapeake Biological Laboratory Publication 90, Solomons, Maryland.

Beck, A. D., and D. C. Miller. 1974. Analysis of inner plant passage of estuarine biota. Paper presented at the American Society of Civil Engineers Power Div. Spec. Conference, Boulder, Colorado. August 12–14, 1974.

Bigelow, H. B., and W. C. Schroeder. 1953. Fishes of the Gulf of Maine. U.S. Fish and Wildlife Service Fishery Bulletin 53.

Breder, C. M., Jr., and D. E. Rosen. 1966. Modes of reproduction in fishes. Natural History Press, Garden City, New York.

Carlson, F. T., and J. A. McCann. 1969. Hudson River fisheries investigations (1965–1968). Report to Consolidated Edison Co., New York, by the Hudson River Policy Commission.

Carpenter, E. J., B. B. Peck and S. J. Anderson. 1974. Survival of copepods passing through a nuclear power station on Northeastern Long Island Sound, USA. Marine Biology 24:49–55.

Childers, W. F., and G. W. Bennett. 1961. Hybridization between three species of sunfish (*Lepomis*). Illinois Natural History Survey Biological Notes No. 46, Urbana.

Cianci, J. M. 1969. Larval development of the alewife, *Alosa pseudoharengus* (Wilson) and the glut herring, *Alosa aestivalis* (Mitchill). M.S. thesis. University of Connecticut, Storrs.

Cooper, R. A. 1961. Early life history of the alewife, *Alosa pseudoharengus*. M.S. thesis, University of Rhode Island, Kingston.

Coutant, C. C. 1970. Biological aspects of thermal pollution I. Entrainment and discharge canal effects. CRC Critical Reviews in Environmental Control 1(3):341–381.

Coutant, C. C. 1971. Effects on organisms of entrainment in cooling water. Steps toward predictability. Nuclear Safety 12:600–607.

Dovel, W. L., J. A. Mihursky, and A. J. McErlean. 1969. Life history aspects of the Hogchoker, *Trinectes maculatus,* in the Patuxent River Estuary, Maryland. Chesapeake Science 10:104–119.

Dovel, W. L., J. A. Mihursky, and A. J. McErlean. 1971. Fish eggs and larvae of the upper Chesapeake Bay. University of Maryland, Natural Resources Institute, Special Report 4.

Draley, J. E. 1972. The treatment of cooling waters with chlorine. Argonne National Laboratory, Environmental Project No. ANL/ES-12, Argonne, Illinois.

Edsall, T. A. 1970. The effect of temperature on the rate of development and survival of alewife eggs and larvae. Transactions of the American Fisheries Society 99(2):376–380.

Edsall, T. A., and T. G. Yocom. 1972. Review of recent technical information concerning the adverse effects of once-through cooling on Lake Michigan, prepared for the Lake Michigan Enforcement Conference, September 19–21, 1972, Chicago, Ill. Great Lakes Fishery Laboratory, Ann Arbor, Michigan.

Flemer, D. A., et al. 1971a. The effects of steam electric station operation on entrained organisms. In J. A. Mihursky and A. J. McErlean (Co-Principal Investigators). Postoperative assessment of the effects of estuarine power plants. University of Maryland, Natural Resources Institute, Reference No. 71-2b.

Flemer, D. A., C. W. Keefe, D. H. Hamilton and J. A. Mihursky. 1971b. The effects of thermal loading and water quality on estuarine primary production. University of Maryland, Natural Resources Institute, Reference No. 70-6b.

Grosse Ile Laboratory. 1972. Lake Michigan entrainment studies, Big Rock nuclear power plant, Escanabe power plant, Nov.–Dec. 1971. Environmental Protection Agency Office of Research and Monitoring, Report 1, Grosse Ile, Michigan.

Hale, J. G. 1960. Some aspects of the life history of the smelt, *Osmerus mordax,* in Western Lake Superior. Minnesota Fish and Game Investigations Fish Series 2:25–41.

Heckman, J. R. 1969. Embryological comparison of *Lepomis macrochirus macrochirus* and *Lepomis macrochirus gibbosus*. Transactions of the American Fisheries Society 98(4):669–675.

Herman, S. S. 1963. Planktonic fish eggs and larvae of Narragansett Bay. Limnology and Oceanography 8:103–109.

Hildebrand, S. F., and L. E. Cable. 1930. Development and life history of fourteen teleostean fishes at Beaufort, North Carolina. U.S. Bureau of Fisheries Bulletin 46:383–488.

Industrial Bio-Test Laboratories, Inc. 1972. Intake-discharge experiments at Waukegan generating stations. Report to Commonwealth Edison Co. Project XI, IBT No. W9861, Biology Section, Chicago.

Kerr, J. E. 1953. Studies of fish preservation at Contra Costa Steam Plant of the Pacific Gas and Electric Company. California Department of Fish and Game, Fish Bulletin 92. Kissil, G. W. 1974. Spawning of the anadromous alewife, *Alosa pseudoharengus* in Bride Lake, Connecticut. Transactions of the American Fisheries Society 103(2):312–317.

Kissil, G. W. 1968. A contribution to the life history of *Alosa pseudoharengus* (Wilson). M.S. thesis. University of Connecticut, Storrs.

Kuntz, A. 1914. The embryology and larval development of *Bairdiella chrysura* and *Anchovia mitchilli*. U.S. Bureau of Fisheries Bulletin (1913) 33:3–19.

Kuntz, A., and L. Radcliffe. 1917. Notes on the embryology and larval development of twelve teleostean fishes. U.S. Bureau of Fisheries Bulletin. 35:87–134.

Lai, C., F. H. Ruggles, Jr. and L. A. Weiss. 1971. Evaluation of flow in tidal reaches of the Connecticut River by mathematical model. U.S. Geological Survey, Hartford, Connecticut.

Lauer, G. J., et al. 1973. Entrainment studies on Hudson River organisms. New York University Medical Center, Laboratory for Environmental Studies, November report, New York.

Leggett, W. C. 1969. Studies on the reproductive biology of the American shad (*Alosa sapidissima,* Wilson). A comparison of populations from four rivers of the Atlantic Seaboard. Ph.D. thesis. McGill University, Montreal.

Leim, A. H. 1924. The life history of the shad, *Alosa sapidissima* (Wilson), with special reference to factors limiting its abundance. Biological Board of Canada, Contrib. Can. Biol. 2:16–284.

Loesch, J. 1968. A contribution to the life history of *Alosa aestivalis* (Mitchill). M.S. thesis. University of Connecticut, Storrs.

Mansueti, A. J. 1964. Early development of the yellow perch, *Perca flavescens*. Chesapeake Science 5:46–66.

Mansueti, A. J., and J.D. Hardy, Jr. 1967. Development of fishes of the Chesapeake Bay Region, Part I. An atlas of egg, larval and juvenile stages. E. E. Deubler, Jr., editor. Natural Resources Institute, University of Maryland, College Park.

Mansueti, R. J. 1964. Eggs, larvae and young of the white perch, *Roccus americanus,* with comments on its ecology in the estuary. Chesapeake Science 5:3–45.

Marcy, B. C., Jr. 1969. Age determinations from scales of *Alosa pseudoharengus* (Wilson) and *Alosa aestivalis* (Mitchill) in Connecticut waters. Transactions of the American Fisheries Society 98(4):622–630.

Marcy, B. C., Jr. 1971. Survival of young fish in the discharge canal of a nuclear power plant. Journal of the Fisheries Research Board of Canada 28:1057–1060.

Marcy, B. C., Jr. 1973. Vulnerability and survival of young Connecticut River fish entrained at a nuclear power plant. Journal of the Fisheries Research Board of Canada 30(8):1195–1203.

Marcy, B. C., Jr., and F. P. Richards. 1975. Entrainment of organisms at power plants, with emphasis on fishes—an overview. Pages 89–106 *in* S. B. Saila, editor. Fisheries and energy production—a symposium. Lexington Books, D.C. Heath and Co., Lexington, Massachusetts.

Markowski, S. 1962. Faunistic and ecological investigations in Cavendish Dock. Barrow-in-Furness. Journal of Animal Ecology 31:43–51.

Massmann, W. H. 1963. The "critical zone" in estuaries. Bulletin of the Sport Fishing Institute, Richmond, British Columbia.

McKenzie, R. A. 1964. Smelt life history and fishery in the Miramichi River, New Brunswick. Bulletin of the Fisheries Research Board of Canada 144.

McNaught, D. C. 1972. The potential effects of condenser

passage on the entrained zooplankton at Zion station. Prepared statement at Lake Michigan Enforcement Conference, September 21, 1972. Chicago.

Morgan, R. P. III, R. E. Ulanowicz, V. J. Rasin, Jr., L. A. Noe, and G. B. Gray. 1973. Effects of water movement on eggs and larvae of striped bass and white perch. University of Maryland, Natural Resources Institute, Chesapeake Biological Laboratory, NRI Ref. No. 73-111, College Park.

Oglesby, R. T., and D. J. Allee. 1969. Ecology of Cayuga Lake and the proposed Bell Station (nuclear powered). Cornell University, Water Res. and Marine Science Center, Publication No. 27, Cornell University, Ithaca, New York.

Pearson, J. C. 1941. The young of some marine fishes taken in lower Chesapeake Bay, Virginia, with special reference to the gray sea trout, *Cynoscion regalis* (Block). U.S. Fish and Wildlife Service, Fishery Bulletin 50:79–102.

Schubel, J. R. 1973. Effects of exposure to time-excess temperature histories typically experienced at power plants on the hatching success of fish eggs. The Johns Hopkins University, Chesapeake Bay Institute, PPRP-4; CBI Special Report 32, Baltimore, Maryland.

Scotton, L. N., R. E. Smith, N. S. Smith, K. S. Price, and D. P. deSylva. 1973. Pictorial guide to fish larvae of Delaware Bay, with information and bibliographies useful for the study of fish larvae. University of Delaware, College of Marine Studies, Delaware Bay Report Series, Volume 7, Newark.

Swee, V.B., and H.R. McCrimmon. 1966. Reproductive biology of the carp, *Cyprinus carpio* L., in Lake St. Lawrence, Ontario. Transactions of the American Fisheries Society 95(4):372–380.

Tarzwell, C. M. 1972. An argument for the open ocean siting of coastal thermal electric plants. Journal of Environmental Quality 1(1):89–91.

Whitney, R. R. 1961. The Susquehanna Fishery Study, 1957–1960. A report of a study on the desirability and feasibility of passing fish at Conowingo Dam. Maryland Department of Research and Education, Contribution No. 169.

Whitworth, W. R., P. L. Berrien and W. T. Keller. 1968. Freshwater fishes of Connecticut. State Geological and Natural History Survey of Connecticut Department of Environmental Protection, Bulletin 101, Hartford.

American Fisheries Society Monograph 9:155–180, 2004

Early Life History Studies of American Shad in the Lower Connecticut River and the Effects of the Connecticut Yankee Plant

BARTON C. MARCY, JR
Ecological Sciences Division, NUS Corporation
1910 Cochran Road, Pittsburgh, Pennsylvania 15220, USA

Because the American shad *Alosa sapidissima* is the most important sport fish in the Connecticut River, all stages of its life in freshwater were scrutinized to determine possible adverse effects of the power plant. Shad lay their eggs in fresh water and the young spend from 4 to 6 months in the river before they migrate out to sea. During this period, they are tiny and delicate and more vulnerable to environmental forces than at any other time in their life history.

Our studies were principally concerned with the distribution and abundance of eggs and young shad, and the downstream migration—all in the vicinity of the plant. To interpret our findings correctly, however, we had to study the young stages throughout the lower river from Massachusetts to Long Island Sound.

Shad Eggs and Larvae in the Lower Connecticut River

Although the American shad is an important fishery resource in the Connecticut River, its early life history and the distribution of its spawning areas in the lower river was not well known, especially in the vicinity of the Connecticut Yankee Atomic Power Company (CYAP) plant. The only pertinent studies were those of Walburg and Nichols (1967) who described some of the physical characteristics of the river that were associated with the presence of shad eggs between river kilometer (rkm) 29 and 119 and Watson (1968, 1970) who conducted a 4-year study of spawning areas between km 119 and 174 located above Holyoke, Massachusetts. This section is based on the work of Marcy (1972) and describes the distribution and relative abundance of shad eggs in the lower Connecticut and Salmon rivers, factors that influence spawning success, and effects of the CYAP plant.

Methods

Collections of shad eggs were made from April to July in 1967 and 1968 at 61 locations ranging from above Enfield Dam (km 85.3) to Essex (km 11.3) (Figure 1). The 10 nets used were modifications of those described by Massmann (1952). They were 0.5-m cone-shaped plankton nets of No. 2 mesh (0.39-mm) nylon with 1-L bottles on the cod ends. Each net was provided with a 23-kg anchor and was rigged to fish about 0.3 m off the bottom and to swing with the tide. One net was set at each of the 10 arbitrarily designated areas throughout the lower river. Each net was fished from 12 to 24 h and then shifted to another location generally twice weekly. Temperature, dissolved oxygen, and depth were recorded before and after each set. Flowmeters could not be used because of net backwash from silt clogging, and relative densities were calculated on the basis of catch per-unit-effort, that is, number of eggs per net-fishing-hour. The cap of each collecting bottle was fitted with a 2-mm-mesh wire screen to reduce the accumulation of silt, but at the same time not permit the eggs to escape. Silt was washed from the samples through a 2-mm-mesh sieve. Eggs were transferred to Stockards' solution (Costello et al. 1957), which turns the protoplasm white while keeping the yolk transparent, thus facilitating the identification of developmental stages.

To determine the rate of development, 200 to 300 eggs stripped from ripe fish (see Marcy 2004 [1976], this volume) were fertilized and

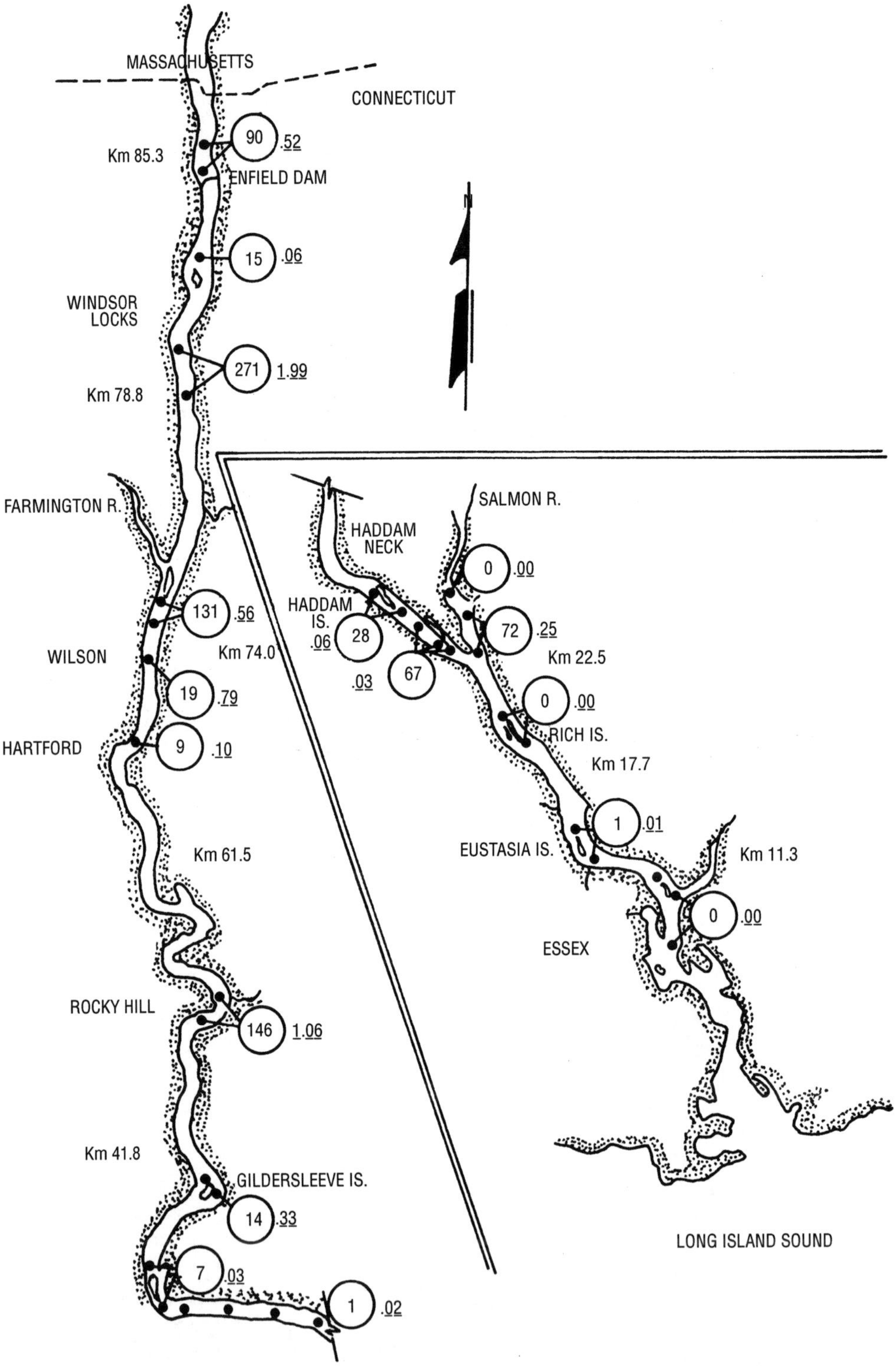

FIGURE 1. Map of lower Connecticut River showing the locations of stations for sampling shad eggs. Circled numbers represent total number of shad eggs collected in 1968 and 1969. The number alongside each circle is the calculated average number of eggs per net-fishing-hour for both sampling years.

reared in the laboratory at temperatures similar to those in the river (13.8–23.0°C). Hatching occurred 71 to 86 h after fertilization, and a developmental timetable was established for aging eggs collected in the river. This allowed calculation of the distance of the eggs from where they were spawned (Table 1).

Several day and overnight boat trips were made from Enfield Dam to Essex in search of spawning sites—generally by drifting and the use of spotlights at night. Whenever spawning activity was suspected, plankton nets were set downstream in order to obtain newly fertilized eggs.

Results and Discussion

Abundance and Development

In the ichthyoplankton survey of the lower Connecticut River (see Marcy 2004 [1976]), the eggs of American shad ranked eighth in abundance and the larvae twelfth. Shad larvae were collected from May through August, more than 80% of them in June, in water that ranged between 9.7°C and 30°C, mostly between 21°C and 23°C. They ranged in length from 4.9 to 14.6 mm and averaged 10.7 mm.

Shad egg size ranged between 1.1 and 3.8 mm, averaging 2.83 mm. The mean egg size in May was 3.1 mm and decreased to 2.39 mm in June. The increase in mean egg size between a nonfertilized and a developing egg due to water hardening was approximately 1.08 mm (Table 1). A regression line was fitted to data obtained on temperature and egg sizes during 1968. The relationship was found to be significant ($P < 0.05$) ($ES = 4.43 - 0.02\,T$; $r = 0.66$; $n = 305$, where ES equals egg size and T equals temperature in °F).

Table 1 describes the developmental sequence of shad from egg fertilization to posthatching. Hatching occurred 71 to 86 h after fertilization. Watson (1968) hatched shad eggs in the field above Holyoke Dam (km 119), Massachusetts in 84 h and provided a regression equation for the number of hours required for hatching at different temperatures ($Y = 729.06 - 8.75X$ ($r = 0.93$; $P < 0.05$), where Y equals the number of hours required for hatching to occur and X equals the mean temperature of development in °F. In the present study, shad larvae were between 5.7 and 6.8 mm long when hatched and absorbed their yolk-sacs in about 88 h. The relationship between shad larval length and yolk-sac size was significant ($P < 0.05$), $YS = 1.4045 - 0.115\,LL$ ($r = .59$), where YS equals yolk sac size in millimeters and LL equals larval length in millimeters. Shad larvae absorb their yolk sacs when they reach 12.2 mm total length (TL).

Spawning Areas

Because of the relative scarcity of shad eggs in our regular collections, special efforts were made to locate shad spawning grounds, particularly in the area around the plant. In 1967 and 1968, 869 eggs of the American shad were collected in 4,341 net-hours (Table 2). Shad spawned throughout the study area but primarily above Rocky Hill (km 51.5). Major spawning areas were found to be at Windsor Locks (km 78.8), Wilson (km 74), and Rocky Hill (km 51.5); eleven times more eggs per hour were collected in these areas than in areas downstream. Although similar collecting efforts were made in the upper and lower Salmon River, no spawning was observed and no eggs were collected more than 0.6 km upstream from its junction with the Connecticut River.

Of the eggs collected during 1967, 63% were 3 h old or less and 91% were 14 h old or less (Table 3). Calculations—based on the age of the eggs when collected and on current velocities, which ranged from 0.15 to 0.6 m/s—indicate that most eggs had traveled only 1.6 to 6.4 km from where they had been spawned. Whitworth and Bennett (1970) followed the movement of shad eggs from the time they were broadcast until they sank or lodged on the bottom and found that this occurred about 5 to 35 m downstream. Barker (1965), Carlson (1968), and Chittenden (1969) have made similar observations in other East Coast rivers.

No shad eggs were found in water with dissolved oxygen concentrations of less than 5 ppm in the present study. The oxygen LC_{50} for Connecticut River shad eggs is 2.0 to 2.5 ppm according to Carlson (1968). Eggs were found at depths of 0.6 to 7.3 m, and as many eggs/h were collected above 3 m as below (0.1981 and 0.1998, respectively). Walburg and Nichols (1967) reported that 40% of the eggs they observed were in water less than 3 m deep. In our

TABLE 1. Developmental sequence of American shad eggs and larvae from fertilization to post-hatching as determined by rearing in the laboratory at 13.8–23.0°C (7–19 June 1967).

Total (h)	Time (min)	Mean egg size (mm)	Mean larval size (mm)	Yolk sac size (mm)	Developmental stage	Developmental features
0	00	2.09			Nonfertilized	Color: light green to clear
0	00	2.14			Fertilized	Many large cells, morula stage
1	45	3.17			Multicellular	Small cells; dome present
3	00	3.14			Gastrulation	Early blastula; dome halfway down yolk
14	20	3.12			Gastrulation	Cells small indistinguishable one-half and three-quarters way down yolk
25	35	3.14			Complete gastrulation	Very early embryo; raised one-half to three-quarters around yolk (shadow)
32	50	3.03			Early embryo to tail free stage	Eye visible; tail pointed; pigmentation on head, body, and yolk; tail separated from yolk; myomeres present
39	05	2.90			Middle embryonic	Extends halfway around yolk; more head pigmentation; eye with fissure
43	35	2.93			Middle embryonic	Otoliths present; heart beating; anus external
49	05	3.05			Late embryonic	Fin development; yolk three-quarters pigmented; pectoral stubs
61	35	3.08			Late embryonic	Extends three-quarters around yolk; eye pigmented; head separate from yolk
70	45	2.99			Late embryonic	Tail extends around yolk; eye fissure closing
71	30		5.73	1.67	Hatched	Hatching
86	45		6.81	2.23	Hatched	
118	00		6.90	1.00	Hatched	Yolk sac three-quarters absorbed; heavy ventral pigmentation
174	45		7.43	0.00	Hatched	Yolk sac absorbed; feeding begun
290	00		7.27	0.00	Hatched	100% mortality

ichthyoplankton survey, shad eggs were distributed almost equally between the surface and the bottom (1.64/tow and 1.58/tow, respectively), while larvae were more than twice as abundant in surface waters (0.78/tow as compared with 0.33/tow), becoming more pelagic in the downstream areas.

Differences in time and place of spawning between 1967 and 1968 may have been caused by differences in temperature. Mean water tem-

TABLE 2. Summary of collections of American shad eggs in the lower Connecticut and Salmon rivers, 1967 and 1968.

Area	River km	Total hours 1967	Total hours 1968	Total eggs 1967	Total eggs 1968	Eggs/h 1967	Eggs/h 1968
Enfield Dam area	87.0	91.50	26.58	88	2	0.9680	0.0759
Enfield boat launch	83.7	120.38	28.58	15	0	0.1247	0.0000
Windsor Locks	80.5	60.55	140.33	216	55	3.5803	0.3922
Farmington River mouth	74.0	99.45	135.33	58	73	0.5842	0.5399
Wilson	72.4	24.25		19		0.7867	
Hartford	70.8	24.17	90.50	3	6	0.1244	0.0664
Rocky Hill	45.0	105.53	133.66	21	125	1.1868	0.9370
Gildersleeve Island	41.8	43.25		14		0.3274	
Middletown	38.6	129.53	80.83	6	1	0.0463	0.0124
Dart Island	35.4	50.91		1		0.0197	
Haddam Island	28.9	275.63	153.66	14	12	0.0508	0.0782
Haddam Neck	25.7	18.75	1,170.00	0	67	0.0000	0.0572
Rich Island	17.7	93.25		0		0.0000	
Eustasia Island	13.3	126.00		1		0.0079	
Essex area	11.3	111.00	95.00	0	0	0.0000	0.0000
Upper Salmon River		19.75	394.33	0	0	0.0000	0.0000
Lower Salmon River		19.66	478.58	7	65	0.3608	0.1358
Totals		1,413.56	2,927.38	463	406		

perature was 11°C in May 1967 and 15°C in May 1968. On the average, shad eggs were collected farther upstream in 1967 than in 1968. The area of mean egg catch was located at rkm 72 in 1967 and at rkm 60 in 1968. This measure of the linear distribution of eggs along the river is based on data from Table 87. The number of eggs per hour for each collecting area was multiplied by its corresponding distance in km from the river mouth. These figures were totaled and then divided by the total catch per hour for all areas.

TABLE 3. Developmental stage and age of eggs collected in the Connecticut and Salmon rivers, 1967.

Developmental stage	Age (h)	Number	Percent
Unfertilized		6	1.65
Fertilized	0.0–1.0	23	6.35
2- to 32-cell	1.0–2.0	63	17.40
Multicellular	2.0–3.0	136	37.56
Blastula	3.0–5.0	49	13.55
Gastrula	5.0–14.0	53	14.66
Primitive streak	14.0–25.0	26	7.18
Early embryo	25.0–33.0	1	0.27
Tail separation	33.0–35.0	0	0.00
Middle embryo	35.0–40.0	5	1.33

Maximum observed spawning occurred from 13 May to 27 June, the peak occurring in June in 1967 and in May in 1968 (Figure 44). The median catch day at Enfield Dam (the date by which half of the total catch had been landed) occurred on 27 May in 1967 and on 13 May in 1968 (Connecticut State Board of Fisheries and Game, personal communication). Mansueti and Kolb (1953) found that shad ovaries developed more slowly at 12.8°C to 18.4°C than at 20°C to 25°C. The lower temperatures of May 1967 may have delayed gonad maturation and the fish may have migrated farther upstream before spawning in 1967 than they did in 1968 when temperatures were higher. Spawning was observed at least 10 different times, but only on dark afternoons and during the evening, confirming observations of Mass-mann (1952), Walburg and Nichols (1967), and Chittenden (1969). Spawning involved pairs of fish, or several males and a single female (cf. Medcof 1957). In group spawning, fish swam with the backs exposed. They splashed about forming a closely packed circle. Several hundred newly fertilized eggs were taken with hand-held plankton nets between 5 and 15 m downstream from this activity in 0.5 to 1 m of water at the confluence of the Farmington and Connecticut rivers.

Spawning Success

Shad eggs were collected in water ranging from 7.5°C to 24°C. Few eggs (3%) were collected below 16°C in 1967, but 65% of them were taken at these low temperatures in 1968 (Figure 2). Peak spawning occurred at 22°C in 1967 and at 14.8°C in 1968. Bradford et al. (1964), Leim (1924), Leach (1925), and Mansueti and Kolb (1953) reported that temperatures below approximately 16°C prolonged the period of development and reduced survival. Watson (1968) provided a regression equation that described the effect of temperature on rate of development. Based on his equation, development, at the time of peak spawning, required 103 h in 1967 and 217 in 1968. It is obvious that most eggs in 1968 experienced prolonged development.

Of the eggs collected in 1967, 18% were dead as compared with 30% of those collected in 1968. Watson (1970) reported that 20% of the eggs collected above Holyoke Dam (km 110) in 1968 were dead, but only 13% in 1969. He attributed the higher egg mortality of 1968 to widely fluctuating water temperatures and stated that many eggs were probably lost from the study area due to high water.

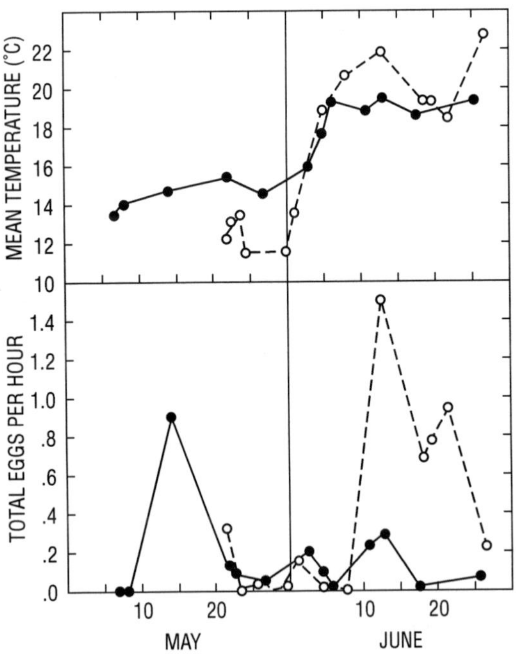

FIGURE 2. Total number of eggs/h and mean water temperatures in May and June. Broken lines 1967, solid lines 1968.

The paucity of juvenile shad collected in 1968 is further evidence of poor spawning success (see page 174). Watson (1970) found juvenile shad scarce above Holyoke Dam in 1968, as compared with 1966, 1967, and 1969. Low water temperatures in June 1968 may have caused poor survival of eggs and larvae and thus weakened this year-class.

Effects of Power Plant on Spawning

An estimate of the number of shad eggs that were entrained in the plant's intake during 1968 and 1969 was attempted by placing two stationary 0.5-m nets at various depths in water that was committed to the plant's intake. These nets were set from 21 May through 23 June in 1968 and 1969. In 1968, 49 shad eggs were collected in 270 net hours. Based on an estimated flow rate through the nets during each collection, the number of eggs per cubic meter of water sampled was 0.0013. The number of m^3 of water taken into the plant over the 29-d period was estimated to be 62,640,000. This figure was multiplied by the number of eggs/m^3 and provides an estimate of 81,432 eggs taken into the intake for 1968. These eggs were assumed to have been destroyed, for continuous sampling in the discharge canal between 1968 and 1972 yielded only a few empty shad egg capsules. In 1969, 87 eggs were collected in 303 net-hours, and the estimated number of eggs/m^3 was 0.0021. The estimated number of shad eggs lost to the intake in 1969 was 131,544.

Leggett (1969) has estimated that approximately one mature adult shad is produced per 100,000 eggs spawned in the Connecticut River. Relating this estimate of natural mortality to the estimated number of shad eggs lost to the intake, only approximately one adult shad would have been lost in 1968 and 1969. The number of shad eggs lost to the intake must be considered minimal when compared with either the number of potential adults lost or the total egg production of the major spawning areas upriver above km 51.5. Moreover, there is no evidence of any change in the location of shad spawning areas in the river as the result of plant operation.

Population Dynamics of Young American Shad in the Lower Connecticut River[1]

The growth, distribution, and feeding habits of young-of-the-year American shad in the lower Connecticut River are not well known. Only Walburg (1960) and Mitchell et al. (1925) have provided data on the food habits and growth of young individuals in the lower part of the river (km 11.3 to 85.3). This section describes the distribution, growth, and feeding habits of young shad in the lower Connecticut River (km 11.3 to 138) and discusses factors that influence seasonal and yearly abundance. The results of studies on the effects of the CYAP plant on young shad are also described.

[1] By Barton C. Marcy, Jr. (Ecological Sciences Division, NUS Corporation, 1910 Cochran Road, Pittsburgh, Pennsylvania 15220, USA) and Paul M. Jacobson (Northeast Utilities Service Co., Post Office Box 270, Hartford, Connecticut 06101, USA).

Methods

Young-of-the-year American shad were sampled biweekly at 12 stations between Essex (rkm 11.3) and just above Enfield Dam (km 85.3) from June through November 1966–1972 and from August through November 1973. Sampling was expanded in 1966 and 1967 to include four stations between Enfield Dam and Northampton, Massachusetts (km 138). The location of each station is described in Table 4 and illustrated in Figure 3. Whenever possible, both sides of the river and the different available types of shoreline at each station were sampled, generally during daylight hours. Temperature and dissolved oxygen were recorded at each location.

Young shad were sampled in 1966 in a 12.5-m nylon bag seine (6-mm bar mesh, 1.2 m deep with a 1.8-m bag). The hauling lines were 23.2 m long. The size of the seine used between 1967 and 1972 was increased to 15.2 m in length, 2.4 m in depth and 1-cm knotless bar mesh. The bag was 2.4 m long and the seine was hauled by 30.5-m lines. From 1967 through 1972, a 4.9-m semi-balloon mid-water trawl was used at stations 8, 6, 5, 3, and 1. The trawl was towed at a speed of one to two knots between the same two points at each

TABLE 4. Location of sampling stations for juvenile shad in the lower Connecticut River, 1966–1973, with fishing gear used.

River kilometer	Station no.	Town	Location	Gear[a] used
138.0	16	Northampton, MA	Below Calvin Coolidge Bridge (east and west)	1
119.0	15	Holyoke	0.4 km above Holyoke Dam (east and west)	1
106.2	14	Chicopee	Across from mouth of Chicopee River (west)	1
101.4	13	Springfield	0.2 km above Memorial Bridge (west)	1
85.3	12	Suffield, CT	Enfield Pool—0.8 km above dam (west)	1
74.0	11	Wilson	Above Bissell Bridge (west)	1
61.0	10	East Hartford	0.4 km upstream from Charter Oak Bridge (east)	1
51.5	9	Glastonbury	0.4 km downstream of Rocky Hill Ferry slip (east)	1
29.8	8	Higganum	Higganum Creek—mouth (east and west)	1, 2
27.4	7	East Haddam	Haddam Island (east)	1
24.9	6	East Haddam	Flats north of CYAP[b] discharge canal (east)	1, 2
24.9	5	Haddam	Opposite CYAP discharge canal (west)	1, 2
24.9	4	East Haddam	Salmon River—middle of cove (east)	1
22.5	3	East Haddam	Mouth of Salmon River—cans "47" to "45" (east)	2
13.7	2	Deep River	Across from Selden Creek (west)	1
11.3	1	Essex	Brockway Island (east)	1, 2

[a] 1 = bag seine; 2 = trawl.

[b] CYAP = Connecticut Yankee Atomic Power Company.

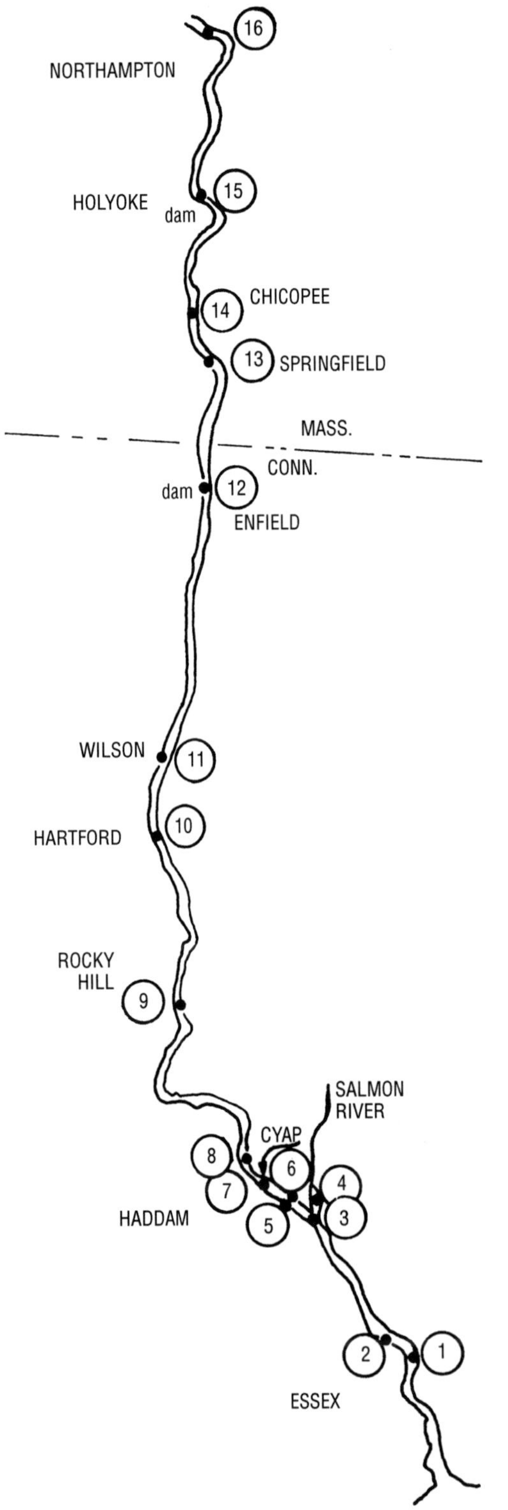

FIGURE 3. Map of lower Connecticut River showing the location of stations for sampling juvenile shad.

station each time. A 4.9-m surface trawl equipped with hydrofoils (Trent 1967) was used at night in the lower river during the late fall to collect large juveniles which had become unavailable to conventional gear.

The criteria used for comparisons of abundance and distribution was catch per unit of effort or the number of juvenile shad collected per net haul. The bag seine and particularly the trawl were always fished in the same fashion and in the same location on each sampling date so that they would represent a fixed standard unit of effort from year to year. Trawl and bag-seine data were pooled at those stations when both were taken concurrently.

Young shad were separated from the two river herrings, alewife *Alosa pseudoharengus* and blueback herring *A. aestivalis*, on the basis of peritoneum color, cheek patch dimensions, and tongue extension. Only the blueback herring has a black peritoneum. The outline of the lower jaw of the shad is less abrupt than that of either the alewife or the blueback herring, and unlike the river herrings, its cheek patch is more deep than wide and its tongue extends beyond the premaxilla when its mouth is held open. A detailed description of the separation of the three species has been provided by Wang (1970). The total length of each shad was measured to the nearest millimeter.

Length–weight relationships, growth rates, and stomach contents were determined from samples of 20 to 50 juveniles collected on each of 22 dates between 13 August and 21 October 1968 at station 12, 0.8 km above Enfield Dam. Fish were measured to the nearest millimeter and weighed to the nearest gram. An analysis was made of the stomach contents of five juveniles collected from late afternoon to early evening on each date. Fish ranged in size between 39 and 141 mm TL. Stomach contents were recorded as the percentage of stomachs in which a particular food item occurred in all stomachs that contained food.

Results and Discussion

Stomach Contents

Analysis of the food items contained in juvenile shad stomachs collected at station 12 (Enfield Dam) in 1968 showed that adult and larval aquatic insects occurred in 92.7% and cladocerans in 68.2% of the stomachs which contained food. Of the 90 stomachs examined, 5 were empty. The majority of the aquatic insects were chironomid larvae, and there were lesser numbers of dipteran adults, Ephemeroptera, and Trichoptera. Invertebrate eggs, fish larvae, copepods, arachnids, terrestrial insects, and 1–3-mm plastic balls occurred in 14.8% of the stomachs. The dominance of aquatic insects and cladocerans in the guts of our sample of juvenile shad agrees not only with the observations of Mitchell et al. (1925) and Levesque and Reed (1972) for the Connecticut River, but also with Massmann (1963), Maxfield (1953), Walburg (1956), Williams and Bruger (1972), Leim (1924), Davis and Cheek (1966), and Chittenden (1969) for other rivers along the Atlantic Coast.

As has been pointed out by several observers, juvenile shad are often opportunistic in their feeding habits and seem to feed on the most available food source. Walburg (1960) examined the stomachs of young shad from above and below Enfield Dam and found that fish taken above the dam had fed primarily on copepods and cladocerans, while the majority of those from below had fed on adult insects. In a study conducted in 1966, Watson (1968) found that juvenile shad from a 57.3-km section between the Holyoke and Turners Falls dams on the Connecticut River fed mostly on copepods and cladocerans. Three years later, Levesque and Reed (1972) studied juvenile shad in the same area and found that aquatic insects made up 46% of the total volume of food and that 34% consisted of crustaceans.

Massmann (1963) and Levesque and Reed (1972) found that the most voluminous stomach contents occurred in early evening. Similarly, juvenile shad that were observed feeding on the surface near shore in the Enfield Pool (above the Enfield Dam) during the summers of 1967 and 1968 dimpled the surface only from late afternoon to early evening. Stomachs from 50 juveniles were analyzed during this surface-feeding activity and all were completely filled with the flying stages of ants (Formicidae). Wind direction and force appeared to be important factors in the regulation of insect availability and consequently the amount of surface activity of the fish. Both Massmann (1963) and Maxfield (1953) have noted orientation toward terrestrial foods when available. Massmann particularly noted that the most important sources of food for young shad did not originate in Virginia rivers but rather in the wooded areas bordering them.

Growth at Enfield Dam (Station 12)

Growth curves in length and weight of juvenile shad collected on 22 occasions between 13 August and 21 October 1968 at station 12 (Enfield Dam) are shown in Figure 4. These measurements were highly variable early in the summer, partly as a result of the variable spawning season, but they became more uniform later in the season.

Growth curves for the shad, from posthatching to emigration in the lower Connecticut River, are shown in Figure 5. Mean total lengths from 22 June through 15 August were plotted from data provided by Mitchell et al. (1925) from the Farmington, Salmon, and Connecticut rivers. Mean lengths from 13 August through 21 October were plotted from data collected at station 12 (Enfield Dam) in 1968. The length–weight relationship for young shad, based on calculations from the present study and from Watson (1968) for the population above Enfield Dam, as well as an estimate of the daily growth in weight for this area are provided in Table 5.

There was a significant increase in weight, but not in length, of the juvenile shad collected at station 12 after 8 September 1968. A significant difference ($P < 0.01$) was found in the slopes of weight, but not of length ($P > 0.05$), when a comparison was made between separate regression lines fitted to the length and weight data from 19 August through 4 September and from 9 September through 21 October. One or more explanations may be responsible for the significant rise in weight: (1) The rise may represent a normal curvilinear function of growth; (2) new recruits from upstream may have entered the population in the Enfield Pool from 4 through 8 September, and although of similar length, may have been significantly heavier; (3) a shift in food preference may have caused a sharp increase in weight. Stomachs analyzed after 8 September were found to be more muscular with elongated posterior ends and fat deposits outside the gut area. Qualitative estimates of the percentage of the total stomach volume occupied by each food item indicated that cladocerans occupied the greater volume before 9 September and aquatic insects the greater volume after 8 September.

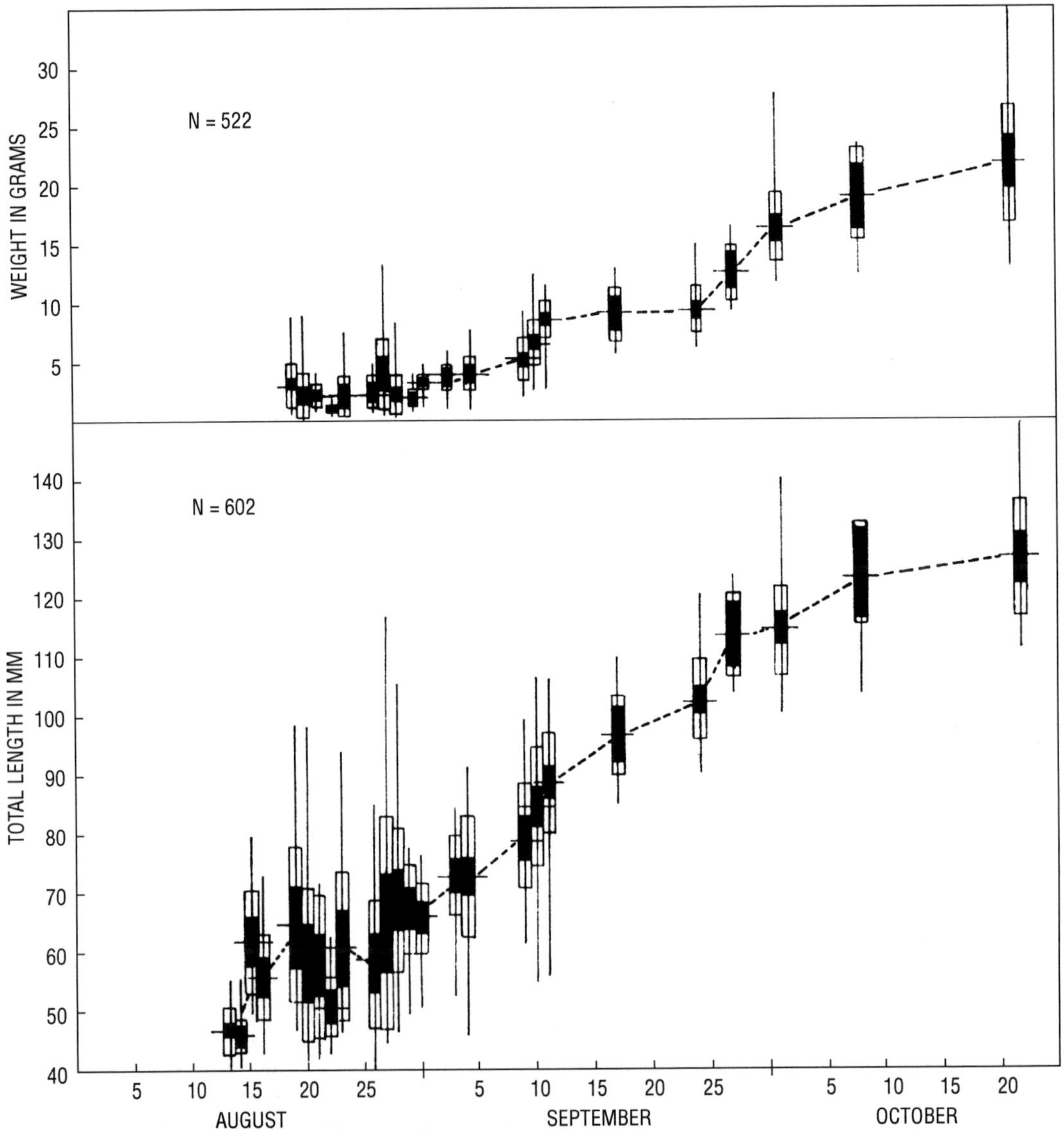

FIGURE 4. Growth curves of Connecticut River shad based on total length and weight of juveniles collected at Station 12 (Enfield Pool) between 13 August and 21 October 1968. Means are indicated by horizontal lines, ranges by vertical ones. Rectangles mark 95% confidence intervals with solid black indicating standard deviation.

Growth at Lower 138 km of the Connecticut River

The total lengths of 8,978 juvenile shad were measured between 1966 and 1973. They ranged in size between 17 and 153 mm. Figure 6 shows the changes in size of these juveniles from June through November. Growth was most rapid early in the season, reaching its peak in October. The decline in mean size in November was probably the result of the movement of larger fish out of the river at this time. Watson (1970) collected juvenile shad up to 182 mm with a scoop net at Holyoke Dam in 1968. Larger fish swim faster, and this was the probable reason the more conventional gear used in the lower river did not capture any specimens longer than 153 mm.

The mean total length of juvenile shad was greater in the downriver areas (Figure 7). Growth is faster in upriver areas (see below), but the larger

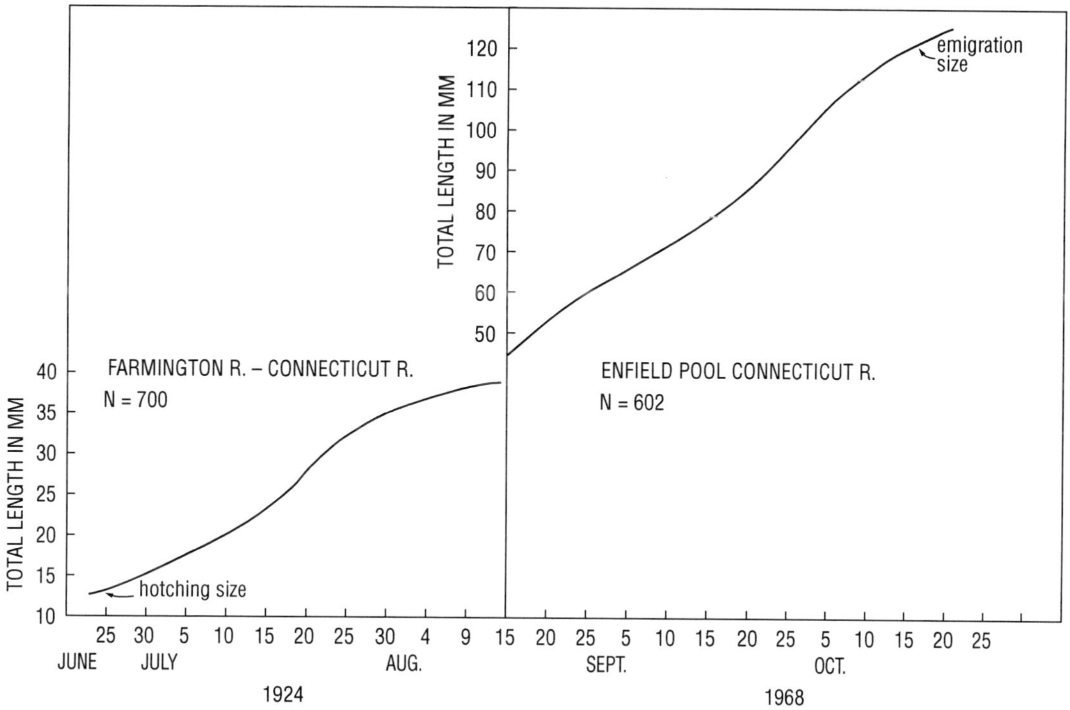

FIGURE 5. Approximate growth curve of the American shad: its early life history in the Connecticut River from hatching to emigration. Fish from 13 to 38 mm were plotted from data provided by Mitchell et al. (1925). Fish from 42 to 121 mm were plotted from data obtained from Enfield Pool (station 12) in 1968.

fish move out of the river first. The majority of adult shad spawn above km 51.5 (station 9) (Marcy 1972), and as the larvae and juveniles grow, they are subject to a cumulative slow downstream transport. This movement is accelerated as time goes on until peak emigration velocities are reached in September or October. In July, the mean length was highest in the Haddam area (station 5) and southward (Table 6). This was perhaps the result of some earlier spawning in the Haddam or lower Salmon River areas and/or rapid downstream transport from upstream spawning areas during the higher flows that occurred early in the season. Juveniles were smaller in the shallower areas below Enfield Dam (stations 10 and 11) where most spawning occurred. Their size apparently increased below Glastonbury (station 9) where the river deepens. By August, juveniles above Enfield Dam were about equal in length to those near the river mouth. The smallest fish were still located between Wilson (station 11) and Hartford (station 10), with an increase in size below Hartford. By September, the mean size had increased steadily below Wilson and the largest fish occurred in the Essex area (station 1). Mean length decreased below the Haddam area in October, indicating the movement of larger fish out of the river.

The growth rate of juvenile shad appears to be faster in upriver than downriver areas (Table 5). The calculated regression equations, showing the increase in length with time, indicate that the most rapid growth occurred at the most upstream station at Northampton. Growth rates, as based on linear approximations, declined rather consistently from upriver to downriver areas and reached the nadir of 0.34 mm per day at the Salmon River (station 4). Mean length (Figure 7) and calculated length intercepts (Table 5) became progressively larger toward the mouth of the river. Young shad grew faster upriver, and the larger sizes emigrated downstream, apparently to leave the river first.

By analyzing the change in size of the members of various populations of young shad as the summer progressed, an average rate of growth for this species could be determined. Measurements from most of the fish collected between km 11.3 and 85.3 (stations 1–12) for 1966–1972 were combined. The mathematical expression of this growth

TABLE 5. Equations of instantaneous (logarithmic) and per-diem (linear) growth[a] of young shad for each station between 1966 and 1972 (Y equals total length in mm and X equals time in days).

River kilometer	Station[b]	Years sampled	Number measured	Instantaneous growth (logarithmic)	Per diem growth (linear)	Length increase per day (mm)
130.4	16	1966–1967	35	$\log Y = -0.7723 + 1.3430 \log X$ (0.98)[c]	$Y = -17.650 + 1.004 X$ (0.97)[c]	1.00[d]
106.2	14	1966–1967	247	$\log Y = -0.4805 + 1.2070 \log X$ (0.98)	$Y = -8.886 + 0.936 X$ (0.97)	0.94
101.4	13	1966–1967	35	$\log Y = -0.1686 + 1.0353 \log X$ (0.94)	$Y = 4.526 + 0.753 X$ (0.91)	0.75
85.3	12[e]	1966–1972	3,889	$\log Y = 0.3560 + 0.7779 \log X$ (0.82)	$Y = 19.856 + 0.623 X$ (0.84)	0.62
74.0	11	1966–1972	981	$\log Y = 0.4846 + 0.6952 \log X$ (0.91)	$Y = 24.419 + 0.501 X$ (0.88)	0.50
61.0	10	1966–1972	365	$\log Y = 0.3872 + 0.7365 \log X$ (0.86)	$Y = 18.930 + 0.536 X$ (0.83)	0.54
51.5	9	1966–1972	347	$\log Y = 0.2067 + 0.8471 \log X$ (0.85)	$Y = 17.028 + 0.624 X$ (0.82)	0.62
29.8	8	1966–1972	340	$\log Y = 0.0288 + 0.9309 \log X$ (0.66)	$Y = 25.794 + 0.517 X$ (0.76)	0.52
27.4	7	1966–1972	282	$\log Y = 0.2135 + 0.8384 \log X$ (0.91)	$Y = 19.152 + 0.574 X$ (0.88)	0.57
24.9	6	1966–1972	288	$\log Y = 0.4269 + 0.7310 \log X$ (0.79)	$Y = 27.196 + 0.504 X$ (0.73)	0.50
24.9	5	1966–1972	1,308	$\log Y = 0.7054 + 0.5887 \log X$ (0.78)	$Y = 35.965 + 0.406 X$ (0.70)	0.40
24.9	4	1966–1972	53	$\log Y = 0.7309 + 0.5558 \log X$ (0.72)	$Y = 34.702 + 0.344 X$ (0.70)	0.34
22.5	3	1966–1972	708	$\log Y = 0.9825 + 0.4713 \log X$ (0.73)	$Y = 46.412 + 0.377 X$ (0.71)	0.38
11.3	1	1966–1972	154	$\log Y = 0.6848 + 0.6247 \log X$ (0.86)	$Y = 37.769 + 0.475 X$ (0.79)	0.47

[a] See Table 92 for explanation of the use of regression equations.
[b] Stations 15 and 2 not included because of insufficient samples.
[c] Correlation coefficient (r).
[d] Slope of linear equation.
[e] The length-weight relationship for 522 young shad collected just above Enfield Dam (km 85.3) is described by the regression $\log W = -5.230 + 3.1094 \log L$ (r = 0.98), W being weight in grams and L total length in millimeters. Watson (1968) provided a length-weight regression for 3,208 young shad collected between the Holyoke and Turners Falls dams of $\log W = -5.210 + 3.09 \log L$. Both growth equations are similar for areas above Enfield Dam. The growth rate of weight, based on 22 sampling days between August 13 and October 21, 1968 at Enfield Dam, is described by the regression $\log W = -0.1461 + 0.7006 \log D$ ($r = 0.77$) or $W = -0.29 + 0.33 D$ ($r = 0.88$), W being weight in grams and D time in days. From the linear expression, the increase in weight was calculated to be 0.33 g per day or 23 g per week.

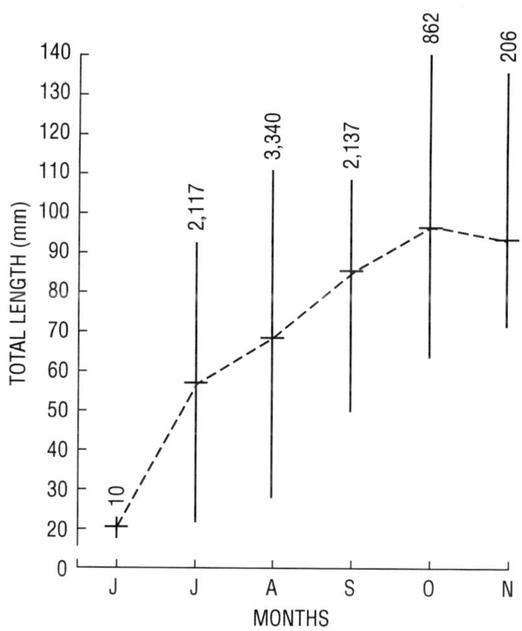

FIGURE 6. Growth of juvenile shad in the lower Connecticut River, 1966–1972. Horizontal lines represent monthly mean total length and vertical lines represent ranges. Numbers are size of sample.

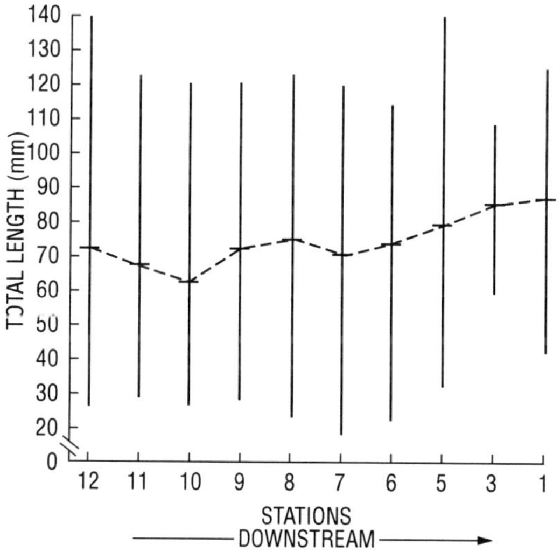

FIGURE 7. Mean total length (horizontal lines) and length range (vertical lines) of juvenile shad collected at each station, 1966–1972.

(instantaneous growth) is expressed by the regression equation $\log Y = 0.3806 + 0.7555 \log X$, where Y is total length in millimeters and X is time. Based on a linear approximation ($Y = 25.88 + 0.52X$) the slope or growth in length per day was 0.52 mm or 3.6 mm per week. To determine if differences existed in the growth rates above and below Enfield Dam, both 1966 and 1967 data were combined for stations 12, 13, 14 and 16 above the dam and stations 1 through 11 below the dam. The regression equations calculated for stations above and below the dam were $\log Y = -0.0048 + 0.9559 \log X$ ($Y = 3.85 + 0.87X$) and $\log Y = -0.3832 + 0.7530 \log X$ ($Y = 8.01 + 0.72X$), respectively. A statistical test (comparing the two linear regression slopes) showed that the growth rate above Enfield Dam was significantly faster than the growth rate below the dam ($t = 4.97$, df = 124, $P < 0.001$).

Earlier upstream spawning and intraspecific competition for food and space are two factors that may bring about the difference in growth rate between the fish in the two areas. Below Enfield Dam juvenile shad must compete with tremendously greater numbers of other plankton-feeding clupeids than they do above it. The number of juvenile clupeids of four species caught per net haul below Enfield Dam is an indication of their comparative abundance. In 1,008 bag seine and trawl collections made at six stations between km 29.8 and 11.3 from 1965 through 1971, there were an average of 70.9 alewives and blueback herring per net haul, 33.8 Atlantic menhaden *Brevoortia tyrannus*, and only 2.4 shad. No menhaden were collected above Enfield Dam, but we have taken young menhaden as far upstream as Wilson (km 74), that is, 50 km above salt water. Ichthyoplankton sampling has shown that alewives and blueback herring together make up 78.8% of all the planktonic eggs and 87% of the larvae collected between Enfield Dam and Essex (see Marcy 2004 [1976]). Enfield Dam is a deterrent for the upstream spawning of both species of river herring, and the total number of their eggs collected per plankton net tow above the dam was only 0.08 between 1965 and 1969, as compared with 871.8 at Windsor Locks just below the dam. The total number of eggs of the 11 species that were collected above Enfield Dam was only 3.6 per tow, while 929.3 per tow were taken at Windsor Locks. This represents a percentage difference of 99.6. Less competition for food and

TABLE 6. Mean length (mm), range in length, and number of juvenile shad measured each month at each station, lower Connecticut River, 1966–1972 combined.

Station	River	June			July			August		
no.	kilometer	*N*	Mean	Range	*N*	Mean	Range	*N*	Mean	Range
12	85.3				1,279	57.8	25–82	1,859	73.7	32–110
11	74.0				170	51.9	28–82	319	63.4	45–92
10	61.0				156	48.1	26–65	79	60.9	40–85
9	51.5				63	45.8	27–70	118	70.7	27–97
8	29.8				81	48.5	22–90	126	72.4	39–111
7	27.4	10	20.7	17–23	89	48.3	29–92	97	67.3	47–100
6	24.1				72	48.4	21–83	82	68.4	57–100
5	24.1				186	73.5	31–90	438	67.5	43–99
3	22.5				5	72.0	60–80	202	73.9	58–105
1	11.3				16	51.6	41–65	20	74.3	45–105
Totals		10	20.7	17–23	2,117	56.8	21–92	3,340	68.4	27–110

Station	River	September			October			November		
no.	kilometer	*N*	Mean	Range	*N*	Mean	Range	*N*	Mean	Range
12	85.3	619	92.4	62–125	132	97.4	70–140	–	–	–
11	74.0	449	74.6	56–123	38	83.9	70–104	5	79.4	70–99
10	61.0	70	79.0	60–120	23	86.5	70–93	37	81.7	70–90
9	51.5	90	80.7	67–103	76	88.3	66–121	–	–	–
8	29.8	42	83.3	51–106	89	97.5	62–123	2	85.5	81–90
7	27.4	22	85.7	77–100	53	99.2	74–120	21	87.6	76–100
6	24.1	58	81.7	49–107	73	98.9	78–114	3	80.7	79–84
5	24.1	438	82.8	58–128	132	102.7	67–140	114	92.2	74–125
3	22.5	275	83.9	63–108	218	95.7	72–134	8	109.1	98–135
1	11.3	74	99.0	76–125	28	86.7	74–120	16	86.7	74–115
Totals		2,137	84.9	49–108	862	96.2	62–140	206	93.05	70–135

space as a result of fewer competitors thus appears to be one of the most important factors influencing the more rapid growth of juvenile shad above Enfield Dam.

Juvenile shad above Holyoke Dam (km 119) attain greater lengths than those below the dam (Watson 1968, 1970). This finding was also supported by a study conducted by the U.S. Fish and Wildlife Service in 1965 (Bureau of Commercial Fisheries Biological Laboratory, Beaufort, North Carolina 1965). In this study, the mean total length of juveniles collected in October above Holyoke Dam (129.8 mm) was significantly greater than below Windsor Locks (123.1 mm).

The growth rates calculated by Watson (1968, 1970) for shad collected between the Holyoke and Turners Falls dams were 7.2, 8.3, and 5.4 mm per week for 1967, 1968, and 1969, respectively. These rates were considerably higher than the growth rate of 3.6 mm per week calculated for the area between Enfield and Essex in the present study. Mitchell et al. (1925), who collected juvenile shad in the Salmon and Farmington rivers (both tributaries of the lower Connecticut River) from June through August, obtained an estimated rate of growth of 3.5 mm per week, which is almost identical with the one obtained in the present study.

The growth of young shad appears to be slower in more southerly rivers along the East Coast. Watson (1968) adapted data provided by Hildebrand and Schroeder (1928) on the length of juvenile shad taken in Chesapeake Bay between 1910 and 1922 and fitted a regression line to these data. The weekly growth rate was 2.0 mm between June and November. Based on a linear regression approximation, Davis et al. (1970) reported weekly growth rates for young shad of 1.9 and 2.9 mm in two creeks off the James River in lower Chesapeake Bay from June to September. The weekly growth rate between May and February in the St. Johns River, Florida was calculated as only 1.7 mm (1969–1970) and 2.4 mm (1970–1971) by Williams and Bruger (1972).

The instantaneous (logarithmic) and per diem (linear) growth in length for each sampling year (1966–1972) are shown in Table 7. Based on linear approximation, the growth rates ranged between 0.27 mm per day in 1971 and 0.83 per day in 1968. The calculated daily growth rates given by Watson (1968, 1970) fluctuated in similar fashion as did those calculated for the same years in the present study. Watson's daily growth rates for the area between Holyoke and Turners Falls dams were 1.02 mm in 1967, 1.18 in 1968, and 0.77 in 1969. Comparing these rates with those calculated for the same three years in the lower river (Table 7), the fastest growth rate in both studies was in 1968, the next fastest in 1967, and the slowest in 1969. The mean monthly lengths of juvenile shad for those years show the same trends (Table 8).

The fastest growth rate (0.83 mm per day) of all the years was in 1968 when juvenile abundance was at one of its lowest levels, while the slowest rate (0.27 mm per day) occurred in 1971 when juveniles reached their greatest abundance (Tables 7 and 9). The faster growth in 1968 may have been influenced by less competition for food and space as a result of reduced numbers of juveniles, perhaps brought about by increased egg mortality and/or loss of eggs downstream following high June flows (see page 159).

Watson (1970) found that the largest juvenile shad were collected above Holyoke Dam in 1968 as also found in the present study. He attributed the lower population and faster growth rate in 1968 to fewer spawning adults entering his study area over Holyoke Dam and higher egg mortality due to widely fluctuating water temperatures during June. He stated that many eggs were probably lost from the study area due to high water. Burgner (1962), studying red salmon smolts, noted that the largest escapement produced the smallest smolts, and Foerster (1954) and Burgner (1962) have shown that large salmon smolts show a higher oceanic survival rate. Shad may behave in a similar manner and Watson (1970) has suggested that the juvenile shad produced in the upper section of the Connecticut River, where growth rates are faster, may suffer less mortality than those produced farther downriver. This, in turn, suggests that when the juvenile production is high in any particular year and a reduction in the growth rate occurs, that a higher oceanic mortality might then follow, thus tending to offset the higher freshwater production. Such feed-back relationships between growth,

TABLE 7. Equations of instantaneous (logarithmic) and per-diem (linear) growth[a] of young shad for each sampling year between 1966 and 1972.

Year	Number[b] measured	Instantaneous growth (logarithmic)	Per-diem growth (linear)	Length increase per day (mm)
1966	231	$\log Y = 0.1408 + 0.8541 \log X$ (0.92)[c]	$Y = 8.885 + 0.629 X$ (0.90)[c]	0.63[d]
1967	1,230	$\log Y = 0.0863 + 0.9170 \log X$ (0.88)	$Y = 13.075 + 0.705 X$ (0.89)	0.70
1968	218	$\log Y = -0.3512 + 1.1388 \log X$ (0.69)	$Y = 1.002 + 0.833 X$ (0.84)	0.83
1969	622	$\log Y = 0.4560 + 0.7227 \log X$ (0.86)	$Y = 16.557 + 0.631 X$ (0.83)	0.63
1970	1,549	$\log Y = 1.7472 + 0.5906 \log X$ (0.81)	$Y = 41.279 + 0.436 X$ (0.81)	0.44
1971	4,477	$\log Y = 1.1050 + 0.3785 \log X$ (0.85)	$Y = 45.459 + 0.267 X$ (0.82)	0.27
1972	361	$\log Y = 0.6771 + 0.6164 \log X$ (0.83)	$Y = 32.915 + 0.489 X$ (0.82)	0.49

[a] Logarithmic equations are given to express instantaneous growth since they more accurately describe its curvilinear nature. Nevertheless, the corresponding linear equations are utilized in discussion about growth in the text for the purpose of comparing estimated rates of growth per interval of time (per diem) to reveal differences between stations and across years, and to compare these rates with growth rates calculated from linear regressions given in the literature.
[b] Stations 1–12 were used (except station 2), since they were sampled in all years.
[c] Y = total length in mm and X = time in days; correlation coefficient is in parentheses (r).
[d] Slope of linear equation.

TABLE 8. Mean length (mm), range in length, and number (*N*) of juvenile shad collected each month and year at all stations between km 85.3 and 11.3, lower Connecticut River, 1966–1973.

Month	1966			1967			1968			1969		
	N	Mean	Range	*N*	Mean	Range	*N*	Mean	Range	*N*	Mean	Range
June	9	20.5	17–22									
July	127	42.7	22–83	109	40.9	27–69	88	38.5	21–92	146	37.0	27–73
August	60	57.2	35–82	599	69.6	27–111	95	73.5	40–100	73	66.9	40–90
September	34	77.4	58–98	365	95.4	62–121	30	97.7	70–123	403	83.5	60–108
October	1	82.0	82	157	106.5	86–140						
November							5	123.6	107–160			
Total	231			1,230			281			622		

Month	1970			1971			1972			1973		
	N	Mean	Range	*N*	Mean	Range	*N*	Mean	Range	*N*	Mean	Range
June	1	23.0	23									
July	379	68.9	40–90	1,274	59.1	25–83						
August	629	61.4	46–110	1,809	70.3	43–95	78	71.7	52–99	70	72.2	56–99
September	248	90.7	68–114	888	79.7	57–123	170	84.1	51–125	150	83.1	62–113
October	253	102.6	76–134	347	82.8	62–125	105	108.5	68–140	54	89.6	78–153
November	39	110.8	98–135	159	84.3	70–115	8	102.5	86–116	16	94.6	79–116
Total	1,549			4,477			361			290		

year-class strength, and mortality in anadromous fishes should be studied further.

Abundance, Distribution, and Downriver Movements

The catch of juvenile shad per net haul for each station for all sampling years combined is shown in Table 9. The largest numbers of juveniles were collected at station 12 (Enfield Dam), followed by station 11 (Wilson) and station 6 (East Haddam). The Salmon River supports a relatively small population of juveniles compared with the lower Connecticut River. This perhaps results from the limited spawning in this area (see page 157). The earliest that juveniles were collected was on

TABLE 9. Number of juvenile shad caught per bag-seine and trawl haul at 11 stations sampled in the lower Connecticut River, 1966–1972.

River kilometer	Station no.	Sampling years: 1966	1967	1968	1969	1970	1971	1972	Mean number per haul
85.3	12	10.14	60.09	1.00	29.00	69.44	210.70	8.44	69.60
74.0	11	1.57	3.81	11.25	18.83	19.33	52.10	8.66	17.54
61.0	10	1.00	6.00	1.00	8.57	1.77	21.00	0.22	6.50
51.5	9	0.28	1.36	8.12	15.00	4.88	9.10	1.00	5.43
29.8	8	0.52	2.82	1.50	0.20	3.81	1.43	0.86	1.65
27.4	7	10.37	0.50	5.66	3.50	5.55	5.20	0.88	4.32
24.9	6	0.16	0.88	0.43	6.56	7.73	22.67	2.68	7.45
24.9	5	1.22	0.82	1.53	0.76	1.57	1.42	0.68	1.17
24.9	4	0.07	1.33	0.75	3.00	3.00	0.90	2.16	1.25
22.5	3	0.66	1.00	1.50	0.75	8.21	17.91	3.04	6.78
11.3	1	1.52	0.38	0.23	0.69	0.55	4.43	0.71	1.38
Mean number per haul		1.84	6.34	2.19	5.63	8.33	21.98	2.18	

29 June 1966, and the latest was on 20 November 1970. Juveniles were fairly abundant by July (8.8 per net haul) and reached their peak in August (13.6 per net haul). The catch per net haul declined only slightly to 12.0 by September but was followed by a rapid decline to 4.45 in October and 1.7 by November. Moreover, no shad were ever collected either at Enfield Dam or in the Salmon River in November (Table 10). The fish must have moved from these areas by the end of October. Most of the juveniles probably leave the river between late September and early October.

The distribution of juvenile shad were variable from one year to the next, but the data from all years can be combined to establish general trends. Figure 8 provides a graphic representation of their seasonal abundance and distribution based on samples taken every other week between June and November 1966–1971. Juveniles became available upstream earlier (June) than downstream (July). This was perhaps the result of earlier spawning upstream. The earliest peak abundance was reached in late July in the middle course of the study area (stations 10, 8 and 7). Some of the young shad may move into the Haddam area at this time causing an earlier peak there. Their numbers peaked at Enfield Dam (station 12) in early August, but not until early September at Wilson (station 11) just below the dam. Peaks occurred in early September at Glastonbury (station 9), East Haddam (station 6), and at Essex (station 1). It appears that the larger upstream juveniles first begin their emigration downstream around the middle of August, as there was a general reduction in the population after this time. It is possible, however, that normal mortality or lowered gear efficiency due to larger and hence faster fish as well as emigration from the river contributed to this decline.

Peaks of abundance occurred progressively later toward the river mouth. Some juveniles probably began to leave the river in the middle of September, as there was a peak in early September at station 1 (Essex). Peak abundance occurred near the end of September in the Salmon River with a decline in early October, all juveniles having disappeared by the end of October. As has already been pointed out, the rapid decrease in abundance at most stations in late September and early October indicates that the bulk of the population leaves the river at this time. The secondary peaks that occurred at down river stations in late October and early November probably indicate the time of final movement out of the Connecticut river. Few juveniles were collected in the lower river after early November despite extensive gill netting in 1966 and subsurface trawling at night in 1968.

In the Connecticut River, young shad apparently begin their downstream movement as early as the middle of August. This downstream movement probably starts as a relatively slow downstream transport of the population as a whole, that

TABLE 10. Monthly summary of catch per net haul of juvenile shad at each station, 1966–1972.

		Sampling months											
River	Station	June		July		August		September		October		November	
kilometer	no.	CPUE[a]	*N*[b]	CPUE	*N*	CPUE	*N*	CPUE	*N*	CPUE	*N*	CPUE	*N*
85.3	12	0.00	3	91.35	14	180.00	10	43.90	11	10.58	12	0.00	3
74.0	11	0.00	3	12.21	14	26.58	12	37.41	12	3.45	11	1.25	4
61.0	10	0.00	3	11.14	14	6.58	12	5.83	12	2.00	11	9.25	4
51.5	9	0.00	3	3.87	16	8.06	15	7.33	12	5.45	11	0.00	4
29.8	8	0.00	14	2.54	33	2.00	30	1.63	19	2.04	21	0.17	17
27.4	7	5.00	2	6.23	13	5.92	14	2.00	10	1.40	10	5.25	4
24.9	6	0.00	16	3.63	38	11.51	37	15.96	27	5.96	31	4.64	25
24.9	5	0.00	16	1.68	38	0.53	39	2.60	25	1.38	26	0.43	23
24.9	4	0.00	10	0.64	14	2.21	14	2.42	7	1.33	6	0.00	6
22.5	3	0.00	13	0.22	22	8.45	22	17.81	16	12.15	19	0.50	16
11.3	1	0.00	5	0.64	31	1.14	28	4.52	17	1.68	22	0.62	16
Totals		0.11	88	8.37	247	13.55	233	12.00	168	4.45	180	1.72	122

[a] CPUE = catch per unit effort, that is, number of shad caught per net haul.
[b] *N* = number of seine and trawl hauls.

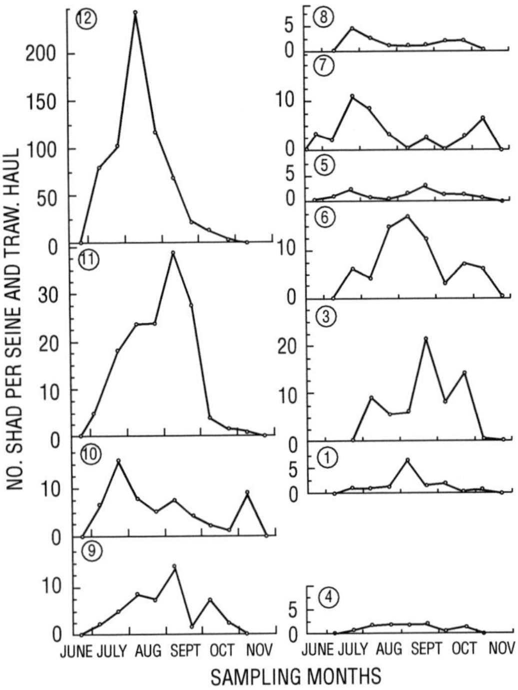

FIGURE 8. Relative seasonal abundance and distribution of juvenile shad. Data points represent the mean number per bag seine and trawl haul, collected biweekly at circled Stations 12 through 1 between June and November 1966–1972.

accelerates with time until peak emigration velocities are reached in September with the bulk of the population leaving the river in late September and early October.

Emigration Rate and Population Size Studies

In the fall of 1965, an attempt was made to determine the rate of downstream emigration by fin-clipping juvenile shad in the Windsor Locks Canal (immediately below and adjacent to the Enfield Dam) and releasing them just below the canal for subsequent recapture. As reported by Moss (1966), the relatively small number of fish marked (18,552), combined with poor survival from seagull predation after release, yielded too small a sample of the total population for a reasonable chance of recapture. Experiments with vital stains in 1966 did not produce a satisfactory mark on the small fish and in some instances had deleterious side effects.

A "mark–recapture" program was carried out from 13 August to 21 October 1968 just above Enfield Dam to determine emigration rates and to make a population estimate of the 1968 year class. Young shad were collected in bag seines, transported to Enfield Dam, marked and released over the dam. The marking procedure involved the embedding of fine fluorescent pigment granules into the scales and mucous layer by means of a pressurized spray gun (7 kg/cm^2). After a short time, the harmless mark was detectable only under ultraviolet light. Fish were never available in sufficient numbers, however, for such an emigration or population estimate. Although two crews seined 18 h each day for 40 d, only 7,000 young shad were collected and marked.

Effects of Temperature on Movement of Juveniles

The largest number of juvenile shad occurred in the Haddam area in early August when the water temperature was 26°C (Figure 9). Juveniles began moving downstream and leaving the river by the middle of August between temperatures of 26°C and 23°C. The bulk of the population emigrated downstream and out of the river from early September through early October at temperatures ranging between 23°C and 17.8°C. A smaller portion

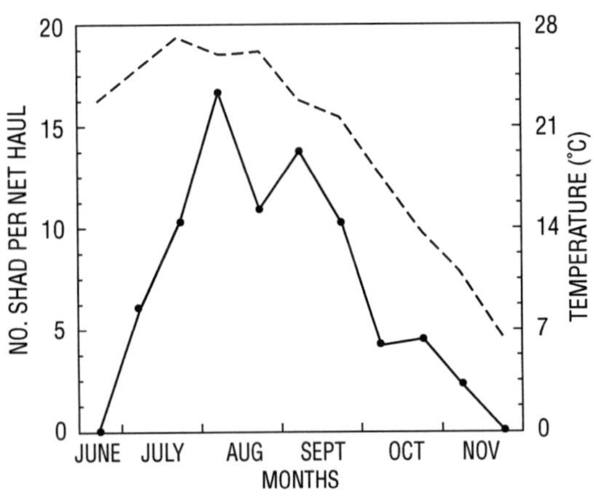

FIGURE 9. Relative abundance of juvenile shad in the Haddam area in relation to water temperature. Solid line represents the mean number of shad collected per net haul and dashed line the mean water temperature at the time of biweekly samples from June through November 1966–1972.

of the population, probably small fish or those from upriver areas, left the river between the middle of October and early November between 17.8°C and 10.9°C. No juveniles were caught in the river by the end of November at temperatures below 6.6°C. The final emigration from the river is associated with the high river flows and low temperatures of late fall.

Watson (1968) indicated that all juvenile shad had left the area above Holyoke Dam when temperatures fell below 10°C. He showed that the larger juveniles started to move downstream in October when temperatures dropped below 18.3°C. In the present study, juveniles reached their largest size in the lower river in October when temperatures were between 17.8°C and 13.8°C. The mean length of shad decreased in November, indicating the movement of larger fish out of the river earlier, that is, in October. The present study supports the suggestion of Watson (1970) that the larger juvenile shad move downstream prior to the time when temperature becomes a critical factor. Sykes and Lehman (1957) noted that juvenile shad move only when temperatures fall below 21°C and that the largest number migrate in temperatures of less than 15.6°C. Chittenden and Westman (1967) reported that juvenile shad movement out of the Delaware River was almost complete when temperatures fell to 8.3°C. Chittenden (1972) determined the lower limits of temperature at which juvenile shad can survive at between 4°C and 6°C. The vast majority of juveniles appear to have left the Connecticut River before a temperature of 6°C was reached. Only one shad was caught in the lower river at a temperature of 5°C—on 11 November 1971.

Hoar (1953) suggested that the downstream movement of anadromous fishes may be initiated by changes in the endocrine system that are in turn induced by environmental factors. As Chittenden (1969) has pointed out and as has been confirmed in the present study, temperature decreases to not far above the lower lethal limit trigger the final movement of juvenile shad from fresh water, while factors associated with size appear to initiate the earlier stages of seaward emigration.

Predictions of Annual Year-Class Strength

In all, 9,190 juvenile shad were collected in 1,485 hauls of the bag seine and trawl between Essex (km 11.3) and Northampton, Massachusetts (km 138) from 1966 through 1973. The average catch of juvenile shad in the lower Connecticut River, using both types of gear, was 6.2 per net haul. The bag seine was more efficient in catching than the trawl. In 883 seine hauls, 7,152 fish were caught, an average catch of 8.1 per haul. In contrast, 602 trawl hauls caught 2,038 juvenile shad, an average of 3.4 per haul. However, trawl collections generally contained fish of larger sizes.

An estimate of the yearly abundance of juvenile shad for each station, expressed as number caught per haul, is shown in Table 9. Since the catch per net haul in 1966 probably represents a low estimate because a smaller seine was used, only the estimates from 1967 through 1973 should be used for comparisons. The mean catch per net haul decreased from 6.34 in 1967 to 2.19 in 1968, and then increased to 5.63 in 1969 and 8.34 in 1970. In 1971, the catch per net haul reached its highest peak (21.98) and then decreased in 1972 to a low similar to that in 1968 (2.18). Because few trawl hauls were taken in 1966 and because the bag seine was found to be more efficient, Table 11 probably provides a more realistic comparison of the yearly abundance of juvenile shad. Based on this table, the lowest number occurred in 1973 and the next lowest in 1972, followed by 1968.

The strength of each annual year class of juvenile shad is dependent upon many factors. Among the factors that were measured were the numbers of adults potentially available to spawn each spring (based on population estimates and escapement calculations), the temperature of the water, and the amount of discharge flow. Table 12 lists the mean catch per bag-seine haul of juvenile shad, mean water temperature for June, mean estimated discharge of the Connecticut River into Long Island Sound for June, and the annual population estimate of adult shad potentially available to spawn for each year. A multiple linear regression model based on these data revealed a highly significant relationship between year-class strength (catch per seine haul of juveniles) and the combination of potential spawning population, water temperature, and the discharge flow: $Y = -233.11 + 0.00021$ (estimated annual adult shad potential spawning population) + 7.4 (river temperature—°C) + 0.057 (river discharge flow—m^3/s); ($r = 0.93$; $R^2 = 0.86$; F (4,3) = 7.90; $P < 0.05$). This equation can explain 86% of the vari-

TABLE 11. Number of juvenile shad caught per bag-seine haul at 10 stations sampled in the lower Connecticut River, 1966–1973.

River kilometer	Station no.	1966	1967	1968	1969	1970	1971	1972	1973
		Sampling year							
85.3	12	10.4	60.09	1.00	29.00	69.44	210.70	8.44	4.10
74.0	11	1.57	3.81	11.25	18.83	19.33	52.10	8.66	2.00
61.0	10	1.00	6.00	1.00	8.57	1.77	21.00	0.22	1.78
51.5	9	0.28	1.36	8.12	15.00	4.88	9.10	1.00	2.07
29.8	8	0.77	6.00	3.00	0.28	7.27	2.36	8.72	1.14
27.4	7	10.37	0.50	5.66	3.50	5.55	5.20	0.88	3.63
24.9	6	0.22	0.87	0.50	0.50	1.45	6.50	1.63	0.94
24.9	5	2.22	1.25	3.00	1.62	4.90	2.45	1.27	3.83
24.9	4	0.07	1.33	0.75	3.00	3.00	0.90	2.16	0.75
11.3	1	3.22	0.66	0.60	1.80	0.90	6.45	1.50	0.60
Mean number per haul		2.71	9.08	3.60	7.90	11.62	30.86	2.55	2.12

ability in year-class strength and therefore can be used to predict the production of juvenile shad in a particular year with a high degree of success.

There was a significant relationship between the number of adults potentially available to spawn (escapement) and the production of juveniles: $Y = -32.65 + 0.00015\ X$ ($r = 0.80$; $R^2 = 0.64$; $P < 0.05$), in which Y equals the number of juveniles per seine haul (year-class strength) and X equals the estimate of the number of adults potentially available to spawn that year. Thus, the number of potential spawning adults, and presumably, egg production, significantly influences the production of juveniles. There was also a significant relationship between river temperature and the production of juveniles: $Y = -83.44 + 4.43\ X$ ($r = 0.73$; $R^2 = 0.54$; $P < 0.05$). Water temperature significantly influences juvenile production because low water temperatures cause prolonged development and reduced survival. The relationship between river discharge flow and juvenile production, however, was nonsignificant: $Y = 23.1 - 0.028\ X$ ($r = -0.68$; $R^2 = 0.46$; $P = 0.06$). This equation indicates only a 46% confidence in pre-

TABLE 12. Abundance of juvenile shad, mean water temperature and discharge flows in June, and size of adult shad population potentially available to spawn (escapement).

	1966	1967	1968	1969	1970	1971	1972	1973	Mean	SE
Juvenile shad catch per bag seine haul[a]	8.7[b]	9.1	3.6	7.9	11.6	30.9	2.6	2.1	9.56	3.28
Mean June water temperature (°C)[c]	22.4	20.4	20.1	22.3	21.6	23.0	18.8	19.4[d]	21.0	0.54
Mean June discharge flow volume (m^3/s)[e]	275	518	711	408	306	244	892	578	492	80.87
Adult shad population available to spawn (escapement)[f]	287,000	292,000	222,000	264,000	348,000	353,000	224,000	263,000	281,625	17,510

[a] Data from present study.
[b] Adjusted by multiplying the catch per seine haul in 1966 (2.71) by the increase (3.2) in area covered by the larger seine used from 1967 to 1973.
[c] Data from W. A. Boyd (personal communication): bottom temperature at hydrology station 3.
[d] Data from CYAP intake temperature records.
[e] Data from U.S. Geological Survey, Hartford, Connecticut.
[f] Data from Leggett (2004 [1976], this volume). Number of fish that escaped the commercial fishery, that is, the estimated number of adult shad entering the river minus the number caught in the commercial fishery.

dicting juvenile production based on river flow alone.

Standard partial regression coefficients have been used as measures of the relative importance of each parameter in a predictive regression model (Snedecor and Cochran 1972). Based on these calculated coefficients, flow (1.40) was the most important contributor or predictor in the regression model, followed by temperature (1.22) and last by potential adult spawners (1.13). Some interesting trends and relationships among the three variables over years are shown in Table 12. For example, in 1972, the catch per net haul of juvenile shad was only slightly higher than the lowest ever recorded, which was in 1973. The mean discharge flow in June of 1972 was the highest of all the years, and the mean water temperature for June was the lowest. Only in 1972 and 1973 were juveniles not available until August; in all the other years, they became available to the sampling gear in June and July. This indicates that spawning occurred later in 1972 and 1973, as compared with the other years. High flows and low temperatures during the spawning season in late May and June undoubtedly prolong the development of the eggs and reduce their survival. Moreover, because of the delay in hatching and high flow rates, many eggs probably were swept away from the spawning grounds to the lower part of the river where increased salinity may have caused mortalities. Leggett (personal communication) observed an unusual number of ripe female shad in the lower river in the spring of 1972 as compared with other years. The 1972 estimate of the size of the adult spawning population was also next to the lowest (Table 12). For both of these reasons, the number of females on the spawning grounds, and therefore egg production, must have been reduced in 1972, and this would have contributed to the paucity of juveniles. The next lowest year in terms of juvenile abundance was 1968. The flow rates in June of 1968 were the second largest, and the mean June temperature was the third lowest of all the years. In addition, the estimated adult spawning population was the lowest recorded.

Juvenile shad were most abundant at station 12 above Enfield Dam in all years except 1968 and 1972. The 1968 and 1972 peak abundance occurred at station 11, downstream at Wilson (Table 9). High June flows and low temperatures probably caused more spawning to occur farther downriver with the consequence that more eggs and larvae were carried downstream in those years. The year of greatest juvenile shad abundance was 1971 (Table 12). In that year, the June mean flow rate was the lowest of all years, and the mean water temperature was the highest. The potential adult spawning population was the highest ever estimated.

In summary, during years of lower discharge flow rates, higher water temperatures, and larger adult shad spawning populations, the production of juvenile shad was higher. The combination of river discharge flow, temperature, and adult shad spawning population estimates in a multiple linear regression model has provided a tool to predict the relative year-class strength of juvenile shad with a high (86%) percentage of success, the predictive ability of which exceeds most expectations concerning natural populations.

Effects of the Connecticut Yankee Plant on Juvenile Shad

In studying the occurrence, distribution, and abundance of juvenile shad in the lower Connecticut River, no evidence was obtained to indicate that the Connecticut Yankee plant exerted any harmful effects on these fish. On the other hand, only circumstantial evidence can be adduced to show that the young shad are unaffected by the heated effluent during their downstream emigration.

The catch per net haul of juvenile shad increased after plant operation began in 1968 at the two stations (3 and 1) below the plant as well as at station 6, which may be covered by the heated effluent on a flood tide (Table 9). The mean catch per net haul for each year was calculated for those stations above the plant not affected by the heated effluent (12, 11, 10, 9, 8, and 7) and for those unaffected stations below the plant (3 and 1). A chi-square analysis was then used to reveal any possible change in catch per net haul before and after plant operation and up and downstream (Table 13). This analysis of the catch both before and after plant operation indicates that the catch of juvenile shad in the lower river has remained relatively unchanged from 1968 through 1972.

There was a possibility that the juvenile shad might be delayed in their seaward movement by preference for the heated water from the discharge canal. Were there any extended delay, ambient river temperatures would eventually fall below

TABLE 13. Comparison of mean catch per net haul of juvenile shad for collection stations above and below the Connecticut Yankee plant for the pre-operation years (1966 and 1967) and post-operation years (1968–1972).[a]

Locality	Before plant operation (1966–1967)	After plant operation (1968–1972)
Stations above plant	11.23	14.66
Stations below plant	2.62	1.32

[a] Chi-square = 0.09, nonsignificant.

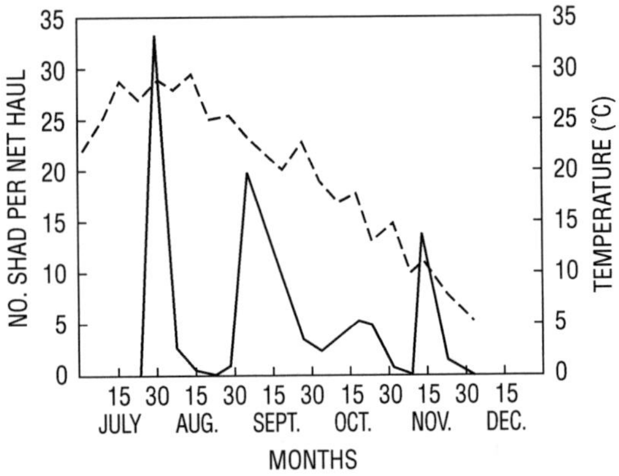

FIGURE 10. Number of young shad collected per net haul near the heated effluent of the Connecticut Yankee Atomic Power Company during 1970. The broken line indicates ambient water temperature.

the lower lethal limit for young shad and cause mortalities. There was no evidence to suggest that juvenile shad overwintered in the heated effluent during the years after plant operation. Only 33 juveniles were collected in the 239 trawl hauls that were taken biweekly in the discharge canal between November and March 1969 through 1972. These 33 juveniles were all taken at the confluence of the discharge canal and the river on 11 and 20 November 1972. The latest date on which a juvenile was collected in the canal was 21 December 1970 when a single live specimen was captured. Yearly observations in the area around the heated effluent during peak downstream emigration have failed to detect any evidence of juvenile shad mortality.

Observations on the Reactions of Young Shad to the Heated Effluent

To obtain definite evidence concerning the effect of the heated effluent on juvenile shad, a series of experiments was conducted in the warm water in 1968, 1970, and 1971. The descriptions and results of these experiments are based on the published work of Marcy et al. (1972).

Young shad begin to emigrate downstream in early August when the water temperatures are highest. The majority of the population is present in the Haddam area when added heat from the plant's effluent could well influence their safe passage (Figure 10). Although the heated effluent does not reach the bank opposite the discharge canal except on slack tide, on such a tide the surface temperature has been recorded at 10ºC above ambient with a maximum rise of 4ºC at a depth of 1.2 m in the channel, which is 4.6 to 6.1 m deep (Boyd 1970; Merriman 1970).

In an experimental tank at our laboratory, Moss (1970) demonstrated that young shad responded minimally to rapid temperature changes of about 1°C, but that they avoided rapid temperature increases of about 4ºC in the experimental system used. Moss concluded that these small fish are behaviorally capable of avoiding temperature changes of a kind that otherwise could prove lethal. We caught young shad in water temperatures ranging between 10ºC and 31ºC, but only one specimen was ever found in water warmer than 30ºC.

Several studies have shown that other species of clupeids respond to small temperature differentials: adult alewife and blueback herring respond to 0.5 differentials (Collins 1952), Pacific herring *Clupea pallasii* to 0.2ºC (Shelford and Power 1915), and dwarf herring *Jenkinsia lamprotaenia* to 0.1ºC (Breder 1951).

Definition of the vertical distribution of juvenile shad in and around the heated effluent was attempted in November 1968 by alternately drifting two 1.3-cm bar-mesh nylon gill nets. One net (30.5 · 4.9 m deep, depth-marked at 1.2-m intervals) was set out at the mouth of the discharge canal and was allowed to drift with the tide up or downstream for 1 h. The second was set at the initial starting point at the time when the first was retrieved. In this fashion, the heated area was continuously fished over a 24-h period. Surface and bottom temperatures were taken before and after each drift. This study indicated that juvenile shad were distributed at the surface and on the bottom when the effluent temperature was 21.5ºC. Of the 71 caught, 87% were collected near the bottom

(3.7–4.9 m) during the day and 100% close to or on the surface (0.24 m) at night.

Observations on the upper lethal temperature and direct mortality were made by drifting young shad and blueback herring (60–95 mm TL) in a floating live-car through the heated effluent on 23 September and 14 October 1970. The tests were made just after the peak abundance of young had occurred in the Haddam area (Figure 10). The live-car was drifted for approximately 1 h during ebbtide to the discharge point at which point the bow window was opened. The live-car was then drifted through the heated effluent. A dual-track-chart temperature recorder registered temperatures at the surface and the bottom of the live-car. Dissolved oxygen was measured before and during each drift. Photographs were taken of the fish at 5-s intervals. Dead fish were counted after a 30-min recovery period in river water of ambient temperature well below the effluent. The drift time between the seining of the fish and entrance into the effluent areas served as a control for each test. In addition, one experiment was performed on 29 October and another on 4 November in water of ambient temperature near the bank opposite the discharge canal. All procedures described for the effluent drifts were duplicated in these river drifts.

Avoidance reactions to effluent of different temperatures were studied within a chamber (2.4 × 4.6 × 2.1 m deep) constructed of 1.3 × 2.5-cm wire mesh, ovate on both ends and with a plywood top and bottom. Submerged vertically in the water column and held in position by a floating platform, the cage was operated at depths between 2.4 and 3.7 m. In September and October 1971, five tests were made in the effluent of the plant and six in ambient water. Shad 70–92 mm TL were caught by bag seine from the shore opposite the discharge canal, and 15 to 77 specimens were immediately transferred into the cage. Underwater observations, extending from 20 min to 2 h, of the thermal stratification and the behavior of the fish were made by two divers using SCUBA gear, metric rules, and hand-held thermometers.

The test drifts of young shad and blueback herring in the live-car through the heated effluent yielded different results (Table 14). Although effluent temperatures were similar, fish tested on 23 September had a lower mortality than on 14 October. The lower mortality was probably related to the higher acclimation temperature and lower ΔT on 23 September. The lower ΔT was a function of a reduced power level. The slight drop in dissolved oxygen concentration encountered by fish entering the effluent was not considered to be a major factor causing the mortalities. There was no evidence of stress or mortality during the 1-h control drifts or in ambient water drifts made near the bank opposite the discharge canal.

A composite of time-sequence photographs of young shad and observations made during the effluent test on 14 October suggested the following reactions as temperatures increased from ambient to maximum effluent temperature: (1) The distribution of the fish in ambient water was maintained throughout the live box with no evidence of schooling (19.0°C); (2) as the live box entered the effluent, the fish immediately formed a tight school in the center as heated water entered from both ends (27.8°C); (3) after 1 min, fish began swimming rapidly in a circular pattern, followed by the break-up of the single school into smaller schools (30.0°C); (4) after 2 min, there was evi-

TABLE 14. Percent mortality of American shad and blueback herring in live-car tests in effluent and ambient waters, 23 September–4 November 1970.

Date	Test water	Ambient river temp. (°C)	Effluent water temp. (°C)	ΔT[a] (°C)	Power level (MW_e)	Dissolved O_2 (ppm) Ambient	Dissolved O_2 (ppm) Effluent	Blueback herring No. tested	Blueback herring Percent mortality	American shad No. tested	American shad Percent mortality
23/9	Effluent	22.7	32.9	10.2	488	7.8	7.3	366	61.7	8	12.5
14/10	Effluent	19.0	32.2	13.2	596	9.0	8.3	3	100.0	68	100.0
29/10	Ambient	11.9			605	10.6		4	0.0	1	0.0
4/11	Ambient	10.8			604	10.0		76	0.0	2	0.0

[a] Difference between ambient and effluent temperature.

dence of disorientation and the smaller schools began to break up (31.2°C); (5) after 3–4 min, there was no schooling, swimming speeds were reduced, and there was continued disorientation with loss of equilibrium (31.8°C); and (6) death of all specimens occurred between 4 and 6 min (32.2°C). These field observations confirmed the laboratory work of Moss (1970) who found that young shad rapidly die when temperatures are suddenly raised from 24–28°C to 32.5°C.

Results and observations from the tests with the submerged cage showed that young shad avoid effluent temperatures above 30°C (Table 15). During the first three effluent tests, shad exhibited vertical movements (±4 cm) that coincided with rapid temperature changes resulting from oscillating levels of heated water. As the temperature rose to approximately 30°C, the fish always moved downward. Shad exhibited no specific depth or temperature preference at ambient (control) temperatures, and at effluent temperatures less than 30°C.

The upper natural temperature limit of young shad may be about 30°C. Young shad avoid potentially lethal temperatures above 30°C and are capable of traversing or avoiding the CYAP heated effluent during their downstream migration.

Acknowledgments

Special appreciation is extended to Dr. James W. Atz for his continuing advice and critical evaluations of the manuscript. Gratitude is expressed to the Ecological Sciences Division, NUS Corp. for its time and support for manuscript preparation, especially to Mr. Henry Hess for statistical advice. Finally, I wish to acknowledge the support and encouragement of my wife Joan.

References

Barker, J. 1965. Observations on some areas of the Delaware River between Belvidere and Scudders Falls,

TABLE 15. Depths and associated water temperatures for each submerged cage test, 9 September–25 October 1971.

Water depth	Dates and Temperature (°C)							
	Effluent tests					Ambient (control) tests[a]		
(cm)	9/9	24/9	29/9	15/10	25/10	7/9	29/9	25/10
0	37.0	30.8	31.5	27.5	26.0	24.0	22.0	16.0
10	37.0	30.8	31.0	27.5	26.0	24.0	22.0	16.0
20	36.0	30.8	30.0	27.0	26.0	24.0	22.0	16.0
30	35.1	30.5	29.0	26.1	25.2	24.0	22.0	16.0
40	35.0	29.6	27.4	25.5	24.2	24.0	22.0	16.0
50	33.0	26.5	27.4	24.5	23.4	24.0	22.0	16.0
60	32.2	26.5	24.6	23.5	22.0	24.0	22.0	16.0
70	31.2	24.2	23.0	21.2	21.0	24.0	22.0	16.0
80	29.0	24.0	22.7	20.0	19.0	24.0	22.0	16.0
90	29.0[b]	23.5	22.5	18.5	17.3	24.0	22.0	16.0
100	29.0	22.6	22.5	17.8	17.0	24.0	22.0	16.0
110	28.2	22.2	22.5	17.4	17.0	24.0	22.0	16.0
120	28.2	21.8	22.5	17.4	17.0	24.0	22.0	16.0
130	28.0	21.8	21.2	17.2	17.0	24.0	22.0	16.0
140	27.2	21.8	21.2	16.9	17.0	24.0	22.0	16.0
150	27.2	21.8	21.2	16.9	17.0	24.0	22.0	16.0
160	27.2	21.8	21.2	16.8	17.0	24.0	22.0	16.0
170	27.0	21.6	21.2	16.8	17.0	24.0	22.0	16.0
180	27.0	21.5	21.2	16.8	17.0	24.0	22.0	16.0
Number of fish	50	19	74	15	15	58	77	20

[a] Two tests conducted on each date.

[b] Numbers in boldface indicate schooling shad observed.

New Jersey in respect to their utilization by American shad, *Alosa sapidissima* (Wilson), for spawning purposes in 1963 and 1964. New Jersey Department of Conservation and Economic Development, Bur. Fish. Lab. Miscellaneous Report 28, Trenton.

Boyd, W. A. 1970. Hydrology report. Pages 2–8 *in* D. Merriman, editor. The Connecticut River investigation. 10th Semiannual Progress Report. Connecticut Water Resources Commission, Hartford.

Bradford, A. D., J. G. Miller and K. Buss. 1964. Progress report summary on phase B-2 to determine by bioassay techniques the inherent tolerance of the shad during its egg and larval stages to specific environmental factors of the Susquehanna River and its tributaries. Pennsylvania Fish Commission, Division of Res. and Fish Management, Benner Springs Fish Research Station (manuscript), Bellefonte.

Breder, C. M., Jr. 1951. Studies on the structure of the fish school. Bulletin of the American Museum of Natural History 98.

Bureau of Commercial Fisheries Biological Laboratory, Beaufort, North Carolina. 1965. Rep. for fiscal year ending June 30, 1966. U.S. Fish and Wildlife Service Circular 264.

Burgner, R. L. 1962. Studies of red salmon smolts from the Wood Lakes, Alaska. Pages 248–314 *in* T. S. Koo, editor. Studies of Alaska red salmon. University of Washington Publications in Fisheries, New Series 1.

Carlson, F. T. 1968. Report on the biological findings. Pages 4–41 *in* Suitability of the Susquehanna River for restoration of shad. U.S. Department of the Interior, Maryland Board of Natural Resources, N.Y. Conservation Department, and Pennsylvania Fish Commission.

Chittenden, M. E., Jr. 1969. Life history and ecology of the American shad, *Alosa sapidissima,* in the Delaware River. Ph.D. thesis. Rutgers University, New Brunswick, New Jersey.

Chittenden, M. E., Jr., and J. R. Westman. 1967. Highlights of the American shad, *Alosa sapidissima,* in the Delaware River. Department of Environmental Science, Rutgers University, New Brunswick, New Jersey.

Chittenden, M. E., Jr., and J. R. Westman. 1972. Responses of young American shad, *Alosa sapidissima* to low temperatures. Transactions of the American Fisheries Society 101(4):680–685.

Collins, G. B. 1952. Factors influencing the orientation of migrating anadromous fishes. U.S. Fish and Wildlife Service Fishery Bulletin 52:373–396.

Costello, D. P., M. E. Davidson, A. Eggers, M. H. Fox, and C. Henley. 1957. Methods for obtaining and handling marine eggs and embryos. Lancaster Press, Inc., Lancaster, Pennsylvania.

Davis, J. R., and R. P. Cheek. 1966. Distribution, food habits, and growth of young clupeids, Cape Fear River System, North Carolina. Southeastern Association of Game and Fish Commissioners, Proceedings of the 20th Annual Conference: 250–260.

Davis, J. R., J. P. Miller, and W. L. Wilson. 1970. Completion report—embodying annual progress report of the anadromous fish project: V. Institute of Marine Science, Project No. AFC-1, 1 October 1969–30 September 1970.

Foerster, R. E. 1954. On the relation of adult sockeye salmon (*Oncoryhnchus nerka*) returns to known smolt seaward migrations. Journal of the Fisheries Research Board of Canada 11:339–350.

Hildebrand, S. F., and W. C. Schroeder. 1928. Fishes of Chesapeake Bay. U.S. Bureau of Fisheries Bulletin 43 (Part I).

Hoar, W. S. 1953. Control and timing of fish migration. Cambridge University Biological Rev. 28:437–452.

Leach, G. C. 1925. Artificial propagation of shad. Pages 459–486 *in* U.S. Bureau of Fisheries Document 981, Appendix VIII, Report U.S. Commission Fish.

Leggett, W. C. 1969. Studies on the reproductive biology of the American shad (*Alosa sapidissima,* Wilson). A comparison of populations from four rivers of the Atlantic Seaboard. Ph.D. thesis. McGill University, Montreal.

Leggett, W. C. 2004. The American shad, with special reference to its migration and population dynamics in the Connecticut River. Pages 181–238 *in* P. M. Jacobson, D. A. Dixon, W. C. Leggett, B. C. Marcy, Jr., and R. R. Massengill, editors. The Connecticut River Ecological Study (1965–1973) revisited: ecology of the lower Connecticut River 1973–2003. American Fisheries Society, Monograph 9, Bethesda, Maryland. (Originally published in 1976)

Leim, A. H. 1924. The life history of the shad, *Alosa sapidissima* (Wilson), with special reference to factors limiting its abundance. Biological Board of Canada, Contrib. Can. Biol. 2:16–284.

Levesque, R. C., and R. J. Reed. 1972. Food availability and consumption by young Connecticut River shad, *Alosa sapidissima*. Journal of the Fisheries Research Board of Canada 29:1495–1499.

Mansueti, R. J., and H. Kolb. 1953. A historical review of the shad fisheries of North America. Chesapeake Biological Laboratory Publication 97, Solomons, Maryland.

Marcy, B. C., Jr. 1972. Spawning of the American shad, *Alosa sapidissima* in the lower Connecticut River. Chesapeake Science 13:116–119.

Marcy, B. C., Jr. 2004. Planktonic fish eggs and larvae

of the lower Connecticut River and the effects of the Connecticut Yankee plant, including entrainment. Pages 125–153 *in* P. M. Jacobson, D. A. Dixon, W. C. Leggett, B. C. Marcy, Jr., and R. R. Massengill, editors. The Connecticut River Ecological Study (1965–1973) revisited: ecology of the lower Connecticut River 1973–2003. American Fisheries Society, Monograph 9, Bethesda, Maryland. (Originally published in 1976)

Marcy, B. C., Jr., P. M. Jacobson, and R. L. Nankee. 1972. Observations on the reactions of young American shad to a heated effluent. Transactions of the American Fisheries Society 101(4):740–743.

Massmann, W. H. 1952. Characteristics of spawning areas of shad, *Alosa sapidissima* (Wilson) in some Virginia streams. Transactions of the American Fisheries Society 81:78–93.

Massmann, W. H. 1963. Summer food of juvenile American shad in Virginia waters. Chesapeake Science 4:167–171.

Maxfield, G. H. 1953. The food habits of hatchery-produced pond-cultured shad (*Alosa sapidissima*) reared to a length of two inches. Chesapeake Biological Laboratory Publication 98, Solomons, Maryland.

Medcof, J. C. 1957. Nuptial or pre-nuptial behavior of the shad, *Alosa sapidissima* (Wilson). Copeia 3:252–253.

Merriman, D. 1970. The calefaction of a river. Scientific American 222:45–52.

Mitchill, P. H., and Staff. 1925. A report of investigations concerning shad in the rivers of Connecticut. (Part I). Connecticut State Board of Fish and Game, Hartford.

Moss, S. A. 1966. Juvenile shad report. Pages 4–6 *in* D. Merriman, editor. The Connecticut River investigation. Connecticut Water Resources Commission, 3rd semi-annual progress report, Hartford.

Moss, S. A. 1970. The response of young American shad to rapid temperature changes. Transactions of the American Fisheries Society 99(2):381–384.

Shelford, V. E., and E. B. Power. 1915. An experimental study of the movements of herring and other marine fish. Biological Bulletin 28:315–334.

Snedecor, G. W., and W. G. Cochran. 1972. Statistical methods. The Iowa State University Press, Ames.

Sykes, J. E., and B. A. Lehman. 1957. Past and present Delaware River shad fishery and considerations for the future. U.S. Fish and Wildlife Service Research Report 46.

Trent, W. L. 1967. Attachment of hydrofoils to otter boards for taking samples of juvenile fish and shrimp. Chesapeake Science 8:130–131.

Walburg, C. H. 1956. Observations on the food and growth of juvenile American shad, *Alosa sapidissima.* Transactions of the American Fisheries Society 86:302–306.

Walburg, C. H. 1960. Abundance and life history of shad, St. Johns River, Florida, U.S. Fish and Wildlife Service, Fishery Bulletin 60:487–501.

Walburg, C. H., and P. R. Nichols. 1967. Biology and management of the American shad and status of the fisheries, Atlantic Coast of the United States, 1960. U.S. Fish and Wildlife Service, Special Scientific Report Fisheries 550.

Wang, Johnson C. S. 1970. Characters of, and keys to, young clupeids collected in the Delaware River in the vicinity of Artificial Island. Pages 154–172 *in* V. J. Schuler, project leader. An ecological study of the Delaware River in the vicinity of Artificial Island. Ichthyological Associates, Delaware Project Report 2 (Part 3), Ithaca, New York.

Watson, J. F. 1968. The early life history of the American shad, *Alosa sapidissima* (Wilson), in the Connecticut River above Holyoke, Massachusetts. M.S. thesis. University of Massachusetts.

Watson, J. F. 1970. Distribution and population dynamics of American shad, *Alosa sapidissima* (Wilson), in the Connecticut River above Holyoke Dam, Massachusetts. Ph.D. thesis. University of Massachusetts.

Whitworth, and D. H. Bennett. 1970. A limnological study of the lower Farmington River with special reference to the ability of the river to support American shad, University of Connecticut, Inst. Water Res, Report 9, Storrs.

Williams, R. O., and G. E. Bruger. 1972. Investigations on American shad in the St. Johns River. Florida Department of Natural Resources, Marine Research Laboratory, Technical Series No. 66, St. Petersburg, Florida.

American Fisheries Society Monograph 9:181–238, 2004

The American Shad, with Special Reference to Its Migration and Population Dynamics in the Connecticut River

WILLIAM C. LEGGETT
Essex Marine Laboratory Inc., Essex, Connecticut 05426, USA
and
Department of Biology, McGill University, Post Office Box 6070, Station A, Montreal, Quebec H3C 3G1, Canada

The American shad *Alosa sapidissima*, an anadromous clupeid, is native to the eastern coast of North America. Its range, encompassing more than 4,800 km of coastline and 21° of latitude, extends from the St. Johns River, Florida to the St. Lawrence River, Canada (Leim 1924; Welander 1940; Bigelow and Schroeder 1953; Walburg and Nichols 1967). Shad are also occasionally reported along the coast of Newfoundland (Hodder 1966; Scott and Crossman 1973), but no spawning populations are known to occur north of the St. Lawrence River. Shad were introduced to the Pacific Coast in 1871 and have since become abundant there. They now occur from Baja California, to Kodiak Island, Alaska (Welander 1940; McHugh and Fitch 1951; Rodel 1953; Neave 1954; Claussen 1959). Historically, shad are believed to have been most abundant in the central portion of their Atlantic Coast range (Leim 1924).

The American shad is most closely related to the allis shad *A. alosa*, which ascends rivers discharging into the Atlantic Coast of Europe from Norway to Portugal (Regan 1916; Borodin 1927). In more southern waters, *A. alosa* is gradually replaced by Twaite shad *A. fianta*, a closely related species (Regan 1916). Of the five major North American species of Atlantic clupeid fishes, *A. sapidissima* is most closely related to blueback herring *A. aestivalis*, followed by alewife *A. pseudoharengus*, Atlantic menhaden *Brevoortia tyrannus*, and Atlantic herring *Clupea harengus* (Mairs and Sinderman 1962). The shad of the Atlantic Coast were originally divided into three species on the basis of number of gill rakers on the lower anterior gill arch. These were named *A. sapidissima*, Alabama shad *A. alabamae*, and Ohio shad *A. ohiensis* (Regan 1917). Their validity has been questioned frequently (D'Ancona 1927; Vladykov and Wallace 1937; Warfel and Olsen 1947), and at present, only two species are recognized, *A. sapidissima* and *A. alabamae* (Hildebrand 1963). A number of local races of *A. sapidissima* are known to occur on the Atlantic Coast; these are distinguishable on the basis of morphological (Fischler 1959; Hill 1959; Carscadden and Leggett 1975a) and reproductive (Leggett 1969; Carscadden and Leggett 1975b) characteristics.

Shad undertake extensive ocean migrations. Fish from all Atlantic Coast populations move together at sea (Vladykov 1936, 1956; Hollis 1948; Talbot and Sykes 1958; Leggett and Whitney 1972). When shad become sexually mature, they return to their native rivers to spawn (Hollis 1948; Vladykov 1950; Judy 1961; Walburg and Nichols 1967). The timing of the river migration varies with latitude, being earliest in the south and progressively later to the north. Peak migrations occur in January in Florida, and in May and June in New Brunswick and Quebec (Walburg and Nichols 1967; Leggett and Whitney 1972). The timing of these migrations appears to be regulated by water temperatures both at sea and in the home river (Leggett and Whitney 1972).

Shad spawn only in freshwater (Leim 1924; Massmann 1952; Walburg 1960), but there does not appear to be any required distance above brackish water (Massmann 1952). Shad prefer areas dominated by broad flats or shallow water with moderate currents (Smith 1907; Bigelow and Welsh 1925; Massmann 1952; Marcy 1972). Time of spawning activity and survival of eggs and larvae are closely related to water temperatures, the temperatures for maximum hatch and survival of eggs and larvae being 15.5–26.5°C (Leim 1924; Massmann 1952; Walburg 1960; Chittenden and Westman 1964; Bradford et al. 1968; Marcy 1972). Juvenile shad migrate to the sea in the fall of the year in which they are spawned (Smith 1899; Walburg and Nichols 1967). The timing of these migrations, too, is apparently triggered by temperature (Sykes and Lehman 1957; Chittenden 1969; Leggett and Whitney 1972).

The length of the ocean residence varies depending on the sex of the fish and the latitude of the home river. On the average, males mature 1 year earlier than females. Most males are mature at age 4, while most females mature at age 5. Average age at maturity is earlier for both males and females in the south and increases with latitude toward the north (Leggett 1969; Carscadden and Leggett 1975a, 1975b). Shad native to rivers south of Cape Fear, North Carolina, die after spawning. In rivers north of this latitude, some shad survive to spawn again (Talbot and Sykes 1958; Leggett 1969). The proportion that spawns more than once increases with latitude and ranges from 3% in the Neuse River, North Carolina, to more than 50% in the St. John River, New Brunswick (Leggett 1969; Carscadden and Leggett 1975a, 1975b).

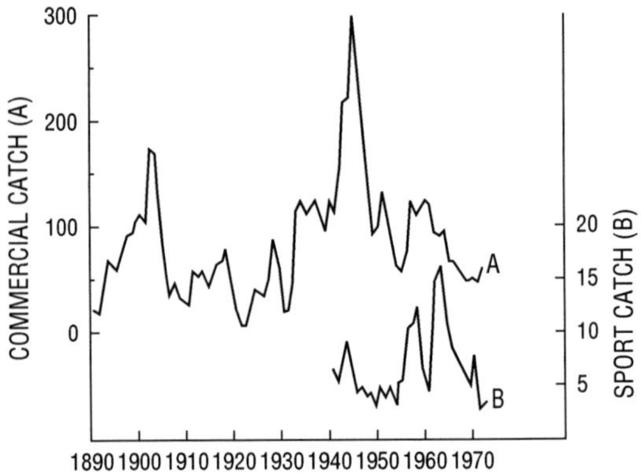

FIGURE 1. Connecticut River commercial shad fishery 1890–1973 (A) and shad sport fishery 1940–1973 (B) catches (in thousands of fish).

Connecticut River Commercial and Sport Fisheries

The Connecticut River commercial catch of American shad in 1880 was estimated by McDonald (1887) to have been 501,415 kg (approximately 250,000 fish). By 1892, the total reported landings had declined to 8,602 kg (Fredin 1954). The trend in Connecticut River commercial landings from 1890 to 1973 is shown in Figure 1. During this interval, reported commercial landings fluctuated between 50,000 and 150,000 fish per year with only periodic excursions beyond these limits. The most notable of these digressions was the large increase in catch between 1940 and 1950; this was associated with a 4-times increase in fishing effort over pre-1940 levels (the major increase in effort occurred between 1943 and 1946, possibly in response to increased prices during the war period), and followed the opening of an additional 33 km of river for spawning between Enfield Dam (constructed in 1880) and the Holyoke Dam (constructed in 1849) as a result of the installation of a fish ladder at Enfield in 1933.

In recent years, improvements in refrigeration and transportation techniques and the development of techniques for deboning the fillets, coupled with the delicate flavor of shad ("sapidissima" means most delicious), have led to a substantial increase in the demand for this fish. In some areas, notably New England, shad is considered a gourmet item. As a result the Connecticut River commercial fishery has achieved considerable economic importance (estimated capitalized value $7.5 million: Merriman 1970).

Sport fishing for shad has also been popular in the Connecticut River for many years (Nichols and Tagatz 1960). Prior to the hurricane and floods that affected the Connecticut River valley in 1938, the sport fishery was largely restricted to the spawning grounds of the Salmon River, a major tributary to the Connecticut at km 22.5. After that date, the fishery moved to the main stem in the area bounded by Wilson (km 74) and Suffield (km 86); Marcy (1972) has recently identified this area as one of the major spawning areas in the Connecticut River. In 1941, the Connecticut State Board of Fisheries and Game began an annual inventory of the numbers of shad taken by fishermen in a state-controlled fishing area at Enfield Dam (Suffield). The number of shad captured annually at this site are shown in Figure 1. In 1960, when 17,900 shad were estimated to have been taken by the entire sport fishery (Nichols and Tagatz 1960), this area alone produced 6,794 shad or approximately 37% of the total sport catch.

The capitalized value of the Connecticut River sport fishery was reported by Merriman (1970) to be $14 million.

The Problem with Regard to the Connecticut Yankee Plant

With the announcement of plans for the construction of the Connecticut Yankee Atomic Power Company's nuclear-fueled electric-generating plant at Haddam Neck (km 26), concern developed over the possible effects of the operation of this facility on the American shad population of the Connecticut River. This concern was based on the belief that alterations in the thermal characteristics of the river might detrimentally affect reproductive success and/or adult survival, thereby leading to reductions in the size of populations entering the river to spawn.

Although no specific information on the thermal tolerances of shad or the effects of temperature changes in their migratory behavior was available in 1964, there was a considerable body of observational data which implied that the timing of the river migrations of the species were to some degree determined by water temperature (McDonald 1884; Leim 1924; Talbot 1953, 1954; Massman and Pacheco 1957). Concern therefore existed that the artificial heating of the river as a result of the discharge of condenser cooling waters into the river would result in significant alterations of the timing of upriver migrations. This could reduce reproductive success in a number of ways. Shad might hold in elevated temperatures near the river mouth and mature too soon; they might proceed upriver too early and spawn where temperatures were below optimum for egg and larval development; or they might halt their upriver migration, spawn in the warmer waters of the lower river, and thus fail to fully utilize the available spawning and nursery areas.

Changes in the temperature characteristics of the river, and in the timing and patterns of the migration, might also result in increased adult mortality. Connecticut River shad may spawn several times. This capability of spawning more than once conveys to populations a greater degree of stability and a reduced probability of extinction in unstable fluctuating environments characteristic of north temperate regions (Ricker 1954, 1958b; Beverton and Holt 1957; Murphy 1966, 1968; Holgate 1967). Leggett (1969) concluded that repeat spawning was an important component of the total reproductive adaptation of shad in the north. It was considered possible that changes in the migration pattern resulting from increased river temperatures might prolong the period of freshwater residence sufficiently to cause excessive utilization of body reserves, for shad do not feed in freshwater and must therefore maintain themselves there on reserves accumulated at sea (Leim 1924; Leach 1925; Nichols 1959). This increased utilization of reserves might lead, in turn, to higher adult mortality. Significant increases in prespawning mortality would materially reduce the reproductive effort of the population. This reduction would be compounded by the associated loss of repeat spawners, which make a large contribution to total reproductive effort (Leggett 1969). Loss of repeat spawners would also reduce the stability of the population and its ability to respond to major environmental fluctuations.

Changes in the level of mortality, operating at any stage in the life cycle, that resulted from exposure to the thermal effluent from Connecticut Yankee would be expected to have direct effects on the number of shad entering the river to spawn each year. The rate at which these changes occurred following startup of the plant would depend directly on the stage of the life cycle affected. Thus if significantly increased mortality operated on the upriver or downriver migrating adults, changes in population size would be expected to become apparent the year after the resulting mortality, since these shad would be prevented from returning to the river to spawn. If, on the other hand, the mortality were restricted to the egg, larval or juvenile stages of the life cycle, pronounced reductions in the number of shad entering the river to spawn would not become apparent until 4 or 5 years following the mortality, at which time these shad would reach maturity and return to the river to spawn. In the case of Connecticut Yankee, which commenced operation late in 1967 after the adult migration has been completed, changes in population size resulting from effects on adults would not have been expected to appear until the spring of 1969, while changes in population size resulting from mortality to preadult stages would not have occurred prior to 1971 or 1972.

The adult shad program of the Connecticut

River Ecological Study was designed to provide the data required to assess the impact, if any, of the heated water discharge from the Connecticut Yankee facility on these aspects of the biology of the American shad.

Migrations of Adult Shad

As early as 1884, Marshall McDonald had developed a theory that shad are restricted to a narrow range of temperatures and that the timing of shad runs depends on when water temperatures in individual rivers warm to the optimum. Leggett and Whitney (1972) examined this proposed correlation between shad runs and river temperatures in four widely separated North American rivers. The populations studied were those of the St. Johns River, Florida, at the southern limit of the Atlantic Coast range, the York River, Virginia, as well as the Connecticut River. The fourth river examined was the Columbia River in the Pacific Northwest. For the Columbia River, 32 years of data were available, and these showed a highly significant correlation of temperature with the timing of the shad run, as well as very little variation in the temperatures at which the peak of the run occurred from year to year. In 78% of the years, the peak appeared within the range of 16.5°C to 19.0°C, and on the average, 90% of any given run was counted when the temperatures ranged between 15.5°C and 19.5°C. Furthermore, in 26 of the 32 years, the temperature varied only 4°C or less during the time when 90% of the shad appeared at the Bonneville fish ladders, and in only 3 years did the range exceed 4.5°C. At the Holyoke Dam in the Connecticut River, where 15 years of data were available, the peak passage occurred at temperatures ranging between 16.5°C and 21.5°C. Similar relationships between shad migration and water temperature were observed in the York River, Virginia, and the St. Johns River, Florida. In the York, analysis of daily catch-per-unit-effort data for the commercial fishery that operates at the river mouth showed that few shad entered the river until late January when river temperatures exceeded 4°C, and that peak numbers were captured in April when temperatures were approximately 14°C. At higher temperatures, the catches declined. Daily catch records of the St. Johns River commercial fishery showed that the shad migrations corresponded closely to the period of lowest annual river temperatures and were timed to occur in mid-January when temperatures were at approximately 15°C. Few, if any, shad entered the river at temperatures in excess of 20°C prior to mid-November, and as temperatures increased after the seasonal low in January, the relative abundance of shad began to decline.

Leggett and Whitney (1972) also examined the relationship between ocean temperatures and the seasonal distribution of shad at sea as indicated by the recovery of tagged individuals. Here, too, a strong correlation between the distribution of shad and the distribution of specific water temperatures was noted, with the seasonal location of shad, in time and space, corresponding closely to the position of the 13°C to 18°C surface isotherms.

The northern and southern limits of the range of shad on the Atlantic Coast also appear to be defined by the same temperature relationship. In June and July, a band of 13°C water typically extends to the mouth of the St. Lawrence River, the northern extent of the fish's reproductive range. In the south, shad appear to be blocked from extending their range into the Gulf of Mexico by a band of water south of Cape Canaveral that rarely cools below 21°C, even in December and January. While little information is available on the seasonal distribution of shad off the Pacific Coast, the occasional reports of shad catches there agree well with the predicted distribution based on a knowledge of the seasonal positioning of the 13–18°C surface waters along the Pacific Coast of North America. Leggett and Whitney concluded that both the marine and the freshwater migrations of American shad are regulated by water temperature. In the four separate rivers considered, temperatures at which shad appeared in the river and at which peak movements occurred were remarkably consistent. The seasonal adjustment in the timing of the entry of shad into the various rivers along the Atlantic and Pacific coasts appears, therefore, to be related to the sequential development of optimum temperatures in rivers from the south to the north with the earliest runs occurring in Florida on the Atlantic Coast and in California on the Pacific Coast. The latest runs in June and July occur in northern rivers such as the St. Lawrence and the Columbia.

On the basis of this analysis, it seems reasonable to conclude that a significant elevation of temperature in northern rivers could alter the migratory behavior of the shad, perhaps causing the fish to enter the rivers earlier than normal, providing that they were in the area when optimum river temperatures developed. In rivers where the heating occurs below the spawning grounds, as is the case in the Connecticut River, this could result in a prolonged freshwater residence if the shad delayed spawning until temperatures in the spawning area were optimum, or alternatively, spawning might commence immediately, but at temperatures below the optimum, causing a reduction in egg and larval survival.

Studies of the rate and pattern of the upriver migration of shad in the Connecticut River were therefore undertaken both before and after plant operation in order to evaluate the environmental factors that normally influence migratory behavior, as well as to assess the effects of plant operation on this behavior. Both conventional and ultrasonic tagging techniques were employed.

Methods

Conventional Tagging

During the years 1965–1973, adult shad were tagged at the time of their entry into the river and were then released to continue their upriver migration. Recoveries of these tagged fish at commercial and sport fishing sites provided an estimate of the individual and average rates of upriver migration. All shad were captured with a 14-cm (stretch measure) monofilament nylon drift gill net. The mark employed was a yellow spaghetti-dart tag (Floy Tag Co. No. FT-1A) 11.5 cm long, serially numbered and bearing the legend "Mail Tag to Fish and Game Hartford Ct. Reward." Tags were placed in the musculature immediately below the dorsal fin. In this position, the barb locked behind the interneural bones and this reduced the probability of tag loss. Tags were inserted with a stainless steel tagging needle into which the tag was fitted. A portable tape recorder was used to record tagging data. For each tagging effort, records were maintained on time, location, tidal state, water temperature, number, sex, and tag number of each shad tagged. Fork length measurements (to the nearest 0.5 cm) were recorded for tagged fish whenever possible. Commercial fishermen were advised of the nature of the tagging program, the number of tags to be applied, the approximate tagging locations, and the location of tag redemption centers by letter prior to the beginning of each tagging season. Posters were placed in the major sport fishing areas to advise sport fishermen of the tagging program and to inform them of the location of redemption centers. A $1.00 reward was paid for each tag returned. When complete details of the location and date of capture were not provided with the returned tag, a follow-up letter was mailed with the reward requesting this information. Commercial fishermen were provided with tag return envelopes prior to each tagging season. These envelopes were printed with instructions for recording the necessary information relating to the recovery of each tag. In an attempt to create interest in the tagging program, and insure return of tags, details on the date and location of tagging, number of days between tagging and recovery, and miles traveled by the fish in the interim were provided to all commercial and sport fishermen for each tag they returned. A small number of fishermen (1–2 per year) obviously took little care in recording tag recoveries because tags returned by them were occasionally reported as recovered on dates prior to their use; all tags returned by these fishermen were excluded from analyses of migration rates. All commercial fishermen were interviewed personally at the end of the fishing season in an effort to ensure as complete a return of recovered tags as possible.

Several analyses of the data were undertaken in an attempt to develop more than one independent estimate of migration rates. Average migration rates (expressed as km/d) were calculated for all fish tagged and released on each day during each tagging season; this was done to determine whether migration rates varied with time of tagging. Average migration rates were also calculated for all shad recovered in a given year, and these data were further analyzed as to average rates of males and females. Estimates of migration rates were also obtained by calculating the average number of days free for all fish recovered at each recovery location and then regressing average number of days against the distance traveled from release to recovery site. In this case, the slope of the regression line provided an estimate of migration rate. Ricker (1973) has noted that regression analysis provides the most reliable analysis for such data. Since 11 of the 14 recovery sites were below the spawning grounds and approximately 92% of all tag recoveries were from these 11 locations, bias resulting from inclusion of shad that had terminated their migration and were milling for some time prior to capture was minimal.

Ultrasonic Tagging

In all, 230 shad were captured and fitted with ultrasonic tags; (Smith Root SR-69B); 27 in 1967, 115 in 1968, and 88 in 1969. All but two of the fish were collected with 14-cm stretched-mesh monofilament gill nets; the two exceptions were captured with a haul seine. Tagging methods were improved throughout the study. The first eleven shad were tagged externally. These tags were secured to the fish immediately anterior to the dorsal fin by a stainless steel wire passing through the dorsal musculature. These fish were held, for varying times, in a floating live-car prior to and after tagging, generally in the company of other shad that were marked with spaghetti-dart tags.

The remaining 219 fish were tagged internally. Insertion was via the mouth, and the tag lodged in the stomach easily. To minimize handling stress, short sections of floating gill net were set and the fish were removed from the net immediately upon capture, as indicated by moving headrope floats. If lightly caught and apparently unharmed, they were immediately tagged and returned gently to the water. Total time from capture to release was generally less than 1 min.

Tag pulse rates, which varied from 20 to 510 pulses per minute for individual tags had been previously established by laboratory tank tests. Tag-use schedules were prepared to achieve the maximum separation in time and space of tags that exhibited similar pulse rates. Pulse rates were rechecked immediately after each fish was released to insure that they had not changed. Laboratory tests confirmed that pulse rates remained sufficiently constant over the first 2 weeks of the transmitter life to allow for the continued identification of individual fish.

Observations of the movements of these tagged fish were accomplished by

(1) Immediate and continuous tracking with boat mounted receivers: 72 shad were so tracked, 25 in 1967, 25 in 1968, and 22 in 1969. The duration of these tracks ranged from 0.5 to 73.5 h. The procedure involved maintaining the boat in line with and directly behind the tagged fish. This was possible because of the directional characteristic of the hydrophone. During periods when no fish movement was observable, the tracking boat was anchored to avoid confusion between fish movement and boat movement. The tonal characteristics of the received signal changed radically when the hydrophone passed directly over the tagged shad, allowing the exact determi nation of fish location. This procedure was seldom employed, however, in order to avoid "herding" or disturbing the fish with the boat. Changes in signal strength due to range en abled experienced trackers to estimate accurately the distance of the fish from the tracking boat, which was normally maintained 25 to 50 m behind the fish. Shore features, navigation aids, and river depths were used to establish the position of the fish, which was plotted at least every 15 min (more frequently when movement was active) on enlargements of navigation charts of the U.S. Coast and Geodetic Survey. Water temperature and river velocity readings coincident with fish position were made wherever possible. Temperatures were determined with an ARA five-thermistor electronic thermometer; river velocity was measured with a Gurley electronic flow meter.

(2) Subsequent pickup: During times when individual fish were being tracked, signals were frequently received from other tagged fish. In each instance, the pulse rate, time, and position of these tagged fish were recorded. At other times, deliberate attempts were made to relocate and track previously tagged fish. Sixty-two shad so located were tracked; 8 in 1967, 21 in 1968 and 33 in 1969. The duration of these tracks ranged from 1.25 to 99 h.

(3) Automatic recording: During the 1968 season, seven shore-based recording stations were obtained on loan from the U.S. National Marine Fisheries Service in Seattle, Washington. These monitors were located as follows: Essex, km 11, just above the saltwater influence; East Haddam Bridge, km 21, just below the influence of the Connecticut Yankee effluent; Connecticut Yankee Atomic Power Company (CYAP) outfall canal, km 24; CYAP intake, km 26; Middletown, km 35; Hartford, km 62; and Enfield Dam, km 84. These dual receiving units passively monitored the noise level in the river. As the tagged fish approached one of two monitor hydrophones,

the recording mechanism was activated. The tag signal was recorded on pressure-sensitive tape as a single mark for each pulse of the tag. Tag identification was accomplished by relating the tag marks to 1-s reference marks placed on the tape during recording. Time holes were punched in the tape every 6 min, and the time of passage of any fish was obtained by adding the time holes in tenths of hours to the starting time of the tape. The direction of movement of the fish was indicated by the track on which the signal was first recorded, the upriver hydrophone and downriver hydrophones being recorded on separate tracks. During 1969, magnetic tape monitors were utilized. These units, of the same basic design as the pressure-sensitive tape monitors, utilized an electronic data-interrogation system. They proved unreliable over the course of the study and data obtained from analyzing magnetic tapes were unusable.

Swimming speed calculations were made by matching boat-speed to the fish's swimming-speed and then measuring the forward velocity of the boat with the flow meter. In other instances, the boat was anchored, current velocities were measured, and swimming speeds were calculated by comparing water velocity and fish displacement with time. All data relating to the movements of conventionally and sonic-tagged shad, and related information on environmental parameters were reduced to digital form and stored on IBM punch cards. Statistical analysis was performed using an IBM 1130 computer.

Results

Rates of Migration—Conventional Tagging

In the interval 1965–1973, 34,086 adult American shad were captured, tagged, and released in the lower 3 km of the Connecticut River. From these fish, 7,778 were returned, 6,120 with complete data on the time and place of recapture, by commercial and sport fishermen and by project personnel (Table 1).

Significant numbers of shad were recovered at the release site a number of days after tagging. The mean time from release to recapture for these was 9.1 d (range 6.5–12.5 d) (Table 2). The average time until recapture for males that were recaptured at the release site appeared to be longer (9.0 d, range 6.9–11.0 d) than for females (8.2 d, range 6.5–9.8 d), but the difference was not statistically significant (t = 1.92; df = 12; 0.5 < P < 0.1). These data suggest that the upriver migration of tagged shad was delayed, perhaps as a result of shock related to capture and tagging.

The mean annual rate of upriver migration of Connecticut River shad, as estimated from the recovery of 6,120 marked fish at various locations upriver, ranged from 1.6 to 2.7 km/d (Table 3). Females moved upriver at a slightly faster rate than did males; the average annual rate for males

TABLE 1. Number of shad tagged (T), number returned (Rt), and number returned with data on place and time of recapture (Rd), 1965–1973.

	Males[a]			Females[a]			Total		
Year	T	Rt	Rd	T	Rt	Rd	T	Rt	Rd
1965							2,200	342	293
1966							3,991	970	766
1967	2,112	447	369	1,519	349	303	4,011	900	762
1968	1,567	362	312	2,253	583	476	4,362	1,032	860
1969	1,594	297	270	2,216	555	478	3,810	852	748
1970	1,793	391	329	2,049	447	368	4,072	883	734
1971	2,123	409	346	1,964	474	387	4,147	900	746
1972	2,177	412	299	2,204	510	377	4,622	992	729
1973	1,240	251	202	1,432	324	261	2,871	607	482
Totals	12,606	2,569	2,127	13,637	3,242	2,650	34,086	7,778	6,120

[a] Number sexed at time of tagging.

TABLE 2. Number of shad recaptured near the release site and mean number of days free before recapture.

Year	Males		Females		Total	
	N	Days free	*N*	Days free	*N*	Days free
1965					56	6.5
1966					121	12.5
1967	112	9.7	74	7.9	201	9.1
1968	61	6.9	95	6.5	174	6.9
1969	71	9.6	89	8.0	160	8.7
1970	77	10.2	50	7.5	136	8.9
1971	75	11.0	59	9.8	135	10.5
1972	76	10.9	64	9.5	153	10.4
1973	52	8.3	54	7.9	108	8.0
Totals	524	9.0	485	8.2	1,244	9.1

TABLE 3. Mean annual upriver migration rates (km/d) of shad based on dart tag recoveries.

Year	Males		Females		Combined	
	N	Rate	*N*	Rate	*N*	Rate
1965					293	2.7
1966					766	1.6
1967	369	1.4	303	1.6	762	1.6
1968	312	2.5	476	2.6	860	2.5
1969	270	2.1	478	2.5	748	2.4
1970	329	2.4	368	3.1	734	2.7
1971	346	2.2	387	2.5	746	2.3
1972	299	2.2	377	2.8	729	2.4
1973	202	2.1	261	2.1	482	2.1

ranged from 1.1–3.5 km/d, while the range for females was 1.6–3.1 km/d (Table 3). Analysis of the average upriver migration rates of fish tagged and released on specific days indicated that shad entering the river late in the run moved upriver faster than those entering early. Grouped data produced significant positive correlations of migration rate vs. date of tagging in 7 of the 9 tagging years (Table 4). Similar trends were observed when the data were separated by sex, males showing significant positive correlations in 5 and females in 6 of the 7 years for which data were available (Table 4).

Rates of upriver migration were also determined by linear regression analyses of the relationship between the mean time for all shad recovered at a specific location to reach that point against the distance from the tagging to the recovery site (9-year mean: 6.0 km/d; SD: 1.4; Table 5); these were significantly greater than those calculated from the analysis of the average daily displacement of individual tagged shad (9-year mean: 2.3 km/d; SD: 0.41; Table 2). The reason for the lower migration rate estimated from displacement of individual shad is that the calculation employs total elapsed time and so incorporates the tagging delay in the estimate. The regression analysis, on the other hand, has the effect of removing this tagging delay before calculating the migration rate. The regression analysis indicated that females moved upriver at faster rates than males, the average rate for females being 7.1 km/d (SD: 0.54) and for males 5.8 km/d (SD: 1.3). In addition, these data also indicated a delay in the initiation of upriver migration following tagging. Extrapolation of fitted regression lines to the *x*-axis indicated an average delay of 9.5 d (SD: 1.5) for males and 8.2 d (SD: 1.5) for females. The estimated average rate of upriver migration in the preoperation years

TABLE 4. Regression analysis of average upriver migration rate of shad (km/d) vs. date of tagging. Asterisk (*) indicates $P < 0.05$.

Year	Males			Females			Sexes combined		
	N	Slope	*F*	*N*	Slope	*F*	*N*	Slope	*F*
1965							19	–0.017	1.057
1966							25	0.021	28.977*
1967	31	0.004	0.199	23	0.034	19.679*	34	0.012	4.233*
1968	30	–0.006	0.473	32	0.014	3.214	33	0.008	1.571
1969	26	0.034	13.467*	30	0.034	30.192*	32	0.045	50.808*
1970	32	0.071	33.275*	31	0.061	41.358*	33	0.065	54.831*
1971	33	0.035	12.866*	31	0.051	39.505*	32	0.041	18.647*
1972	31	0.048	39.183*	29	0.075	62.881*	33	0.060	91.073*
1973	26	0.088	27.317*	33	0.106	31.613*	35	0.073	47.614*

TABLE 5. Regression analysis of distance traveled vs. days free based on mean time to recovery for all shad recovered at specific locations. All regressions are significant at $P = 0.05$.

	Males		Females		Sexes combined	
Year	b[a]	X intercept	b[a]	X intercept	b[a]	X intercept
1965					7.9	6.2
1966					4.2	9.2
1967	5.1	10.2	5.7	9.2	6.8	10.8
1968	5.8	7.4	5.0	5.8	5.6	5.8
1969	5.9	9.3	7.0	6.7	5.5	7.4
1970	5.8	9.4	8.0	8.7	4.4	6.6
1971	6.9	11.7	8.2	10.2	7.9	10.9
1972	5.6	10.7	6.8	8.6	6.6	10.2
1973	5.7	7.8	8.2	8.2	5.1	6.8

[a] b = km/d upriver migration rate.

1965–1967 (6.3 km/d) did not differ significantly from the postoperation years (5.9 km/d). ($t = 0.40$; df = 8; $P > 0.9$).

Estimates of the rate of upriver migration were also obtained by observing the time required for the peak abundance of shad to pass from Old Saybrook (estimated from catch-per-unit-effort data from the tagging net), at the river mouth, to Holyoke Dam, a distance of 119 km. These rates were lower than the ones developed by the regression analysis described above, the 9-year average being 4.3 km/d (range 3.1–5.9) (Table 6).

Rates of Migration—Ultrasonic Tagging

Migration rates of ultrasonic-tagged shad were determined by recording their progress up or down the river during continuous tracking, and also by combining data from initial tracks with subsequent relocation of tagged fish and, in 1968, with data on fish movements obtained by means of shore-based receivers.

In general, rates based on data obtained exclusively from continuous tracks (Tables 7 and 8) were higher than those determined by grouping data from all available sources (Table 9). An inverse relationship was observed between the duration of the track and the rate of upriver migration; longer tracks gave lower estimates (Table 7).

Estimates of migration rates based on tracks exceeding 20 h in duration (10.6 km/d) agreed most closely with estimates based on long-term periodic position data (6.2 km/d). Considerable individual variation was observed in migration rates. For continuous observations of more than 20 h, the range was 0 to 32.3 km/d; for rates based on long-term periodic position data the range was 0 to 40.8 km/d .

Statistically significant differences were noted in the migration rates of shad tracked continuously from the time of tagging and those re-

TABLE 6. Date of peak catch/unit effort at Saybrook, date when median shad was moved through the Holyoke fish lift, and resultant estimate of upriver migration rate.

Year	Saybrook	Holyoke	Rate (km/d)
1965	6 May	26 May	5.9
1966	2 May	8 June	3.1
1967	15 May	11 June	4.4
1968	10 May	1 June	5.4
1969	5 May	5 June	3.8
1970	6 May	4 June	4.1
1971	12 May	7 June	4.6
1972	9 May	15 June	3.2
1973	12 May	11 June	4.0

TABLE 7. Mean upriver rates of migration based on data obtained from continuous tracking (internally tagged shad only).

Year	Duration of track	No. of fish	Total hours	Mean km/d	Range km/d
1967	All data	5	117.1	13.7	5.6–22.4
	>5 h	5	117.1	13.7	5.6–22.4
	>10 h	5	117.1	13.7	5.6–22.4
	>15 h	4	104.8	11.7	5.6–15.6
	>20 h	3	88.8	10.4	5.6–15.6
1968	All data	29	360.5	22.4	1.3–54.7
	>5 h	21	340.6	19.5	1.3–59.9
	>10 h	11	273.8	13.5	1.3–32.3
	>15 h	9	251.1	13.7	1.3–32.3
	>20 h	6	201.6	14.7	6.0–32.3
1969	All data	27	375.6	38.2	0–136.5
	>5 h	17	344.0	32.6	0–136.5
	>10 h	12	302.6	25.4	0–136.5
	>15 h	7	247.3	25.2	0–136.5
	>20 h	6	230.6	6.8	0–17.4
1967–1969	All data	61	853.2	28.7	0–136.5
	>5 h	43	801.7	22.8	0–136.5
	>10 h	28	693.5	18.6	0–136.5
	>15 h	20	603.2	17.4	0–136.5
	>20 h	15	521.0	10.6	0–32.3

TABLE 8. Mean downriver rates of migration based on data obtained from continuous tracking (internally tagged shad only).

Year	Duration of track	No. of fish	Total hours	Mean km/d	Range km/d
1967	All data	7	53.2	27.8	5.9–84.4
	>5 h	5	48.7	20.9	9.3–39.3
	>10 h	2	27.8	9.6	9.3–10.0
	>15 h	1	15.5	9.3	
	>20 h	0			
1968	All data	17	149.8	13.5	0–53.4
	>5 h	9	125.3	8.0	0–25.8
	>10 h	6	114.5	7.8	1.8–25.8
	>15 h	4	93.0	3.7	1.8–4.3
	>20 h	3	77.5	2.9	1.8–4.3
1969	All data	20	180.2	44.0	6.9–106.0
	>5 h	11	148.9	45.1	6.9–91.7
	>10 h	6	114.4	23.6	6.9–49.1
	>15 h	1	53.4	10.6	
	>20 h	1	53.4	10.6	
1967–1969	All data	44	383.2	29.7	0–106.0
	>5 h	25	322.9	26.0	0–91.7
	>10 h	14	256.7	15.0	1.8–49.1
	>15 h	6	161.9	4.8	1.8–4.3
	>20 h	4	130.9	6.0	1.8–4.3

located and tracked several hours or days after tagging. Shad tracked immediately after tagging moved upriver more slowly and downriver more rapidly than did shad tracked after an interval of time. The mean rate of upriver migration of initially tracked shad was 20 km/d, as compared with 27.2 km/d for subsequently tracked shad. Corresponding rates for downriver migrations were 33 and 21.5 km/d.

Migration rates were reduced in the area of the salt–freshwater interface. Data from the shore-based recording stations operated in 1968, augmented by records of tracked and subsequently relocated individuals confirm these findings. The mean rate of upriver migration for 31 shad tagged at km 3.5, in brackish water, was 4.6 km/d (SD: 4.8). For 13 shad tagged upriver at km 24.0 in freshwater, the mean rate was three times this value or 13.9 km/d (SD: 13.7). The average rate of movement of shad tagged at km 3.5 to the first monitor located at km 11.0, just above the salt–freshwater interface, was 4.1 km/d (SD: 4.5). The mean migration rate of these same fish in freshwater was 12.4 km/d (SD: 12.8).

TABLE 9. Mean annual upriver rates of migration based on data from tracking, subsequent pickup, and monitors (1968 only).

Year	No. fish	Total hours	Mean km/d	Range km/d
1967	4	1,138.8	3.1	0.7–5.0
1968	44	9,110.4	7.4	0–40.8
1969	16	3,040.5	3.6	0–12.6
1967–1969	64	13,289.7	6.2	0–40.8

Swimming Speeds

Swimming speed data for internally tagged shad released in 1967, 1968, and 1969 were combined and analyzed in relation to all possible combinations of the direction of fish movement, water current, and salinity (Table 10), as well as fish position relative to channeled and shallow areas of the river. Swimming speeds were generally greater in fresh than brackish water regardless of flow direction or salinity conditions. In freshwa-

TABLE 10. Swim speeds of shad under varying environmental conditions and different orientation in the Connecticut River.

Condition	N	Mean swim speed cm/s	SD	Mean flow cm/s
Brackish water observations				
Fish down, flow out	13	26	24	29
Fish down, flow in	28	30	22	06
Fish down, slack	2	47	40	0
Fish up, flow out	86	46	31	21
Fish up, flow in	22	35	25	10
Fish up, slack	8	40	40	0
Channeled areas	137	39	31	29
Shallow areas	19	48	20	44
All conditions	159	39	30	32
Freshwater observations				
Fish down, flow out	31	45	33	37
Fish down, flow in	25	26	25	04
Fish down, slack	12	39	43	0
Fish up, flow out	274	71	39	37
Fish up, flow in	27	65	49	04
Fish up, slack	30	55	37	0
Channeled areas	248	64	41	35
Shallow areas	97	63	40	39
All conditions	401	63	41	38

ter and in brackish water, no significant differences were found between swim speeds in channeled and shallow areas. The mean swim speed in brackish water (all conditions combined) was 39 cm/s; in freshwater, it was 63 cm/s. This difference was highly significant (t = 8.94; df = 558; P < 0.001). The frequencies of occurrence of various swim speeds in salt- and freshwater are shown in Figures 2 and 3.

The influence of varying current velocities, resulting from change in tides, on swim speeds was examined by regression analysis (Table 11). With one exception, significant regressions were obtained only for those analyses relating outgoing tide with upriver migration, or when all observations, regardless of conditions, were grouped in the analysis. In these cases, a significant regression was found between swim speed and flow velocity. Gross examination of all values of t in the t-test (without reference to levels of significance) revealed much higher values when fish were oriented against the current than when oriented with it (Table 11).

The relationship between flow velocity and swim speed under the most frequently observed conditions (fish migration upriver, current out) were plotted for freshwater (Figure 4) and saltwater (Figure 5). Speeds over the bottom (the differential between swim speed and current speed) were also plotted. These figures show that shad migrating in freshwater adjusted their swim speed in relation to current speed in a manner that produced an approximately uniform rate of upriver progress under all flow conditions. In brackish water, however, the increase in swim speed was

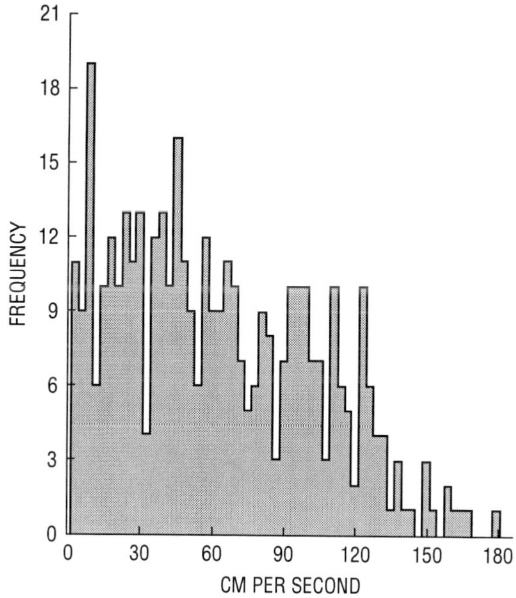

FIGURE 3. Frequency of occurrence of various swim speeds in freshwater.

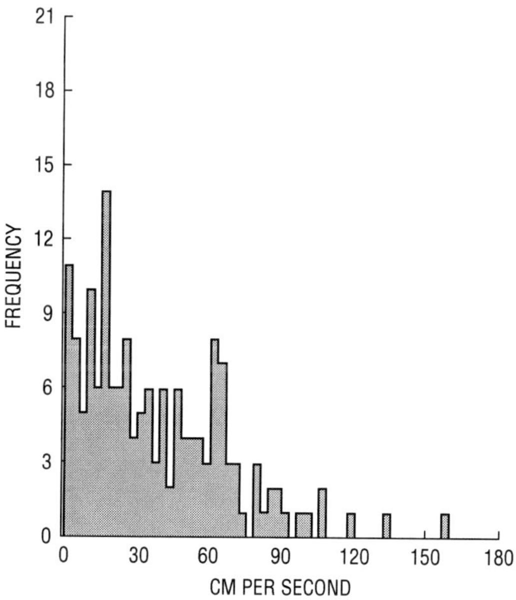

FIGURE 2. Frequency of occurrence of various swim speeds in brackish water conditions (lower estuary).

TABLE 11. Summary of regression analysis of relationship between swim speed and current velocity. Asterisk (*) indicates statistical significance at P = 0.05 level.

Condition	N	Regression	Intercept	t
Brackish water observations				
Fish down, flow out	13	0.052	0.24	0.15
Fish down, flow in	28	0.616	0.14	1.66
Fish up, flow out	86	0.435	0.28	4.03*
Fish up, flow in	22	0.202	0.39	0.37
All conditions combined	159	0.397	0.27	4.50*
Freshwater observations				
Fish down, flow out	33	0.359	0.31	1.13
Fish down, flow in	25	1.454	0.03	2.23*
Fish up, flow out	274	0.934	0.26	11.04*
Fish up, flow in	27	0.521	0.58	0.42
All conditions	401	0.727	0.36	10.70*

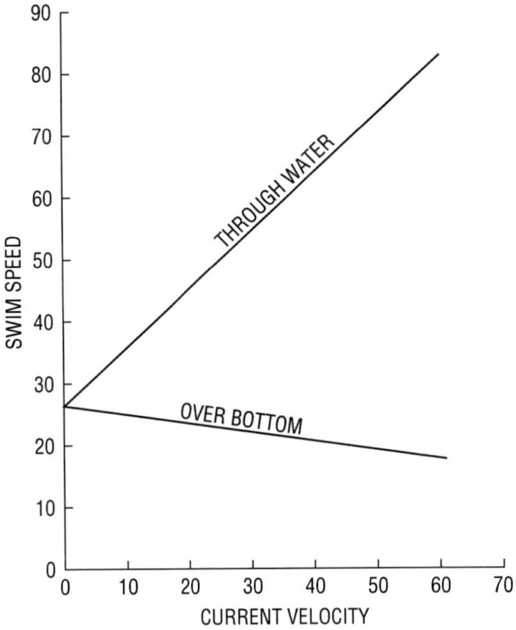

FIGURE 4. Least squares regression of swim speed through water vs. current velocity, and resultant progress over the bottom. Fish oriented upriver, fresh water. Units are centimeters per second.

inadequate to compensate for the increase in current velocity, and the rate of upriver movement declined as current velocity increased.

The possibilities of changes in swim speed associated with increasing distance upriver from the river mouth or increasing water temperature during the spawning migration were also examined by regression analysis. No significant regression was found between swim speed and river kilometer [rkm]. This was true both during upriver (N = 447; b = 0.0039; t = 1.69) and downriver (N = 113; b = 0.0019; t = 0.63) movements. Similarly, there was no significant regression between swim speed and water temperature (N = 560; b = 0.0281; t = 1.56).

An apparent diel cycle of swimming speed was observed when all freshwater observations were grouped (Figure 6). Maximum values occurred around 0400 and 1800 hours. These peaks were associated with sunrise and sunset, which occurred between 0415 and 0510 hours and between 1830 and 1930 hours, respectively, during the three tracking seasons. Minimum values occurred around 1100 and 2400 hours. Diel cycles of swimming activity were less evident in the brackish-water data (Figure 7), with swim speeds

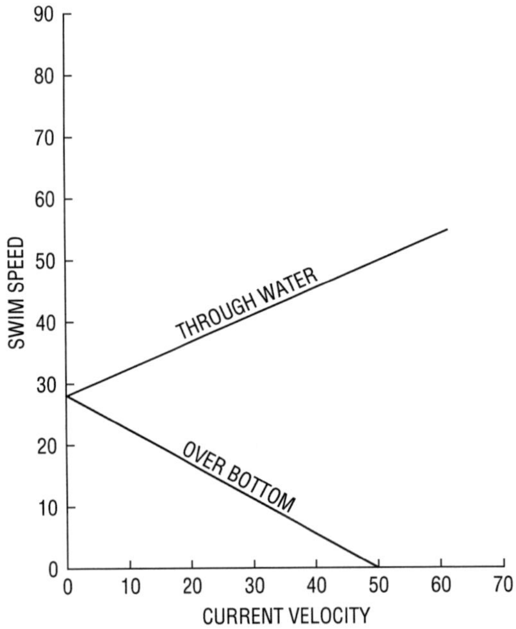

FIGURE 5. Least-squares regression of swim speed through water vs. current velocity, and resultant progress over the bottom. Fish oriented upriver, brackish water conditions. Units are centimeters per second.

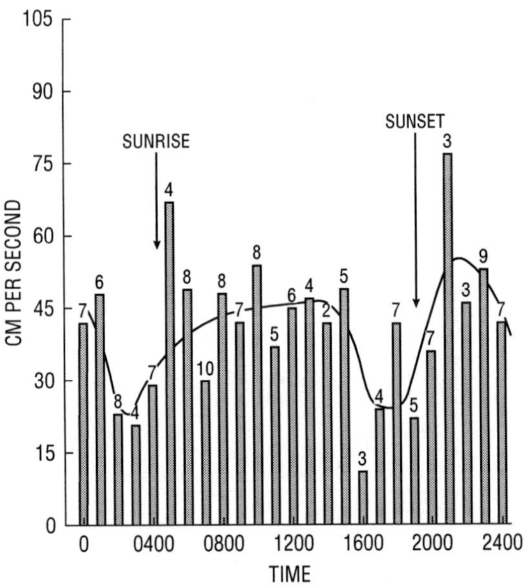

FIGURE 6. Diel periodicity in swim speeds in freshwater. Numbers over vertical bars represent number of observations; data smoothed by a moving average of threes for the linear representation.

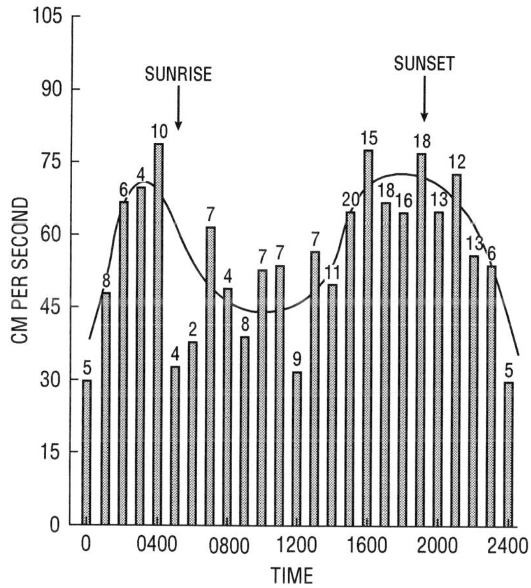

FIGURE 7. Diel periodicity in swim speeds in brackish water conditions. Numbers over vertical bars represent number of observations; data smoothed by a moving average of threes for the linear representation.

being more uniform throughout the 24-h period except for distinct reductions at 0300 and 1600 hours, approximately 1–2 h before sunrise and sunset, respectively.

Discussion

The differences among estimates of upriver migration rates derived from analyses of individual tag recoveries (2.3 km/d), from regression analyses of conventional tag-return data (6.0 km/d), from the timing of the movement of peak numbers of shad from the river mouth to Holyoke (4.3 km/d), and from long-term ultrasonic tracking records (6.2 km/d) appear to result mainly from the delay in upriver migration following conventional tagging. The average time until recovery at the tagging site (9.1 d) and the x-axis intercept (8.2 d) of regression lines describing the average upriver migration rates indicate an 8–9-d delay. The average distance traveled by tagged shad between release and recovery sites (1965–1973) was 32.6 km (SD: 7.06). At an average rate of 2.3 km/d, 14.2 d would be required to migrate this distance. At 6 km/d, 5.4 d would be required. The difference, 8.8 d, does not differ significantly from either estimate of the delay associated with tagging. The estimate based on the timing of the movement of the peak abundance of shad is thought to be biased downward by delays in upriver migration occurring because of operating conditions at the Holyoke fish lift. This 8–9-d delay may have resulted, at least in part, from the effects of handling and/or tagging. Physiological shock and irritation resulting from the tagging wound could have altered both the ability and the motivation of shad to resume their upriver migration following release. It is also very likely that a portion of this delay resulted from normal behavioral responses associated with transition to freshwater. Dodson et al. (1972) observed delays in upriver migration ranging from 17 to 71 h near the upper limit of the saltwater intrusion. All shad tagged in this study were released at or below this point (see Patterns of Migration, p. 194). The lower swim speeds noted for shad migrating in the region of the salt–freshwater interface, as compared with shad migrating in freshwater, may also have resulted from the physiological stresses caused by movement into freshwater. This reduced swim speed resulted in the fact that shad migrating in the lower river failed to compensate for increases in tidal current velocity. As current speed increased in this area, the rate of progress over the bottom declined until, at current velocities of 50 cm/s or more, little or no net upriver movement occurred. This swimming behavior must have contributed significantly to the reduced rates of upriver migration observed in the lower river.

In sharp contrast to this pattern, shad fully adapted to freshwater had higher mean swim speeds and altered their speed to offset changes in current velocity, thereby maintaining a relatively constant rate of progress over the bottom. This behavior resulted in faster rates of upriver migration in freshwater as reflected in both the ultrasonic tracking data and in regression of conventional tagging data.

The reduced swim speeds associated with migration with the current (generally downriver) may be indicative of an attempt by the shad to minimize the rate of movement away from the spawning area, or it may be a reaction to increased speeds relative to the bottom resulting from moving with rather than against the current, or both.

The data on migration rates suggest that the upriver migration in both brackish and freshwater is not direct. Rates determined from individual ultrasonic tracks were generally higher than those developed from all available sources of data. In addition, an inverse relationship was found between the duration of individual tracks and the resulting estimate of migration rates: longer tracks gave consistently lower estimates. This probably occurred because shorter continuous tracks failed to incorporate the full range of behavior, including meandering, holding, and downriver movements.

The observation that swim speeds did not increase with water temperature, even though individual migration rates of both males and females increased as the spawning run progressed, indicates that the increased rate of upriver migration later in the spawning run results from a more direct course upriver rather than from adjustments in swim speed with time. Females enter the river later than males, and this might also explain the faster upriver migration rate of females, as compared with males.

The most reliable estimates of upriver migration rates appear to be those provided by regression analyses of conventional tagging data and by long-term ultrasonic tracks. These results indicate an average rate of upriver migration rate of 5–8 km per day in freshwater. At this rate, the average time to migrate from the salt–freshwater interface to the main spawning area at Enfield Dam would be 10–16 d.

Patterns of Migration

Conventional Tagging

Conventional tagging experiments provided only limited data on the pattern of migration between the release and recovery sites. However, analysis of the distribution of recoveries during the nine study years indicated that there had been no significant changes in the pattern of recoveries from the commercial or sport fishery during that period. In addition, there was no significant change in the distribution of tag recoveries above and below the Connecticut Yankee plant in pre- (1965–1967) and post- (1968–1973) operative years (Table 12). In the years 1970–1972, there were increased tag returns from the Holyoke fish lift due to a more thorough surveillance of fish being lifted in connection with the extensive studies conducted at that facility by personnel of the Massachusetts Cooperative Fishery Unit based at the University of Massachusetts, Amherst.

TABLE 12. Distribution (percent) of tag recoveries of shad in the Connecticut River 1965–1973.

	Commercial fishery			
Year	Below CYAP	Above CYAP	Sport fishery	Holyoke lift
1965	57%	26%	12%	5%
1966	72	16	9	3
1967	56	34	7	3
1968	58	32	7	3
1969	47	43	5	5
1970	43	34	8	15
1971	52	30	8	10
1972	57	28	5	10
1973	57	36	6	1

Ultrasonic Tracking

With the initial use of external sonic tags in 1967, 10 of 11 shad at first made significant downriver movements. Two shad tagged in the area of Essex, Connecticut, moved downriver more than 8 km and returned to the salt waters of Long Island Sound; shad no. 1 completed this return to saltwater in 20 h; no. 2, in 56 h. Both fish were tracked continuously during these intervals. Shad no. 1 was recovered 29 d later by sport fishermen at Windsor Locks, Connecticut (80 km upriver from the river mouth). Three other externally tagged shad made downriver movements of at least 24-h duration. These movements were at rates in excess of 9 km per day and resulted from a combination of active downriver migration and drift. These three fish were subsequently relocated several kilometers upriver.

With the change to internal tags, a greater proportion of the tracked shad moved upriver after release. Fourteen internally tagged shad were tracked in 1967. The first of these was subjected to considerable handling and is not included in the discussion. Eleven internally tagged shad were released in the area of East Haddam, approximately 20 km from the river mouth and 3 km below the area of the Connecticut Yankee outfall. Five of the 11 shad released in this area moved upriver after initial downriver movements that lasted from a few minutes to 2.5 h. In all but one instance, the rates of these downriver movements were equal to or slightly less than the velocity of the water current. In each instance, lateral movements occurred during this period and upriver migration was resumed when the fish passed over the edge of the navigation channel or bottom drop-off. These five fish were tracked continuously for periods ranging from 12 to 41 h; during this time, they continued their upriver migration. The remaining six shad in this group were tracked for periods ranging up to 15.5 h; all made downriver movements characteristic of the externally tagged shad. One of these shad made an initial downriver movement of 4.5 km in 12.5 h. Contact with this fish was re-established 12 h later, an additional 3 km downriver, at which time it was moving slowly upriver. Two internally tagged shad were released at Wilson, Connecticut, in the shallow unchanneled area of the river. The first, released 13 June drifted downriver 4.5 km, at which point it apparently died. The second, released 15 June died at the tagging site within minutes of its release. Mean water temperatures on these dates were 22.8°C and 22.6°C respectively. Shad respond poorly to handling at these high temperatures, and death presumably resulted from handling stress.

In 1968, 25 fish were individually tracked from the time of release. Of these, 15 moved upriver and 10 downriver. Sixteen were tracked for more than 5 h, of which 13 began upriver migrations within 1 h of release and 1 within 4 h of release. Short downriver movements generally followed release. These movements took place at rates equal to or slightly less than current velocities, indicating that they resulted from drift rather than active migration. Two shad, however, moved downriver for more than 5 h following release.

In 1969, of 22 fish tracked from the time of release, 6 moved upriver and 16 downstream. Fourteen were tracked initially for more than 5 h. Only 4 of the 14 moved upriver. The other 10 moved actively downriver until tracking was discontinued. The duration of these downriver tracks ranged from 5.2 to 53.5 h, the average being 12.8 h. In contrast to shad tracked immediately after tagging, those relocated and tracked at varying times after release exhibited a greater frequency of upriver migration. In 1967, 4 of 8 moved upriver; in 1968, 14 of 21 moved upriver, 5 moved downriver, and 2 made no net movement; in 1969, 23 of 33 moved upriver, 6 moved downriver, and 4 made no net movement during tracking.

The behavior of adult shad during the transition from salt- to freshwater was investigated by Dodson et al. (1972). The Connecticut River has a relatively narrow mouth and large discharge, and as a result, the salinity distribution in the lower river forms a well-defined "salt-wedge" (Meade 1966). This wedge consists of a short zone located several kilometers upstream of the river mouth where Connecticut River water mixes with saltwater. The salinity wedge moves up- and downriver in a tidal rhythm. Water on the river bottom may contain 15% chloride at high tide but be almost completely fresh in the same place at low tide. The position of the salinity wedge is also sensitive to changes in upland discharge, being depressed downriver by increasing flow. This also results in a steepening of the salinity gradient in both the horizontal (downriver) and vertical directions, the most saline end of the gradient being respectively downriver and on the bottom. Migrating adult shad enter the Connecticut River in April and May when the river discharge generally varies from 18,000 to 4,000 m^3/s. At high slack tide with a discharge of 18,000 m^3/s, the salt wedge extends less than 1.8 km upriver, and at a discharge of 4,000 m^3/s, the salt wedge extends 11 km upriver.

In addition to salinity changes, the fish also experience a temperature change in their transition from salt to freshwater. In early May, 1970, a 5°C increase in temperature coincided with reduction in salinity from 26‰ to 0‰. In mid-June of the same year, a 10°C increase in temperature coincided with a reduction in salinity from 27‰ to 0‰. The observed increase in ΔT with time across the salinity wedge is brought about by the more rapid warming of the river water than the water from Long Island Sound.

The behavior of five of seven shad tracked in the lower estuary consisted of extensive meandering in the area of the upriver extent of saltwater. This movement resulted from a combination of passive drift and leisurely swimming. The fish were rarely observed to outswim the tide when oriented into it. Thus, they were passively maintained near the leading edge of the salt wedge. Fish no. 1 was tagged 2.8 km above the leading edge of the salt wedge and meandered for 53 h. During this interval, the fish maintained itself close to the upper limit of the saltwater wedge. It then began a direct upriver migration covering 18 km in 7.5 h. Similar behavior was observed in four other shad. Fish no. 2 was captured at the leading edge of the salt wedge. After 57 h of meandering in the lower estuary (Figure 8), a direct upriver migration of 24 km in 8 h occurred. Fish

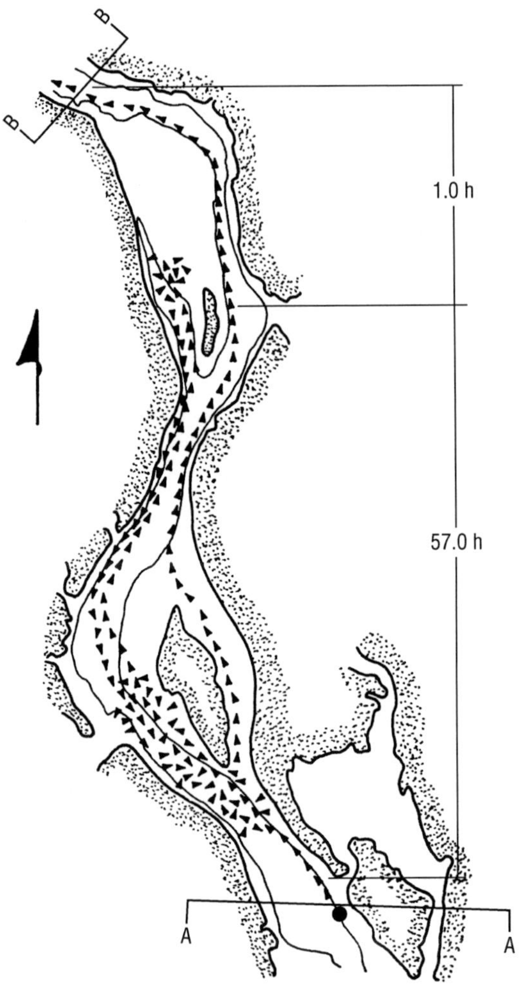

FIGURE 8. Behavior of shad no. 2 at the leading edge of the saltwater–freshwater interface. Arrows indicate direction of travel. Where no clear direction is evident, meandering behavior was observed.

no. 3 was captured 2.3 km below the upriver extent of saltwater. It immediately migrated 5.6 km upriver into freshwater where meandering behavior began and continued for 17 h. Following this behavior, a direct upriver migration of 11 km in 6 h occurred. In all three of these cases, direct upriver migration occurred in freshwater and was continuous during periods of zero and reverse flow resulting from tidal action. Fish nos. 4 and 5 showed similar meandering behavior in the vicinity of the upriver extent of saltwater but were lost before upriver migration occurred. The subsequent upriver migration of these two fish was revealed by data obtained from an automatic shore-based ultrasonic tag monitor located at rkm 11. Fish no. 4 moved upriver past the monitor 71 h after tagging. Fish no. 5 moved upriver past the monitor 64 h after tagging. The distance of these 5 fish, during meandering behavior, from the mouth of the river at the time of high and low slack water was positively correlated with the distance of the upriver extent of the salt wedge from the river mouth at these times ($r = 0.682$; $N = 27$; $P < .001$). Of the remaining two shad, Fish no. 6 was tagged 2.3 km above the leading edge of the salinity wedge and migrated 26 km continuously upriver in 20 h. Fish no. 7 was tagged 5.6 km below the upriver extent of saltwater; after holding its position for 10 h, it migrated downriver into Long Island Sound.

In contrast to the meandering nature of fish movements in the lower estuary, the upriver movement of individually tracked fish in freshwater tended to be more direct. Periodic meandering, holding of position, and downriver movement occurred during most tracks, but these were more or less random events, showing no obvious pattern in time or space. The precise path taken by all shad was remarkably consistent.

Visual inspection of the route taken by shad during 1967, 1968, and 1969 suggested that fish migrating in freshwater followed the natural and dredged river channel, with only periodic movements into shallower water. To quantify these tendencies, a chi-square analysis of the relationship of the observed distribution of the tracks between channel and shallow areas and the expected distribution based on the assumption of random distribution across the river was undertaken. Expected distributions were calculated for 47 transects arbitrarily established at 1-km intervals starting from the river mouth. For each transect, the relative proportion of channel and shallow areas was determined from U.S. Coast and Geodetic Survey navigation charts, and this proportion was used to calculate the expected distribution of tracks at this transect. Ultrasonic tracking data from 1967, 1968, and 1969 were then analyzed for the relative position of tracks at each transect. In each of the 3 years, grouped data produced a significant deviation from that expected, the trend in each case being toward the channel ($\chi^2_{1967} = 30.9$, df = 19, $P < 0.05$; $\chi^2_{1968} = 152.8$, df = 47, $P < 0.001$; $\chi^2_{1969} = 176.5$, df = 39, $P < 0.001$). A second chi-square analysis of the position of tracks at individual transects showed that at 37 of 42 individual transects (88%) the fish were in the channel more often than would be expected by chance ($P < 0.05$).

Four of the transects where fish were more uniformly distributed between channel and shallow areas were located at the approximate leading edge of saltwater intrusion, between km 5 and 11, where extensive milling was observed. Two others were adjacent to the mouth of the Salmon River (km 24), the only large tributary to the Connecticut River within the study area. At this point, the river channel is situated near the east bank, in close proximity to the mouth of the Salmon River; west of the channel lies a broad shallow area. In most instances, shad migrating through this area left the channel approximately 0.8 km below the mouth of the Salmon River and moved over this shallow region to the west bank before continuing their upriver migration. The remaining nine transects were located between km 35 and 45; these suffered from inadequate observations, the maximum number of observations for any one transect being six.

The response of shad to changes in current direction was also investigated. River flows during April, May, and June of 1967 were generally sufficient to overcome the effect of tide in most areas of the river. Consequently, the frequency of occurrence of reverse and zero flow conditions in the tracking area was low, and observations on the response of shad to such tidal phenomena were limited. However, reversed river flows resulting from tidal action were recorded five times. The total tracking time during such conditions was 17 h. Definite changes in migration patterns were noted for each reversal of flow. Direct upriver migration was replaced by meandering, holding, and, in two instances, by downriver migrations against the current. Slack tide (zero flow) conditions were recorded on an additional five occasions lasting a total of 6 h. Upriver migration of four shad was again replaced by meandering and holding behavior, but one shad continued to migrate upstream.

In 1968, as in 1967, the occurrence of reverse and zero flow conditions was minimized by spring runoff. A total of 10 observations was nevertheless recorded. Four responses to zero and reverse flow were observed in the saline waters of the lower estuary. On two occasions, one shad ceased its upriver progress, began the milling behavior often associated with slack-tide conditions, and oriented into the incoming tide; in both cases, its swimming speed was less than the opposing current velocities and the fish was carried upriver by the current. The controlling factor in these movements is not clear, however, since the milling behavior occurred at the approximate upper limit of the saltwater intrusion, and milling may have been associated with the fish's acclimation to freshwater. The second shad made no noticeable response to these conditions and continued its upriver movements through periods of slack and incoming flow on two separate occasions. Six responses to zero and reverse flow were observed in freshwater areas of the river. These responses involved five different shad. Two of them exhibited meandering and downriver movements during slack-tide periods, but one of these fish continued its upriver movement without hesitation through the next period of slack, reverse flow, slack current. Three other shad continued their upriver movements through slack and reverse-flow conditions with no noticeable change in rate or pattern of migration.

In 1969 a total of 14 changes in water current direction occurred while fish movements were being monitored. No change in migratory behavior was observed during nine of these periods. In the other five cases, distinct alterations in behavior associated with changes in current direction were noted. One of these occurred at the approximate upper limit of saltwater intrusion, however, and may have been related to the meandering behavior previously described in this area.

Movements in the Area of the Heated Effluent

The initial discharge of heated water from the Connecticut Yankee facility into the Connecticut River occurred in October 1967, 5 months after the termination of that year's tracking program. The average daily operating load in 1968, during the period when shad were being tracked through the discharge area, was 480 MW_e. In 1969, plant operation during the tracking season averaged 480 MW_e on all but 4 d, and the minimum operating level was 240 MW_e (see Figure 5 in Boyd 2004 [1976], this volume).

At an operating level of 480 MW_e, 25 m^3 of water heated approximately 10°C above ambient, are discharged from the plant each second (see

Figure 3 in Boyd 2004 [1976]). This ΔT is reduced approximately 1°C by heat loss to the atmosphere during passage down the 1.8-km discharge canal, resulting in a 9°C ΔT at the point of discharge into the river. Because of its lesser density, the heated water discharged from the canal slides over the surface of the cooler river water and a marked stratification rapidly develops. The heated water is restricted to the upper levels of the river and is seldom in contact with the bottom except in the immediate area of the discharge point (see Boyd 2004 [1976]).

During the spawning migration of the shad, river discharge rates are generally sufficient to overcome the influence of tide, and reversals of flow in this area are infrequent. When slack tide or reversals of flow do occur, they are much shorter in duration than in late summer and fall. In the spring, the upriver distribution of temperatures tends to be compressed toward the upstream edge of the discharge point, and the downriver finger of the plume is more confined to the discharge side of the river, with heated water reaching the opposite shore only occasionally and for brief periods. At no time does the heated water extend to the bottom in the channeled area of the river, which, in this sector, lies near the west bank on the opposite side of the river from the discharge point.

Five shad were tracked through the discharge area in 1967 prior to plant operation. All five had been tagged and released at rkm 20, and each had resumed its upriver migration after a brief initial downriver movement. In all cases, the migration through the area was direct and channel-oriented. All five shad avoided the mouth of Salmon River by crossing to the west bank approximately 1 km below its confluence with the Connecticut and then continuing upriver along the western edge of the navigation channel until approximately 1 km above the discharge point. No hesitation or meandering was observed between km 20 and km 27, which includes the entire portion of the river affected by plant discharge under maximum operating conditions. The average time taken to move through the area was 7.6 h (SD: 4.8). Three of the five shad continued their direct upriver movements to km 30 where tracking was discontinued. The remaining two terminated their upriver migration at Haddam Island (km 27) where their movements were recorded for more than 48 h; their behavior during this period consisted of meandering movements in water less than 4 m in depth, and one fish remained in this area for 7 d. The migratory paths of these shad are illustrated in Figure 9.

In 1968 and 1969, 27 shad were tracked through the discharge area during plant operation, 12 in 1968 and 15 in 1969. During nine of these tracks (33%) no elevation in temperature was recorded along the path taken by the shad. This occurred because of the tendency of shad to favor the west side of the river, while the plume

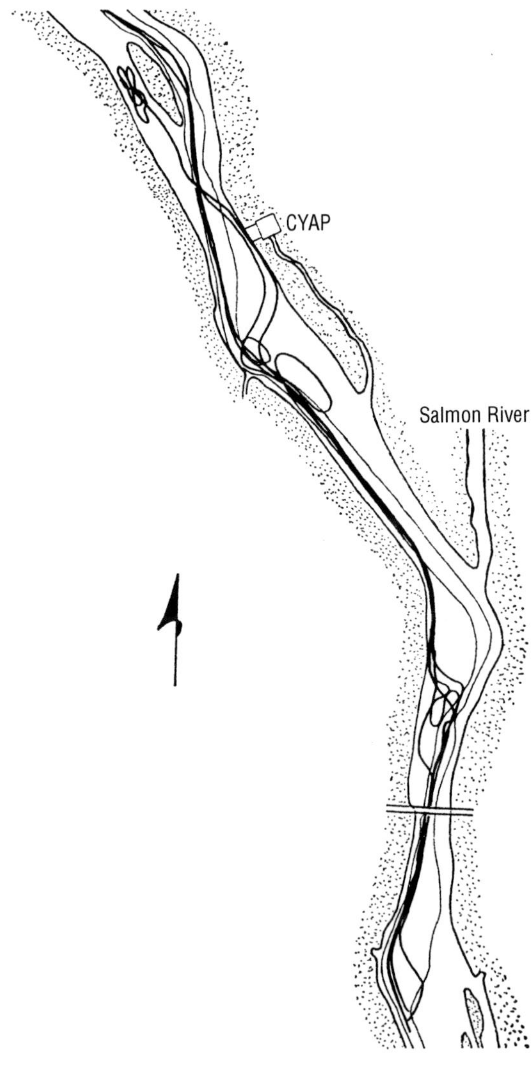

FIGURE 9. Migration routes of five ultrasonic tagged shad tracked past Connecticut Yankee in 1967 prior to plant operation.

was generally restricted to its eastern half by the strong outgoing current. Seven of these shad (78%) made direct upriver migrations through the area and exhibited no significant alterations in orientation or swim speed. One halted its upriver migration approximately 700 m below the discharge point, moved downriver approximately 1,000 m, and then resumed its upriver migration immediately without deviation or delay. One shad halted its upriver migration 300 m north of the discharge point and moved downriver 4 km where it meandered for 24 h; this behavior was followed by a direct upriver migration through the area with no unusual alterations in orientation or swim speed. The average time required for these fish to move from km 20 to km 27 was 3.9 h (SD: 1.8). The paths of these nine shad are presented in Figure 10.

Distinct temperature elevations resulting from the presence of the heated water plume were recorded along the path taken by the remaining 18 shad. These elevated temperatures were generally restricted to the top 2 m, although temperature increases were occasionally recorded as deep as 3 m along the east bank and close to the discharge point. Because the depth at which shad were swimming during the tracks was not known, it was not possible to determine whether the shad actually came in contact with these higher temperatures or simply swam under them in the underlying water of ambient temperature. Seven of these 18 shad (39%) made direct upriver migrations through the area, during which there were no changes in orientation or swim speed that could be related to the observed changes in water temperature along the track. The average time required to move through the area was 2.3 h (SD: 1.2). The tracks of these shad are presented in Figure 11.

Six of the 18 shad (33%) tracked through the area made minor alterations in swim speed or orientation in the vicinity of the heated water plume. This behavior consisted of brief meandering in the immediate area of the discharge and/or movements out of the channel toward the discharge during upriver progress. The maximum duration of this meandering behavior was 2 h. The average migration time through the area was 3.1 h (SD: 1.2). The behavior of these shad is summarized in Table 13. Three additional shad were observed to make minor changes in orientation in the area of the plume, but in these cases, tracking was terminated because of adverse weather conditions before they moved out of the area of influence.

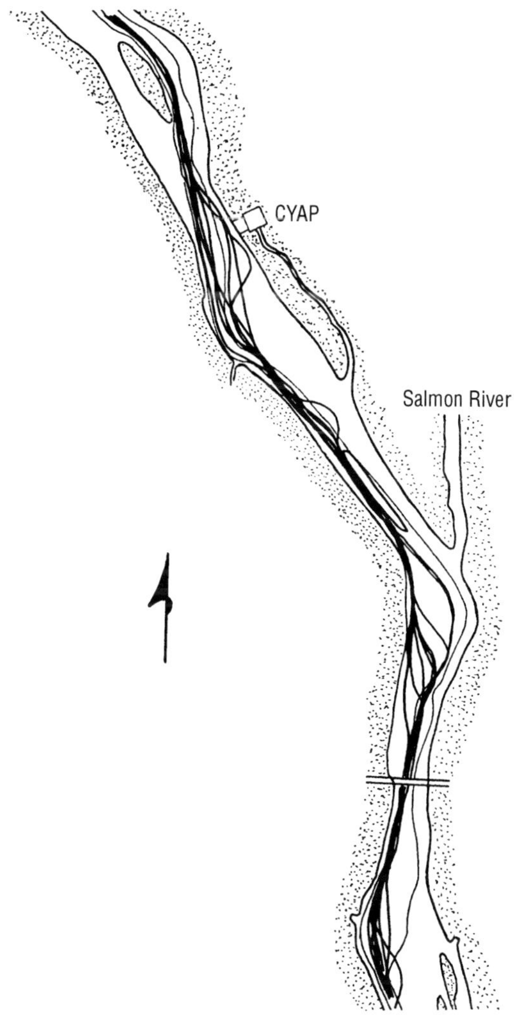

FIGURE 10. Migration routes of nine ultrasonic tagged shad tracked past Connecticut Yankee in 1968 and 1969 during plant operation. No elevation in water temperature was recorded during these tracks.

The remaining two shad tracked through the plume exhibited prolonged meandering behavior in the general area of the discharge. The first of these (Figure 12) moved directly upriver into the plume and initiated a period of extensive milling on the east side of the river in the area between the discharge canal and the mouth of the Salmon River. This behavior continued for 11 h, and the shad even swam directly into the mouth of the discharge canal on four occasions; however, the maximum ΔT was only 5°C since the

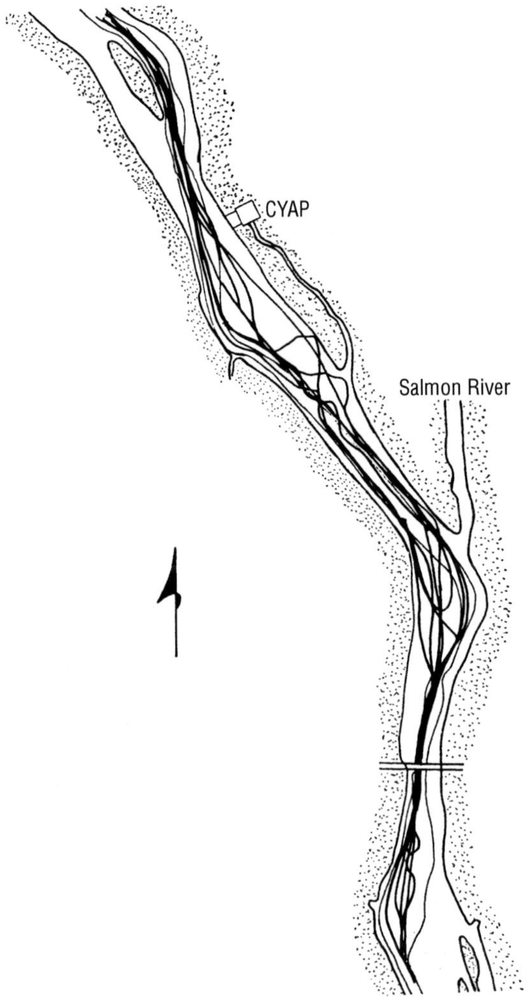

FIGURE 11. Migration routes of seven ultrasonic tagged shad tracked past Connecticut Yankee in 1968 and 1969. Elevated temperatures (max. 8°C ΔT) were recorded along the tracks, but no significant changes in behavior were observed.

operating level of the plant had been reduced to 240 MW_e on the date of the track. Following this period of meandering, the shad moved upriver to a point approximately 3 km above the plume where it began a second period of milling lasting approximately 2 h. It then migrated downriver into the plume where it meandered an additional 2 h before continuing downriver. The second shad (Figure 13) moved directly up the west shore to a point approximately 1,000 m north of the discharge where it halted its upriver migration and commenced an up- and downriver movement along approximately 1 km of the west side of the river, opposite the discharge point. This behavior continued for 22 h. Elevated surface temperatures to a maximum of 8°C above ambient (19.0°C) were recorded during two separate intervals of 4.5 and 2.5 h during this 22-h period; no temperature increase was noted during the balance of this period. Following this period of meandering, direct upriver migration was resumed and continued for 4.5 h, during which time the shad moved 3 km; then, upriver migration halted and meandering behavior continued throughout the remaining 20 h of the track.

During 1970, 1971, and 1972, attempts were made to study the behavior of shad during the seaward migration following spawning. Great difficulty was experienced in ultrasonic tagging spent shad because of their extremely weakened condition and the high water temperatures (>24°C) prevailing during the downstream migration. Several hundred shad were captured, but only 15 were sufficiently lively to tag. Of these only 7 survived tagging and were successfully tracked, 2 in 1970, 3 in 1971, and 2 in 1972. All were captured and tagged approximately 1.5 km above the Connecticut Yankee intake structure (km 26). Six were tracked downstream past the Connecticut Yankee discharge area and four were tracked until they entered Long Island Sound.

The behavior of downstream migrant shad differed little from that of upstream migrants tracked during 1967–1969. The movements of all seven were highly channel-oriented, and this resulted in similar downstream migration routes being followed by the lot. In the area of the Connecticut Yankee facility, the tendency, previously reported, of shad to migrate along the west side of the river was repeated. As a result, no increase in water temperature was recorded during the downriver movement of these fish past the discharge area. One of the six shad tracked downriver past Connecticut Yankee meandered briefly along the east shore between the plant's intake and discharge structures but soon returned to the channel on the west side and continued downriver. The majority of the downstream migration was with the current. As was the case with upstream migrants, the response of downstream migrants to reversals in the direction of the current was mixed. Two shad moved actively upriver during an incoming tidal current, one briefly and one for 2 h. Both then resumed their downriver migration through the balance of the flood and the following ebb tide. The

TABLE 13. Environmental conditions, and behavior of six shad exhibiting minor alterations in swim speed and orientation, in the area of thermal plume.

Shad no.	Temperature at 1 m depth: Ambient	Temperature at 1 m depth: Max ΔT during track	Path taken	Time through area (km 20–27)	Behavior
1	19°C	5°C	Channel	2 h	Left channel and moved briefly toward discharge. Returned immediately to channel. No meandering or delay in upriver progress.
2	15°C	3°C	Channel	4 h	Moved upriver directly avoiding mouth of Salmon River. Moved toward plume when first ΔT was recorded and continued upriver on discharge (east) side. Brief meandering occurred below discharge where 1°C ΔT at 1M was recorded. Upriver migration halted 2 km above discharge and fish moved directly downriver through area without delay. No ΔT on downriver portion of track.
3	16°C	2°C	Channel, shallow	5 h	Meandered briefly before reaching plume. Then migrated directly upriver through plume.
4	20.5°C	4°C	Channel	2 h	Moved upriver directly with minor slowing of upriver rate opposite discharge area.
5	16°C	4°C	Channel	3 h	Meandered at mouth of discharge for 2 h before resuming upriver movement.
6	17°C	3°C	Channel	2.5 h	Moved directly through plume, meandered 15 min at upstream side in unaffected water, then continued upriver progress.

remaining two shad tracked during reversals of current did not alter their orientation.

Downstream migrant shad appear to be able to withstand a rapid transition from fresh- to saltwater, although two of the four spent shad tracked into Long Island Sound did display brief (<2 h) meandering behavior at the approximate upper limit of saltwater intrusion. The remaining two showed no signs of hesitation on entering brackish water or the saltwater of Long Island Sound. This is in agreement with Tagatz' (1961) observation that adult shad could withstand rapid direct transfers from freshwater to saltwater.

Discussion

The high frequency of initial downriver movements displayed by shad tagged both externally (19%) and internally (66%) with ultrasonic transmitters suggests that these movements may result from physiological shock and/or disorientation related to capture and tagging. Internally tagged shad released and relocated one or more days after tagging, however, also showed a relatively high frequency (35%) of prolonged downriver movements after relocation. Nevertheless, a chi-square analysis of the frequency of downriver movement of shad tracked immediately after release, as compared with shad that were relocated after one or more days and then tracked, revealed that shad tracked immediately after tagging performed a significantly greater proportion of downriver movements (χ^2 = 9.47; df = 1; $P < 0.005$). Internally tagged shad tracked continuously for more than 5 h after release generally exhibited initial downriver movements, but the frequency of downriver movement lasting more than 2 h was less (46%) in this group as compared with those tracked for shorter intervals. Chi-square analysis indicated that the frequency of downriver movement in this group did not differ significantly from the subsequent recovery group (χ^2 = 2.97; df = 1; $P > 0.05$). These findings suggest that the stress effects of internal tagging are minimal, and that in most cases recovery occurs within the first 5 h after tagging. The relatively high frequency of

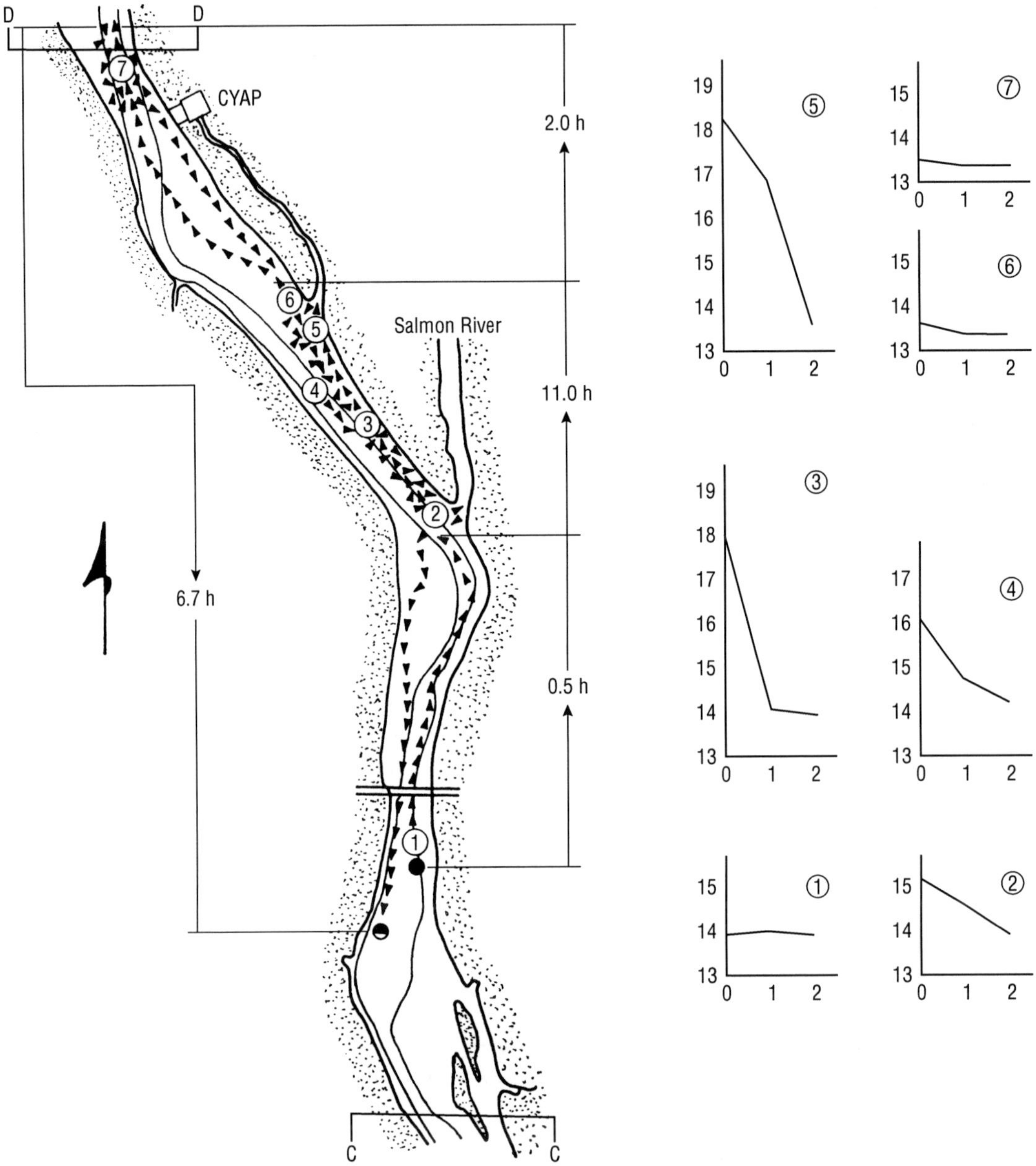

FIGURE 12. Migrations of shad showing marked alterations in behavior and rates of upriver progress in the area of the CYAP discharge. Temperature profiles to a depth of 2 m are shown for selected points along the track. Arrows indicate direction of travel. Where no clear direction is evident, meandering was observed.

downriver movement demonstrated both by subsequently relocated shad and by shad tracked continuously for 5 or more hours after tagging, suggests that the migration from the sea to the spawning grounds is not direct.

The extensive meandering exhibited by five of the seven shad tracked in the lower estuary could have resulted from three factors: (a) atypical behavior due to physiological shock resulting from handling, (b) normal behavioral response to the tidal cycle, or (c) behavioral response to the transition from salt to freshwater.

The observed meandering does not appear to be a result of handling. Fish no. 3 was tagged 2.4

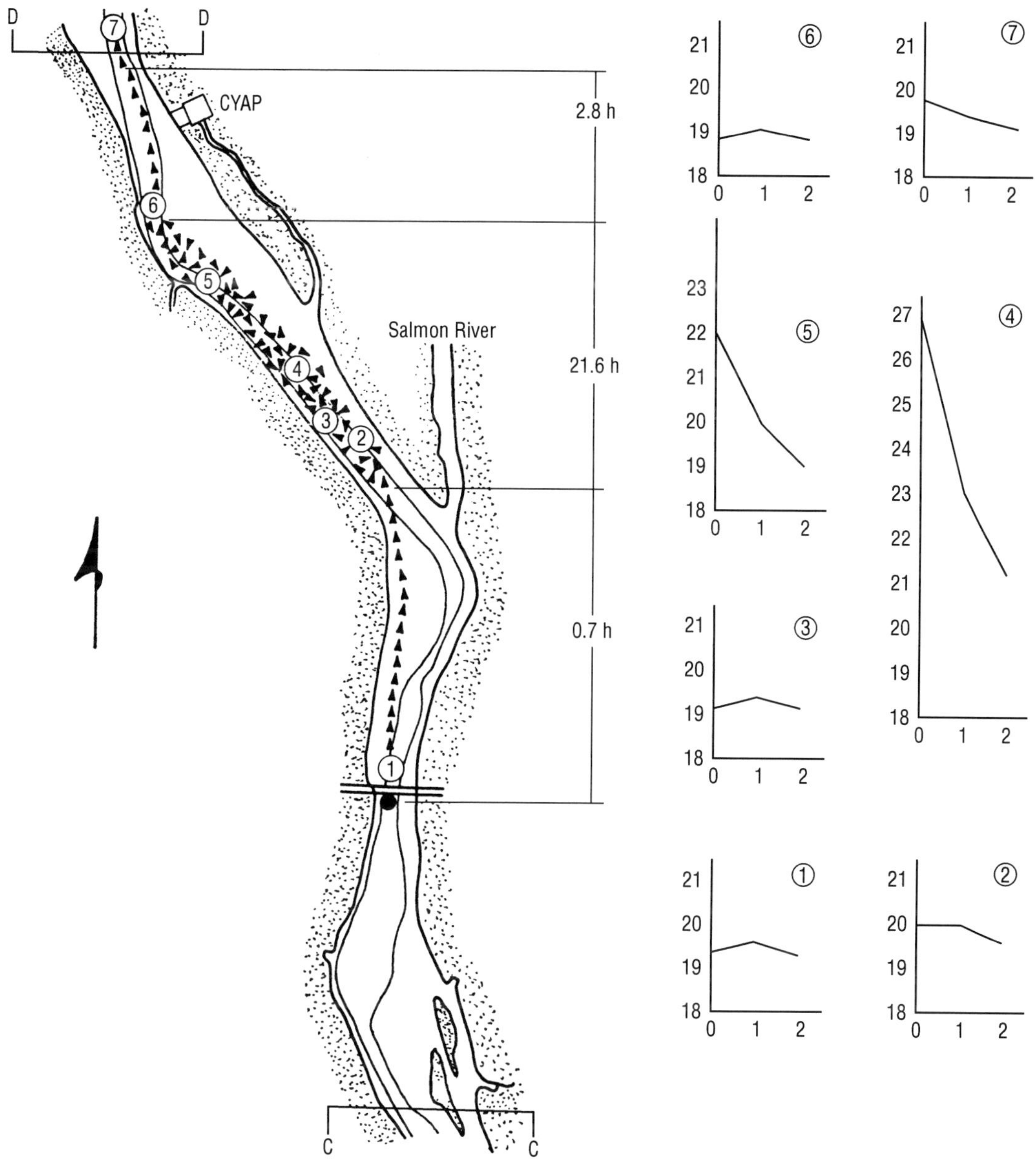

FIGURE 13. Migrations of shad showing marked alterations in behavior and rates of upriver progress in the area of the CYAP (Connecticut Yankee Atomic Power Company) discharge. Temperature profiles to a depth of 2 m are shown for selected points along the track. Arrows indicate direction of travel. Where no clear direction is evident, meandering was observed.

km below the leading edge of the salt wedge and migrated 4.8 km upstream in 5 h. Only then did it begin to meander in the vicinity of the upriver extent of the salt wedge. If the meandering activity had been related to handling effects, it seems reasonable to suppose it would have occurred shortly after release. As previously noted, such effects appear to be of relatively short duration, whereas the meandering behavior in the vicinity of the leading edge of the salinity wedge exceeded 12 h in five of the seven shad tracked there.

Similarly, meandering in the lower estuary does not appear to be a response to tidal cycle. Three fish (nos. 1, 2, 3) were tracked both in the lower estuary and in freshwater. In each case, extensive meandering occurred in the lower estuary, while upriver migration in freshwater was continuous through conditions of zero and reverse flow due to tidal action. In addition, the 31 observations of the movements of shad during slack and reverse flow conditions failed to demonstrate any clear behavior patterns associated with such events.

I conclude that the observed meandering behavior of five of the seven shad tracked in the lower estuary of the Connecticut River is a consequence of the transition from salt- to freshwater. The reactions of shad to specific salinities are not known as it was not possible to acquire accurate data as to what osmotic condition the fish were exposed to during the course of their tracks. The prediction of the upriver extent of saltwater was approximate and did not permit exact determination of salinity values coincident with fish position. Furthermore, the position of fish in the vertical salinity gradient was unknown since the ultrasonic tracking technique did not furnish information on the depth of the fish.

Tagatz (1961) showed that abrupt transfers of adult American shad from salt- to freshwater of similar temperature caused some mortality. Apparently, shad must adapt slowly to freshwater. Those fish that were tracked into freshwater meandered from 24 to 71 h at the leading edge of salinity. These time differences may have resulted from differences in the stage of freshwater adaptation that had been attained by each fish at the time of capture.

These findings also suggest that the delay in upriver migration noted in the conventional tagging studies may not be totally due to the effects of tagging, but may also result, at least in part, from meandering behavior related to entry into freshwater.

The means by which shad locate and orient themselves to the channeled areas of the river are not known. At least two possibilities exist: shad may simply seek the deepest parts of the river in which to migrate and thereby maintain themselves in channeled areas, or alternately, they may detect some difference in the environmental conditions such as temperature or flow velocities in the channel, as compared with shallower areas, and this may allow them to locate and orient to the channel. No precise knowledge of the depths at which shad migrate is available at present, so the first possibility cannot be evaluated. However, the results of a series of flow measurements made at 34 stations along 8 transects ranging from km 6 to km 24 indicated that there was a definite trend in flow velocities. Flow velocities were recorded at 2.5 m depth in shallow, channel edge, and channel areas of the river at each transect; in all cases, the highest average velocities occurred in the channel (mean 75 cm/s), intermediate velocities at the edge of the channel (mean 61 cm/s), and lowest velocities in shallow areas (mean 51 cm/s). A vertical velocity gradient in which the flow velocity decreased with increasing depth was also noted. For example the average flow at 5.0 m depth at the channel edge and in the channel at these same stations was 54 and 72 cm/s. The greatest absolute reduction in velocity (V) from surface to bottom was noted in the channel (ΔV = 34 cm/s) followed by the channel edge (ΔV = 23 cm/s) and the shallows (ΔV = 18 cm/s). These data suggest that the channel is distinguishable from other areas of the river with respect to flow velocities. The precise adjustment of swim speed to changes in current velocity both at sea (Dodson and Leggett 1973, 1974) and in freshwater indicate that shad are able to detect and respond to slight changes in current velocity and may employ them to orient to the channel.

Why shad select the channel is also unknown. There was no significant difference in the mean bottom velocity in shallow (40 cm/s), channel edge (42 cm/s), or channel (41 cm/s) areas of the river at the transects studied. However, if shad swim more than one meter off the bottom in the channel, they would have to swim harder than they would if they were migrating through shallow areas in order to achieve the same rate of displacement over the bottom. Perhaps the channel provides a better directional clue and thereby reduces the incidence of unproductive wandering from the desired upriver route. If so, this could partially offset the added energy of migrating in the channel.

These observations, plus the finding that swimming speeds were consistently higher when shad were oriented into the current, and especially so when oriented upriver, suggest that a rheotaxis directs the upriver migration of shad. This may be an extension of the olfactory-rheotaxis response that Dodson and Leggett (1974) consider to be the prime orienting mechanism during homing to the Connecticut River from Long Island Sound. However, the 16 instances of unaltered upriver migration during slack water and during conditions of upriver currents indicate that shad are capable of orienting their upriver migrations even during complete reversals in the direction of the rheotaxis clue. How this is accomplished is not known. Perhaps olfactory, inertial, or visual mechanism are substituted or become more important at such times. The fact that olfactory clues are important in locating the river initially, and the frequent departure from strongly channel-oriented behavior immediately below the confluence of the Salmon River with the Connecticut, suggests that olfaction may be important if only in guiding shad past tributaries to the main spawning sites in the Connecticut River.

The similarity of the overall rates and patterns of upriver migration from year to year, as revealed by the conventional tagging studies, the absence of changes in the distribution of conventional tag recoveries relative to the Connecticut Yankee discharge over the 9 years investigated, and the absence of significant changes in the behavior or migration times of 22 of 24 shad (92%) in the immediate area of the thermal discharge as illustrated by ultrasonic tracking studies, all clearly indicate that under present levels of plant operation no thermal block to the upriver passage of shad exists. This results, at least in part, from the fortuitous location of the channel, the preferred avenue of migration of most of the shad, on the opposite, west bank of the river away from the discharge point. The rapid thermal stratification in the area of the discharge, with the heated water largely restricted to the surface 1–2 m, also provides a pathway under the plume. Downstream migrant shad also appear to be unaffected by the thermal discharge.

Population Size 1965–1973

Estimates of the numbers of adult shad that enter the Connecticut River to spawn were obtained annually from 1965 to 1973 using the standard Petersen mark and recapture method (Ricker 1958a). Confidence limits for the estimates were calculated using the method of Chapman (1948). Tagging and recovery methods are described in the section dealing with the rate and pattern of upriver migrations (page 185). Only those tags recovered from shad captured in the Connecticut River during the upriver phase of the migration (unspent fish) during the year of tagging have been included in the analysis, and only tag recoveries from the Connecticut commercial and sport fisheries were used in the population estimates. Additional tags were recovered from fish being transported over the Holyoke Dam, from experimental netting conducted by conservation agencies of Connecticut and Massachusetts and from the sport fishery in Massachusetts. However, because of difficulties in accurately determining the size of the samples from which these tag recoveries were obtained and questions as to the detection of tags by operators of the Holyoke fish lift (see Patterns of Migration, page 194), these tags were excluded from the analysis.

All possible efforts were taken to minimize mortality resulting from capture, handling, and tagging procedures. The gill net used to capture shad for tagging was manned by commercial fishermen who were experts in net handling and were able to lift and clear the net quickly and with minimum harm to the fish. The net was fished so that the end set first was lifted first, in order to minimize the duration of confinement of captured fish. In the early phases of the run when water temperatures were low (4–10°C), the net was lifted approximately every half hour. Later in the season when water temperatures were higher, the frequency of lifting was increased because it was noted that half-hour sets produced more fish that appeared to be in a weakened condition. At that time, it was normal procedure to set the net and then begin lifting the end set first almost immediately. One half the net was lifted, cleared, and reset; then, the other half was lifted, cleared, and reset. This procedure was repeated for the duration of the drift. All tagging was done in the net boat, each fish being tagged and released immediately after being freed from the net. The average time out of water was 10 to 15 s. When fish appeared noticeably weak after capture and tagging, this condition was recorded. In 1965, the first year of the study, 123 fish were so noted. Recoveries from this group totaled 9 (7.2%) as compared with 334 recoveries (16.1%) from 2,077 fish judged to be in good condition at the time of release. This reduction in recovery was statistically significant ($\chi^2 = 5.74$; $P < 0.05$) and clearly indicated differential mortality resulting from handling and tagging. In all subsequent years, care was taken not to tag fish that appeared to have been seriously weakened by the capture process.

Additional studies were undertaken in 1966 and 1967 in an attempt to document the precise level of mortality resulting from tagging. Shad were captured, 50% of the captured fish were tagged, and these together with the unmarked fish were placed in holding pens anchored in the river (1966) or in a circular wading pool (3 m diameter, 1 m deep) supplied with a continuous flow of river water and having a counterclockwise current (1967). These experiments proved unsuccessful as all tagged and untagged fish died within 24 h with no clear pattern in the timing of death of tagged and untagged individuals. It was concluded from these experiments that mortality was due to the inadequacy of the containment apparatus rather than the stress related specifically to capture, handling, and tagging. Tagging per se appears to have little effect on mortality of tagged shad. Returns of double-tagged shad released in 1972 and 1973 (Table 14) did not differ significantly from returns of single-tagged shad (1972 $\chi^2 = 2.37$, df = 1, $P = 0.1$; 1973 $\chi^2 = 3.08$, df = 1, $P = 0.05$). These double-tagged shad bore two identical marks, which were both applied to the back musculature as previously described. Approximately every third shad was double-tagged in order to distribute the double marks as uniformly as possible. If tagging had been a significant factor, the application of two tags might have been expected to lead to increased mortality and a reduced rate of recovery of double-tagged fish.

No evidence of differential mortality with time of tagging was evident from recovery rates during the shad runs of 1965–1973 (Table 15). The slightly lower recovery rates during the last 5 d of the season resulted from the fact that fishing pressure was reduced, many fishermen having ceased fishing because of relatively poor success at this time.

While the tags themselves, or their insertion, had no effects, the effects of handling associated with tagging remain unmeasured. That handling had little effect is shown by ultrasonic tracking results. As previously described, 72 shad were tracked immediately following tagging and release. Of these, 2 (2.7%) are thought to have died; it should be noted, in this regard, that lack of movement may also have resulted from tag regurgitation, which would be incorrectly interpreted as resulting from death. Another two of these shad were released on the spawning grounds when water temperatures were above 20°C and are known to have died within 1 h of tagging; their death was not unexpected and should not be regarded as typical of the majority of shad released. In a similar study conducted in Long Island Sound

TABLE 14. Rate of tag shedding, 1972–1973.

Year	No. shad double-tagged	No. recovered with both tags	No. recovered with only one tag	Percent losing one tag
1972	1,844	422	12	2.8
1973	981	222	7	3.1
Total	2,825	644	19	

TABLE 15. Average recovery rate of shad per 5-d interval, 1965–1973.

Year	Day 0–5	5–10	10–15	15–20	20–25	25–30	30–35	35–40
1965	15.7	17.4	20.4	7.9	4.8(1)[a]			
1966	27.1	23.2	27.0	25.8	23.2(4)			
1967	21.8	28.9	23.2	21.4	22.3	23.3	20.4	
1968	26.4	22.5	21.4	28.0	27.6	20.1	12.7(4)	
1969	12.8	16.6	23.1	24.5	24.1	22.5	19.5(3)	
1970	36.6	19.5	28.0	24.8	21.6	19.8	13.8	
1971	26.7	23.4	21.7	18.5	23.7	21.6	18.5	0.5(2)
1972	25.9	16.2	23.7	21.6	19.5	22.3	25.1	
1973	20.7	24.8	22.1	20.9	26.1	20.8(4)	10.5(2)	16.2(4)
Mean	23.7	21.4	23.4	21.5	21.4	21.4	17.2	8.4

[a] Numbers in parentheses show number of days per interval when less than five.

during 1970–1973, Dodson et al. (1973) reported one death among 65 shad tracked (1.5%); here again, regurgitation may have been the cause of lack of movement.

While the capture and tagging procedures used in the ultrasonic tagging and conventional tagging operations were not identical, both groups of fish were subjected to gill-net capture, handling, and in the case of the shad tracked in Long Island Sound, 22 fish were also subjected to blinding, olfactory occlusion, or both. On the basis of this evidence, it was concluded that while some mortality occurred as a result of handling and tagging, the magnitude of this mortality was no more than 2.2% (3 out of 72 + 65) and the bias in population estimates from this factor was small. Moreover, the use of identical fishing and tagging techniques in all years would tend to standardize this error and result in accurate data on the trends occurring in population size. Interyear analyses based on the population estimates therefore can be considered to be unaffected by handling mortality as long as it is assumed to be similar in all years.

Differences in vulnerability of marked shad relative to unmarked members of the population could also bias population estimates. Such differences could result from differences in behavior resulting from tagging, or from the effects of the tag itself, which might increase the vulnerability of the fish to capture by commercial gear. The use of spaghetti dart tags in this study greatly reduced the possibility of the latter factor being a serious source of error, since these tags are not subject to tangling in the nets and therefore should not increase the probability of capture. Changes in behavior are more difficult to evaluate. Analysis of upriver migration rates as determined from conventional tagging data suggests that after being tagged with a spaghetti-dart tag, shad delay their upriver migration for approximately 10 d. However, as some tracking data show, shad may normally delay their upriver migration at the salt–freshwater interface for up to 5 d. The added delay by dart-tagged shad, possibly as a result of shock related to handling and tagging, could cause them to be more vulnerable to the fishery because they would be exposed to it for a greater period of time. However, analysis of the distribution of tag recoveries within three areas of the commercial fishery reveals no evidence of higher-than-average rates of recovery from the river mouth relative to points farther upriver (Table 16).

TABLE 16. Ratio of tagged to total shad in commercial catches from three segments of the commercial fishery.

Year	River mouth (km 0–11)	Middle stretch (km 20–29)	Upriver area (km 42–47)
1965	0.0015	0.0028	0.0018
1966	0.0047	0.0112	0.0096
1967	0.0103	0.0015	0.0094
1968	0.0141	0.0144	0.0098
1969	0.0049	0.0116	0.0124
1970	0.0047	0.0156	0.0046
1971	0.0047	0.0049	0.0092
1972	0.0153	0.0176	0.0150
1973	0.0049	0.0130	0.0044
Means	0.0072	0.0103	0.0085
SD	0.0048	0.0059	0.0042

The observation that sonic-tagged shad avoid gill nets with considerable facility (Leggett and Jones 1971) could be explained on the basis that once shad have encountered a gill net, they are wary and avoid similar conditions in their subsequent movements for a time at least. If such is the case, a reduction in the vulnerability of tagged shad, relative to untagged shad, would occur. We have no direct evidence with which to evaluate this possibility. However, the literature pertaining to the behavior of other species of fishes when confronted with nets (Blaxter and Parrish 1965; Hester and Taylor 1965; Beamish 1969; Hemmings 1969; Mohr 1969; Parrish 1969; Parrish et al. 1969) suggest that such avoidance behavior is normal even among unmarked members of the population. Differential gear selectivity between the tagging and recovery nets could also result in differences in the vulnerability of marked and unmarked shad. In this study, the nets used to capture fish for tagging were identical in mesh size and design with those used in the commercial fishery. They were fabricated, however, from monofilament nylon, which, at the time of this study, was outlawed in the commercial fishery except in the estuary of the river below the Penn Central Railroad bridge (km 3). The possibility that the composition of the tagging catch and the catch of the commercial fishery, as a whole, differed significantly was tested by comparing the mean size of all fish tagged with the mean size of tagged fish that were recovered in the commercial fishery. No significant difference was found between the mean lengths of shad that were tagged and tagged shad that were taken from commercial nets (Table 17). This suggests that differential vulnerability due to differences in fishing gear did not exist.

The extent of tag shedding during the fishing season was investigated by double-tagging experiments conducted in 1972 and 1973. Of the shad tagged in 1972 and 1973, 40% and 34% were double-tagged. Recovery of these double-tagged fish indicated that 2.8% and 3.1% of the double-tagged shad in 1972 and 1973 lost one tag during the interval between tagging and recovery (Table 14); these two rates did not differ significantly (χ^2 = 0.95; df = 1, $P > 0.75$). For the purpose of calculating population estimates, an annual rate of shedding of 3% was assumed, and the number of tags applied was reduced by 3% to eliminate the error associated with tag shedding.

With the exception of 1965, attempts to capture fish for tagging began prior to, or very close to, the beginning of the annual shad run as indicated by the fact that in 7 of the 8 years, no shad were captured on the first day of fishing. Subsequent to the initial capture of shad, fishing continued on a regular basis until later in the season when either fishing success declined drastically with the onset of the end of the run or water temperature became a significant factor in obtaining active fish for tagging (1965). The distribution of tagged shad in the population was therefore approximately proportional to the number of shad entering at any given time during the run. Some alteration in the proportionality of the distribution may have occurred in the upriver areas because of the estimated 10-d delay in upriver migration resulting from tagging and/or adaptation to freshwater; however, this was not considered to be a significant source of error in the population estimates, since the distribution of fishing effort in the commercial and sport fisheries was approximately proportional to the abundance of fish in that it was largely governed by fishing success.

TABLE 17. Mean fork lengths of tagged shad released and mean lengths of tagged shad recaptured in the commercial fishery.

Year	Tagged N^a	Tagged Mean length (cm)	Tagged SD	Recaptured N^b	Recaptured Mean length (cm)	Recaptured SD
			Males			
1967	55	47.6	2.6	7	48.1	1.4
1968	237	48.4	2.3	41	48.4	1.8
1969	564	47.2	3.7	107	47.9	2.8
1970	1,425	47.5	2.3	306	47.5	2.3
1971	1,130	47.3	2.2	233	48.1	2.1
1972	1,841	47.9	2.2	235	47.9	2.0
1973	1,132	47.8	2.3	176	47.9	2.4
			Females			
1967	74	50.1	3.2	15	50.2	2.3
1968	520	50.2		87	49.7	1.8
1969	801	49.6	2.8	180	49.3	2.6
1970	1,433	50.6	2.5	360	50.5	2.4
1971	917	50.5	2.5	222	50.5	2.4
1972	1,712	51.0	2.2	280	51.0	2.2
1973	1,386	50.8	2.3	247	51.4	2.2

[a] Number measured at time of tagging.

[b] Number of tagged fish recovered for which length at time of tagging was known.

Advertising of the tagging program, personal contacts with commercial fishermen prior to and after the fishing season, and contact with sport fishermen during the creel census that was conducted in the sport fishing areas served to alert most fishermen as to the nature of the tagging program, the type of tag being employed, its position on the fish, and the reward offered for its return. In spite of these efforts, however, some tags were not observed by commercial fishermen. A number of tags were returned from the Fulton fish market in New York and from individuals who had purchased shad from local markets; however, the number of these returns was less than 1% of the total recoveries in all years. Although some bias must have resulted from missed tags, it was small and furthermore it was approximately the same throughout the 8 years of the study.

Since tagging and recaptures took place over a short time period, and only fully recruited (mature) fish enter the river, recruitment did not have to be considered in these population estimates.

The Connecticut River Commercial Shad Fishery, 1965–1973

Walburg and Nichols (1967) have reviewed the history of the commercial shad fishery in the Connecticut River. They noted that seines were traditionally used for the commercial capture of shad in the river, but that more recently, they have been superceded by drifting gill nets. In 1960, only one seine was fished commercially for shad, and in subsequent years, no significant numbers of shad have been taken by this method. The physical locations of the main commercial fishing sites have changed little since 1896. Then, as now, the fishery was largely restricted to the lower 64 km of the river with most of the fishing occurring in the lower 30 km. The nets, which average 180–275 m in length and 30–50 meshes in depth, are predominantly of 14 cm mesh (stretch measure), although a small number of fishermen employ 14.6 cm mesh in an attempt to capture larger fish. The nets are generally tailored to the area fished and extend close to the bottom. Very light footlines are used and the nets are weighted with wire rings, which tear free easily, thus minimizing damage to the nets as a result of frequent contact with the bottom and with snags. Fishing occurs mainly at night, generally on the last of the ebb tide. A minor amount of daytime fishing occurs in the estuary (km 0–4) where monofilament gill nets are allowed. Fishing is permitted 5 nights per week with Friday and Saturday nights closed to allow a portion of the run to proceed upriver unmolested. Fishing normally begins in the second week of April and continues until fishing success becomes marginal (generally during the last week of May or first week of June) as the run proceeds upriver beyond the main commercial fishing areas. After 1966, the commercial fishermen largely adopted the policy of allowing our study tagging crews the dubious honor of determining when the run into the river had begun and then initiating their efforts in earnest. Connecticut River commercial fishermen are required by law to report to the Connecticut Department of Environmental Protection at the close of each fishing season. These reports were used as the basis estimating the size of the catch from which tag recoveries from the commercial fishery were derived. Bampton (1964) speculated that the size of the true catch might be 50% to 200% higher than the reported catch. This occurs for a number of reasons such as lack of vigorous law enforcement, failure to maintain adequate records at the time of fishing, and attempts to avoid the declaration of income from fishing. This problem was reduced in 1973 when new reporting laws were vigorously enforced. In an effort to detect those fishermen whose reports might not be reliable and to correct for the bias that would result in our estimates of population size, a test was developed to examine the ratio of untagged and tagged fish in reported catches of individual fishermen and to relate this ratio to that of all other fishermen reporting catches in a given year. The method assumed that the catchability of tagged shad was the same as untagged fish and that, subject to normal variability, a uniform recovery ratio applied to all fishermen. As noted previously, the use of spaghetti dart tags supports the initial assumption because these tags do not become entangled in gill nets and therefore do not increase the probability of capture. But since no data were obtained on the relative catchability of tagged and untagged shad because of possible changes in behavior resulting from tagging, the possibility remains that the first criterion may not have been completely met. The second assumption dealing with the uniform probability of capture of tagged fish among fishermen appears to be valid (Table 16).

The method that was developed employed the chi-square test to examine the relationship between the number of tags expected on the basis of the reported catch and the number of tags actually returned:

$T_{ej} = AC_{fj}/C_t$, where T_{ej} = number of tags expected from individual fisherman j; A = total number of tags returned from the fishery; C_{fj} = total reported catch for fisherman j; and C_t = total reported catch for the fishery.

$\chi^2_{total} = \sum_{j=1}^{n} (T_{oj} - T_{ej})^2/T_{ej}$, with $(n - 1)$ degrees of freedom, where T_{oj} = observed number of tags returned by fisherman j.

If χ^2_{total} exceeded the tabular value ($P < 0.01$) at the appropriate number of degrees of freedom, the catches of those fishermen whose reported catches represented the greatest cell contribution to χ^2_{total} were removed from the list of reliable fishermen's catches until χ^2_{total} approached the expected value. The annual minimum, mean, and maximum individual chi-square values for rejected catches are given in Table 18. At this point, the reliability of the records of the fisherman who remained in the list was assumed, and the mean ratio of tagged to total fish in their catches was accepted as the standard for the fishery as a whole (Table 19). The commercial catch for a given year was then estimated as: $C_e = R_t/R_r$, where C_e = estimated (corrected) commercial catch; R_t = total commercial tags recovered; R_r = mean ratio tagged/total fish in catches of reliable fishermen.

The reported and estimated commercial shad catches for 1965–1973 are presented in Table 19. The greatest differences between reported and

TABLE 18. Annual minimum, mean, and maximum individual chi-square values for catches of commercial fishermen rejected for inaccurate reporting.

Year	Chi-square		
	Minimum	Mean	Maximum
1965	7.0	33.1	128.3
1966	9.2	39.9	109.9
1967	7.2	86.5	630.5
1968	11.5	62.9	404.3
1969	7.0	45.3	214.7
1970	7.1	59.8	388.5
1971	8.1	51.1	185.2
1972	9.8	21.7	72.6
1973	12.0	49.3	104.0

estimated catches occurred in years when the fishing was good as judged by the size of the catch (Spearman rank test: $r = -0.76$, $N = 9$, $P < 0.05$; Seigel 1957). This pattern of reporting might be expected to occur on intuitive grounds since large catches, if accurately reported, would lead to high income tax liabilities.

The Connecticut River Sport Fishery for Shad, 1965–1973

The history and character of the Connecticut River sport fishery has been reviewed by Nichols and Tagatz (1960). The fishery is presently localized in about eight main areas from the mouth of the Farmington River to South Hadley Bridge, Holyoke, Massachusetts (Figure 14). Most of the sport fishing for shad in Connecticut is shore-based, with boat fishing limited to areas at the base of the Enfield Dam and south of the Windsor Locks Bridge. The legal sport fishing season for shad in Connecticut begins on the third Saturday in April and continues until the end of June, the exact date of closure being determined by the state on the basis of the success at its controlled fishing area at Enfield Dam and on the general conditions in the river. In some years, when river discharge is very low in June, shad become susceptible to foul hooking and other outlawed practices, and the season may be closed early to prevent such infractions.

The total number of fish captured each year in the sport fishery in the state of Connecticut was estimated by combining the results of two creel surveys, one a total census conducted by the Connecticut Department of Environmental Protection at their controlled access fishing site at Enfield Dam (km 85), and the other an incomplete survey conducted by personnel of the River Study. The latter included seven fishing sites representing the major sport fishing areas (apart from the state-censused area) in Connecticut (Figure 14); it was initiated on opening day, and the termination date was arbitrarily set at 30 June for the purpose of establishing a sampling schedule. The maximum deviation between this date and the actual date of termination occurred in 1969 when the season officially ended on 22 June.

The official fishing season, as defined above, was partitioned into fortnightly periods with four sampling dates per period. These dates were selected at random with the restriction that two weekdays and two weekend days be sampled in each period (Carlander et al. 1958). The daily survey began at 0700 hours on weekdays and 0800 hours on weekends and terminated after dark at 2100 hours. The number of census takers was adjusted through the season in accord with changes in fishing pressure. Early and late in the season when fishermen were scarce, one biologist surveyed all seven areas. At the height of the season as many as seven biologists were employed on each census date, with one individual at each location throughout the day. Each area was surveyed approximately every one and a half hours. At these times, the number of rods fished, number of shad taken, and sex of the shad caught was recorded for each fisherman present. When the same fisherman was interviewed several times in succession, only those shad caught in the interval between interviews were recorded, and the time between interviews was taken as the fishing time. Boat (or trailer) registrations were recorded at all boat-launching sites during each survey, and these were checked off when the parties returned. When boats were occasionally missed, the daily average catch for the area was substituted to correct for the resulting error. With this exception, no fish was recorded as caught unless seen by the census taker. Extrapolation of the census data to an estimate of the total catch of the sport fishery and the related effort was achieved by first conducting a hypothetical census employing catch-effort data from the state fishing area. The method used was that of Best and Boles (1955). State census-area data were reduced to dates corresponding to the study survey dates. Catch effort values for intervening dates were estimated by substituting a combina-

TABLE 19. Total commercial tags recovered, ratio of tags/total catch (T/C) for all fishermen reporting catches (unadjusted), for fishermen judged reliable in their reporting (adjusted), and corrected (estimated) commercial catch.

		Unadjusted					Adjusted						
											Chi-square		
Year	Total commercial tags recovered[a]	Fishermen reporting catches	Tags[a] returned	Catch reported	T/C	Chi-square	Fish-ermen judged reliable	Tags returned	Catch reported	T/C	Calculated	$P = 0.01$	Total estimated catch
1965	251	41	197	84,426	0.00233	328.3	33	104	71,675	0.00145	57.6	53.4	173,103
1966	853	38	687	67,422	0.01019	444.9	27	549	52,032	0.01055	46.7	45.6	80,853
1967	803	40	727	67,695	0.01074	1191.0	27	431	40,767	0.01057	45.6	45.6	75,969
1968	879	39	822	61,092	0.01346	758.8	28	582	38.528	0.01511	51.0	46.9	58,173
1969	764	38	577	49,772	0.01159	760.0	22	264	23,765	0.01111	38.9	38.9	68,767
1970	673	33	526	48,484	0.01085	785.8	20	211	22,385	0.00943	43.1	36.2	71,367
1971	730	38	701	48,772	0.01437	405.2	33	333	34,073	0.00977	59.0	53.4	74,718
1972	834	37	806	48,968	0.01646	361.9	23	510	31,463	0.01629	47.6	40.3	51,197
1973	575	54	575	60,665	0.00948	484.4	44	383	45,644	0.00839	72.9	66.2	68,534

[a] Differences result from failure of some known fishermen to report catches as required by law.

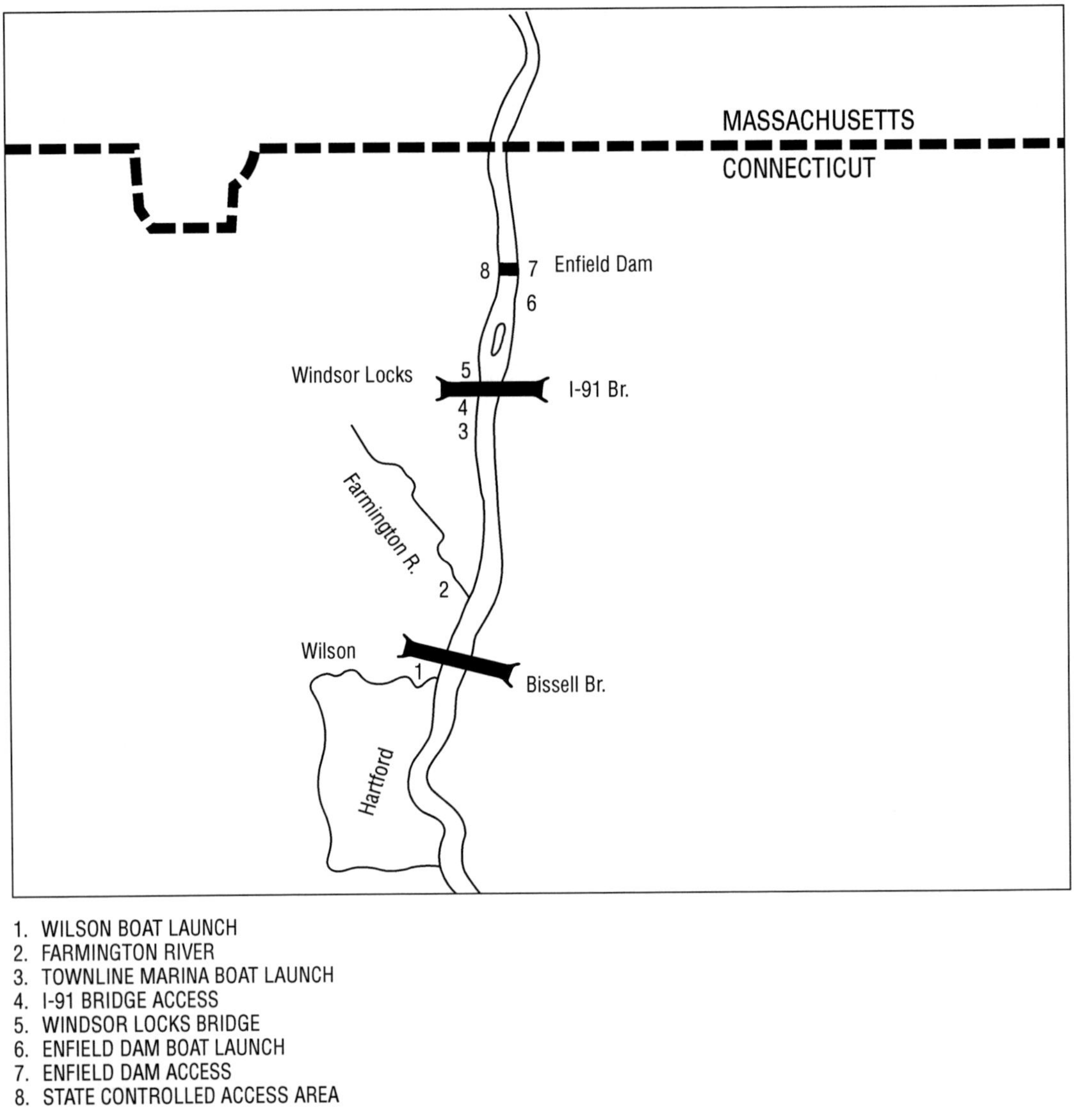

FIGURE 14. Connecticut River creel survey sites.

tion of seasonal, weekly, and weekend averages for the uncensused dates. Several possible combinations were evaluated until a method was derived that produced a difference of less than 5% between the actual state census counts and the estimates resulting from the hypothetical census. This method was then used to estimate the total catch and effort values for the study survey area. The study and state catch and effort data were then combined to form an estimate of the catch and effort in the Connecticut State sport fishery. This procedure was considered valid because the state-censused area lay within the area surveyed by the study personnel and therefore was similarly affected by changes in river flow, temperature, atmospheric conditions, and fishing pressure.

Creel surveys were conducted by the River Study each year from 1965 to 1969. During that period, various modifications of the initial sampling design were tested by conducting further hypothetical manipulations of assembled data. None proved to be more accurate. In 1970, the annual census was discontinued because the 1965–1969 data had revealed a relatively constant relationship between the total effort and the catch per unit effort in the state and study census areas

(Table 20). The catch per unit effort and fishing pressure data for 1965–1969 was therefore averaged, and the 5-year average ratio between the state and study areas was calculated for each year. In the remaining years of the study (1970–1973), the total catch per unit effort and fishing effort in the Connecticut portion of the sport fishery was determined by the following formulae:

$$R_c = R_s \times A,$$
$$S_c = S_s \times B,$$
$$C_c = R_c \times S_c, \text{ and}$$
$$C_t = C_s + C_c,$$

where R_c = rod hours in study census area, R_s = rod hours in state census area, A = 1965–1969 ratio R_c/R_s = 1/0.75, B = 1965–1969 ratio S_c/S_s = 1/0.95, S_c = Catch/unit effort in study area, S_s = Catch/unit effort in state area, C_c = estimated shad catch in study area, C_s = counted shad catch in state area, and C_t = total shad catch in Connecticut sport fishing areas.

Effort, catch and catch-per-unit-effort data for the state and study survey areas and for the total sport fishery in the Connecticut River, 1965–1973, are given in Table 20. The largest catch and the greatest effort occurred in 1966. The lowest catch recorded was in 1972. Catch-per-unit-effort data correspond very closely to the commercial catch estimates with the greatest success occurring in 1965, followed by a downward trend to 1968. This, in turn, was followed by a gradual increase in fishing success through 1971. In 1972, a significant decline in fishing success and total catch was recorded making it the poorest sport fishing year of the nine investigated. Fishing success improved marginally in 1973.

Population Estimates

The number of shad entering the Connecticut River to spawn each year from 1965 to 1973 was estimated by means of tag returns and catch estimates from the commercial fishery and from the sport fishery. The resulting estimates are given in Table 21. Estimates based on the commercial catch data were significantly correlated with three independent estimates of abundance (Figure 15) (i.e., catch-per-unit effort in the sport fishery [Table 120; Spearman Rank Test: $r = 0.83$, $N = 9$, $P < 0.01$; Siegel 1957], catch per unit effort when capturing shad for tagging [Table 22; $r = 0.63$, $N = 9$, $P < 0.05$] and catch per unit effort in the commercial fishery [Table 23; $r = 0.7$, $N = 9$, $P < 0.05$]). Population estimates derived from sport-fishing data were much larger and, because of the smaller number of tags recovered from this source, had much broader confidence limits. These estimates, too, were significantly correlated with catch per unit effort in the sport fishery (Spearman Rank Test: $r = 0.73$, $N = 9$, $P < 0.05$), in the tagging catch ($r = 0.69$, $N = 9$, $P < 0.05$), and in the commercial fishery ($r = 0.74$, $N = 9$, $P < 0.05$). Several factors may contribute to the higher population estimates based on the sport fishery. In spite of advertisements and interviews with sport fishermen, the tagging program was less well known to them than to commercial fishermen. This, coupled with the inability to contact all licensed sport fishermen to ensure that tags were returned, as well as the desire expressed by some of them to retain tags as souvenirs, may have produced a higher frequency of nonreporting of tags in the sport fishery. This would lead to higher population estimates. For these reasons, the population esti-

TABLE 20. Summary of sport fishery statistics for shad in the Connecticut River, 1965–1973. C/E = catch per effort.

	State area			Study area			Total		
Year	Rod-hours	Catch	C/E	Rod-hours	Catch	C/E	Rod-hours	Catch	C/E
1965	29,216	12,384	0.424	21,367	7,501	0.351	50,583	19,885	0.393
1966	30,732	8,661	0.282	52,449	14,023	0.268	83,181	22,693	0.273
1967	29,712	7,817	0.263	39,512	9,709	0.246	69,224	17,526	0.253
1968	34,808	5,684	0.163	40,516	8,593	0.212	75,324	14,277	0.190
1969	27,444	5,632	0.205	32,017	7,186	0.224	59,461	12,818	0.216
1970	24,580	4,953	0.201	32,773	6,981	0.213	57,353	11,934	0.208
1971	31,952	6,942	0.217	42,496	9,689	0.228	74,448	16,631	0.223
1972	22,304	2,838	0.127	29,664	3,945	0.133	51,968	6,783	0.131
1973	24,180	3,247	0.134	32,159	4,534	0.141	56,339	7,781	0.138

TABLE 21. Connecticut River shad population estimates based on commercial and sport fishery statistics, 1965–1973.

	Commercial				Sport			
	Tags				Tags			
Year	Available	Recovered	Catch	Estimated population	Available	Recovered	Catch	Estimated population
1965	2,200	251	107,103	1,470,000 ± 148,000	1,949	36	19,885	1,076,000 ± 299,000
1966	3,991	853	80,853	367,000 ± 25,000	3,138	69	22,693	1,023,000 ± 215,000
1967	4,011	803	75,969	368,000 ± 24,000	3,208	62	17,526	905,000 ± 199,000
1968	4,362	879	58,173	280,000 ± 19,000	3,483	92	14,277	539,000 ± 99,000
1969	3,810	764	68,767	333,000 ± 22,000	3,046	46	12,818	848,000 ± 216,000
1970	4,072	673	71,367	419,000 ± 37,000	3,399	73	11,934	556,000 ± 114,000
1971	4,147	730	74,718	428,000 ± 31,000	3,417	77	16,631	738,000 ± 147,000
1972	4,622	834	51,197	275,000 ± 17,000	3,785	37	6,783	694,000 ± 190,000
1973	2,871	575	68,534	332,000 ± 28,000	2,296	20	7,781	893,000 ± 314,000

[a] Total tagged less number recovered in commercial fishery.

mates based on the commercial catch are thought to reflect more accurately the size of the annual populations of shad that enter the river to spawn. The smaller confidence intervals associated with these estimates also make them more reliable for detecting annual trends in population size.

Population Estimates 1935–1964

Fredin (1954) estimated the size of the shad populations entering the Connecticut River annually from 1936 to 1951. These estimates were derived from a single Petersen mark–recapture experiment, conducted in 1951, in which 633 shad were marked with Petersen-disc tags at the mouth of the river and released to continue their upriver migration. Subsequently, 359 tags (56.7%) were recovered from the commercial fishery, in which a catch of 100,967 shad was reported for the year. Fredin estimated the 1951 population to be 178,072 shad.

Population estimates for 1935–1950 were calculated by means of the formula $N_x = C_x/(1 - q^n)$, where N_x = estimated population year x, C_x = reported commercial catch year x, n = number of standard fishing unit (SFU) days[2] in year x, $q = (1 - p)$, and p = portion of total population removed per standard fishing unit day.

From the 1951 tagging data, q was estimated as follows: $q^nN = E$, where N = population size and E = escapement. Thus, q^{2589} (178,072) = (178,072 − 100,967); $q^{2589} = 0.433$; and $q = 0.999675$ (Fredin 1954).

Since neither the location of the fishery, the nature of the gear, nor the number of legal fishing days have changed significantly since 1951, it would be reasonable to expect that if the assumptions inherent in Fredin's methods are valid, a reasonable approximation of our 1965–1972 population estimates should be obtained by applying his formula, $N_x = C_x/(1 - 0.999675^n)$, to the catch and effort data for those years. The population estimates for 1965–1972 based on this method are given in Table 24. These estimates are based on corrected commercial catches. Total SFU days (n) per season were calculated by assuming that all fishermen fished an average of 26 d (Fredin 1954: Table 2, 1948–1951). The number of SFU days was calculated on the basis of the number of fishermen known to have fished, as indicated by tag recoveries, rather than on the number who reported catches. This was done because in recent years, the state of Connecticut has not vigorously enforced its reporting laws, with the result that several fishermen failed to report their catches, or reported catching no shad, in years when they returned one or more (frequently more than 10) tags for rewards. This situation was improved in 1973 when new reporting requirements were strongly enforced. As Table 24 clearly shows, the Fredin calculations underestimate the size of the 1965–1973 populations, averaging 40% lower (range 23–64%) than those of the present study. This results from Fredin's 1951 estimate of p (325×10^{-6}) being larger than the values for 1965–

[2] One SFU day = 1 gill net fished 1 d = 0.625 seine days = 0.769 pound-net days.

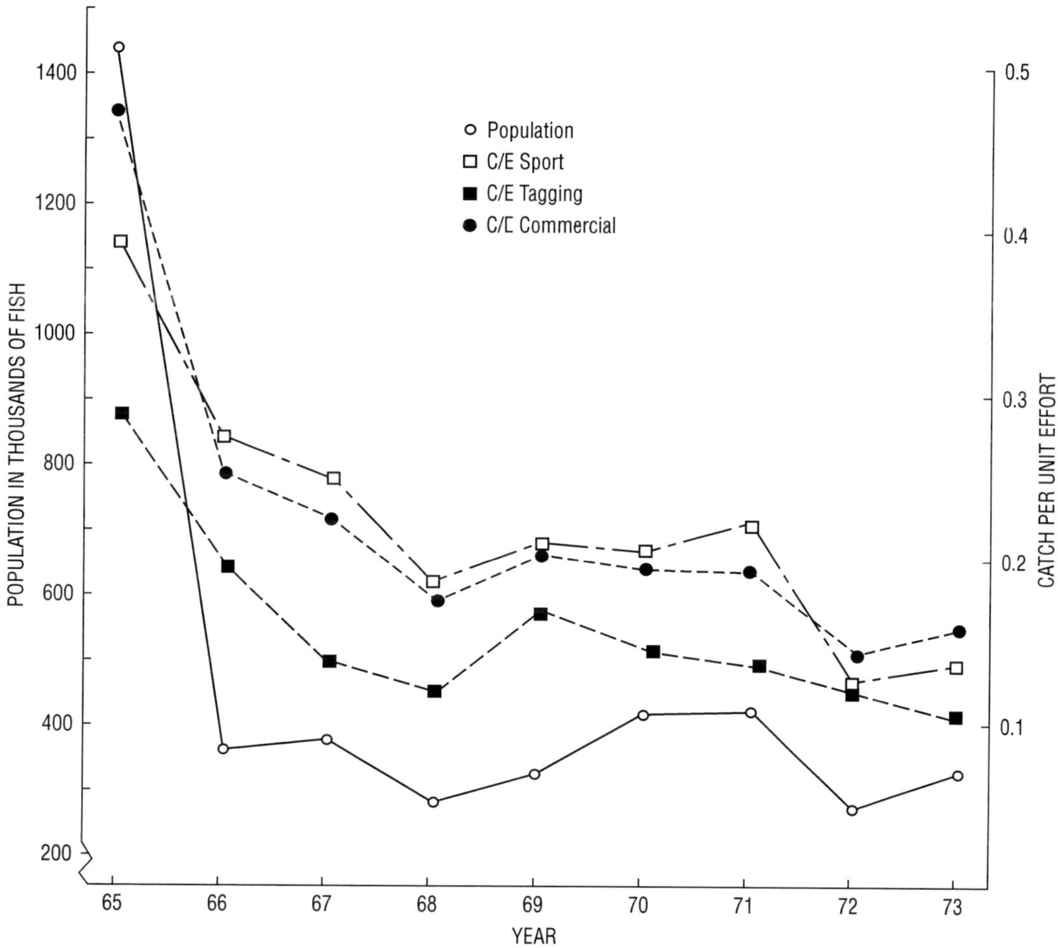

FIGURE 15. Population size and related fishing success in the tagging, commercial, and sport fisheries, 1965–1973.

1973 calculated in the present study (mean = 180 × 10^{-6}, range 103 × 10^{-6} to 239 × 10^{-6}; Table 24).

Several factors may account for this difference. The first is that Petersen-disc tags, inserted as they were through the dorsal musculature near the origin of the dorsal fin, may have significantly increased the susceptibility of capture of the shad tagged in Fredin's study. This would be especially true for the smaller shad that might normally escape through the net mesh. In addition, shad moving along the nets in an attempt to avoid them (Leggett and Jones 1971) might become tangled by catching the disc in the mesh. Any shad struggling to become free after touching the net would also almost certainly tangle the disc in the mesh and reduce its chances of escape. According to Fredin (1954), there was no evidence that the Petersen-disc tags increased the susceptibility of the fish to capture, but evidence to the contrary does exist. Fredin noted that there was no difference in the proportion of returns of fish that were tagged from gill nets and those tagged from pound nets. Leggett (1969) analyzed unpublished age-frequency data for the 1957 Connecticut River

TABLE 22. Catch per standard-net-minute, of the shad-tagging catch 1965–1973.

1965	1966	1967	1968	1969	1970	1971	1972	1973
0.2925	0.1975	0.1406	0.1222	0.1725	0.1500	0.1400	0.1302	0.1051

TABLE 23. Percentage of the total shad population (p) removed during one standard fishing unit day and number of standard fishing unit days (n).

Year	n	p
1965	1,222	103×10^{-6}
1966	1,040	239×10^{-6}
1967	1,118	207×10^{-6}
1968	1,040	224×10^{-6}
1969	1,092	212×10^{-6}
1970	1,222	153×10^{-6}
1971	1,274	151×10^{-6}
1972	1,248	165×10^{-6}
1973	1,404	165×10^{-6}

commercial fishery (provided by the National Marine Fisheries Service, Beaufort, North Carolina) and found that pound-net catches had a higher proportion of virgin (first-time spawners), younger, and smaller fish than did the gill net catches (67% and 59%, respectively). In addition, the proportion of virgins in the sport fishery catch (84%) suggested that the commercial fishery was more effective against larger members of the population. Assuming no effect of tagging on subsequent recapture, one would expect a lower recovery rate among shad tagged from the pound nets, since that group would contain a higher proportion of smaller, less vulnerable fish than the group tagged from gill nets. The absence of any significant difference in recovery rates may therefore have been due to increased susceptibility of small fish to capture in gill nets due to the tag. When questioned about the "button" tag used in 1951, commercial fishermen expressed considerable dislike for the tag because it became so badly tangled in the net and made removal of the fish difficult. They also reported finding "significant" numbers of fishless tags entwined in their nets. These observations suggest that the shad marked with Petersen-disc tags were, in fact, selected for, and that the resulting rate of tag recovery in 1951 was too great for use in population estimates.

As the data in Table 23 show, there is some annual variability in p (proportion of total population captured per SFU day). The range of observed values, 1965–1973, was 103×10^{-6} to 239×10^{-6} (mean 180×10^{-6}). There was, however, no significant correlation between annual estimates of p and total effort ($r = -0.26$, df = 7, $P > 0.1$), population size ($r = -0.08$, df = 7, $P > 0.1$) or mean river discharge in May ($r = 0.32$, df = 7, $P > 0.1$). Furthermore, there was no significant partial correlation between p and any one of these variables when the possible influence of the others was excluded: $r_{p,\ SFU} = -0.27$, df = 5, $P > 0.1$; $r_{p,\ population} = -0.02$, df = 5, $P > 0.1$; $r_{p,\ discharge} = 0.31$, df = 5, $P > 0.1$.

This indicates that the observed annual differences in p are independent of changes in effort, population size, and river discharge, and are an expression of the normal range of variability in this parameter. It should be noted, however, that the range of total effort, 1965–1973, was not large (1,040–1,404 SFU). A significant relationship between p and these variables might exist over a greater range of effort.

Estimates of annual population sizes 1935–1964 were developed from Fredin's equation, N_x

TABLE 24. Comparison of population estimates, 1965–1973 based on the present study or calculated by the method of Fredin (1954).[a]

Year	Catch	Standard fishing units (SFU)	Estimated population (Fredin)	Estimated population (Leggett)	Fishing rate Fredin	Fishing rate Leggett
1965	173,103	1,222	528,000	1,470,000	32.8	11.8
1966	80,853	1,040	282,000	367,000	28.7	22.0
1967	75,969	1,118	249,000	368,000	30.5	20.6
1968	58,173	1,040	203,000	280,000	28.7	20.8
1969	68,767	1,092	230,000	333,000	29.9	20.7
1970	71,367	1,222	218,000	419,000	32.7	17.0
1971	74,718	1,274	220,000	428,000	33.9	17.4
1972	51,197	1,248	154,000	275,000	33.2	18.6
1973	68,534	1,404	187,000	332,000	36.6	20.6

[a] $q = 0.999675$.

$= C_x/(1 - q^n)$, where the 9-year (1965–1973) mean value of $q = (1 - p) = (1 - 0.000180) = 0.999820$. For use in this analysis, reported annual catches (1935–1951: Fredin 1954; 1952–1964: Connecticut Department of Environmental Protection, unpublished data) were adjusted by substitution into the equation $Y = -32742 + 1.876\ X$, which describes the relationship between reported and actual catches (page 209). This was done to eliminate the error in reported catches resulting from nonreporting and underreporting of catches by commercial fishermen. Values of n (SFU days) for the years 1935–1951 were taken from Fredin (1954). For the years 1952–1964, SFU was estimated by multiplying the number of nets licensed by 26 (average number of days fished/year; page 216).

The population estimates for the years 1935–1964 (Table 25) indicate that the numbers of Connecticut River shad have oscillated between 990,000 (1944) and 267,000 (1956) during the 30-year interval. This is of the same order of magnitude as population fluctuations during the 1965–1973 period (1,470,000–275,000). Estimated fishing mortality rates for 1935–1964 (mean 31.3%, SD: 11.2) were higher than the levels observed in 1965–1973 (mean 18.8%, SD: 3.1). This is consistent with the heavier fishing pressure of the earlier era. These higher fishing rates are especially apparent in the period 1945–1953 (mean 47.0%, SD: 10.3) and coincide with both the greatly increased effort in those years and the dramatic increase in catch which occurred after 1942 (Figures 1 and 18).

TABLE 25. Size of Connecticut River shad population, 1935–1964.

Year	Catch	Standard fishing units (SFU)	Estimated population size	Fishing rate
1935	200,375	1,617	739,000	27.1
1936	166,975	1,504	704,000	23.7
1937	183,459	1,730	686,000	26.7
1938	202,425	1,918	693,000	29.2
1939	174,691	1,767	641,000	27.3
1940	141,410	1,265	694,000	20.4
1941	195,014	1,231	981,000	19.9
1942	182,662	1,298	877,000	20.8
1943	266,210	1,845	942,000	28.3
1944	364,901	2,554	990,000	36.9
1945	374,136	3,764	760,000	49.2
1946	529,657	5,309	861,000	61.5
1947	410,825	5,140	681,000	60.3
1948	299,745	4,118	573,000	52.3
1949	212,254	3,692	437,000	48.6
1950	111,879	2,749	287,000	39.0
1951	156,672	2,589	420,000	37.3
1952	223,148	2,808	714,000	31.3
1953	185,206	3,146	428,000	43.3
1954	124,519	2,418	353,000	35.3
1955	79,520	2,158	247,000	32.2
1956	70,973	1,716	267,000	26.6
1957	120,737	1,482	516,000	23.4
1958	208,101	2,002	688,000	30.2
1959	174,113	1,742	647,000	26.9
1960	184,825	1,768	678,000	27.3
1961	203,045	1,950	686,000	29.6
1962	194,640	2,054	630,000	30.9
1963	117,100	1,872	409,000	28.6
1964	109,078	1,612	433,000	25.2

Mortality

In an effort to detect whether there were changes in the mortality rates of shad related to the operation of the Connecticut Yankee plant, and whether such changes could be related to changes in population size, estimates were obtained of the fishing and the natural mortality of the two sexes of each age-class comprising the adult shad populations that entered the Connecticut River to spawn during the years 1965–1973.

The relative proportions of shad of various ages in the commercial and sport fisheries of the Connecticut River were determined by scale analysis. Approximately 100 samples were collected weekly from each fishery throughout the spawning migration. Age and spawning histories were read from the scales, using the method developed for shad by Cating (1953) and validated by Judy (1961). All scales were read by two biologists without reference to any data on sex, length, or age. The results of these independent readings were compared; when differences occurred (approximately 25%), the scales were reread. If differences persisted on the second reading, the sample was discarded (approximately 2%).

Leggett (1969) concluded that accurate estimates of the proportion of various age-classes of shad present in annual populations can be obtained by determining their contribution to the sport and commercial fisheries and then weighting these two estimates equally to provide an estimate for the river as a whole. This manipulation is required because the commercial fishery, which utilizes drift gill nets almost exclusively, is selective for larger fish, and samples drawn from it, therefore, contain a higher proportion of older age-classes than is characteristic of the entire population. In contrast, the average age and size of the fish reaching the sport fishing areas, which are located upriver from the commercial fishery, are smaller and younger than the average for the population as a whole (Fredin 1954; Leggett 1969). The estimated percentage contributions of the 1957–1970 year-classes to the structure of the 1965–1973 Connecticut River shad populations are given in Table 26. Estimates of the abundance of males and females in annual populations were obtained by assuming a 50:50 sex ratio and applying this to the annual population estimates (developed from commercial catch statistics). The assumption of a 50:50 sex ratio is approximate, a more reliable means of establishing the true ratio of males to females in the annual populations being unavailable. The ratio of males to females in the tagging catches 1967–1972 (Table 27) indicates that this is a reasonable approximation, although gear selectivity probably caused the proportion of males in these catches to be slightly low because males are smaller at any given age and young males may therefore escape the nets more readily.

TABLE 27. Proportion of males and females in tagging catches, 1967–1973.

Year	% males	% females
1967	58	42
1968	41	59
1969	42	58
1970	47	53
1971	52	48
1972	50	50
1973	46	54
Mean	48	52

The contribution of various age and sex categories to the structure of the 1965–1973 shad populations was determined (Table 28) from $A_j = NB_j$, where A_j = estimated number of shad of the particular sex and of age j in the population, N = estimated number of shad of that sex in the population, and B_j = proportion of shad of that sex and of age j in the sample.

The contribution of recruitment to the various age-classes was assessed by determining the proportion of virgin shad (by age and sex) in the annual populations 1965–1973 by scale-reading techniques (Table 29). These estimates were then used to calculate the contribution of the 1957–

TABLE 26. Contribution of year-classes 1957–1970 to structure of Connecticut River shad population 1965–1973 (in percent).

Sample			Year-class													
year	Sex	No.	1970	1969	1968	1967	1966	1965	1964	1963	1962	1961	1960	1959	1958	1957
1965	M	59									No	15.0	57.9	22.5	3.7	0.9
	F	149									data	11.9	69.0	15.6	2.5	1.0
1966	M	328								9.2	22.0	34.7	30.9	3.1	0.1	
	F	311									7.0	43.1	39.8	9.3	0.8	
1967	M	188							9.3	24.0	27.8	20.6	15.3	3.0		
	F	322								8.6	37.9	30.7	17.2	5.2	0.4	
1968	M	343						19.8	30.5	23.6	7.8	9.2	8.3	0.8		
	F	489							27.4	48.9	12.3	6.1	4.6	0.7		
1969	M	259					13.5	40.5	20.5	17.8	4.2	3.1	0.4			
	F	453					0.2	19.0	50.3	19.0	5.1	3.1	2.9	0.4		
1970	M	409				9.9	45.0	26.3	15.0	2.8	1.0					
	F	540				0.9	22.3	48.3	21.4	5.7	0.8	0.6				
1971	M	315			2.0	22.3	40.7	20.3	9.9	4.6	0.2					
	F	435				14.8	60.6	16.6	6.6	1.2	0.2					
1972	M	285		8.4	21.5	40.0	19.6	7.4	2.5	0.6						
	F	492			16.4	61.7	15.3	3.6	3.0							
1973	M	399	1.6	43.8	23.8	16.9	8.5	3.8	1.6							
	F	494		18.1	34.7	33.6	8.7	4.3	0.6							

TABLE 128. Contribution of year-classes, 1957–1970 (absolute values) to the structure of Connecticut River shad populations, 1965–1973 (absolute values).

Sample year	Sex	Number	Year-class 1970	1969	1968	1967	1966	1965	1964	1963	1962	1961	1960	1959	1958	1957
1965	M	735,000									No	110,250	425,565	165,375	27,195	6,615
	F	735,000									data	87,465	507,150	114,660	18,375	7,350
1966	M	183,500								16,882	40,370	63,675	56,700	5,689	184	
	F	183,500									12,845	79,089	73,033	17,066	1,467	
1967	M	184,000							17,112	44,160	51,152	37,904	28,152	5,520		
	F	184,000								15,824	69,736	56,488	31,648	9,568	736	
1968	M	140,000						27,720	42,700	33,040	10,920	12,880	11,620	1,120		
	F	140,000							38,360	68,460	17,220	8,540	6,440	980		
1969	M	166,500					22,477	67,432	34,133	29,637	6,993	5,162	666			
	F	166,500					333	31,634	83,750	31,634	8,492	5,162	4,829	666		
1970	M	209,500				20,741	94,274	55,099	31,425	5,866	2,095					
	F	209,500				1,886	46,719	101,189	44,833	11,941	1,676	1,256				
1971	M	214,000			4,280	47,722	87,098	43,442	21,186	9,844	428					
	F	214,000				31,672	129,684	35,524	14,124	2,568	428					
1972	M	137,500		11,550	29,563	55,000	26,950	10,175	3,437	825						
	F	137,500			22,550	84,837	21,038	4,950	4,125							
1973	M	166,000	2,656	72,708	39,508	28,054	14,110	6,308	2,656							
	F	166,000		30,046	57,602	55,776	14,442	7,138	996							

1970 year-classes to the recruitment of the 1965–1973 populations, according to the relationship (Table 30): $R_j = A_j V_j$, where R_j = estimated number of recruits of the particular sex and of age j in the population, and V_j = percent virgin shad of that sex and of age j in the sample.

Annual total mortality by age and sex was calculated from (Table 31): $D_{tj} = A_j - (A_{j+1} - R_{j+1})$, where D_{tj} = total mortality (absolute values) of the particular sex and of age j.

Annual fishing mortality by age and sex was estimated by determining the number of shad of each sex in the combined commercial, sport, and experimental catches and then using these data to estimate the distribution of the annual catches by age and sex as follows (Table 32): $D_{fj} = CB_j$, where D_{fj} = fishing mortality of the particular sex and of age j, and C = catch of shad of the particular sex.

Two assumptions are implicit in this calculation. The first is that the distribution of sexes in the combined catches was proportional to the distribution of sexes in the tagging catch. This appears reasonable because the commercial fishery, which employed nets identical in mesh size and design with the tagging nets and which showed no significant differences in size selection (Table 17), accounts for approximately 85% of the combined commercial and sport catch in the river. The sport fishery as a whole does not appear to be highly sex-selective in spite of the fact that fishermen favor females for their roe and will retain them preferentially when fishing success allows. The creel surveys noted no significant deviation from a 50:50 sex ratio for the whole fishery over the entire season. At the most successful fishing area (the state-controlled access area at Enfield Dam), the 12-year average (1962–1973) did reveal a bias in favor of females (41%/59%). However, our observation here indicates that this bias was achieved not by the selective release of live fish, but rather by the release of individual dead males from stringers and their replacement by females, as they are caught later in the day in an effort by the fisherman to improve his catch within the legal limits. Thus, even here, the sex ratio of the kill appears to approximate 50:50, although the estimate of total kill may be low by 500–1,000 fish a year as a result of this practice.

The second assumption is that the distribution of the age of the fish in the combined sport and commercial catches was proportional to the estimated distribution in the population as a whole. Leggett (1969) has shown this to be true.

Annual natural mortality (Table 33) was calculated as follows: $D_j = D_{tj} - D_{fj}$, where D_j = natural mortality of shad of the particular sex and of age

TABLE 29. Percent virgin shad and distribution by age class in annual populations, 1965–1973.

Sample year	Sex	% virgin	Year-class											
			1970	1969	1968	1967	1966	1965	1964	1963	1962	1961	1960	1959
1965	M	27										82	24	5
	F	69										100	80	14
1966	M	45								100	95	38	4	
	F	49									100	84	11	14
1967	M	39							100	97	22			
	F	54								100	93	27	2	
1968	M	56						100	93	34	3			
	F	77							100	94	29			
1969	M	55					100	86	28	6				
	F	70					100	99	89	31	4			
1970	M	57				100	95	18						
	F	66				100	100	87	8					
1971	M	38			100	90	36	5						
	F	72				100	86	32						
1972	M	45		100	94	37	6							
	F	72			100	88	5							
1973	M	59	100	99	57	4								
	F	54		100	87	18								

TABLE 30. Contribution of the 1959–1970 year-classes to recruitment in the 1965–1973 Connecticut River shad population.

Sample year	Sex	Number	Year-class 1970	1969	1968	1967	1966	1965	1964	1963	1962	1961	1960	1959
1965	M	200,810										90,405	102,136	8,269
	F	509,237										87,465	405,720	16,052
1966	M	81,699								16,882	38,352	24,197	2,268	
	F	89,703									12,845	66,435	8,034	2,389
1967	M	71,200							17,112	42,835	11,253			
	F	96,563								15,824	64,854	15,252	633	
1968	M	78,993						27,720	39,711	11,234	328			
	F	107,706							38,360	64,352	4,994			
1969	M	91,804					22,477	57,992	9,557	1,778				
	F	116,337					333	31,319	74,538	9,807	340			
1970	M	120,219				20,741	89,560	9,918						
	F	140,226				1,886	46,719	88,034	3,587					
1971	M	80,757			4,280	42,950	31,355	2,172						
	F	154,568				31,672	111,528	11,368						
1972	M	61,306		11,550	27,789	20,350	1,617							
	F	98,209			22,550	74,657	1,052							
1973	M	98,279	2,656	71,981	22,520	1,122								
	F	90,200		30,046	50,114	10,040								

TABLE 31. Number of shad dying (total mortality, absolute values) by year-class.

Sample year	Sex	Year-class 1969	1968	1967	1966	1965	1964	1963	1962	1961	1960	1959	1958	1957
1965/1966	M									70,772	371,133	159,686	27,011	6,615
	F									74,811	442,151	99,983	16,908	7,350
1966/1967	M							15,557	471	25,771	28,548	160	184	
	F								7,963	37,853	42,018	7,498	731	
1967/1968	M						14,123	22,354	40,560	25,024	16,532	4,400		
	F							11,716	57,510	47,948	25,208	8,588	736	
1968/1969	M					18,280	18,124	5,181	3,927	7,718	10,954	1,120		
	F						29,148	46,632	9,068	3,378	1,611	314		
1969/1970	M				17,763	22,251	2,708	23,771	4,898	5,162	666			
	F				333	18,480	42,504	19,693	6,816	3,905	4,829	666		
1970/1971	M			15,969	38,531	13,829	10,239		1,667					
	F			1,886	28,563	77,033	30,709	9,374	1,248	1,257				
1971/1972	M		2,506	13,072	61,765	33,267	17,748	9,019	428					
	F			21,591	109,698	30,574	9,999	2,568	428					
1972/1973	M	10,823	12,575	28,068	12,840	3,867	782	825						
	F		15,062	39,102	6,596		3,129							

TABLE 32. Distribution of annual shad catches by age-class, Connecticut River, 1965–1973.

Sample year	Sex	Number	Year-class 1970	1969	1968	1967	1966	1965	1964	1963	1962	1961	1960	1959	1958	1957
1965	M	96,494										14,474	55,871	21,711	3,570	868
	F	96,494										11,483	66,581	15,053	2,412	965
1966	M	51,773								4,763	11,390	17,965	15,998	1,605	52	
	F	51,773									3,624	22,314	20,606	4,815	414	
1967	M	54,227							5,043	13,014	15,075	11,171	8,297	1,627		
	F	39,268								3,377	14,883	12,055	6,754	2,042	157	
1968	M	29,704						5,881	9,060	7,010	2,317	2,733	2,465	238		
	F	42,746							11,712	20,903	5,258	2,608	1,966	299		
1969	M	34,266					4,626	13,878	7,025	6,099	1,439	1,062	137			
	F	47,319					95	8,991	23,801	8,991	2,413	1,467	1,372	189		
1970	M	39,151				3,876	17,618	10,296	5,873	1,096	392					
	F	44,150				397	9,845	21,324	9,449	2,517	353	265				
1971	M	47,501			950	10,593	19,333	9,643	4,703	2,184	95					
	F	43,848				6,490	26,571	7,279	2,894	526	88					
1972	M	28,990		2,435	6,233	11,596	5,682	2,145	725	174						
	F	28,990			4,754	17,887	4,435	1,044	870							
1973	M	36,631	586	16,044	8,718	6,191	3,114	1,392	586							
	F	39,684		7,183	13,770	13,334	3,453	1,706	238							

TABLE 33. Number of shad dying (natural mortality, absolute values) by year-class.

Sample year	Sex	Year-class 1969	1968	1967	1966	1965	1964	1963	1962	1961	1960	1959	1958	1957
1965/1966	M									56,298	315,262	137,975	23,441	5,747
	F									63,328	375,570	84,930	14,496	6,385
1966/1967	M							10,794		7,806	12,550		132	
	F								4,339	15,539	21,412	2,683	317	
1967/1968	M						9,080	89,340	25,485	13,853	8,235	2,773		
	F							8,339	42,627	35,893	18,454	6,546	579	
1968/1969	M					12,399	9,064		1,610	4,985	8,489	882		
	F						17,436	25,729	6,751	770		15		
1969/1970	M				13,137	8,373		17,672	3,459	4,100	529			
	F				238	9,489	18,703	10,702	4,403	2,438	3,457	477		
1970/1971	M			12,093	20,913	3,532	4,366		1,275					
	F			1,489	18,718	55,709	21,261	6,857	895	992				
1971/1972	M		1,556	2,479	42,432	23,624	13,045	6,834	333					
	F			15,101	83,126	23,295	7,105	2,042	340					
1972/1973	M	8,388	6,342	16,472	7,158	1,722	57	651						
	F		10,308	21,215	2,161		2,259							

j. Instantaneous total (Z), natural (M), and fishing mortality (F) coefficients (Tables 34–36) were calculated using the method of Royce (1972: pp. 234–235).

Annual natural mortality estimates were also calculated from tag-recapture data using Ricker's (1958a) formula 5.2, modified to remove the effects of the recovery of marked fish during the year of tagging (Table 37). The equation used was

$$M_n = 1 - \frac{R_{12}M_2}{(M_1 \quad R_{11})(R_{22}+1)},$$

where M_n = estimated natural mortality;
M_1 = number of fish tagged in year 1;
M_2 = number of fish tagged in year 2;
R_{11} = number of fish tagged in year 1 and recovered in year 1;
R_{12} = number of fish tagged in year 1 and recovered in year 2;
R_{22} = number of fish tagged in year 2 and recovered in year 2.

In these calculations, R_{12} was adjusted for an estimated 63% annual tag loss. The method used to calculate this rate of tag loss was that of Chapman et al. (1965):

Number single tagged at large (Ns) = 3603;
Number double tagged at large (Nd) = 1810;
Number single tagged recovered (ns) = 19;
Number double tagged recovered with two tags (ndd) = 6);
Number double tagged recovered with one tag (nds) = 7.

$$\text{Annual loss} = 1 - \left(\frac{\text{nds} \times \text{Ns}}{2\text{ns} \times \text{Nd}}\right) = 63\%$$

Considerable age-specific and sex-specific, year-to-year variation in the levels of total and natural mortality occurred during the 8 years investigated (Tables 34 and 35). The largest total mortalities occurred in 1965/1966 and 1971/1972 and coincided with the lowest (1972) and third lowest (1966) levels of recruitment (Table 29) during the period. This contributed to the large declines in population observed between 1965/1966 and between 1971/1972. The average instantaneous total mortality coefficients for males and females over the 8-year period were 1.19 (N = 43, SD = 1.02) and 1.24 (N = 38, SD = 0.59), respectively. These rates were not significantly different (t = 0.31, df = 79, $0.4 < P < 0.5$). They correspond to annual mortality rates of 70% for males and 71%

TABLE 34. Instantaneous total mortality (Z).

Year	Sex	Age III/IV	IV/V	V/VI	VI/VII	VII/VIII	VIII/IX	IX/X	Mean (SD)
1965/1966	M		1.03	2.06	3.37	5.00			2.86 (1.71)
	F		1.93	2.05	2.06	2.53			2.14 (0.26)
1966/1967	M	2.54	0.01	0.90	0.70	0.03			0.84 (1.03)
	F		0.97	0.65	0.86	0.58	0.69		0.75 (0.16)
1967/1968	M	1.74	0.71	1.57	1.08	0.88	1.60		1.26 (0.43)
	F		1.35	1.74	1.89	1.59	2.28		1.77 (0.35)
1968/1969	M	1.08	0.55	0.17	0.45	0.91	2.86		1.00 (0.97)
	F		1.43	1.14	0.75	0.50	0.29	0.39	0.75 (0.45)
1969/1970	M	1.56	0.40	0.08	1.62	0.36			0.80 (0.73)
	F		0.88	0.71	0.97	1.62	1.41		1.12 (0.38)
1970/1971	M	1.47	0.53	0.29	0.39		1.59		0.85 (0.62)
	F		0.95	1.43	1.16	1.54	1.37		1.29 (0.24)
1971/1972	M	0.88	0.32	1.23	1.45	1.82	2.48		1.36 (0.75)
	F		1.14	1.87	1.97	1.23			1.55 (0.43)
1972/1973	M	2.77	0.55	0.71	0.65	0.48	0.26		0.90 (0.93)
	F		1.10	0.62	0.38	–[a]	1.42		0.88 (0.47)
		Grand mean (SD)							
Mean (SD)	M	1.72 (0.70)	0.45 (0.33)	0.88 (0.71)	1.21 (0.98)	1.35 (1.70)	1.76 (1.00)		1.19 (1.02)
	F		1.22 (0.35)	1.28 (0.58)	1.26 (0.64)	1.37 (0.69)	1.24 (0.69)	0.39	1.24 (0.59)

[a] Dash (–) indicates negative value.

TABLE 35. Instantaneous natural mortality (*M*).

Year	Sex	Age							
		III/IV	IV/V	V/VI	VI/VII	VII/VIII	VIII/IX	IX/X	Mean (SD)
1965/1966	M		0.82	1.75	2.91	4.34			2.46 (1.52)
	F		1.63	1.74	1.75	2.17			1.82 (0.24)
1966/1967	M	1.76	–[a]	0.89	0.20	–			0.95 (0.78)
	F		–	0.21	0.48	0.21	0.30		0.30 (0.13)
1967/1968	M	1.12	0.30	0.99	0.60	0.08	1.01		0.68 (0.43)
	F		0.96	1.29	1.41	1.16	1.74		1.31 (0.29)
1968/1969	M	0.73	0.26	–	0.18	0.59	2.22		0.80 (0.83)
	F		0.86	0.58	0.32	0.11		0.02	0.38 (0.35)
1969/1970	M	1.15	0.15	–	1.20	0.25			0.69 (0.56)
	F		0.37	0.31	0.53	1.05	0.88		0.63 (0.32)
1970/1971	M	1.11	0.29	0.07	0.17	–	1.22		0.57 (0.55)
	F		0.62	1.03	0.80	1.13	0.98		0.91 (0.20)
1971/1972	M	0.55	0.06	0.84	1.03	1.49	1.88		0.98 (0.65)
	F		0.80	1.42	1.50	0.87			1.15 (0.36)
1972/1973	M	2.15	0.28	0.42	0.48	0.21	0.02		0.59 (0.78)
	F		0.75	0.34	0.12	–	1.03		0.66 (0.57)
					Grand mean (SD)				
Mean (SD)	M	1.22 (0.56)	0.31 (0.24)	0.83 (0.57)	0.85 (0.92)	1.16 (1.64)	1.27 (0.85)		0.92 (0.90)
	F		0.86 (0.39)	0.87 (0.58)	0.86 (0.61)	0.96 (0.69)	0.99 (0.51)	0.02	0.87 (0.55)

[a] Dash (–) indicates negative value.

TABLE 36. Instantaneous fishing mortality (*F*).

Year	Sex	Age							
		III/IV	IV/V	V/VI	VI/VII	VII/VIII	VII/IX	IX/X	Mean (SD)
1965/1966	M		0.21	0.31	0.46	0.66			0.41 (0.20)
	F		0.30	0.31	0.31	0.36			0.32 (0.03)
1966/1967	M	0.78	–[a]	0.01	0.50	–	0.01		0.32 (0.38)
	F		–	0.44	0.38	0.37	0.39		0.40 (0.03)
1967/1968	M	0.62	0.41	0.58	0.48	0.80	0.59		0.58 (0.13)
	F		0.39	0.45	0.48	0.43	0.54		0.46 (0.06)
1968/1969	M	0.35	0.25	–	0.27	0.32	0.64		0.37 (0.16)
	F		0.57	0.51	0.43	0.39		0.37	0.46 (0.09)
1969/1970	M	0.41	0.25	–	0.42	0.11			0.29 (0.13)
	F		0.51	0.40	0.44	0.57	0.53		0.49 (0.07)
1970/1971	M	0.36	0.24	0.22	0.22	–	0.37		0.28 (0.08)
	F		0.33	0.40	0.36	0.41	0.39		0.38 (0.03)
1971/1972	M	0.33	0.26	0.39	0.42	0.33	0.60		0.39 (0.12)
	F		0.34	0.45	0.47	0.36			0.41 (0.07)
1972/1973	M	0.62	0.27	0.29	0.17	0.27	0.24		0.31 (0.16)
	F		0.35	0.28	0.26	–	0.39		0.32 (0.06)
					Grand mean (SD)				
Mean (SD)	M	0.50 (0.18)	0.27 (0.06)	0.30 (0.19)	0.37 (0.12)	0.42 (0.26)	0.41 (0.25)		0.38 (0.19)
	F		0.40 (0.10)	0.41 (0.08)	0.39 (0.08)	0.41 (0.07)	0.45 (0.08)	0.37	0.41 (0.08)

[a] Dash (–) indicates negative value.

TABLE 37. Summary of tagging and recovery data and related mortality estimates.

Years (*X*)	Number tagged	Recovered[a] year (*X*)	Recovered[b] year (*X* + 1)	Natural mortality %
		Males		
1965				
1966				
1967	2,112	447	108	72
1968	1,567	364	97	57
1969	1,594	300	124	56
1970	1,793	393	76	73
1971	2,123	430	38	89
1972	2,177	426	32	91
1973	1,240	251		
		Females		
1965				
1966				
1967	1,519	349	78	75
1968	2,253	590	213	49
1969	2,216	558	130	64
1970	2,049	449	49	88
1971	1,964	490	51	85
1972	2,204	521	46	88
1973	1,432	324		
		Sexes combined		
1965	2,200	346	27	94
1966	3,991	976	103	85
1967	4,011	904	208	72
1968	4,362	1,039	362	52
1969	3,810	858	254	61
1970	4,072	887	122	83
1971	4,147	937	92	87
1972	4,622	1,019	78	90
1973	2,871	607		

[a] Includes river and coastal recoveries.

[b] Adjusted for estimated 63% annual rate of tag shedding.

for females. Average instantaneous natural mor tality coefficients for males and females were 0.92 (N = 39; SD = 0.90) and 0.87 (N = 36, SD = 0.55), corresponding to annual rates of 54 and 50% respectively, and these rates also did not differ significantly (t = 0.05, df = 75, $P > 0.9$). The average instantaneous fishing mortality for males (0.38, SD = 0.19, N = 40) and females (0.41, SD = 0.08, N = 36) did not differ significantly (t = 0.81, df = 75, $0.4 < P = 0.5$). This represents an average fishing mortality of 22.4% in males and 23.5% in females. The mean fishing mortality (sexes combined) of 22.7% (SD = 8.7, N = 76) does not differ significantly from the estimate of 19.8% (SD = 3.0, N = 9), which was derived from the analysis of commercial and sport fishery tag recoveries.

The levels of total and natural mortality determined in this study are in agreement with previous estimates of the Connecticut River population of shad. A natural mortality rate (based on frequencies of age and spawning groups) of 58% was reported by Walburg (1961) for the 1956–1959 period; this is in close agreement with the present estimate (52%) for 1965–1972. Fredin (1954) reported natural mortality for 6-year-old shad in the 1946 population to be 36%; this is within the range of values observed in this study. Walburg (1961) estimated total mortality (1956–1959) to be 73%, as compared with 70% in the present study. These data suggest that average natural and total mortality have not changed significantly during the last decade.

Estimates of average annual natural mortality based on tag recoveries were consistently much higher (Table 37), the average for males being 73%, for females 75% (1967–1972), and for sexes combined (1965–1972) 78%. One cause for this might be an above-average mortality at sea among tagged fish, possibly resulting from increased susceptibility to predation or infection of the tagging wound. Another cause might be an inaccurate estimate of the annual rate of tag shedding. For Susquehanna River shad, Whitney (1961) reported annual mortalities, determined from tag recovery records, of 80.7% (1959) and 90.8% (1960). These values also appear to be too high since the repeat-spawning component of this population, as based on an 85% annual mortality rate from time of maturity, would be 15%, but Lapointe (1957) has reported the frequency of repeat spawning in the Susquehanna River to be 37%. Whitney made no attempt to compensate for tag loss; this may be the cause of his overestimate of mortality.

Natural mortality rates were significantly lower in postoperation (1968–1973) than in preoperation (1965–1967) years. There does not appear to be any reasonable mechanism that would explain a relationship between increased survival and plant operation, and I conclude that there is no relationship of this kind (Table 38).

TABLE 38. Average instantaneous natural mortality rates (Z) prior to and during CYAP operation.

Sex	Pre-operation (1965–1967) Z	SD	Operation (1968–1973) Z	SD	t	
Males	1.50	1.24	0.73	0.65	2.56	$P < 0.05$
Females	1.16	0.67	0.71	0.40	2.53	$P < 0.05$

Parent–Progeny Relationship and Maximum Equilibrium Yield

Walburg (1963) attempted to define the parent–progeny relationship of the Connecticut River shad population. His relationship was developed from Fredin's (1954) analysis of population sizes for 1935–1951 and additional population estimates, based on Fredin's method, for 1952–1961. These estimates, as has been previously shown, were too low, and this led to an incorrect estimate of stock-recruitment relationships.

For the present purposes, the contribution of spawners in the 1935–1967 population to recruitment in subsequent populations was calculated by determining the mean proportion of 3-, 4-, 5-, and 6-year-old virgin fish in annual populations 1965–1973 (Table 26) and then calculating recruitment from individual parent populations as follows: virgins in year x = 0.07 P_{x3} + 0.23 P_{x4} + 0.29 P_{x5} + 0.03 P_{x6}, where P_{x3} = total population year x + 3, P_{x4} = total population year x + 4, P_{x5} = total population year x + 5, and P_{x6} = total population year x + 6.

This method assumes that the proportion of 3-, 4-, 5- and 6-year-old virgin shad in annual populations is constant from year to year. The deviations from the 9-year average were generally small (Table 39).

Connecticut River shad are susceptible to the fishery more than once because some of them are repeat spawners. It was therefore necessary to convert parent stock and recruitment to equivalent units (Ricker 1958a, p. 237). This was achieved by converting number of recruits and adult stock to number of eggs either spawned or potentially available for spawning. The figures for parent stock and recruitment were divided by two on the assumption that the sex ratio in each approximates 50:50 (Table 27), and the quotient was multiplied by the appropriate fecundity factor. Leggett (1969) reported the mean lifetime fecundity (number of eggs laid during the life of an average female) of Connecticut River shad as 353,000 eggs and the mean population fecundity (average number of eggs laid per female in any given year) as 269,000 eggs. It therefore follows that

$$F = 1/2 \text{ recruitment} \times 353{,}00($$
$$P = 1/2 \text{ parentstock} \times 269{,}00(\text{ ,}$$

where F is recruitment and P is parent stock in terms of egg numbers (Table 40).

The parent–progeny relationship (recruitment curve) was calculated following the method outlined in Ricker (1958a) viz.

$$F = \mathrm{P}e^{a(1 - P/P_r)} \text{ ,}$$

where P_r = replacement size of parental generation; $a = P_r/P_m$; P_m = level of parental stock that produces the maximum recruitment. P_r and P_m are estimated from the relationship

$$\log_{10} F - \log_{10} P = \frac{0.4343\, P_r}{P_m} -$$

The calculated regression relationship based on this analysis was $F = Pe^{0.7118(1 - P/86.59)}$ (Figure 16). Data related to the analysis are presented in Table 40. Fitting the curve significantly reduced the sum of squares from 0.86 (for deviations from the mean log F) to 0.71 (for deviations for the log curve). Variance of squared deviations of observed-calculated log F (0.0111) was also significantly reduced by fitting (as compared with the simple variance of log F values about the mean log F (0.0279) (Table 40). This justified the use of the above equation in developing the parent–progeny relationship (Ricker 1958a).

A polynomial regression of degree 2 fitted to the data did not differ significantly from the Ricker function for values of P between 0 and P_r (Figure 16). The equation for this curve was $F = 2.5745 + 1.6784P - 0.0084P^2$. The regression was significant at the $P = 0.01$ level (F statistic = 40.6).

TABLE 39. Nine-year mean and standard deviation of proportion of total population composed of virgin shad aged III, IV, V, VI, and VII years.

Age	III	IV	V	VI	VII
Mean	0.07	0.23	0.29	0.03	0.00
SD	0.02	0.06	0.08	0.02	

TABLE 40. Population size, escapement, and resulting recruitment, 1935–1967.[a]

Year	Pop. (1000s)	Esc. (1000s)	P	Rec. (1000s)	F	$\log P$	$\log F$	$\log F -$ $\log P$	d_1	d_1^2	Comp. $\log F$	d_2	d_2^2
1935	739	539	73	604	107	1.86	2.03	0.17	−0.25	0.0625	1.91	0.12	0.0144
1936	704	537	72	515	91	1.86	1.96	0.10	−0.18	0.0324	1.91	0.05	0.0025
1937	686	503	68	557	98	1.83	1.99	0.16	−0.21	0.0441	1.90	0.09	0.0081
1938	693	491	66	573	101	1.82	2.00	0.18	−0.22	0.0484	1.89	0.11	0.0121
1939	641	466	63	588	104	1.80	2.02	0.22	−0.24	0.0576	1.88	0.14	0.0196
1940	694	553	74	540	95	1.87	1.98	0.11	−0.20	0.0400	1.91	0.07	0.0049
1941	981	786	106	514	91	2.03	1.96	−0.07	−0.18	0.0324	1.96	0.00	0.0000
1942	877	694	93	466	82	1.97	1.91	−0.06	−0.13	0.0169	1.95	0.04	0.0016
1943	942	676	91	396	70	1.96	1.85	−0.11	−0.07	0.0049	1.94	−0.09	0.0081
1944	990	625	84	315	56	1.92	1.75	−0.17	0.03	0.0009	1.93	−0.18	0.0324
1945	760	386	52	236	42	1.72	1.62	−0.10	0.16	0.0256	1.84	−0.22	0.0484
1946	861	331	45	238	42	1.65	1.62	−0.03	0.16	0.0256	1.80	−0.18	0.0324
1947	681	270	36	337	59	1.56	1.77	0.21	0.01	0.0001	1.74	0.03	0.0009
1948	573	273	37	328	58	1.57	1.76	0.20	0.02	0.0004	1.75	0.01	0.0001
1949	437	225	30	258	46	1.48	1.66	0.18	0.12	0.0144	1.68	−0.02	0.0004
1950	287	175	24	191	34	1.38	1.53	0.15	0.25	0.0625	1.60	−0.07	0.0049
1951	420	263	35	174	31	1.54	1.49	−0.05	0.29	0.0841	1.72	−0.23	0.0529
1952	714	419	56	249	44	1.75	1.64	−0.11	0.14	0.0196	1.86	−0.22	0.0484
1953	428	243	33	356	63	1.52	1.80	0.28	−0.02	0.0004	1.71	0.09	0.0081
1954	353	303	41	402	71	1.61	1.85	0.24	−0.07	0.0049	1.77	0.08	0.0064
1955	247	167	22	414	73	1.34	1.86	0.52	−0.08	0.0064	1.57	0.29	0.0841
1956	267	196	26	419	74	1.41	1.87	0.46	−0.09	0.0081	1.63	0.24	0.0576
1957	516	395	53	421	74	1.72	1.87	0.15	−0.09	0.0081	1.84	0.03	0.0009
1958	688	480	65	325	57	1.81	1.76	−0.05	0.02	0.0004	1.89	−0.13	0.0169
1959	647	473	64	308	54	1.81	1.73	0.08	0.05	0.0025	1.89	−0.16	0.0256
1960	678	493	66	566	100	1.82	2.00	0.18	−0.22	0.0484	1.89	0.11	0.0121
1961	686	483	65	486	86	1.81	1.93	0.12	−0.15	0.0225	1.89	0.04	0.0016
1962	630	435	59	302	53	1.77	1.72	−0.05	0.06	0.0036	1.87	−0.15	0.0225
1963	409	290	39	201	35	1.59	1.54	−0.05	0.24	0.0576	1.76	−0.22	0.0484
1964	433	324	44	199	35	1.64	1.54	−0.10	0.24	0.0576	1.79	−0.25	0.0625
1965	1,470	1,297	174	229	40	2.24	1.60	−0.64	0.18	0.0324	1.93	−0.33	0.1089
1966	367	286	38	251	44	1.58	1.64	−0.06	0.14	0.0196	1.75	−0.11	0.0121
1967	368	292	39	202	36	1.59	1.56	−0.03	0.22	0.0484	1.76	−0.20	0.0400

[a] P and F defined in text, d_1 = mean $\log F - \log F$, $d_2 = \log F -$ Comp. $\log F$. Comp. $\log F$ determined as $\log F = \log P - 0.003570P + 0.309139$ (method of Ricker 1958a).

Both recruitment curves indicate P_r (replacement size of parental stock) to be approximately 85×10^9 eggs. This level of egg production is achieved by a spawning stock of 630,000 shad and results in a recruitment of 480,000 shad. Maximum surplus egg production occurs at a parent stock of approximately 40×10^9 eggs (300,000 shad) and results in a potential surplus of $16–18 \times 10^9$ eggs. At this level of escapement, annual recruitment is approximately 332,000 shad.

Recruitment, measured in numbers of fish, is less than the parent stock at P_r. This occurs because of the substantial number of fish in the population that spawn more than once (Table 29). In repeat spawning stocks, the level of recruitment required to maintain replacement = $(1 - p)$ parent stock. This level may vary seasonally with annual fluctuations in survival of fully recruited adults. Only when $s = 0$ (single reproduction) is it necessary for recruitment to equal parent stock to maintain the spawning stock at P_r (Ricker 1958a).

Because of the repeat-spawning nature of the population, the optimum parent stock (escapement) is not that which provides the maximum

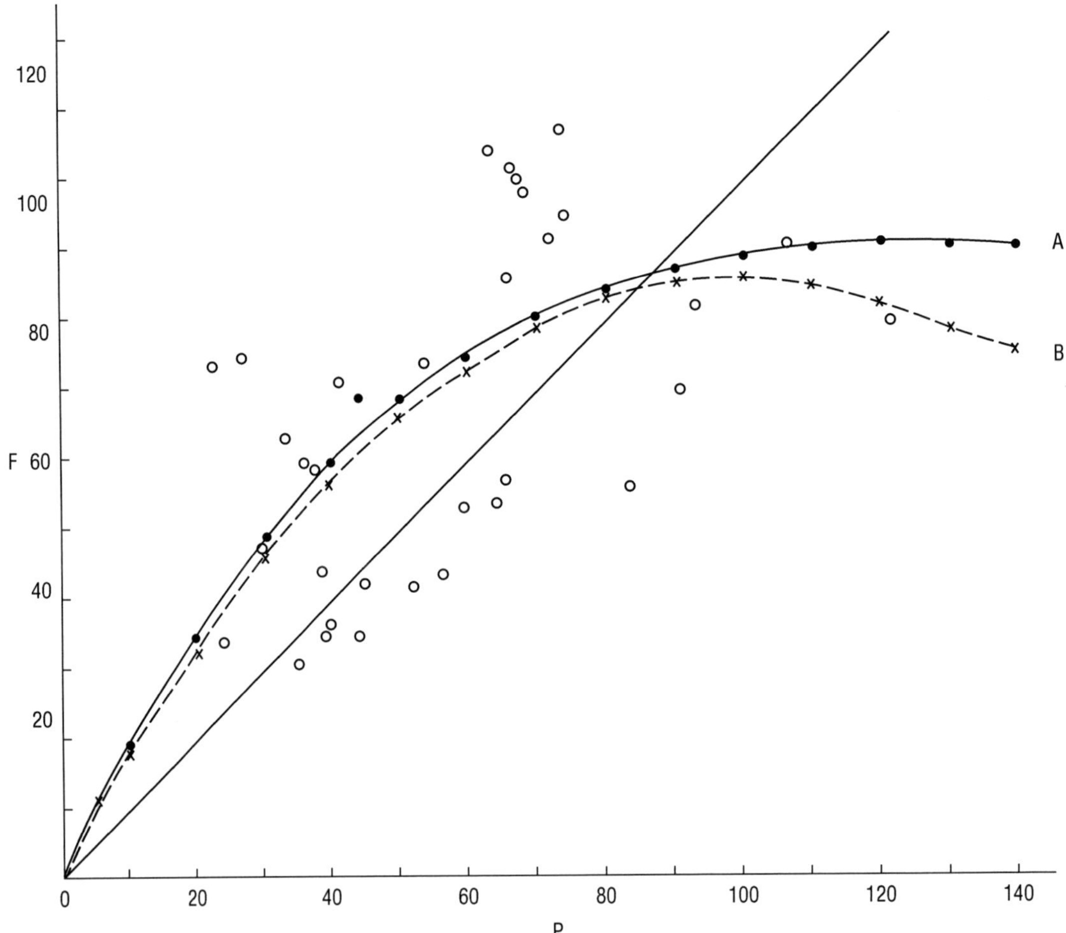

FIGURE 16. Parent (P) – progeny (F) (recruitment) curves for Connecticut River shad. A.$F = Pe^{0.7118(1 - P/86.59)}$ B.$F = 2.5745 + 1.6784P - 0.0084P^2$. Axes indicate relative egg production in millions.

surplus recruitment, but rather that which provides the maximum number of catchable recruits *and* repeat spawners. The proportion of repeat spawners in annual populations is quite stable, and hence, an approximation of the optimum escapement can be made by adopting the average natural mortality rate of 52%, as observed in this study (Table 35). Optimum escapement for the Connecticut River stock under prevailing conditions is estimated by this means to be approximately 450,000 shad (Table 41). At this level of escapement, the maximum equilibrium yield is estimated to be 193,000 shad and is achieved at an equilibrium population size of 643,000 shad (Figure 17). The average exploitation rate required to maintain this yield is 30%. This would produce a population containing, on the average, 66% virgin and 34% repeat spawners.

Discussion

The shad entering the Connecticut River each year include fish that will spawn for the first time (virgins) and fish that have escaped the fishery in previous seasons, have spawned, and have returned to spawn again (repeaters). The absolute numbers and relative proportions of these two categories exhibit modest annual variations, both as a result of changes in the natural mortality of the repeat-spawning segment of the population and of changes in recruitment. As has been shown, the number of virgin shad entering the river from year to year is related in part to the size of the parental stock, which in turn means the number of shad that escaped the fishery 3, 4, 5, and 6 years previously. As the data in Figure 16 plainly show, however, the annual variation in this par-

TABLE 41. Potential yield from various average escapements of shad.[a]

Escapement (average) (A)	Recruitment (B)	Repeat spawners (C = 0.48A)	Population size (D = B + C)	Potential yield (E = D – A)	Exploitation rate (F/D)	% repeat spawners (C/D × 100)
10	15	5	20	10	0.50	25
50	73	24	97	47	0.48	25
100	139	48	187	87	0.47	26
150	196	72	268	118	0.44	27
200	250	96	346	146	0.42	28
250	297	120	417	167	0.40	29
300	332	144	476	176	0.37	30
350	369	168	537	187	0.35	31
400	400	192	592	192	0.32	32
450	427	216	643	193	0.30	34
500	446	240	686	186	0.27	35
550	465	264	729	179	0.24	36
600	480	288	768	168	0.22	38
650	490	312	802	152	0.19	39
700	500	336	836	136	0.16	40
800	513	384	897	97	0.11	43
900	516	432	948	48	0.05	46

[a] Recruitment based on $F = Pe^{0.7118(1 - P/86.59)}$, with repeat spawners estimated by assuming an average survival rate after maturity of 0.48. All values in thousands of fish.

ent–progeny relationship is high. This implies that density-independent factors, perhaps environmental, are operating to influence the number of recruits that return to the Connecticut River each year. Marcy (2004 [1976], this volume) shows that 64.0% of the annual variation in the production of juvenile shad is directly related to the size of the spawning stock, but that an additional 22% can be attributed to conditions of water temperature and flow during the spawning season. Changes in the levels of natural mortality operating on immature shad during their ocean life could also significantly influence recruitment. Nothing is presently known about the mortality experienced by these age-groups (I–III) or of the factors determining it. It thus appears that the influence of the environment in determining recruitment may be almost as important as the size of the parent stock.

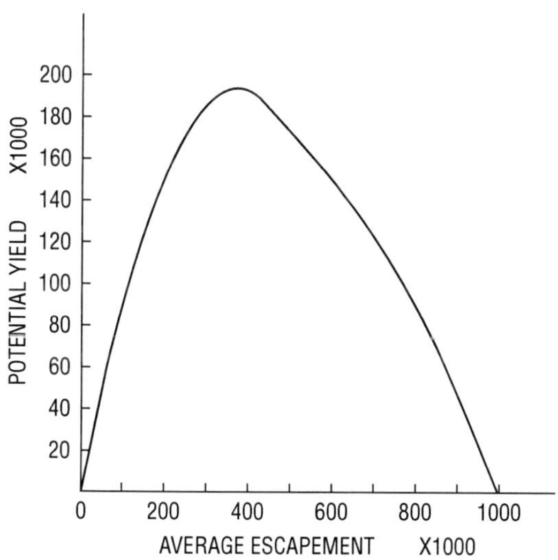

FIGURE 17. Potential equilibrium yield for the Connecticut River shad population at various levels of escapement. (All values in thousands of fish.)

The importance of escapement one (potential repeaters) and 3, 4, 5, and 6 (parent stock) years previously in determining the size of annual populations 1940–1973 was investigated by stepwise multiple regression analysis. Of the total variation in population size over this period, 35% was accounted for by these variables. Of this, 25% was the result of escapement 1 year before and 10% was the result of escapement 3, 4, 5, and 6 years earlier. This suggests that in addition to environmental influences, exploitation is a major factor influencing changes in population size. This is supported by the record of commercial catches 1890–1973 (Figures 1 and 18) and popu-

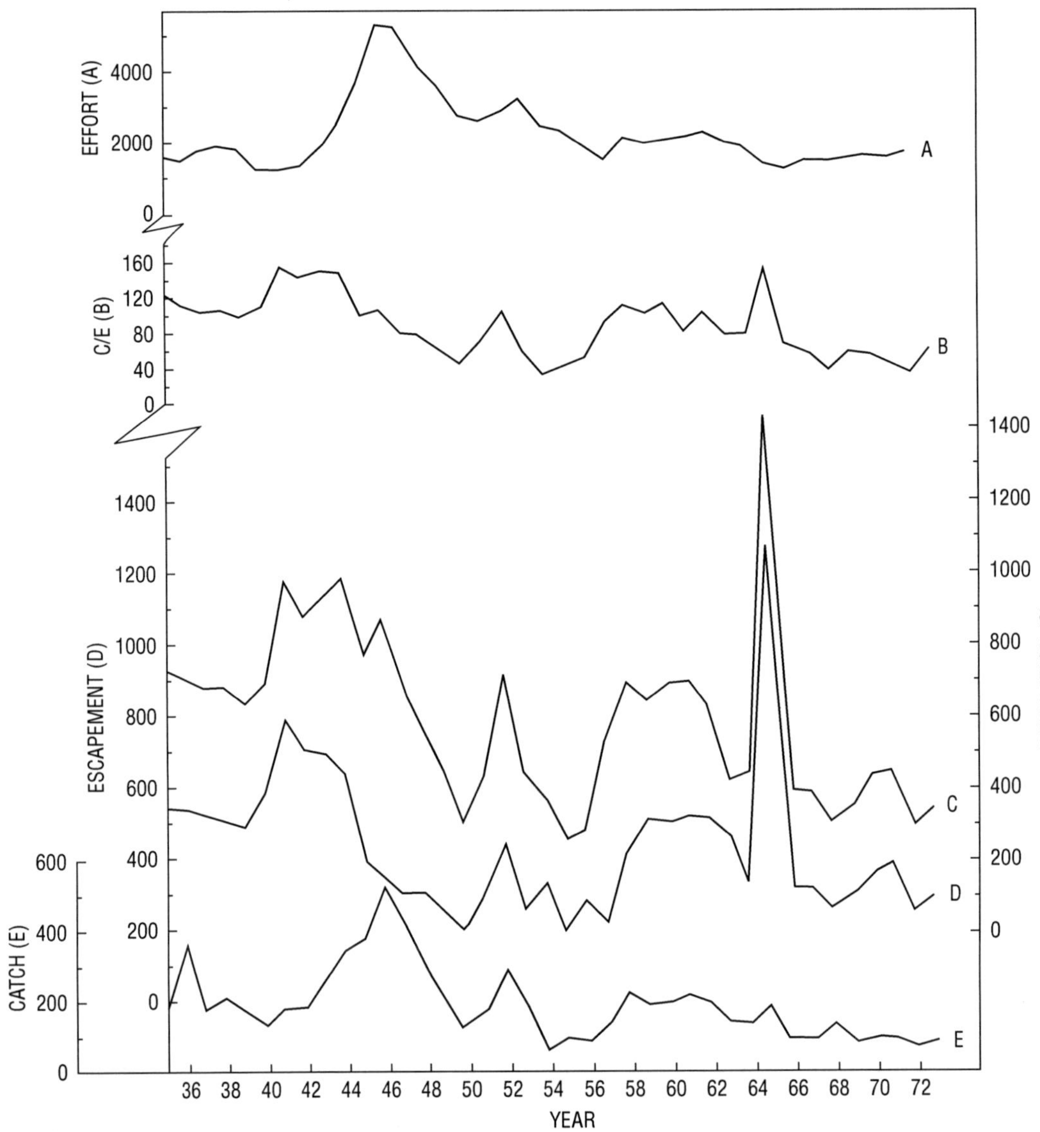

FIGURE 18. Effort, catch-per-unit effort, population size, escapement, and catch, 1935–1973.

lation size 1935–1973 (Figure 18), which show that both have declined substantially over these periods. The population appears to have been largely underexploitated from 1890 to 1934—reported annual catches ranging from 18,000 to 175,000 shad—and to have remained well below 100,000 in all but six of these years. These catches were below the maximum equilibrium yield (193,000 shad) indicated in Table 41. Reference to Table 41 suggests that the average population size during this period was in the order of 800,000 to 1,000,000 shad. This is based, of course, on the assumption that spawning success and/or survival at sea have not changed sufficiently since that time to alter seriously the stock–recruitment relationship.

The major changes in population size that have occurred since 1935 appear to have been primarily due to the effects of fishing, although all changes obviously cannot be explained on this basis. During 1935–1940 population size, catch, effort, and catch per unit effort remained quite stable. However, in 1940, a significant increase in population size began. The reason for this increase is not known, but it is possible that it resulted from the construction of a fish ladder at Enfield Dam in 1933, which opened an additional 41 km of river to spawning. These higher populations prevailed until 1944 and resulted in substantially increased fishing success (measured in terms of shad/SFU day) also beginning in 1941 and continuing until 1945. Fishermen appear to have responded positively to these improved catches, for in 1943, after a 2-year lag, fishing effort began to increase dramatically, and it more than quadrupled, from 1,298 SFU days in 1942 to 5,309 in 1946. In response to this increased effort, catches also increased sharply from 141,000 shad in 1940 to 530,000 shad in 1946 (Figure 18).

These increased catches appear to have resulted from the so-called fishing-up phenomenon (Ricker 1958a) in which surplus adults are removed from the stock and the total population is reduced to a lower equilibrium level of population size. This conclusion is based on the observation that in 1944, the population size and catch per unit effort began to decline under the pressure of this increased effort. These declines continued unchecked until 1950. In 1946, again after a 2-year lag, effort began to decline, probably as a result of less successful or opportunistic fishermen leaving the fishery as the catch per unit effort declined (Figure 18). This reduced fishing pressure, beginning in 1946, coupled with the declining population size that began 2 years earlier, in 1944, led to the decline in catches, which occurred between 1946 and 1950. It is important to note, however, that escapement, which is the major determinant of the contribution of repeat spawners to subsequent populations, and thus significantly affects population size, began a rapid decline in 1944 exactly coincident with the major increase in fishing effort. This decline in escapement also continued unchecked until 1950 (Figure 18).

It thus appears that the major decline in population size that occurred between 1944 and 1950 resulted primarily from overfishing. In this connection, it appears that escapement first declined below the level required for maximum equilibrium yield in 1949 and that it has since remained well below this level with the exception of the 4 years 1959–1962. This appears to have resulted from the fact that while fishing effort has declined to pre-1940 levels, the reduction in population size 1944–1950 was sufficiently large to prevent recovery of the population to pre-1944 levels as long as fishing pressure remains at present levels.

It must also be noted that this analysis is based exclusively on the commercial catch data. An additional 10,000 to 20,000 shad are angled annually in Connecticut and Massachusetts, and hence, actual escapements are lower than those shown by this amount. Limited escapement thus appears to be the major factor restricting the size of present day shad populations in the Connecticut River and was the reason for its failure to provide catches at the level of maximum sustained yield in more than 3 years (1952, 1958, and 1961) over the period 1950–1973.

The reason for the sharp increase in population size between 1956 and 1958 cannot be firmly established; however, it may have resulted from recruitment of a very successful year class spawned in 1952 or 1953. This conclusion is highly speculative, however, since it is based exclusively on the observation (Leggett 1969) that the proportion of virgin male shad in the 1956 population (93%) was the highest recorded during the years 1956–1958 and 1965–1973, and the proportion of virgin males in 1957 (60%) was the second highest observed in this period. Similarly, the proportion of virgin females in the 1956, 1957, and 1958 populations (78%, 81%, and 71% respectively) were the highest observed during this same period. Unfortunately, age and spawning history data are unavailable for the years 1959–1964, and consequently, the progress of this suspected strong year-class cannot be confirmed by continued evaluation of its contribution to subsequent populations. However, the 5-year duration of these population sizes is in good agreement with the difference between the mean age of maturity (4.4 years, sexes combined: Leggett 1969) and the maximum age of shad in the population (approximately 8–9 years).

One very interesting feature of the population trends 1935–1973 is the approximate 7–10 year periodicity in peaks and troughs. This is in good agreement with Ricker's (1954) conclusion, based on a theoretical analysis of stock–recruitment relationships, that in multiple-age stocks, the periodicity of population cycles would be twice the mean generation time (interval from egg to egg). This interval is approximately 4–5 years for shad.

The variability in population size during

1965–1973 is typical of trends in earlier years (Figure 18). The large population observed in 1965 appears to have resulted from recruitment of a very successful year-class spawned in 1960. The contribution of this year class to subsequent populations is apparent in Table 26. The high natural mortality experienced by shad in 1965 reduced the contribution of repeat spawning shad to 1966 and subsequent populations. This, coupled with lower than average recruitment in 1966 and 1967, led to the marked difference in population size between 1965 and subsequent years. The 1968 population, the lowest recorded during 1965–1973, appears to have resulted primarily from poor recruitment and from higher than average mortality of females spawning in 1967. The 1969, 1970, and 1971 populations, the first to be potentially influenced by Connecticut Yankee operation, were approximately equal to or larger than three of the four preoperation populations (1966–1968) and exhibited both good recruitment and a strong repeat spawning compo- nent. Recruitment in 1972 was the lowest recorded during 1965–1973; however, it appears to have been related, at least in part, to very low juvenile production in 1968 (Table 9 in Marcy 2004 [1976]), which followed a highly predictable pattern in relation to flow and temperature conditions on the spawning grounds located well above the influence of the Connecticut Yankee facility. Recruitment in 1973, while ranking only fifth of the 9 years studied, improved over 1972 levels. This, coupled with lower-than-average adult mortality in 1972–1973, resulted in the small increase in population size noted in 1973. Based on the success of the commercial fishery in 1974 (Robert Jones, personal communication), this improving trend appears to have continued in 1974. This pattern of population changes does not indicate any meaningful influence of the Connecticut Yankee on the size or structure of annual shad populations to date.

General Discussion

Migration of Adults

The results of our investigation of the relationship between temperature and the timing of shad migrations, as reported here and in Leggett and Whitney (1972), have clearly demonstrated a strong correlation between specific water temperatures and the movements of shad, both into rivers and while at sea. The similarity of the preferred temperatures in populations widely separated geographically suggest that this characteristic is conservative and that it has considerable adaptive value for the species. The precise significance of this adherence to specific water temperatures while at sea is far from understood; it may, for example, relate to the maintenance of the population, during all seasons, in waters that have high food production, or alternately, it may result from a narrow range of temperature tolerance. In any event, it is clear that the tendency does have considerable significance for reproduction. It brings shad to the mouth of their home river each year at a time when river conditions are approaching the optimum for reproduction. Several investigators (Leim 1924; Bradford et al. 1968; Marcy 1969) have demonstrated that the optimum temperatures for egg and larval development are in the range 15.5–26.5°C. Timing the entry of peak numbers of shad into the river when water temperatures are in the range of 13–18°C thus brings the maximum number of adults to the spawning grounds under thermal conditions that are optimum for the survival of eggs and larvae. The importance of this is clearly demonstrated by the strong relationship between temperature and flow conditions and juvenile production and by the significant contribution of temperature at time of spawning to recruitment in subsequent populations.

This timing may also have survival value for the adults. Leggett (1969, 1972) suggested that the energy cost of the freshwater migration may be directly related to the temperatures experienced en route and that this could explain differences in the frequency of repeat spawning (mortality) in populations at different latitudes along the Atlantic Coast. Leggett (1972) and Glebe and Leggett (1973) reported that somatic weight losses of shad during the freshwater migration in the Connecticut River range from 27% to 51% of total body weight. This corresponds to a utilization of up to 60% of the fish's total available energy (fat and protein) during the migration, and it falls within the lethal range reported for other species (Belding 1934). Glebe and Leggett have also shown that shad that migrate early in the season, when water temperatures are low, experience significantly smaller weight losses and energy expenditures than do shad that migrate during the latter half of the run when water temperatures are in the range 15–20°C.

In addition, Dodson and Leggett (1973, 1974) have suggested that the annual return to the mouth of the natal river as a result of adherence to specific temperature conditions at sea may be a critical feature of the homing process. This, coupled with the development of sensory receptiveness to home-river clues, perhaps related to the annual reproductive cycle, results in home-seeking behavior and, ultimately, in entry into the river to spawn.

It thus appears that the precise timing of the marine and freshwater migrations of shad to correspond with temperatures in the range of 13–18°C has adaptive significance for both juveniles and adults. Significant changes in river temperature profiles as a result of human activities could therefore present a serious threat both to the timing of shad migrations and to the continued vitality of the affected populations.

The freshwater migration of American shad native to the Connecticut River appears to have at least three phases, each having distinct behavior correlates. Dodson and Leggett (1973, 1974) have shown that approach to the home river from Long Island Sound is not followed by immediate entry. Rather, there appears to be a temporal restriction, perhaps related to gonadal development and modulated by the endocrine system, that delays entry into the river until conditions, both physiological and hydrographic, are optimum.

Movements in the saline portion of the estuary were found to be characterized by reduced swimming speeds compared with those observed both at sea- and in freshwater. This, in addition to the prolonged meandering behavior observed while tracking shad near the upper limit of the salt–freshwater interface, and to the delay in upriver migration following conventional tagging, would appear to indicate that adaptation to freshwater may require considerable time and may result in reduced swimming speeds and upriver orientation, perhaps as a result of physiological stresses related to osmoregulation. This was most clearly demonstrated in the swim-speed response to changes in flow velocity. Tagatz (1961) demonstrated that shad are highly sensitive to rapid transfers between salt and fresh water. The meandering behavior may be an attempt to minimize this stress by reducing the rate of transition and perhaps by selecting the optimum salinity conditions for each stage of an adaptation sequence.

The freshwater phase of the migration is clearly more direct and is executed at higher average swimming speeds. However, both ultrasonic and conventional tagging studies indicate that a considerable amount of meandering, holding, and even downriver migration occurs en route to the spawning grounds. The behavior during this period is interesting in that the shad maintains a pattern of adjusting its swim speeds to current velocities, and this in turn facilitates the maintenance of an approximately uniform rate of movement relative to the bottom. This behavior is also characteristic of the saltwater migration (Dodson and Leggett 1973, 1974). The strict counter-current response observed in saltwater is abandoned in freshwater, however, and is replaced by an overriding upriver orientation that is independent of the direction of current flow. This change in basic behavior pattern following entry into freshwater may be analogous to changes in the preferred direction of orientation in saltwater as a result of sign stimuli related to the presence or absence of home river clues (Dodson and Leggett 1973, 1974). How this shift in behavior is triggered is not known, but one obvious possibility involves hormonal changes related to oso-adaptation to freshwater.

Estimates of the rates of upriver migration based on regression analysis of spaghetti tag recoveries and on long-term ultrasonic tracking records are in close agreement and indicate an average upriver movement of approximately 6 km/d. This rate is consistent with observations of the average time required for the peak abundance of shad to proceed upriver. This rate compares with a theoretical rate of 34 km/d based on a calculated average swimming speed, relative to the bottom, in freshwater, of 40 cm/s, providing that the fish continuously maintains its upriver orientation. This is in reasonable agreement with the fastest observed rate of movement from the river mouth to the spawning grounds (23 km/d) based on dart tag recoveries. However, the much slower rate characteristic of the population as a whole reinforces the conclusion that the upriver migration, in most cases, is not direct.

The observed increase in the upriver migration rate as the season progressed, the faster rate among females, which enter the river later than the males, and the lesser delay in upriver migration in females all suggest an increase in the motivation or drive of shad to migrate as the season progresses. This appears to be due to the approach of optimum spawning conditions, since Glebe and Leggett have found that the gonads are essentially mature at the time the shad enters the river and that there is no significant augmentation of fat or protein content of the eggs during the upriver migration.

Population Trends

Analysis of the annual variations in population size in the Connecticut River, both during the period of intensive study (1965–1973) and over the longer interval (1935–1973), clearly demonstrate that large-scale, year-to-year variations in population size are common. These fluctuations in population size result from harvesting and from the interaction between recruitment and the survival of the mature shad following its initial and subsequent spawning acts. Recruitment has been shown to be dependent on the size of the parental stock; however, environmental variables, notably temperature and flow conditions during the spawning interval, have also been demonstrated to influence juvenile production and subsequent recruitment significantly. River temperatures during the spawning migration have also been implicated as a significant factor in determining spawning and postspawning survival, even though the role of temperature in this relationship is not yet well quantified.

Over the longer term, the pattern of changes in population size and related commercial catches follows closely the responses predicted for exploited populations by Ricker (1954, 1958a). The higher population sizes and larger catches per unit effort that were characteristic of the 1930s and early 1940s appear to have resulted from the operation of the fishery at the level of maximum sustainable yield. Between 1943 and 1946, the total fishing effort quadrupled and this led to catches well in excess of the maximum sustainable level. Escapement and population size declined sharply from 1946 to 1950 and has since remained at a lower average level (approximately 50%) of the pre-1942 level. Failure of the population to recover following reductions in fishing pressure appears to have resulted from the fact that under prevailing conditions escapement, and hence both recruitment and repeat spawning contributions to subsequent populations, is now considerably less than that required to produce the optimum equilibrium population size and maximum sustainable yields.

The oscillations in population size since 1950 have been more modest, and their amplitude is not atypical of oscillations predicted for multiple age stocks by Ricker (1954, 1958a, 1958b) and Allen and Basasibwaki (1974). The periodicity of these oscillations is also in reasonable agreement with predicted ranges, approximately twice mean generation time (Ricker 1954) and approximately twice mean age (Basasibwaki 1972). The amplitude of these oscillations also appears to be declining as times goes on. This is to be expected in situations where the number of age-groups in the spawning stock is declining (Allen and Basasibwaki 1974). Reduction in the number of age-groups and in the relative importance of older age-groups are a direct consequence of "fishing up" and can be assumed to have occurred since the early 1930s.

The periodicity and amplitude of oscillations in fish populations is also significantly affected by random environmental factors that operate on survival (Ricker 1954, 1958a, 1958b; Allen and Basasibwaki 1974). The significant influence of river temperature and flow on juvenile production and recruitment, and the implied effect of temperature on adult survival, could thus be expected to alter the direction and amplitude of population oscillations significantly. This may well serve to explain the low predictability of recruitment and population size based on escapement (parental stock) data alone.

The evidence available suggests that the current size of annual shad populations in the Connecticut River is well below the level producing maximum equilibrium yield. This does not imply that the current populations are being overfished; in fact, tag return data suggest that the current fishing rate is only 20% to 25%. However, the average escapement from the fishery has been below the optimum level almost continuously since 1945. This is because the population declined rapidly under very heavy fishing pressure until 1950, when the major decline in population size stopped. The size of annual populations has since oscillated between approximately 250,000 and 650,000 shad with the exception of the very large (1,470,000) population in 1965.

Effects of Thermal Additions

The absence of changes in the timing and pattern of shad migrations and in population trends associated with the pre- and postoperation phases of the Connecticut Yankee facility indicate that with its present operating mode, the discharge of heated water has no meaningful effect on the structure or behavior of the Connecticut River shad population.

As noted on page 183, full assessment of the impact of the heated water discharges, which began in October 1967, will not be possible until the progeny from several post-1967 year-classes have made their appearance in the Connecticut River. Progeny from the 1968 year-class, the first produced during conditions of near normal plant operation, returned to the Connecticut River in 1972 and 1973. There is no evidence that recruitment in these years was significantly affected by plant operation. Recruitment in 1972, as previously noted, was the lowest recorded during the period from 1965 to 1973; it appears to have been related, at least in part, to below-average juvenile production in 1968, which followed a highly predictable pattern in relation to flow and temperature conditions on the spawning grounds located well above the influence of the Connecticut Yankee facility. Recruitment in 1973, on the other hand, was above the average for the years 1966–1973 and exceeded that observed in 1966, 1967, and 1968, which were unaffected by plant operation. These data, coupled with above-average juvenile production in 1969, 1970, and 1971 (Table 9 in Marcy 2004 [1976]) and the behavioral data reported by Moss (1970) and Marcy, which showed that juvenile shad are able to detect and avoid artificially induced temperature increases of as little as 1°C, argue against any significant impact by the heated effluent from Connecticut Yankee plant on the survival of juvenile shad.

In considering the possible effects of future alterations in the heat budget of the Connecticut River, it is important to remember that several strong relationships have been demonstrated between water temperature and the migratory behavior and population dynamics of the shad. First, the timing of its migrations, both in the sea and in freshwater, is strongly influenced by water temperatures. This characteristic of shad behavior appears to be remarkably conservative, since the temperature at which peak numbers of shad enter freshwater to spawn (13–18°C) is similar in all Atlantic and Pacific Coast rivers studied to date (Leggett and Whitney 1972).

Second, the production of juveniles and subsequent recruitment is strongly related to water temperatures and discharge conditions in the spawning area. Significant alteration of the temperature characteristics of the river, either at its mouth or in the spawning area, could, therefore, lead to a lack of synchrony between the timing of the entry of shad into the river and the optimum conditions for egg and larval development. In turn, this could lead to reduced juvenile production and recruitment.

Third, Glebe and Leggett have data that suggest a relationship between temperatures experienced during the freshwater migration and the energy cost of the migration. This increased energy cost may be related to lower adult survival following spawning in southern rivers and perhaps also to differences in annual mortality rates in single populations; the high mortality experienced by Connecticut River shad in 1965 was associated with one of the lowest flow and highest water temperature years on record.

Significant elevations of river temperature could result in the migration being made at higher average temperatures. This would occur if river warming became advanced with respect to the warming of the ocean, since the movements of shad in the ocean are governed by the same temperature criteria that influence their entry into rivers. Such a development could lead to a reduction in the repeat spawning component of the stock.

Finally, the channel-oriented behavior of adult shad migrating in freshwater should be considered in the siting of future thermal discharges. The location of the Connecticut Yankee discharge on the opposite side of the river from the main river channel significantly reduces the exposure of shad to the thermal plume. Similar strategies should be employed in future sites if thermal discharges are deemed acceptable on other grounds.

Acknowledgments

I am deeply indebted to Dr. Richard R. Whitney for his continuing interest and creative criticism over the past 9 years; I have benefited greatly from his untiring efforts. The very significant contribution of Mr. Robert A. Jones to the design and execution of the ultrasonic tracking program needs also be acknowledged. Thanks are also due to Drs. Daniel Merriman and James W. Atz for editorial assistance and the entire Connecticut River Ecological Study staff for active support.

References

Allen, R. L., and P. Basasibwaki. 1974. Properties of age structure models for fish populations. Journal of the Fisheries Research Board of Canada 31:1119–1125.

Bampton, T. 1964. Testimony to State of Connecticut Water Resources Commission, Second Public Hearing on Application of Connecticut Yankee Atomic Power Co.

Basasibwaki, P. 1972. Characteristics of cyclic fluctuations generated by stock-recruit systems. M.Sc. thesis. University of British Columbia, Vancouver.

Beamish, F. W. H. 1969. Photographic observations on reactions of fish ahead of otter trawls. F.A.O. (Food and Agriculture Organization of the United Nations) Fisheries Reports 2:511–521.

Belding, D. L. 1934. The cause of high mortality in the Atlantic salmon after spawning. Transactions of the American Fisheries Society 64:219–224.

Best, E. A., and H. D. Boles. 1955. An evaluation of creel census methods. California Fish and Game 42(2):109–115.

Beverton, R. J. H., and S. J. Holt. 1957. On the dynamics of exploited fish populations. Fishery Investigation Min. Agri. Fish. Food (G.B.) Series II. Volume XIX.

Bigelow, H. B., and W. C. Schroeder. 1953. Fishes of the Gulf of Maine. U.S. Fish and Wildlife Service Fishery Bulletin. 53:1–577.

Bigelow, H. G., and W. W. Welsh. 1925. Fishes of the Gulf of Maine. Bulletin of the U.S. Bureau of Fisheries 40(II):1–509.

Blaxter, J. H. S., and B. B. Parrish. 1965. The importance of light in shoaling, avoidance of nets and vertical migration by herring (*Clupea harengus*). Journal du Conseil Conseil International pour L'Exploration de la Mer 30:40–57.

Borodin, M. A. 1927. Revision of the shad and shad like fishes of North America. Zoologische Jahrbuecher Abteilung fuer Systematik Oekologie und Geographie der Tiere 53(1/3):1–22.

Bradford, A. D., J. G. Miller and K. Buss. 1968. Bioassays of eggs and larval stages of American shad, *Alosa sapidissima*. In Suitability of the Susquehanna River for restoration of shad. U.S. Department of the Interior, Washington, D.C.

Carlander, K. D., C. J. DiCostanzo, and R. J. Jessen. 1958. Sampling problems in Creel Census. Progressive Fish-Culturist 20(2):73–81.

Carscadden, J. E., and W. C. Leggett. 1975a. Meristic differences in spawning populations of American shad, *Alosa sapidissima*: evidence for homing to tributaries in the St. John River, New Brunswick. Journal of the Fisheries Research Board of Canada 32:653–660.

Carscadden, J. E., and W. C. Leggett. 1975b. Life History variations in populations of American shad, *Alosa sapidissima* (Wilson), spawning in tributaries of the St. John River, New Brunswick. Journal of Fish Biology:595–609.

Cating, J. P. 1953. Determining age of Atlantic shad from their scales. U.S. Fish and Wildlife Service Fishery Bulletin 54(85):187–199.

Chapman, D. G. 1948. Problems in enumeration of populations of spawning sockeye salmon. 2. A mathematical study of confidence limits of salmon populations from sample tag ratios. International Pacific Salmon Fisheries Commission Bulletin II:69–85.

Chapman, D. G., B. D. Fink, and E. B. Bennett. 1965. A method for estimating the rate of shedding of tags from yellow fin tuna. Inter-American Tropical Tuna Commission Bulletin 10(5):335–352.

Chittenden, M. E. 1969. Life history and ecology of the American shad, *Alosa sapidissima,* in the Delaware River. Ph.D. thesis. Rutgers University, New Brunswick, New Jersey.

Chittenden, M. E., and J. Westman. 1964. Progress report of shad research on the Delaware River (Manuscript).

Claussen, L. G. 1959. A southern range extension of the American shad to Todos Santos Bay, Baja, California. California Fish and Game 45(3):217–218.

D'Ancona, U. 1927. Sal valore sistematico delle unita morfologiche de numero variabile ed in particolare suela sistematica delle Alsoe mediterranee. Monitoire Zoologico Italiano 38(8):187–191.

Dodson, J. J., and W. C. Leggett. 1973. Behavior of adult American shad (*Alosa sapidissima*) homing to the Connecticut River from Long Island Sound. Journal of the Fisheries Research Board of Canada 30:1847–1860.

Dodson, J. J., and W. C. Leggett. 1974. The role of olfaction and vision in the behavior of American shad (*Alosa sapidissima*) homing to the Connecticut River from Long Island Sound. Journal of the Fisheries Research Board of Canada 31:1607–1619.

Dodson, J. J., W. C. Leggett, and R. A. Jones. 1972. The behavior of adult American shad (*Alosa sapidissima*) during migration from salt to fresh water as observed by ultrasonic tracking techniques. Journal of the Fisheries Research Board of Canada 29:1445–1449.

Dodson, J. J., W. C. Leggett, and R. A. Jones. 1973. The behavior of American shad (*Alosa sapidissima*) during the homing migration to the Connecticut River. Essex Marine Laboratory, Report no. 5, Essex, Connecticut.

Fischler, K. J. 1959. Contributions of Hudson and Connecticut Rivers to New York–New Jersey shad catch of 1956. U.S. Fish and Wildlife Service Fishery Bulletin 60:161–174.

Fredin, R. A. 1954. Causes of fluctuations in abundance of Connecticut River shad. U.S. Fish and Wildlife Service Fishery Bulletin 54:247–259.

Glebe, B. D., and W. C. Leggett. 1973. A study of weight loss of American shad during the freshwater migration. U.S. National Marine Fisheries Service, Progress Report Project AFC-8-1 (Connecticut),.

Hemmings, C. C. 1969. Observations on the behavior of fish during capture by Danish seine net, and their relation to herding by trawl bridles. FAO (Food and Agriculture Organization of the United Nations) Fisheries Reports 2:645–655.

Hester, F., and J. H. Taylor. 1965. How tuna see a net. Commercial Fisheries Review 27:11–16.

Hildebrand, S. F. 1963. Family Clupeidae. Pages 257–454 in H. B. Bigelow et al., editors. Fishes of the western North Atlantic. Memoir 1, part 3. Sears Foundation for Marine Research, Yale University, New Haven, Connecticut.

Hill, D. R. 1959. Some uses of statistical analysis in classifying races of American shad (*Alosa sapidissima*). U.S. Fish and Wildlife Service Fisheries Bulletin 59:269–286.

Hodder, V. M. 1966. Two further records of the American shad in Newfoundland waters. Transactions of the American Fisheries Society 95(2):228–229.

Holgate, P. 1967. Population survival and life history phenomena. Journal of Theoretical Biology 14:1–10.

Hollis, E. H. 1948. The homing tendency of shad. Science 108:332–333.

Judy, M. H. 1961. Validity of age determination from scales of marked American shad. U.S. Fish and Wildlife Service Fishery Bulletin 61:161–170.

Lapointe, D. F. 1957. Age and growth of the American shad from three Atlantic Coast rivers. Transactions of the American Fisheries Society 87:139–150.

Leach, G. C. 1925. Artificial propagation of shad. Report U.S. Comm. Fish., 1924, appendix 8:459–486 (Doc. 981).

Leggett, W. C. 1969. Studies on the reproductive biology of the American shad (*Alosa sapidissima,* Wilson). A comparison of populations from four rivers of the Atlantic seaboard. Ph.D. thesis. McGill University, Montreal.

Leggett, W. C. 1972. Weight loss in American shad (*Alosa sapidissima,* Wilson) during the freshwater migration. Transactions of the American Fisheries Society 101(3):549–552.

Leggett, W. C, and R. A. Jones. 1971. Net avoidance behavior in American shad (*Alosa sapidissima*) as observed by ultrasonic tracking techniques. Journal of the Fisheries Research Board of Canada 28:1167–1171.

Leggett, W. C., and R. R. Whitney. 1972. Water temperature and the migrations of American shad. U.S. Fish and Wildlife Service Fisheries Bulletin 70(3):659–670.

Leim, A. H. 1924. The life history of the shad, *Alosa sapidissima* (Wilson) with special reference to factors limiting its abundance. Biological Board of Canada Contrib. Can. Biol. Fish 2(11):163–284.

Mairs, D. F., and C. J. Sinderman. 1962. A serological comparison of five species of Atlantic clupeoid fishes. Biological Bulletin 123(2):330–343.

Marcy, B. C. 1969. Shad early life histories. Pages 19–36 *in* The Connecticut River investigation. 8th semi-annual progress report.

Marcy, B. C. 1972. Spawning of the American shad, *Alosa sapidissima,* in the lower Connecticut River. Chesapeake Science 13(2):116–119.

Marcy, B. C., Jr. 2004. Early life history studies of American shad in the lower Connecticut River and the effects of the Connecticut Yankee plant. Pages 155–180 *in* P. M. Jacobson, D. A. Dixon, W. C. Leggett, B. C. Marcy, Jr., and R. R. Massengill, editors. The Connecticut River Ecological Study (1965–1973) revisited: ecology of the lower Connecticut River 1973–2003. American Fisheries Society, Monograph 9, Bethesda, Maryland. (Originally published in 1976)

Massmann, W. H. 1952. Characteristics of spawning areas of shad (*Alosa sapidissima,* Wilson) in some Virginia streams. Transactions of the American Fisheries Society 81:78–93.

Massmann, W. H., and A. L. Pacheco. 1957. Shad catches and water temperature in Virginia. Journal of Wildlife Management 21(3):351–352.

McDonald, M. 1884. The shad—*Clupea sapidissima.* Pages 594–607 *in* G. B. Goode and Associates. The fisheries and fishery industries of the United States. Section I. Natural history of useful aquatic animals. U.S. 47th Congress, 1st Session, Senate Miscellaneous Document 124.

McDonald, M. 1887. The Connecticut and Housatonic rivers and minor tributaries of Long Island Sound. In G. B. Goode. The fisheries and fishery industries of the United States. Section 5. 1:659–667. U.S. Government Printing Office, Washington, D.C.

McHugh, J. L., and L. E. Fitch. 1951. An annotated list of the clupeoid fishes of the Pacific Coast from Alaska to Cape San Lucas, Baja California. California Fish and Game 37(4):491–495.

Meade, R. H. 1966. Salinity variations in the Connecticut River. Water Resources Research 2:567–579.

Merriman, D. 1970. The calefaction of a river. Scientific American 222(5):42–52.

Mohr, H. 1969. Observations on the Atlanto-Scandian herring with respect to schooling and reactions to fishing gear. FAO (Food and Agriculture Organization of the United States) Fisheries Reports 2:567–577.

Moss, S., A. 1970. The response of young American shad to rapid temperature changes. Transactions of the American Fisheries Society 99(2):381–384.

Murphy, G. I. 1966. Population biology of the Pacific sardine (*Sardinops caerulea*). California Academy of Sciences, Fourth Series 34(1):1–84.

Murphy, G. I. 1968. Vital statistics of the Pacific sardine (*Sardinops caerulea*) and the population consequences. Ecology 48(5):731–736.

Neave, F. 1954. Introduction of anadromous fishes on the Pacific Coast. Canadian Fish Culturist 16:25–27.

Nichols, P. R. 1959. Extreme loss in body weight of an American shad (*Alosa sapidissima*). Copeia 1959(4):343–344.

Nichols, P. R., and M. Tagatz. 1960. Creel census Connecticut River shad sport fishery, 1957–1958, and estimate of the catch, 1941–1956. U.S. Fish and Wildlife Service Special Scientific Report Fisheries 351.

American Fisheries Society Monograph 9:239–246, 2004

Recurrent Mass Mortalities of the Blueback Herring in the Lower Connecticut River

SANFORD A. MOSS, WILLIAM C. LEGGETT, AND WILLIAM A. BOYD
Essex Marine Laboratory, Inc., Essex, Connecticut 06426, USA

The sudden mortality of large numbers of fish in natural waters is of concern to biologists and nonbiologists alike. Mass mortalities are conspicuous and produce unsightly public nuisances. They may also have important consequences with respect to the population structures and the maintenance of adequate stocks of food or forage fish. Moreover, in most cases, they signal a change, usually a deterioration, in water quality.

Mass mortalities of fish result from a variety of causes. Rapid changes in physiologically important parameters, such as temperature, dissolved oxygen, carbon dioxide, and salinity, may be lethal. Mortalities may also result from unusually heavy parasitic or pathogenic infestations (Spotte 1970). Other causes include supersaturation of gasses and the introduction into waterways of toxic substances, including pesticides and herbicides, industrial wastes, and petroleum products. Unfortunately, it is often difficult to satisfactorily identify the specific cause of mass mortality. Workers frequently arrive several hours or even days after a mortality has occurred when the causal conditions no longer exist. Detailed autopsies of the victims frequently fail to provide requisite information. A more subtle difficulty is that mass mortalities of fish (except in the case of acutely toxic substances) are rarely the result of a single factor. Stressful situations imposed by poor, but sublethal conditions may facilitate the unfavorable action of other environmental parameters. This point has been stressed by Mihursky and Kennedy (1967), Doudoroff and Shumway (1967), and Hoff et al. (1968) for sublethal extremes of temperature, dissolved oxygen, and physiological stress.

Since 1965, we have had the opportunity to study five mass mortalities of blueback herring *Alosa aestivalis* (Mitchill). This is a report of our observations and an analysis of the variables that appear to be involved in these mortalities.

Methods

The study area consisted of the lower reaches of the Connecticut River from Saybrook, Connecticut, at the mouth, to Canel, approximately 32.2 km upstream (Figure 1). In this region, the Connecticut is a broad, tidal river, with silt and sand bottoms. A navigation channel is maintained to a minimum depth of 3.65 m. Measurable salinities may intrude upstream to Haddam (km 21) during extreme low flow, but normally, in summer, they occur only as far north as Hadlyme (km 17). Barge traffic is moderate and recreational boating in the area is seasonally moderate to heavy. The resident fish fauna is dominated by white catfish *Ictalurus catus*, brown bullhead *I. nebulosus*, white perch *Morone americana*, common carp *Cyprinus carpio*, American eel *Anguilla rostrata*, and spottail shiner *Notropis hudsonius*. Important anadromous fish using the river and its major tributaries for spawning and rearing include American shad *Alosa sapidissima*, alewife *A. pseudoharengus*, and blueback herring. Of the anadromous species, the blueback spawns latest; spawning does not commence until water temperatures reach approximately 20°C (Bigelow and Schroeder 1953).

Various aspects of our survey (finfish, plankton, benthic, and hydrographic sampling) called for nearly daily visits by boat to the study area. During these trips, temperature and dissolved oxygen were routinely recorded from several sampling stations. In addition, five continuously recording hydrographic stations were operated in the study area from June of 1966 to November 1970. These stations were located approximately 17, 22, 24, 28, and 32 km from the river mouth

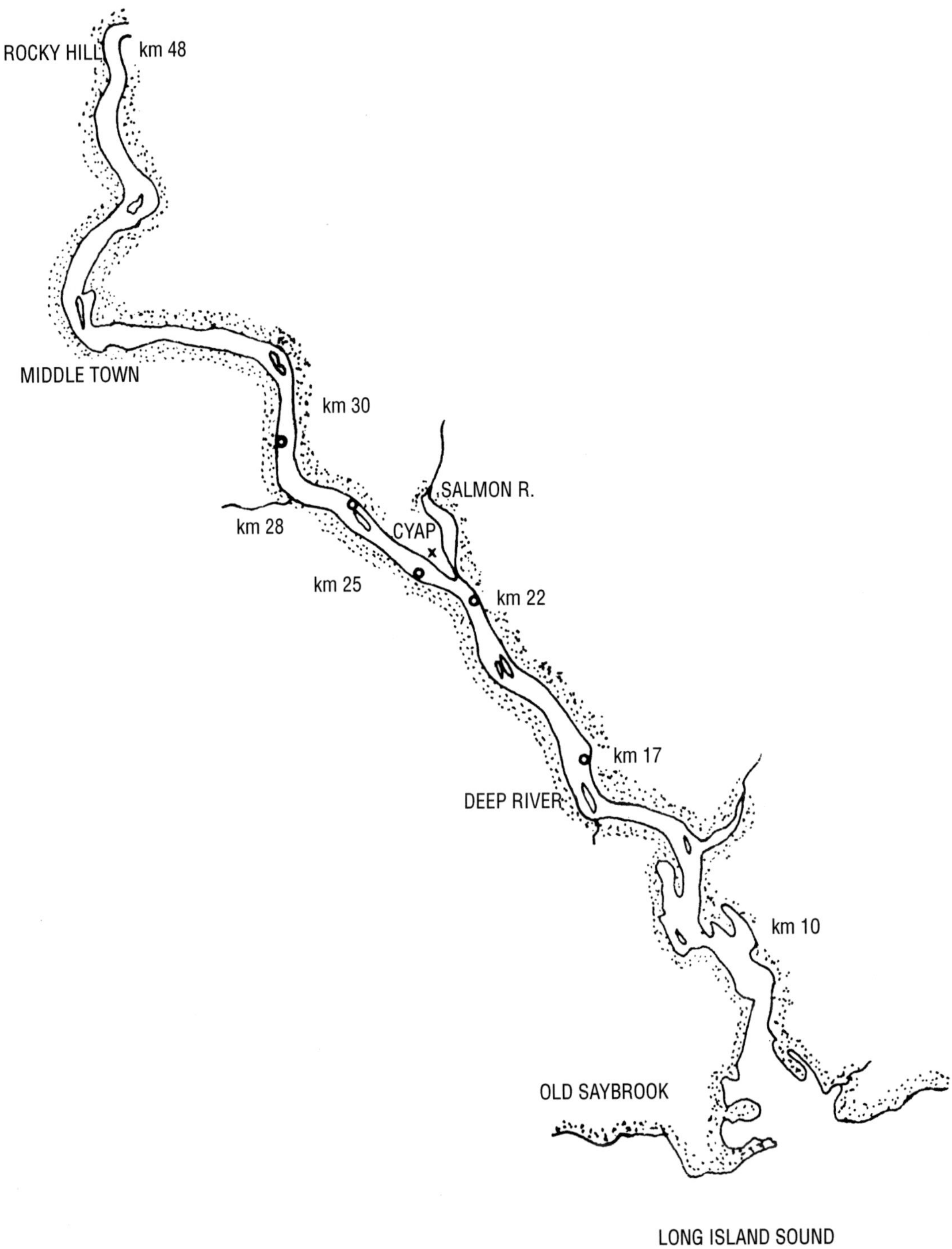

FIGURE 1. Lower Connecticut River including regions of recorded fish kills involving blueback herring. Locations of fixed, continuously recording hydrographic stations are indicated (o); CYAP refers to the location of the Connecticut Yankee Atomic Power Company plant.

(Figure 1). In November 1970, the number of these hydrographic stations was reduced to two, located at km 14.4 and 24. Each hydrographic station consisted of thermistors at surface and bottom, polarographic oxygen electrodes, and conductivity probes. One station, km 24, was equipped with pH electrodes, and a river stage transducer (Boyd 2004 [1976], this volume).

Beginning in the summer of 1966, when our normal sampling program indicated that water quality conditions were marginal, or when reports of dying fish reached us, early-morning transects of the study area were made and additional measurements of temperature, dissolved oxygen, and pH were taken. Counts and identification of dead and dying fish were made, and autopsies were performed at the laboratory. Dissolved oxygen was determined by oxygen electrodes (YSI and Precision Scientific) supplemented and calibrated by standard Winkler determinations. pH was determined colorimetrically in the field or was measured in the laboratory using refrigerated samples drawn an hour or two earlier.

Observations

Two substantial mortalities of fish occurred in the lower Connecticut River during the summer of 1965. The first, on 2 July, occurred in the Haddam region of the river, near the Connecticut Yankee Atomic Power Company (CYAP) plant construction site (Figure 1). Examination of the area on the morning of 3 July showed that dead fish were confined to the region between km 29 (Higganum Creek) and km 20 (3 km downstream from the Haddam Bridge). The dead fish consisted almost exclusively (99.9%) of *A. aestivalis*. A few shad were noted, but no more than is normal for that time of year. 3 July midday surface temperatures in this area ranged between 24°C and 25°C. Several days of dry, hot weather had preceded the mortality. Approximately 15 to 20 thousand fish were estimated to have been included in this event.

On 13 July a second major mortality occurred. It extended from km 48 (Rocky Hill) to km 27 (Haddam Island). Dying fish were first reported by Middletown police department at 0730 hours. State biologists estimated the kill to have included approximately 100,000 fish (Hartford [Connecticut] Courant, 13 July 1965) and considered such a mortality to be "normal" for the Connecticut River at this time of year (Hartford Courant, 14 July 1965). Nearly all the fish examined by us were again blueback herring, although a few alewives and shad were also seen. 14 July midday temperatures in the region of the kill ranged between 25.6°C and 26.5°C, dissolved oxygen 3.4 to 4.1 mg/L, and pH ranged between 6.7 and 7.1.

In 1966 another large fish kill occurred, beginning in the early morning of 4 July. At midday 4 July, most dead fish were found from km 20 downstream to km 9 (Brockway Island). The kill was calculated by random counts to have included in excess of 50,000 fish, composed mainly (99%) of adult herring. At dawn on 5 July, a smaller mortality was observed by us from km 29 (Higganum Creek) to km 17 (Hadlyme). A similar mortality occurred on the morning of 6 July. Blueback herring were observed spawning in the main river at Essex, below the region of the kill, and in an unaffected tributary (Salmon River) on the morning of 5 July. Adult blueback herring, in good condition, as well as large numbers of young-of-the-year shad and alewives were collected by bag seine in the affected area on 6 July. A sample of 24 dead and dying (swimming weakly into the current at the surface) fish collected at dawn on 5 July consisted of 13 males and 11 females. All were fully ripe with the exception of one half-spent female. None of the fish was obviously infested with internal or external parasites; 33% were clean fish with no abrasions or signs of hemorrhage; 42% were lightly marked with minor abrasions or epidermal irritations; 24% were moderately marked with dermal hemorrhages, particularly near the bases of the pectoral fins and occasionally on the operculum. Temperature, oxygen, and pH values obtained during the morning hours while fish were dying on 5 July are given in Table 1. These may be compared with temperature and oxygen values collected outside the affected area in transects of the lower 160 km of the Connecticut River prior to (29, 30 June) and after (7, 8 July) the 1966 kill (Figure 2). In general, dissolved oxygen was lower and temperatures were higher in the affected area on both transects; oxygen readings were lowest and temperatures highest on 7 and 8 July.

In 1967, a moderate fish kill occurred in the same general region of the river. On 13 July, a few

TABLE 1. Water conditions in the lower Connecticut River during blueback herring mortality, 6 July 1966.

Time (EDT)	Location (km upstream)	Dis-oxygen (mg/L)	Tem-perature °C	pH	Dead fish (*) or none (0)
0550	8.0	5.7	22.5	6.2	0
0600	9.5	6.1	22.7	6.2	0
0608	11.5	5.0	25.1	6.2	0
0625	16.0	4.8	26.6	6.2	*
0645	17.5	4.7	27.0	6.5	*
0655	20.5	4.4	27.0	6.7	*
0728	22.0	5.1	26.9		*
0740	23.5	5.0	27.2	7.1	*
0750	26.0	4.4	27.2	6.4	*
0800	28.0	4.6	27.3	6.2	few

dead fish were noted in the Haddam to Hadlyme section of the river. At dawn on 14 July, blueback herring were found in moderate numbers both freshly killed and swimming weakly at the surface from km 27 (Haddam Island) to km 9 (Brockway Island), with the largest concentration between km 22.5 (Haddam Bridge) and km 18 (Hadlyme). Dissolved oxygen, temperature, and pH data collected while these fish were dying are presented in Table 2. In general, dissolved oxygen values ranged from 2.6 to 3.0 mg/L, and the temperatures were in the neighborhood of 26°C. A sample of 13 fish (most still showing opercular movements) were dipnetted and examined. The sample was composed of nine males and four females. All were ripe, but little spawning had yet occurred. The fish were clean without any signs of abrasions or hemorrhage. Freshly dead blueback herring were also observed on the morning of 15 July, but further mortality beyond this date was not noted. The total number of fish involved

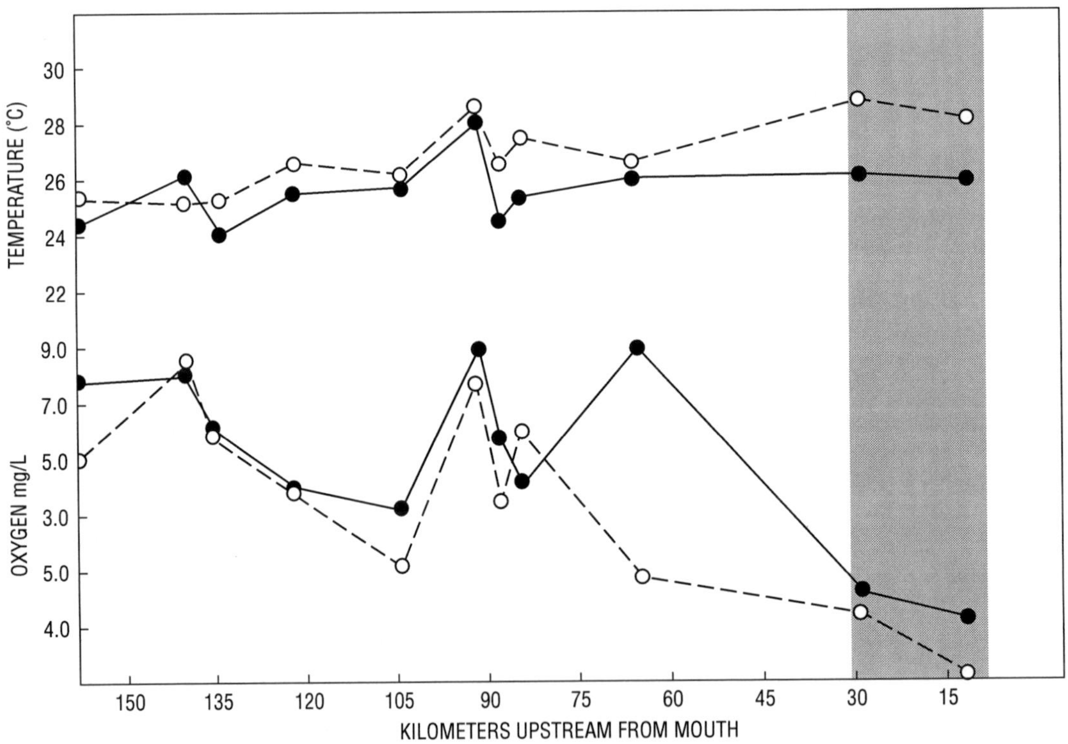

FIGURE 2. Dissolved oxygen and temperature readings taken prior to the 4–6 July 1966 blueback herring mortality (29 and 30 June, solid lines) and afterward (7 and 9 July, broken lines) along the lower 160 km of the Connecticut River. The vertical distance between the lines indicates the change that occurred between the sampling periods. The area in which the mortality occurred (km 16 to 30) suffered the greatest increase in temperature and a substantial decline in dissolved oxygen during the week the mortality occurred.

TABLE 2. Water conditions in the lower Connecticut River during blueback herring mortality, 14 July 1967.

Time (EDT)	Location (km upstream)	Dis-oxygen (mg/L)	Tem-perature °C	pH	Dead fish (*) or none (0)
0515	7.3	3.2	24.7	6.5	0
0540	9.5	2.6	25.4	6.5	*
0600	11.5	3.0	25.4		*
0620	16.0	2.7	25.7	6.5	*
0640	17.5	2.7	26.1	6.5	*
0655	19.0	2.9	26.3	6.6	*
0715	22.0	2.6	26.8		few
0730	23.5	2.6	26.2	6.5	few
0745	26.0	3.5	26.2	6.5	0

over the three mornings of 13, 14, and 15 July 1967 probably did not exceed 15,000 fish.

No substantial mortalities of blueback herring were reported from the lower Connecticut River in 1968, 1969, and 1970.

In 1971, another mass mortality of blueback herring occurred. This kill, which began 21 June and continued through 25 June was preceded by several days of very hot, humid weather. It was investigated by us on the mornings of 24 and 25 June at the peak of the kill. On both days, large numbers of freshly dead blueback herring were observed between km 23 and km 30. Very few fresh specimens were observed in other areas investigated, although large numbers of badly decomposed fish were evident from km 40 (Middletown) to Long Island Sound. On both 24 and 25 June, dissolved oxygen was depressed (Tables 3 and 4). Thirty specimens collected on 24 June showed no signs of disease or physical harm. All fish in the sample, which contained 17 males and 13 females, were ripe but unspawned. Nevertheless, large numbers of blueback herring were observed spawning at Haddam Island (km 27) on 24 June and at Higganum Creek (km 29) on 25 June.

TABLE 3. Water conditions in the lower Connecticut River during blueback herring mortality, 24 June 1971.

Time (EDT)	Location (km upstream)	Dis-oxygen (mg/L)	Tem-perature °C	pH	Dead fish (*) or none (0)
0930	7.3	6.6	26.8	6.5	0
0950	13.0	7.0	27.0	7.0	0
1015	18.3	5.6	26.0	6.5	0
1040	25.5	2.8	25.5	6.5	*
1100	27.0	3.0	24.9	6.5	*

TABLE 4. Water conditions in the lower Connecticut River during blueback herring mortality, 25 June 1971.

Time (EDT)	Location (km upstream)	Dis-oxygen (mg/L)	Tem-perature °C	pH	Dead fish (*) or none (0)
0600	9.5	6.6	25.0	7.0	0
0620	12.0	5.2	26.2	6.5	few
0640	16.0	5.1	26.0	6.5	0
0700	20.0	4.3	27.0	6.5	few
0715	22.0	3.3	25.3	6.5	*
0730	24.0	3.0	25.5	6.5	*
0745	25.5	3.4	25.5	6.5	*
0800	27.0	3.3	24.5	6.5	*
0840	33.5	3.5	25.5	6.5	few
0930	40.0	3.7	25.8	6.5	few

Discussion

The specificity of the several mortalities observed by us (greater than 99% blueback herring) tends to minimize the possibility of toxic pollutants being a causal factor. Possible points of origin of such pollutants were considered, but the variation of the mortality locus in the river, the early-morning timing of the mortalities, and the usual mortality redundancy over several days all suggest that such a cause is unlikely. pH records obtained from samples collected in the survey showed no apparent abnormalities coincident with the observed mortalities. In general, pH ranged from 6.2 to 7.3 in the river.

Facilities were not available for detailed bacterial or epizooite examination of the affected fish. Some of the fish autopsied in the 1966 kill showed dermal hemorrhages at the base of the pectoral fins and on the operculum. Such marks have been reported from young shad suffering from high temperature, low oxygen, and infection of *Aeromonas liquefaciens* (Haley et al. 1967; Pippy and Hare 1969). Such marks were absent, however, from many of the fish examined in 1966 and all of those examined in 1967 and 1971. Moreover, none of the fish examined showed the inflamed lower digestive tract typical of infestations of this epizootic. The possibility remains, however, that epizootic proliferations were contributory to these mortalities.

The dissolved oxygen, temperature, and pH data collected during our surveys of the fish mortalities were taken in a sequential fashion from downstream up. Since the surveys covered several kilometers of river, there is an appreciable time difference, 2.5 to 3.5 h, between the first and last samples. Also, an imperfectly known time lag occurred between the data collection at any station and the actual events that resulted in mortality. For these reasons, the data presented in Tables 1 through 4 are probably biased in at least two ways. First, the upstream stations, sampled later than the downstream stations, probably had relatively higher dissolved oxygen and higher temperatures than the downstream ones. Second, all stations were made sometime after the mortalities commenced, resulting again in the measurement of higher temperatures and higher dissolved oxygen values than were probably present at the initiation of the mortality. For these reasons, we feel that data collected by the continuously recording hydrographic stations are more reliable in assessing the actual dissolved oxygen and temperature values that accompanied the mortalities.

Consequently, the early morning, low oxygen, and temperature data from the fixed hydrographic stations were analyzed for years when mortalities were recorded (1966, 1967, and 1971) as well as for similar periods (late June and all of July) in years when no mortalities were observed (1968, 1969, 1970, and 1972). These data, plotted on a temperature–oxygen scatter diagram (Figure 3), clearly show that mortalities occurred only during periods of low oxygen in the presence of high temperatures. It is also apparent that the lethal condition is complex, being a function of both oxygen and temperature. The data-point distribution, however, is such that a straight line may be drawn between those oxygen–temperature conditions that coincided with mortality and those that did not. This line, by inspection, was found to fit the equation $T = 23.5 + 1.25\ O$, where T = early morning temperature (°C) and O = morning (low) dissolved oxygen in mg O_2/L.

The importance of this equation is that it allows the prediction of conditions under which mortalities of blueback herring can be expected to occur in the Connecticut River. By rearrangement, the relationship can be expressed as $(T - 23.5)\ O = 1.25$. Therefore, we suggest that when this value, the early-morning temperature minus 23.5 and divided by the early-morning oxygen level, exceeds a value of 1.25, mortality of these fish can be expected.

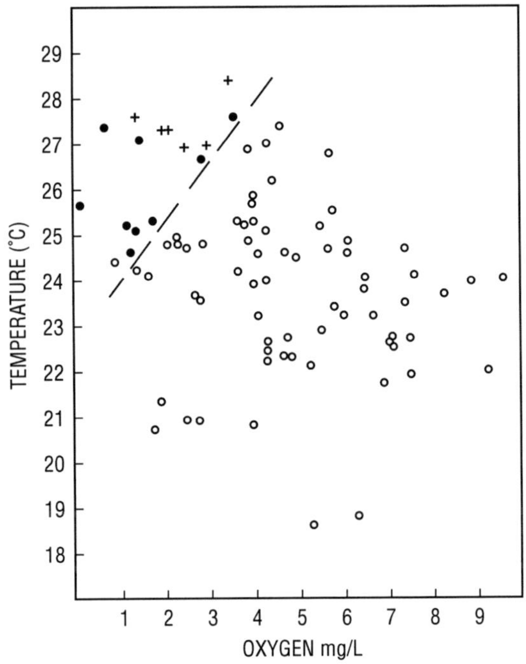

FIGURE 3. Scatter diagram of early-morning low temperature and low oxygen values recorded by a fixed hydrographic station located at river km 22. Data shown were taken from the record of the first 15 d of July in the years 1966 and 1972. Open circles indicate conditions on days when no mortality occurred. Closed circles indicate conditions on days when mortalities occurred. Crosses indicate conditions in 1966 when no mortality occurred beyond an earlier three days of mortalities. See text for further discussion.

A final factor must be considered in analyzing the mortalities described here. Blueback herring in the Connecticut River are known to spawn from the middle of June into August, with the spawning peak occurring in the first half of July. Peak spawning activity, as in the alewife and American shad, occurs during dawn and dusk hours. The appearance of dying fish at dawn corresponds with a period of great activity associated with spawning, as well as with a period of low oxygen. The recurrent observation that nearly all dead fish examined were ripe supports the idea that activity associated with courtship or spawning may have been related to the mortalities reported here. Such activity would, of course, increase the animal's oxygen demand. Resultant oxygen debts not promptly repaid in the face of continued low environmental oxygen and high temperature could contribute to death directly or encourage damaging epizootic proliferation. Higher evening dissolved oxygen in spite of generally higher temperature may allow the repayment of an oxygen debt incurred while spawning.

The fact that sustained poor water quality was not associated with mortality beyond a 3-d period may be due to acclimatization to low oxygen and high temperatures by the survivors. It is also possible that virtually all the fish in a spawning mode were eliminated early in the lethal period and that the survivors were those fish that had either finished spawning or had not yet begun to spawn. In the latter case, a protracted spawning period (such as that shown by *Alosa aestivalis* in the Connecticut River) may be adaptive in the sense that it allows a large, presumably nonspawning, percentage of the population to escape transient lethal conditions.

Water conditions in the Connecticut River during June and July of 1968, 1969, and 1970, when no mortalities occurred, differed from those of previous years in that morning oxygen–temperature relationships always fell outside the lethal zone, as illustrated in Figure 3. This observation again reinforces the conclusion that although the causes of blueback herring mortality in the Connecticut River may be multivariate in nature, low dissolved oxygen with concurrent elevated environmental temperatures are instrumental in initiating them. Frey (1955) and Colby and Brooke (1969) reported similar temperature–oxygen relationships to be instrumental in large summer kills of cisco *Coregonus artedi* in Indiana and Michigan.

The reach of the Connecticut River surveyed by us and in which blueback herring mortalities were observed includes the region thermally influenced by cooling water effluents of the Connecticut Yankee atomic power plant (Figure 1). This warmwater effluent has not been related to catastrophic blueback herring mortalities to the present time. The mortalities observed by us in 1965, 1966, and 1967 occurred before the plant became operational. Those years of no abnormal mortalities, 1968, 1969, and 1970, coincided with periods of near-maximum generating output by this power plant (Figure 4). Only in June 1971 did power plant operation coincide with a mass mortality of blueback herring. Nevertheless, the occurrence of blueback herring mortalities in 4

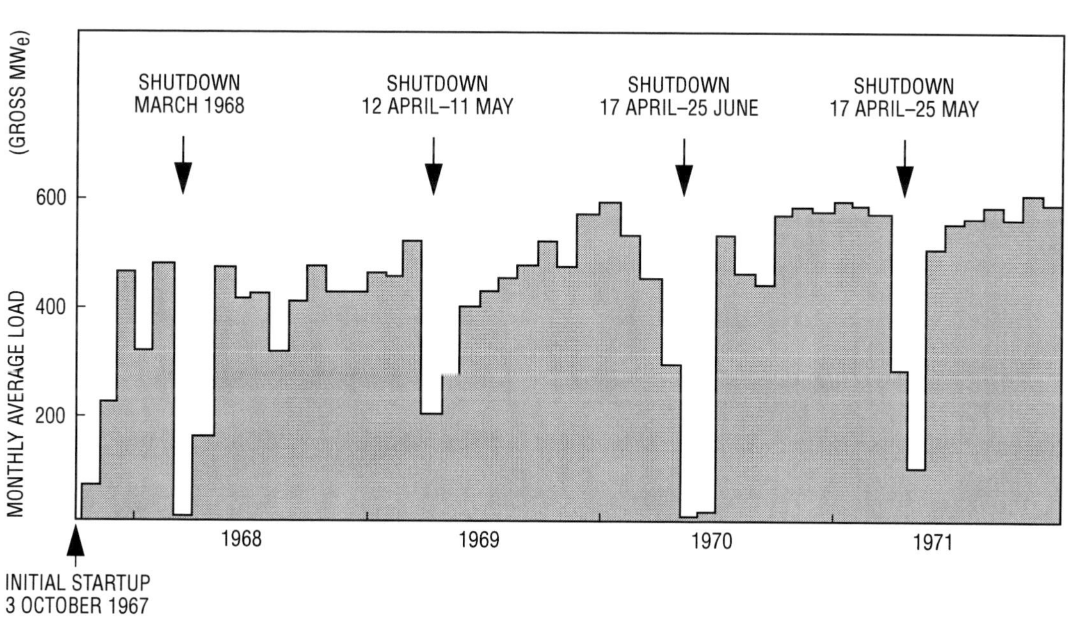

FIGURE 4. Monthly average loads (MW_e) of the Connecticut Yankee atomic power plant during the study period. Blueback herring mortalities in this period occurred only on 21–25 June 1971.

years out of the 7 observed by us suggests that further deterioration of water quality in the lower Connecticut River may threaten the continued spawning success of this fish.

References

Bigelow, H. B., and W. C. Schroeder. 1953. Fishes of the Gulf of Maine. U.S. Fish and Wildlife Service Fishery Bulletin 53:1–577.

Boyd, W. A. 2004. Hydrogaphy. Pages 27–36 *in* P. M. Jacobson, D. A. Dixon, W. C. Leggett, B. C. Marcy, Jr., and R. R. Massengill, editors. The Connecticut River Ecological Study (1965–1973) revisited: ecology of the lower Connecticut River 1973–2003. American Fisheries Society, Monograph 9, Bethesda, Maryland. (Originally published in 1976)

Colby, P. J., and L. T. Brooke. 1969. Cisco (*Coregonus artedii*) mortalities in a southern Michigan lake, July, 1968. Limnology and Oceanography 14(6):950–960.

Doudoroff, P., and D. L. Shumway. 1967. Dissolved oxygen criteria for the protection of fish. Transactions of the American Fisheries Society, Special Supplement 4:13–19.

Frey, D. G. 1955. Distributional ecology of the cisco, *Coregonus artedii,* in Indiana. Invest. Indiana Lakes Streams 4:177–228.

Haley, R., S. P. Davis, and J. Hyde. 1967. Environmental stress and *Aeromonas liquefaciens* in American and threadfin shad mortalities. Progressive Fish-Culturist 29(4):193.

Hoff, J. G., M. E. Chittenden, and J. R. Westman. 1968. Oxygen requirements of some marine and anadromous fishes, with particular reference to problems of measurement. Proceedings of the 21st Ind. Water Conference Part I, Series No. 121:125–140.

Mihursky, J. A., and V. S. Kennedy. 1967. Water temperature criteria to protect aquatic life. Pages 20–32 *in* E. L. Cooper, editor. A Symposium on Water Quality Criteria to Protect Aquatic Life. American Fisheries Society, Special Publication 4, Bethesda, Maryland.

Pippy, J. H. C., and G. M. Hare. 1969. Relationship of river pollution to bacterial infection in salmon (*Salmo salar*) and suckers (*Catostomus commersoni*). Transactions of the American Fisheries Society (98)4:685–690.

Spotte, S. H. 1970. Fish and invertebrate culture. Wiley-Interscience, New York

American Fisheries Society Monograph 9:247–260, 2004

Summary and Conclusions

DANIEL MERRIMAN AND LYLE M. THORPE
Connecticut River Ecological Study, Connecticut Yankee Atomic Power Company
Box 270, Hartford, Connecticut 06101, USA

1. The Connecticut River Ecological Study, 1965–1973, was originally undertaken as a condition of a dredging permit (21.X.'64) granted by the State of Connecticut Water Resources Commission to the Connecticut Yankee Atomic Power Company (CYAP). At that time, Connecticut Yankee's nuclear-powered electric-generating plant at Haddam Neck on the lower reaches of the Connecticut River was under construction. The commission did not then have direct jurisdiction over thermal discharges as such; however, its responsibilities "for the protection of ... fin-fish habitats" clearly gave it reason for concern. In short, what would be the effects of the operation of the plant on the biota of the river after the dredging was done? Accordingly, the Water Resources Commission (incorporated into the Department of Environmental Protection 1.X.'71) specified that there should be a

    > thorough and comprehensive study of the river conditions and environment undertaken under the auspices and with the support of the applicant ... throughout the starting-up and initial operating stages of the plant, for a period of five (5) years after the plant is placed in full operation ... etc.

    This monograph presents the results of the prescribed study, 1965–1967, before the plant went into action and 1968–1973, while the plant was in commercial operation—for the most part at the licensed level of 600 MW_e with a discharge of 23.5 m^3/s of condenser cooling water, and a ΔT of 12.5°C.
2. The Connecticut Yankee plant site (212 ha) is on the east side of the Connecticut River, 26 km upstream from Long Island Sound, 38 km below Hartford. In this region, the river is 5–600 m wide; its depth ranges from 1 m or less on the east shore flats to 9 m in the navigable channel adjacent to the west bank. Over the past half century, the average daily rate of flow has been about 450 m^3/s, with an additional minimum daily average tidal flow of 425 m^3/s (vertical range 0.75 m, no salinity).
3. Connecticut Yankee's once-through condenser-cooling system operates to pump river water into the plant via four adjacent bays at 0.3–0.6 m/s and to pass it through two condensers where the heat exchange takes place; passage through the plant requires approximately 1.5 min. The heated water flows out over a rocky weir with considerable turbulence, then down a canal (18 m in width, 3 m in depth) to enter the river 1.8 km below the intake via a broad (180 m) mouth. The drop in temperature of the thermal effluent during its passage down the canal is approximately 1°C.
4. The Introduction includes a brief account of the nuclear-generated power industry in the United States and the associated problems of "thermal pollution," as well as a general review of pertinent literature and an account of the physiography and history of the Connecticut River. It details the origin and distinctive character of the Connecticut River Ecological Study, namely that (a) it was undertaken before the advent of the Water Quality Act (1965), the National Environmental Policy Act (1969), and the requirements for environmental impact studies; (b) it has been a long-term "controlled experiment" in nature because of its before-and-after character—preplant operation 1965–1967, postplant 1968–1973; (c) it has demonstrated the fallacy in attempting to make all-inclusive regulations for thermal discharges from electric generating plants without regard for lo-

cal site conditions; and (d) apart from fulfilling its obligation "to collect relevant data and formulate sound conclusions," it has provided a broad basis for future comparison and further elaboration as needed.

5. The Introduction also includes summary recitals of associated research projects and monitoring programs dealing with microbiology, impingement of fish on the intake screens of the plant, and environmental radiation.

Hydrography and Chemistry

6. The section on hydrography outlines the operating history of the Connecticut Yankee plant 1967–1972, discusses the Connecticut River discharge and climatological conditions and describes the hydrographic monitoring system.
7. The normal upstream limit of significant salinity resulting from the seawater incursion from Long Island Sound is well below the maximum downstream extent of the thermal plume from the Connecticut Yankee plant. The sole bearing of salinity on the subject of this study is that the proximity of brackish water permits the occasional presence of holeuryhaline species as well as the anadromous fishes.
8. The results of both long- and short-term studies of oxygen concentrations in the Haddam reach of the Connecticut River show no measurable effect attributable to the operation of the Connecticut Yankee plant.
9. The area and extent of the thermal plume in the river varies so widely under the broad spectrum of conditions imposed by river discharge and tidal flow that averages are mere indicants (e.g., up- and downriver surface ranges average < 1 km). Under normal or low-flow conditions, however, surface temperatures usually showed elevations up to 8°C at least once each day near the west bank of the river opposite the mouth of the effluent canal. Nonetheless, that portion of Connecticut Yankee's effluent plume, which retained a temperature of 2°C or more above ambient river temperature, was largely found at depths of less than 1.2 m. It is concluded that there are no temperature distributions resulting from the thermal effluent that constitute a thermal barrier to the passage of migratory or resident fishes, assuming their ability to sense the heated plume and to swim under or around it.
10. Measurements of dissolved oxygen at the point of intake of the condenser cooling water system, the outfall from the plant above the weir, and near the top and bottom of the effluent canal, as well as estimates of the production and consumption of oxygen by the plants and animals in the canal, provide further evidence that plant operation has no meaningful effect on dissolved oxygen levels in the heated water discharged to the river.

Benthic Fauna

11. More than 4,700 samples and 550,000 organisms involving 72 species were collected in the course of the study in order to assess changes in the populations of benthic invertebrates in the vicinity of the power plant during the years 1965–1972.
12. Four general benthic biotopes were encountered in the vicinity of the power plant: gravel, rock, sand, and silt. Each possessed distinctive biological characteristics. Depth of the channel of the Connecticut River in the Haddam reach off the power plant averaged 5 or more m, and the bottom was composed of gravel. Light transmission in the channel was markedly reduced at about 1 m off the bottom by a layer of suspended moving particulate matter, and the bottom was at all times sparsely populated.
13. The rocky area extended along the entire west shore across the river from the power plant. Heated water sometimes reached the rocky stations briefly in the area opposite the mouth of the canal at the surface, but no biological changes were noted.
14. At the water intake to Connecticut Yankee in an area of about 0.5 ha, high current velocities completely depopulated invertebrates from the substratum when the plant went into operation. Heat did not reach the bottom. During intervals when the plant was shut down, the bottom was repopulated.
15. Scattered beds of *Vallisneria* occupied approximately 20 ha of a sandy flat upstream of the opening of the discharge canal into the river before the plant went into operation. Heated water in the plume came in contact

with the sediment for an average distance of 0.9 km upstream of the discharge point and as much as another 1.5 km on the surface. During early operation of the power plant in 1968, as siltation began and temperature increased, the number of species in the area increased markedly. Thereafter, as silt continued to accumulate, there was a decrease in the density of organisms, and *Vallisneria* disappeared. *Pisidium* in particular was markedly reduced; by contrast, the dipteran larva, *Glyptotendipes*, increased sharply. Diversity indices reflected an increase in the number of species, but a reduction in the number of individuals, as operation of the plant continued.

16. The bottom area adjacent to and directly below the mouth of the canal for 100 m had temperature increases up to 10°C. The sand substratum became overlain by silt and detritus by 1968, and the kinds and numbers of species decreased from 1968 to 1972—from 22 to 10 species and from 1,470 to 828/m^2 in mean numbers or organisms. Annelids (*Limnodrilus hoffmeisteri*) comprised more than two-thirds of the animals collected, insects were abundant, and clams disappeared entirely.
17. The bottom area starting some 300 m below the mouth of the canal and continuing downstream to the Salmon River was never in contact with the thermal plume. It had a typical silt substratum and remained relatively unchanged throughout the 91 months of the study, with a rich population of annelids dominant and insects plentiful.
18. Approximately 2.2 million organisms was the estimated population decrease from the area in direct contact with the plume (10.8 ha), during the periods 1965–1967 and 1968–1972; this estimate does not include the population that had been associated with *Vallisneria*. In contrast, all stations downstream from the plant and beyond the influence of the plume reflected increases in populations of benthic invertebrates during this period.
19. In the heated discharge canal itself, the density of invertebrates decreased as temperatures rose to 37–40°C. During shutdown of the plant when temperatures returned to ambient, populations increased markedly. This was particularly apparent in April–May 1971 and June–July 1972. In 1972, populations reached maximum density during shutdown, and minimum density in August and September when temperatures again rose to 35–40°C. During the cold months of the year, resident bottom feeding fishes moved into the canal, and during these periods, the abundance of benthic invertebrates was reduced. In the summer, when the plant was in operation and water temperatures reached maximum levels, relatively few fish remained in the canal. Fly larvae and the annelid *L. hoffmeisteri* were collected alive in plankton nets and grab samples during the full range of thermal fluctuations in the canal.
20. At all control stations in the river, that is, those not affected by the plume, the number of species and number of individuals increased continuously throughout the entire study from June 1965 to December 1972.
21. At all stations affected by the plume, decreases in abundance and biomass, and changes in percentage composition, of dominant taxa were moderate. It is suggested that the principal effects of the Connecticut Yankee power plant on the benthos were mediated primarily by changes in the composition of the sediment rather than by direct thermal stress.

Entrainment of Zooplankton

22. The zooplankton population in the Haddam reach of the Connecticut River is dominated by entomostracan crustacea and includes insect and fish larvae. The zooplankton were generally more abundant in the region of the channel and showed a strong diurnal trend in vertical distribution.
23. Of the zooplankton in the vicinity of Connecticut Yankee, the numbers entrained at the intake of the plant averaged about 4%. Mortalities of cladocerans and cyclopoid copepods (the dominant forms in the warmer months) attributable to mechanical damage in passing through the condenser-cooling system were negligible. Mortalities of these organisms due to heat were essentially 100% when temperatures were above 31°C, a period of about 120 d a year; however, during cooler periods most of the population appeared to survive.

Fishes of the Lower Connecticut River

24. A survey of the fish population in the lower Connecticut River was conducted for the years 1965–1972. In this period, some 229,933 fish representing 44 species and 21 families were collected in 1,398 bag-seine and trawl hauls between Higganum (km 29.8) and Essex (km 11.3). Including six additional species that have been reported in the commercial and sport fisheries, a total of 50 fish species is now known to inhabit this area.
25. The composition and relative and seasonal abundance of fish species regularly sampled at frequent intervals at eight stations for both young and adults are detailed. Life-history data are presented for the four most abundant species: white perch *Morone americana*, spottail shiner *Notropis hudsonius*, brown bullhead *Ictalurus nebulosus*, and white catfish *I. catus*—together representing 88% of the total "resident" fishes collected.
26. Six species of fishes, involving 9,817 individuals, were tagged in an effort to obtain information on mobility and growth; more than 90% of these were brown bullheads and white catfish, each of which yielded an approximate return of 10%. The data demonstrate that both the catfishes are far from stationary, that in fact they traverse the entire 85 km from Enfield Dam to Old Saybrook.
27. Before the plant started to operate, the number of fish species was higher at the "heated" stations than at the control stations, and it remained higher through 1969, but then dropped below the number found at the control stations. As the environment stabilized after the addition of heat to the river, fewer species were found in the peripheral areas that had become warmed, but the number of species in the discharge canal increased.
28. After the plant began to operate, there was a significant reduction in the number of white perch, brown bullheads, white catfish, and spottail shiners at the heated stations. There was also a significant reduction in the number of white catfish and spottail shiners at the control stations; however, a reduction in the numbers of the white perch and brown bullheads did not occur at the control stations.
29. The percentage decrease, in the catch per net haul, for 21 major fish species after plant operation was greater at the heated stations (18%) than at the control stations (6.6%). Much of the decrease observed at the heated stations in the vicinity of the discharge canal was most likely caused by the movement of these fishes from the peripheral heated areas to the preferred higher temperatures within the canal. Species abundant at the control stations before the beginning of plant operation were probably also attracted to the discharge canal during their normal seasonal movements, and this may explain the decreases that occurred at those stations.
30. In the discharge canal, 38,093 fish of 25 species were collected in 441 trawl hauls between 1968 and 1972; brown bullheads dominated the catch (67.3%) and, together with white catfish, made up 86.5% of the total catch. About one-third of the fishes caught were young of the year; brown bullheads dominated (86.6%) and, with white perch and blueback herring *Alosa aestivalis*, constituted 95.5% of the aggregate. The average number of fish per trawl haul in 1968, the first full year of plant operation, was double that recorded in any other year. After the initial surge of fish into the canal in late 1968, the number per trawl haul was approximately halved in 1969, was further reduced in 1970 and 1971, and then increased slightly in 1972.
31. Two distinct seasonal peaks occurred in the fish population in the canal, one in March and April and another, smaller one in November and December. The number of fishes remained relatively constant from May through October. Brown bullheads and white catfish were most abundant in March and April, golden shiners *Notemigonus crysoleucas* in June, white perch and spottail shiners in August, and common carp *Cyprinus carpio* in December. These six species made up 97.7% of the total catch. Golden shiners, white catfish, brown bullheads, common carp, and spottail shiners were the only species found in the canal during all months. More species were collected in June than in any other month. Five species (pumpkinseed *Lepomis gibbosus*, white perch, golden shiner, white catfish, and brown bullhead) have been col-

lected in the discharge canal in water as warm as 40°C. The majority of the fishes left the canal when the water reached about 35°C, but on occasion returned immediately after as little as a one-degree drop in temperature. Generally, the smaller or younger fish remained in the CYAP canal at higher temperatures than did adults of the same species. There was no evidence to suggest that the species of fishes that inhabit the CYAP discharge canal during the warmer season suffered any mortality because of increased water temperature; the fishes either avoid the canal or leave it as the rising temperature approaches their upper limits of tolerance.

32. Of the various species of fishes that were attracted to the warm waters of the discharge canal, 15 overwintered there during the cooler months (November through March). During these months, 11,173 bullheads made up 68% and 4,562 white catfish 28% of the total adult catch. Both species consistently entered the canal when the water temperature there dropped to about 20°C (range, 19–23°C). The number of catfishes collected by trawl in the Connecticut River in the vicinity of the plant decreased considerably whenever ambient river temperatures fell below 10°C. Both species of catfishes reached their greatest abundance in the canal between mid-February and early April. With the exception of periods of plant shutdown for refueling purposes in the spring, both species generally remained in the canal for about 8 months, or through June, when discharge temperatures rose above about 35°C. With few exceptions, neither species was found in the canal again until the following October.
33. Power plant shutdowns had a striking effect on the abundance of catfishes in the canal. Except in 1968, the first full year of plant operation, a major shutdown in the spring or early summer resulted in both species leaving the canal almost immediately. After each of four shutdowns, the two species were observed at the surface exhibiting lethargic behavior, as if they had been stunned, while being carried out of the canal with the 0.3–0.9 m/s current. However, no evidence was found to suggest that either species of catfish suffered any mortality as a result of these sudden drops in temperature.
34. Estimates of the number of brown bullhead and white catfish inhabiting the discharge canal during the winter were made from 1969 through 1972—generally in March, when ambient river temperatures were below 4°C and trawl collections showed that the largest number of fish were present in the heated discharge. In this study, 5,158 brown bullheads and 1,335 white catfish caught in the canal were tagged, and the population estimates were based on a return of 4.1% of the bullheads and 1.1% of the white catfish. The yearly population estimates of both species combined ranged between 17,000 and 39,700, and it was estimated that some 74,800 bullheads and 38,900 white catfish spent their winters in the canal in the 4 years under consideration. A reversal in the number of the two species occurred in the 1969–1970 and 1970–1971 winter seasons; although the total population remained about the same, the brown bullhead became the dominant species in 1970–1971 in about the same ratio that the white catfish had dominated during the previous winter. The carrying capacity of the canal seems to have remained at about 40,000 individuals, and the ratio between the species may have depended upon which one first entered the canal in considerable numbers.
35. The estimated percentage of the total population of white catfish and brown bullheads in the 386-ha area in the vicinity of the plant that was subject to overwintering stress in the CYAP canal in 1968–1969 was 14.3%. The plant's intake was estimated to have entrapped and subsequently killed about 5% of the total population of the two species of catfishes in that area in 1969. (This entrapment loss was substantially reduced in subsequent years by the installation of a protective device at the plant intake.) In 1969, it was estimated that sport fishermen harvested 4% of the local catfish population. Hence, in that year, intake fatality and sport fishing accounted for a mortality of approximately 9% of the total adult catfish population, with an additional 14% of the population subject to overwintering stress in the discharge canal.
36. The members of both catfish species that were living in the canal during the winter weighed significantly less and were in significantly poorer condition than those in two cooler

overwintering areas outside the canal. Weight and condition of the canal fish declined continuously in contrast to the fish outside. The average weight loss of brown bullheads from tagging to recapture for a 141-d period over one winter was about 30%. The average weight loss of white catfish was about 20% for a winter season. Evidence from tagged individuals indicates that brown bullheads that had spent one winter in the discharge canal made up the loss in weight and condition after leaving the canal in the spring and spending the summer outside the canal. Twenty brown bullheads that had been subjected to overwintering stress for two winters in the canal (tagged in the canal in March and recaptured in the canal the following March), however, gained only 175 g or 8.8 g/fish over a period of 351 d. Granted that these fish were reweighed at a time of maximum emaciation, this is a much lower yearly increase than that represented by the difference in weight between any of the age-classes of this species. Four white catfish, recaptured after spending two winters in the canal, lost 14 g/fish over a period of 363 d. However, we have no data to show how much weight these fish might have made up after spending a summer back in the river. In all events, the effect of this poor growth rate on the total population of catfishes must have been minimal because only 2.5% and 0.5% of the tagged population of brown bullheads and white catfish returned to the canal the next winter. Possible reasons for the loss in weight and condition of the catfishes that overwinter in the heated discharge canal are (a) metabolic rate must be higher in the canal than in the cove areas; (b) greater expenditure of energy is necessary for the fishes to maintain their positions in the canal with its continuous current of 0.3 to 0.9 m/s; (c) high population densities in the canal result in overcrowding with increased competition for a limited amount of food.

37. The ovaries of female brown bullheads and white catfish develop significantly faster, that is, unseasonably earlier, during the winter in the discharge canal as compared with females overwintering in the cooler waters outside. Several factors indicate that this gonadal phenomenon probably has a minimal effect on the entire population of catfishes in the lower Connecticut River: (a) If fish did spawn earlier after leaving the canal, this could not occur before April, at which time sufficient numbers of zooplankton would undoubtedly be available in the river to support the young. (b) In 1972, there was no evidence to suggest that catfishes spawned in the canal earlier than the plant shutdown in June; in all other years, both species generally left the canal in early April immediately after shutdown. It is probable that the development of the ova slowed when the fish entered ambient water after plant shutdowns in the spring and that this would have caused most fish from the canal to spawn at a time closer to their normal spawning season. (c) The possibility exists that bullheads spawn twice yearly in the Connecticut River, and this would further minimize the effects of earlier maturation. (d) No damage to the ovaries as the result of overwintering was ever observed in the canal fish. Fish that had been tagged in the canal and 1 and 2 years later returned to the canal exhibited ovarian weights and egg sizes similar to fish of comparable lengths taken in the canal in December 1971 and March 1972. This suggests that the ovaries functioned and developed normally after a previous winter's sojourn in the canal.

38. From 1965 through 1969, a creel census was conducted from April through November in the vicinity of the CYAP plant and in 1972 and 1973 during the winter and spring months in the heated discharge canal. The purpose of this survey was to determine the effect of the operation of the Connecticut Yankee plant on the sport fishery in the lower Connecticut and Salmon rivers and to evaluate the newly established winter–spring sport fishery in the plant's heated discharge.

39. The 5-year creel survey in the vicinity of the plant showed that the average yearly catch rate remained stable both before and after the operation of the plant, ranging between 1.02 and 1.18 fish/rod-hour. The highest catch of fish per rod-hour was 1.41 in the lower Salmon River, as compared with 1.23 in the Connecticut River and 0.81 in the upper Salmon River. The catch of 25 different species was dominated by the brown bullhead (27.0%), white perch (26.0%), and white catfish (16.0%); the

American eel *Anguilla rostrata*, pumpkinseed, and yellow perch *Perca flavescens* made up 22.0%. Together, these two groups constituted more than 90% of all the fish caught by sport fishermen in the Haddam area. Analysis of the data clearly indicates that the sport fishing in the vicinity of the plant did not change significantly as a result of the plant's operation.

40. A creel census of the sport fishery was conducted in the discharge canal from January through June in 1972 and 1973. During the first year of the survey, the average number of fish caught per hour of effort averaged 2.37 over the 6-month period. During the second year of the winter-spring fishery survey, the average number of fish caught per hour of effort averaged 1.51. During 1973, the fishermen expended four times the effort but caught only twice as many fish as those who fished in 1972. It may be that increased fishing effort (especially for northern pike) and competition among the more numerous fishermen, as well as the decrease in the number of available fish brought about the decline in catch per rod and man-hour. It is reckoned that in 6,912 trips, fishermen caught an estimated 27,044 fish in the canal during the two seasons, or almost four fish for each fishing trip. In 1972, 18 species of fishes were caught by anglers in the canal, and in 1973, 16 were taken. White catfish, white perch, and brown bullheads made up more than three-quarters of the catch in both years. White perch dominated the catch in 1972 and white catfish in 1973. Other species that were caught in relatively high numbers were common carp, American eel, striped bass *Morone saxatilis*, pumpkinseed, and northern pike *Esox lucius*. The species most sought after and caught with reasonable frequency were the northern pike and striped bass; pike fisherman averaged 0.56/man-hour from January through March, the months when this species was most available. No clearcut relationship between water temperature and overall catch rate was evident, except that (a) on those days when the power level of the plant was reduced enough to lower the water temperature sharply, the catch rate declined markedly; and (b) that the catch rate increased sharply in April, when water temperatures approached 17ºC, but generally dropped when temperatures exceeded 22ºC. It appears that one of the most successful of all winter–spring fisheries has developed in the discharge canal of the Connecticut Yankee plant. Fishermen who utilize its heated discharge during the cooler months are able to catch twice as many fish per-unit-effort as those who fish in the Haddam area of the river from April through November. These anglers also caught as many as 10 times more fish per hour-of-effort than did the fishermen in another creel survey that was carried out 141 km upstream, and as many as four times more fish per hour compared with winter estimates in this upstream area. The northern pike provides a striking example of their success. Five times as many pike were caught in 1973 as in the previous year (one pike for every two fishing hours), and this species has become a major attraction for fishermen during the winter months.

Ichthyoplankton and Entrainment

41. From 1965 to 1969, eggs and larvae (112,896) representing 16 species of fish were collected in 690 biweekly plankton tows at 16 sampling stations on the lower Connecticut River and at 9 stations in 5 of its tributaries or coves, all of them between km 11.3 and 85.3 from the mouth of the river.

42. The alewife, blueback herring, white perch, and yellow perch made up 98% of the total egg production. Enfield Dam (km 83.7) is a deterrent to upstream migration of these species, as exemplified by the great difference between the total number of eggs per tow collected below the dam at Windsor Locks (929/tow) and above the dam (less than 4/tow). The eggs of these species hatch upstream and the larvae are carried downstream in a planktonic state by freshwater runoff. These larvae possess yolk-sac food reserves and presumably do not depend on the nutritional capacity of their spawning grounds; active feeding probably begins when they move downstream toward estuarine waters of low salinity. Larvae taken southward of the Haddam area had generally passed through

the yolk-sac stage, indicating that this is a major nursery area.

43. Spawning occurred from April through August and peaked in June in the river and in May in the tributaries and coves. Egg production in the tributaries and coves was more than twice that found in the river, the Salmon River being the most productive. In the river, 87% of the egg production occurred above Middletown (km 35.4), the greatest quantity in the Windsor Locks area (km 78.8). Larval abundance in the river was more than twice that found in the tributaries or coves; as a result of drift from freshwater upstream spawning areas, the majority of the larvae become concentrated in a nonsaline or low-salinity nursery area between Higganum (km 29.8) and Essex (km 11.3).
44. More than three-quarters of the eggs occurred in water that fell within a temperature range of 19–25°C; 15% occurred in a lower range of 7–11°C. The larvae of 16 species of fishes were found in water between 7.5°C and 31°C; of these, 95% occurred within a temperature range of 19–29°C and 77% between 21°C and 27°C. Eggs were 4 times and larvae 11 times more abundant near the bottom than near the surface in the upper part of this section of the river; however, the larvae of most species became more pelagic as they drifted downstream.
45. At Essex (km 11.3), the young of 10 marine species were collected by seine and trawl from 1965 through 1972. None of these species was ever collected farther upstream. The area from km 11.3 to the river mouth appears to be an important nursery area for marine species.
46. In the lower Connecticut River and its tributaries and coves, 77% of the egg production occurred above the Connecticut Yankee plant (km 25.7). The number of eggs collected near the Connecticut Yankee intake represented only 0.2% of the egg production in the lower Connecticut River and its tributaries and coves. For this reason, the potential effect of entrainment on fish eggs at the intake of the plant appears to be minimal. On the other hand, nearly half of the fish larvae in the lower Connecticut River and its tributaries and coves were collected between Higganum and the mouth of the Salmon River. Excluding the tributaries and coves, 61% of all larvae collected in the lower Connecticut River were found in this region near the plant site.
47. Entrainment studies conducted between 1970 and 1972 showed that all fish larvae and early juveniles of nine fish species that passed through the plant were dead by the time they reached the end of the discharge canal. Sampling during June and July, when 95% of the nonscreenable fish were present near the intake, showed that approximately 80% of the total mortality in the canal had been caused by mechanical damage and 20% by heat shock and prolonged exposure to temperatures above 28°C. There was no measurable mortality from the injection of sodium hypochlorite as a biocide. A conservative estimate of the number of nonscreenable young living fish entrained by the intake averaged about 4% of all those passing by the plant under conditions of unidirectional net tidal flow.
48. Any estimate of the effect of mortality from entrainment on the total fish population must take into account the natural mortality that occurs during specific life stages of each species, since entrainment mortality adds to the rate of natural mortality. Until information on natural mortality on the life stages of the various species encountered in this study is available and more is known about the carrying capacity of the river system, no inferences can be made as to the impact that the 4% entrainment has on the population of adult fishes.

Juvenile Shad

49. Because the American shad *Alosa sapidissima* is the most important commercial and sport fish in the Connecticut River, all stages of its life in freshwater were scrutinized to determine possible adverse effects of the power plant. The studies were principally concerned with the distribution and abundance of eggs and young shad, and with the downstream migration—all in the vicinity of the plant. However, studies on the location of spawning and nursery areas as well as the relative abundance of the young stages were made throughout the lower Connecticut River from Massachusetts to Long Island Sound.
50. To determine the rate of development, 200 to 300 eggs were stripped from ripe fish, fertilized, and then reared in the laboratory at tem-

peratures similar to those in the river. This provided a chronological description of the early developmental stages of the shad. Hatching occurred 71 to 86 h after fertilization, and a developmental timetable was established to make possible the aging of eggs collected in the river. This permitted the calculation of the distance of the place where eggs were collected from the place where they had been spawned. Calculations—based on the age of the eggs when collected and on current velocities of the river (range from 0.15 to 0.6 m/s)—indicated that most eggs had traveled only 1.6 to 6.4 km from the point where they had been spawned. Eggs collected in the river were, therefore, not far from the actual spawning sites.

51. Collections of shad eggs were made generally twice weekly from April to July in 1967 and 1968 at 61 locations from above Enfield Dam (km 85.3) to Essex (km 11.3). Spawning was observed at least 10 different times, but only on dark afternoons and during the evening. Spawning involved pairs of fish, or several males and a single female. In group spawning, fish swam with the backs exposed and splashed about in a closely packed circle. Shad spawned throughout the study area, but they did so primarily above Rocky Hill (km 51.5). Major spawning areas were located at Windsor Locks (km 78.8), Wilson (km 74), and Rocky Hill (km 51.5); 11 times more eggs per hour were collected in these areas than in areas downstream. Although similar collecting efforts were made in the upper and lower Salmon River, no spawning was observed and no eggs were collected farther than 0.6 km upstream from its junction with the Connecticut River. The major portion of the spawning of the shad occurs well above the CYAP plant.
52. Differences in time and place of spawning between 1967 and 1968 may have been caused by differences in temperature. On the average, shad eggs were collected farther upstream in 1967 than in 1968. The water temperatures of May 1967 were 4°C lower than those of May 1968; this may have delayed gonad maturation, and the fish apparently migrated farther upstream before spawning in 1967 than they did in 1968. Peak spawning occurred at 22°C in June 1967 and at 14.8°C in May 1968. Few eggs (3%) were collected at temperatures below 16°C in 1967, but 65% of them were taken at such low temperatures in 1968. Temperatures below approximately 16°C prolonged the period of development and reduced survival. Embryonic development, at the time of peak spawning, required 103 h in 1967 and 217 h in 1968. Of the eggs collected in 1967, 18% were dead as compared with 30% of those collected in 1968. The paucity of juvenile shad collected in 1968 is further evidence of poor spawning success in that year.
53. In the ichthyoplankton survey of the lower Connecticut River (km 11.3–85.3), the eggs of American shad ranked eighth in abundance and the larvae twelfth. Shad egg size ranged between 1.1 and 3.8 mm and averaged 2.83 mm; the mean egg size in May was 3.1 mm and decreased to 2.39 in June. Shad larvae were collected from May through August, more than 80% of them in June in water that ranged between 9.7°C and 30°C, mostly between 21°C and 23°C. The larvae ranged from 4.9 to 14.6 mm in length with an average length of 10.7 mm; they were generally between 5.7 and 6.8 mm long when hatched and absorbed their yolk sacs in about 88 h at 12.0 mm.
54. An attempt to estimate the number of shad eggs that were entrained in the plant's intake from 21 May through 23 June during 1968 and 1969 was made by placing two stationary 0.5-m nets at various depths in the water that had become committed to the plant's intake. The number of shad eggs estimated to have been lost to the intake was 81,432 in 1968 and 131,544 in 1969. When an estimate of natural mortality is taken into consideration in relation to the estimated number of shad eggs lost to the intake, only approximately one adult shad would have been lost by this means in 1968 and 1969. The number of shad eggs lost to the intake must be considered minimal when compared with either the total number of potential adults lost or the total egg production of the major spawning areas upriver above km 51.5. It should also be noted that no evidence was found of any change in the location of shad spawning areas in the river as a result of plant operation.
55. Young-of-the-year American shad were sampled biweekly at 12 stations between

Essex (river kilometer 11.3) and just above Enfield Dam (km 85.3) from June through November 1966–1972 and from August through November 1973. Sampling was expanded in 1966 and 1967 to include four stations between Enfield Dam and Northampton, Massachusetts (km 138). These data were used to describe the distribution, growth, and feeding habits of young shad in the Connecticut River and factors that influence their seasonal and yearly abundance. Between 1966 and 1973, the total lengths of 8,978 juvenile shad were measured (17–153 mm). Growth was most rapid in the earlier juvenile stages, reaching its peak in October. The apparent decline in mean size in November was probably the result of the movement of larger fish out of the river at this time. Young shad grew faster upriver, and the larger individuals then emigrated downstream. Measurements of fish collected between km 11.3 and 85.3 showed a general daily growth in length of 0.52 mm or 3.6 mm per week. The fastest growth (0.83 mm per day) took place in 1968 when juvenile abundance was at one of its lowest levels, while the slowest rate (0.27 mm per day) occurred in 1971 when juveniles reached their greatest abundance. The faster growth in 1968 may have been influenced by less competition for for food and space as a result of reduced numbers of juveniles, perhaps brought about by increased egg mortality and/or loss of eggs downstream following high June flows. It may be that when the juvenile production is high in any particular year a higher oceanic mortality may follow, thus tending to offset the higher freshwater production. Such feedback relationships between growth, year-class strength, and mortality in anadromous fishes merits detailed study.

56. In the Connecticut River, young shad apparently begin their downstream movement as early as the middle of August. This movement probably starts with the relatively slow downstream transport of the population as a whole. As time goes on, it accelerates, and peak emigration velocities are reached in September. The bulk of the population leaves the river in late September and early October. The largest numbers of juvenile shad were found in the Haddam area in early August when the water temperature was 26°C. Juveniles began moving downstream and leaving the river by the middle of August between temperatures of 26°C and 23°C; the bulk of the population emigrated from early September through early October at temperatures ranging from 23°C to 17.8°C. A smaller portion of the population, probably small fish or those from upriver areas, left the river between the middle of October and early November between 17.8°C and 10.9°C. No juveniles were caught in the river by the end of November at temperatures below 6.6°C. The decrease in water temperature almost to the lower lethal limit (6°C) apparently triggers the final movement of juvenile shad from freshwater, while factors associated with size appear to initiate the earlier stages of seaward emigration.

57. The year-class strength of juvenile shad depends upon many factors. The construction of a multiple linear regression model established a highly significant relationship between the catch per seine-haul of juveniles (a measure of year-class strength) and the combination of the estimated number of adult shad potentially available to spawn that year, the water temperature, and the discharge flow. This equation represents an extremely useful predictive tool, since it can be used effectively to estimate the relative production or year-class strength of juvenile shad in any particular year. There was a significant relationship between the number of adults potentially available to spawn (and, hence, egg production) and the production of juveniles. There was also a significant relationship between river temperature and the production of juveniles; low water temperatures caused prolonged development and reduced survival. However, the relationship between river discharge flow and juvenile production was nonsignificant and showed a 46% confidence in predicting juvenile production based on river flow alone. In summary, during years of lower discharge flow rates, higher water temperatures, and larger adult shad spawning populations, the production of juvenile shad was higher. The combination of river discharge flow, temperature, and the number of potentially spawning adult shad in a multiple linear regression model provides a tool

to predict the relative year-class strength of juvenile shad with a high (86%) percentage of success; this predictive ability exceeds most expectations concerning natural populations.

58. In studying the occurrence, distribution, and abundance of juvenile shad in the lower Connecticut River, no evidence was obtained to indicate that the Connecticut Yankee plant exerted any harmful effects on these fish. On the other hand, only circumstantial evidence can be adduced from these sources to show that the young shad are unaffected by the heated effluent during their downstream emigration. A chi-square analysis was used to reveal any possible change in catch per net haul before and after plant operation and up and downstream. This analysis indicated that the catch of juvenile shad in the lower river remained relatively unchanged from 1968 through 1972. There was no evidence to suggest that juvenile shad overwintered in the heated effluent during the years after plant operation. Yearly observations in the area around the heated effluent during peak downstream emigration failed to detect any evidence of juvenile shad mortality or of delay in passing the plant site.
59. To obtain definite evidence about the effect of the heated effluent on juvenile shad, a series of experiments was conducted in the warm water in 1968, 1970, and 1971. It had been demonstrated in an experimental tank that young shad responded only minimally to rapid temperature changes of about 1°C, but that they avoided rapid temperature increases of about 4°C. It was concluded from these experiments that small shad are behaviorally capable of avoiding temperature changes of a kind that otherwise might prove lethal. Young shad have been collected in our study in water temperatures ranging between 10°C and 31°C, but only one specimen was ever found in water warmer than 30°C. Sampling by gill net in and around the thermal plume when the effluent temperature was 21.5°C indicated that 100% of the juvenile shad were distributed at or near the surface (0.24 m) at night and 87% near the bottom (3.7–4.9 m) during the day.
60. Observations on the upper lethal temperature and direct mortality were made by placing young shad in a shallow floating live-car and drifting it through the heated effluent on 23 September and 14 October 1970, just after the peak abundance of young had occurred in the Haddam area. Reactions of the caged fish as temperatures increased from ambient to that of the maximum effluent were recorded by a composite of time-sequence photographs on 14 October with the following results: (a) In ambient water (19.0°C), an even distribution of the fish was maintained throughout the live-car with no evidence of schooling; (b) as the live-car entered the effluent, the fish immediately formed a tight school in the center as heated water (27.8°C) entered from both ends; (c) after 1 min at 30.0°C, the fish began swimming rapidly in a circular pattern, followed by the break-up of the single school into smaller schools; (d) after 2 min at 31.2°C, there was evidence of disorientation and the smaller schools began to break up; (e) after 3–4 min at 31.8°C, there was no schooling, swimming speeds were reduced, and there was continued disorientation with loss of equilibrium; and (f) death of all specimens occurred between 4 and 6 min at 32.2°C. These field observations confirmed our laboratory work which showed that young shad die rapidly when temperatures were suddenly raised from 24–28°C to 32.5°C.
61. Avoidance reactions to effluent of different temperatures were observed within a cage (2.4 × 4.6 × 2.1 m deep) constructed of 1.3 × 2.5 cm wire-mesh. The cage was submerged vertically in the water column and held in position by a floating platform; it was operated in September and October 1971, five tests being made in the effluent of the plant and six in ambient water. Underwater observations during these tests showed that young shad avoid effluent temperatures above 30°C. Shad exhibited vertical movement (±40 cm) that coincided with the rapid temperature changes resulting from oscillating levels of heated water. As the temperature rose to approximately 30°C, the fish always moved downward. Shad exhibited no specific depth or temperature preference at ambient (control) temperatures and at effluent temperatures less than 30°C. It is concluded that young shad avoid potentially lethal temperatures above 30°C and are capable of traversing or

avoiding the CYAP heated effluent during their downstream migration.

Adult Shad

62. The introductory sections on the adult American shad deal with its geographical range, relationship to other members of the genus *Alosa*, and life history as reported in the literature. There follows an account of the Connecticut River commercial and sport fisheries for shad, past and present, and a discussion of the problems posed by the construction and operation of the Connecticut Yankee plant at Haddam Neck—especially the close relationship of the migratory behavior of shad to temperature.
63. From 1965 to 1973, more than 34,000 adult shad were tagged with yellow spaghetti darts in the lower 3 km of the Connecticut River. Of these, more than 20% of the tags were returned with full data on the time and place of recapture. There was no significant change in the distribution of spaghetti tag recoveries above and below the Connecticut Yankee plant in the pre- (1965–1967) and post- (1968–1973) operative years.
64. In 1967, 1968, and 1969, an additional 230 adult shad were fitted with ultrasonic tags, 219 of which carried the device in their stomachs. These fish were tracked continuously for as much as 3 d with boat-mounted receivers, and additional information was derived from shore-based recording stations, which monitored the movements of individual fish between Essex (km 11) and Enfield Dam (km 84). The tracking of shad by ultrasonic means revealed a definite meandering behavior in the area of the saltwater wedge several kilometers upstream of the mouth of the Connecticut River. The upriver movement of individually tracked fish once they reached freshwater tended to be more direct, although periodic meandering, holding of position, and even downriver movement occurred during most tracks; however, these were more or less random events, showing no obvious pattern in time or space. The shock of the tagging process apparently modified to some extent the normal migratory behavior of meandering or downriver movement; this resulted in a delay of upriver migration that averaged 8–9 d. Following this initial delay, shad accomplished their upriver migration at an average rate of 5–8 km per day in freshwater. At this rate, the average time to migrate from the mouth of the river to the main spawning area at Enfield Dam would be 10–16 d. Shad migrating upstream followed the natural and dredged river channel, with only periodic movements into shallower water. The path taken by all shad was remarkably consistent. The pattern of downstream (postspawning) migrant shad was similarly highly channel-oriented.
65. Clear evidence that no thermal block to the upriver passage of shad exists as a result of the Connecticut Yankee plant operation at past or present levels is provided by (a) The similarity of rates and patterns of upriver migration from year to year, as revealed by conventional (spaghetti) tagging; (b) the similarity in the annual percentage of tag recoveries above the Connecticut Yankee discharge from fish that had been released in the lower 3 km of the river over a 9-year span before and after plant operation; and (c) the absence of significant changes in behavior or migration times of shad tracked by ultrasonic-means in the immediate area of the thermal discharge before and after the onset of plant operation. Downstream migrant shad also appear to be unaffected by the thermal discharge.
66. The second half of the study of the adult American shad involves a penetrating discussion of population trends in the Connecticut River, not only for the immediate period (1965–1973), but also for the preceding three decades. Detailed analysis of the annual variations from 1935 to 1973 in the size of the Connecticut River run of shad, which return to their native river to spawn after 4 (most males) to 5 (most females) years at sea, clearly demonstrate that large scale, year-to-year fluctuations in population size are common. Thus, the estimates for the years 1935–1964 taken from the data of earlier studies indicate that the numbers of Connecticut River returners oscillated between 267,000 (1956) and 990,000 (1944). The population fluctuations during the period of our study (1965–1973) were of the same order of magnitude—275,000 in 1972 and 1,470,000 in 1965.

67. It is concluded that the pattern of variation in the yearly run of adult shad since the Connecticut Yankee plant has been in operation is in no way indicative of any meaningful influence by the thermal discharge on the size or structure of the annual population. In considering the possible effects of future alterations in the heat budget of the Connecticut River, the following points have emerged from this study: (a) The timing of shad migrations, both in the sea and in freshwater, is strongly influenced by water temperatures; (b) the production of juveniles and subsequent recruitment is closely related to water temperatures and discharge conditions in the spawning area; (c) data are available that suggest a relationship between the temperature experienced during freshwater migration and the energy cost of the migration; and (d) the channel-oriented behavior of adult shad migrating in freshwater should be a primary consideration in the siting of future thermal discharges.

Fish Kills

68. Five mass mortalities of blueback herring occurred in the lower 48 km of the Connecticut River in the latter part of June and/or the first half of July in 1965, 1966, 1967, and 1971. These mortalities each involved 10,000 to 100,000 fish. Three of these, in the general area of the Connecticut Yankee facility, occurred before the plant went into operation. The mortalities occurred in the early morning when dissolved oxygen fell in the ranges below 1.3 mg/L at 24.6°C and 3.6 mg/L at 27.6°C.
69. No substantial mortality of blueback herring occurred in the lower Connecticut River in 1968, 1969, 1970, and 1972—all years that the plant was in operation, but years in which water temperatures and dissolved oxygen values remained outside the apparent zone of lethality during the blueback spawning run.

Summation

Connecticut Yankee was one of the first large nuclear baseload units to go into operation in the United States, and towards the specified termination of the main body of our study (December 1972), it had produced more power from nuclear fission for commercial use than any other single unit in the world. In view of the magnitude of this operation with its 23.5 m^3/s of condenser cooling water being returned to its source at a ΔT of 12.5°C, there was widespread concern that there might be substantial changes in the aquatic faunal regime of the lower reaches of the Connecticut River. And "change," in this context, by implication, was most commonly taken to mean for the worse.

The most obvious harmful effect of the warmwater plume in the river might have been to change the migratory pattern of the anadromous shad, leading to a reduction or even elimination of the spawning population. But such was not the case. The operation of the Connecticut Yankee plant has had no effect on the spawning habits, the migratory pattern, or the size of the population of the shad, which annually returns to spawn in its native river.

Then there was possibility that the warmwater effluent might cause massive fish kills, but such has proven not to be so under the existing operational conditions. And there was justifiable concern over the entrainment of fish larvae and early juveniles with resultant mortality at levels that might affect the size of the adult populations of fishes. We estimate that the average number of young living fish killed in their passage through the entire condenser cooling system was about 4% of all those passing by the plant under conditions of unidirectional net tidal flow. Of this 4%, more than 97% were alewives and blueback herring. In the present state of knowledge, few would argue that this is a matter for serious concern.

With regard to the resident fishes, clearly there have been changes in the relative abundance and distribution of the various species in the river in the surrounding area of the Connecticut Yankee plant, with definite, though in no way portentious, decreases. These have been offset by the development of a remarkably successful winter–spring fishery in the 1.8-km discharge canal leading from the plant to the river.

The benthic fauna has undergone considerable change along the several kilometers on the east side of the river where the shallower waters are most directly influenced by the warmwater plume. At the intake to the plant, an area of 0.5 ha has become depopulated as a result of the scouring of the bottom by high current velocities. In the 10.8-ha area in most direct contact with the plume, there have been notable changes in the composition of the benthic fauna, with overall decreases in the numbers of organisms, and scattered beds of *Vallisneria* growing in an area of some 20 ha have disappeared. By contrast, the areas downstream from the Connecticut Yankee plant beyond the influence of the plume reflected increases in the populations of benthic invertebrates. There is also a new and abundant benthic fauna established by the creation of the discharge canal, which supports the winter–spring fishery.

The details of these and other changes effected by the operation of the Connecticut Yankee plant are recounted in the text of this monograph. In sum, it is clear that such alterations to the ecosystem that might be regarded as deleterious appear to have been remarkably limited and, under present conditions, well within the carrying capacity of the river. In the words of Spilhaus (1973) (p. 4), "... there is an acceptable burden of man's wastes of the proper kind that these components [air, land and water] can carry," and the thermal discharge from the Connecticut Yankee plant appears in no way to exceed the limits of acceptability as judged by any reasonable means. We conclude that in the case of the Connecticut Yankee Atomic Power Company plant, such ecological alterations as have been observed as a result of the local warming of the Connecticut River in the general region of the Haddams are of relatively small moment—especially when weighed against the utilitarian benefits.

Part II: The Connecticut River Ecological Study Revisited

American Fisheries Society Monograph 9:263–272, 2004

Introduction: Connecticut River Watershed Management—Past, Present, and Future

PAUL M. JACOBSON*

Ecological Assessments, LLC, Post Office Box 376, Old Saybrook, Connecticut 06475, USA

CHARLES G. FREDETTE[1]

Connecticut Department of Environmental Protection, 79 Elm Street Hartford, Connecticut 06106-5127, USA

NELS BARRETT

U.S. Department of Agriculture, Natural Resources Conservation Service 344 Merrow Road (Route 195), Suite A , Tolland, Connecticut 06084-3917, USA

Abstract.—The 1965–1973 Connecticut River Ecological Study (the "Study") is revisited in the context of ecological information collected in the intervening 30 years and the ongoing national dialog regarding the potential impact of cooling water intake flow withdrawals. This historic Study examined the ecological impact of the Connecticut Yankee Haddam Neck (CY) nuclear power station on the lower Connecticut River. The Study predated environmental regulations governing cooling water intake and discharge and provided a scientific foundation for impact assessments to follow. Subsequent and ongoing conservation and watershed management initiatives, efforts to preserve the unique features of the lower Connecticut River and enhance water quality, coupled with more recent ecological data following completion of the Study, combine to provide an extraordinary opportunity to revisit the ecology of this portion of New England's largest river, both from the perspective of evaluating longer-term impacts of CY and also with respect to future resource management. The principal investigators of the Study collaborated with investigators now involved in the protection and study of the lower Connecticut River to produce a compendium of ecological research relevant to improvements in the ecological assessment process and the development of new analytical tools and approaches (Part II, this volume). Homage is paid to D. Merriman and L. Thorpe, directors of the original Study and editors of the first monograph of the American Fisheries Society (Part I, this volume), whose insights provide a foundation for this retrospective.

Introduction

This monograph examines, retrospectively, one of the first "grand experiments" in power plant ecological impact assessment, bearing on power plant construction and operation undertaken during the advent of nuclear powered electric generation in the United States. The 1965–1973 Connecticut River Ecological Study (the "Study"), conducted in association with the construction and operation of the Connecticut Yankee Haddam Neck power plant (CY), was a seminal effort in that it established the theoretical and scientific basis for power plant ecological studies that followed and were later implemented under Clear Water Act (CWA) §316(a), thermal discharge, and §316(b), entrainment and impingement impact assessment requirements. Since the completion of the Study in 1973, there has been no comprehensive review of its conclusions regarding the long-term operation, closure, and subsequent decommissioning of CY. The Connecticut Yankee Haddam Neck power plant began commercial operation in 1968 and was shut down in

* *Corresponding author*: jacobsonpaul@msn.com
[1] Present address: 186 Francis Street, New Britain, Connecticut 06053, USA

July 1996. The decommissioning process began in December 1996 and is ongoing.

The context for, and relevance of, this review is to be found in the ongoing national dialog regarding the ecological impact of industrial cooling water withdrawals. The U.S. Environmental Protection Agency (EPA) is developing Clean Water Act §316(b) regulations that are intended to provide a framework for assessing and limiting those impacts (Dixon and Zammit 2000; Nagle and Morgan 2000; EPRI 2002). The Study methodology, results, and conclusions augmented by information collected during the post-1973 operation and postclosure phases of CY are relevant to this dialog. In that regard, no better introduction to this "revisitation" of the Study could be articulated than that already provided by Merriman and Thorpe (2004c [1976], this volume) in their summary remarks found in the original monograph. They concluded that "apart from fulfilling its obligation to collect relevant data and formulate sound conclusions, it [the Connecticut River Ecological Study] has provided a broad basis for future comparison and further elaboration as needed."

In this "revisited" account of the Study (Part II, this volume), we examine the original 8 years of research in relation to what we now know about the current post-1973 state of the ecology of the lower Connecticut River. Thirty years of new information on various aspects of its ecology are presented by way of comparison. Water quality improvements resulting from requirements of the Clean Water Act combined with regional conservation efforts are discussed. It is clear, too, that this compilation of ecological information on the lower Connecticut River, in addition to informing the regulatory debate on cooling water intake withdrawals, should also serve to inform and assist watershed managers and policy makers in their continuing efforts to preserve and enhance the natural resource values unique to the Connecticut River and its watershed and, by extension, to similar challenges being faced elsewhere.

Of particular interest is whether the original conclusions regarding the potential impact of CY on the lower Connecticut River are consistent with the body of ecological information collected in the three decades that have passed since the Study results were first published. Current regulatory and scientific issues bearing on the management of thermal effluents and cooling water withdrawals clearly depend on the validity of the inferences that can be drawn from this seminal study regarding key factors influencing potential impact. Furthermore, while the Connecticut River Ecological Study established a scientific foundation for impact assessments that followed, a substantial body of new analytical tools, approaches, and information has developed in the intervening decades. This information is reviewed and assessed (EPRI 1999; Dey et al. 2000; Van Winkle et al. 2003; Saila and Lorda 2004, this volume).

Regulatory History

From an historical regulatory view point, it is noteworthy that the Study was initiated prior to enactment of federal and state environmental laws and regulations that established the current framework for evaluating impacts of once-through cooling water systems in the United States. Planning for the Study, and the field work (conducted between 1965 and 1973) that led to its conclusions, began several years before the National Environmental Policy Act of 1969 and enactment of the 1972 Federal Water Pollution Control Act Amendments, which under §316(a and b) and §402 of the National Pollutant Discharge Elimination System (NPDES) permit program provided the statutory basis for assessment of the ecological effects of power plant cooling water intake systems. The Study also predated the preparation of EPA's Development Document (EPA 1973), which compared alternative cooling water intake technologies, and prior to EPA's development of guidance on the conduct of ecological impact assessments (EPA 1977). The Study also predated the Connecticut Water Diversion Policy Act of 1982, which gave the Connecticut Department of Environmental Protection (CTDEP) latitude to regulate and assess the potential ecological impacts of proposed source water diversions. In fact, the Study was initiated as a result of the Connecticut Water Resources Commission requirements related to approvals to proceed with dredging and construction of the cooling water intake structure. Moreover, rather than developed based on regulatory standards for ecological impact assessment, the studies were designed and guided by an Ecological Advisory Committee of prominent aquatic and marine scientists who provided critical independent peer review (Merriman and Thorpe 2004b [1976], this volume).

The first NPDES permit for CY was issued in December 1974 (CTDEP 1974). Upon issuance, the newly formed CTDEP determined, on the basis of the Study findings that the thermal component of the discharge met appropriate water quality criteria. At the same time, the CTDEP required that CY undertake additional investigations to determine whether the cooling water intake system represented "best technology available" (BTA) for minimizing adverse impacts, which may be associated with entrainment and impingement (CTDEP 1974). Reports submitted to CTDEP (CYAPCO 1974a) and the Nuclear Regulatory Commission pursuant to the Environmental Technical Specifications (CYAPCO 1974b) from 1976 through 1983 (CYAPCO 1976–1983) were based on results from the Study and from studies that examined impingement, the efficacy of fish deterrent devices, and the installation of a fish return system (CYAPCO 1980). These submittals led the CTDEP to conclude in 1986 that the CY intake system represented BTA pursuant to §316(b) of the federal Clean Water Act (CTDEP 1986). No further studies or technology requirements were imposed. By way of contrast, a review of the recently promulgated standards for cooling water intakes at new facilities (EPA 2001) suggests that CY, as built, would not meet the a priori Track I technology criteria (i.e., closed cycle cooling) established in the new rule.

Retrospective Assessment

The genesis for this monograph stems from a workshop sponsored by the Electric Power Research Institute (EPRI) and held at the University of Connecticut Marine Sciences Program, Groton, Connecticut in November 2001 (EPRI 2001). The workshop provided a forum for the original principle investigators of the Connecticut River Ecological Study, namely William C. Leggett, Ronald R. Massengill, and Barton C. Marcy, Jr., to share the results of their investigations performed during 1965 through 1973. Related ecological research performed over the intervening 30-year period was presented for comparison. These results suggested that the Connecticut River and CY experience covering the period 1965 through 2003 was highly relevant to the science of ecological impact assessment as well as to the ongoing EPA effort (Nagle and Morgan 2000) to develop rules to implement §316(b) of the Clean Water Act (i.e., regulation of cooling water withdrawal impacts at power plants and other industrial facilities) and, therefore, warranted further examination.

No finer starting point for Part II of the Connecticut River Ecological Study could be sought than the insightful preface from the original American Fisheries Society Monograph 1 (Merriman and Thorpe 2004b [1976]). Those readers steeped in the science and evolution of ecological impact assessment and related regulatory matters will find this preface, written some 30 years ago by Dr. Dan Merriman and Lyle Thorpe, as relevant and timely today as it was then. The reader is advised by Merriman and Thorpe that "It is not always easy to determine in advance the precise confines of the various areas of an ecological investigation, perhaps especially so when one seeks a clear answer to a pragmatic problem." For example, the Study design was based on what is today called the BACI sampling design for ecological impact assessments (Green 1979). This classic approach focused on developing ecological characterizations [B] before and [A] after (during) operation at [C] control and [I] impacted sampling stations and was particularly suited for the study of thermal impacts. However, in the very early power plant operational phases of the Study, it became clear that in addition to the postulated effects of thermal additions to the Connecticut River, entrainment and impingement impacts associated with the CY cooling water intake structure were of equal and perhaps even greater ecological importance. Not only were entrainment and impingement important effects, but their complexity required even more sophisticated assessment tools, and the related population-level information requirements were greater than had been initially assumed. At that time, no other studies were available in the United States on entrainment and impingement study design or impact assessment and mitigation alternatives. This led Merriman and Thorpe (2004b [1976]) to highlight the difficulties associated with making population-level assessments of entrainment and impingement in relation to power plant siting and operation. This retrospective review examines this important question by drawing upon a considerable body of information on the life history parameters of several key species, most notably the American shad *Alosa sapidissima*, collected during and subsequent to the Study.

In looking back on the Study and in interpreting its results in the context of ecological information collected subsequently, we are once again counseled by Merriman and Thorpe (2004b [1976]) that "Because of the complexity of the interrelationship between the various elements in a natural aquatic ecosystem, it is next to impossible to provide a masterwork of the whole. It is not as though the system were stable as with an artist in the delineation of a still life; it is a constantly changing scene from season to season and year to year." While this seems obvious, the summary of ecological information on resident fishes, anadromous fishes, and benthic organisms examined in this retrospective assessment confirms that, at least in the context of the Connecticut River, there have been dramatic shifts in species dominance, some man-induced and some natural, since 1973, that appear to totally overwhelm any real or implied power plant impacts that did or might have occurred. As a result, the relative importance of such changes to the ecology of the lower river are considered here both as an aid to our understanding and future management of this and other natural resources and as a guide to our ability to assess and interpret the impact of power plant operations, given the reality of the dynamic character of this and presumably all other ecosystems subjected to human intervention, whether positive or negative in character.

Ecological Significance of the Lower Connecticut River: A Story of Change, Success, and Future Challenges

The Connecticut River is the longest and largest river system in New England. From its headwaters in Canada at the border with New Hampshire, it flows some 660 km to its terminus in Long Island Sound at Old Saybrook, Connecticut (Figure 1). The watershed encompasses approximately 2.9 million ha or more than 28,500 km^2 located in four states (Vermont, New Hampshire, Massachusetts, and Connecticut). The character of the Connecticut River watershed is approximately 77% forested, 9% agricultural, 7% wetlands and water, and 7% developed. Over much of its length, river flow is subject to modification and regulation by power generation and flood control reservoirs. In fact, there are 16 main-stem dams above the Connecticut–Massachusetts border and nearly 1,000 dams in the entire watershed (USFWS 1995). The Connecticut River main stem is unregulated from Windsor Locks near the Connecticut–Massachusetts border to Long Island Sound. It is tidally influenced from its mouth northward approximately 90 km (Merriman and Thorpe 2004a [1976], this volume; Garabedian et al. 1998; Caplis 2001; Rozsa 2001). The primary focus of the Study and of this retrospective assessment is this lower tidal reach.

The history of the Connecticut River, in many respects, mirrors the settlement and development of many of the major river systems across the United States. It is a chronicle of early colonial settlement, industrial development, and urbanization (Delaney 1996; Maloney et al. 2001; Wetherell 2002). Yet, in large measure, the Connecticut River retains much of its early charm and beauty. This is particularly true in the lower reaches of the Connecticut River below the city of Hartford, Connecticut (Figure 1). "The Connecticut River is the only principal river in the northeastern United States without a major port, harbor, or urban area at its mouth. This is the result of shifting sandbars in Long Island Sound that impede navigation. This situation has served to preserve the largely rural characteristics of the regional landscape and maintain the river's extraordinary assemblage of natural and relatively undisturbed biotic communities" (Caplis 2001). As a result, the naturally occurring freshwater tidal wetlands in this reach and the large remaining undeveloped areas of shoreline have been the beneficiary of aggressive efforts to preserve these natural values.

The state of Connecticut was a pioneer in water resource management when, in 1925, it established a state water commission to assist municipalities in improving treatment of domestic sewage. Other landmark legislation has included the Connecticut Clean Water Act in 1967, a precursor of the federal Water Pollution Control Act amendments of 1972 that established the NPDES permit program. Soon after followed protection for tidal and inland wetlands, development of water quality standards, increased funding for sewerage infrastructure improvements, non-point-source pollution management, and finally watershed programs to coordinate and fund pollution abatement activitics in targeted river basins. The adopted use goals for the Connecticut

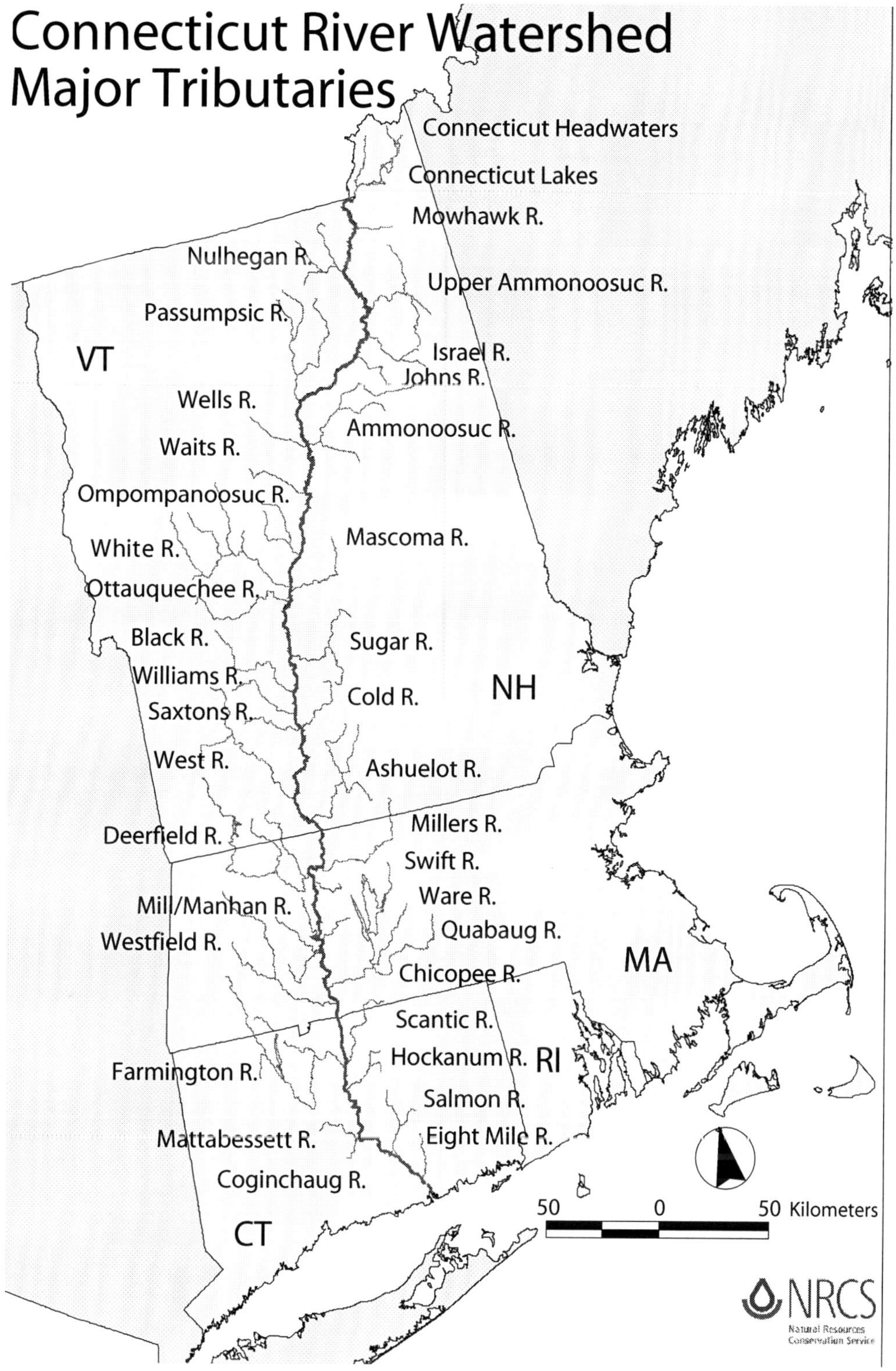

FIGURE 1. The Connecticut River watershed and its major tributaries.

River include recreation, fish and wildlife habitat, navigation, and assimilation of treated wastewaters. These goals are achieved in most respects, although water contact recreation is partially impaired by episodes of elevated concentrations of indictor bacteria, and fish consumption is partially impaired by PCB contamination in some species. Water quality management priorities within the watershed include the control of combined sewer overflows in the Hartford, Connecticut and Springfield, Massachusetts areas; urban stormwater management; stream bank stabilization; remediation of contaminated sediments; and control of invasive species of aquatic organisms (CTDEP 2002).

The state and federal water resource regulatory programs have established the underpinning for current preservation efforts. Water quality improvements (Mullaney 2004, this volume) have, at least in part, led to the renewed interest in the river's recreational and esthetic values. In 1991, federal legislation established the Silvio O. Conte National Fish and Wildlife Refuge, encompassing the entire 7.2 million acre watershed within U.S. borders, acknowledging that protecting its natural heritage was to be both a social challenge and a scientific endeavor. The Nature Conservancy (TNC) recognized the ecological significance of the lower river and environs as one of its "Last Great Places" in 1993. Known locally as the "Tidelands," and understanding that ensuring natural landscape integrity is the key to protecting natural diversity, the TNC has been aggressively acquiring lands and agreements to set aside conservation lands. In 1994, the Connecticut River Estuary and Tidal River Wetlands Complex was selected as the nation's fourteenth Wetlands of International Importance under the Ramsar convention (Dreyer and Caplis 2001), due largely to the unique wetlands complexes located there. More recently, the Connecticut River was designated as one of only 14 American Heritage Rivers recognizing its ecological and cultural significance. That the unique features of the lower Connecticut River are now better understood and managed can be attributed, in part, to an array of natural resource inventories performed since the late 1980s that informed decision-makers. These inventories can be found in the U.S. Fish and Wildlife Action Plan and Environmental Impact Statement related to the Silvio O. Conte National Fish and Wildlife Refuge (USFWS 1995), the CTDEP Nomination Report to Ramsar Convention (CTDEP 1994), Dreyer and Caplis (2001), and Maloney et al. (2001).

Content of the Revised Monograph

The monograph "revisited" is divided into two parts. The American Fisheries Society Monograph 1 is republished in its entirety as Part I. This is intended not only to reprint the original monograph, but also aid in comparison with later work found in Part II. The "Connecticut River Ecological Study Revisited," Part II, includes papers summarizing information collected since the Connecticut River Ecological Study. Where appropriate, the papers update specific topical areas such as the migration, spawning, and population dynamics of American shad. Species such as the shortnose sturgeon *Acipenser brevirostrum*, Atlantic salmon *Salmo salar*, striped bass *Morone saxatilis*, and the nonindigenous bivalve Asian clam *Corbicula fluminea* that were not addressed during the Connecticut River study (Part I, this volume) are also discussed.

For example, the Connecticut River shortnose sturgeon population is the focus of current study, as it is a listed endangered species and exists in the lower river (Savoy 1999 and 2004, this volume). No specimens were reported to have been collected during the 1965–1973 CY studies. Restoration of the Atlantic salmon to the Connecticut River has been underway since the late 1970s and is of great historical and recreational interest (Gephard and McMenemy 2004, this volume). American shad migrating upstream now have access to their original natal areas, due to removal of the first upstream barrier in Enfield and significant improvements in fish passage facilities at the Holyoke Dam in southern Massachusetts and at dams further upstream. Striped bass abundance has increased dramatically, representing an important ecological dynamic that has resulted in major shifts in abundance of certain prey species (Savoy and Crecco 2004, this volume). The exotic Asian clam now dominates the benthic sand infauna but did not exist in the lower Connecticut River during the earlier study period (Morgan et al. 2004, this volume). Demonstrable improvements in water quality, largely due to the

Clean Water Act, have occurred (Mullaney 2004), while at the same time, there has been a continuing trend toward increased water temperature (Figure 2).

While Part II of the monograph focuses on aquatic benthic and fish species potentially impacted by CY, it does not address other habitats and biotic values, although ecologically and recreationally important. The freshwater tidal wetlands that occur south of CY in coves and tributaries have been previously described and are being aggressively protected. Vegetation of the freshwater and brackish tidal wetlands of the lower Connecticut River were characterized by Metzler and Rosza (1982), Barrett (1989), Barrett et al. (1997), Barrett and Prisloe (2001), and Rosza et al. (2001). Barrett (1994) provided an understanding of the factors affecting ecological succession and community structure of these habitats, while Barrett et al. (1997) provided a comprehensive inventory of the submerged aquatic vegetation in the lower Connecticut River. Craig (1990), Craig and Beal (1992), and Benoit (2001) described the estuarine marsh birds of these unique wetlands and other rare bird species that inhabit or migrate along the river corridor. The impact of *Phragmites australis* invasion on fish food resources in the lower Connecticut River has been investigated by Fell et al. (1998). Generalized summaries of the richness and diversity of the biota and habitats of the lower Connecticut River are found in Dreyer and Caplis (2001) and Maloney et al. (2001). Papers in Part II also elaborate on advances in analytical impact assessment tools that were not available during the original Connecticut River Ecological Study.

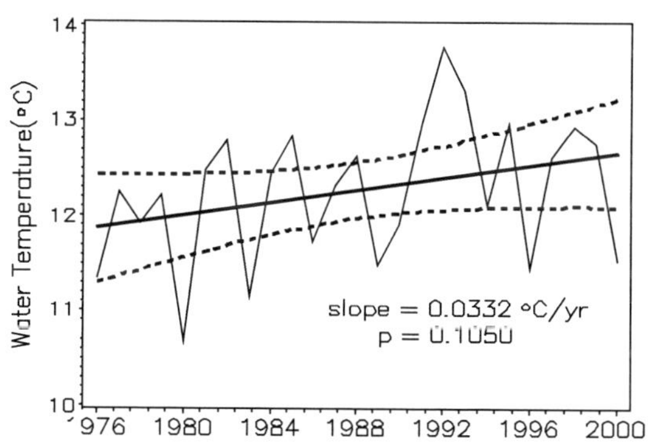

FIGURE 2. Annual mean water temperature (°C; thin solid line) measured at the cooling-water intake at the Connecticut Yankee atomic power plant (CY) and a linear regression of the annual mean values (thick solid line) with 95% confidence interval (dashed lines). Water temperature recorded at 15-min intervals at the CY intake was used to calculate daily and then yearly averages.

Finally, Part II examines questions relevant to our understanding of the ecology of the lower Connecticut River and provides that "future comparison" to which Merriman and Thorpe spoke. To what extent can we draw inferences from the additional data through 2003? What are the limitations, strengths of these inferences? To what extent have analytical methods improved? Can we identify information key to sustaining improvements in habitat observed over the last four decades? What would we do differently today?

With these questions as a framework, it is our hope that the information presented in this monograph provides valuable insights for natural resource managers faced with developing regulatory strategies and programs that preserve our unique ecological systems, such as the lower Connecticut River.

Acknowledgement

We thank Beth Goettle, Jan Rowan, Robert Wilce, John Tietjen, Robert Whitlatch, Barton C. Marcy, Jr., and William Leggett for their helpful review comments and suggestion on the manuscript. We also wish to thank our editor R. Massengill for his help and guidance during the development and review process. The editors are particularly grateful to Douglas Dixon of EPRI, whose conceptual guidance, encouragement, and financial support made possible this revisit of the Connecticut River Ecological Study. We are indebted to his foresight and commitment to this undertaking.

References

Barrett, J. M., M. Prisloe, L. Giannotti and N. Barrett. 1997. Distribution and abundance of submerged aquatic vegetation in the lower tidal Connecticut River. Submitted to the Connecticut Department of Environmental Protection, Hartford.

Barrett, N. 1989. Vegetation of the tidal wetlands of the lower Connecticut River. Ecological relationships of plant community-types with respect to flooding and habitat. Master's thesis. University of Connecticut, Storrs.

Barrett, N. 1994. Vegetation patch dynamics in freshwater tidal wetlands. Doctoral dissertation. University of Connecticut, Storrs.

Barrett, N., and S. Prisloe. 2001. Spatial distribution of *Phragmites australis* (CAV.) Trin. Ex Stued. within the tidelands of the Connecticut River, 1994. Final Report to the Connecticut Department of Environmental Protection, Office of Long Island Sound Programs, Hartford.

Benoit, L. 2001. Birds of the lower Connecticut River. Pages 48–55 *in* G. D. Dreyer and M. Caplis, editors. Living resources and habitats of the lower Connecticut River. Bulletin 37. The Connecticut College Arboretum, New London.

Caplis, M. 2001. Introduction. Pages 7–10 *in* G. D. Dreyer and M. Caplis, editors. Living resources and habitats of the lower Connecticut River. Bulletin 37. The Connecticut College Arboretum, New London.

Craig, R. J. 1990. Historic trends in the distribution and populations of estuarine marsh birds of the Connecticut River. Connecticut Agricultural Experiment Research Station, Storrs.

Craig, R. J., and K. G. Beal. 1992. The influence of habitat variables on marsh bird communities of the Connecticut River Estuary. Wilson Bulletin 104:295–311.

CTDEP (Connecticut Department of Environmental Protection). 1974. NPDES Permit. Connecticut Yankee Atomic Power Company, Hartford, Connecticut.

CTDEP (Connecticut Department of Environmental Protection). 1986. NPDES Permit. Connecticut Yankee Atomic Power Company, Hartford, Connecticut.

CTDEP (Connecticut Department of Environmental Protection). 1994. Nomination report to the Convention on Wetlands of International Importance (RAMSAR): Connecticut River Estuary and Tidal Wetlands Complex. Hartford, Connecticut.

CTDEP (Connecticut Department of Environmental Protection). 2002. 2002 water quality report to Congress – prepared pursuant to Clean Water Act 305(b). Hartford, Connecticut.

CYAPCO (Connecticut Yankee Atomic Power Company). 1974a. Summary of the environmental assessment of the once-through condenser cooling system. Submitted to the State of Connecticut Department of Environmental Protection by Northeast Utilities Service Company. Hartford, Connecticut.

CYAPCO (Connecticut Yankee Atomic Power Company). 1974b. Environmental Technical Specifications. Appendix B to the Facility Operating License DPR-61, Docket 50-213. Haddam Neck plant. Haddam Neck, Connecticut.

CYAPCO (Connecticut Yankee Atomic Power Company). 1976–1983. Haddam Neck plant, Annual Environmental Operating Reports, Part A: Non-radiological Report. Operating License No. DPR-61, Docket No. 50-213. Haddam Neck, Connecticut.

CYAPCO (Connecticut Yankee Atomic Power Company). 1980. Feasibility of bypass facilities to return impinged fish to the source water, Haddam Neck station. Submitted to the Connecticut Department of Environmental Protection by Northeast Utilities Services Company. Hartford, Connecticut.

Delaney, D. 1996. The Connecticut River, New England's historic waterway. The Connecticut River Foundation at Steamboat Dock, Essex, Connecticut.

Dey, W. P., S. M. Jinks and G. L. Lauer. 2000. Pages S15–23 *in* J. Wisniewski, editor. Power plants & aquatic resources: issues and assessment. Environmental Science & Policy 3(Supplement 1), Elsevier, New York.

Dixon, D. A., and K. D. Zammit. 2000. Preface. Pages vii–viii *in* J. Wisniewski, editor. Power plants & aquatic resources: issues and assessment. Environmental Science & Policy 3(Supplement 1), Elsevier, New York.

Dreyer, G. D., and M. Caplis. 2001. Living resources and habitats of the lower Connecticut River. Bulletin 37. The Connecticut College Arboretum, New London.

EPA (Environmental Protection Agency) 1973. Development document for proposed best technology available for minimizing adverse environmental impact of cooling water intake structures. Report 440/1-76/015, Washington, D.C.

EPA (Environmental Protection Agency). 1977. Draft guidance for evaluating the adverse impact of cooling water intake structures on the aquatic environment: Section 316(b) P. L. 92-500. Washington, D.C.

EPA (Environmental Protection Agency). 2001. National pollutant discharge elimination system: regulations addressing cooling water intake structures for new facilities. Federal Register, Volume 66, No. 243. Washington, D.C.

EPRI (Electric Power Research Institute). 1999. Catalog of assessment methods for evaluating the effects of power plant operations on aquatic communities. Electric Power Research Institute, Report TR-112013, Palo Alto, California.

EPRI (Electric Power Research Institute). 2001. Connecticut River Ecological Study Workshop: revisiting the impact of a power plant: November 15–16, 2001, the University of Connecticut Marine Sciences Institute at Avery Point. Electric Power Research Institute, Report (Workshop presentations and notes) 1006900, Palo Alto, California.

EPRI (Electric Power Research Institute). 2002. Evaluating the effects of power plants on aquatic communities: guidelines for selection of assessment methods. Electric Power Research Institute, Report 1005176, Palo Alto, California.

Fell, P. E., S. P. Weissbach, D. A. Jones, M. A. Fallon, J. A. Zeppieri, E. K. Faison, K. A. Lennon, K. J. Newberry, and L. K. Reddington. 1998. Does invasion of oligohaline tidal marshes by reed grass, *Phragmites australis* (CAV.) Trin. Ex Steud., affect the availability of prey resources for the mummichog, *Fundulus heteroclitus*, L. Journal of Experimental Marine Biology and Ecology 222:59–77.

Garabedian, S. G., F. G. Coles, S. J. Grady, E. C. Todd Trench, and M. J. Zimmerman. 1998. Water quality in the Connecticut, Housatonic, and Thames river basins, Connecticut, Massachusetts, New Hampshire, New York and Vermont, 1992-95. U.S. Geological Survey Circular 1155. U.S. Geological Survey, Marlborough, Massachusetts.

Gephard, S., and J. McMenemy. 2004. An overview of the program to restore Atlantic salmon and other diadromous fishes to the Connecticut River with notes on the current status of these species in the river. Pages 287–317 *in* P. M. Jacobson, D. A. Dixon, W. C. Leggett, B. C. Marcy, Jr., and R. R. Massengill, editors. The Connecticut River Ecological Study (1965–1973) revisited: ecology of the lower Connecticut River 1973–2003. American Fisheries Society, Monograph 9, Bethesda, Maryland.

Green, R. H. 1979, Sampling, design and statistical methods for environmental biologists. John Wiley and Sons, New York.

Maloney, T. and eight coauthors. 2001. Tidewaters of the Connecticut River, an explorers guide to hidden coves and marches. River's End Press, LLC. Old Saybrook, Connecticut.

Merriman, D., and L. M. Thorpe. 2004a. Introduction. Pages 3–25 *in* P. M. Jacobson, D. A. Dixon, W. C. Leggett, B. C. Marcy, Jr., and R. R. Massengill, editors. The Connecticut River Ecological Study (1965–1973) revisited: ecology of the lower Connecticut River 1973–2003. American Fisheries Society, Monograph 9, Bethesda, Maryland. (Originally published in 1976)

Merriman, D., and L. M. Thorpe. 2004b. Original preface. Pages xiii–xv *in* P. M. Jacobson, D. A. Dixon, W. C. Leggett, B. C. Marcy, Jr., and R. R. Massengill, editors. The Connecticut River Ecological Study (1965–1973) revisited: ecology of the lower Connecticut River 1973–2003. American Fisheries Society, Monograph 9, Bethesda, Maryland. (Originally published in 1976)

Merriman, D., and L. M. Thorpe. 2004c. Summary and conclusions. Pages 247–260 *in* P. M. Jacobson, D. A. Dixon, W. C. Leggett, B. C. Marcy, Jr., and R. R. Massengill, editors. The Connecticut River Ecological Study (1965–1973) revisited: ecology of the lower Connecticut River 1973–2003. American Fisheries Society, Monograph 9, Bethesda, Maryland. (Originally published in 1976)

Metzler, K. M., and R. Rozsa. 1982. Vegetation of fresh and brackish tidal marshes in Connecticut. Connecticut Botanical Society Newsletter 10(1):1–3.

Morgan, D. E., M. Keser, J. T. Swenarton, and J. F. Foertch. 2004. Effect of Connecticut Yankee power plant on population dynamics of Asiatic clams and their interaction with native bivalves. Pages 419–439 *in* P. M. Jacobson, D. A. Dixon, W. C. Leggett, B. C. Marcy, Jr., and R. R. Massengill, editors. The Connecticut River Ecological Study (1965–1973) revisited: ecology of the lower Connecticut River 1973–2003. American Fisheries Society, Monograph 9, Bethesda, Maryland.

Mullaney, J. 2004. Summary of water quality trends in the Connecticut River, 1968–1998. Pages 273–286 *in* P. M. Jacobson, D. A. Dixon, W. C. Leggett, B. C. Marcy, Jr., and R. R. Massengill, editors. The Connecticut River Ecological Study (1965–1973) revisited: ecology of the lower Connecticut River 1973–2003. American Fisheries Society, Monograph 9, Bethesda, Maryland.

Nagle, D. G., and J. T. Morgan, Jr. 2000. A draft regulatory framework for analyzing potential adverse environmental impact from cooling water intake structures. Pages ix–xiv *in* J. Wisniewski, editor. Power plants & aquatic resources: issues and assessment. Environmental Science & Policy 3(Supplement 1), Elsevier, New York.

Rozsa R. 2001. Hydrology of the lower Connecticut River. Pages 18–22 *in* G. Dreyer and M. Caplis, editors. Living resources and habitats of the lower Connecticut River. Bulletin 37. The Connecticut College Arboretum, New London.

Rosza, R., K. J. Metzler, and P. Fell. 2001. Ecology of the lower Connecticut River: plants, animals, and their habitats. Pages 29–47 *in* G. D. Dreyer and M. Caplis, editors. Living resources and habitats of the lower Connecticut River. Bulletin 37. The Connecticut College Arboretum, New London.

Saila, S. B., and E. Lorda. 2004. Developments in fishery science with special reference to power plant impact assessment methods and the Connecticut River Ecological Study: 1973–present. Pages 475–500 *in* P. M. Jacobson, D. A. Dixon, W. C. Leggett, B. C. Marcy, Jr., and R. R. Massengill, editors. The Connecticut River Ecological Study (1965–1973) revisited: ecology of the lower Connecticut

River 1973–2003. American Fisheries Society, Monograph 9, Bethesda, Maryland.

Savoy, T. 1999. Final report, anadromous fish studies in Connecticut waters, 1996-1999. Project AFC-25. Submitted to National Marine Fisheries Service. Connecticut Department of Environmental Protection, Hartford.

Savoy, T. F. 2004. Population estimate and utilization of the lower Connecticut River by shortnose sturgeon. Pages 345–352 *in* P. M. Jacobson, D. A. Dixon, W. C. Leggett, B. C. Marcy, Jr., and R. R. Massengill, editors. The Connecticut River Ecological Study (1965–1973) revisited: ecology of the lower Connecticut River 1973–2003. American Fisheries Society, Monograph 9, Bethesda, Maryland.

Savoy, T. F., and V. Crecco. 2004. Factors affecting the recent decline of blueback herring and American shad in the Connecticut River. Pages 361–377 *in* P. M. Jacobson, D. A. Dixon, W. C. Leggett, B. C. Marcy, Jr., and R. R. Massengill, editors. The Connecticut River Ecological Study (1965–1973) revisited: ecology of the lower Connecticut River 1973–2003. American Fisheries Society, Monograph 9, Bethesda, Maryland.

USFWS (U. S. Fish and Wildlife Service). 1995. Final action plan and environmental impact statement, October 1995, the Silvio O. Conte National Fish and Wildlife Refuge. Department of the Interior, Turners Falls, Massachusetts.

Van Winkle, W., W. P. Dey, S. M. Jinks, M. S. Bevelhimer, and C. C. Coutant. 2003. A blueprint for the problem formulation phase of EPA-type ecological risk assessments for 316(b) determinations. Pages 264–290 *in* D. A. Dixon, J. A. Veil and J. Wisniewski, editors. Defining and assessing adverse environmental impact from power plant impingement and entrainment of aquatic organisms. A. A. Balkema Publishers, Lisse, The Netherlands.

Wetherell, W. D., editor. 2002. This American river, five centuries of writing about the Connecticut. University Press of New England, Lebanon, New Hampshire.

American Fisheries Society Monograph 9:273–286, 2004

Summary of Water Quality Trends in the Connecticut River, 1968–1998

JOHN R. MULLANEY*

U.S. Geological Survey, 101 Pitkin Street, East Hartford, Connecticut 06108, USA

Abstract.—The Connecticut River has a long history of water quality impairment. From the 1800s to the late 1960s, untreated or minimally treated waste discharges from population centers and industries have caused serious water quality problems. Trend analysis of selected water quality data in Connecticut from 1968 to 1998, collected by the U.S. Geological Survey in cooperation with the Connecticut Department of Environmental Protection, shows that water quality has improved, benefiting aquatic plants and animals, recreation, and the esthetics of the river. Many of the trends detected—including downward trends in total phosphorus, total nitrogen, and indicator bacteria, and upward trends in pH and dissolved oxygen—can be attributed primarily to improvements in wastewater treatment following the Federal Clean Water Act of 1972. Some uncertainty remains in evaluating the environmental significance of trends in dissolved oxygen, pH, and other constituents with diurnal fluctuations caused by plant metabolism. Downward trends in sulfate concentrations likely are attributable to reductions in sulfur dioxide emissions mandated by the Clean Air Act of 1970 and amendments made in 1990. Upward trends in chloride concentrations illustrate some effects of increasing urbanization and nonpoint-source pollution. Current (2003) and future water quality challenges for the Connecticut River include reducing nitrogen loads from point and nonpoint sources, reducing bacteria and other contaminant concentrations in urban stormwater runoff, and separating stormwater and sanitary sewers at some locations to prevent combined-sewer overflows.

Introduction

The Connecticut River basin, with its rich natural resources, has attracted people since precolonial times. European settlers and associated industries came to the region in the early 1600s. The region was favored for its plentiful waters, which were used for agriculture, industrial processing, navigation, and waste disposal. By 1897, the problem of pollution of Connecticut's waterways was recognized by the Connecticut General Assembly. A commission was formed to investigate sewage-disposal practices because the quantity of raw sewage being discharged into some streams had become unbearable (Hupfer 1965). By the 1930s, after the creation of the State Water Resources Commission, wastewater discharges to the Connecticut part of the basin had reached 154,000 m^3/d. About 35% of this wastewater had some form of primary treatment, and a number of facilities, including one in Hartford, were under construction. The reported municipal wastewater discharge to the Massachusetts part of the basin was about 112,000 m^3/d; only about 5% of this discharge received any form of treatment (New England Regional Planning Commission 1937). By 1962, 95 municipalities in four states were discharging wastewater to the Connecticut River Basin; of these, 41 provided primary treatment and 26 provided secondary treatment. Water samples collected in Agawam, Massachusetts, during September 1963, contained concentrations of fecal coliform bacteria as high as 550,000 colonies/100 mL. Industrial discharges of untreated wastes from paper, chemical, metal, plating, dyeing, and other industries were recognized as a serious problem (U.S. Department of Health, Education and Welfare 1963). These industrial discharges included pollutants that contributed to high biological oxygen demand (BOD), causing reduced concentrations of dissolved oxygen in the Connecticut River. Concentrations of

* *Corresponding author*: jmullane@usgs.gov

dissolved oxygen in an area below the Holyoke Dam in Massachusetts were nearly zero during the drought of 1966 and less than 2 mg/L in 1971 (EPA 2000).

In 1965–1972, the original Connecticut River Ecological Study was conducted to evaluate the ecological effects of the construction and operation of the Connecticut Yankee Haddam Neck nuclear power station (Merriman and Thorpe 1976 and Part 1 of this volume). This study provided only limited information on water quality, but collected a great deal of information on the ecology of the Connecticut River, and provides a baseline for current comparisons.

A number of initiatives, including the Connecticut Clean Water Act of 1967 (State of Connecticut 1967) and the Federal Water Pollution Control Act of 1972, have led to improvements in the water quality of the Connecticut River in Connecticut. There, the water quality currently (2003) is classified using standards adopted in 2002. The upper part of the Connecticut River between the Massachusetts border and just below the confluence with the Farmington River (the nontidal section of the river in Connecticut) is classified as C/B. The tidal section of the river (from the segment described above to Middle Haddam) is classified as SC/SB. The last reach from Middle Haddam to Long Island Sound is classified as SB. These classifications are subsequently defined.

The goal for water quality in the Connecticut River is to achieve standards for class B, or SB in coastal areas. Class B waters are designated for habitat for fish and other aquatic life and wildlife, recreation, navigation, and industrial and agricultural water supply. Class SB has a similar designated use to class B, except it includes habitat for marine fish and commercial harvesting of shellfish. Criteria for attainment of class B includes factors such as good esthetics, dissolved oxygen concentrations not less than 5 mg/L, and bacteria levels within established ranges.[1] Criteria for class B limit the color and amount of sludge, oils/grease, suspended solids, silt and sand deposits, turbidity, pH, allowable temperature increase, and concentrations of chemical constituents (CTDEP 2002). Classes C/B, or SC/SB indicate that the waters currently are not meeting class B criteria for one or more reasons (CTDEP 2002), but the established goal for water quality is B or SB conditions.

The U.S. Geological Survey (USGS) has been collecting water quality samples in the Connecticut River since passage of the state and federal Clean Water Acts. An expanded cooperative effort between the USGS and the Connecticut Department of Environmental Protection (CTDEP) began in 1973 and has continued to the present. Water quality data also have been collected as part of various federal initiatives, including the National Stream Accounting Network (NASQAN) (Ficke and Hawkinson 1975) and the National Water-Quality Assessment (NAWQA) program (Gilliom et al. 1995).

This paper summarizes selected water quality trends and constituent loads in the main stem of the Connecticut River. Information has been summarized from analyses of trends in water quality in Connecticut during 1969–1988 (Trench 2000), 1989–1998 (Colombo and Trench 2002), and Trench and Vecchia (2002). Information on loads in phosphorus and nitrogen has been summarized from Trench (2000) and Mullaney et al. (2002). Most of the information summarized herein postdates the Connecticut River Ecological Study (Merriman and Thorpe and Part 1 of this volume 1976).

Study Area

The five water quality monitoring stations discussed herein are on the Connecticut River in Connecticut (Figure 1; Table 1). Four of the stations are in tidally affected sections of the Connecticut River (all but 01184000, Thompsonville); therefore, no streamflow data are available. Water quality sampling frequency at these stations has varied from monthly during the earlier part of the sampling record to eight or six times per year at some stations in the latter part of the record.

The monitoring station on the Connecticut River at Thompsonville is the farthest downstream station that is unaffected by tidal conditions. The Thompsonville station is near the Connecticut border with Massachusetts, just downstream from Springfield. Annual mean flow of the river at this station during water years 1928–2001 was 476 m^3/s (Morrison et al. 2002). During 1968–2001, the highest annual peak flow was measured on 31

[1] Current standards based on enterococci concentration data. an older standard for fecal coliform has been referenced in this paper (CTDEP 1996) because the collection of primarily fecal coliform data during the time period described in this paper.

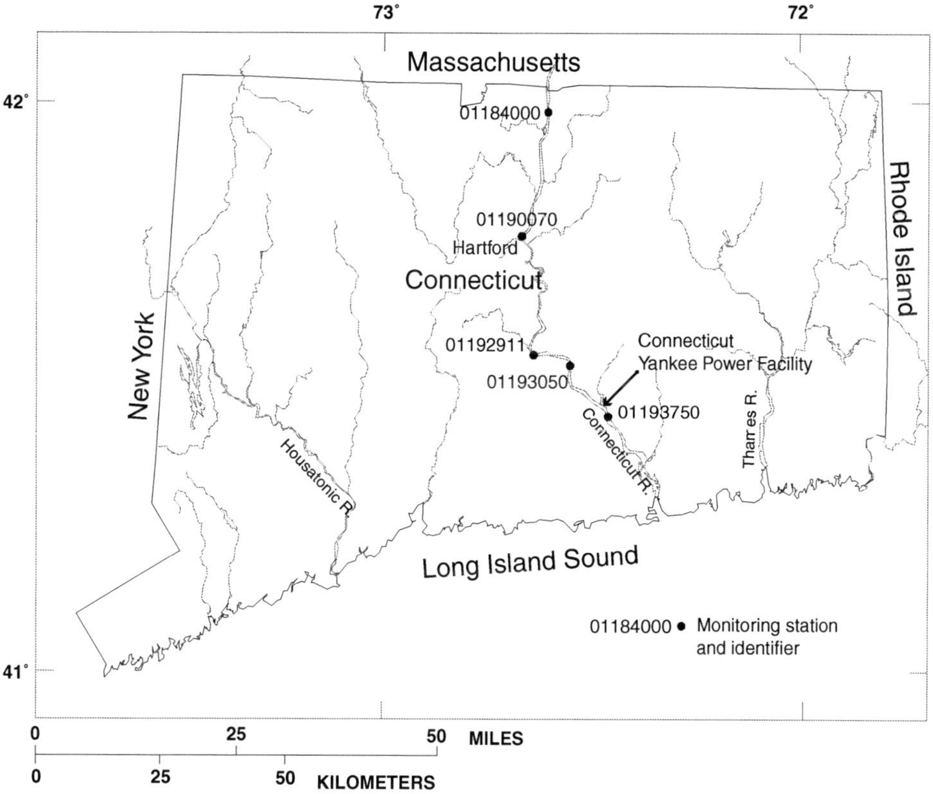

FIGURE 1. Selected monitoring stations on the Connecticut River, Connecticut, 1960s–1990s.

May 1984 (5,270 m^3/s), and the lowest mean daily flow was measured on 2 September 1968 (29.7 m^3/s). The average flow at this station during the Connecticut River Ecological Study (Merriman and Thorpe 1976 and Part 1 of this volume) for water years 1965–1972 was 411 m^3/s. Monthly flow statistics are shown in Figure 2.

The only station located downstream from the Connecticut Yankee nuclear power facility is the East Haddam station (01193750). The stations at Middle Haddam and Middletown are downstream of the Hartford Connecticut metropolitan area.

Methods

The nonparametric seasonal Kendall test was used to analyze trends in selected water quality constituents in two statewide studies (Trench 1996; Colombo and Trench 2002). The seasonal Kendall test can be applied to actual constituent concentrations or to constituent concentrations that have been adjusted for streamflow-related variability. In the seasonal Kendall test, the trends are assumed to be monotonic for the period of analysis. The use of the seasonal Kendall test for analysis

TABLE 1. Water-quality monitoring stations on the Connecticut River, Connecticut, 1960s–present.

Water-quality monitoring station	U.S. Geological Survey identifier	Drainage area (km^2)	Period of sampling
Connecticut River at Thompsonville, Connecticut	01184000	25,019	1966–present
Connecticut River at Hartford, Connecticut	01190070	27,161	1976–present
Connecticut River at Middletown, Connecticut	01192911	28,151	1974–1991
Connecticut River at Middle Haddam, Connecticut	01193050	28,223	1966–present
Connecticut River at East Haddam, Connecticut	01193750	28,728	1968–present

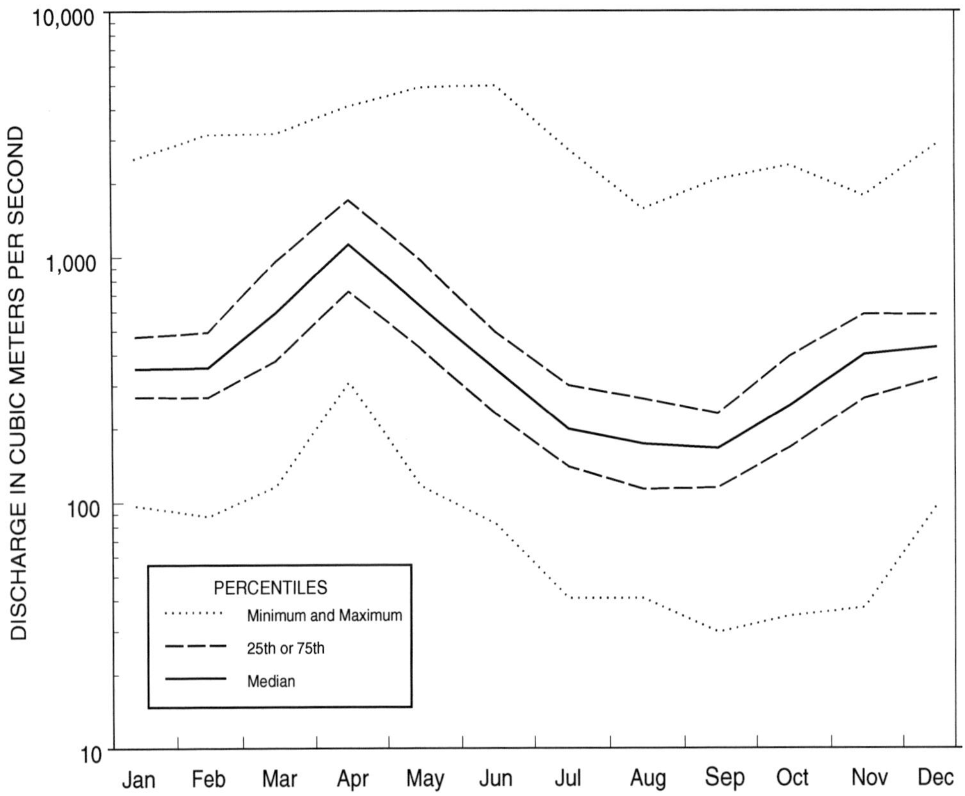

FIGURE 2. Monthly flow statistics Connecticut River at Thompsonville, Connecticut (01184000), 1968–2001.

of trends in water quality is described by Hirsch et al. (1982), Smith et al. (1982), Schertz (1990), and Schertz et al. (1991).

Trench and Vecchia (2002) used a time-series analysis of river discharge and concentration at five water quality monitoring stations in Connecticut to understand and remove the effects of annual and seasonal varibility to produce flow-adjusted concentration data. Variation in the data set that could not be explained by annual and seasonal variability was filtered using a periodic autoregressive moving average (PARMA) model. The application of this method to streamflow was described by Vecchia (1985) and for concentration by Vecchia (2000). The time-series method was used to identify linear trends and step trends, and also to understand both long-term (10 years or more) and short-term (1–2 years) water quality trends. Step trends were used to evaluate periods where changes or biases in analytical methods may have affected measured constituent concentrations (Trench and Vecchia 2002). Results were used to design optimum sampling schedules for trend detection at specific stations.

Annual loads of total phosphorus and total nitrogen were estimated for the Connecticut River at Thompsonville as part of two regional studies (Trench 2000; Mullaney et al. 2002). Loads were estimated using a log-linear regression model that incorporates algorithms to deal with censored water quality data and to account for retransformation bias. The regression models relate the natural log of load to flow, time, and seasonal variability (Cohn et al. 1989, 1992).

Trends in Selected Water-Quality Constituents

Phosphorus

Total phosphorus concentrations in the Connecticut River at Thompsonville have been decreasing steadily since 1969 (Figure 3). Concentrations of total phosphorus at Thompsonville have decreased

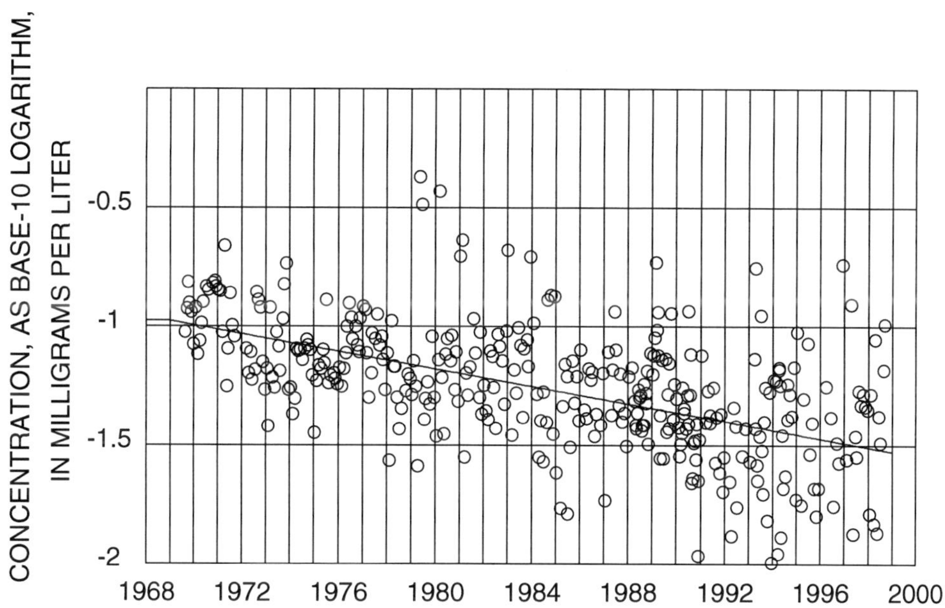

FIGURE 3. Trends in flow-adjusted concentrations of total phosphorus, Connecticut River at Thompsonville, Connecticut, 1969–1998 (from Trench and Vecchia [2002]: Figure 12A).

from an average of 0.15 mg/L in the late 1960s to about 0.05 mg/L in 2001. Trench and Vecchia (2002) report a decline in median total phosphorus concentrations at this station of 4.1% per year. The load of total phosphorus at Thompsonville has decreased from about 1.5 million kg/year in the early 1970s to less than 0.7 million kg/year during the mid-1990s (Figure 4) (Trench 2000). Significant downward trends in total phosphorus concentrations (not flow adjusted) also were reported at the four tidally affected Connecticut River monitoring stations during 1975–1988 and 1981–1988 (Trench 1996) (Table 2).

Nitrogen

Trends in nitrogen concentrations in the Connecticut River have varied during 1971–1998 (Figure 5) at the Thompsonville station. Total nitrogen increased during 1975–1988 and began to decline thereafter (Trench and Vecchia 2002). Colombo and Trench (2002) report no significant trends in total nitrogen at Thompsonville and three other stations for 1992–1998. Annual loads of total nitrogen calculated for the Connecticut River at Thompsonville ranged from about 7 million kg in 1995 (a dry year) to more than 18 million kg in 1984 (Figure 6).

Analysis of data from 1975 to 1988 for the Connecticut River at Thompsonville and some downstream stations indicated that significant upward trends were present in concentrations of total organic nitrogen, nitrite-plus-nitrate nitrogen, and dissolved ammonia nitrogen. Downward trends in the concentrations of dissolved ammonia nitrogen were detected for three stations on the Connecticut River for 1989–1998 (Colombo and Trench 2002).

Indicator Bacteria

Downward trends in concentrations of fecal coliform and fecal streptococcus bacteria (Table 2) were detected at some Connecticut River stations for the periods between 1975 and 1988 (Trench 1996). Trends in fecal coliform and enterococcus bacteria were analyzed during 1989–1998 (Colombo and Trench 2002). Concentrations of these indicator bacteria showed downward trends during 1975–1988 and 1981–1988 (Table 2). No significant trends were detected during 1989–1998.

Median and maximum measured concentrations of fecal coliform bacteria generally decreased in the Connecticut River at Thompsonville by an order of magnitude from 1970 to the mid-1980s

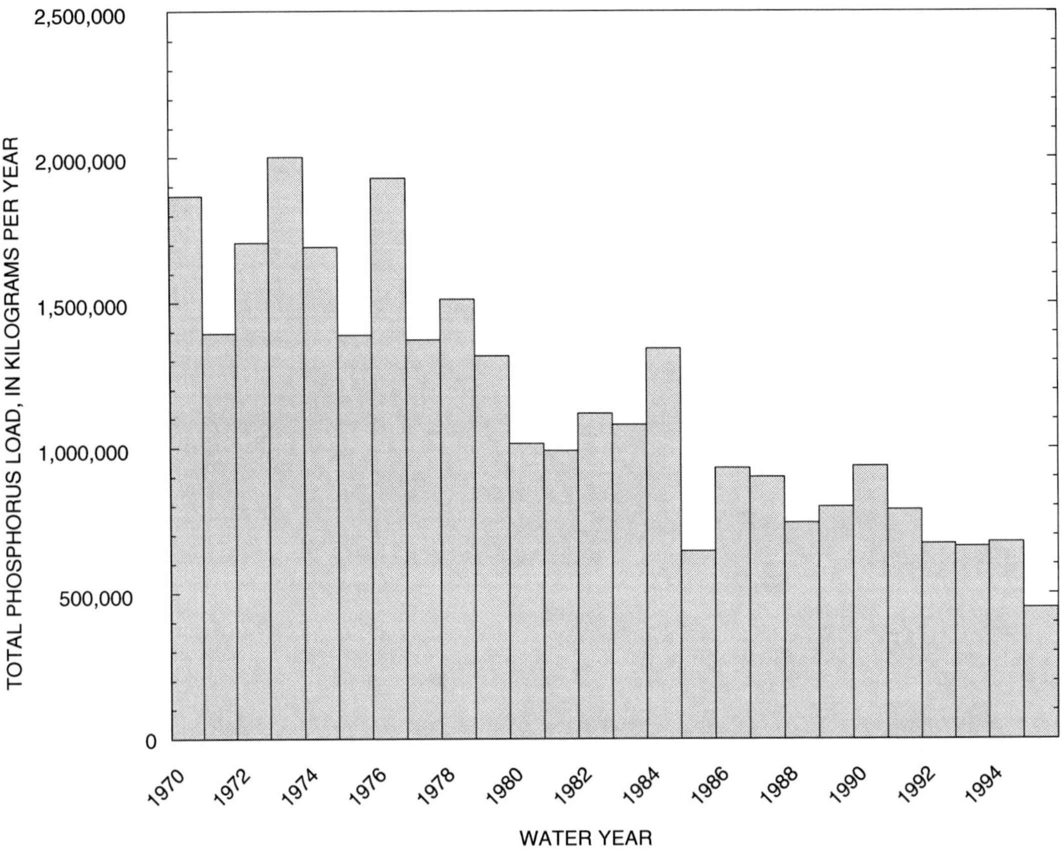

FIGURE 4. Annual loads of total phosphorus, Connecticut River at Thompsonville, Connecticut, water years 1970–1995 (from Trench [2000]: Figure 17).

TABLE 2. Summary of selected water-quality trends reported for monitoring stations on the Connecticut River, Connecticut, 1968–1998. (Data compiled from Trench [2002], Colombo and Trench [2002], Trench [1996]; *, insufficient data for trend analysis)

U.S. Geological Survey identifier	Total phosphorus		Total nitrogen		Fecal coliform		Chloride	
	Trend	Period	Trend	Period	Trend	Period	Trend	Period
01184000	–	1969–1998	–	1971–1975	–	1975–1988	–	1968–1973
			+	1976–1982			+	1974–1998
			–	1988–1998				
01190070	–	1981–1988	*		–	1981–1988	+	1981–1988
01192911	–	1975–1988	+	1975–1988	–	1981–1988	+	1975–1988
01193050	–	1975–1988	+	1975–1988		no significant trends	+	1981–1988
01193750	–	1975–1988	+	1975–1988		no significant trends		no significant trends

+ indicates upward trend.

– indicates downward trend.

Table 2. Continued

U.S. Geological Survey identifier	Sulfate		pH		Dissolved Oxygen	
	Trend	Period	Trend	Period	Trend	Period
01184000		1968–1976	+	1969–1988	+	1969–1988
	+	1977–1981				
	–	1988–1998				
01190070	–	1989–1998	+	1981–1988	+	1981–1988
01192911	*	*	+	1975–1988		no significant trends
01193050	–	1969–1998	+	1969–1988	+	1969–1988
01193750	*		+	1975–1998		no significant trends

(Figure 7). The number of years with more than 50% of the samples higher than the state standard (400 colonies/100 mL in greater than 10% of samples) has decreased with time (CTDEP 1996).

No significant trends in indicator bacteria were reported for the Connecticut River at East Haddam, Connecticut, during 1975–1998. Annual maximum concentrations and many annual median concentrations of fecal coliform have exceeded the state standard throughout the period of record (Figure 8).

Inorganic Compounds

Upward trends in chloride concentrations have been detected at all four Connecticut River monitoring stations (Table 2). Upward trends in chloride concentrations may be attributable to nonpoint sources, including deicing chemicals.

Sulfate concentrations have declined in the Connecticut River during the past 30 years (Figure 9). Significant trends in flow-adjusted concentrations of sulfate were detected at the

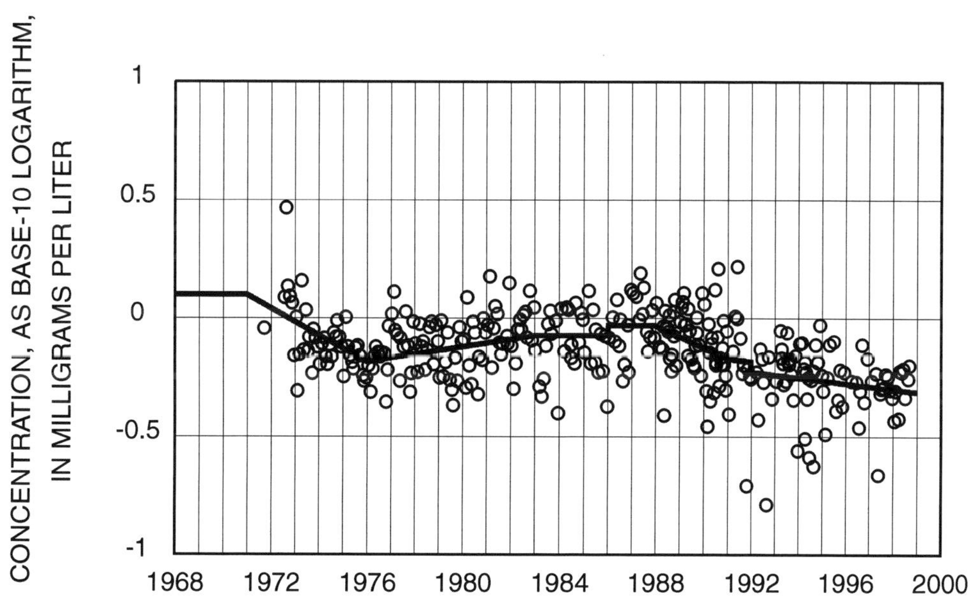

FIGURE 5. Trends in flow-adjusted concentrations of total nitrogen, Connecticut River at Thompsonville, Connecticut, 1971–1998 (from Trench and Vecchia [2002]: Figure 11A).

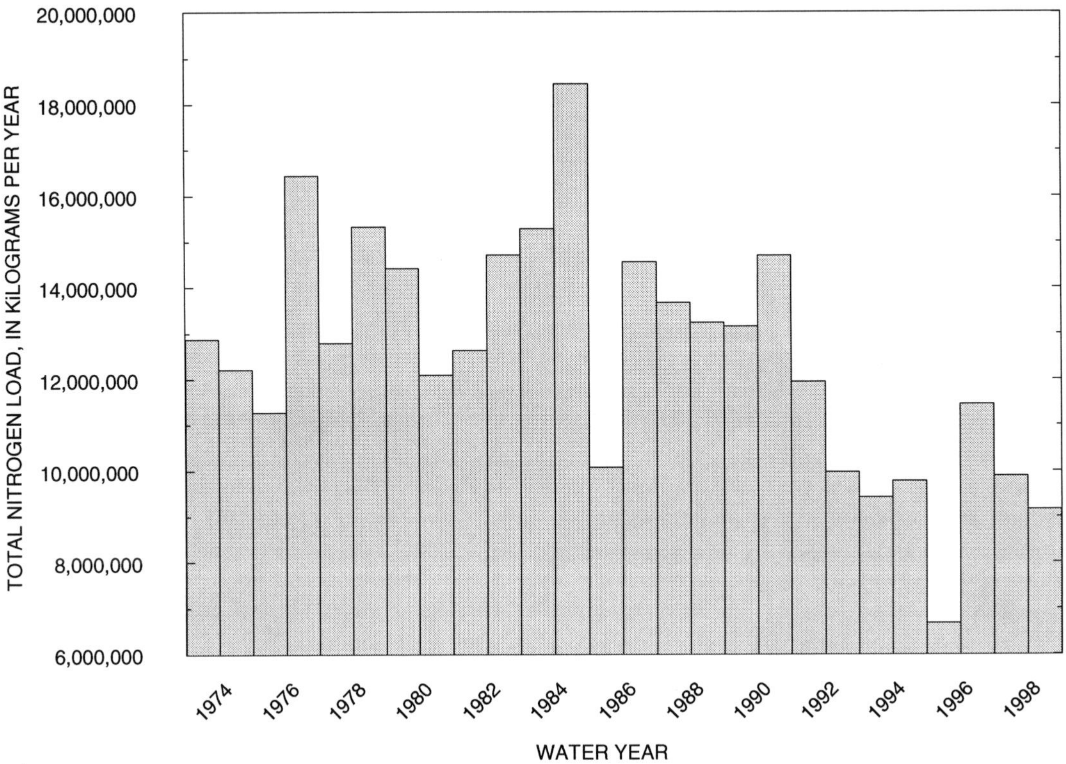

FIGURE 6. Annual loads of total nitrogen, Connecticut River at Thompsonville, Connecticut, water years 1973–1998.

Connecticut River at Thompsonville for samples collected from 1968 to 1998 (Trench 2002) (Figure 9). Downward trends were present during 1968–1976 and 1988–1998. Upward trends were detected for sulfate concentration data collected during 1983–1989 (step trend possibly related to analytical methods) and 1977–1981. Downward trends in sulfate concentrations have been detected for the monitoring stations at Hartford and at Middletown (Table 2). Colombo and Trench (2002) reported downward trends in sulfate from 1989 to 1998 as a regional trend in Connecticut.

Dissolved Oxygen and pH

Upward trends in the concentration of dissolved oxygen were detected for the Connecticut River from the late 1960s to the late 1980s. Upward trends in flow-adjusted dissolved oxygen concentrations were detected in data from 1969 to 1988 at Thompsonville (Trench 1996) (Table 2). Significant upward trends were reported for Hartford during 1981–1988 and for Middle Haddam (Table 2) during 1969–1988. No significant trends in dissolved oxygen concentrations were detected in data for East Haddam. No significant trends in dissolved oxygen concentrations were detected for Connecticut River stations during 1989–1998 (Colombo and Trench 2002). Trends in concentrations of dissolved oxygen may be attributed primarily to improvements in wastewater-treatment practices and a reduction in the discharge of oxygen-demanding effluent; however, trends detected prior to 1974 may have resulted because of changes in sampling methods (Trench 1996). The trend also was detected in basins with no point-source discharges. Dissolved oxygen concentrations with time for the Middle Haddam station are shown in Figure 10.

Upward trends in pH were detected at all five stations during various periods from 1969 to 1988 (Table 2). Significant upward trends in pH also were detected for Thompsonville and East Haddam

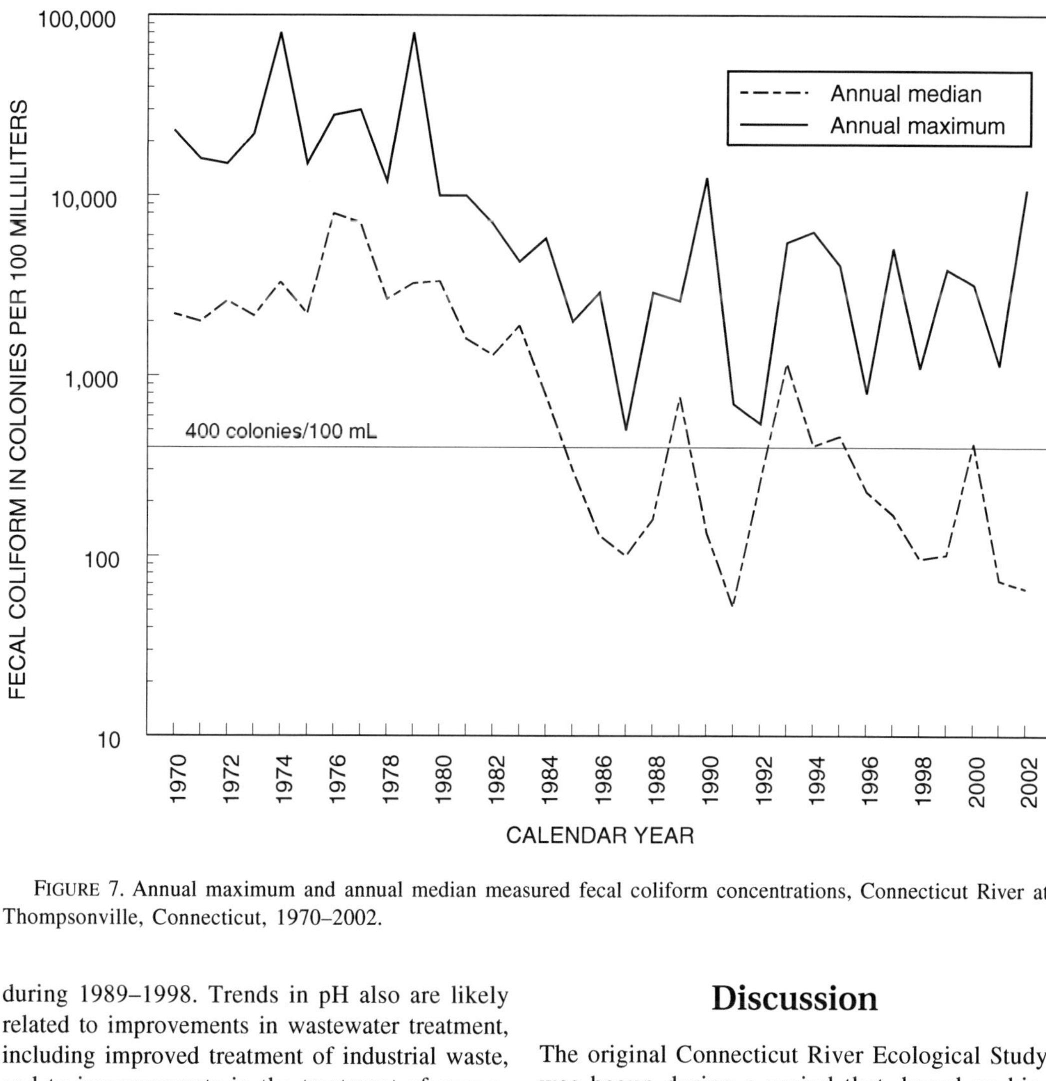

FIGURE 7. Annual maximum and annual median measured fecal coliform concentrations, Connecticut River at Thompsonville, Connecticut, 1970–2002.

during 1989–1998. Trends in pH also are likely related to improvements in wastewater treatment, including improved treatment of industrial waste, and to improvements in the treatment of ammonia at municipal wastewater-treatment facilities (Trench 1996).

An additional complexity in evaluating detected trends in dissolved oxygen concentrations, pH, and temperature results from the diurnal fluctuations that typically occur instream during the growing season. In addition to the seasonal variability that is accounted for in the seasonal Kendall test, the time of day can affect the measured value. There has been a tendency, for example, towards collecting samples later in the day during the latter part of the record. This change in sampling time complicates the interpretation of environmental trends because pH, temperature, and dissolved oxygen may increase during the day during the growing season.

Discussion

The original Connecticut River Ecological Study was begun during a period that, based on historical observations, likely represents the most degraded water quality conditions for the Connecticut River. Many positive changes in water quality began during this study period and have continued to the present. The data from the Connecticut River Ecological Study provide a baseline of the aquatic habitat that existed on the lower river prior to the implementation of management action.

Water quality trend analysis for the main stem of the Connecticut River in Connecticut during 1968–1998 shows substantial improvement in the concentrations of many water quality constituents; therefore, improvement in the support of aquatic life, esthetics, and the use of the river for recreation. Many of these improvements can be

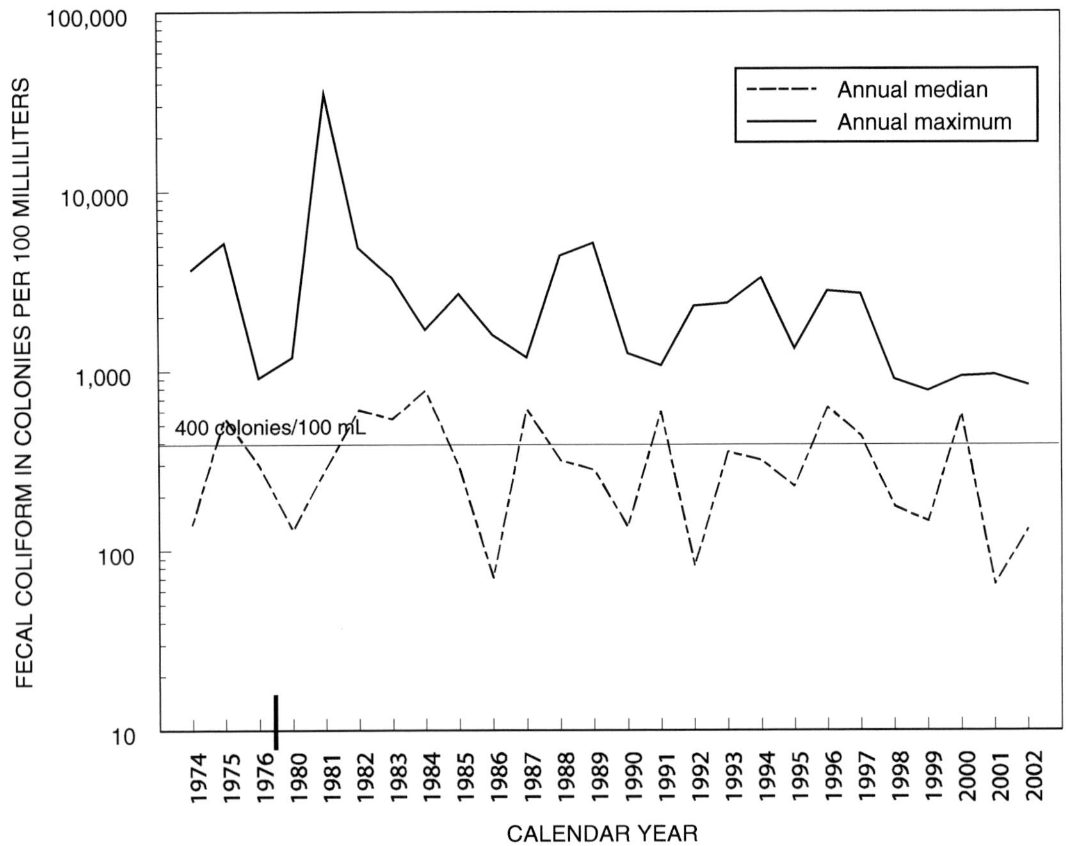

FIGURE 8. Annual median and annual maximum measured fecal coliform concentrations, Connecticut River at East Haddam, Connecticut, 1974–2002.

attributed to the successful management actions following the Connecticut Clean Water Act of 1967 and the Federal Water Pollution Control Act of 1972.

Downward trends in the concentrations of total phosphorus primarily are due to improvements in municipal wastewater treatment and to a reduction in the use of phosphate detergents since the 1970s (Litke 1999). These improvements also could be caused in small part by the loss of agricultural land and by the implementation of Best Management Practices (BMPs) on some farms.

Downward trends in total nitrogen since 1988 at Thompsonville are most likely the result of improved nitrogen removal at municipal wastewater-treatment facilities, but could relate to changes in land use (i.e., agricultural to residential or forest). Reductions in nitrogen concentrations in the Connecticut River are likely not related to atmospheric sources because wet deposition of nitrogen oxides in precipitation has remained relatively unchanged since the 1980s (Driscoll et al. 2001). Additional reductions in nitrogen load are mandated by the Long Island Sound Total Maximum Daily Load (TMDL) calculation, which is designed to improve dissolved oxygen conditions in Long Island Sound (NYDEC/CTDEP 2000). A reduction in nitrogen load of 58.5% by 2014 has been mandated by this TMDL; the reduction has been allocated between municipal wastewater discharges and nonpoint contributions of nitrogen. This amounts to a reduction of 1.7 million kg/year for the Connecticut part of the Connecticut River basin (NYDEC/CTDEP 2000).

Concentrations of indicator bacteria have decreased in the Connecticut River at four of the monitoring stations upstream from East Haddam (Table 3); yet, concentrations in samples fre-

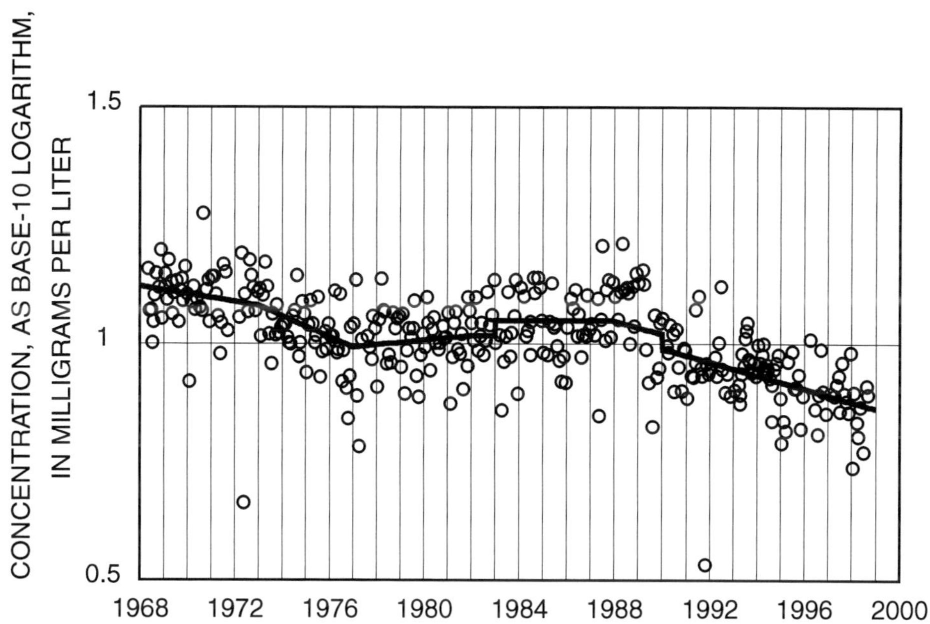

FIGURE 9. Trends in flow-adjusted concentrations of sulfate, Connecticut River at Thompsonville, Connecticut, 1968–1998 (from Trench and Vecchia [2002]: Figure 10A)

quently exceeded water quality standards for class B waters. The reductions noted in these concentrations likely result because of wastewater-treatment improvements in the watershed during the past 30 years. The continued presence of elevated bacteria concentrations may be attributable to combined sewer overflows that still are present in Connecticut and Massachusetts and to nonpoint stormwater runoff.

The upward trends in chloride concentration may be indicative of the effects of the increasing nonpoint runoff of road-deicing chemicals from the expansion of impervious areas through new land development during the last 30 years.

Upward trends in pH and dissolved oxygen are primarily attributed to improvements in wastewater-treatment, both municipal and industrial. Trend analyses for these constituents and for other measurements, such as temperature (not studied), are complicated by the large diurnal fluctuations in values coupled with an apparent trend in sample collection time. A possible bias in dissolved oxygen concentrations was discussed by Trench (1996).

Trends in sulfate concentrations are consistent with declines in sulfate concentration in precipitation that have been linked to reductions in sulfur dioxide emissions from power plants in the northeast (Driscoll et al. 2001). Sulfur dioxide emissions were highest during 1973 and began to decline thereafter (EPA 2003). The Title IV Amendments to the Clean Air Act of 1990 have accelerated the downward trend in emissions beginning in 1995. The importance of this trend to aquatic life in the Connecticut River is uncertain; however, reductions in emissions of sulfur dioxide may be a partial cause for the upward trends that were detected in pH, and ultimately may improve the quality of habitat in acid-sensitive lakes and headwater streams in parts of the Connecticut River basin.

Acknowledgments

The author would like to thank the Connecticut Department of Environmental Protection, the major cooperator with the USGS for the Connecticut statewide ambient monitoring network. Samples also were collected under the funding of the USGS National Stream Accounting Network (NASQAN) and the National Water-Quality Assessment (NAWQA) program. The author also thanks Elaine Trench of the USGS in Connecticut, Paul Jacobson of Ecological Assessments

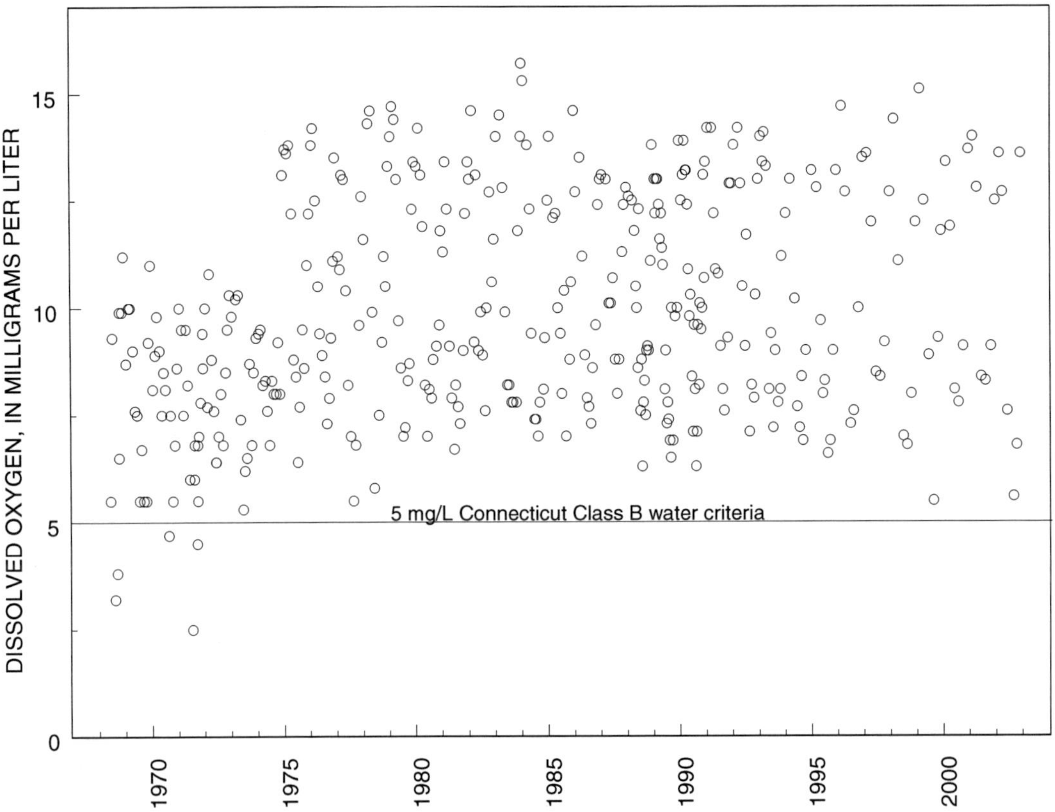

FIGURE 10. Concentrations of dissolved oxygen at the Connecticut River at Middle Haddam, Connecticut, 1968–2002.

TABLE 3. Significant trends (95% confidence level) in colonies of indicator bacteria in the Connecticut River in Connecticut, 1975–1988. (Data from Trench [1996]; values in percent per year; negative sign denotes downward trend; na, not analyzed or insufficient data; ns, no significant trend; F, flow adjusted; *, significant at 90 % confidence level)

Water-quality monitoring station	1975–1988 Fecal coliform	1975–1988 Fecal streptococcus	1981–1988 Fecal coliform	1981–1988 Fecal streptococcus
Connecticut River at Thompsonville Connecticut 01184000	–28.6 F	–18.1 F	–37.5	–34.7 F
Connecticut River at Hartford Connecticut 01190070	na	–14.9	–29.0	–35.3
Connecticut River at Middletown Connecticut 01192911	–3.9*	ns	–17.6	–16.8
Connecticut River at Middle Haddam Connecticut 01193050	ns	ns	ns	–16.9*
Connecticut River at East Haddam Connecticut 01193750	ns	ns	ns	ns

LLC, and Paul Stacey of the Connecticut Department of Environmental Protection for peer reviews, and Barbara Korzendorfer and Ron Massengill for editorial assistance.

References

Cohn, T. A., D. L. Caulder, E. J. Gilroy, L. D. Zynjuk, and R. M. Summers. 1992. The validity of a simple statistical model for estimating fluvial constituent loads—an empirical study involving nutrient loads entering Chesapeake Bay. Water Resources Research 28(9):2353–2363.

Cohn, T. A., L. L. DeLong, E. J. Gilroy, R. M. Hirsch, and D. K. Wells. 1989. Estimating constituent loads. Water Resources Research 25(5):937–942.

Colombo, M. J., and E. C. T. Trench. 2002. Trends in surface-water quality in Connecticut. U.S. Geological Survey Water-Resources Investigations Report 02–4012, East Hartford, Connecticut.

CTDEP (Connecticut Department of Environmental Protection). 1996. Water quality standards. Bureau of Water Management, Hartford, Connecticut.

CTDEP (Connecticut Department of Environmental Protection). 2002. Water quality standards. Bureau of Water Management, Hartford, Connecticut.

Driscoll, C. T., G. B. Lawrence, A. J. Bulger, T. J. Butler, C. S. Cronan, C. Eagar, K. F. Lamber, G. E. Likens, J. L. Stoddard, and K. C. Weathers. 2001. Acid rain revisited: advances in scientific understanding since the passage of the 1970 and 1990 Clean Air Act amendments. Hubbard Brook Research Foundation, Science Links™ Publication, volume 1, no. 1, Hanover, New Hampshire.

EPA (Environmental Protection Agency). 2000. Progress in water quality—an evaluation of the national investment in municipal wastewater treatment, chapter 5, Connecticut River case study. U.S. Environmental Protection Agency, EPA 832/R-00–008, Washington D.C.

EPA (Environmental Protection Agency). 2003. Response of surface water chemistry to the Clean Air Act Amendments of 1990. U.S. Environmental Protection Agency, EPA 620/R-03/001, Washington D.C.

Ficke, J. F., and R. O. Hawkinson. 1975. The national stream quality accounting network (NASQAN)—some questions and answers. U.S. Geological Survey Circular 719, Bismarck, North Dakota.

Gilliom, R. J., W. M. Alley, and M. E. Gurtz. 1995. Design of the National Water Quality Assessment Program—occurrence and distribution of water-quality conditions. U.S. Geological Survey Circular 1112, Reston, Virginia.

Hirsch, R. M., J. R. Slack, and R. A. Smith. 1982. Techniques of trend analysis for monthly water-quality data. Water Resources Research 18(1):107–121.

Hupfer, M. E. 1965. Forty years of water pollution control in Connecticut. Presented at the 81st Annual Meeting Connecticut Society of Civil Engineers, Inc. Cheshire, Connecticut, April 22, 1965.

Litke, D. W. 1999. Review of phosphorus control measures in the United States and their effects on water quality. U.S. Geological Survey Water-Resources Investigations Report 99–4007, Denver.

Merriman, D., and L. M. Thorpe, editors. 1976. The Connecticut River Ecological Study: the impact of a nuclear power plant. American Fisheries Society, Monograph 1, Bethesda, Maryland.

Morrison, J., P. L. Provencher, J. W. Martin, and J. R. Norris. 2002. Water resources data, Connecticut, water year 2001. U.S. Geological Survey Water-Resources Data Report CT-01–1, East Hartford, Connecticut.

Mullaney, J. R., G. E. Schwarz, and E. C. T. Trench. 2002. Estimation of nitrogen yields and loads from basins draining to Long Island Sound, 1988–98. U.S. Geological Survey Water-Resources Investigations Report 02–4044, East Hartford, Connecticut.

New England Regional Planning Commission. 1937. Water resources of New England. New England Regional Planning Commission, National Resources Committee, Region One, publication number 51, Boston.

NYDEC/CTDEP (New York State Department of Environmental Conservation/ Connecticut Department of Environmental Protection). 2000. A total maximum daily load analysis to achieve water-quality standards for dissolved oxygen in Long Island Sound: NYDEC, Albany, New York and CTDEP, Hartford, Connecticut.

Schertz, T. L. 1990. Trends in water-quality data in Texas. U.S. Geological Survey Water-Resources Investigations Report 89–4178, Austin, Texas.

Schertz, T. L., R. B. Alexander, and D. J. Ohe. 1991. The computer program Estimate Trend (ESTREND), a system for the detection of trends in water-quality data. U.S. Geological Survey Water-Resources Investigations Report 91–4040, Reston, Virginia.

Smith, R. A., R. N. Hirsch, and J. R. Slack. 1982. A study of trends in total phosphorus measurements at NASQAN stations. U.S. Geological Survey Water-Supply Paper 2190, Reston, Virginia.

State of Connecticut. 1967. Connecticut's Clean Water Act of 1967, an analysis of Public Act 57. Clean Water Task Force, Hartford, Connecticut.

Trench, E. C. T. 1996. Trends in surface-water quality in

Connecticut, 1969–88. U.S. Geological Survey Water-Resources Investigations Report 96–4161, Hartford, Connecticut.

Trench, E. C. T. 2000. Nutrient sources and loads in the Connecticut, Housatonic, and Thames river basins. U.S. Geological Survey Water-Resources Investigations Report 99–4236, East Hartford, Connecticut.

Trench, E. C. T., and A. V. Vecchia. 2002. Water-quality trend analysis and sampling design for streams in Connecticut, 1968–98. U.S. Geological Survey Water-Resources Investigations Report 02–4011, East Hartford, Connecticut.

U.S. Department of Health, Education and Welfare. 1963. Statement on water quality management, states of Connecticut and Massachusetts portion of Thames, Connecticut, and Housatonic river basins. Presented before the Natural Resources and Power Subcommittee of the House Committee on Government, 4 October 1963, U.S. Department of Health, Education and Welfare, Region I, Boston.

Vecchia, A. V. 1985. Periodic autoregressive-moving average (PARMA) modeling with applications to water resources. Water Resources Bulletin 21(5):721–730.

Vecchia, A. V. 2000. Water-quality trend analysis and sampling design for the Souris River, Saskatchewan, North Dakota, and Manitoba. U.S. Geological Survey Water- Resources Investigations Report 00–4019, Bismarck, North Dakota.

American Fisheries Society Monograph 9:287–317, 2004

An Overview of the Program to Restore Atlantic Salmon and Other Diadromous Fishes to the Connecticut River with Notes on the Current Status of these Species in the River

STEPHEN GEPHARD*

Connecticut Department of Environmental Protection, Inland Fisheries Division
Post Office Box 719, Old Lyme, Connecticut 06371, USA

JAMES MCMENEMY

Vermont Department of Fish and Wildlife, Springfield Regional Office
100 Mineral Street, Suite 302, Springfield, Vermont 05156, USA

Abstract.—A federal and multi-state cooperative program to restore American shad *Alosa sapidissima* and Atlantic salmon *Salmo salar* to the Connecticut River basin was begun in 1967 and has evolved to include many other species. The program began in the last years of the Connecticut River Ecological Study, but most of its activities have occurred since the study ended. The Connecticut River Atlantic Salmon Commission manages the program. Emphasis has been placed on the provision of fish passage at barrier dams. Early fishways were justified on the basis of existing American shad runs, and later upriver fishways were built to support future salmon runs. Fishways exist at five main-stem dams and eight tributary dams, with facilities for downstream fish passage provided at many additional dams. Salmon restoration has been pursued with stocking of hatchery-reared fry and smolts, catch prohibitions, kelt reconditioning, fish health management, and various genetic management and marking schemes. Annual runs typically have numbered in the hundreds but recently have declined to less than 100 at the same time runs elsewhere through the species' range have also declined. Annual runs of American shad, blueback herring *A. aestivalis*, and alewife *A. pseudoharengus* increased but recently experienced declines, for which stock recovery of the striped bass *Morone saxatilis* is thought to be at least partially responsible. Gizzard shad *Dorosoma cepedianum* and hickory shad *A. mediocris* experienced significant range extensions into the Connecticut River basin since the 1990s, and the number of nonspawning striped bass that enter the river annually has increased dramatically during the same time period. Brief reviews of the status in the Connecticut River basin of these anadromous species as well as of the shortnose sturgeon *Acipenser brevirostrum*, Atlantic sturgeon *A. oxyrinchus*, white perch *Morone americana*, rainbow smelt *Osmerus mordax*, sea lamprey *Petromyzon marinus*, and sea-run brown trout *Salmo trutta* and the catadromous American eel *Anguilla rostrata* are provided.

Introduction

Diadromous fish species were very abundant throughouft the Connecticut River basin prior to European contact (Moffitt et al. 1982). Atlantic salmon *Salmo salar* are reported to have ascended the river approximately 615 km to the present day site of Beechers Falls, Vermont (Atkins 1874; Kendall 1935; Moffitt et al. 1982), as well ascending many tributaries of the river until generally stopped by waterfalls (Atkins 1874). Runs were extirpated in southern tributaries beginning in the early to mid-1700s due to the construction of dams to power mills and factories. The first dam that completely blocked the main-stem river was built in 1798 near the present day site of Turners Falls, Massachusetts, and it resulted in the extirpation of the last run of Atlantic salmon to the river (Atkins 1874; CRASC 1998). Every other diadromous fish species was able to survive

* *Corresponding author:* steve.gephard@po.state.ct.us

but apparently in greatly reduced numbers, even as downstream dams were built in Holyoke, Massachusetts and Enfield, Connecticut (Figure 1) because some spawning habitat for these species remained accessible downstream of all of the dams.

Early History of Restoration Efforts

The first effort to restore runs of fish to the Connecticut River was initiated in 1867 when the newly created Fish Commissions of the four Connecticut River states (Connecticut, Massachusetts, New Hampshire, Vermont) met to organize the effort (CRASC 1998). The restoration program resulted in hundreds of salmon returning to the Connecticut, but the effort was abandoned after 25 years due to the lack of effective fish passage at the dams and the failure to protect the returning salmon from harvesting (Foster 1991). There is no indication that the program increased runs of other species to the river.

The Connecticut River suffered from severe water pollution and heavy exploitation of remnant runs of fish throughout much of the 1900s, but much of the original habitat for salmon and other diadromous fish species remained, due to the rural character of much of the basin. In 1965, the U.S. Congress passed the Anadromous Fish Conservation Act (Public Law 89–304) that provided federal funds to states that joined cooperatively to restore anadromous fish runs to their rivers. There had been growing interest since the 1940s to attempt to restore Atlantic salmon to the river. The availability of federal funds in 1967 prompted the four states to once again unite to create an anadromous fish restoration program for the Connecticut River (Foster 1991). The original governing body of the restoration program was named "The Policy Committee for Fisheries Management of the Connecticut River Basin" (the Policy Committee). It consisted of fish and game commissioners from each of the four states as well as high level administrators from the U.S. Bureau of Sport Fisheries (later to become the U.S. Fish and Wildlife Service) and the U.S. Bureau of Commercial Fisheries (later to become the NOAA—Fisheries). The Policy Committee received scientific advice from the "Technical Committee for Fisheries Management of the Connecticut River Basin" (Technical Committee), which was comprised of a fisheries biologist from each state and federal agency. It was charged with carrying out the necessary field activities (culture and stocking of juvenile salmon, adult capture, transport, holding, and spawning, egg incubation, etc.). The states typically funded their activities through Sport Fish Restoration Funds (authorized by the "Dingell-Johnson Act" and later the "Wallop-Breaux Act").

The U.S. Fish and Wildlife Service (USFWS) provided a river program coordinator to expedite communication among the partners and coordinate their efforts. In a sense, the coordinator was a one-person staff for the program. The four states helped fund the coordinator's office.

During the final years of the Connecticut River Ecological Study, the restoration program was getting organized, and the Policy and Technical committees were in negotiation with the power companies about the provision of fish passage at the Holyoke, Turners Falls, Vernon, Bellows Falls, and Wilder dams (dams 5 through 9 in Figure 1) (Foster 1991). However, by the time the study concluded in 1972, only token numbers of salmon had actually been stocked, no adult returns had been documented, and virtually no efforts had been expended on behalf of other diadromous fish species. Fisheries biologists from the USFWS and individual state agencies along with experienced salmon anglers assessed the habitat upstream of the dams to confirm that suitable habitat for Atlantic salmon persisted after nearly 200 years of human impacts. This effort resulted in the targeting of five major tributaries for the program: Salmon River (#2 in Figure 1), Farmington River (#3), Westfield River (#4), Deerfield River (#10), West River (#16), White River (#26), and Ammonoosuc River (#30). The locations of these targeted tributaries determined which dams were targeted for fish passage. Other tributaries were targeted for restoration later in the program.

Creation of the "Commission"

The program partners sought congressional authorization for an interstate commission, which they received in 1983 when Congress passed P.L. 98–138 and the states passed related State Acts, creating the Connecticut River Atlantic Salmon Commission. The commission met for the first time in January of 1984 (Foster 1991). The commission, commonly referred to by its acronym CRASC, replaced the Policy Committee, but its

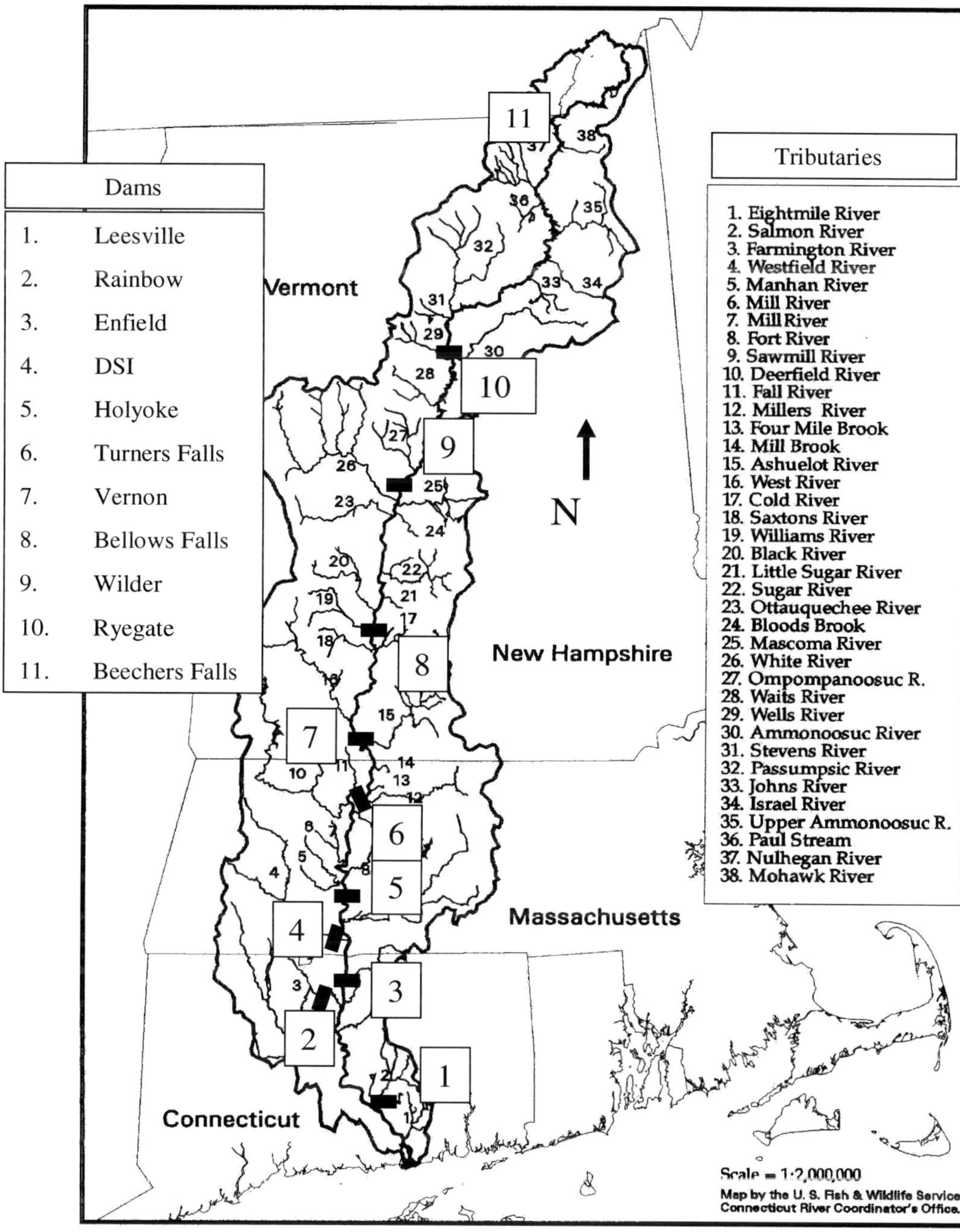

FIGURE 1. Connecticut River basin with key dams and tributary watersheds.

make-up was similar, adding one private sector member from each state (appointed by the state's governor). The current membership of the CRASC and agency abbreviations used in this paper are listed in Table 1. Atlantic salmon remains the prime focus of CRASC, but management decisions concerning other diadromous species are routinely made by CRASC. The original act authorized CRASC to exist for 20 years. In 2002, Congress passed P.L. 107–171, which re-autho-

TABLE 1. Members of the Connecticut River Atlantic Salmon Commission and active partners of the commission in the restoration program. (Abbreviations in parenthesis are used throughout the text.)

Voting members	Partners
State of Connecticut, Department of Environmental Protection (CTDEP)	United States Department of Agriculture, U.S. Forest Service (USFS)
State of Massachusetts, Division of Fisheries and Wildlife (MADFW)	United States Department of Interior, U.S. Geological Survey, Biological Resources Division (USGS-BRD)
State of Massachusetts, Division of Marine Fisheries	United States Department of Defense, U.S. Army Corps of Engineers
State of Vermont, Department of Fish and Wildlife (VTDFW)	Northeast Utilities
State of New Hampshire, Fish and Game Department (NHFGD)	Pacific Gas and Electric Company
United States, Department of Interior, U.S. Fish & Wildlife Service (USFWS)	Connecticut River Watershed Council
United States, Department of Commerce, National Oceanic Atmospheric Administration, National Marine Fisheries Service (NOAA-Fisheries)	Connecticut River Salmon Association
	Trout Unlimited
	Various watershed councils focused on tributary watersheds

rized CRASC for another 20 years. During its first 20 years, CRASC received no federal appropriations and had no budget. It accomplished necessary tasks of the restoration program through the funded activities of its member agencies. Not all restoration activities for diadromous species in the watershed are conducted by CRASC.

The Technical Committee continues its mandate within CRASC, working with a number of subcommittees (Salmon Studies, Shad Studies, Genetics, Smolt Advisory, and Fish Passage) staffed by committee members as well as colleagues from their agencies and others, including researchers. CRASC typically meets twice a year and the Technical Committee three or four times a year, and the workgroups regularly, as needed.

The U.S. Fish & Wildlife Service (USFWS) program coordinator now serves as executive secretary for CRASC, works with the Technical Committee, coordinates the activities of all of the partners, acts as the primary means of communication among the partners, and serves as the contact person for the restoration program.

The Atlantic salmon restoration effort has been guided by a strategic plan, developed in 1980 and revised in 1982 and 1998 (CRASC 1998). It identified the habitat within the basin targeted for salmon restoration, identified the dams requiring fish passage, set the annual Atlantic salmon stocking target of 10 million fry and 100,000 hatchery-reared smolts, and expected future adult returns in the range of 900–15,000 (CRASC 1998). Figure 1 shows the tributary streams targeted for salmon restoration as well as the dams that currently have fish passage and dams targeted for future fish passage facilities (CRASC 1998). Other fisheries management plans were adopted for other diadromous species and will be discussed later.

Fish Passage Facilities

Provision of fish passage is the foundation of the restoration program. The basin contains more than 1,000 dams, many which are located within the historical range of diadromous species. The first fishways on the river were built prior to the initiation of the restoration program. A ramp was installed in the 2-m-high Enfield Dam (river kilometer [rkm] 110, #3 in Figure 1) in 1933 (Moffitt et al. 1982). This timber crib–stone dam has always been overtopped by water and some fish were always able to surmount it. The effectiveness of the ramp was never clearly documented. The dam continually deteriorated during the 1950s–1970s (Robert A. Jones, Director, Connecticut Department of Environmental Protection/ Fisheries Division–retired, personal communication) and a large section of the dam washed out in 1978, at which time the dam ceased to be an impediment to fish migration. No further management actions have been taken in regards to this dam other than CRASC resolutions opposing various informal proposals to rebuild it.

A fish elevator installed at the 10-m-high Holyoke Dam (rkm 139, #5 in Figure 1) began operation in 1955 primarily to provide passage of American shad *Alosa sapidissima* (Moffitt et al. 1982). By modern standards, the elevator was rudimentary. Workers were required to push lifted shad in handcarts to the headpond for release. It allowed successful spawning of shad above the dam for the first time since 1849. The Holyoke facility was expanded in 1975 to a two-lift facility and modified so that the lift hoppers emptied directly into an exit flume, allowing fish to swim into the headpond of their own volition. This upgrade was a result of negotiations by the Policy Committee (Stephen Rideout, U.S. Geological Survey [USGS], Director- Conte Anadromous Fish Research Center, personal communication). The presence of an observation window in the flume allowed enumeration. Other fishways were built between the late 1970s and early 1990s (Table 2). These opened the lower 440 km of the main-stem river and portions of targeted lower tributaries to fish passage. Most of these facilities were built by the power companies as a result of project relicensing requirements of the Federal Energy Regulatory Commission (FERC) following fishway prescriptions by the USFWS or a result of settlement agreements between the companies and the states/federal government. Decisions to build most of these fishways were made before any adult Atlantic salmon had returned to the river. Justification for all fishways downstream of Bellows Falls, Vermont (rkm 280) was based on the need to pass the sizeable existing run of American shad (Moffitt et al. 1982). Fishways at the dams at Bellows Falls (#8 in Figure 1) and Wilder

TABLE 2. Status of upstream fish passage facilities at key dams in the Connecticut River basin.

Dam name	River	Town	Dam use	FERC#	Style	Year installed
Holyoke	Connecticut	Holyoke, MA	Hydro	2004A	Lift	1955,1976, 2003 modifications
Turners Falls	Connecticut	Turners Falls, MA	Hydro	1889A	Ice Harbor	1980
Vernon	Connecticut	Vernon, VT	Hydro	1904A	Ice Harbor/ vertical slot	1981
Bellows Falls	Connecticut	Rockingham, VT	Hydro	1855A	vertical slot	1984
Wilder	Connecticut	Hartford, VT	Hydro	1892A	modified Ice Harbor	1987
Ryegate	Connecticut	Ryegate, VT	Hydro	8011A	n.a.	Scheduled for future[a]
McColloch	Rowland Brk	Old Lyme, CT	Aesthetics	n.a.	pool and weir	
Mary Steube	Mill Brook	Old Lyme, CT	Aesthetics	n.a.	steeppass	
Moulson Pond	Eightmile	Lyme, CT	Hydro	n.a.	steeppass	1997
Lower Pond	Joshua Creek	Lyme, CT	Aesthetics	n.a.	pool and weir/ steeppass	
Leesville	Salmon	East Haddam, CT	Fisheries	n.a.	Denil	1980
Rainbow	Farmington	Windsor, CT	Hydro	n.a.	vertical slot	1976
DSI	Westfield	Westfield, MA	Hydro	2608A	Denil	1996
No. 2	Deerfield	Shelburne, MA	Hydro	2323A	n.a.	Scheduled for future[a]
Townshend	West	Townshend, VT	Flood control	n.a.	trap & truck	1993
Woodsville	Ammonoosuc	Woodsville, NH	Hydro	5307A	n.a.	Deferred

[a] Will be built when agreed upon numbers of fish are passed at downstream fishways.

(rkm 350, #9 in Figure 1) were justified by the need to pass adult Atlantic salmon, which had begun to return when these decisions were made. These dams are above the historic range of American shad in the river. Additional fishways at main-stem dams upstream of the Wilder Dam are planned if and when Atlantic salmon run sizes increase to a point that justifies this action.

Fishways were justified on the basis of these anadromous species but it was recognized that nonanadromous (riverine) species would benefit from these facilities and large numbers of white sucker *Catastomus commersonii*, common carp *Cyprinus carpio*, centrarchids, and other fishes are passed annually and were considered in the design capacity of the facilities.

It was initially assumed that downstream migrants would either use the fishways designed for upstream migrants or safely pass over the spillways. Years of operating experience and data analysis have shown that many downstream migrants passed through the turbines (Bell and Kynard 1985; Taylor and Kynard 1985; Stier and Kynard 1986; McMenemy and Kynard 1988). Estimated turbine mortality rates at Holyoke were 12–14% for Atlantic salmon smolts (Stier and Kynard 1986), 22% for adult American shad (Bell and Kynard 1985), and 62–82% for juvenile clupeids (Taylor and Kynard 1985). Downstream fish passage facilities were subsequently installed at all dams on the main stem downstream of and including Ryegate and many dams on tributaries. The delay in providing effective downstream fish passage significantly slowed the pace of the restoration program. Some downstream passage facilities were effective from the beginning (e.g., Rainbow Dam), but most required modification to achieve satisfactory performance (e.g., Holyoke Dam) (Kynard and O'Leary 1993). Downstream passage for shortnose sturgeon *Acipenser brevirostrum* and American eel *Anguilla rostrata* remains problematic. Facilities are opened, operated, and closed according to an annual operating schedule developed by the CRASC.

The construction in 1990 of the Silvio Conte Anadromous Fish Research Center (SCAFRC) (adjacent to the Turners Falls Dam in Turners Falls, Massachusetts) by the USFWS added research capability for fish passage on the Connecticut River. Its engineering facilities that allow full-scale models of fishways and testing with actively migrating diadromous fishes are unique in the world. The laboratory is now part of the USGS–Biological Resource Discipline (BRD) and has a national focus, but its location on the Connecticut River has resulted in its intimate involvement with many fish passage challenges that CRASC faces on the Connecticut River.

Status of Diadromous Fishes in the Basin with Summaries of Restoration Activities

Atlantic Salmon

Status Prior to Restoration Program

Atlantic salmon was native to the Connecticut River, one of the southernmost rivers within North America in which the species was found (Atkins 1874; MacCrimmon and Gots 1979). The species was extirpated from the basin after the first dam across the main-stem river was constructed in 1798 near the present day site of Turners Falls, Massachusetts (Atkins 1874; Moffitt et al. 1982). An earlier attempt to restore the species during the latter half of the 1800s failed (Foster 1991), and only occasional stray salmon were reported in the river during the 1900s prior to the present-day restoration program (Merriman and Jean 1949).

Restoration Strategies

1. *Targeted Habitat.* It seems likely that little of the main-stem Connecticut River downstream of the present-day site of the Ryegate Dam (rkm 440, #10 in Figure 1) historically supported Atlantic salmon rearing habitat. Most of the rearing habitat was located in the river's many tributaries. Some historical Atlantic salmon habitat has been permanently lost due to human activities such as dam construction and water diversions. Most of the remaining salmon habitat in the basin is currently being stocked as part of the restoration program. Stocked habitat ranges from the Eightmile River in southern Connecticut to the Mohawk River in northern New Hampshire (#1 and # 38, respectively, in Figure 1). Much of this habitat is not currently accessible to adult salmon because of the presence of many dams.
2. *Fish Passage.*
   a. *Upstream passage*—All main-stem dams from Vernon (rkm 228, #7 in Figure 1) down-

stream have fishways that were designed primarily for upstream passage of American shad. These can easily accommodate Atlantic salmon. Fish ladders at Bellows Falls (rkm 280, #8) and Wilder (rkm 350, #9) dams were built specifically for Atlantic salmon. All fishways planned for main-stem dams upstream of Wilder and tributary dams upstream of Bellows Falls and on the Deerfield River are primarily for salmon. Tributary fishways on the Eightmile, Salmon, Farmington, and Westfield rivers were constructed for a variety of anadromous fish including salmon. A "trap and truck" facility has provided upstream passage for salmon at the Townshend Dam on the West River. Collectively, these upstream passage facilities allow salmon to ascend the main-stem river to Ryegate Dam (rkm 440) and allow access to much or all of the Eightmile, Salmon, Farmington, Westfield, West, and White river systems and portions of many other tributaries.

b. *Downstream passage*—Most of the existing salmon habitat in the Connecticut River basin is located above dams, often many dams. Downstream passage, particularly at hydroelectric dams, is therefore essential to successful smolt emigration. The Connecticut River Atlantic Salmon Commission signed memoranda of agreement with two utility companies in 1990 to provide passage at the five lowermost main-stem dams and the Northfield Mountain pumped storage plant (Massachusetts). The need for downstream passage on tributaries and on the main stem upstream of the Wilder Dam increased with the dramatic expansion of fry stocking in the early 1990s. Currently, downstream passage facilities exist at the lowermost seven dams on the main stem and at 48 tributary dams. The Connecticut River Atlantic Salmon Commission has stated that some of these facilities require improvements and an additional 15 dams require downstream passage. There are no significant diversions of water other than for hydroelectricity that pose a risk for down- stream migrating fishes in the Connecticut River system.

3. *Reintroduction into Habitat*. The native stocks of Connecticut River Atlantic salmon had been extinct for more than 150 years when the restoration program began. There were few sources of Atlantic salmon eggs available to the program, and out of necessity, eggs were imported from wherever they were available, usually in small numbers. Since salmon imprint on the chemical signature of the water in which they smoltify (Hasler and Scholz 1983), salmon stocked into the Connecticut River were expected to return to the Connecticut rather than their river of genetic origin. Small numbers of salmon fry, parr, and smolts were stocked beginning in 1967 (Table 3).

a. *Hatchery smolt stocking*—Hatchery smolts have high egg-to-smolt survival rates and can be released below dams. The USFWS had previous experience growing Atlantic salmon smolts at two salmon hatcheries in Maine and began raising smolts for the Connecticut River program at the Pittsford National Fish Hatchery (PNFH) in Vermont. The states produced some smolts at existing trout hatcheries. Initially, most hatchery smolts were 2 years old. In 1981, the USFWS constructed the White River National Fish Hatchery (WRNFH), a large facility in Bethel, Vermont dedicated to salmon production. When WRNFH began operation, PNFH reverted to production for the Great Lakes programs. A decision to increase the numbers of smolts switched the production from 2-year smolts to yearlings in 1983. This accelerated rearing regime also resulted in the stocking of large numbers of parr that were graded out and stocked because they were too small to become smolts in 1 year. These fish experienced poor survival and produced very few adult returns. The White River National Fish Hatchery, Kensington State Salmon Hatchery (KSSH) in Connecticut, and the Roger Reed State Fish Hatchery (RRSFH) in Massachusetts provided yearling smolt production for the program from 1983 to 1995. The Roger Reed State Fish Hatchery was converted from smolt to broodstock production in 1987 because of its limited capacity and the program's greater need for eggs for fry stocking. Poor adult returns from KSSH smolts (likely due to the constant water temperature of 10°C that impeded full smolti-fication) led to the conversion of KSSH from smolt to captive broodstock production in 1993 to supply eggs for fry stocking. The White River Na-

TABLE 3. Summary of the stocking and adult return rates of Atlantic salmon fry and hatchery-reared smolts.

Year	Atlantic salmon smolts Stocked (in 1,000s)	Atlantic salmon smolts Returned (per 1,000)	Atlantic salmon fry Stocked (in 1,000s)	Atlantic salmon fry Returned (per 1,000)
1967	0	0	3	0
1968	5	0	50	0
1969	11	0	0	0
1970	0	0	0	0
1971	18	0	60	0
1972	18	0.06	0	0
1973	33	0.09	0	0
1974	54	0.04	16	0
1975	73	0.10	32	0
1976	35	2.57	27	0
1977	99	0.61	50	0
1978	94	1.89	50	1.40
1979	145	3.57	54	0.56
1980	52	1.21	286	0.63
1981	79	0.53	168	1.13
1982	209	0.31	294	1.57
1983	98	3.06	226	0.09
1984	312	0.90	584	0.05
1985	255	1.35	422	1.11
1986	291	0.32	176	1.59
1987	206	0.29	1,169	0.44
1988	395	0.58	1,310	0.83
1989	218	0.78	1,243	0.54
1990	476	0.74	1,346	0.51
1991	351	0.40	1,724	0.20
1992	313	0.84	2,009	0.59
1993	383	0.42	4,147	0.45
1994	375	0.39	5,979	0.49
1995	1	0	6,818	0.21
1996	12	0	6,675	0.15
1997	1	0	8,526	0.04[a]
1998	2	0	9,119	0.05[a]
1999	23	0	6,428	0[a]
2000	49	0.08[1]	9,328	–[b]
2001	0	0	9,585	–[b]
2002	0	0	7,278	–[b]

[a] Additional adult returns are possible for the 2000 year-class of hatchery smolts and for the 1997 year-class of fry. Additional adult returns are likely for all fry year-classes subsequent to 1998.
[b] Adult returns have not yet begun for fry year-classes 2000–2002.

tional Fish Hatchery initiated captive broodstock production in addition to smolt production in 1992 also to support the program's expanding need for fry (see below). Smolt production ceased at WRNFH in 1994 due to disease and budgetary issues. No hatchery smolts were reared between 1994 and 1999. During this period, CRASC searched for alternative facilities for smolt production.

In 1998, the USFWS agreed to raise 100,000 smolts (primarily 2 year old) for the program at PNFH. In 1999, the larger members of that cohort were stocked as yearlings. The remaining fish were stocked as 2 year olds in 2000. This smolt program was interrupted in 2000–2002 by an outbreak of furunculosis at the facility. Smolt production resumed in 2002 with 90,000 smolts scheduled for stocking in 2003 and a goal of 100,000 smolts stocked annually, thereafter.

b. *Fry stocking*—Salmon stocked as fry have lower egg-to-smolt survival than hatchery smolts because of high natural mortality experienced in the wild before smolt migration (McMenemy 1995). However, they also have much higher smolt-to-adult survival than hatchery smolts. Early studies in the Farmington River, Connecticut (Orciari et al. 1994) revealed that fry stocking was an effective means of producing smolts in the extensive nursery habitat in the basin. The 1982 Restoration Plan was modified to highlight fry stocking throughout the basin with a goal of maximizing wild smolt production. However, fry were stocked into only minor portions of the available habitat prior to 1987 because of limited numbers of available eggs. Fry stocking exceeded one million fry for the first time in 1987 and included major portions of most tributaries in the lower basin. Fry stocking continued to expand as egg supplies from Connecticut River captive broodstock increased and by 1998, close to 10 million fry were being stocked annually (Table 3). Most of the fry are stocked when their yolk sacs are nearly absorbed and they are about to begin exogenous feeding. Fry produced at KSSH are fed because warm wellwater results in yolk sac absorption before stream conditions allow for successful stocking. Studies have shown little difference in survival between fed and unfed fry (McMenemy 1995; Whalen and LaBar 1998). Rates of production and probably overwinter survival vary greatly over time and space within the basin; however, stocked fry have generally performed well with survival rates and density estimates similar to those reported in the literature from other regions (Orciari et al. 1994; McMenemy 1995; CRASC, unpublished data). Wild smolt production increased in response to increased fry stocking (Figure 2). The numbers shown in Figure 2 summarize estimates derived from electrofishing as-

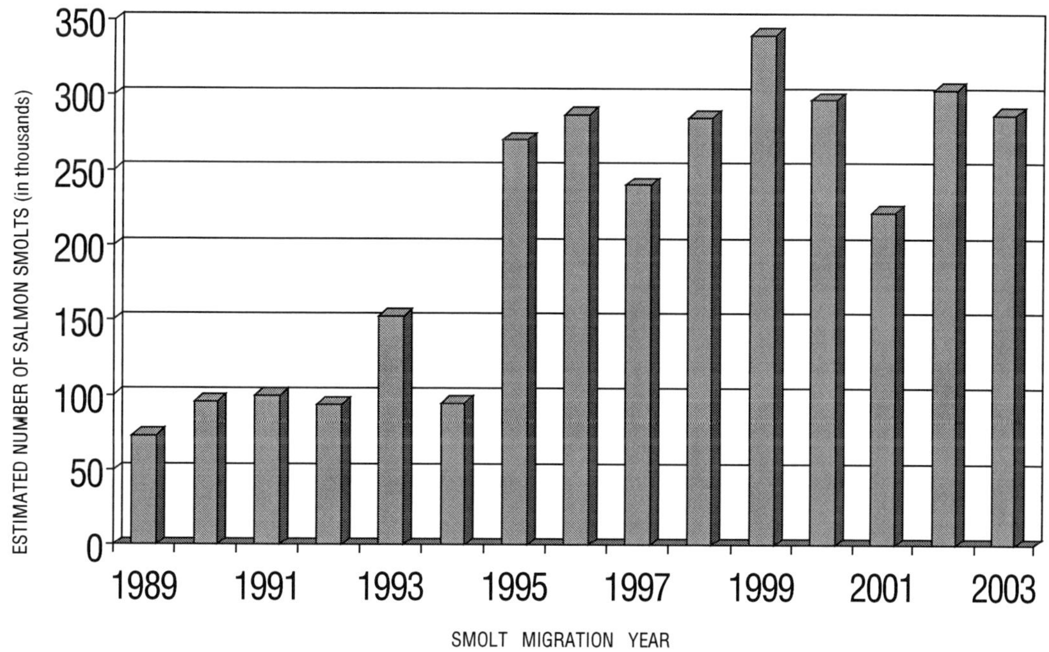

FIGURE 2. Index of Atlantic salmon smolts produced from stocking fry in the Connecticut River basin (CRASC, unpublished data).

sessments by the four states because it is not possible to enumerate all smolts departing the river, but increases in production have been obvious. Increased numbers of fry were used to stock habitat previously unstocked, and the increase in production depicted in Figure 2 reflects the increase in the amount of productive habitat.

4. *Sea Returns.*
   a. *Broodstock capture strategy*—The rearing program was designed to develop a strain of salmon that is adapted to the particular conditions of the Connecticut River and nearshore environment. Salmon reared in the wild from fry and returned to the river as adults have survived these conditions and presumably have more adaptive genes than did stocked fish that did not survive to adulthood. The use of these returning fish as broodstock over generations will lead to the development of a new strain of salmon adapted to the Connecticut River. Returning adult salmon captured at fishways at the Holyoke Dam and Salmon, Farmington, and Westfield rivers are routinely used for broodstock.
   b. *Spawning escapement*—The importance of returning salmon for broodstock and the limited numbers of returns have led to the decision to release few adult salmon upstream from the fishways. Beginning in 1987, 10% of returning salmon reaching Holyoke Dam were released after capture and data and tissue sampling and allowed to migrate upstream. These "releases" at Holyoke have resulted in as few as 4 to as many as 38 salmon released annually. Monitoring the movement of these released salmon was limited to observations at upstream fishways and limited snorkeling surveys through 1997. Beginning in 1998, most adult salmon released at Holyoke have been radio-tagged. These have been tracked both upstream and downstream into tributary systems in all four states. Successful spawning has been confirmed in the West River (#16 in Figure 1), the Westfield River (#4), and the Salmon River (#2).
   c. *Kelt reconditioning*—The high value of sea-return adult salmon combined with the extremely low return rate of kelts (post-spawned adult salmon), particularly in southern New England, led to the decision to recondition the kelts for egg production. This process was pioneered in Nova Scotia in the late 1970s (Ron Gray, Canada DFO, personal communication) and was recommended to CRASC early in the program (Calaprice and Hogsett 1975). The Connecticut Department of Environmental Protection (CTDEP) initiated a pilot project at the Quinebaug Valley Trout Hatchery in 1980. The program was transferred to the Whittemore Salmon station in 1981 and has maintained between 20 and 100 kelts. The USFWS began reconditioning kelts at the Berkshire National Fish Hatchery in Great Barrington, Massachusetts in 1982. When that facility was closed in 1994, the program was transferred to the North Attleboro National Fish Hatchery in North Attleboro, Massachusetts. It has maintained a kelt population between 100 and 200. These kelt facilities have annually produced between 310,000 and 670,000 eggs.

      Male kelts were rarely successfully reconditioned and, as a result, few males were retained. Efforts to synchronize spawning time by use of hormone implants begun in 2001 show promise for increasing the success rate of male reconditioning.
   d. *Adult return rates*—Adult salmon return from the ocean as a result from stocking either hatchery-produced fry or smolts into the habitat. Return rates are calculated by dividing the adults from each source (as determined by scale analysis and/or tag recovery) by the number of juveniles of each source stocked in the appropriate previous year. Return rates of hatchery smolts have been lower than those observed in the Penobscot River, Maine, but follow the same trends from year to year (Friedland 1994). Return rates peaked at over three per thousand smolts stocked in 1979 and 1983 but have not exceeded one smolt per thousand since 1985. In general, return rates have declined over the last 30 years both in the Connecticut River and throughout much of the range of Atlantic salmon (Friedland 1994). Return rates for fry are lower, but marine survival of smolts produced from stocked fry is about 10 times higher than hatchery smolts (Rideout and Stolte 1988;

Meyers 1994). Fry-to-adult return rates have exceeded one per 10 thousand in early years when numbers of fry released were relatively low, but have not exceeded one per 10 thousand since 1986. Fry to adult survival has declined precipitously since 1995 despite high survival of fry and parr in freshwater and continual improvements in downstream passage. This decline is believed to be due to marine conditions since similar trends have been observed elsewhere in the range of Atlantic salmon. Comparison of return rates for smolts and fry is difficult since there is much natural mortality of fish between the fry and smolt stages and there is no reliable way of counting smolts produced by fry stocking as they leave the river.

5. *Genetic Considerations.* Managing the genetic resources of a salmon population may be the most important task for managers seeking to reestablish a run of salmon into a river that lost its native run. The Connecticut River presents great challenges for genetic management because it is a large, diverse watershed. The environments of the southern tributaries (e.g., Salmon River at rkm 26, latitude 41°N) are considerably different than those of the northern tributaries (e.g., Ammonoosuc River at rkm 428, latitude 44°N). The native run of salmon most likely included multiple stocks. In addition, the Connecticut River is near the southern edge of the species' North American range and requires all migrants to pass through marine waters that reach higher temperatures than anywhere else in the species' global distribution. It is possible that the native stocks possessed unique genes that are now lost. Saunders (1981) recognized the special requirements of a stock adapted to the Connecticut River.

The Policy Committee commissioned a report to make recommendations on genetic policies for the restoration. Calaprice and Hogsett (1975) recommended that eggs used to generate fry and smolts be imported: (1) from stocks originating from the geographically closest rivers, (2) from multiple stocks to allow natural selection to act upon a wide diversity of genes. Following the first recommendation was difficult because at that time Atlantic salmon runs had been extirpated from all nearby rivers in southern New England and all but seven small rivers in eastern Maine. None of those rivers had effective broodstock collection infrastructure (Ed Baum, Maine Atlantic Salmon Commission–retired, personal communication). The restoration program on the Penobscot River had begun only 2 years previous to the start of the Connecticut River program, and there were no spare eggs to share with the Connecticut River program (R. A. Jones, Director, CTDEP/Fisheries Division–retired, personal communication). The first salmon eggs used by the program originated from rivers in Newfoundland and the Gulf of St. Lawrence (Rideout and Stolte 1988), which are geographically distant from the Connecticut River. Only 11 adult salmon returned to the Connecticut River during all of the years prior to the use of the Penobscot River stock (PRS), which first returned to the Connecticut River in 1978, when 90 salmon returned.

A thorough discussion of the genetic background and management of the restoration program is beyond the scope of this paper. However, several important points can be made:

- Many of the early returns from nonPenobscot River stocks (NPRS) did not survive until spawning due to fish health problems in hatcheries, and therefore their genes did not contribute to future generations of the Connecticut River Stock (CRS).
- We assume that a large percentage of returns to the river after 1977 were from imported Maine stocks. In addition to the PRS, fish from the Union River stock (URS) from Maine were used.
- In years when PRS, URS and NPRS stocks would have been expected to return, the marking of individual stocks was ineffective, and it is impossible to conclude how many representatives of each group contributed to future generations of the CRS.
- All contributions from outside river stocks ceased after 1994 and all subsequent stockings have been CRS-based (i.e., all parents returned to the Connecticut River).
- The genetic backgrounds of the PRS and URS are complicated. Both were stocked with fish from the Machias and Narraguagus rivers in Maine as well as each other (Baum 1997; Michael Hendrix, USFWS-Craig Brook NFH, unpublished data). The Penobscot, Machias, and Narraguagus riv-

ers were all stocked with New Brunswick stocks (NBS) such as the Miramichi River prior to this mixing of stocks (Baum 1997) and it is impossible to know which stockings succeeded and which stocks contributed significantly to the future identity of the stocks. Spidle et al. (2003) found statistically significant genetic variation among all anadromous Atlantic salmon populations in Maine and of those populations, the Penobscot, Machias, and Narraguagus rivers were genetically the most similar. Martinez et al. (2001) and Spidle et al. (in press) concluded that the CRS is now significantly different than the PRS, which is often considered as the largest contributor to the founding of the CRS although this assumption may underestimate the contributions of the URS. More research is needed to determine whether this difference is due to adaptation or to different initial contributions to the two founding populations.

Several genetic management strategies have been considered in recent years with the goal of improving return rates. These include starting over with a completely new donor stock(s), adding contributions from different stocks (including southern Europe) to the CRS, and continuing with the existing CRS. The addition of more stocks was rejected to avoid outbreeding depression. The addition of southern European stocks may have contributed some genes adaptive to southern latitudes but also some genes maladaptive to North America. The decision to continue using the existing CRS was based on the fact that

- it has succeeded in generating adult returns;
- the closest salmon river to the Connecticut is the Penobscot;
- there is evidence that the CRS has differentiated from the PRS and if such differences are adaptive, re-injecting present day PRS would equate to going backwards;
- knowledge of Atlantic salmon genetics is inadequate to allow informed choice of stocks that would perform well in the Connecticut River; and
- state, federal, and international fish health and genetic concerns and agreements limit which stocks may now be transferred across political boundaries (NASCO 2002).

In 1988, a decision was made to improve the management of the existing CRS by

- using all available sea-returns, including grilse, for mating;
- randomizing and equalizing the use of all available broodstock;
- using only fish that returned from the sea as parents for future domestic broodstock;
- maximizing the number of crosses possible, given the number of available broodstock;
- increasing the number of matings that cross year classes (e.g., cross sea-returns with kelts, mature parr, and younger and older year-classes of domestic broodstock); and
- maintaining a sea-run broodstock population equal to or greater than 50 pairs.

The randomized cross protocol (second bullet above) was terminated in 1997 when DNA fingerprinting of sea-run broodstock made possible the determination of the genetic relationship between captive adults (Letcher and King 1999, 2001). Currently in use is a nonrandomized (nonblind) mating scheme in which only broodstock that are not closely related (sharing a relatively large number of common indicator genes) are mated.

6. *Control of Harvest.* Initially, rewards were offered to anyone capturing and reporting an adult Atlantic salmon in the Connecticut River. This was done to document any sea returns. Once sea returns became common in the 1980s, all four basin states and the state of New York enacted regulations that prohibited the taking of any life phase of Atlantic salmon by any means within the Connecticut River basin and Long Island Sound. Occasional rod catches and catches in the in-river American shad drift gillnet commercial fishery have occurred, but many of these fish were released alive and the resulting mortality is believed to have been low. In 1987, the New England Fisheries Management Council adopted an Atlantic salmon management plan that prohibited the taking of Atlantic salmon in U.S. territorial waters (NEFMC 1987).

In 1983, the U.S. signed an international treaty as part of the creation of the North Atlantic Salmon Conservation Organization (NASCO). This regulated the taking of salmon in international waters and negotiated harvest levels in key interceptor fisheries (NASCO

2002). The North Atlantic Salmon Conservation Organization has prohibited the taking of Atlantic salmon in international waters and reduced the annual harvest of Atlantic salmon in the West Greenland fishery from 870 tons in 1984 to 22–55 tons (depending upon early CPUE) in 2002. In some years (including 2002), the harvest at West Greenland has been zero due to the "buy out" of the NASCO quota by private conservation organizations. Interceptor fisheries in Canada (e.g., Nova Scotia, Newfoundland, and Labrador) have all been closed during the lifetime of the restoration program. Tagging studies reveal that Connecticut River salmon were caught in the West Greenland fishery (Friedland 1994), but the current role of any distant interceptor fishery in reducing the returns to the Connecticut River is unclear.

Current Status

Atlantic salmon is currently very abundant in juvenile life stages in at least 33 tributary subdrainages throughout the basin due to fry stocking. Adults return to the river from the sea annually in varying numbers (Table 4), but the long-term survival of this run is not assured.

Discussion

1. *Major Ecological Changes in the River since the Time of Extirpation.* Two major challenges confront the task of restoring Atlantic salmon to the Connecticut River. First, the Connecticut River is at the southern extent of the range, and the saltwater off of its mouth is warmer than that of all other Atlantic salmon rivers in the world. Moreover, the mouth of the river is geographically isolated and distant from the mouths of all other salmon rivers in North America. The native strain of Atlantic salmon clearly possessed genes adapted to these unique conditions, but it is not known whether those genes (or other suitable ones) persist in the North American gene pool, much less the Penobscot River gene pool on which the restoration program relies so heavily. Second, there have been many changes in the Connecticut River basin since the extirpation of Atlantic salmon in 1810, including massive deforestation, urbanization, loss of habitat, warming of fresh, estuarine, and salt waters, the loss of many native species, and the addition of many nonnative species. Had a viable run of Atlantic salmon persisted to the present day, it may well have adapted to these changes, as have runs in Spain and France at the southern extent of the species' European range. The capacity to "genetically engineer" a new strain of salmon adapted to current conditions is extremely limited. It remains to be seen whether the traditional selective breeding strategy of animal husbandry will meet the challenge.
2. *Restoration during a Period of Global Decline in Atlantic Salmon Stocks.* The causes of the extirpation of Atlantic salmon are to be found in its freshwater habitat. However, as these problems in the basin are addressed, it is clear that there are new problems for the species in its marine habitat. Atlantic salmon runs are declining worldwide, including runs from pristine rivers (Parrish et al. 1998). There is much speculation that the source of the problem is marine in origin (Friedland et al. 1993; Friedland 1998). This makes the effort to restore Atlantic salmon to the Connecticut River even more challenging because these marine factors are beyond the control of CRASC.
3. *Impacts to Salmon Restoration by the Connecticut Yankee Atomic Power Plant.* No studies were undertaken to determine the impact of the Connecticut Yankee Atomic power plant (CY) on the Atlantic salmon in the Connecticut River. No rearing habitat exists in the main-stem Connecticut River anywhere near CY, and therefore any potential impact to salmon would be limited primarily to effects on the outgoing smolts. The original study documented that the warmwater discharge of the plant created a limited and discreet thermal plume that adult American shad could easily avoid (Leggett 2004 [1976], this volume). It is likely that adult Atlantic salmon can do likewise. Atlantic salmon smolts are known to migrate in the main river channel (Fried et al. 1978), which in the area of CY is located on the opposite side of the river from the plant, causing most to avoid the plant's intake. During the years of 1968–1982 when CY impingement data were regularly reported, only eight Atlantic salmon smolts were reported from the plant (Jacobson et al. 2004b, this volume) despite in excess of half a million Atlantic salmon smolts migrating downstream during those years.

Table 4. Documented adult returns of Atlantic salmon to the Connecticut River during the restoration program.

	Capture locations					
Year	Holyoke[a]	Rainbow[b]	Leesville[c]	DSI[d]	Misc.[e]	Total
1967–1973	0	*	*	*	0	0
1974	1	*	*	*	0	1
1975	1	*	*	*	2	3
1976	2	*	*	*	0	2
1977	2	0	*	*	5	7
1978	23	56	*	*	11	90
1979	19	32	*	*	7	58
1980	126	26	1	*	22	175
1981	319	62	118	*	30	529
1982	11	41	11	*	7	70
1983	25	14	0	*	0	39
1984	66	6	11	*	9	92
1985	285	9	5	*	11	310
1986	260	39	12	*	7	318
1987	208	126	10	*	9	353
1988	72	14	5	*	4	95
1989	80	24	3	*	2	109
1990	188	37	36	*	2	263
1991	152	33	11	*	7	203
1992	370	97	18	2	3	490
1993	169	14	0	10	5	198
1994	262	42	12	7	3	326
1995	151	22	7	6	2	188
1996	202	29	4	21	4	260
1997	96	60	3	39	1	199
1998	197	50	3	47	3	300
1999	91	36	9	18	0	154
2000	52	6	8	11	0	77
2001	24	6	2	8	0	40
2002	34	4	0	5	1	44

[*] Trapping facilities not yet constructed.
[a] Holyoke Dam Fishlift, Connecticut River, Holyoke, MA, rkm 139.
[b] Rainbow Dam Fishway, Farmington River, Windsor, CT, rkm 101.
[c] Leesville Dam Fishway, Salmon River, East Haddam, CT, rkm 32.
[d] Decorative Specialties Industries Dam, Westfield River, Westfield, MA, rkm 140.
[e] Misc. reflects angling catches and live and dead fish observed at various locations.

Atlantic salmon is a coolwater species and any systematic warming of the river, which is on the southern edge of the species' range, could negatively impact the species' ability to adapt to the river. However, Jacobson et al. (2004a, this volume) showed that the river experienced a general warming upstream of the plant during the past 30 years. Therefore, it is difficult to conclude that CY caused a warming that was deleterious to Atlantic salmon.

American Shad

Status Prior to the Restoration Program

This species was very abundant in the river at the time of European contact, and by the American Revolution (1770s), a significant commercial fishery became established in the Connecticut portion of the river (Judd 1905). The species experienced declines in the early 1800s after the first main-stem dams were constructed and impressions of population size between the 1880s and 1950s are based on fluctuating commercial landings (Mansueti and Kolb 1953). The species remained common in the river at the time the restoration program was initiated. For more information on the history of the run, see Savoy et al. (2004, this volume) and Leggett et al. (2004, this volume).

Restoration Strategies

1. *Targeted Habitat.* The entire native range of American shad within the main-stem river (the mouth up to Bellows Falls, Vermont) is targeted for restoration as are significant portions of larger tributaries such as Eightmile, Mattabesett, Hockanum, and Farmington rivers (Connecticut), Westfield River (Massachusetts), West River (Vermont), and Ashuelot River (New Hampshire).
2. *Fish Passage.* American shad is the primary target species for most fishways built in the Connecticut River basin, downstream of Bellows Falls, Vermont. American shad is a strong swimming species that does not readily ascend many types of fishways. Fishways in the basin that effectively pass American shad always pass all other targeted anadromous fish species. Therefore, shad is the focus of much fish passage research. The greatest challenge in passing American shad upstream has been at the 12-m-high Turners Falls Dam (Massachusetts, rkm 198), where there are three fishways. Two of the three are "Ice Harbor style" pool-and-weir fishways that have been modified over the years to improve shad passage. More studies and possible modifications are proposed.
3. *Reintroduction of Species into Habitat.* Hatcheries were operated for the culture of American shad beginning during the restoration program of the late 1800s and extending to 1938 (Moss 1952). There have been no aquaculture activities for American shad associated with the current restoration program because there has been no shortage of shad in the lower river.. Connecticut River Atlantic Salmon Commission member agencies have employed the transplantation of prespawned adult American shad from the Holyoke Dam fishlift to upstream portions of the basin targeted for shad restoration. This allows juvenile production in unutilized or underutilized habitat and creates a subpopulation of fish that is imprinted to that portion of the basin. Adult shad have been trucked into the main-stem Connecticut River between the Vernon and Bellows Falls dams, the Ashuelot River (New Hampshire), and the Farmington, Hockanum, and Eightmile rivers (Connecticut).

Current Status

American shad is very abundant in the basin and continues to support popular sport and commercial fisheries, although both fisheries have declined in recent years for reasons beyond stock abundance. A general trend of abundance over time is reflected in the passage data from the Holyoke Dam Fishlift (Table 5).

Discussion

American shad have migrated to the base of Bellows Falls (the extent of its historical range) and beyond (due to the Bellows Falls Fishway, built for Atlantic salmon), but the numbers that have passed the Turners Falls, Vernon, and Bellows Falls fishways are so small that the habitat in the pools upstream of these dams is very much underutilized by the species. The restoration of American shad to the Connecticut River has been delayed by the inability to provide satisfactory passage around the Turners Falls Dam. Only 2.45% (recent 6-year average) of the shad passed at Holyoke Dam passed the Turners Falls Dam whereas 73.3% (same 6-year average) of those fish passed the next upstream dam at Vernon, Vermont.

It is difficult to assess the role of fish passage in the population trends of American shad in the river because there have been other factors occurring during the same time period. See Savoy et al. 2004 and Leggett et al. 2004 for a more complete discussion of these issues. Savoy and Crecco (2004, this volume) discuss the role of the recovery of the stocks of striped bass *Morone saxatilis* in the population trends of the American shad.

There have been no investigations into the role recent environmental trends in the Connecticut River have played in the recent declines of American shad. These include a general warming trend in water temperature (Jacobson et al. 2004a), the shift in ictalurid dominance in the river from white catfish *Ameiurus catus* to channel catfish

TABLE 5. Adult returns of anadromous fish to the Holyoke Dam Fishlift, Holyoke, Massachusetts: 1955–2002 (0–999 to the nearest individual; 1,000–9,999 to the nearest 100; 10,000–99,9999 to the nearest 1,000; and greater than or equal to 100,000 to the nearest 10,000; all data except 2003 and for shortnose sturgeon are from Caleb Slater, Massachusetts Division of Fish and Wildlife; sturgeon data for 1955–1974 are from Robert Stira, Northeast Generation Services; sturgeon data for 1975–1996 are from Kynard (1998); sturgeon data for 1997–2002 and all data for 2003 are from Jan Rowan, USFWS; and sturgeon counts for 1955–1957 and 1968–1974 are not available).

Year	American shad	Blueback herring[a]	Atlantic salmon	Striped bass	Sea lamprey[b]	Gizzard shad	Shortnose sturgeon
1955	4,900	–	–	–	–	–	*
1956	7,700	–	–	–	–	–	*
1957	8,800	16	1	–	–	–	*
1958	5,700	29	1	–	2	–	0
1959	15,000	20	–	–	73	–	0
1960	15,000	796	2	–	17	–	0
1961	23,000	1,200	–	–	42	–	0
1962	21,000	19	–	–	209	–	1
1963	31,000	32	–	–	64	–	0
1964	35,000	13	–	–	537	–	1
1965	34,000	53	–	–	26	–	2
1966	16,000	54	–	–	2	–	1
1967	19,000	356	–	–	46	–	0
1968	25,000	–	–	–	–	–	*
1969	45,000	10,000	–	–	–	–	*
1970	66,000	1900	–	–	–	–	*
1971	53,000	302	–	–	–	–	*
1972	26,000	188	–	–	–	–	*
1973	25,000	302	–	–	–	–	*
1974	53,000	504	–	–	–	–	*
1975	110,000	1,600	–	–	23,000	–	5
1976	350,000	4,700	–	–	32,000	–	3
1977	200,000	33,000	2	–	52,000	–	0
1978	140,000	38,000	23	–	43,000	–	1
1979	260,000	40,000	19	103	31,000	–	3
1980	380,000	200,000	126	148	34,000	–	0
1981	380,000	420,000	319	510	53,000	–	4
1982	290,000	590,000	11	231	26,000	–	4
1983	530,000	450,000	25	346	29,000	–	4
1984	500,000	480,000	66	110	22,000	–	10
1985	480,000	630,000	285	369	40,000	–	6
1986	350,000	520,000	260	187	20,000	27	13
1987	280,000	360,000	208	521	23,000	94	3
1988	290,000	340,000	72	256	16,000	95	4
1989	350,000	290,000	80	923	15,000	294	4
1990	360,000	390,000	188	1,000	22,000	956	5
1991	520,000	410,000	152	1,200	41,000	486	0
1992	720,000	310,000	368	327	28,000	1,100	4
1993	340,000	100,000	167	194	23,000	341	6
1994	181,000	32,000	256	159	30,000	165	1
1995	190,000	110,000	150	1,300	18,000	2,000	1
1996	276,000	55,000	202	537	45,000	1,100	16
1997	299,000	64,000	94	679	32,000	2,100	0
1998	316,000	11,000	196	492	97,000	1,100	25
1999	194,000	2,700	91	859	20,000	35,000	1

TABLE 5.Continued

Year	American shad	Blueback herring	Atlantic salmon	Striped bass	Sea lamprey	Gizzard shad	Shortnose sturgeon
2000	225,000	11,000	50	489	21,000	38,000	0
2001	273,000	11,000	25	1,200	49,000	5,500	4
2002	375,000	2,000	34	1,100	74,000	3,100	0
Total	9,688,100	5,921,784	3,473	13,240	960,018	91,458	132

[a] *Alosa aestivalis*
[b] *Petromyzon marinus*

Ictalurus punctatus (Jacobs et al. 2004, this volume), the increase in numbers of other introduced predators such as northern pike *Esox lucius*, walleye *Sander vitreus*, and smallmouth bass *Micropterus dolomieu* (Jacobs and Hyatt 2004), and the dramatic increase in numbers in the river by gizzard shad *Dorosoma cepedianum* and hickory shad *Alosa mediocris* (see below). It is possible that the recent decline of anadromous clupeids in the Connecticut River may be due to the synergistic effects of marine factors such as striped bass recovery, the colonization of the river by gizzard shad, and fundamental changes in the river's food web.

Alewife

Status Prior to the Restoration Program

Alewife *Alosa pseudoharengus* was common in the main-stem river south of the Enfield Dam (#3 in Figure 1). Early fishermen and fisheries agency officials tended to lump both alewife and blueback herring together as "alewife" or "river herring," so it is difficult to distinguish historical abundance of alewife and blueback herring.

Restoration Strategies

1. *Targeted Habitat.* The CTDEP has targeted portions of tributaries within Connecticut (including many small brooks) for alewife restoration. The Farmington River above the Rainbow Dam is the largest tributary targeted. The only tributary of the Connecticut River outside of the state of Connecticut where alewife is known to spawn is Raspberry Brook, just north of the Connecticut–Massachusetts border (Bruce Kindseth, Fanny Stebbins Nature Preserve, East Long-meadow, Massachusetts; personal communication). CRASC has not targeted additional waters for alewife restoration since there is no evidence that the species is being denied access to any waters within its historic range within Massachusetts, Vermont, and New Hampshire.
2. *Fish Passage.* Seven of the eight fishways built at dams on Connecticut tributaries are intended to pass alewife. Many more are planned. The "steeppass" design is commonly used at small dams.
3. *Reintroduction of Species into Habitat.* The state of Connecticut is transplanting prespawned adult alewife from the Pattaganset River in southeastern Connecticut into five Connecticut River tributaries. Fish are seined and dipped from the donor stream, loaded into a transport truck at loads of 400, and released above dams targeted for future fish passage. Fall visual surveys at dusk for surface-feeding juveniles usually confirm the presence of young-of-the-year alewives in the recipient habitat, confirming successful reproduction. Transplants are conducted to allow upstream production in advance of fish passage projects and to accelerate restoration programs in streams with fishways.

Current Status

While alewife are common, the population may be declining. No population estimate is available for alewife in the Connecticut River since the species does not ascend as far as the Holyoke Dam. The number of alewives entering the river annually probably increased between 1967 and 1993, based on personal observations, but likely decreased subsequent to 1993. Alewife has been able to expand its range in four tributaries where fishways have been constructed. However, these fishways lack counting facilities so changes in population size cannot be documented.

Discussion

Alewife is one of the more difficult diadromous species in the river to track because significant numbers do not use any of the primary fish passage facilities (Holyoke Dam Fishlift, DSI Dam Fishway, Rainbow Dam Fishway, and Leesville Dam Fishway) that are used to monitor the size of runs. The species' historical range extended upstream only from the Rainbow Dam on the Farmington River, but the passage of alewife at this facility is so low and inconsistent (due to improper fishway design) that it cannot be used to monitor population trends. Alewife spawn in the many coves and backwaters of the main-stem river (e.g., Chapmans Pond, Deadman's Swamp, and Hamburg, Salmon, Keeney, and Wethersfield coves) and does not rely as much on tributary stream habitat as other species.

Despite the paucity of data with which to monitor population trends, qualitative visual observations and angler reports to the CTDEP indicate that the population size of alewife has probably followed the same general trend that has been documented for American shad and blueback herring: a general increase from the late 1960s to the early 1990s and then a marked decrease thereafter (see Table 5 for annual counts of the other two species).

Some of the same issues that were listed as possible factors for the decline in numbers of American shad are also likely factors in the decline in alewife numbers. Alewife is also targeted by anglers, who use dip nets to capture alewife mostly for bait in the striped bass sport fishery. Alewife run up the river from very late March to very early May and striped bass commonly enter the river in early May. Not only does the alewife population appear to overlap with the striped bass run to a lesser extent than does the blueback herring run, but the alewife run appears to have been exploited less by the dipnetters than the blueback herring run. It appears from the little data available that the alewife run has not experienced as great of a decline as the blueback herring run. The general trend of alewife populations size summarized for the Connecticut River is also apparent throughout the rest of the state of Connecticut, the Hudson River (K. A. Hattala, NYSDEC-Bureau of Marine Resources, personal communications) and parts of Rhode Island (Phil Edwards, RIDEC- Fish and Wildlife Division, personal communication). The CTDEP imposed a total statewide prohibition on the taking of either species of river herring, effective 1 March 2002, to conserve the dwindling stocks of river herring.

Blueback Herring

Status Prior to the Restoration Program

Blueback herring were common in the main-stem river and tributaries up to natural falls. The native range on the main stem extended to Bellows Falls, Vermont, but at the beginning of the restoration program, it was restricted to south of the Holyoke Dam (Figure 1). Early fishermen and fisheries agency officials tended to lump both alewife and blueback herring together as "alewife" or "river herring," so it is difficult to distinguish historical abundance of alewife and blueback herring.

Restoration Strategies

1. *Targeted Habitat.* CRASC has targeted much of the historic range of blueback herring for restoration, including the main-stem river as far upstream as Bellows Falls, Vermont and significant portions of the Eightmile, Farmington, Westfield, West, and Ashuelot rivers. Lower portions of smaller streams in Connecticut (e.g., Roaring, Pewterpot, and Salmon brooks) and Massachusetts (e.g., Fort River) have also been targeted.
2. *Fish Passage.* Fishways at the lowest three dams on the main stem and three of the eight fishways built at dams on Connecticut tributaries are intended to pass blueback herring. Many more tributary fishways are planned. At small dams, "steeppass" fishways are commonly used. The Rainbow Dam Fishway on the Farmington River is a vertical slot design but only passes

small numbers of blueback herring due to the design. There are plans to replace the vertical slot fishway with a Denil fishway, which should pass blueback herring more effectively.

3. *Reintroduction of Species into Habitat.* The Massachusetts Division of Fisheries and Wildlife (MADFW) and the New Hampshire Fish and Game Department, in cooperation with the USFWS, are transplanting prespawned adult blueback herring from the lower Chicopee River (Massachusetts tributary) into the Westfield River (#4, Figure 1) and the Ashuelot River (#15, Figure 1). This is done to accelerate restoration programs in upstream tributaries by creating subpopulations that are imprinted to these waters and will likely have a stronger urge to continue upstream.

Current Status

Blueback herring is common downstream of the Turners Falls Dam and the first dam on most tributaries, but in serious decline (Table 5). It is present in low numbers between the Turners Falls Dam and the Bellows Falls Dam.

Discussion

It is apparent that, historically, the blueback herring penetrated much farther inland on all streams, utilizing more habitat than alewife. It is our opinion that the native blueback herring population in the river greatly exceeded that of the alewife. During the mid-1960s when the water pollution of the river was severe, large numbers of adult blueback herring died regularly during summer nights. This was attributed to the regular depletion of dissolved oxygen in the river (Moss et al. 2004 [1976], this volume). Water pollution was greatly reduced beginning in the 1970s (Anonymous 1984) and such mass mortalities ceased. Fish passage was provided at several dams during the 1970s and 1980s, resulting in an increase in the size of the annual blueback herring run. The number lifted over the Holyoke Dam peaked in 1985 at 630,000. Since that time, the run has steadily declined, and in 2002, the number lifted was 1,939, a reduction of 99.7% (Table 5).

Some of the same issues that were listed as possible factors for the decline in numbers of American shad (see Savoy and Crecco 2004) are also likely factors in the decline in blueback herring numbers. In addition, blueback herring is also targeted by anglers who use dip nets to capture herring, mostly for bait in the striped bass sport fishery. The blueback herring run appears to overlap with the striped bass run to a greater extent than does the alewife run, and the species appears to have been exploited more by the dipnetters than the alewife run. It appears that the blueback herring run has experienced a greater decline than the alewife run.

Commercial harvest in the Atlantic Ocean is another possible cause of blueback herring decline. Since the recovery of Atlantic herring *Clupea harengus* stocks, both effort and landings of Atlantic herring have increased (Anonymous 1998). The bycatch of blueback herring in this fishery is generally low and likely not significant to the coastwide population of blueback herring, but the impact of bycatch by nearshore boats on homewater returns of nearby rivers is unknown (Bill Overholtz, NOAA Fisheries- NEFC, personal communication). There has also been an increase in the landings of blueback herring from offshore fisheries in Connecticut during the 1990s (CTDEP/Marine Fisheries Division, unpublished data) as well as anecdotal reports of the presence of blueback herring in the Atlantic herring catch.

The general trend of blueback herring populations described for the Connecticut River is similar for runs throughout the rest of the state of Connecticut (CTDEP/Inland Fisheries Division, unpublished data). The CTDEP/Inland Fisheries Division imposed a total statewide prohibition on the taking of either species of river herring (inland and marine waters), effective 1 March 2002, to conserve the dwindling stocks of river herring.

Gizzard Shad

Status Prior to the Restoration Program

This species was not present prior to the restoration program. Long Island Sound and southern New England was north of the historical range of the coastal populations.

Restoration Strategies

1. *Targeted Habitat.* None.
2. *Fish Passage.* No fish passage facility within the basin has been designed to specifically pass

gizzard shad since the species was not present when most of the fishways were planned. Gizzard shad currently use the Holyoke Dam Fishlift in large numbers (Table 5) and have used the fishways at Turners Falls, Vernon, and Bellows Falls, ascending more than 280 km up the main-stem river. On tributaries, the species has used the DSI Dam Fishway on the Westfield River in relatively small numbers and ascended the Rainbow Dam Fishway only up as far as the fish trap, where staff remove them to prevent colonization of the upstream portions of the Farmington River. Future fish passage projects may have to take gizzard shad into account during the design process. Even though the species is not targeted for restoration, it will be difficult to exclude them from some fishways, and their abundance needs to be accommodated.

3. *Introduction of Species into Habitat.* There have been no efforts to deliberately introduce this species into habitat, other than allowing it to use fishways constructed for the benefit of other anadromous species.

Current Status

This species is now present in the Connecticut River basin due to a natural range extension during the time period of the restoration program (O'Leary and Smith 1987). It is now abundant in the main stem and the lower reaches of many tributaries in the basin below the Bellows Falls Dam. Juvenile gizzard shad have been documented in the Vernon Pool, attesting to successful reproduction.

Discussion

Gizzard shad were first reported in the tidal portion of the Connecticut River in 1976 by commercial American shad netters, who referred to them as "Geechee River shad" (Whitworth et al. 1980; Peter Minta, CTDEP/Marine Fisheries Division–retired, personal communication). Reports increased during the mid-1980s, and the first gizzard shad was lifted at the Holyoke Dam Fishlift in 1985 (O'Leary and Smith 1987). The species appears to have a very adaptable life history, and each year's run appears to consist of anadromous individuals entering the river from the ocean and potamodromous individuals that overwintered in freshwater. Backwater areas such as Wethersfield Cove in Connecticut and "the Oxbow" in Massachusetts support large, multi-year-class, year-round populations of gizzard shad (O'Leary and Smith 1987; R. P. Jacobs, CTDEP/Inland Fisheries Division, personal communication). It appears that the coastal populations of the species may not be fully adapted to the winters of New England. It experienced large mortalities throughout Connecticut during the cold winter of 2000–2001, and the spring runs of 2001 were much smaller than the preceding 2 years (Table 5).

No overt benefits of gizzard shad have been observed. No user group on the river is advocating its active management. However, since the species is abundant, overlaps with targeted anadromous species in time and space and is difficult to exclude from fishways, it has become a de facto restoration species. Furthermore, CRASC does not have any evidence that the presence of the species is detrimental to the targeted species. Gizzard shad are sympatric with American shad, alewife, and blueback herring in the Mid-Atlantic states (Rohde et al. 1994). Therefore, there may not be any reason to exclude the nonnative gizzard shad from colonizing additional habitat while other species are restored to the same habitat.

Hickory Shad

Status Prior to the Restoration Program

Hickory shad appear to have been sporadically present in Long Island Sound prior to the restoration program (Collette and Klein-MacPhee 2002), but there was no evidence of the species in the Connecticut River during the 1960s and 1970s.

Restoration Strategies

1. *Targeted Habitat.* None.
2. *Fish Passage.* There is no evidence that hickory shad extend inland as far as any dam in the watershed. There is no perceived need to provide fish passage to this species.
3. *Introduction of Species into Habitat.* None.

Current Status

Hickory shad increased in abundance in the lower Connecticut River during the 1990s. It is currently common in the lower 8 km of the river and is commonly recreationally fished as far upstream as the East Haddam Bridge (rkm 26). It has been documented in low densities as far upstream as Hartford (rkm 80) (Tom Savoy, CTDEP/Marine Fisheries Division, personal communication). It is most common between August and November, but there have been unconfirmed reports of hickory shad ascending the river in the spring.

Discussion

It appears that hickory shad is experiencing a northerly range extension similar to that exhibited by gizzard shad. It is possible that hickory shad have been present in Long Island Sound and the lower Connecticut River in the past, but the species became more abundant during the 1990s than at any other time in recent memory. Batsavage (1997) reported large population increases in North Carolina, where the species is native, during the 1990s, indicating that there may be more to the trend than range extension. If the species is present in the Connecticut River during the spring, it raises the possibility that the species is spawning in the river. Richkus and DiNardo (1984) reported that the species did not spawn north of Maryland, and Collette and Klein-MacPhee (2002) reported that the species does not spawn in the New England region. That may no longer be true after the northerly shift of the species during the 1990s. No juvenile hickory shad have been sampled in the river during the annual juvenile clupeid index surveys (Tom Savoy, CTDEP/Marine Fisheries Division, personal communication). Batsavage (1997) speculated that young-of-year hickory shad depart rivers quickly and utilize the ocean as nursery habitat and therefore may not be available to summer juvenile surveys.

Hickory shad have not been reported in the river upstream of Connecticut and does not warrant the interest of CRASC. The only management activity taken on behalf of hickory shad has been the establishment of a daily creel limit of six for conservation purposes within Connecticut waters. The colonization of the Connecticut River by hickory shad is currently seen as a positive development since the species is highly regarded by anglers. However, there have been no studies of the ecological relationships between hickory shad and other clupeids. The increase in hickory shad occurred during the same time as the increase in gizzard shad and the decrease in American shad and river herring.

Shortnose Sturgeon

Status Prior to the Restoration Program

This native sturgeon was present prior to the restoration program but it has likely not been abundant during anytime in the past 50 years. The historical abundance of this species is not clear due to the lack of a long-term scientific database and the likely confusion by the general public of the shortnose sturgeon with the Atlantic sturgeon *Acipenser oxyrinchus*.

Restoration Strategies

1. *Targeted Habitat.* The historic range of this species within the river appears to have extended from the mouth of the river upstream as far as Turners Falls, Massachusetts (rkm 198). Shortnose sturgeon also likely entered most large tributaries (e.g., Farmington, Westfield, Deerfield rivers) as far upstream as the first falls or heavy rapids. There has been no formal adoption of a restoration plan for this species by CRASC, but the area of the basin described above is considered the de facto targeted habitat.
2. *Fish Passage.* The only fishway that has been targeted for upstream and downstream passage of shortnose sturgeon is at the Holyoke Dam, which is located in between two known spawning areas for the species. Shortnose sturgeon seem to safely pass upstream via the Holyoke lifts, but there is concern that a large percentage of the spawning migrants cannot locate the entrance to the lifts. Downstream passage of sturgeon at Holyoke is thought to be inadequate. It is possible that the species historically ascended above the present-day location of the DSI fishway (Westfield River), but this fishway was not designed for sturgeon passage and the species cannot use this facility.
3. *Reintroduction of Species into Habitat.* There have been no efforts to culture or transplant shortnose sturgeon within the basin.

Current Status

The current population of shortnose sturgeon in the river is an estimated 1,800 individuals (Savoy

2004, this volume). The species is currently listed as "Endangered" under the U.S. Endangered Species Act and under the Connecticut Endangered Species Act.

Discussion

This species is generally referred to as "anadromous" yet evidence indicates that relatively few individuals may venture outside of the mouth of the river (Tom Savoy, CTDEP/Marine Fisheries Division, personal communication). The species engages in long riverine migrations between the salt wedge in Old Saybrook (rkm 13) to the Enfield (rkm 110) and Holyoke (rkm 139) dams (Savoy 2004) that might better be referred to as pota-modromous. Shortnose sturgeon is known to spawn above the Holyoke Dam, near Turners Falls, Massachusetts (Kynard 1998) and are thought to spawn below the Holyoke Dam (Tom Savoy, CTDEP/Marine Fisheries Division, personal communication). However, the relationship between the two apparent subpopulations and the nature of recruitment from one area to the other is not understood.

Commercial shad netters now report that they commonly catch sturgeon in their gill nets. The Connecticut River Ecological Study (1965–1972) conducted numerous net sets for American shad using similar techniques to those of the current shad netters, yet the study never collected any sturgeon (Marcy 2003a [1976], this volume). It suggests that the species may have been very uncommon during those years, and the population has since experienced a recovery to the present-day level. For a more complete discussion of the biology of Connecticut River shortnose sturgeon and results of recent studies, see Savoy (2004) and Savoy and Benway (2004, this volume).

Atlantic Sturgeon

Status Prior to the Restoration Program

The Atlantic sturgeon is native to the Connecticut River and at one time supported a large fishery (Galligan 1960). However, the population may have already been extirpated upon the beginning of the restoration program.

Restoration Strategies

1. *Targeted habitat.* None.
2. *Fish passage.* None. It is assumed that the fish passage requirements of Atlantic sturgeon would be similar to those of shortnose sturgeon. Shortnose sturgeon use the Holyoke Fish Lift, but no Atlantic sturgeon has been documented using it.
3. *Reintroduction of Species into Habitat.* None.

Current Status

It is believed that this species no longer spawns in Connecticut River, but Atlantic sturgeon from other river systems seasonally visit the lower Connecticut River (Savoy and Pacileo 2003).

Discussion

Atlantic sturgeon was once very common in the Connecticut River and the sturgeon fishery in the state of Connecticut was centered in the Cromwell reach of the river (~rkm 42), although some fish were taken near the Enfield Dam (Galligan 1960). No Atlantic sturgeon were collected during the Connecticut River Ecological Study (Marcy 2004a [1976]). The CTDEP has conducted sturgeon studies in the Connecticut River for the past 18 years and collected 98 immature Atlantic sturgeon. All but one Atlantic sturgeon were collected in the estuary (lower 26 km), and all are believed to be from the Hudson River (Savoy and Pacileo 2003).

Striped Bass

Status Prior to the Restoration Program

Striped bass were abundant upon European Contact (Judd 1905) and have been present in fluctuating numbers ever since. The population in the Connecticut River was not very abundant during the late 1960s and 1970s when the restoration program began (Marcy 2004a [1976]).

Restoration Strategies

1. *Targeted Habitat.* None.
2. *Fish Passage.* No fishway was designed to deliberately pass striped bass and few use fishways. Moderate numbers of striped bass are

passed annually at the Holyoke Dam Fishway (Table 5), and very small numbers are passed annually at the Turners Falls, Vernon, DSI, and Rainbow Dam fishways. Striped bass are not known to spawn in the Connecticut River basin, and therefore the species does not have to access upstream habitat to flourish.

3. *Reintroduction of Species into Habitat.* There have been no deliberate efforts to introduce striped bass into Connecticut River habitats.

Current Status

Striped bass are extremely abundant in the Connecticut River, perhaps numbering in the hundreds of thousands annually. For a more thorough discussion of the species' abundance, see Savoy and Crecco (2004). The increase of abundance of striped bass in the river (Table 5) is due to the coast-wide stock recovery of the species in 1995 (ASMFC 1995).

Discussion

Striped bass supports a very popular sport fishery in the Connecticut River. Commercial fishing of striped bass is prohibited in Connecticut and in the Massachusetts portion of the Connecticut River. Numerous striped bass in the river possess mature gonads yet there has never been any direct evidence of the species spawning in the Connecticut River. Less than 10 young of year have been collected in river by the annual clupeid juvenile survey and those were collected in the lower 16 km and attributed to the Hudson River (Tom Savoy, CTDEP/Marine Fisheries Division, personal communication). Striped bass did not typically overwinter in the river prior to the late 1990s. They generally entered the river in low numbers in April during the river herring runs and departed in late June/early July when the blueback herring run left the river. While this general trend is still apparent, anglers report catching striped bass in the river year-round.

White Perch

Status Prior to the Restoration Program

White perch *Morone americana* was very common and widely distributed from the mouth of the river up to the Holyoke Dam and the first dam on most tributaries below the Holyoke Dam.

Restoration Strategies

1. *Targeted Habitat.* No specific range targeted.
2. *Fish Passage.* Not specifically targeted for fish passage but significant numbers of the species ascend some fishways.
3. *Reintroduction of Species into Habitat.* There have been no deliberate efforts by CRASC to reintroduce/introduce white perch into upstream habitat. However, humans have transplanted the species widely within the watershed (Whitworth 1996).

Current Status

White perch are abundant and widespread throughout the lower basin. Howell and Molnar (2004, this volume) estimated the lower Connecticut River population at 1.6 million. The species is common in the middle portion of the main stem (Holyoke to Wilder dams) and is routinely seen in the Vernon Dam fishway. The species has been introduced into many inland lakes in New Hampshire (Scarola 1973) and the fish in the main-stem river in New Hampshire likely originated from fish dropping downstream from the lakes.

Discussion

White perch that overwinter in Hamburg Cove (rkm 12.8) greatly contribute to spring populations in upstream coves (e.g., rkm 78) and the mouth of the river (Maltezos et al. 1979; Howell and Molnar 2004). It is unclear what proportion, if any, of white perch migrate upstream from Long Island Sound and the species may be more accurately described as potamodromous than anadromous. Fishways have added additional complexity to the white perch community in the basin by allowing migratory fish to access historical and nonhistorical habitat upstream of dams. Howell and Molnar (2004) concluded that fishing pressure on the population has remained stable during the past 30 years, but the population has been recently depressed by natural mortality that is likely caused by striped bass predation. The CTDEP implemented creel and minimum size limits for the sport fishery for the first time in 2003, but the species does not require any additional management strategies by CRASC partners. For more information about the white perch population in the lower Connecticut River, see Howell and Molnar (2004).

Rainbow Smelt

Status Prior to the Restoration Program

Anadromous rainbow smelt *Osmerus mordax* was widespread in the lower river (Marcy 2004b [1976], this volume).

Restoration Strategies

1. *Targeted Habitat.* None.
2. *Fish Passage.* Most dams are upstream of the native range of rainbow smelt, which is not able to surmount falls of greater than 0.5 m (Collette and Klein-MacPhee 2002). Furthermore, smelt are not known to use any fishway design (Alex Haro, USGS-BRD, Conte Anadromous Fish Research Center, personal communication), and no effort has been made to pass smelt above dams.
3. *Reintroduction of Species into Habitat.* There have been no efforts to reintroduce rainbow smelt into historical habitat.

Current Status

CTDEP has collected very small numbers of rainbow smelt. It is assumed that the species is still present in the lower river but there are no data on abundance or distribution of spawning runs into lower river tributaries. A recent effort to document spawning runs in two lower river tributaries known to support smelt in the past failed to observe any smelt (Heather Fried, University of Connecticut, Department of Life Sciences, Storrs, Connecticut; personal communication). The distribution of rainbow smelt in the upper portion of the basin is similar to that of white perch. It was not native to the upper basin but was introduced extensively into inland lakes for forage for sport fish (Scarola 1973) and the species is now found throughout the basin, including nonanadromous populations in the main stem.

Discussion

There is anecdotal evidence of the loss or decline of anadromous rainbow smelt runs in coastal streams along the entire Connecticut shoreline. There are data to indicate recent stock collapses of rainbow smelt in the Hudson River (Daniels et al., in press). The CTDEP and the University of Connecticut began a study in 2003 to assess the status of anadromous rainbow smelt in the lower Connecticut River.

Sea Lamprey

Status Prior to the Restoration Program

Sea lamprey was abundant in the Connecticut River and tributaries as far upstream as the first barrier falls. There are reports that the species may have been able to surmount the falls at Bellows Falls, Vermont (Scarola 1973) and we suspect that its ability to ascend the falls may have been sporadic and linked to favorable flow conditions. When the restoration program began, the species was common in the main stem upstream to the base of the Holyoke Dam and to the base of the first dams on most tributaries downstream of Holyoke. Some of these tributary runs may have had reduced numbers due to poor water quality and nonconsump-tive killing by humans.

Restoration Strategies

1. *Targeted Habitat.* The Connecticut River Atlantic Salmon Commission and its partners have targeted for restoration the extent of the native distribution of sea lampreys that lies within the targeted habitat for Atlantic salmon, American shad, and blueback herring. This means that fish passage would not be implemented for the sole purpose of passing sea lamprey. However, when fish passage is targeted for other species, sea lamprey would be part of the suite of species targeted for restoration.
2. *Fish Passage.* Sea lamprey use fishways very effectively and can be expected to utilize any style of fishway that other species of anadromous fish utilize. No fishway has been designed specifically for sea lamprey. There have been times when heavy use of the vertical slot

Rainbow Dam fishway have deterred passage of American shad because resting (attached) lamprey clogged the 25-cm slot openings.

3. *Reintroduction of Species into Habitat.* There have been no efforts other than fish passage to reintroduce sea lamprey into historical habitat.

Current Status

Adult sea lampreys were not consistently counted at the Holyoke Dam Fishlift prior to 1975, but the counts have subsequently ranged between 15,000 and 100,000 adults annually (Table 5). Annual sea lamprey counts at the Vernon Dam fishway are usually several hundred, but have been as high as 16,000 (VTDFW, unpublished data). The largest number passed at the Bellows Falls fishway was 198 in 1998 (VTDFW, unpublished data). No lampreys were observed using the Wilder Dam fishway, 1987–1994, but the fishway has not been monitored for fish counts since 1994. Sea lampreys may have utilized it during subsequent years (e.g., 1998). The sizes of runs in the Westfield, Farmington, and Salmon rivers have experienced increases similar to those seen at Holyoke during the period of the restoration program (CRASC, unpublished data). Sea lamprey reproduction has been documented in the Cold River, New Hampshire (#17, Figure 1) (Ken Sprankle, USFWS, personal communication), the West River, Vermont (#16), and the White River, Vermont (#26, Figure 1) (Rich Kirn, VTDFW, personal communication).

Discussion

There has been some public questioning of restoring sea lamprey runs within the basin due to the widespread knowledge of the ecological problems created by the introduction of sea lamprey into the Great Lakes (Scott and Crossman 1973). However, sea lamprey has been present in the basin since prehistoric times with no problems of fish parasitism noted. Juvenile sea lampreys are preyed upon by striped bass (Tom Savoy, CTDEP/ Marine Fisheries, personal communication) and presumably other species. Stonefly larvae are frequently observed feeding on sea lamprey carcasses in Connecticut River tributaries. CRASC believes that the restoration of sea lamprey runs to the basin increases the basin's biodiversity and results in ecosystem benefits. It is believed that the reduction of public vandalism of nesting lampreys is due to public outreach efforts by the CRASC partners.

The abundant runs of sea lamprey provide research opportunities. CRASC partners have collected adult lampreys at fishways and juvenile lampreys in streams for the USFWS to support research associated with the international effort to control nuisance lamprey populations in the Great Lakes. The CTDEP has cooperated with academic institutions in the collection of sea lamprey ammocoetes to support neurological research.

Sea-Run Brown Trout

Status Prior to the Restoration Program

Brown trout *Salmo trutta* is not native and was not present in the basin prior to the introduction of freshwater forms from Europe in the late 1800s (Scott and Crossman 1973). It is not known when anadromous forms of brown trout were first reported but they were common in the lower Connecticut River immediately prior to the beginning of the restoration program.

Introduction Strategies

1. *Targeted Habitat.* CRASC has not targeted any habitat for sea-run brown trout introduction. The CTDEP has targeted free-flowing stretches above dams on six tributaries in Connecticut.
2. *Fish Passage.* No fishway outside of Connecticut has been designed for sea-run brown trout. Several fishways on Connecticut tributaries have been planned for sea-run brown trout but the species is an extremely strong migrant and can use fishways designed for other species. Therefore, sea-run brown trout are not given special consideration for fish passage.
3. *Introduction of Species into habitat.* Brown trout are currently being stocked in the lowest section of the Eightmile River in an experimental manner to increase the returns of sea-run brown trout.

Current Status

Sea-run brown trout generally do not reach the Holyoke Dam, although one large individual was captured at the fishlift in 2001 (Mickey Novak, USFWS, Cronin National Salmon Station; personal communication). Some sea-run brown trout may be passing up the DSI Dam Fishway (#4, Figure 1) on the Westfield River (Caleb Slater, MADFW, personal communication). Annual adult returns to the Leesville Dam Fishway and the Rainbow Dam Fishway have ranged from 2 to 39.

Discussion

The state of Connecticut has stocked large numbers of hatchery-reared brown trout annually since before the 1950s and imported anadromous strains from Denmark and Tasmania during the 1950s and 1960s. Sea-run brown trout continue to enter the Connecticut River and its tributaries in small numbers and are believed to be the result of the state hatchery trout stocking program. The fish are highly prized by anglers, prompting the recent effort to increase their numbers. Sea-run brown trout are allowed to ascend fishways and spawn upstream, but it is not known whether spawning by sea-run parents produce anadromous progeny or whether increasing the density of wild brown trout parr will increase the tendency for some juveniles to drop downstream to tidewater.

Historically, the only stream-dwelling salmonid native to the Connecticut River other than Atlantic salmon was the brook trout *Salvelinus fontinalis*. There are no good data on anadromous populations of brook trout in the Connecticut River, but it is likely that there were anadromous brook trout into the Connecticut River upon European Contact. The last known report of a "sea-run" brook trout was a catch from a gillnet at the mouth of the Connecticut River in 1979, and at that time, veteran biologists considered it to be extremely unusual. Sea-run brown trout are stocked to provide a replacement for the native sea-run brook trout that appears to no longer be capable of living in Long Island Sound.

American Eel

Status Prior to the Restoration Program

American eel is one of the most widely distributed fish species in the Connecticut River basin. Historically, it was found as far upstream as the Connecticut Lakes (rkm 642) near the Quebec border (Scarola 1973). The species was common and abundant in the tidewater portions of the river prior to the restoration program (Marcy 2004a [1976]).

Restoration Strategies

1. *Targeted Habitat.* The Connecticut River Atlantic Salmon Commission has not targeted any habitat for American eel restoration, but individual member state agencies have. The CTDEP has targeted habitat upstream of the Rainbow Dam on the Farmington River, upstream of the Leesville dam on the Salmon River, and habitat in many smaller tributaries such as Scantic, Hockanum, Mattabesett, Coginchaug, and Eightmile rivers. The MADFW has targeted habitat above the Holyoke Dam on the main stem, above the DSI Dam on the Westfield River, above the Dwight Station Dam on the Chicopee River, and above the Advocate Dam on the Mill River (Caleb Slater, MADFW, personal communication).
2. *Fish Passage.* Some pool-and-weir fishways (e.g., Rainbow Dam and Turners Falls fishways) allow the passage of American eels, but many of the other fishways already constructed do not pass American eels (e.g., Holyoke Dam, Leesville Dam, and DSI fishways). In most cases, separate, specially-designed passage facilities are needed for American eel. The design of eel passes will vary among dams, but the most common design for tall dams will likely be a bristle or pegboard substrate in a trough carrying small amounts of water and dead-ends into a holding tank/trap that allows the enumeration and hand transportation of the eels above the dam (Haro et al. 2002). Downstream passage of "silver" American eel is a problem since these individuals may be up to 1 m long (Facey and Van Den Avyle 1987) and are particularly vulnerable to mutilation in hydroelectric turbines (McCleave 2001; EPRI 1999). Silver eels will sometimes use facilities constructed for the downstream passage of juvenile anadromous fish or the upstream passage of adult anadromous fish (such as at the Rainbow Dam on the Farmington River). More research is needed to design effective downstream passageways for eels, and SCAFRC is currently conduct-

ing studies at the Turners Falls and Rainbow dams (Alex Haro, USGS-SCAFRC, personal communication).

3. *Reintroduction of Species into Habitat.* There have been no efforts to reintroduce American eel into historical habitat by any other means other than passage over barrier dams.

Current Status

CRASC member agencies have only recently begun to collect American eel data within the basin, but it seems safe to conclude that the species is less abundant in the watershed than it was historically and is less abundant in some portions of the basin than it was prior to the beginning of the restoration program. Eels were routinely captured in the West River in southern Vermont in the 1980s and early 1990s but have not been observed since. The species has not been captured from central (Rich Kirn, VTDFW, personal communication) or northern (Len Gerardi, VTDFW, personal communication) Vermont during the last 10 years. The Atlantic States Marine Fisheries Commission has reported coast-wide declines in American eel stock abundance since 1967 (ASMFC 2000) due to loss of habitat and mortality of downstream migrants. These factors are known to exist in the Connecticut River basin and it is reasonable to conclude that the distribution and abundance of American eel have declined during the period of time of the restoration program.

Discussion

Undocumented increases in population densities of American eels have likely occurred on portions of the Westfield, Salmon, and Farmington rivers above dams where effective upstream eel passage currently exists. The effort to provide upstream passage for American eel in the Connecticut River basin is about 25 years behind that for anadromous fish. Many eel passage projects are expected during the next 10 years. Effective downstream passage of American eel may lag behind while directed research is conducted. Population increases in anadromous fish species have generally followed successful passage at dams. It is not clear how improved eel passage at dams will affect overall eel population sizes in the Connecticut River basin. Improved upstream passage should increase upstream distribution and population densities. Improved downstream passage will improve spawning escapement to the ocean. However, neither trend may result in increased glass eel recruitment to the mouth of the Connecticut River since the glass eels do not return to the river in which their parents resided and any increases in the number of eggs and larvae in the Sargasso Sea will likely be shared among all North American rivers that receive glass eel runs (Moriarty 1978). Increases of the number of glass eels to the mouth of the Connecticut River may occur only after similar restoration plans have been implemented in many North American rivers and significant increases in spawning escapement to the Sargasso Sea occur. CRASC expects to adopt an American eel management plan in 2003.

Summary

Restoring runs of migratory fish that cross many national jurisdictions to a large watershed that spans many U.S. political jurisdictions is extremely challenging. The Connecticut River watershed contains some of the most natural, undeveloped land in the Northeastern United States and some of the most urban, densely-populated areas in the entire nation. The watershed is very dynamic: it is recovering from past industrialization; it has been experiencing widespread reforestation, but the forests are now being fragmented due to increased residential development; it is impacted by global influences such as climate change and acid rain; and is experiencing the effects of coast-wide shifts in fish species distribution. Restoring native species in this context is analogous to "aiming at a moving target." The program has achieved great successes and has experienced disappointments. It is clear that successful restoration of sustained fish runs will require many more years of work.

The diadromous fish restoration program for the river would be impossible without the close partnerships between the state and federal agencies and many nongovernmental organizations. The spirit of cooperation and willingness of all groups to make long-term commitments to fish restoration as a watershed-based activity have been crucial to the program's success. In this paper, we have reviewed the restoration activities undertaken for each diadromous species. Some species only migrate short distances up the river (e.g., alewife, rainbow smelt, hickory shad) and therefore management activities are conducted only by the most downstream state (Connecticut). However, when CRASC considers activities to restore selected diadromous fish runs to a large portion of the watershed, all four states share equal responsibility, authority, and commitment.

The program has benefited from applied research conducted at many institutions, most notably the Silvio O. Conte Anadromous Fish Research Center (USGS/BRD), the USGS/BRD Cooperative Fish and Wildlife Research Unit, the U.S. Forest Service Cooperative Research Unit, and the Department of the Forestry and Wildlife at the University of Massachusetts (Amherst), USGS/BRD Cooperative Fish and Wildlife Research Unit at the University of Vermont (Burlington), Dartmouth College, and the Northeast Fisheries Center at Woods Hole (NMFS). The Connecticut River Ecological Study provided encouragement to the nascent restoration program and provided critical scientific information that became the foundation of much future work with anadromous species on the river.

The program has succeeded with broad public support throughout all four states and at all levels of government. It remains to be seen if such support can be sustained during periods of poor economic growth, shrinking government budgets, and inconsistent fish runs, particularly in respect to Atlantic salmon, which receives the most attention from the public despite the many successes associated with the other species.

Acknowledgments

The authors would like to thank the other members of the CRASC Technical Committee: Caleb Slater, MADFW; Gabe Gries, New Hampshire Department of Fish and Game; Steve Roy, USFS; and John Warner, USFWS and the Connecticut River Coordinator, Janice Rowan, USFWS, for their help in compiling the information for this paper. The assistance of Ben Letcher, Tim King, Tom Savoy, Joe Ravita, Paul Jacobson, and Penny Howell is gratefully acknowledged. We appreciate the helpful reviews of William Leggett, E. J. Crossman, and Stephen Rideout.

References

Anonymous. 1984. The Connecticut River—worth the cost! Connecticut Department of Environmental Protection, Water Compliance Unit, Hartford.

Anonymous. 1998. Report of the 27th Northeast Regional Stock Assessment Workshop, July 1998: section G: Atlantic herring. Pages 281–309 *in* Woods Hole ref. doc. 98–15. National Marine Fisheries Service, Northeast Fisheries Center, Woods Hole, Massachusetts.

Atkins, C. G. 1874. On the salmon of eastern North America, and its artificial culture. Pages 226–335 *in* Report of the Commissioner, for 1872 and 1873, Part II. U.S. Commission of Fish and Fisheries, GPO, Washington, D.C.

ASMFC (Atlantic States Marine Fisheries Commission). 1995. Amendment #5 to the interstate fishery management plan for striped bass. Atlantic States Marine Fisheries Commission, Fisheries Management Report 24, Washington, D.C.

ASMFC (Atlantic States Marine Fisheries Commission). 2000. Interstate fishery management plan for American eel. Atlantic States Marine Fisheries Commission, Fisheries Management Report 36, Washington, D.C.

Batsavage, C. F. 1997. Life history aspects of the hickory shad (*Alosa mediocris*) in the Albemarle Sound/ Roanoke River watershed, North Carolina. Master's thesis. East Carolina University, Greeneville, North Carolina.

Baum, E. T. 1997. Maine Atlantic salmon- a national treasure. Atlantic Salmon Unlimited, Hermon, Maine.

Bell, C. E., and B. Kynard. 1985. Mortality of adult American shad passing through a 17-megawatt Kaplan turbine at a low-head hydroelectric dam. North American Journal of Fisheries Management 5:33–38.

Calaprice, J. R., and M. L. Hogsett. 1975. Genetic engineering in the restoration of Atlantic salmon to the Connecticut River. Final report to the Technical Committee for Fisheries Management, Connecticut River Basin. BCS, Seattle.

Collette, B. B., and G. Klein-MacPhee. 2002. Bigelow and Shroeder's fishes of the Gulf of Maine. Smithsonian Institution Press, Washington, D.C.

CRASC (Connecticut River Atlantic Salmon Commission). 1998. Strategic plan for the restoration of Atlantic salmon to the Connecticut River. Connecticut River Atlantic Salmon Commission, Sunderland, Massachusetts.

Daniels, R. A., K. E. Limburg, R. E. Schmidt, D. L. Strayer, and R. C. Chambers. In press. Changes in fish assemblages in the tidal Hudson River, New York. Transactions of the American Fisheries Society.

EPRI (Electric Power Research Institute). 1999. American eel (*Anguilla rostrata*) scoping study: a literature and data review of life history, stock status, population dynamics, and hydroelectric impacts. Electric Power Research Institute, Technical Report TR-111873, Palo Alto, California.

Facey, D. E., and J. J. Van Den Avyle. 1987. American eel. Species Profiles: life histories and environmental requirements of coastal fishes and invertebrates (North Atlantic). U.S. Fish and Wildlife Service National Wetlands Research Center, Biological Report 82(11–74), Washington, D.C.

Foster, C. H. W. 1991. Yankee salmon—the Atlantic salmon of the Connecticut River. CIS, Cambridge, Massachusetts.

Fried, S. M., J. D. McCleave, and G. W. LaBar. 1978. Seaward migration of hatchery-reared Atlantic salmon, *Salmo salar*, smolts in the Penobscot River Estuary: riverine movements. Journal of the Fisheries Research Board of Canada 35:76–87.

Friedland, K. D. 1994. Marine survival of restoration stocks. Pages 223–239 *in* S. Calabi and A. Stout, editors. A hard look at some tough issues. New England Atlantic Salmon Management Conference. New England Salmon Association, Newburyport, Massachusetts.

Friedland, K. D. 1998. Ocean climate influences on critical Atlantic salmon (*Salmo salar*) life history events. Canadian Journal of Fisheries and Aquatic Sciences 55(Supplement 1):119–130.

Friedland, K. D., D. G. Reddin, and J. F. Kocik. 1993. Marine survival of North American and European Atlantic salmon: effects of growth and environment. ICES Journal of Marine Science 50:481–492.

Galligan, J. P. 1960. History of the Connecticut River sturgeon fishery. The Connecticut Wildlife Conservation Bulletin 6(1):1–6.

Haro, A., S. Gephard, and T. Wildman. 2002. Design and performance of upstream eel passes in the northeast. Fourth Bioengineering Symposium. American Fisheries Society, 132nd Annual Meeting, Baltimore, Maryland.

Hasler, A. D., and A. T. Scholz. 1983. Olfactory imprinting and homing in salmon. Springer-Verlag, New York.

Howell, P., and D. Molnar. 2004. Stock assessment of white perch in the lower Connecticut River. Pages 379–390 *in* P. M. Jacobson, D. A. Dixon, W. C. Leggett, B. C. Marcy, Jr., and R. R. Massengill, editors. The Connecticut River Ecological Study (1965–1973) revisited: ecology of the lower Connecticut River 1973–2003. American Fisheries Society, Monograph 9, Bethesda, Maryland.

Jacobs, R. P., W. A. Hyatt, N. T. Hagstrom, E. B. O'Donnell, E. C. Schluntz, P. Howell, and D. R. Molnar. 2004. Trends in abundance, distribution, and growth of freshwater fishes from the Connecticut River in Connecticut (1988–2002). Pages 319–343 *in* P. M. Jacobson, D. A. Dixon, W. C. Leggett, B. C. Marcy, Jr., and R. R. Massengill, editors. The Connecticut River Ecological Study (1965–1973) revisited: ecology of the lower Connecticut River 1973–2003. American Fisheries Society, Monograph 9, Bethesda, Maryland.

Jacobson, P. M., C. G. Fredette, and N. Barrett. 2004. Introduction: Connecticut River watershed management—past, present, and future. Pages 263–272 *in* P. M. Jacobson, D. A. Dixon, W. C. Leggett, B. C. Marcy, Jr., and R. R. Massengill, editors. The Connecticut River Ecological Study (1965–1973) revisited: ecology of the lower Connecticut River 1973–2003. American Fisheries Society, Monograph 9, Bethesda, Maryland.

Jacobson, P. M., C. Tomichek, and D. J. Danila. 2004. Twenty years of impingement history: Connecticut Yankee Haddam Neck nuclear power plant. Pages 455–473 *in* P. M. Jacobson, D. A. Dixon, W. C. Leggett, B. C. Marcy, Jr., and R. R. Massengill, editors. The Connecticut River Ecological Study (1965–1973) revisited: ecology of the lower Connecticut River 1973–2003. American Fisheries Society, Monograph 9, Bethesda, Maryland.

Judd, S. 1905. History of Hadley, Massachusetts. H. R. Hunting and company, Springfield, Massachusetts.

Kendall, W. C. 1935. The salmons. Pages 1–116 *in* The fishes of New England- the salmon family, part 2. Monograph of the natural history of New England. Memoirs of the Boston Society of Natural History 9(1).

Kynard, B. 1998. Twenty-two years of passing shortnose sturgeon in fishlifts on the Connecticut River: what has been learned? Pages 255–264 *in* M. Jungwirth, S. Schmutz and S. Weiss, editors. Fish migration and fish bypasses. Fish News Books, Oxford, England.

Kynard, B., and J. A. O'Leary. 1993. Evaluation of a bypass system for spent American shad at Holyoke Dam, Massachusetts. North American Journal of Fisheries Management 13:782–789.

Leggett, W. C. 2004. The American shad, with special reference to it migration and population dynamics in the Connecticut River. Pages 181–238 *in* P. M. Jacobson, D. A. Dixon, W. C. Leggett, B. C. Marcy, Jr., and R. R. Massengill, editors. The Connecticut River Ecological Study (1965–1973) revisited: ecology of the lower Connecticut River 1973–2003. American Fisheries Society, Monograph 9, Bethesda, Maryland. (Originally published in 1976)

Leggett, W. C., T. F. Savoy, and C. A. Tomichek. 2004.

The impact of enhancement initiatives on the structure and dynamics of the Connecticut River population of American shad. Pages 391–405 *in* P. M. Jacobson, D. A. Dixon, W. C. Leggett, B. C. Marcy, Jr., and R. R. Massengill, editors. The Connecticut River Ecological Study (1965–1973) revisited: ecology of the lower Connecticut River 1973–2003. American Fisheries Society, Monograph 9, Bethesda, Maryland.

Letcher, B. H., and T. L. King. 1999. Targeted stock identification using multilocus genotype 'family-printing.' Fisheries Research 43:99–111.

Letcher, B. H., and T. L. King. 2001. Parentage and grandparentage assignment with known and unknown matings: application to Connecticut River Atlantic salmon restoration. Canadian Journal of Fisheries and Aquatic Sciences 58:1812–1821.

MacCrimmon, H. R., and B. L. Gots. 1979. World distribution of Atlantic salmon, *Salmo salar*. Journal of the Fisheries Research Board of Canada 36:422–457.

Maltezos, G., R. McMahon, and M. Leslie. 1979. Movements of white perch in the lower Connecticut River. Connecticut Department of Environmental Protection, Fisheries Division, 88–309 Program Final Report, Hartford.

Mansueti, R., and H. Kolb. 1953. A historical review of the shad fisheries of North America. Chesapeake Biological Laboratory, Publication No. 97, Solomons, Maryland.

Marcy, B. C., Jr. 2004a. Fishes of the lower Connecticut River and the effects of the Connecticut Yankee plant. Pages 65–124 *in* P. M. Jacobson, D. A. Dixon, W. C. Leggett, B. C. Marcy, Jr., and R. R. Massengill, editors. The Connecticut River Ecological Study (1965–1973) revisited: ecology of the lower Connecticut River 1973–2003. American Fisheries Society, Monograph 9, Bethesda, Maryland. (Originally published in 1976)

Marcy, B. C., Jr. 2004b. Planktonic fish eggs and larvae of the lower Connecticut River and the effects of the Connecticut Yankee plant, including entrainment. Pages 125–153 *in* P. M. Jacobson, D. A. Dixon, W. C. Leggett, B. C. Marcy, Jr., and R. R. Massengill, editors. The Connecticut River Ecological Study (1965–1973) revisited: ecology of the lower Connecticut River 1973–2003. American Fisheries Society, Monograph 9, Bethesda, Maryland. (Originally published in 1976)

Martinez, J. L., S. Gephard, F. Juanes, J. Pezez, and E. Garcia-Vazquez. 2001. Genetic and life history differentiation between donor and derivative populations of Atlantic salmon. Transactions of the American Fisheries Society 130:508–515.

McCleave, J. D. 2001. Simulation of the impacts of dams and fishing weirs on reproductive potential of silver phase American eels in the Kennebec River basin, Maine. North American Journal of Fisheries Management 21:592–605.

McMenemy, J. R. 1995. Survival of Atlantic salmon fry stocked at low density in the West River, Vermont. North American Journal of Fisheries Management 15(2):366–374.

McMenemy, J. R., and B. Kynard. 1988. Use of inclined-plane traps to study movements and survival of Atlantic salmon smolts in the Connecticut River. North American Journal of Fisheries Management 8:481–488.

Merriman, D., and Y. Jean. 1949. The capture of an Atlantic salmon in the Connecticut River. Copeia 1949(4):210–§121.

Meyers, T. F. 1994. The program to restore Atlantic salmon to the Connecticut River. Pages 11–21 *in* S. Calabi and A. Stout, editors. A hard look at some tough issues. New England Atlantic Salmon Management Conference. New England Salmon Association, Newburyport, Massachusetts.

Moffitt, C. M., B. Kynard, and S. G. Rideout. 1982. Fish passage facilities and anadromous fish restoration in the Connecticut River basin. Fisheries 7(6):2–11.

Moriarty, C. 1978. Eels - a natural and unnatural history. Universe Books, New York.

Moss, D. D. 1952. A history of the Connecticut River and its fisheries. Connecticut Board of Fisheries and Game, Hartford.

Moss, S. A., W. C. Leggett, and W. A. Boyd. 2004. Recurrent mass mortalities of the blueback herring in the lower Connecticut River. Pages 239–246 *in* P. M. Jacobson, D. A. Dixon, W. C. Leggett, B. C. Marcy, Jr., and R. R. Massengill, editors. The Connecticut River Ecological Study (1965–1973) revisited: ecology of the lower Connecticut River 1973–2003. American Fisheries Society, Monograph 9, Bethesda, Maryland. (Originally published in 1976)

NASCO (North Atlantic Salmon Conservation Organization). 2002. Manual of resolutions, agreements, and guidelines. NASCO, Edinburgh, Scotland.

NEFMC (New England Fishery Management Council). 1987. Fishery management plan for Atlantic salmon. NEFMC, Saugus, Massachusetts.

O'Leary, J., and D. G. Smith. 1987. Occurrence of the first freshwater migration of the gizzard shad, *Dorosoma cepedianum*, in the Connecticut River, Massachusetts. Fishery Bulletin 85(2):380–383.

Orciari, R. D., G. H. Leonard, D. J. Mysling, and E. C. Schluntz. 1994. Survival, growth, and smolt production of Atlantic salmon stocked as fry in a southern New England stream. North American Journal of Fisheries Management 14:588–606.

Parrish, D. L., R. J. Behnke, S. R. Gephard, S. D. McCormick, and G. H. Reeves. 1998. Why aren't there more Atlantic salmon (*Salmo salar*). Canadian Journal of Fisheries and Aquatic Sciences 55(Supplement1):281–287.

Richkus, W. A., and G. DiNardo. 1984. Current status and biological characteristics of the anadromous alosid stocks of eastern United States: American shad, hickory shad, alewife, and blueback herring. Interstate Fisheries Management Program, Atlantic States Marine Fisheries Commission, Washington, D.C.

Rideout, S. G., and L. W. Stolte. 1988. Restoration of Atlantic salmon to the Connecticut and Merrimack rivers. Pages 67–81 *in* Richard H. Stroud, editor. Present and future Atlantic salmon management-measuring progress toward international cooperation. Marine Recreational Fisheries 12. Atlantic Salmon Federation & National Coalition for Marine Conservation, Inc., Ipswich, Massachusetts.

Rohde, F. C., R. G. Arndt, D. G. Linquist, and J. F. Parnell. 1994. Freshwater fishes of the Carolinas, Virginia, Maryland, and Delaware. University of North Carolina Press, Chapel Hill.

Saunders, R. L. 1981. Atlantic salmon (*Salmo salar*) stocks and management implications in the Canadian Atlantic provinces and New England, USA. Canadian Journal of Fisheries and Aquatic Sciences 38:1612–1625.

Savoy, T., and D. Pacileo. 2003. Movements and important habitats of subadult Atlantic sturgeon in Connecticut waters. Transactions of the American Fisheries Society 132:1–8.

Savoy, T. F. 2004. Population estimate and ultization of the lower Connecticut River by shortnose sturgeon. Pages 345–352 *in* P. M. Jacobson, D. A. Dixon, W. C. Leggett, B. C. Marcy, Jr., and R. R. Massengill, editors. The Connecticut River Ecological Study (1965–1973) revisited: ecology of the lower Connecticut River 1973–2003. American Fisheries Society, Monograph 9, Bethesda, Maryland.

Savoy, T. F., and J. Benway. 2004. Food habits of shortnose sturgeon collected in the lower Connecticut River from 2000 through 2002. Pages 353–360 *in* P. M. Jacobson, D. A. Dixon, W. C. Leggett, B. C. Marcy, Jr., and R. R. Massengill, editors. The Connecticut River Ecological Study (1965–1973) revisited: ecology of the lower Connecticut River 1973–2003. American Fisheries Society, Monograph 9, Bethesda, Maryland.

Savoy, T. F., and V. A. Crecco. 2004. Factors affecting the recent decline of blueback herring and American shad in the Connecticut River. Pages 361–377 *in* P. M. Jacobson, D. A. Dixon, W. C. Leggett, B. C. Marcy, Jr., and R. R. Massengill, editors. The Connecticut River Ecological Study (1965–1973) revisited: ecology of the lower Connecticut River 1973–2003. American Fisheries Society, Monograph 9, Bethesda, Maryland.

Savoy, T. F., V. A. Crecco, and B. C. Marcy, Jr. 2004. American shad early life history and recruitment in the Connecticut River: a 40-year summary. Pages 407–417 *in* P. M. Jacobson, D. A. Dixon, W. C. Leggett, B. C. Marcy, Jr., and R. R. Massengill, editors. The Connecticut River Ecological Study (1965–1973) revisited: ecology of the lower Connecticut River 1973–2003. American Fisheries Society, Monograph 9, Bethesda, Maryland.

Scarola, J. F. 1973. Freshwater fishes of New Hampshire. New Hampshire Department of Fish and Game, Concord.

Scott, W. E., and E. J. Crossman. 1973. Freshwater fishes of Canada. Fisheries Research Board of Canada Bulletin 184.

Spidle, A. P., S. T. Kalinowski, B. A. Lubinski, D. L. Perkins, K. F. Beland, Kocik, J. F., and T. L. King. 2003. Population structure of Atlantic salmon in Maine with reference to populations from Atlantic Canada. Transactions of the American Fisheries Society 132(2):196–209.

Spidle, A. P., T. L. King, and B. H. Letcher. In press. Genetic diversity in Connecticut River Atlantic salmon (*Salmo salar* L.). Aquaculture.

Stier, D. J., and B. Kynard. 1986. Use of radio telemetry to determine the mortality of Atlantic salmon smolts passed through a 17-MW Kaplan turbine at a low-head hydroelectric dam. Transactions of the American Fisheries Society 115:771–775.

Taylor, R. E., and B. Kynard. 1985. Mortality of juvenile American shad and blueback herring passed through a low-head hydroelectric turbine. Transactions of the American Fisheries Society 114:430–435.

Whalen, K. G., and G. W. LaBar. 1998. Survival and growth of unfed and fed Atlantic salmon fry stocked in a Vermont tributary of the Connecticut River. North American Journal of Fisheries Management 18:931–935.

Whitworth, W. R. 1996. Freshwater fishes of Connecticut. CTDEP/ State Geological and Natural History Survey of Connecticut, Bulletin 114 (2nd edition), Hartford.

Whitworth, W. R., P. Minta, and R. Orciari. 1980. Further additions to, and notes on, the freshwater ichthyofauna of Connecticut. Pages 27–29 *in* P. M. Jacobson, editor. Studies of the ichthyofauna of Connecticut. University of Connecticut, Storrs Agricultural Experiment Station, Bulletin 457, Storrs.

American Fisheries Society Monograph 9:319–343, 2004

Trends in Abundance, Distribution, and Growth of Freshwater Fishes from the Connecticut River in Connecticut (1988–2002)

ROBERT P. JACOBS, WILLIAM A. HYATT, NEAL T. HAGSTROM, EILEEN B. O'DONNELL, AND ERIC C. SCHLUNTZ

Connecticut Department of Environmental Protection
Bureau of Natural Resources, Inland Fisheries Division
79 Elm Street, Hartford, Connecticut 06106, USA

PENELOPE HOWELL AND DAVID R. MOLNAR

Connecticut Department of Environmental Protection
Bureau of Natural Resources, Marine Fisheries Division
333 Ferry Road, Old Lyme, Connecticut 06371, USA

Abstract.—This paper presents population statistics for freshwater fishes collected by boat electrofishing in three main-stem sections and eight coves of the Connecticut River within the state of Connecticut, and by stream electrofishing in 19 of the river's tributaries. In addition, results are presented from a creel survey conducted from Hartford to the mouth of the river. Forty-six species in 18 families were collected by all sampling methods combined. The most ubiquitous species in the river was the American eel *Anguilla rostrata*, which was found among 97% of all sites sampled. The most abundant species in the river was the spottail shiner *Notropis hudsonius*. Many fish species apparently preferred different habitat types and several exhibited pronounced north–south trends in abundance. Other species were found in tributary streams, but were not sampled in the main stem or its coves. Growth of most Connecticut River fishes was faster than average for the state's lakes and ponds and, for many species, was fastest in the southern section of the river. Anglers fished more than 264,000 h and caught more than 411,000 fish annually during 1997 and 1998. The most numerous fishes caught and harvested were catfish (white catfish *Ameiurus catus*, channel catfish *Ictalurus punctatus*, and brown bullhead *A. nebulosus*) and white perch *Morone americana*. More than 63,000 catfish and white perch were caught annually and greater than 60% harvested. Greater numbers of striped bass *M. saxatilis* were caught during 1 year of the survey (99,000 in 1998), but less than 6% were harvested. The river also supported large sport fisheries for largemouth bass *Micropterus salmoides* and smallmouth bass *M. dolomieu* and panfish. Rock bass *Ambloplites rupestris*, pumpkinseed *Lepomis gibbosus*, redbreast sunfish *L. auritus*, hickory shad *Alosa mediocris*, gizzard shad *Dorosoma cepedianum*, and channel catfish *Ictalurus punctatus* all experienced significant population expansions in the river since the 1970s. We conclude that additional sampling is needed to better quantify the status of the river's fisheries, that stock assessment data on the important catfish fishery are lacking, that a creel survey of the fishery north of Hartford should be conducted, and that the effects of relocating bass by tournament anglers needs to be determined.

Introduction

The Connecticut River is the state's largest freshwater fishery (~4,000 ha) and supports the greatest diversity of fish species of any freshwater ecosystem in Connecticut (See Appendix 1, this volume). Effectively sampling all habitats and fish populations within a large and diverse system requires a substantial commitment of effort and resources. Historically, the Connecticut Department of Environmental Protection (CTDEP) Inland Fisheries Division and its predecessors

(Board of Fish and Game, Bureau of Fisheries) were responsible for managing fisheries in a large number of water bodies and lacked the resources needed to undertake a comprehensive inventory of the Connecticut River. Consequently, the river was largely ignored by biologists throughout most of the 20th century. Those data gathered were mostly the result of studies that targeted a specific species or species group. Examples include efforts to assess the river's northern pike *Esox lucius* population in the 1970s and early 1980s (Phillips and Bender 1984) and shad population in the 1970s and 1980s (Leggett 2004 [1976]; Crecco and Savoy 1987). A general survey of fish populations was done by Marcy (2004 [1976], this volume); however, this study was confined to one section of the river located between Higganum and Essex (encompassing 18.5 river kilometer). Only recently has the CTDEP Inland Fisheries Division begun collecting information on the distribution, abundance, and growth rates of freshwater fishes inhabiting all regions and habitat types found throughout the Connecticut portion of the Connecticut River.

Information in this paper on present-day fish populations comes from a variety of relatively recent sources. A statewide lake and pond electrofishing survey was conducted during 1988–1992 to gather baseline data on all of the state's warmwater fish populations (Jacobs and O'Donnell 1996). A number of cove and main-stem sites on the Connecticut River were included in this effort. These survey data provided a valuable, but imprecise snapshot of the river's fish populations. Variability among samples was high due to the large number of species, presence of diadromous species, and the open environment that allowed for migration. The CTDEP Inland Fisheries Division has conducted ongoing electrofishing since 1992 to increase our understanding of the river's fish populations and trends over time.

A statewide stream survey was conducted during 1988–1995 to gather baseline data on fish populations inhabiting Connecticut's brooks, streams, and moderate-sized rivers (Hagstrom et al. 1996). Many sample sites were located on tributaries of the Connecticut River, some of which were close to the main stem or its coves. The boundary between the main stem and its tributary waters is a blurry distinction at best. By including data collected from tributary streams near their confluence with the Connecticut River, we are able to document the presence of many species that are difficult to collect in the main stem, but which occasionally frequent the river, nonetheless. These species may play an important part in the ecology of shoal areas and freshwater marshes adjacent to the mouths of tributary streams. They also interact with many of the larger species that enter these habitats during some phase of their life cycle.

A creel survey was conducted on the Connecticut River from Hartford south in 1997 and 1998 to gather data on angler effort and harvest needed to manage targeted sport fisheries (Howell and Molnar 1999). In the process, additional information on the distribution and relative abundance of many of the larger fish species was obtained. Data collected during the creel survey also provided valuable insight into annual and intra-annual variation in abundance, the potential effects of fishing mortality, and the importance of various species to recreational anglers.

Our presentation of these data are also influenced by insight gained while working on a variety of species specific projects (e.g., annual collection of broodstock northern pike and operation of spawning marshes) and directed searches for rare species (e.g., American brook lamprey *Lampetra appendix* and burbot *Lota lota*). Information pulled from all of these divergent sources provides a valuable record of our present-day knowledge of the freshwater fishes inhabiting the Connecticut River and enables us to better identify those areas where additional data are most needed.

Study Area

For the purposes of this paper, the Connecticut River was divided into three sections (Figure 1). The "North" section includes all sites from the Massachusetts State line to the East Windsor/South Windsor town line. The "Central" section includes those sites south of this point to the East Hampton/East Haddam town line. The "South" section includes those sites south of the Central section to Long Island Sound. Tidal influence on the river extends to an area just north of the demarcation between our North and Central sections (tidal fluctuation around Hartford is approximately 30 cm), and the salt wedge from Long Island Sound can penetrate as far north as the Route

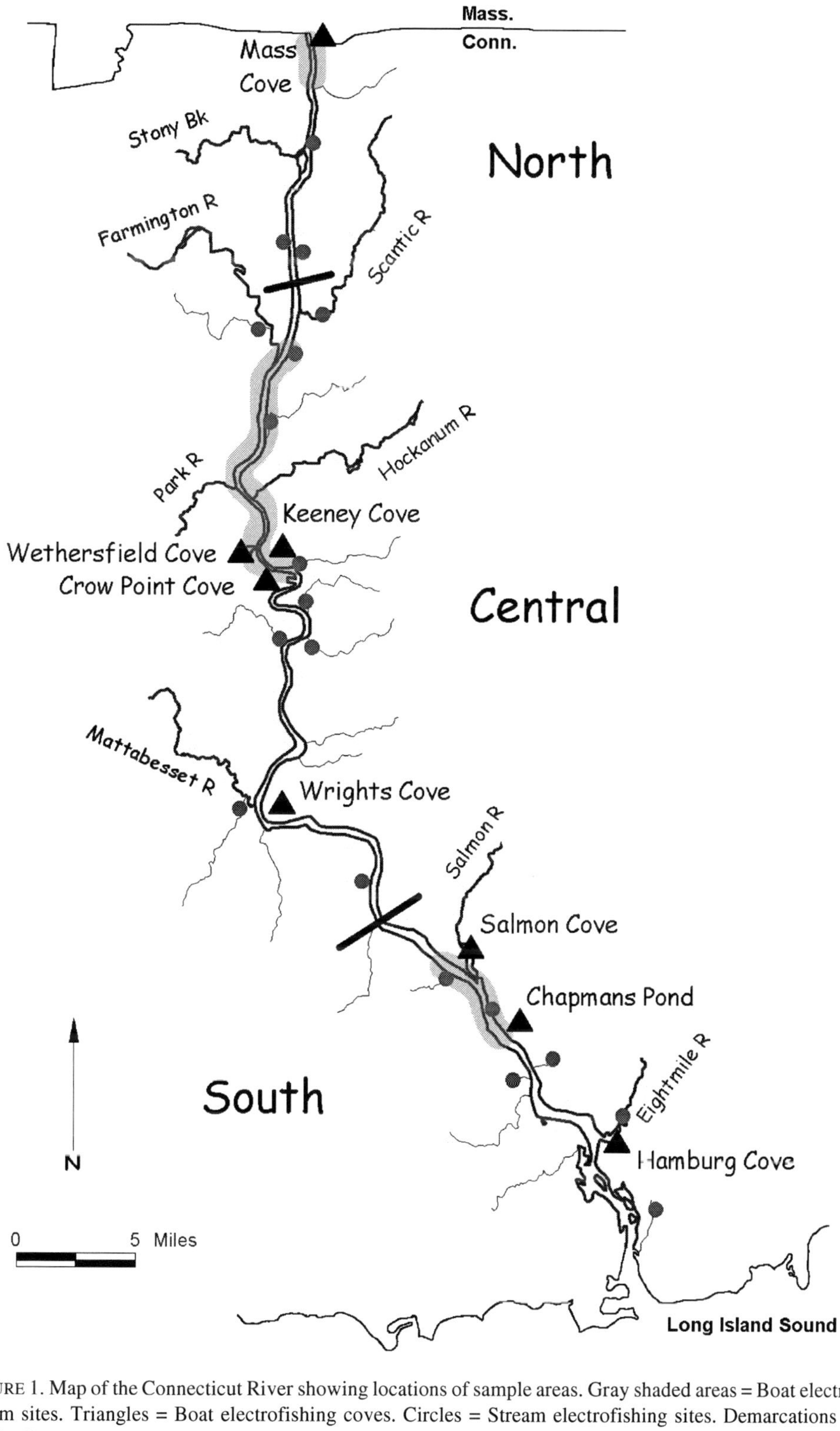

FIGURE 1. Map of the Connecticut River showing locations of sample areas. Gray shaded areas = Boat electrofishing mainstem sites. Triangles = Boat electrofishing coves. Circles = Stream electrofishing sites. Demarcations between "North," "Central," and "South" River sections are marked by bold lines intersecting the main stem.

82 bridge in East Haddam (between Salmon Cove and Chapmans Pond, see Figure 1). Therefore, our Central section can be roughly characterized as "tidally influenced" and our Southern section as "estuarine."

Three main-stem sections and eight coves of the Connecticut River were sampled by boat electrofishing (1988–2002) (Figure 1). The three sections of main stem sampled were the Massachusetts line to the Enfield Dam (North section), the mouth of the Farmington River in Windsor to the Route 3 bridge in Wethersfield (Central section), and the Connecticut Yankee atomic power plant to the Route 82 bridge in East Haddam (South section). Only one cove was sampled in the northern part of the river. This is a small, presumably unnamed backwater cove (2.6 ha) on the east side of the river near the Massachusetts border and, hence, was dubbed the "Mass. Cove." Four coves (Wethersfield, Keeney, Crow Point, and Wrights) ranging in size from 19.8 to 35.2 ha were sampled in the Central section. Three coves were also sampled from the southern part of the river: Chapmans Pond, a small, backwater cove, and Salmon River and Hamburg coves (72.0 and 69.6 ha), which are the mouths of the Salmon and Eight Mile rivers.

Stream electrofishing sites chosen for inclusion in this paper were any tributary sites sampled that were below the first migratory barrier on the stream and no more than 6 m in elevation above that of the Connecticut River at the stream's mouth (Figure 1). Small tributary streams (<10 m width) sampled were North—Beeman Brook (Enfield), Waterworks Brook (Windsor Locks), and Namerick Brook (East Windsor); Central—Mill Brook (Windsor), Stoughton Brook (South Windsor), Podunk River (South Windsor), Salmon Brook (Glastonbury), Hubbard Brook (Glastonbury), Goff Brook (Rocky Hill), and Hubbard Brook (Middletown); South—Mill Creek (Haddam), Succor Brook (East Haddam), Roaring Brook (Lyme), and Great Brook (Ches-ter). Larger tributary streams (≥10 m width) sampled were Central—Scantic River (East Windsor), Roaring Brook (Glastonbury), and Coginchaug River (Middletown); South—Eightmile River (Lyme) and Mill River (Old Lyme). No large tributaries were sampled in the North section of the river.

The creel survey covered the river's main stem and larger coves from Hartford south to the river's mouth (i.e., the area covered by the creel survey did not coincide with the entire area surveyed by electrofishing) (Figure 2). Interview sites included the most popular shore access areas and all public boat launches downstream of Riverside Park in Hartford. The entire area was divided into three sections or "bus routes." Route 1 extended from the river mouth at Old Saybrook/Old Lyme north to the Haddam Bridge and included 12 access sites. Route 2 extended from the Haddam Bridge north to the Portland Bridge and included 10 sites. Route 3 extended from the Portland Bridge north to Riverside Park in Hartford and included 10 sites. Routes 1, 2, and 3 used in the creel survey were different from the North, Central, and South distinctions applied to the electrofishing data.

Methods

Boat Electrofishing

The Connecticut River and its coves were sampled by boat electrofishing between 1988 and 1992 as part of a larger statewide lake and pond electro-fishing survey (Jacobs and O'Donnell 1996). A few sites on the river were sampled more recently (2000–2002), and these data are also included. Boat electrofishing was conducted at night during late April–early June and October–early November. Each site was sampled on one to four different dates. The boats used were constructed by Coffelt Electronics Inc. and were 5.0–5.8-m aluminum johnboats with model VVP-15 control units, powered by 6,500-W generators. The electrode arrays consisted of six 2-m-long anodes (64-mm braided stainless cables) mounted from two booms, which extended approximately 2 m ahead of the bow. The boat hull served as the cathode. Electrofishing was done in a pulsed DC mode (80 pps, 60% pulse width) at 200–400 V and 7–9 A. Electrofishing technique consisted of slowly moving along the shoreline in water generally less than 1.5 m in depth, with the current being controlled via a foot pedal on the boat's bow. A technique was used where an area would be shocked for approximately 5 s; then the foot pedal would be released for a similar amount of time in order to move forward into a new area. This system was employed in an effort to reduce the number of fish that escape by moving ahead of the field. The resulting "on time," used as our unit of sam-

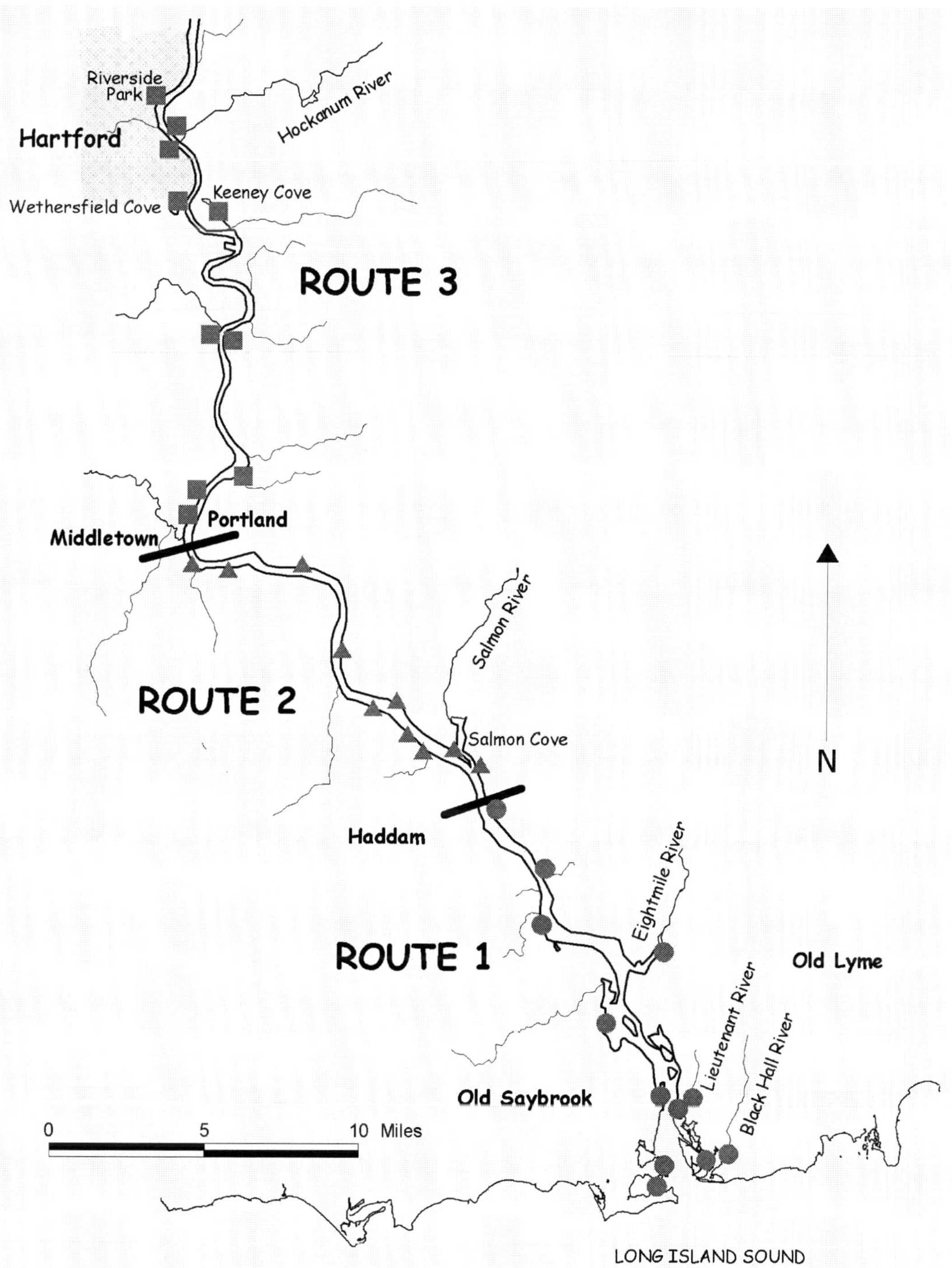

FIGURE 2. Map of the Connecticut River showing the locations of the creel survey interview sites. Circles = Route 1 interview sites, Triangles = Route 2 interview sites, Squares = Route 3 interview sites. Demarcations between Routes 1, 2, and 3 (The Haddam Bridge between Routes 1 and 2 and the Portland Bridge between Routes 2 and 3) are marked by bold lines intersecting the main stem.

pling effort, was recorded in seconds by a control unit meter and is approximately equal to 40% of actual time spent electrofishing.

During a night's electrofishing, abundant species such as sunfish and spottail shiners *Notropis hudsonius* were subsampled for about half the total time spent shocking, whereas all less common species were sampled throughout. All sampled fish were placed in a livewell and were subsequently identified to species and measured (total length [TL] in centimeters). Additionally, length (mm) and scales were taken from a subsample (a maximum of two per centimeter size-class) of each fish species. Electrofishing data included in this paper are from 34 nights on the Connecticut River between 1988 and 2002, including 4 nights in each of the main-stem North, main-stem Central, and main-stem South, and 22 nights in the river's coves.

Stream Electrofishing

Tributaries of the Connecticut River were sampled once each by stream electrofishing during 1989 as part of a larger statewide stream survey (Hagstrom et al. 1996). Some tributaries were sampled more recently (1995–1999) and those data are also included. Stream electrofishing was conducted during the day from July through early September. Different gear types were employed depending on stream size. In smaller streams (<5 m wide), a Coffelt BP-4 dual electrode backpack shocker was used. Two backpacks were used side by side in 5–10-m-wide streams. For streams wider than 10 m, Coffelt VVP-2 stream shocker units (towed in canoes) with 3-m dual electrodes were used side by side, one for each 10 m increment of stream width. The length of each stream sample site depended on stream width as follows: 50 m long for 0–1.5-m-wide streams, 100 m long for 1.5–3.0-m-wide streams and 150 m long for streams wider than 3.0 m. A single upstream run was conducted at each site. One to four netters collected stunned fish, and all fish were identified and measured with the exception of young-of-year centrachids (<4.5 cm), catostomids, and cyprinids (<3.5 cm). The length of time taken to complete each site was recorded to calculate catch per unit effort. Data included in this paper are from 19 tributary streams sampled by electro-fishing on 1 d each between 1988 and 1999.

Creel Survey

An Access Area Survey of the Connecticut River was conducted from March through October in 1997 and 1998. The survey employed a stratified "bus-stop" design (Pollock et al. 1994) that included three sections or "bus routes" (Figure 2). Interview sites included the most popular shore access areas and all public boat launches within the boundaries of the study area. Sampling strata included month (grouped into bimonthly seasons), route, and day type (weekday or weekend/holiday). Initially, sampling of each river section, or bus route, took 7 h to complete. Start times were equally divided between morning and afternoon, and the site within each route where sampling began was randomly chosen each sampling day. Beginning with the start site, sampling proceeded from site-to-site in an established pattern with fixed "wait" times at each site. The allotted wait time at each site was assigned based on angler use, with more popular sites given longer wait times. Wait times were shortened to create a 6-h sampling day in the fall so that the survey could be completed during daylight hours.

At each site, the creel agent counted the shore anglers and boat trailers present and then interviewed and noted the arrival and departure times of all anglers during the assigned period. Total time for shore and boat anglers were recorded and expanded separately. Shore counts were adjusted to account for anglers fishing in areas other than the interview sites based on percentages determined from boat based surveys of entire sample sections. Trailer counts were adjusted to reflect the relative percentage of boaters that were fishing based on interviews done at each site. All recorded anglers' time at the site was then summed and expanded to produce estimates of total angler-hours and associated precision (Howell and Molnar 1999; Pollock et al. 1994). A mean catch/h for each species on each sample day was computed from all anglers interviewed. Total catch and associated precision were estimated by expanding daily catch rate data to produce estimates for each river section, season, and year. All procedures followed those described by Pollock et al. (1994) except that catch-per-hour data from incomplete trips were included.

Sampling effort was similar between the 2

years. In 1997, creel agents interviewed 2,105 anglers during 112 sampling days, and in 1998, agents interviewed 1,873 during 113 d. In addition to the creel survey, boaters at individual boat-launch sites were interviewed on 42 d (to determine the percentage of boaters who were fishing), and the total number of shore anglers fishing entire sections were counted on 48 d (to determine the percentage of total shore fishing effort occurring at sites used in the bus-stop survey).

Gear Selectivity

No direct quantitative comparisons can be made between this study and those of previous authors because of inherent differences in efficiency and fish size selectivity of gear types used. Historical work by Marcy (2004 [1976]) employed beach seines and otter trawls. Beach seines select for smaller fish and work well only in areas with little structure. Otter trawls are best for pelagic fish species, but undersample larger, stronger-swimming fish. Trap nets as used by Howell and Molnar (2004, this volume) and Maltezos et al. (1980) select for larger fish and more mobile species, and because it is a passive gear type, catches can vary considerably depending on where the nets are set. Night boat electrofishing, while good for detecting a wide variety of species, nonetheless often underestimates abundance of smaller fishes, fishes that are nocturnal and/or demersal, and fishes that tend toward open or deeper water habitats (Nielsen and Johnson 1983; Reynolds 1996). Equally important is the inability of electrofishing to sample water with any significant salt content; therefore, species that prefer brackish water are poorly represented here. Data collected by creel survey agents are obviously limited to those species and sizes caught by anglers. However, harvest and length-frequency data gathered from angler catches can provide insight into the distribution, abundance and growth of many species, including some that are not readily sampled by boat electrofishing. In general, stream electrofishing is probably the least selective of the gears discussed in this paper, primarily because it is a mobile gear type used in a confined environment. It nonetheless has its own difficulties that can result in undersampling small, negatively buoyant species and larger, fast-swimming fish like smallmouth bass *Micropterus dolomieu*.

Results

Distribution and Trends in Abundance of Fishes

Forty-six species in 18 different families were identified from our electrofishing samples of the Connecticut River (Table 1). Of these, 34 were strictly freshwater species and the remaining 12 were either estuarine or diadromous. Only 50% of the strictly freshwater species sampled are native to Connecticut. The paucity of native species is primarily due to the state's recent glaciation (Bond 1979), whereas the abundance of naturalized species is primarily the result of widespread introductions done during the last century to increase fishing opportunities (Wilde 1959).

The most ubiquitous species in the Connecticut River was the American eel *Anguilla rostrata*, which was found among 97% of all sites sampled, followed by white sucker *Catostomus commersonii* (87% of sites), pumpkinseed *Lepomis gibbosus* (83%), tessellated darter *Etheostoma olmstedi* (83%), and redbreast sunfish *L. auritus* (77%) (Table 1). The most abundant species overall was the spottail shiner, which greatly outnumbered other species in most main-stem and cove areas (Tables 2 and 3). The American eel was the dominant species in almost all of the river's tributary streams, being nearly four times as abundant on average than any other species (Table 4).

A number of species were probably underrepresented in our main-stem and cove samples due to the selectivity of the boat electrofishing gear (Tables 2 and 3). Most notably among freshwater species, channel catfish *Ictalurus punctatus*, white catfish *Ameiurus catus*, and American eels were probably undersampled because they are active at night, and thus can more readily avoid our gear. They are also more likely to inhabit deepwater habitat than many other species. The high abundance of catfish in the Connecticut River and the large fisheries they support are well documented in the creel survey data (Table 7). Anadromous fishes such as blueback herring *Alosa aestivalis*, alewife *A. pseudoharengus*, and American shad *Alosa sapidissima* are also poorly represented in our samples because they tend to be pelagic, mobile, and most importantly, their abundance is highly seasonal.

Relative abundance of fish species in the river is greatly influenced by habitat type. Specific

TABLE 1. Percent occurrence among main-stem, cove, and tributary sites of the Connecticut River for all fish taxa sampled by Connecticut Inland Fisheries Division boat and stream electrofishing (1988–2002). Native species denoted by asterisks (*). Blank cells indicate that the species was not detected at any site within that category.

		Habitat			
Scientific name	Common name	Main stem	Coves	Tributaries	All sites
	Number of sites =	3	8	19	30
Amiidae					
Amia calva	Bowfin		13		3
Anguillidae					
**Anguilla rostrata*	American eel	100	100	95	97
Atherinidae					
**Menidia beryllina*	Inland silverside	33			3
Catostomidae					
**Catostomus commersonii*	White sucker	100	100	79	87
Centrarchidae					
Ambloplites rupestris	Rock bass	100	38	37	43
**Lepomis auritus*	Redbreast sunfish	100	88	68	77
**Lepomis gibbosus*	Pumpkinseed	100	100	74	83
Lepomis macrochirus	Bluegill	100	88	58	70
L. macrochirus × *L. gibbosus*	Hybrid sunfish	33	38		13
L. macrochirus × *L. auritus*	Hybrid sunfish		13		3
Micropterus dolomieu	Smallmouth bass	100	63	11	33
Micropterus salmoides	Largemouth bass	100	88	63	73
Pomoxis annularis	White crappie	33	38		13
Pomoxis nigromaculatus	Black crappie	100	100	11	43
Clupeidae					
**Alosa aestivalis*	Blueback herring	100	100		37
**Alosa pseudoharengus*	Alewife	33	50	5	20
**Alosa sapidissima*	American shad	100	38	5	23
**Brevoortia tyrannus*	Atlantic menhaden		25		7
**Dorosoma cepedianum*	Gizzard shad	67	63		23
Cyprinidae					
Carassius auratus	Goldfish		13		3
Cyprinus carpio	Common carp	100	100	11	43
**Luxilus cornutus*	Common shiner			32	20
**Notemigonus crysoleucas*	Golden shiner	100	100	47	67
**Notropis bifrenatus*	Bridle shiner			5	3
**Notropis hudsonius*	Spottail shiner	100	100	47	67
Pimephales promelas	Fathead minnow			5	3
**Rhinichthys atratulus*	Eastern blacknose dace			63	40
**Rhinichthys cataractae*	Longnose dace			21	13
**Semotilus corporalis*	Fallfish	68		53	40
Esocidae					
**Esox americanus*	Redfin pickerel	33		58	40
Esox lucius	Northern pike	68	63	11	30
**Esox niger*	Chain pickerel		50	11	20
Fundulidae					
**Fundulus diaphanus*	Banded killifish	68	25	11	20
Gadidae					
**Lota lota*	Burbot			5	3
Gasterosteidae					
**Apeltes quadracus*	Fourspine stickleback			5	3

TABLE 1. Continued

Scientific name	Common name	Habitat			All sites
		Main stem	Coves	Tributaries	
Ictaluridae					
Ameiurus catus	White catfish	68	50		20
**Ameiurus nebulosus*	Brown bullhead		88	11	30
Ictalurus punctatus	Channel catfish	100	75		30
Moronidae					
**Morone americana*	White perch	68	100		33
**Morone saxatilis*	Striped bass	68	50		20
Percidae					
**Etheostoma olmstedi*	Tessellated darter	68	75	89	83
**Perca flavescens*	Yellow perch	100	100	32	57
Sander vitreus	Walleye		13		3
Petromyzontidae					
**Petromyzon marinus*	Sea lamprey	33	13	47	37
Salmonidae					
Salmo trutta	Brown trout	33		42	30
**Salvelinus fontinalis*	Brook trout			16	10
Soleidae					
**Trinectes maculatus*	Hogchoker		25		7
Umbridae					
Umbra limi	Central mudminnow			32	20

habitat types found in the lower Connecticut River include the main stem, coves, tributary streams, and the interface between the tributaries and the main stem (tidal streams and marshes, etc.). The relative abundance of all species combined was greater in the main-stem South than in other main-stem sections of the river (Table 2). This is in large part due to relatively higher abundance of white perch and redbreast sunfish in the south. Interestingly, however, total fish biomass (electrofishing kg/h) did not follow the same pattern and was nearly equal for all sections of river (100.4 kg/h in the North, 96.5 kg/h in the Central and 91.2 kg/h in the South). This difference is due almost exclusively to there being a greater abundance of large common carp in the central and northern sections. Although these fish were relatively few in number, they accounted for a large portion of the total biomass of fish sampled by electrofishing.

Three species that exhibited a pronounced north–south increasing trend in abundance were white perch, redbreast sunfish, and pumpkinseed (Tables 2, 3, and 4). Conversely, rock bass, smallmouth bass, common carp, and bluegill tended to be more abundant in the North section of the river. The reasons for some of these differences in distribution are not readily apparent. For example, the South section appears to contain more lentic habitat and to be better suited to support most sunfish species. As such, it is no surprise that pumpkinseed sunfish were more abundant in this section, yet bluegill, whose relative abundance should parallel pumpkinseeds, were more abundant in the North section.

Some species were more abundant in the coves than in the main stem. These included largemouth bass, bluegill, black crappie, and golden shiner (Tables 2 and 3). This is not surprising since each of these species is more typically associated with lake and pond type environments than with flowing waters (Whitworth et al. 1968; Hartel et al. 2002). Species found more often in main-stem habitats than coves included smallmouth bass, redbreast sunfish, and rock bass.

A number of fish species were in higher abundance or were only found in tributary streams of the river. These include eastern blacknose dace, longnose dace, common shiner, brook trout, fallfish, redfin pickerel, and sea lamprey (Table 4).

TABLE 2. Catch/effort (fish/h) for all fish species captured by boat electrofishing in three sections (North, Central, and South) of the Connecticut River main stem (1988–2002).

	River location			
Species	Main stem NORTH	Main stem CENTRAL	Main stem SOUTH	Main stem Mean
Spottail shiner	287.6	342.1	336.8	322.2
Redbreast sunfish	10.9	45.1	301.7	119.2
White perch		14.4	336.1	116.8
Yellow perch	52.3	31.5	58.0	47.3
White sucker	16.7	68.7	34.2	39.9
Smallmouth bass	55.2	21.5	27.5	34.7
Pumpkinseed	2.5	0.5	96.0	33.0
Rock bass	51.9	8.5	31.2	30.5
American eel	17.1	26.9	21.4	21.8
Blueback herring	4.1	4.7	44.2	17.7
American shad	8.3	9.4	18.5	12.1
Common carp	11.6	13.3	3.1	9.3
Largemouth bass	17.8	2.3	7.3	9.1
Bluegill	18.1	8.5	0.8	9.1
Golden shiner	1.1	1.3	13.4	5.3
Channel catfish	1.1	7.4	7.3	5.3
Black crappie	6.9	4.4	0.6	4.0
White catfish		4.9	4.9	3.3
Fallfish	2.9	0.3		1.1
Tessellated darter	1.8	1.0		0.9
Gizzard shad	0.4	1.5		0.6
Northern pike		0.5	1.2	0.6
Banded killifish	0.4	1.2		0.5
Striped bass		0.8	0.6	0.5
Alewife			1.0	0.3
Sea lamprey		0.8		0.3
Inland silverside			0.7	0.2
Brown trout		0.5		0.2
Redfin pickerel	0.4			0.1
White crappie		0.3		0.1
Bluegill × pump. hybrid		0.3		0.1
Total catch per hour	569.1	622.6	1346.5	846.1
Hours sampled	2.8	3.9	1.6	

Species that were common in both the main-stem river and tributary streams species include American eel, white sucker, golden shiner, redbreast sunfish, and pumpkinseed (Tables 3 and 4). American eel and white sucker use both habitats and migrate between them for spawning and overwintering, whereas sunfish species are more likely to be resident in either location.

A greater number of species (32) were sampled from smaller (<10 m wide) tributary streams than from larger ones (21). This may have been an artifact of sampling intensity because 14 small streams were sampled versus only 5 large streams. The greater apparent abundance of many fish species in smaller streams may similarly have been due to decreasing sampling efficiency with increasing stream size. Larger and more mobile fish, such as common carp, can more easily evade the stream electrofishing gear in the less confined areas of larger streams. Despite possible sampling bias, it is apparent that some species were more associated with small stream habitat (eastern blacknose dace, brook trout, fallfish, redfin pickerel, sea lamprey ammocetes, and

TABLE 3. Catch/effort (fish/h) for all fish species captured by boat electrofishing in coves within North, Central, and South sections of the Connecticut River (1988–2002).

	River location								
	NORTH	CENTRAL				SOUTH			
Species	Mass. Cove	Wethersfield Cove	Keeney Cove	Crow Point Cove	Wrights Cove	Salmon Cove	Chapmans Pond	Hamburg Cove	Mean All coves
Spottail shiner	118.3	350.3	76.6	745.1	11.0	506.2	355.2	42.3	275.6
Yellow perch	78.2	89.2	19.2	127.7	63.4	84.8	93.1	167.3	90.4
Pumpkinseed	49.8	45.8	6.6	30.3	88.2	24.2	199.9	142.1	73.4
Largemouth bass	148.0	35.1	5.9	52.1	74.4		67.4	25.3	51.0
Bluegill	121.2	61.0	14.0	43.2	79.9		20.3	28.9	46.1
White sucker	3.8	33.4	26.5	72.1	11.0	115.3	10.3	51.5	40.5
Golden shiner	6.7	4.0	2.9	15.4	148.9	10.1	70.4	60.8	39.9
White perch	2.6	16.1	14.0	42.3	8.3	28.3	74.6	81.7	33.5
American eel	5.1	24.9	21.4	18.2	52.4	50.5	23.1	36.0	29.0
Blueback herring	2.1	56.9	5.2	99.7	13.8	8.1	22.0	11.8	27.5
Redbreast sunfish	1.3	7.6	19.9	32.0		42.4	1.3	40.0	18.1
Black crappie	37.3	15.8	1.5	32.5	35.8	4.0	9.1	0.3	17.0
Common carp	19.5	6.6	13.3	4.4	2.8	6.1	8.9	5.3	8.4
Atlantic menhaden				45.8			1.3		5.9
Gizzard shad	12.4	11.6		4.6	8.3		0.8		4.7
American shad		6.1		16.5			3.2		3.2
Brown bullhead	0.6	0.7		1.3	8.3	2.0	9.9	1.9	3.1
Northern pike	0.3	1.5		1.8	11.0		2.4		2.1
Alewife				1.5	8.3		4.1	1.9	2.0
Tessellated darter	0.6	1.2		3.1		8.1	0.3	0.3	1.7
Rock bass	1.3	0.7				10.1			1.5
Channel catfish	0.0	0.7	1.5	0.4	5.5		1.7	0.5	1.3
White crappie	3.2	0.2		4.4					1.0
Banded killifish							6.9	0.3	0.9
Smallmouth bass	0.6	3.2		0.4			0.7	0.8	0.7
White catfish		0.7	0.7				1.7	2.4	0.7
Striped bass		1.5		2.2			0.2	1.1	0.6
Chain pickerel		0.5		0.9			0.5	1.9	0.5
Hogchoker							1.7	0.8	0.3
Goldfish		1.2							0.2
Bluegill × pump. hybrid	0.3			0.4				0.3	0.1
Walleye		0.7							0.1
Sea lamprey	0.6								0.1
Bowfin	0.3								<0.1
Bluegill × redbreast hybrid							0.3		<0.1
Total catch per hour	614.1	777.2	229.2	1398.3	631.3	900.2	991.3	705.5	780.9
Hours sampled	3.1	4.1	1.4	2.3	0.4	0.5	4.2	3.8	

common shiner), whereas others (American eel, redbreast sunfish, and largemouth bass) were more common in larger streams. Tributary streams provide spawning habitat for blueback herring and shad and spawning and rearing habitat for sea lamprey. Brown trout, fallfish, and sea lamprey ammocetes that were found in the main stem were near the mouths of tributary streams, presumably having emigrated from these streams.

The interface between main-stem and tributary streams typically support high species richness (Minshall et al. 1985) and provide critical

TABLE 4. Mean catch/effort (fish/h) for fish species captured by stream electrofishing in the lower reaches of Connecticut River tributary streams for two size-classes of streams (small = <10 m wide and large = ≥10 m wide) in three sections (North, Central and South) of the river (1988–1999).

	Small streams (<10 m)			Large streams (≥10 m)		
Species	NORTH	CENTRAL	SOUTH	CENTRAL	SOUTH	Mean
American eel	35.4	77.3	40.5	109.4	144.9	81.5
White sucker	17.7	44.2	25.6	12.5	5.0	21.0
Blacknose dace	84.7	7.7	6.6	2.2		20.2
Tessellated darter	15.7	42.7	19.7	9.4	13.1	20.1
Brook trout	7.6	55.4	0.2			12.6
Redbreast sunfish	0.4	1.0	8.4	16.4	34.7	12.2
Pumpkinseed	3.6	6.9	15.7	0.9	21.2	9.7
Fallfish		14.0	26.0	3.1	2.8	9.2
Spottail shiner	16.1	13.5	8.1	3.4	1.6	8.5
Redfin pickerel	1.3	32.0	1.5	0.5	0.9	7.2
Sea lamprey	22.1	5.2	1.9	3.3	1.8	6.9
Banded killifish	1.4	0.4	23.1			5.0
Bluegill	1.3	5.5	4.1	9.1	1.4	4.3
Golden shiner	9.4	3.6	5.7	0.2		3.8
Longnose dace			7.2	10.5		3.5
Rock bass	7.0	0.1	0.2	7.0		2.9
Fourspine stickleback		11.0				2.2
Largemouth bass	0.6	1.7	<0.1	3.3	1.8	1.9
Central mudminnow		5.8	0.2			1.2
Brown trout	0.7	0.2	0.4	3.7	0.5	1.1
Common carp		5.4	0.1			1.1
Common shiner	0.2	3.8		0.8		1.0
Brown bullhead		0.8	2.9			0.7
Yellow perch		0.7		0.2	1.3	0.4
American shad				1.9		0.4
Fathead minnow	1.4					0.3
Chain pickerel			0.4		0.4	0.2
Bridle shiner		0.7				0.1
Smallmouth bass	0.2			0.1		0.1
Northern pike		0.2				<0.1
Alewife			0.1			<0.1
Black crappie		0.1				<0.1
Burbot		0.1				<0.1
Total catch per hour	226.8	340.0	198.6	197.9	231.4	239.3
Hours sampled	1.8	3.7	2.1	2.4	0.8	

habitat for a number of fish species (e.g., black bass, fourspine stickleback *Apeltes quadracus*; Green et al. 1993; Carlson 1990; Smith 1985). For the Connecticut River, this interface includes tidal freshwater marshes and the lower tidal reaches of the tributary streams themselves. Fourspine stickleback, central mudminnow, redfin pickerel, juvenile northern pike, and burbot, a state-listed endangered species, were all found predominantly in these areas.

Age and Growth

Mean growth rates of most fish species in the river are relatively fast compared to state means for Connecticut lakes and ponds (Table 5). This is probably because the river is a very large and productive system that supports a greater diversity of species as well as greater densities of small prey fish than are typically found in most lakes and ponds (e.g., spottail shiners, see Tables 2 and 3). Growth rates of most fish species are faster in

TABLE 5. Mean backcalculated length at age (mm) for selected Connecticut fish species sampled from the Connecticut River (North, Central, and South sections) (1988–1995). State means are from public lakes and ponds where Ns = the number of sites included in the mean. For Connecticut River growth rates, Nf = the number of fish aged.

		AGE										
SPECIES		1	2	3	4	5	6	7	8	9	10	11
Largemouth bass												
Conn. R. (North)	Nf = 61	116	211	290	339	370	392	417	444	471		
Conn. R. (Central)	Nf = 113	126	217	294	339	387						
Conn. R. (South)	Nf = 114	143	261	328	379	416	438	453	464	478		
State mean	Ns = 72	94	192	267	322	364	399	429	454	475	493	
Smallmouth bass												
Conn. R. (North)	Nf = 46	110	196	266	323	366	394					
Conn. R. (Central)	Nf = 65	110	201	280	340	382	438					
Conn. R. (South)	Nf = 29	130	237	284	327							
State mean	Ns = 30	92	158	222	280	318	360	406	444	465		
Northern pike												
Conn. R. (Central)	Nf = 169	419	597	702	739	804						
Conn. R. (South)	Nf = 904	445	619	703	765	819						
Bantam L.*	Nf = 1048	319	551	679	753	824						
Mansfield Hollow Res.*	Nf = 586	338	502	622	739	818						
Black crappie												
Conn. R. (North)	Nf = 25	76	158	214	258	293	307					
Conn. R. (Central)	Nf = 55	79	167	222	264	294	320					
Conn. R. (South)	Nf = 12	83	187	247								
State mean	Ns = 48	78	150	207	240	263	279	301	320			
White crappie												
Conn. R. (All sites)	Nf = 18	75	187	232	262	280						
Yellow perch												
Conn. R. (North)	Nf = 33	114	197	238	283	299	312	325				
Conn. R. (Central)	Nf = 131	96	145	181	215	232						
Conn. R. (South)	Nf = 73	105	161	200	230	255	286					
State mean	Ns = 71	88	139	177	204	224	240	256	270	284	295	310
White perch												
Conn. R. (Central)	Nf = 75	98	174	217	249	273	285	305				
Conn. R. (South)	Nf = 65	114	193	237	268	284	297	308	315			
State mean	Ns = 18	96	161	207	234	251	266	283	298	310	319	
Bluegill												
Conn. R. (North)	Nf = 24	52	100	135	161	183	205					
Conn. R. (Central)	Nf = 80	48	93	134	156	165	182					
Conn. R. (South)	Nf = 25	57	109	151	180	190						
State mean	Ns= 72	47	91	132	162	180	191	198	206			
Pumpkinseed												
Conn. R. (North)	Nf = 15	55	108	135	158							
Conn. R. (Central)	Nf = 36	55	103	132	151	163	169					
Conn. R. (South)	Nf = 32	65	119	147	170							
State mean	Ns = 69	51	89	124	147	162	171	180				
Redbreast sunfish												
Conn. R. (North)	Nf = 10	56	103	134	159							
Conn. R. (Central)	Nf = 69	53	90	125	155	168						
Conn. R. (South)	Nf = 32	59	106	137	159	169						
State mean	Ns = 18	50	84	114	136	150	165					
Rock bass												
Conn. R. (North)	Nf = 25	63	115	159	192	218						

TABLE 5. Continued

SPECIES		AGE										
		1	2	3	4	5	6	7	8	9	10	11
Conn. R. (Central)	Nf = 19	57	107	152								
State mean	Ns = 25	56	95	133	161	178	197	207	223			
Golden shiner												
Conn. R. (Central)	Nf = 11	91	146	182	198	202						
Conn. R. (South)	Nf = 40	102	143	174	199	213	219	236				
State mean	Ns = 52	99	147	182	204	217	227	229				
Gizzard shad												
Conn. R. (All sites)	Nf = 18	199	335	371	381	404						

* CTDEP Inland Fisheries Division unpublished data.

the southern section of the Connecticut River than in either the northern or central sites (Table 5). Fish growth may be faster in the southern part of the river for a number of reasons, including slower flow rates and higher water temperatures. More importantly, the southern site encompasses the Connecticut River Estuary, which increases the productivity of the area. Levels of total phosphorous, nitrate, and alkalinity were all reported to be highest in the southern part of the river (U.S. Geological Survey 1994). It also appears likely that this section contains a more diverse and dense forage base. Electrofishing catch/effort in the southern main-stem site for all fish species combined was more than twice that of either the northern or the central sites (Table 2); however, this was the result of higher densities of only two species, redbreast sunfish and white perch. Also contributing to the diversity and density of the forage base are species that were not sampled or were poorly sampled by our gear, including many marine and estuarine transients (amphidromous species, e.g., killifish) and seasonal influxes of juveniles of various herring species (Whitworth 1996; MacBroom 1998).

Creel Survey Results

Anglers fished between 264,000 and 279,000 h and caught an estimated 411,000–469,000 fish (all species combined) each year during 1997 and 1998 (Tables 6 and 7). Total catch was 50% greater for shore anglers than boat anglers and was greatest in the southern region of the river (Table 8). Approximately half (50%) of the fish caught were harvested and the remainder released (Table 7). The lower harvest rate in 1998 (38% versus 53% in 1997) was a result of a threefold increase in the

TABLE 6. Estimated total fishing effort (angler-hours) on the Connecticut River (from Hartford south), as determined by creel surveys in 1997 and 1998 (Route 1: mouth of the river upstream to Haddam, Route 2: Haddam upstream to Middletown, Route 3: Middletown upstream through Hartford). Standard errors in parentheses.

Year/Mode	Route			Total
	1*	2*	3	
1997				
Shore	61,020 (16,580)	36,240 (8,885)	75,299 (12,519)	172,559 (37,983)
Boat	33,819 (10,569)	33,796 (11,713)	38,305 (12,099)	105,960 (34,381)
1998				
Shore	44,505 (8,473)	39,040 (11,576)	60,386 (16,334)	143,931 (36,383)
Boat	36,953 (11,333)	37,935 (10,382)	45,394 (13,517)	120,282 (35,231)
Combined	81,458 (19,806)	76,975 (21,958)	105,780 (29,851)	264,213 (71,614)

* Routes 1 and 2 not sampled during March and April.

TABLE 7. Estimated total angler catch (standard error [SE] in parentheses) for the Connecticut River (from Hartford south), as determined by creel surveys in 1997 and 1998. Fishery types are listed in rank order of total catch for the 2 years. The percentage of the total catch harvested (kept) is also listed.

Species name	1997 Total catch	1997 (SE)	1997 Percent harvested	1998 Total catch	1998 (SE)	1998 Percent harvested
All species	410,954	(128,600)	53	468,669	(87,924)	38
Catfish*	78,465	(17,322)	62	66,908	(12,554)	60
Striped bass	37,182	(14,869)	6	99,824	(20,051)	2
White perch	72,339	(26,622)	80	63,321	(11,698)	63
Panfish**	38,873	(3,045)	56	59,954	(2,509)	40
Bluefish *Pomatomus saltatrix*	63,217	(33653)	96	19,579	(11,737)	73
LM & SM bass	33,529	(19,225)	9	48,614	(16,189)	11
Yellow perch	31,289	(8,860)	44	27,988	(3,026)	26
River herring***	16,949	(387)	83	19,287	(7,214)	100
Hickory shad *Alosa mediocris*	20,144	(3,521)	<1	14,117	(1,418)	43
Winter flounder *Pseudopleuronectes americanus*	5,877	na[1]	55	19,265	na[a]	68
Summer flounder *Paralichthys dentatus*	2,307	na[1]	15	12,441	(515)	48
Northern pike	3,078	(534)	0	2,385	(1,152)	7

Also taken in the catch: American shad, American eel, Atlantic tomcod *Microgadus tomcod*, brook trout, brown trout, common carp, goldfish, little skate *Leucoraja erinacea*, northern pike, rainbow smelt *Osmerus mordax*, spottail shiner, white sucker, windowpane *Scophthalmus aquosus*. Individual catch was not calculated because fewer than five anglers were interviewed who had caught each species. Estimated catch for all 12 fishery types was 20,500 fish over the 2 years.

[a] Catch was not distributed sufficiently for calculation of a standard error.
* Includes white and channel catfish, and brown bullhead.
** Includes bluegill, pumpkinseed, rock bass *Ambloplites rupestris*, black and white crappie, and other sunfish.
*** Includes alewife and blueback herring.

catch of striped bass, which is, historically, a catch-and-release fishery (2–6% harvest).

Total catch included 26 and 30 species in 1997 and 1998, respectively. More than 95% of the catch consisted of 21 species grouped into 12 fishery types (Table 7). The most common species caught and harvested were catfishes (white catfish, channel catfish, and brown bullhead combined) and white perch. An estimated 66,000–78,000 catfishes and 63,000–72,000 white perch were caught each year by anglers fishing the surveyed portion of the Connecticut River during 1997 and 1998. Each of these fisheries had harvest rates of 60% or higher and were taken principally by shore anglers (Tables 7 and 8). Catfishes and white perch were caught in all three surveyed river sections (Routes 1, 2, and 3). The catfish catch was equally distributed among the three routes, whereas white perch were taken in larger numbers downstream of Haddam in Route 1 (Table 8).

Striped bass were ranked fifth among fishery types in total catch in 1997 and first in 1998; however, harvest never exceeded 2,000 fish either year (only 2–6% harvested). The increased catch in 1998 came from boat-based anglers catching and releasing large numbers of small fish at the mouth of the river (Route 1) during the early season (March–April). The threefold increase in catch matched increases in the abundance of 2–5-year-old striped bass throughout Long Island Sound. The striped bass fishery in the Connecti-

TABLE 8. Distribution of Connecticut River angler catch by mode, location, and season as determined by creel surveys in 1997–1998. Mean percentage of total catch by mode, route, and season is listed by species or species group in rank order of total catch for the 2 years.

	PERCENT OF TOTAL CATCH								
	By mode		By route			By season			
Species name	Boat	Shore	1	2	3	M-A	M-J	J-A	S-O
All species	39	61	48	19	32	17	35	47	11
Catfish*	21	79	33	33	33	6	40	49	5
Striped bass	46	54	56	30	14	41	50	6	3
White perch	27	73	71	16	13	30	30	26	14
Panfish**	16	84	12	12	76	<1	34	51	15
Bluefish	48	52	90	10	0	0	3	88	9
LM & SM bass	73	27	4	43	53	0	41	46	13
Yellow perch	54	46	47	8	46	28	15	27	30
River herring***	45	55	0	8	92	0	99	1	0
Hickory shad	4	96	82	4	14	0	30	28	41
Winter flounder	96	4	100	0	0	90	6	4	0
Summer flounder	96	4	62	38	0	0	45	47	8
Northern pike	54	46	1	11	88	0	34	42	14

* Includes white and channel catfish, and brown bullhead.

** Includes bluegill, pumpkinseed, rock bass, black and white crappie, and other sunfish.

*** Includes alewife and blueback herring.

cut River was widespread during both survey years with fish being caught by boat and shore anglers in all areas of the river. A large fishery for striped bass also takes place each spring upriver from Hartford to the Massachusetts border (outside the range of this creel survey) (Jacobs and O'Donnell 2002).

Panfish and yellow perch were caught and harvested in large numbers, with panfish (bluegill, pumpkinseed, rock bass, white crappie, black crappie, and redbreast sunfish) being largely taken upriver of the Portland Bridge (Route 3) (Table 8). The catch of yellow perch was largest downstream of Haddam (Route 1) and upriver from Portland (Route 3) and was considerably less between Portland and Haddam (Route 2) (Table 8). The fishery for panfish is principally shore-based, whereas yellow perch were taken in equal numbers by shore and boat anglers (Table 8). The estimated combined catch of panfish and yellow perch for 1997–1998 is 70,000–88,000 fish per year with a harvest rate of approximately 50%.

Largemouth and smallmouth bass were taken principally by boat anglers in the northern two routes from May to August (Table 8). This fishery is mostly catch and release with harvest rates of only 9–11% (Table 7). This section of the Connecticut River is a popular tournament fishing site, attracting an average of 108 tournaments per year that account for approximately 5,547 angler-days and 44,378 angler-hours annually (CTDEP Inland Fisheries Division unpublished data). Still, the estimated total catch from these tournaments (5,325 bass annually assuming an average catch rate of 0.12 per hour) is only a small percentage of the total catch of 30,000–50,000 bass per year.

Bluefish ranked high in the catch in 1997, but dropped to seventh in 1998 (Table 7). This fishery consists almost entirely of "snappers." or age-0 (89% of total bluefish harvest), taken at the mouth of the river as part of a coastwide summer fishery. Fluctuations in snapper abundance and availability from year to year make this fishery highly variable. The catch of bluefish, winter flounder, and summer flounder from the lower river (Route 1, Table 7) is a small fraction of the total annual catch of each of these species occurring in Connecticut waters; the majority occurs in Long Island Sound.

Catch of alewife and blueback herring (river herring) was consistent between years and ranked 7–8 overall (Table 7). River herring were taken by both shore and boat anglers fishing the upriver section (Route 3) during the spring season (Table 8). Nearly all of these fish were harvested (83–100%), presumably for use as bait for striped bass. The total catch of river herring from the Connecticut River drainage was greatly underestimated by this survey because catch from the tributaries and the main-stem river north of Hartford are not included. Similarly, there is no targeted recreational fishery for American shad in the main-stem Connecticut River south of Hartford, and hence, this species did not show up in our creel survey. A large recreational shad fishery does take place north of Hartford and a commercial fishery occurs south of Rocky Hill (Crecco and Savoy 1987; Gephard and McMenemy 2004, this volume). Among other herring species, hickory shad support a recently growing shore-based fishery that is largest downstream of Haddam (Route 1) (Table 8), and gizzard shad are occasionally caught by anglers fishing throughout the river. The growing recreational catch and abundance of hickory shad and the presence of gizzard shad are probably reflective of recent range expansions of these species (Dr. Robert E. Schmidt, Simon's Rock College, Great Barring-ton, Massachusetts, personal communication).

The Connecticut River and its coves support a recreational fishery for northern pike (Tables 7 and 8). This fishery is underrepresented in our data as much of the annual fishing activity occurred prior to the onset of the creel survey. There is a directed fishery for pike during the winter, both through the ice and in open water during March prior to spawning (CTDEP Inland Fisheries Division, unpublished data). Anglers targeting other species take most of the pike caught during the remainder of the year. Data collected in this creel survey indicate that the catch of northern pike is largest north of Middletown (Route 3); however, much of the unrecorded late winter/early spring fishery is known to occur in the Haddam area (Route 2).

The diversity of the Connecticut River fishery is evident in both the variety of game fish and panfish targeted by anglers and in the variety of species that add to their catch (Figure 3). A third to one-half of shore anglers said they were not fishing for any particular species (37% in 1997, 47% in 1998). In contrast, only 7–20% of boat anglers had no particular catch in mind. Of those who were targeting a specific type of fish, a third (30–42%) of both boat and shore anglers were pursuing striped bass (Figure 2). A second third of boat anglers were pursuing largemouth and smallmouth bass. Only 6–10% of shore anglers targeted freshwater bass. More popular targets for shore anglers were white perch, bluefish, and catfish, each sought after by 11–18% of anglers. White perch were targeted by 8% of boat anglers, whereas all other species were each targeted by less than 6% of boat or shore anglers.

The percentage of anglers catching striped bass, largemouth, and smallmouth bass was considerably less than the percentage that said they were seeking these species (Figure 3). In contrast, greater numbers of anglers caught catfish, white perch, and panfish than were targeting them. Fish in the "Other species" category (i.e., other than those listed in Figure 3) were well represented among the species caught by both boat and shore anglers.

Comments on Selected Fishes of the Lower Connecticut River

Largemouth and Smallmouth Bass

Largemouth and smallmouth bass collectively are Connecticut's most ubiquitous game fish, with one or both being present in almost all lakes, ponds, and larger streams (Jacobs and O'Donnell 1996). They are also the principal fish predators in most lakes and large river systems and are certainly among the most influential resident fish predators in the Connecticut River. Largemouth bass were most numerous in coves, whereas smallmouths were predominantly found in main-stem sites (Tables 2 and 3). The combined electrofishing catch/effort for largemouth and smallmouth bass (43.8 fish/h) was far greater than for any other large predator (i.e., northern pike, channel catfish, brown trout, striped bass) (Table 2). While this is due in large part to these species being more vulnerable to electrofishing gear (Reynolds and Simpson 1978), it also reflects their greater availability among all habitats (both main stem and coves) during both spring and fall. Channel catfish were more abundant in the main stem (based on the creel survey data) than in the coves, and striped bass were seasonally abundant in the main stem but are absent much of the year. Northern pike, walleye, and brown trout were considerably less abundant than largemouth bass or smallmouth bass in all areas of the river.

The Connecticut River supports a large directed fishery for largemouth and smallmouth bass (Table 7), including many of the largest competitive fishing tournaments in New England (CTDEP

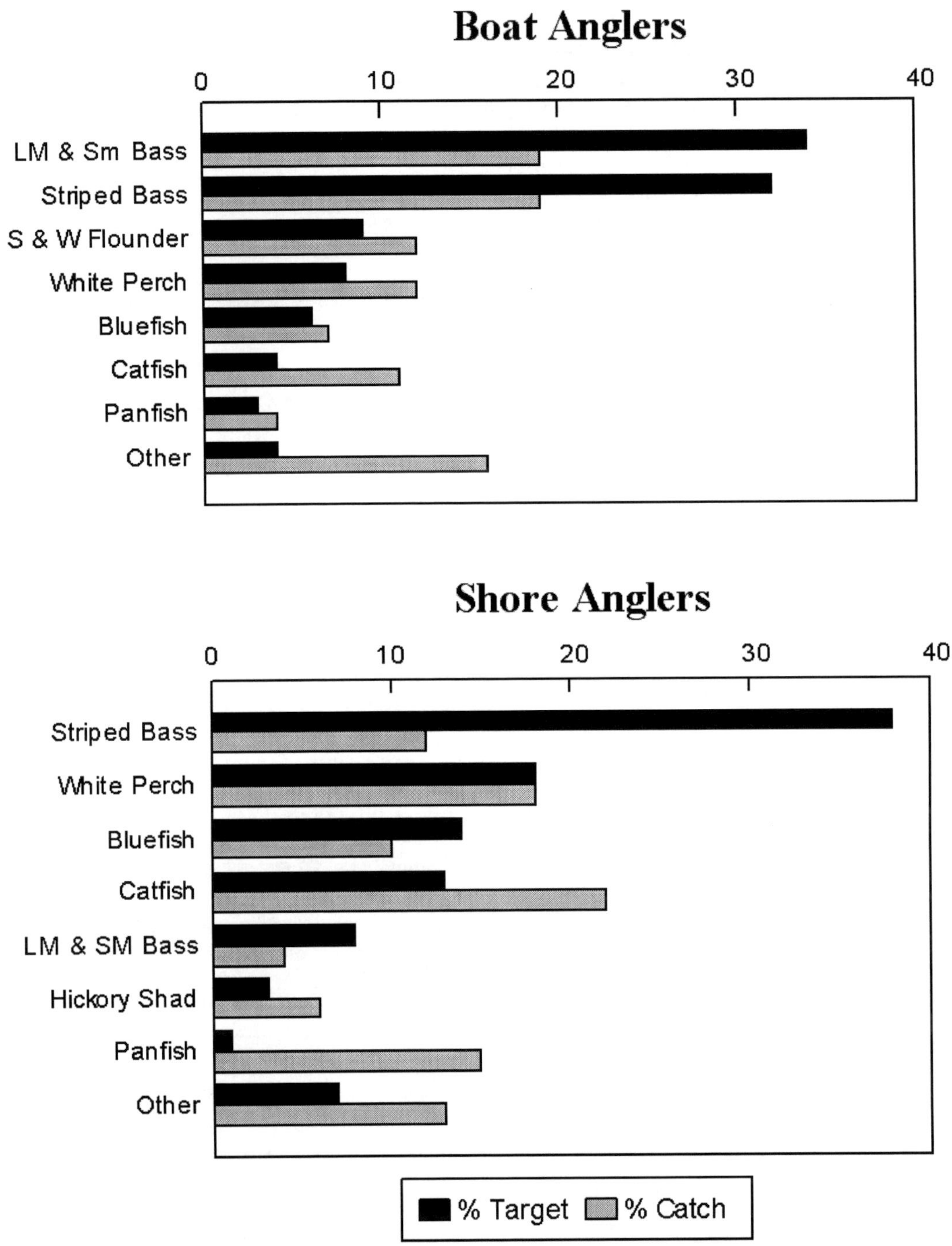

FIGURE 3. Comparison of the frequency (%) a species was targeted by boat and shore based anglers and percent of total catch by mode. Catfish include white and channel catfish, and brown bullhead; panfish include bluegill, pumkinseed, rock bass, white and black crappie, and redbreast sunfish. "LM & SM Bass" = largemouth and smallmouth bass and "S & W Flounder" = summer and winter flounder.

Inland Fisheries Division, unpublished data). In the course of conducting these tournament events, bass are routinely captured in one section of river and transported 30 km or more before being released. The effects on the fish populations and on the fishery of the constant relocation of apex predators among habitats (cove and main stem) and over distance are unknown.

Northern Pike, Chain Pickerel, and Walleye

The northern pike is the river's largest resident game fish and fish predator. Northern pike in the Connecticut River are relatively fast-growing and short-lived, reaching 60 cm by age 2 and 80–100 cm at their maximum age of 5–6 (Phillips and Bender 1984). Relative abundance of northern pike was greater in coves than in the main stem (Tables 2 and 3). Northern pike sampled by boat electrofishing in the main stem tended to be in backwaters and associated with structure or vegetation. Northern pike captured during stream electrofishing were predominantly juveniles and were taken in the lower sections of tributaries near their junctions with the river (Table 4). Northern pike are probably underrepresented in our electrofishing samples because they seem to be sensitive to the electric field and more likely to avoid our gear rather than display the typical electrotaxis of many other fish species.

Pike spawn during the early spring (late February through March), usually over terrestrial vegetation in seasonal marshes created by spring floodwaters (Barry and Machowski 1996). Since the early 1980s, the CTDEP Inland Fisheries Division has managed two pike spawning marshes located adjacent to the river at Haddam Meadows State Park (Phillips and Bender 1992). In these marshes, water levels are manipulated to provide optimal spawning conditions. Adult prespawn pike are captured during March, either by trap net or by a weir trap located at the marsh, and stocked into the marshes to spawn. Adults are allowed to spawn naturally and the young pike grow for 3 months within the marshes. During June, the marshes are drained and fingerling (100–200 mm TL) and adult pike are captured in weir traps as they attempt to emigrate. Adults are returned to the river and fingerlings are used to enhance the Connecticut River population and to maintain several other pike fisheries in Connecticut lakes.

The pike population in the lower river appears to have increased over the past 15 years. During the late 1980s to early 1990s, an average of less than three adult pike per day were captured at the Haddam Meadows weir trap during marsh management operations; whereas, by the late 1990s, catch rates at this trap averaged more than 15 pike per day. Anecdotal reports by anglers have also indicated an increase in pike catches in this area. Although chain pickerel are very common in Connecticut lakes and ponds, they are relatively uncommon in the Connecticut River. Among river sites, they were encountered only in a few coves and at low densities (Table 3).

A few walleye were sampled in Wethersfield Cove during the Statewide Lake and Pond Electrofishing Survey (Table 3). Other adult walleye have occasionally been taken in trap nets or gill nets from the northern main stem, Wethersfield Cove and Wrights Cove (CTDEP Inland Fisheries Division, unpublished data). Additionally, there have been scattered angler reports of walleye caught from the central and northern river. Little is known about the extent of the river population in Connecticut. These fish could be residents, utilizing the riffle areas below the Enfield Dam to spawn; however, they could also be seasonal migrants from a population that spawns below the Holyoke Dam in Massachusetts (Boyd Kynard, U.S.G.S Conte Anadromous Fish Research Laboratory, Turner Falls, Massachusetts, personal communication).

Salmonids

Brown and brook trout were commonly encountered in most of the Connecticut River tributaries (Table 4). Brook trout were limited to smaller tributaries (<10 m wide) and were almost exclusively wild populations. Brown trout were present in many of the tributaries where there was typically a mix of wild and holdover hatchery-raised fish (Hagstrom et al. 1996). Four streams (Beeman Brook, Waterworks Brook, Mill Creek, and Hubbard Brook in Glastonbury) held predominantly wild brown trout populations. Only one brown trout was sampled by boat electrofishing, and it was near the mouth of a tributary of the main stem (Table 2). A small number of large (36–40 cm) holdover or wild brown trout were taken by stream electrofishing in the mouths of tributaries. Scale aging revealed that these fish grew rapidly, and we believe that they may have used the main stem to forage. Little is known of abundance and habits of potamodromous or sea-run brown trout in the Connecticut River. Restoration of Atlantic salmon *Salmo salar* to the Connecticut River system began in 1967 (Stolte 1980). Returning adults continue to be rare (Gephard and McMenemy 2004, this volume) and were not encountered during our sampling. Additionally, none were collected in the tributary streams that were sampled because these waters were not among those stocked with salmon parr or smolts.

Panfish

White perch are native to Connecticut and, although not encountered by electrofishing in the northern river, are the most abundant panfish species in the lower Connecticut River main stem (Table 2). Marcy (2004 [1976]) similarly reported that they were the dominant resident fish species in the lower river during his 1965–1972 survey. They overwinter and spawn during early spring in large numbers in the southern river (notably Hamburg Cove), after which some of the population migrates north for the remainder of the season (Howell and Molnar 2004, this volume). No white perch were encountered during our summer stream electrofishing. The extent to which white perch enter tributary streams of the river during other seasons is unclear; however, they are known to make large spring spawning runs into tributaries of the nearby Hudson River (Schmidt and Lake, in press).

Populations of both black crappie and white crappie occur in the Connecticut River. Black crappie were sampled from all main-stem and cove sites, whereas white crappie occurred at much lower abundance and were only found in central and northern sections (Wethersfield Cove, Crow Point Cove, Mass. Cove and the main stem near Hartford, Tables 2 and 3). Crappie were rarely encountered in the river's tributary streams (Table 4). There are popular ice fishing and spring openwater sport fisheries for crappie in the river, especially in the central coves (notably Wethersfield and Crow Point coves) (Jacobs and O'Donnell 2002; CTDEP Inland Fisheries Division, unpublished data). The only known population of white crappie in Connecticut occurs in the river. There is no stocking record for white crappie, and it is unclear when or how they became established. They may have been mistakenly stocked along with black crappie.

Four species of sunfish were identified during our survey; bluegill, pumpkinseed, redbreast, and rock bass (Tables 2–4). The most abundant sunfish species overall was the redbreast sunfish. This species prefers flowing water (Whitworth et al. 1968) and were most numerous in main-stem and stream samples. As mentioned previously, redbreast sunfish become increasingly more abundant from north to south throughout the river system, whereas bluegill displayed the opposite geographic trend. Bluegill and pumpkinseed are species that prefer lentic water (Hartel et al. 2002) and were the most common sunfish species in the river's coves (Table 3). Rock bass were more abundant in main-stem than cove samples, probably in part due to their preference for rocky, nonvegetated habitat (Whitworth et al. 1968).

It is noteworthy that rock bass, pumpkinseed, and redbreast sunfish have been abundant in our river samples since the late 1980s. Yet Marcy's (2004 [1976]) exhaustive survey indicates that rock bass were rare and pumpkinseed and redbreast sunfish absent in the Haddam area of the river during 1965–1972. Whitworth et al. (1968) indicated that all three species were present in at least some Connecticut River tributaries during the late 1960s. Range distribution maps from the 1988–1994 Connecticut statewide stream survey (Hagstrom et al. 1996) revealed that all three species have generally extended their ranges both among the river's tributaries and statewide relative to Whitworth et al. (1968). Although Marcy employed different gear types (bottom trawl and beach seine) than our study, it seems unlikely that these species would have gone undetected if they were common at the time. Therefore, it is apparent that all three species (especially redbreast sunfish) experienced significant population expansions in a period of less than 20 years. It is unclear whether this is a continuing trend.

Catfishes

Three catfish species, channel catfish, white catfish, and brown bullhead, are found in the Connecticut River (Tables 1–4). Electrofishing catch rates for all catfish species were characteristically low because the electrofishing methods used in this study do not sample catfish as well as many other species (Vokoun and Rabeni

1999). However, high catch rates in both trap nets (Howell and Molnar 2004) and the creel survey (Table 7) attest to catfishes being abundant in the river.

Both channel catfish and white catfish were more commonly encountered by electrofishing in main-stem sites than in coves (Tables 2 and 3), but were not present in the tributaries during late summer stream shocking (Table 4). Early spring trap netting indicates that both species use the coves seasonally, possibly as wintering and/or spawning areas (Howell and Molnar 2004). Brown bullhead, on the other hand, were only found in cove and small stream habitats and were absent in main-stem samples (Tables 2–4). Channel catfish and brown bullhead were found in all three sections of the river (North, Central, and South), whereas white catfish were not found in the northern section. White catfish only outnumbered channel catfish in extreme southern samples (Hamburg Cove and North Cove, Table 3 and Howell and Molnar 2004), indicating that white catfish probably have a higher tolerance for saline conditions than channel catfish.

The relative abundance and distribution of the three catfish species have undergone a dramatic change in the river over the past 30 years. Channel catfish were relatively rare throughout the Connecticut River during the late 1960s (Whitworth et al. 1968). Marcy (2004 [1976]) indicates that brown bullheads and white catfish were abundant in main-stem areas near Haddam during 1965–1972 and that channel catfish were rare. Interestingly, the warmwater discharge of the Connecticut Yankee atomic power plant was one of the early aggregating areas for channel catfish during those years (i.e., they were more common there than elsewhere in the river). Since the 1970s, channel catfish have come to dominate the main-stem areas of the river (Whitworth 1996), possibly outcompeting and displacing bullhead and white catfish from their historic habitats. This may be a regional trend because Jordan et al. (in review) also believe that channel catfish may have displaced white catfish in the Hudson River.

Rare Fishes

There are a number of noteworthy species in the Connecticut River drainage that are either rare or have a limited distribution. As previously mentioned, white crappie and walleye occur sporadically in the northern and central sections of the river. The central mudminnow can be locally abundant, but is only found in Connecticut in marshes and near the mouths of small tributaries of the Connecticut River between Windsor and Haddam (Hagstrom et al. 1996). Fourspine sticklebacks are typically estuarine species, although strictly freshwater populations do occur (Whitworth 1996). They are locally abundant in just a few freshwater areas of the river around Hartford (Hagstrom et al. 1996).

Two juvenile burbot, a state endangered species, were sampled in Salmon Brook, Glastonbury, within approximately 100 m of Keeney Cove and the Connecticut River. No more is known of this population except that anglers have periodically reported catching fish in the Hartford area that fit the description of burbot. The only other known population of burbot in the state is in the Hollenbeck River in the Housatonic River drainage of northwestern Connecticut (Hagstrom et al. 1996). Another Connecticut State endangered species, the American brook lamprey, was found in Hubbard Brook, Glastonbury and Salmon Brook, Glastonbury, upstream of the areas indicated in Figure 1. Only a few isolated American brook lamprey populations exist in Connecticut, typically upstream of a barrier to migration of sea lamprey.

Two bowfin *Amia calva* have been captured by CTDEP Inland Fisheries Division (one during this survey in the Mass. Cove, see Table 3 and one in Chapmans Pond during sampling on another project). Why bowfin are uncommon in the river given their reputation for being hardy and invasive is unclear. Perhaps the river has insufficient weedy backwater areas to provide the spawning habitat needed to sustain an abundant population. The only other known self-sustaining population of bowfin in Connecticut is in Scoville Reservoir in Wolcott where they are common (Whitworth 1996).

Other species that were rare in our samples from the Connecticut River, but can be found elsewhere in the state are bridle shiner, fathead minnow *Pimephales promelas*, and goldfish.

Anadromous and Estuarine Fishes

Our electrofishing samples do not reliably represent the relative abundance of most anadromous fishes because their spawning runs did not necessarily coincide with our sampling dates. Nonetheless, many anadromous and brackish water species were taken by our gear (Tables 1–4). Juvenile blueback herring were the most numerous of these species, especially in the central cove sites (Table 3). Anadromous alewives were found only in the central and southern sites at relatively low densities. Gizzard shad were most common in the northern and central coves, whereas menhaden, hogchoker, and inland silversides *Menidia beryllina* were primarily found in the southern sites (Tables 3 and 4). Sea lamprey ammocetes were sampled at low densities by boat electrofishing in only two river sites (Mass. Cove and central main stem, Tables 3 and 4), but were common in many tributary sites (Table 4).

American shad and striped bass are both present in the Connecticut River (Tables 3 and 4) (Savoy et al. 2004; Leggett et al. 2004; Savoy and Crecco 2004; all this volume). The river supports an annual spring shad run of 250,000 to more than 1 million adults with juveniles remaining in the river until they emigrate to Long Island Sound in the fall (Crecco and Savoy 1987; Savoy et al. 2004). Most of the shad we encountered were juveniles; however, a number of adults (>42 cm) were sampled during June in the main stem from the Massachusetts line to Wethersfield Cove. Striped bass are not known to spawn in the Connecticut River, but use the river during much of the year to forage (Savoy and Crecco 2004). Low numbers of young striped bass (19–39 cm) were sampled by electrofishing in main-stem areas from Hartford south and in Crow Point, Wethersfield, and Hamburg coves.

Rainbow smelt and Atlantic tomcod were both among the species sampled by Marcy (2004 [1976]) in the 1970s; however, none were collected in our studies. The most recent rainbow smelt collected from the Connecticut River by CTDEP Inland Fisheries staff was taken in a trap net set in Salmon Cove in the mid-1990s. We know that smelt were once abundant in the state's coastal waters and tributary streams, including the Connecticut River and its tributaries (Walcott et al. 1926). However, extensive data from the Hudson River and anecdotal information on local populations indicates that smelt populations crashed in the mid-1990s (Schultz et al. 2002a; Daniels et al., in press). Rainbow smelt are presently being proposed for listing as a state threatened species. Similar anecdotal information points to a less dramatic decline for tomcod in Long Island Sound and natal streams, including the Connecticut River and its tributaries (Kulle et al. 1930; Schultz et al. 2002b).

Future Research Needs

A number of areas where data on Connecticut River fish populations were missing or insufficient were identified in the process of gathering and coalescing information for this paper. Ongoing monitoring to assess seasonal variability, sampling error, and to establish baseline data for tracking trends over time is clearly needed. In addition to baseline monitoring, the following research would add to our collective knowledge and enhance conservation and management of the river's fish populations.

Channel catfish are widespread, support a popular fishery, and are harvested in large numbers by recreational anglers. This species is relatively long-lived, slow-growing, and has the potential to reach large size. As such, catfish populations are susceptible to declines in fishing quality resulting from overfishing. The collection of stock assessment data on channel catfish would enable biologists to determine if harvest levels and fishing regulations are appropriate for maximizing the value of this large sport fishery.

A creel survey of the Connecticut River north of Hartford would provide data similar to that presented in this paper for areas south of Hartford. These data would help to quantify the full magnitude of the white perch and catfish fisheries and would provide data on important fisheries for striped bass, shad, and smallmouth bass. Both smallmouth bass and catfish are subject to intense fishing pressure that may be impacting angling quality. The fishery for American shad takes place entirely north of Hartford and appears to have changed significantly since data were last collected in the 1980s. Deterioration of the Enfield Dam and a growing striped bass fishery have both influenced the shad fishery (Savoy and Crecco 2004).

Consideration should be given to collecting periodic riverwide creel survey data to monitor overall trends in Connecticut River sport fisheries. Creel surveys targeting specific fisheries may be needed if changes are observed in the relative abundance or size distribution of highly sought after game fish (e.g., catfish) or prey species (e.g., river herring).

The effects of heavy fishing pressure and relocation of bass by tournament anglers needs to be determined. Largemouth and smallmouth bass are the most sought after freshwater species inhabiting the river. They are also subject to heavy fishing pressure from tournament anglers. Consequently, large numbers of bass are captured and relocated long distances on the river. The response of individual fish and the cumulative effects on the bass populations and the fishery are unknown. The relocation of smallmouth bass from the northern river to areas south of Hartford is of particular concern.

The migratory patterns of important "resident" freshwater species such as crappie, white perch, and channel catfish warrant further study. Additionally, the status of rare or declining species such as rainbow smelt, tomcod, American brook lamprey (in tributary streams), and river herring needs to be accurately determined and critical habitats identified.

Acknowledgments

The U.S. Fish and Wildlife Service Federal Aid in Sport Fish Restoration (F-57R) provided most of the funding for this project. We thank all the permanent and seasonal personnel of the Connecticut Department of Environmental Protection Bureau of Natural Resources who assisted with field sampling. We also thank our editors, Bart Marcy, Rob Neumann, Michael Paller, and Bob Schmidt, for sage advice and patient review of the manuscript. Special thanks go to Paul Jacobson, Bart Marcy, Ron Massengill, Bill Leggett, and Doug Dixon for having the vision to conceive this revisited ecological study and revised monograph and the ambition to see it through to its end.

References

Barry, T., and E. Machowski. 1996. A preliminary plan for statewide management of northern pike in Connecticut. Connecticut Department of Environmental Protection, Federal Aid in Sport Fish Restoration, Final Report F-57-R, Hartford.

Bond, C. E. 1979. Biology of fishes. Saunders Co., Philadelphia.

Carlson, D. M. 1990. Assessment of the largemouth bass population and fishery in the Hudson River Estuary. New York State Department of Environmental Conservation, Albany.

Crecco, V., and T. Savoy. 1987. Fishery management plan for American shad in the Connecticut River. Connecticut Department of Environmental Protection, Hartford.

Daniels, R. A., K. E. Limburg, R. E. Schmidt, D. L. Strayer, and R. C. Chambers. In press. Changes in fish assemblages in the tidal Hudson River, New York. North American Journal of Fisheries Management.

Gephard, S., and J. McMenemy. 2004. An overview of the program to restore Atlantic salmon and other diadromous fishes to the Connecticut River with notes on the current status of these species in the river. Pages 287–317 *in* P. M. Jacobson, D. A. Dixon, W. C. Leggett, B. C. Marcy, Jr., and R. R. Massengill, editors. The Connecticut River Ecological Study (1965–1973) revisited: ecology of the lower Connecticut River 1973–2003. American Fisheries Society, Monograph 9, Bethesda, Maryland.

Green, D. M., S. E. Landsberger, S. B. Nack, D. Bunnell, and J. L. Forney. 1993. Abundance and winter distribution of Hudson River black bass. Hudson River Foundation for Science and Environmental Research, New York.

Hagstrom, N. T., M. Humphreys, W. A. Hyatt, and W. B. Gerrish. 1996. A survey of Connecticut streams and rivers. Connecticut Department of Environmental Protection, Federal Aid in Sport Fish Restoration Final Report, F-66-R, Hartford.

Hartel, K. E., D. B. Halliwell, and A. E. Launer. 2002. Inland fishes of Massachusetts. Massachusetts Audubon Society, Lincoln.

Howell, P., and D. R. Molnar. 1999. An angler survey of the Connecticut portion of the Connecticut River. Pages 190–205 *in* A study of marine recreational fisheries in Connecticut. Connecticut Department of Environmental Protection, Hartford.

Howell, P., and D. R. Molnar. 2004. Stock assessment of white perch in the lower Connecticut River. Pages 379–390 *in* P. M. Jacobson, D. A. Dixon, W. C. Leggett, B. C. Marcy, Jr., and R. R. Massengill, editors. The Connecticut River Ecological Study (1965–1973) revisited: ecology of the lower Connecticut River 1973–2003. American Fisheries Society, Monograph 9, Bethesda, Maryland.

Jacobs, R. P., and E. B. O'Donnell. 1996. An electrofishing survey of selected Connecticut lakes. Connecticut Department of Environmental Protection, Federal Aid in Sport Fish Restoration, Final Report F-57-R-14, Hartford.

Jacobs, R. P., and E. B. O'Donnell. 2002. A fisheries guide to lakes and ponds of Connecticut including the Connecticut River and its coves. Connecticut Department of Environmental Protection, Bulletin 35, Hartford.

Jordan, S. M., R. M. Neumann, and E. T. Schultz. In review. Distribution, habitat use, growth, and condition of a native and introduced catfish species in the Hudson River Estuary. Journal Freshwater Ecology.

Kulle, K. C., C. E. Wheeler and W. A. True. 1930. Eighteenth biennial report of the state board of fisheries and game for the fiscal year ended June 30, 1930. Connecticut Board of Fisheries and Game, Hartford.

Leggett, W. C. 2004. The American shad, with special reference to its migration and population dynamics in the Connecticut River. Pages 181–238 *in* P. M. Jacobson, D. A. Dixon, W. C. Leggett, B. C. Marcy, Jr., and R. R. Massengill, editors. The Connecticut River Ecological Study (1965–1973) revisited: ecology of the lower Connecticut River 1973–2003. American Fisheries Society, Monograph 9, Bethesda, Maryland. (Originally published in 1976).

Leggett, W. C., T. F. Savoy, and C. A. Tomichek. 2004. The impact of enhancement initiatives on the structure and dynamics of the Connecticut River population of American shad. Pages 391–405 *in* P. M. Jacobson, D. A. Dixon, W. C. Leggett, B. C. Marcy, Jr., and R. R. Massengill, editors. The Connecticut River Ecological Study (1965–1973) revisited: ecology of the lower Connecticut River 1973–2003. American Fisheries Society, Monograph 9, Bethesda, Maryland.

MacBroom, J. C. 1998. The river book. Connecticut Department of Environmental Protection, Hartford.

Maltezos, G. C., R. McMahon, and M. Leslie. 1980. Movements of white perch in the lower Connecticut River 1980. Connecticut Department of Environmental Protection, Hartford.

Marcy, B.C., Jr. 2004. Fishes of the lower Connecticut River and the effects of the Connecticut Yankee plant. Pages 65–124 *in* P. M. Jacobson, D. A. Dixon, W. C. Leggett, B. C. Marcy, Jr., and R. R. Massengill, editors. The Connecticut River Ecological Study (1965–1973) revisited: ecology of the lower Connecticut River 1973–2003. American Fisheries Society, Monograph 9, Bethesda, Maryland. (Originally published in 1976).

Minshall, G. W., K. W. Cummins, R. C. Peterson, C. E. Cushing, D. A. Bruns, J. R. Sedell, and R. L. Vannote. 1985. Developments in stream ecosystem theory. Canadian Journal of Fisheries and Aquatic Sciences 42:1045–1055.

Nielsen, L. A., and D. L. Johnson. 1983. Fisheries techniques. American Fisheries Society. Bethesda, Maryland.

Phillips, C. L. and J. J. Bender. 1984. Northern pike, *Esox lucius* research and management in the Connecticut River. Connecticut Department of Environmental Protection, Final report, Hartford.

Phillips, C. L. and J. J. Bender. 1992. Development and maintenance of a pike spawning marsh at Haddam Meadows. Connecticut Department of Environmental Protection, Federal Aid in Sport Fish Restoration, Final Report F-57-R, Hartford.

Pollock, K. H., C. M. Jones, and T. L. Brown. 1994. Angler survey methods and their application in fisheries management. American Fisheries Society, Special Publication 25, Bethesda, Maryland.

Reynolds, J. B. 1996. Electrofishing. Pages 221–253 *in* B. R. Murphy and D. W. Willis, editors. Fisheries techniques, 2nd edition. American Fisheries Society, Bethesda, Maryland.

Reynolds, J. B., and D. E. Simpson. 1978. Evaluation of fish sampling methods and rotenone census. Pages 11–24 *in* G. D. Novinger and J. G. Dillard, editors. New approaches to management of small impoundments. American Fisheries Society, North Central Division, Special Publication 5, Bethesda, Maryland.

Savoy, T. F., and V. A. Crecco. 2004. Factors affecting the recent decline of blueback herring and American shad in the Connecticut River. Pages 361–377 *in* P. M. Jacobson, D. A. Dixon, W. C. Leggett, B. C. Marcy, Jr., and R. R. Massengill, editors. The Connecticut River Ecological Study (1965–1973) revisited: ecology of the lower Connecticut River 1973–2003. American Fisheries Society, Monograph 9, Bethesda, Maryland.

Savoy, T. F., V. A. Crecco, and B. C. Marcy, Jr. 2004. American shad early life history and recruitment in the Connecticut River: a 40-year summary. Pages 407–417 *in* P. M. Jacobson, D. A. Dixon, W. C. Leggett, B. C. Marcy, Jr., and R. R. Massengill, editors. The Connecticut River Ecological Study (1965–1973) revisited: ecology of the lower Connecticut River 1973–2003. American Fisheries Society, Monograph 9, Bethesda, Maryland.

Schmidt, R. E., and T. R. Lake. In press. The role of tributaries in the biology of Hudson River fishes. In J. Levinton and J. R. Waldman, editors. The Hudson River ecosystem. Cambridge University Press.

Schultz, E., H. Fried, and R. Neumann. 2002a. Anadromous rainbow smelt in Connecticut: assessment of populations, conservation status, and need for restoration plan. University of Connecticut, Fiscal year 2002 State Wildlife Grant proposal, Storrs.

Schultz, E., H. Fried, and R. Neumann. 2002b. Tomcod in Connecticut: assessment of populations, conservation status, and need for restoration plan. University of Connecticut, Fiscal year 2002 State Wildlife Grant proposal, Storrs.

Smith, C. L. 1985. The inland fishes of New York State. New York State Department of Environmental Conservation, Albany.

Stolte, L. W. 1980. A strategic plan for the restoration of Atlantic salmon to the Connecticut River basin. United States Department of the Interior, Fish and Wildlife Service, Laconia, New Hampshire.

U.S. Geological Survey. 1994. Water resources data Connecticut water year 1994. U.S. Geological Survey, USGS Water-data report, CT-94–1, East Hartford, Connecticut.

Vokoun, J. C., and C. F. Rabeni. 1999. Catfish sampling in rivers and streams: a review of strategies, gears, and methods. Pages 271–286 *in* B. R. Murphy and D. W. Willis, editors. Fisheries techniques, 2nd edition. American Fisheries Society, Bethesda, Maryland.

Walcott, F. C., K. C. Kulle, and C. E. Wheeler. 1926. Sixteenth biennial report of the state board of fisheries and game for the years 1924–1926. Connecticut State Board of Fisheries and Game, Hartford.

Whitworth, W. R., P. L. Berrien, and W. T. Keller. 1968. Freshwater fishes of Connecticut. University of Connecticut, Department of Agriculture and Natural Resources, State Geological and Natural History Survey of Connecticut, Bulletin 101, Storrs.

Whitworth, W. R. 1996. Freshwater fishes of Connecticut. 1996. 2nd edition. Department of Environmental Protection, State Geological and Natural History Survey of Connecticut, Bulletin 114, Hartford.

Wilde, C. W. 1959. A fishery survey of the lakes and ponds of Connecticut. Connecticut State Board of Fish and Game, Report Number 1, Hartford.

American Fisheries Society Monograph 9:345–352, 2004

Population Estimate and Utilization of the Lower Connecticut River by Shortnose Sturgeon

THOMAS F. SAVOY*

Connecticut Department of Environmental Protection, Marine Fisheries Division
Post Office Box 719, Old Lyme, Connecticut 06371-0719, USA

Abstract.—From 1 July 1988 through 1 December 2002, 896 shortnose sturgeon *Acipenser brevirostrum* have been collected in the Connecticut River within Connecticut State waters with 668 fish being tagged and 160 recaptured. Long-term average population estimates of numbers of adult fish in the Connecticut area of the Connecticut River range from 1,042 to 1,580. Grouping population estimates by 1989–1994 and 1996–2002 yields an increase in population size of 65–138%. Sixty shortnose sturgeon have had ultrasonic transmitters surgically implanted to determine timing of movements and habitat preferences. Telemetry efforts and conventional tagging document repeated seasonal use of the estuary coincident with the spring freshet that displaces the salt wedge and mixing zone to outside of the river. Residence times varied from 45 to 90 d with gradual movement northward with re-establishment of the salt wedge into the Connecticut River. Shortnose sturgeon were tolerant of salinities up to 28 parts per thousand (ppt) and changes of 15 ppt over a 6-h tidal cycle.

Introduction

The Connecticut River is the longest river in New England and second only to the Susquehanna in volume of all rivers on the East Coast of the United States (Merriman and Thorpe 2004 [1976], this volume). Although the Connecticut River extends some 658 km from the First Connecticut Lake in northern New Hampshire to the mouth, which empties into Long Island Sound (Figure 1), the river is effectively subdivided into numerous parts by various dams (Moffit et al. 1982). Interestingly, despite the enormous volume of field work carried out in what has been described as the seminal work on the Connecticut River (Merriman and Thorpe 2004 [1976]), no shortnose sturgeon *Acipenser brevirostrum* were reported. Few historical accounts of sturgeon presence/abundance exist, and these are generally limited to anecdotal stories that do not provide clear evidence of abundance. McCabe (1942) listed European sturgeon *A. sturio* in his description of fishes of western Massachusetts, and Galligan (1960) described the fishery for Atlantic sturgeon *A. oxyrinchus*, but shortnose sturgeon were virtually unknown in the Connecticut River until a 1978 fish survey (Buckley and Reed 1978). Only two shortnose sturgeon were reported impinged at Connecticut Yankee during the operational life of the plant (1966–1996), one in 1978 and the second in 1982 (Jacobson et al. 2004, this volume). Since then, the Connecticut River population of shortnose sturgeon has become one of the best studied populations on the East Coast (Buckley 1982; Buckley and Kynard 1975b; 1985a; Kynard 1997, 1998; Kynard et al. 2000). Presently, only the lower most two sections of the Connecticut River are known to contain shortnose sturgeon. Controversy exists over whether these two groups are natural or the result ot one subgroup being landlocked after the construction of the Holyoke Dam in 1849 (Green 1939). This paper provides information on seasonal use of the estuary by shortnose sturgeon in the lower portion of the Connecticut River, here defined as the section from the Holyoke Dam in Holyoke, Massachusetts to its mouth in Old Saybrook, Connecticut, and documents an increase in the population of shortnose sturgeon.

* Tom.Savoy@po.state.ct.us

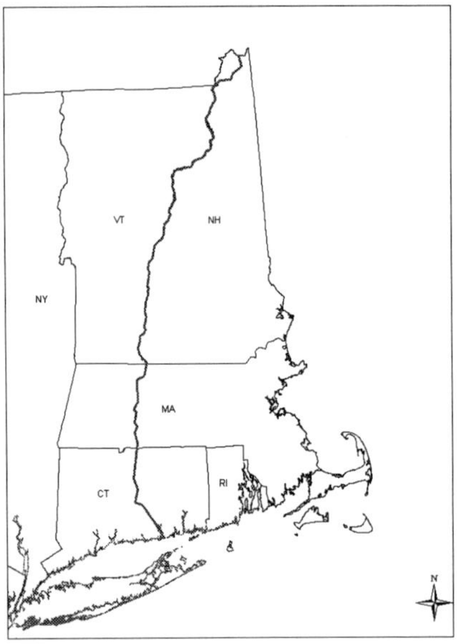

FIGURE 1. Map of the Connecticut River showing extent of the river from northern New Hampshire to Long Island Sound.

Methods

Shortnose sturgeon were captured with anchored multifilament nylon gill nets (single mesh size per net of 10.0, 12.7, 15.2, or 17.8-cm stretch mesh, 30.5 m long by 1.8 m deep) and otter trawls. Gill nets were set parallel to the river flow with anchors to fish the bottom 1.8 m of the water column throughout the main-stem Connecticut River in Connecticut State waters. One to five nets were set for 1–26 h per location. Trawls (10-m headrope) were towed at approximately 2.5 knots for 5–15 min. Tow duration was variable, but the trawl was fished as long as practical in various reaches.

Upon capture, all shortnose sturgeon were removed from the net and temporarily placed in a floating live car (1 m × 1 m × 0.5 m deep) constructed of 2.5-cm netting. Fish were then individually transferred into a water filled examination box in the boat for processing. Each sturgeon was examined for tags, marks, and damage; measured to length (total length [TL] and fork length [FL], cm); and tagged in this box. From 1988 through 1992, uniquely numbered orange color Carlin dangler tags (6 mm × 16 mm) were wired to the base of the left side of the dorsal fin for estimating population size and identifying individual sturgeon. Beginning in 1993, external tag attachment was discontinued and Passive Integrated Transponder (PIT) tags were injected below the skin on the left side of the fish under the dorsal fin. From 1988 through 1997 and 2000 through 2002, sturgeon were weighed (kg) in a mesh sling suspended from a spring scale (Jennings 1989) before being released. Condition factor, C, was calculated as C = (weight in g/FL^3 in mm) × 10^5 for all fish weighed.

The number of shortnose sturgeon in the lower Connecticut River was calculated with the basic Petersen technique and several of its variations, the Schumacher/Eschmeyer, the Schnabel method, and the Chapman technique (Ricker 1975).

Information on movements and preferred habitats utilized by shortnose sturgeon were determined through telemetry efforts. Individual coded ultrasonic transmitters (USTs) were surgically implanted into the body cavity in up to 10 shortnose sturgeon per year. Transmitters were 18 mm diameter by 96 mm long and had expected battery lives of 13 months. Individual shortnose sturgeon were selected for transmitter implants on the basis of overall health and size. Utilizing only large, robust sturgeon limited understanding of movements to adult members of the population but reduced the problems associated with relatively large transmitter size and small fish size. Internal implants were used to ensure the transmitters would be retained throughout their effective battery life. Internal implants were also thought to be less detrimental to the fish than external attachment. Surgical implantations were made by a single incision on the ventral side slightly behind the pectoral girdle and just off the midline of the fish. The incision was kept as small possible (approximately 40 mm) to allow insertion of the UST. After insertion, the transmitter was pushed toward the vent and to the opposite side of the body from the incision. The wound was closed with several individual stitches of a 3/0 nonabsorbable monofilament suture. After 1989, when several sturgeon had been suspected of shedding USTs, this procedure was modified by paraffin wax coating the USTs (Summerfelt and Mosier 1984) prior to implantation.

Tracking efforts to relocate sturgeon bearing ultrasonic transmitters were made at least weekly when water temperatures were above 10°C and bimonthly when below 10°C. After determining the identity of the sturgeon, the position of each sturgeon was plotted onto National Oceanic and Atmospheric Administration charts or U.S. Geological Survey topographic sheets by triangulating signals with a directional hydrophone against landmarks and physical features. Date, time, weather conditions, river state, and other conditions were noted as appropriate. Occasionally, fish were noted to be moving while tracking was taking place, and several breachings were also recorded during tracking efforts.

Results

A total of 896 shortnose sturgeon were collected from 01/07/88 through 01/12/02, of which 668 were tagged and released (Table 1). Additionally, 105 Atlantic sturgeon were also collected in the Connecticut River during shortnose sturgeon sampling (Savoy and Pacileo 2003). Of the 896 shortnose sturgeon collected, 210 were recaptures of sturgeon previously tagged, although only 160 recaptures were sturgeon originally tagged by Connecticut Department of Environmental Protection (CTDEP). Forty-seven of the 50 non-CTDEP-tagged sturgeon were originally tagged in the Connecticut River by U.S. Geological Survey/Biological Resources Division (USGS/BRD) staff working out of the Conte Anadromous Fish Research Center. Notably, two shortnose sturgeon recaptures were of fish originally tagged in the Hudson River by Cornell University researchers, and one of these was captured twice. USGS/BRD staff reported recapture of 65 shortnose sturgeon tagged by CTDEP. Information on sturgeon tagged/recaptured by other parties was not used in the calculation of population size. Thirteen shortnose sturgeon were killed in gill nets, although these mortalities predominately took place prior to 1991 (Table 1). Adoption of netting protocols limiting gill net soak times to less than 12 h when water temperatures exceeded 20°C were generally effective at reducing mortalities.

Gill nets accounted for 49.0% (439/896) of the shortnose sturgeon collected, and trawling accounted for 49.2% (441/896). All other gear types collectively accounted for only 16 sturgeon (6–trammel nets, 8–electrofishing, 1–seine, 1–angler). Catch per unit of effort as an index of relative importance of various habitats or changes in abundance over time was not calculated, since nets were not set randomly but rather concentrated in selected areas to maximize catch, and effort varied among years.

Shortnose sturgeon captured by all gear types ranged from 28.0 to 112.0 cm fork length (FL; Figure 2; total length [TL] 32.5–126.0 cm) in length and from 0.1 to 15.4 kg in weight. One fish collected exceeded the capacity of the scale and was estimated at 24.0 kg. Condition factor of

Table 1. Number of shortnose sturgeon collected and tagged in the Connecticut River and number of recaptures, mortalities, and fish implanted with ultrasonic transmitters (USTs) by year from 1988 to 2002.

Year	Collected	Tagged	Recaptures	Mortalities	UST
1988	48	45	1	2	12
1989	80	64	11	5	7
1990	60	49	7	3	4
1991	74	60	4	0	8
1992	50	37	13	0	6
1993	14	5	7	1	2
1994	20	12	6	0	2
1995	11	9	2	0	2
1996	22	14	6	0	3
1997	34	25	6	0	5
1998	52	35	11	2	6
1999	35	32	2	0	1
2000	117	93	17	0	2
2001	126	85	26	0	0
2002	153	103	31	0	0
Totals	896	668	60	13	60

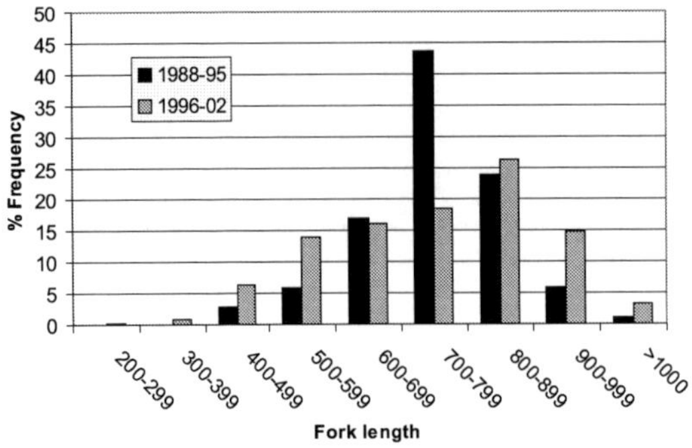

FIGURE 2. Length frequency of shortnose sturgeon captured by CTDEP in the lower Connecticut River, 1998–2002, by all gear types.

shortnose sturgeon collected in the lower river ranged from 0.40 to 1.71 (Table 2). The mean condition factor of all fish weighed (n = 763) was 0.96 (SD = 0.20), which is among the highest reported (Table 2) from nearby river systems (Dadswell 1979; Taubert 1980; Hastings 1983; Squiers 1998). No trends or patterns with condition factor were apparent with length of fish; fish with high condition factors ranged from 55 to 90 cm FL. Condition factor does change seasonally (Table 2), with lowest values during winter months and highest values associated with heavy feeding in the estuary region during the spring months.

The 160 recaptures of CTDEP-tagged shortnose sturgeon (Table 1) are further subdivided as they represent only 121 unique sturgeon. Ninety-six sturgeon were recaptured once, 19 sturgeon were recaptured twice, 3 sturgeon were recaptured three times, 1 sturgeon was recaptured four times, and 2 sturgeon were recaptured five times. Recapture times varied widely with time at liberty ranging from 6 to 4,786 d.

Population size was variable among years and methods (Table 3) but demonstrates a general agreement among techniques within a year and yields long term average population estimates of 1,042–1,580. Data for 1988, 1995, and 1999 were dropped for not meeting the criteria of Marked times Capture exceeding Four times N (Ricker 1975), in large part due to the low number of recaptures in those years. All techniques suggest an increase in size of the population. When grouped into 1989–1995 and 1996–2002, population size increased from 65% to 138%. The mean long-term population estimate from all years combined suggests that the numbers of sturgeon are still modest at 1,100–1,600 for the lower Connecticut River.

Whereas shortnose sturgeon were not aged due to lack of hard parts removed, growth analyses were based on recapture data. One hundred and sixty pairs of data were available from recaptures where all necessary information was available. Limiting growth analyses to those recaptures taking place greater than 30 d apart, and only those fish exhibiting positive growth in length

TABLE 2. Condition factor (mean, standard deviation [SD], minimum, and maximum) of all shortnose sturgeon weighed in the Connecticut River from 1988 to 2002. Seasonal values and selected other river systems for comparison including values determined from shortnose sturgeon collected above the Holyoke Dam.

Location/time	Number	Mean	SD	Minimum	Maximum
CT River – all	763	0.96	0.20	0.40	1.71
Winter	58	0.80	0.13	0.55	1.12
Spring	92	1.05	0.19	0.59	1.59
Summer	337	1.00	0.20	0.40	1.71
Fall	276	0.92	0.19	0.56	1.56
Upper CT River[a]	271	0.73	0.10	0.43	1.34
Saint John River[b]	2890	0.97			
Kennebec River[c]	15	0.86	0.15	0.71	1.19
Delaware River[d]	373	0.91	0.15	0.53	1.80

[a] Taubert (1980)
[b] Dadswell et al. (1984)
[c] Squiers (1998)
[d] Hastings (1983)

TABLE 3. Annual population estimates of shortnose sturgeon greater than 50 cm FL in the Connecticut portion of the Connecticut River from 1989 to 2002, as estimated using four techniques, and average population size from 1989 through 1994 and 1996 through 2002.

Year	Peterson	Chapman	Schnabel	Schumacher
1989	793	743	579	495
1990	1,354	1,212	748	1,337
1991	1,152	1,095	864	729
1992	981	933	875	714
1993	520	489	827	699
1994	907	819	832	1,400
1995				
1996	1,082	973	872	1,602
1997	1,813	1,605	943	2,618
1998	1,420	1,348	1,002	993
1999				
2000	3,304	3,153	1,422	1,972
2001	2,966	2,875	1,667	1,512
2002	2,671	2,618	1,874	1,054
Average	1,580	1,489	1,042	1,261
Ave. 89–94	951	882	788	896
Ave. 96–02	2,209	2,095	1,297	1,625
% increase	132	138	65	81

yield 132 pairs of data. Since immature fish experience higher growth rates, the data were partitioned into two groups, less than and greater than 550 mm FL. Juvenile sturgeon had growth rates (mean = 0.1923 mm/d; SD = 0.2055) three times that of adult sturgeon (mean = 0.0664 mm/d; SD = 0.0739). Maximum observed growth rate of juveniles was 0.74 mm/d, more than twice the maximum exhibited by an adult of 0.31 mm/d. Maximum observed growth rates of juvenile sturgeon (0.7426 and 0.3972 mm/d) were 75 mm in 101 d and 170 mm in 428 d, respectively. Three juvenile fish and four adult sturgeon had growth rates in excess of 0.25 mm/d. Increase in FL observed ranged up to 325 mm, with a maximum annual increase in length of 100 mm in 329 d (growth rate = 0.3034 mm/d). Weight increases ranged up to 6.8 kg in 1376 d and a high of 7.9 g/d (5 fish had weight gain rates of 5.0 g/d or higher), although the mean increase was 0.6604 g/d. As expected, weight gains of juvenile sturgeon were significantly lower than values seen for adult sturgeon (mean growth in weight 300 g versus 866 g), yet rate of weight increase remained higher for juveniles (mean growth rate of 0.9097 g/d versus 0.6605 g/d, juveniles versus adults, respectively). The mean growth rate of recaptured sturgeon collected in the lower river with unrestricted access to the estuary was 2.8 cm/year, which compares to 1.8 cm/year for sturgeon collected in the Holyoke Pool (Taubert 1980).

Sixty shortnose sturgeon had USTs surgically implanted from 1988 through 2000 (Table 1). Forty five of these have provided useful long-term (>60 d) information. As many as 15 transmitters were active and tracked at one time, as during the spring of 1991. Although transmitters had expected battery lives of 13 months, several transmitters provided information over two or more consecutive spring seasons, and one transmitter lasted for 5 years.

Whereas the information presented on location use was predominately taken from telemetry observations, there was some concern whether sturgeon with implanted USTs were acting normally. Nets set in the vicinity of UST locations were successful in capturing nonsonic tagged sturgeon in seven of seven occurrences when UST-implanted sturgeon were recaptured. On two of these occasions, a total of 14 sturgeon were captured in the same net, suggesting that the UST-bearing sturgeon are behaving similarly to untagged fish and that they can be found in close proximity to each other.

General seasonal movement patterns as noted from ultrasonic telemetry were similar among all years. Sturgeon movements downriver into the estuary (rkm 6–20) were rapid and directed with individual fish moving up to 30 km/d. Shortnose sturgeon utilize the estuary predominately during times of high freshwater outflow, essentially when there is little saline water present. Sturgeon move into this reach from the end of April to the first weeks of May and typically remain in the area until mid-June. Some sturgeon stay into July, well after the saltwedge has re-established itself in the river. Of the 23 shortnose sturgeon with USTs that lasted for two or more consecutive spring seasons, 21 moved to the estuary each spring. One sturgeon moved down to this area in each of five consecutive springs and another for 6 years in a row. Only 3 of the 45 sturgeon providing good long-term information did not move to the estuary region during the spring when their transmitters were active. These sturgeon moved only as far south as rkm 25.

Departure from the estuary is a much slower process than immigration into the estuary. Emigration of shortnose sturgeon northward out of the estuary begins when saline waters began to move back into the river; however, adult shortnose sturgeon can tolerate both high salinities and rapid changes in salinity. Shortnose sturgeon were found in salinities up to 28 parts per thousand (ppt) in the river, and in several instances, fish remained in one area over a tide change where salinity changed from 0.0 to 14.5 ppt. Moreover, at least two shortnose sturgeon have moved through Long Island Sound from the Hudson River to the Connecticut River, a minimum straight line distance of more than 100 km. Salinities in Long Island Sound typically range from 24 to 32 ppt with the higher salinities in the western end (Reid et al. 1979).

Discussion

Little quantative information is available on the historical numbers of shortnose sturgeon, and no estimates of abundance in the Connecticut River are available prior to 1984 (Dadswell et al. 1984). The commercial Atlantic sturgeon, prosecuted around the turn of the century, and the American shad fishery, operating continuously from the Colonial period to present time in the Connecticut River, likely contributed to reducing the numbers of sturgeon. Water quality also deteriorated (DO levels of 2–4 mg/L), prompting an investigation into the ability of the Connecticut River to support shad. Mitchell et al. (1925) measured dissolved oxygen levels in the river in 1923 and 1924 and recorded percent saturation as low as 23% in the deep holes below Hartford and Middletown. For a while the river earned the designation of being "the best landscaped sewer in America." All of the above factors may have reduced shortnose sturgeon to very low numbers in the Connecticut River. Kynard (1997) speculated that the Connecticut River population has been stable from 1975 to 1996, with approximately 1,000 individuals. However, the directed tagging efforts described here suggest that significant improvements in water quality and decreases in commercial fishing effort in the Connecticut River over the last 50 years may have resulted in an increase in population size. Population growth would account for the lack of sturgeon in the catches and records of the original Connecticut River Ecological Study, despite the intensive field work conducted.

Movement patterns of UST bearing shortnose sturgeon were very complex. In addition, our inability to externally sex sturgeon, determine stage of sexual maturity, or age the fish without harm (Kohlhorst 1978) complicates matters and makes definitive statements or understanding specific movement patterns difficult. Similar to that reported in Dadswell et al. (1984), we found that tagged shortnose sturgeon have repeatedly been shown to be capable of rapid and directed movements to discrete areas. Rates of movement up to 40 km/d seen in the Connecticut River are among the fastest recorded movements for shortnose sturgeon. In contrast to behavior seen in other river systems (McCleave et al. 1977; Dadswell 1979; Brundage and Meadows 1982; O'Herron et al. 1993), we did not find extensive use of the estuary region during the winter months. We note that the preponderance of adult fish were found in upriver areas, well above any influence of salt water during the winter months. One winter concentration of sturgeon could be found in a deepwater site (depths 6–15 m), yet another group was typically found in a relatively shallow (3–5 m) area with moderate to strong river currents. Despite lack of overwintering in the estuary, we speculate that free access to the estuary must be very important for feeding based on rapid and directed movement into this region by both postspawning and nonspawning fish.

The high condition factor and general robustness of shortnose sturgeon collected in the lower Connecticut River indicates that sturgeon are not food limited (Savoy and Benway 2004, this volume) and additional population growth may be possible. Research efforts need to investigate the age structure of this stock to determine variability in frequency and strength of reproductive success and recruitment and examine the role of environmental factors on these processes.

Acknowledgments

I would like to thank Mark Collins and Webb Van Winkle for reviewing the manuscript and their constructive comments. Thanks to all former employees and coworkers who assisted over the years (Technicians Kurt Gottschall, Deborah Pacileo, and Jacqueline Benway and a host of resource assistants). While fieldwork was not always easy, it was the best of times. Funding was provided by the Connecticut Department of Environmental Protection and from NOAA/NMFS State Federal Aid Office through the Anadromous Fish Conservation Act.

References

Brundage, H. M., III, and R. E. Meadows. 1982. Occurrence of the endangered shortnose sturgeon, *Acipenser brevirostrum*, in the Delaware River Estuary. Estuaries 5:203–208.

Buckley, J., and B. Kynard. 1985a. Habitat use and behavior of pre-spawning and spawning shortnose sturgeon, (*Acipenser brevirostrum*), in the Connecticut River. Developments in Environmental Biology of Fishes 6:111–117.

Buckley, J., and B. Kynard. 1985b. Yearly movements of shortnose sturgeons in the Connecticut River. Transactions of the American Fisheries Society 114(6):813–820.

Buckley, J., and R. J. Reed. 1978. Survey of the Connecticut River for shortnose sturgeon, *Acipenser brevirostrum*, below Holyoke Dam, Holyoke, Massachusetts. Final report to Northeast Utilities, Massachusetts Cooperative Fishery Research Unit, University of Massachusetts, Amherst.

Buckley, J. L. 1982. Seasonal movement, reproduction, and artificial spawning of shornose sturgeon (*Acipenser brevirostrum*) from the Connecticut River. Master's thesis. University of Massachusetts, Amherst.

Dadswell, M. J. 1979. Biology and population characteristics of the shortnose sturgeon, *Acipenser brevirostrum* LeSueur 1818(Osteichthyes: Acipenseridae), in the Saint John River Estuary, New Brunswick, Canada. Canadian Journal of Zoology 57:2186–2210.

Dadswell, M. J., B. D. Taubert, T. S. Squires, D. Marchette, and J. Buckley. 1984. Synopsis of biological data on shortnose sturgeon, *Acipenser brevirostrum* LeSueur 1818. NOAA Technical Report, NMFS 14, Silver Spring, Maryland

Galligan, J. P. 1960. History of the Connecticut River sturgeon fishery. Connecticut Wildlife Conservation Bulletin 6(1):1:5–6.

Green, C. M. 1939. Holyoke Massachusetts. A case history of the industrial revolution in America. Yale University Press, New Haven, Connecticut.

Hastings, W. R. 1983. A study of the shortnose sturgeon (*Acipenser brevirostrum*) population in the upper tidal Delaware River: assessment of impacts of maintenance dredging. Prepared for U.S. Army Corps of Engineers, Philadelphia.

Jennings, M. R. 1989. Use of spring scales for weighing live fish in the field. North American Journal of Fish Management 9:509–511.

Jacobson, P. M., C. Tomichek, and D. Danila. 2004. Twenty years of impingement history: Connecticut Yankee Haddam Neck nuclear power plant. Pages 455–473 *in* P. M. Jacobson, D. A. Dixon, W. C. Leggett, B. C. Marcy, Jr., and R. R. Massengill, editors. The Connecticut River Ecological Study (1965–1973) revisited: ecology of the lower Connecticut River 1973–2003. American Fisheries Society, Monograph 9, Bethesda, Maryland.

Kohlhorst, D. W. 1978. Effect of first pectoral fin ray removal on survival and estimated harvest rate of white sturgeon in the Sacramento-San Joaquin Estuary. California Fish and Game 64:173–177.

Kynard, B. 1997. Life history, latitudinal patterns, and status of shortnose sturgeon, *Acipenser brevirostrum*. Environmental Biology of Fishes 48:319–334.

Kynard, B. 1998. Twenty-two years of passing shortnose sturgeon in fish lifts on the Connecticut River: what has been learned? Pages 255–264 *in* M. Jungwirth, S. Schmutz, and S. Weiss, editors. Fish migration and fish bypasses. Fishing News Books, London.

Kynard, B., M. Horgan, M. Kieffer, and D. Seibel. 2000. Habitats used by shortnose sturgeon in two Masssachusetts rivers, with notes on estuarine Atlantic sturgeon: a hicrarchical approach. Transaction of the American Fisheries Society 129:487–503.

McCabe, B. C. 1942. Distribution of the fishes of the streams of western Massachusetts. Doctoral dissertation. Cornell University, Ithaca, New York

McCleave, J. D., S. M. Fried, and A. K. Towt. 1977.

Daily movements of shortnose sturgeon, *Acipenser brevirostrum*, in a Maine estuary. Copeia 1977:149–157.

Merriman, D., and L. M. Thorpe. 2004. Introduction. Pages 3–25 *in* P. M. Jacobson, D. A. Dixon, W. C. Leggett, B. C. Marcy, Jr., and R. R. Massengill, editors. The Connecticut River Ecological Study (1965–1973) revisited: ecology of the lower Connecticut River 1973–2003. American Fisheries Society, Monograph 9, Bethesda, Maryland. (Originally published in 1976)

Mitchell, P. H., and staff. 1925. A report of investigations concerning shad in the rivers of Connecticut, part 1:7–44. Connecticut State Board of Fish and Game, Hartford.

Moffit, C. M., B. Kynard, and S. G. Rideout. 1982. Fish passage facilities and anadromous fish restoration in the Connecticut River basin. Fisheries 7(6):2–11.

O'Herron, J. C., K. Able, and R. W. Hastings. 1993. Movements of shortnose sturgeon (*Acipenser brevirostrum*) in the Delaware River. Estuaries 16(2):235–240.

Ricker, W. E. 1975. Computation and interpretation of biological statistics of fish populations. Fisheries Research Board of Canada Bulletin 191.

Reid, R. N., A. B. Frame, and A. F. Draxler. 1979. Environmental baselines in Long Island Sound, 1972–1973. NOAA Technical Report, NMFS SSRF-738, Silver Spring, Maryland.

Savoy, T., and D. Pacileo. 2003. Movements and important habitats of sub-adult Altantic sturgeon in Connecticut waters. Transactions of the American Fisheries Society 132:1–8.

Savoy, T. F., and J. Benway. 2004. Food habits of shortnose sturgeon collected in the lower Conecticut River from 2000 through 2002. Pages 353–360 *in* P. M. Jacobson, D. A. Dixon, W. C. Leggett, B. C. Marcy, Jr., and R. R. Massengill, editors. The Connecticut River Ecological Study (1965–1973) revisited: ecology of the lower Connecticut River 1973–2003. American Fisheries Society, Monograph 9, Bethesda, Maryland.

Squiers, T. S. Jr. 1998. Progress Report Kennebec River shortnose sturgeon population study August – December, 1998. Federal Aid Reports, NMFS Contract 40EANF800053, Augusta, Maine.

Summerfelt, R. C., and D. Mosier. 1984. Transintestinal expulsion of surgically-implanted dummy transmitter by channel catfish. Transactions of the American Fisheries Society 113:760–766.

Taubert, B. D. 1980. Biology of shortnose sturgeon (*Acipenser brevirostrum*) in Holyoke Pool, Connecticut River, Massachusetts. Doctoral dissertation. University of Massachusetts, Amherst.

American Fisheries Society Monograph 9:353–360, 2004

Food Habits of Shortnose Sturgeon Collected in the Lower Connecticut River from 2000 through 2002

THOMAS F. SAVOY* AND JACQUELINE BENWAY

Connecticut Department of Environmental Protection, Marine Fisheries Division
Post Office Box 719, Old Lyme, Connecticut 06371-0719, USA

Abstract.—Prey preference of shortnose sturgeon *Acipenser brevirostrum* at an upriver freshwater site and a downriver estuarine site in the Connecticut River was examined through gastric lavage. A total of 327 shortnose sturgeon were collected, and 246 were pumped from 15 May 2000 through 1 November 2002. Bottom grabs were made with a petite ponar grab to estimate prey resource abundance. Overall, 76% of sturgeon examined had evidence of recent feeding. Invertebrates had much higher abundance in the estuary region, but diversity was lower than the upriver site. Gammarid amphipods, chironomids, and polychaetes were the most abundant macroinvertebrate taxa in the estuary. Clams, chironomids, and insects were the most abundant taxa found at the upriver site. Prey utilization in the estuary was 10-fold greater than the upriver site, with estuarine organisms also being larger and heavier. Sturgeon preyed heavily on amphipods and infaunal worms in the estuary versus insects and chironomids at the upriver site.

Introduction

Sturgeons are among the oldest living representatives of the Osteichthyes class of fishes. Shortnose sturgeon *Acipenser brevirostrum* occur in eastern North America with 19 distinct populations distributed from the St. John River, Canada to the St. Johns River, Florida (Dadswell et al. 1984; NMFS 1998). Shortnose sturgeon are sympatric with Atlantic sturgeon *A. oxyrinchus* along the East Coast of the United States. Historically, the Connecticut River was thought to host both Atlantic sturgeon and shortnose sturgeon; however, Atlantic sturgeon native to the river are believed to be extirpated (Savoy 1991). Although shortnose sturgeon are the smallest representative of the five species of sturgeon in North America (NMFS 1998), they are among the larger resident fish in many rivers where they occur.

Shortnose sturgeon were listed as endangered rangewide by the Fish and Wildlife Service as early as 1967 (U.S. Office of the Federal Register 1967) as a result of disappearing from their known range due to habitat degradation, dam construction, and overfishing in the American shad *Alosa sapidissima* and Atlantic sturgeon net fisheries (NMFS 1998). Recent evidence (Savoy 2004, this volume) suggests that the population in the lower river may be slowly increasing. However, even a population of 2,500 shortnose sturgeon in the Connecticut River would yield an average density among the lowest reported for all river systems on the East Coast of the United States (Dadswell et al. 1984).

One of the key findings of research by Connecticut Department of Environmental Protection (CTDEP) staff conducted on shortnose sturgeon in the Connecticut River from 1988 to 1994 was the springtime use of the estuary region (Savoy and Shake 1993; Savoy 1999). Rapid and directed movement of postspawning fish to this area suggests that it must be an important foraging area. Given the low density of sturgeon in the Connecticut River and concerns over the long-term survival of the stock due to the low number of potential spawners, we present information on food habits and prey availability to examine whether shortnose sturgeon in the Connecticut River are food limited.

* *Corresponding author:* Tom.Savoy@po.state.ct.us

Methods

Shortnose sturgeon were collected from 15 May 2000 to 1 November 2002 for food habit studies using a 9-m skiff trawl with 10-cm mesh, and a 6-mm-mesh cod end liner. Trawling took place during daylight hours throughout the Connecticut River, but effort was focused in two areas: the estuary (river kilometer [rkm] 5–10) and an upriver area (rkm 85–90), for comparison (Figure 1). Tows were 5–15 min (mean = 8 min) with duration based on bottom topography, obstructions, and vessel traffic. All sturgeon were handled according to protocols developed for sampling shortnose and Atlantic sturgeon, and all precautions for safeguarding the health of individuals were taken (Moser et al. 2000).

All fish collected in a single tow were removed from the trawl, sorted, and counted. Sturgeon were placed into a 1.5 m × 0.9 m × 0.5 m plastic tub filled with river water. Other species in the trawl were identified (Thomson et al. 1978; Whitworth 1996), counted, and released. All sturgeon from a single net haul were completely processed before expending additional netting effort. Sturgeon were individually moved into a tub containing a buffered solution of MS-222® (tricaine methanesulfonate, 100-mg/L concentration) for tranquilization. Sturgeon remained in the anesthetic for 5–10 min, depending on their size, and were then moved to a water-filled wooden measuring box. Each sturgeon was examined for tags, damage/deformities, and general condition. Each sturgeon was measured and scanned for Passive Integrated Transponder (PIT) tags. All untagged sturgeon had a PIT tag implanted into the flesh at the base of the dorsal fin with a hypodermic needle as part of CTDEP's long-term shortnose sturgeon monitoring program. Sturgeon were weighed (0.1 kg) in a mesh sling suspended from a spring scale (Jennings 1989).

Following examination/tagging, sturgeon

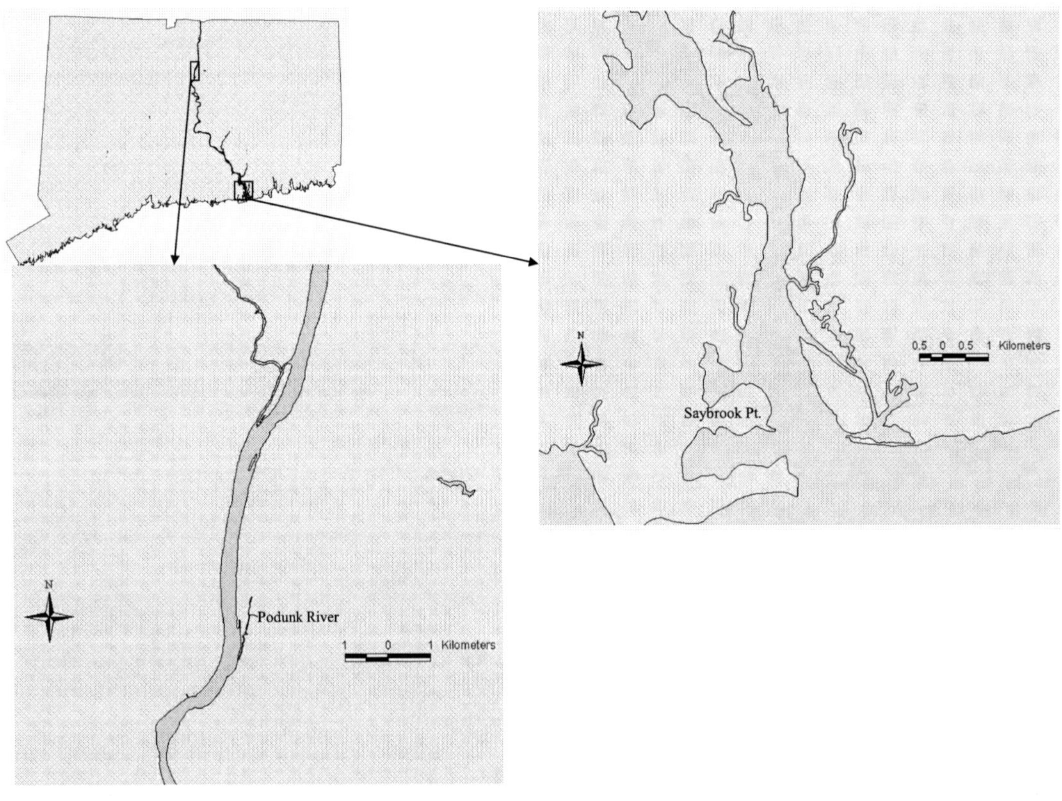

FIGURE 1. Location of study areas in Connecticut River, Estuary (rkm 5–10, map on right) and upriver (rkm 85–90, map on left).

were transferred to a v-board where the stomach contents were flushed out using a water lavage. We modified the technique developed by Haley (1998) specifically for sturgeon, by using a hand-powered pump (Black and Decker "Jackrabbit pump®") instead of individual syringes. A small diameter piece of flexible tubing was inserted into the stomach through the mouth and esophagus, and continuous water flow was used to gently flush water through the tube into the stomach of the sturgeon, expelling gut content. Water and food flushed out through the mouth and/or the anus. The volume of water flushed through the fish (up to 2.0 L) was subjective based on fish size and quantity of food regurgitated. If a fish continued to expel food, additional water was used. If the fish regurgitated little or no food, less water was pumped. All contents were collected in 3.8-L sample jars. River water was added to dilute any stomach acid. After lavaging, the sturgeon was placed in a recovery tub with Stress Coat® (water conditioner used to replace natural slime coat) and released as soon as the fish was active and alert. Sturgeon were held next to the side of the boat until they were vigorously swimming and then released. Samples of stomach contents were brought to the laboratory, fixed with 10% formalin for 24 h, and then preserved with 95% propanol.

Quantitative sediment and benthic invertebrate samples were taken from the estuary and upriver areas using a Petite Ponar grab (Wildco®). Each grab measured 0.02 m^2, and three grabs per site were pooled to form one sample. A total of 69 grabs were collected randomly within discrete areas where sturgeon were typically collected. Fifty-two benthic grabs were collected in the estuary and 17 from the upriver site. Samples were preserved in a solution of 10% formalin and processed in the laboratory. In the laboratory, samples were hand-sorted and invertebrates were counted and removed under 10× dissecting scope and preserved in 95% ethanol. Percent relative abundance (*%N*) was determined for each taxa.

Thirteen of the 69 sediment samples considered representative of the two areas were dried, shaken through a nested sieve set (U.S. standard mesh numbers 5 [pebble], 10 [gravel], 35 [coarse sand], 60 [medium sand], 120 [fine sand], 230 [very fine sand], and >230 [silt]), and individual fractions were weighed to the nearest 0.01 g (Leeder 1982). Percent sediment composition and mean percent grain size were determined for the two sites.

Stomach samples were examined under a 10× dissecting microscope and identified to lowest feasible taxon. Partially digested organisms were individually counted using unique hard parts. Polychaete heads were used to identify individual worms, since they were often in fragments. Invertebrates were keyed to the lowest possible taxon (Marine Biological Laboratory at Woods Hole 1964; Ruppert and Fox 1988; Ruppert and Barnes 1994; Weiss 1995; Merritt and Cummins 1996). For each stomach sample, taxon groups were blotted and wet weighed to the nearest 0.01 g. Volume was measured for each taxon by water displacement to the nearest 0.2 mL.

To evaluate the importance of individual food items in the diet, frequency of occurrence (*%F*) and relative abundance (*%N*) were calculated (Hyslop 1980). Frequency of occurrence (*%F*) was defined as the number of stomachs that contain one or more of the food groups (Hyslop 1980; Bowen 1983). Relative abundance (*%N*), was defined as the number of individuals of one group compared to the total number of all organisms in all stomach samples (Hyslop 1980; Bowen 1983).

Results and Discussion

A total of 663 tows were made from 31 May 2000 to 1 November 2002. Within the estuary region, 443 tows made over 103 d resulted in the collection of 4,303 fish of 34 different species. Estuarine collections were dominated by white perch *Morone americana* (37.3%), striped bass *M. saxatilis* (19.4%), and catfish (white catfish *Ameiurus catus*, brown bullhead *A. nebulosus*, and channel catfish *Ictalurus punctatus* 11.4%), collectively representing 68.1% of the total catch. Shortnose sturgeon comprised only 3.6% of the total catch. Fish abundance was much lower at the upriver site, with only 700 fish of 13 species collected in the 230 tows made on 34 sampling trips. Catfish were numerically dominant upriver, collectively comprising 41.6% of the catch. Shortnose sturgeon and white suckers *Catostomus commersonii* were the other dominant species, comprising 24.9% and 23.0%, respectively. Distribution of sampling effort was unequal between sites with 433 (65%) tows made in the estuary and 230 (35%) tows in the upriver area; however, similar numbers of shortnose sturgeon were collected (estuary = 153, upriver = 174) and pumped (estuary = 125, upriver = 121) from both locations.

Fork lengths (FL) of all sturgeon collected from the upriver area ranged from 300 to 1,086 mm (mean = 670 mm), and weights ranged from 0.15 to 12.7 kg. Shortnose sturgeon collected from the estuary were larger (FL 437–1,120 mm [mean = 834 mm]), and heavier (0.8–15.4 kg) than those captured at the upriver site. The estuary had a higher percentage of sturgeon more than a meter in fork length (Figure 2), while more juvenile shortnose sturgeon (≤50 cm FL) were collected upriver (*n* = 22) than the estuary (*n* = 2).

A total of 246 of the 327 shortnose sturgeon collected were sampled for food habits. Overall, 76% (187 of 246 pumped) of the sturgeon lavaged had evidence of recent feeding. Of these, 30 fish only had trace items (here defined as 1–5 individual prey items). No food was recovered from 24% (59/246) of fish sampled. Feeding incidence was 98% (123/125) in the estuary versus just better than 50% of the sturgeon (65/121) lavaged from upriver. The lower feeding incidence of fish upriver is at least partly a result of the low water temperature (<12°C) from November to March, when 52% of the upriver fish were collected versus 1% of the estuary fish. Despite numerous collection attempts, we were only successful in collecting a single fish in the estuary from October to March. Eighty-five percent of sturgeon collected during the cold water period were empty or had only trace food items. The 15% of fish that had fed during the coldwater period contrasts with 81% that had more than trace food items during the warmwater period.

The estuarine area had orders of magnitude greater abundance of prey items but reduced diversity compared to the upriver area. Benthic sampling yielded 8 taxa from the estuary, 3 of which were restricted to the estuary region, in contrast

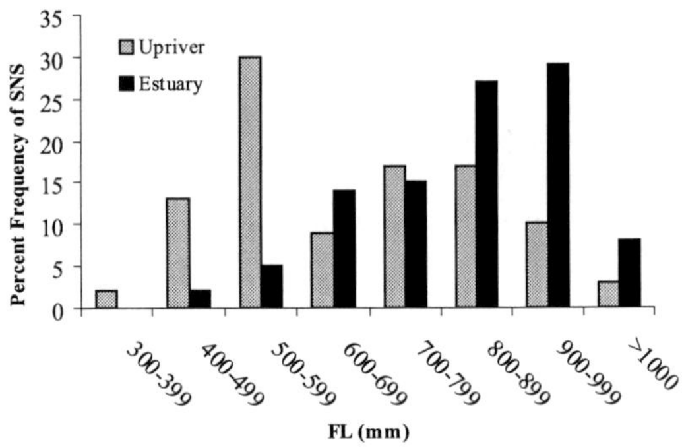

FIGURE 2. Frequency histogram of fork lengths (FL) of shortnose sturgeon (SNS) collected at the estuary and upriver sites.

to 13 upriver. Within the estuary, the three most abundant taxa/m^2 were amphipods (Family Gammaridae), isopods (Family Idoteidae), and palp worms (Family Spionidae). The upriver benthic samples were dominated by fingernail clams (Family Sphaeriidae), Family Chironomidae, and Order Diptera (not including Family Chironomidae) as the three most abundant taxa/m^2 (Table 1).

Sediment composition was very similar between the two areas, with a slightly greater percentage of both fines and coarse sediments in the estuary compared to the upriver site. The dominant grain size throughout the river was primarily course sand (Figure 3).

Shortnose sturgeon utilize a variety of epibenthic and benthic resources (Table 2) in the Connecticut River. Fifteen taxa and 35,137 individuals were enumerated from 125 stomach samples of sturgeon collected in the estuary versus nine taxa and 5,954 individuals identified from 121 stomach samples collected at the upriver

TABLE 1. Relative abundance (%N) of food items in shortnose sturgeon stomachs and in grab samples from the estuary and upriver sites. Invertebrates are broken down by groups of taxa.

	Sturgeon		Bottom grabs	
	Estuary	Upriver	Estuary	Upriver
Clams	0.004	10.62	0.67	44.23
Amphipods	51.75	5.58	47.18	4.12
Isopods	1.37		42.97	
Mayflies		68.34		2.47
Chironomids	<0.1	15.00	4.07	24.69
Other insects	<0.1	0.57	0.65	20.79
Polychaetes	46.81		4.35	3.29

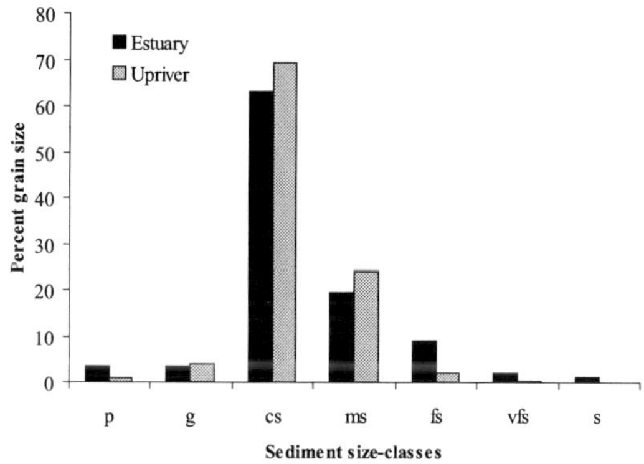

FIGURE 3. Substrate composition based on dominant grain size within sites estuary and upriver. p = pebble, g = gravel, cs = coarse sand, ms = medium sand, fs = fine sand, vfs = very fine sand, and s = silt.

area. Leeches (Hirudinea) and fish lice (*Argulus* spp.) (%*N* = 0.02) were also occasionally collected but were discounted as food items as they are known ectoparasites. Not only were estuarine organisms consumed in higher numbers, but the prey items were also larger with wet weight of prey consumed ranging up to 324.4 g and volumes to 750 mL per fish in contrast to 52 g and 93.8 mL per fish in those collected upriver. Maximum prey count from a single estuarine fish was 10-fold greater (11,099 items in an 825 mm FL, 7.0 kg) than an upriver fish (843 items, 800 mm FL, 3.6 kg).

Shortnose sturgeon collected in the estuary fed on a wider variety and larger quantity of taxa than sturgeon collected upriver. Eighty-six percent of the total relative abundance, as well as 86% of the total wet weight of all taxa consumed, were eaten in the estuary. Fish in the estuary consumed almost six times the total wet weight and eight times the total volume of food compared to fish in the upriver area. Moreover, of the 15 different taxa found in the stomach samples, 9 of these were only utilized in the estuary, indicating that both greater numbers and a greater variety of prey are available in the estuary. Amphipods (*Gammarus* spp.) and palp worms (*Scolecolepides* spp.) had the highest relative abundances (Table 1), frequency of occurrence (82.69 and 74.04), total wet weights (99.5 g and 188.3 g), and total volumes (261.10 mL and 381.40 mL). *Gammarus* spp. ranked number one in abundance in both the stomach samples and the benthic grabs from this area (Figure 4A). While a Petite ponar grab is not the ideal instrument for sampling these mobile epibenthic organisms, that they ranked highest in relative abundance gives some indication of their absolute abundance in the lower river. *Gammarus* spp. were often so abundant that they covered the deck when the trawl was pulled out of the river. The isopods *Cyathura* spp. and *Chiridotea* spp. also had a high frequency of occurrence but were not abundant in any of the stomachs. Similarly, although Chironomids were ubiquitous throughout the Connecticut River and were found in the estuary, sturgeon rarely con-

TABLE 2. Taxa in stomach samples from shortnose sturgeon collected at the estuary (rkm 1–10) and upriver (rkm 85–90) sites from May 2000 to July 2002.

Group	Class	Order	Family	Taxon
Fingernail clams	Bivalvia	Veneroida	Sphaeridae	*Pisidium* spp.
Gemma clams			Veneridae	*Gemma* spp.
Freshwater mussels		Unionoida	Unionidae	*Elliptio* spp.
Amphipods	Malacostraca	Amphipoda	Corophiidae	*Unicola* spp.
			Haustoridae	
			Aoridae	
Isopods		Isopoda	Idoteidae	*Chiridotea* spp.
				Cyathura spp
Mysids		Mysidacea		
Nereid worms	Polychaeta		Nereididae	*Nereis* spp.
Palp worms			Spionidae	*Scolecolepides* spp.
Insects	Insecta	Ephemeroptera		
		Trichoptera		
		Diptera	Chironomidae	

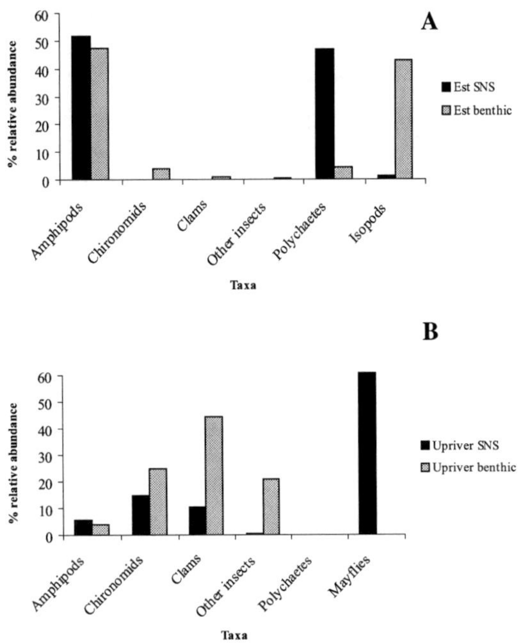

FIGURE 4. (A) Percent relative abundance of taxa food items in shortnose sturgeon (SNS) compared to invertebrates collected in benthic samples from the estuary site. (B) Percent relative abundance of taxa food items in shortnose sturgeon compared to invertebrates collected in benthic samples from the upriver site.

sumed them there. *Scolecolepides* spp. ranked third in abundance in benthic grabs. Total wet weight and volume of sediment was also much higher from estuary stomach samples (744.8 g and 646.2 mL), likely a result of heavy feeding on infaunal worms.

Upriver, 8 of the 13 taxa collected in the benthic grabs were aquatic insect larvae. Mayflies (Order Ephemeroptera) comprised 68% (Table 2; Figure 4B) of the prey ingested upriver, but had a low total wet weight. Many times, a group of several insects recovered from a stomach sample were less than a fraction of a gram. Chironomidae were found in 30% of fish with food but comprised only 15% of contents eaten. *Gammarus* spp. was also consumed in the upriver area ranking third in frequency of occurrence (22.94) and fourth in relative importance. Fingernail clams (*Pisidium* spp.) were fourth in frequency of occurrence, but third in relative abundance because of the high total wet weight (46.6 g) and total volume (62.2 mL) due to the shell. We found no utilization of the invasive Asiatic clam (*Corbicula* spp.), despite *Corbicula* being the most abundant mollusk in middle sections of the river (Massengill et al. 2004; Morgan and Keser 2004; both this volume). The two species are closely related (Order Veneroida); however, *Corbicula* grows to a larger size and has a thicker shell.

The gastric lavage technique is an effective, nonlethal way to retrieve food items from shortnose sturgeon. Seventeen shortnose sturgeon were pumped twice, and two sturgeon were pumped three times without apparent ill effects. The shortest interval between lavages for a single fish was 4 d, although the average time between events was 138 d. The technique appears to extract all of the food from sturgeon stomachs, with prey items retrieved including soft-bodied polychaetes, aquatic insect larvae, amphipods, and hard shelled mollusks. The lavage technique is sufficiently mild that recently ingested prey items often remained alive until fixed in the laboratory. Retrieval of live prey organisms is also notable in that it indicates that sturgeon were actively feeding in both areas and not importing food items from other areas.

The diet of shortnose sturgeon in the Connecticut River is similar to that described in the Hudson River (Haley 1999) and Saint John River Estuary (Dadswell 1979). It differs slightly from the Hudson River where shortnose sturgeon preyed primarily on Gammarid amphipods, while Atlantic sturgeon preyed most heavily on spionid polychaetes (Haley 1999). Dadswell (1979) reported that adult shortnose sturgeon in the St. John River, Canada, fed mainly on small mollusks in estuarine waters and gastropods in freshwater, while juvenile shortnose sturgeon fed on insect larvae and small crustaceans (amphipods and isopods). We found consumption of clams only at the upriver freshwater site, utilization of amphipods at both sites, and a much higher utilization of polychaetes in the Connecticut River Estuary compared to other river systems.

The estuary site is a productive benthic community, providing sturgeon with larger quantities of food compared to the upriver area. Although sturgeon and the other riverine species examined (*Morone saxatilis*, *M. americana*, and *Ictalurus punctatus*) are all feeding primarily on *Scolecolepides* spp. and *Gammarus* spp. in this area, it does not appear that shortnose sturgeon examined were limited in either number or volume of prey consumed. We conclude that unrestricted access to the estuary is important to maintaining the high condition factors seen in the population of shortnose sturgeon in the section of the Connecticut River below the Holyoke dam (Savoy 2004). This may be particularly important in the persistence of this stock, despite the low numbers of potential adult spawners.

Without exception, those sturgeon examined that contained more than trace food items and were actively feeding during the cold water period, were less than 600 mm in FL. Feeding throughout the winter months may be a significant factor in maintaining the high growth rates seen among juvenile sturgeon (Savoy 2004).

Acknowledgments

We would like to thank Dr. Webb Van Winkle and Dr. Douglas Dixon for their helpful review of the manuscript. Thanks to Dr. Steven Burian for assistance with invertebrate identification and to Kevin Nebiolo and Marisa Guillot for all their hard work. Funding support was obtained from The Nature Conservancy Small Grants Program and Connecticut's Income Tax Checkoff.

References

Bowen, S. H. 1983. Quantitative description of the diet. Pages 325–336 *in* L. A. Nielsen and D. L. Johnson, editors. Fisheries techniques. American Fisheries Society, Bethesda, Maryland.

Dadswell, M. J. 1979. Biology and population characteristics of the shortnose sturgeon, *Acipenser brevirostrum* LeSueur 1818 (Osteichthyes: Acipenseridae), in the Saint John River Estuary, New Brunswick, Canada. Canadian Journal of Zoology 57:2186–2210.

Dadswell, M. J., B. D. Taubert, T. S. Squires, D. Marchette, and J. Buckley. 1984. Synopsis of biological data on shortnose sturgeon, *Acipenser brevirostrum* LeSueur 1818. NOAA Technical Report. National Marine Fisheries Service 14, Silver Spring, Maryland.

Haley, N. J. 1998. A gastric lavage technique for characterizing diets of sturgeons. North American Journal Fisheries Management 18:978–981.

Haley, N. J. 1999. Habitat characteristics and resource use patterns of sympatric sturgeons in the Hudson River Estuary. Master's thesis. University of Massachusetts, Amherst.

Hyslop, E. J. 1980. Stomach contents analysis- a review of methods and their application. Journal of Fish Biology 17:411–429.

Jennings, M. R. 1989. Use of spring scales for weighing live fish in the field. North American Journal Fisheries Management 9:509–511.

Leeder, M. R. 1982. Sedimentology-process and product. Allen and Unwin, London.

Marine Biological Laboratory at Woods Hole. 1964. Keys to marine invertebrates of the Woods Hole Region. Spaulding Company, Massachusetts.

Massengill, R. R., J. M. Vozarik, D. E. Morgan, and M. Keser. 2004. Comparison of benthic community structure in the lower Connecticut River before, during, and after operation of the Connecticut Yankee power plant. Pages 441–454 *in* P. M. Jacobson, D. A. Dixon, W. C. Leggett, B. C. Marcy, Jr., and R. R. Massengill, editors. The Connecticut River Ecological Study (1965–1973) revisited: ecology of the lower Connecticut River 1973–2003. American Fisheries Society, Monograph 9, Bethesda, Maryland.

Merritt, R. W., and K. W. Cummins. 1996. An introduction to the aquatic insects of North America. Kendall/Hunt Publishing, Iowa.

Morgan, D. E., M. Keser, J. T. Swenarton, and J. F. Foertch. 2004. Effect of Connecticut Yankee power plant on population dynamics of Asiatic clamsand their interactions with native bivalves. Pages 419–439 *in* P. M. Jacobson, D. A. Dixon, W. C. Leggett, B. C. Marcy, Jr., and R. R. Massengill, editors. The Connecticut River Ecological Study (1965–1973) revisited: ecology of the lower Connecticut River 1973–2003. American Fisheries Society, Monograph 9, Bethesda, Maryland.

Moser, M. L., M. Bain, M. R. Collins, N. Haley, B. Kynard, J. C. O'Herron, G. Rogers, and T. S. Squiers. 2000. A protocol for use of shortnose and Atlantic sturgeons. NOAA Technical Memorandum, National Marine Fisheries Service, NMFS-OPR-18, Silver Spring, Maryland

NMFS (National Marine Fisheries Service). 1998. Recovery plan for the shortnose sturgeon (*Acipenser brevirostrum*). Prepared by the Shortnose Sturgeon Recovery Team for the National Marine Fisheries Service, Silver Spring, Maryland.

Ruppert, E. E., and R. D. Barnes1994. Invertebrate zoology. Saunders College Publishing, Florida.

Ruppert, E. E., and R. S. Fox. 1988. Seashore animals of the southeast. University of South Carolina Press, Columbia.

Savoy, T. 1991. Sturgeon status in Connecticut waters. Completion report for Connecticut Department of Environmental Protection, Old Lyme.

Savoy, T. 1999. Anadromous fish studies in Connecticut waters. Final report for Connecticut Department of Environmental Protection, Project AFC-25, Old Lyme.

Savoy, T. F. 2004. Population estimate and utilization of the lower Connecticut River by shortnose sturgeon. Pages 345–352 *in* P. M. Jacobson, D. A. Dixon, W. C. Leggett, B. C. Marcy, Jr., and R. R. Massengill, editors. The Connecticut River Ecological Study (1965–1973) revisited: ecology of the lower Connecticut River 1973–2003. American Fisheries Society, Monograph 9, Bethesda, Maryland.

Savoy, T., and D. Shake. 1993. Anadromous fish studies in Connecticut waters. Annual Progress report for Connecticut Department of Environmental Protection, Project AFC-21–1, Hartford.

Thomson, K. S., W. H. Weed, A. G. Taruski, and D. E. Simanek. 1978. Saltwater fishes of Connecticut. State Geographical and Natural History Survey of Connecticut, Bulletin No. 105, Hartford.

U.S. Office of the Federal Register. 1967. Endangered and threatened wildlife and plants; endangered status of the shortnose sturgeon. Federal Register 32:48(11 March 1967):4001.

Weiss, H. W. 1995. Marine animals of southern New England and New York. State Geographical and Natural History Survey of Connecticut, Bulletin No. 115, Hartford.

Whitworth, W. R. 1996. Freshwater fishes of Connecticut. State Geographical and Natural History Survey of Connecticut, Bulletin No. 114, Hartford.

American Fisheries Society Monograph 9:361–377, 2004

Factors Affecting the Recent Decline of Blueback Herring and American Shad in the Connecticut River

THOMAS F. SAVOY* AND VICTOR A. CRECCO

Connecticut Department of Environmental Protection, Marine Fisheries Division
Post Office Box 719, Old Lyme, Connecticut 0637, USA

Abstract.—A dramatic and unexpected decline in American shad *Alosa sapidissima* and blueback herring *A. aestivalis* abundance occurred in the Connecticut River since 1992. Several hypotheses were tested in an effort to explain the alosine stock declines, including (1) overharvest by natal (in-river) and coastal intercept commercial fisheries, (2) stock displacement or enhanced ocean mortality rates due to colder than normal ocean water temperatures (winter and/or spring), and (3) increased predation mortality on adult and juvenile alosines. The possibility that coastal and in-river overharvest could have caused the recent shad decline was not consistent with trends in in-river (FR) and coastal instantaneous fishing mortality (FC) rates from 1980 to 2001. The total fishing mortality rates (FT = FR + FC) fell steadily after 1994, and the current average fishing mortality (F) rate (1996–2001) of 0.12 was well below the estimated overfishing definition (F_{30} = 0.43) for Connecticut River shad. Moreover, in-river commercial fishing effort (number of fishing days), as well as in-river and coastal commercial landing of Connecticut River shad, have declined steadily from 1992 to 2002 consistent with the observed decline in fishing mortality. The parallel decline in adult blueback herring and juvenile production cannot be attributed to coastal and in-river commercial fisheries, since no directed commercial fisheries currently exist on Connecticut River blueback herring. The temperature anomaly hypothesis does not provide a satisfactory explanation for the recent alosine decline for several reasons. By contrast, the combination of statistical (regression analyses) and empirical evidence (food habits, striped bass *Morone saxatilis* size frequency, and population abundance in the Connecticut River) supports the predation hypothesis as a direct linkage between increased striped bass abundance and the recent dramatic drop in both adult shad and blueback herring in the Connecticut River.

Introduction

The anadromous restoration program in the Connecticut River involved the construction of upstream passage facilities at the Holyoke and Turners Falls dams after 1975 to restore American shad *Alosa sapidissima* and Atlantic salmon *Salmo salar* to their historic upriver spawning areas (Gephard and McMenemy 2004, this volume). Expansion of spawning areas for American shad, resulting from up-river passage, was expected to raise the annual shad population size to an average of 2 million fish by 1990 (Stolte 1980). Although the number of adult shad passed at Holyoke and Turners Fall dams rose sixfold since 1975, shad population size (Crecco and Savoy 1987; Savoy 1998) has varied erratically from 360,000–1.6 million during the years 1966–1992 (Figure 1). A steady decline in shad stock size (Figure 1) took place after 1992 and was accompanied by an even more severe and persistent drop in the population of blueback herring *A. aestivalis* (Figure 2).

The variation in the adult shad population from 1993 to 2002 does not seem particularly noteworthy compared to historic run sizes from 1965 to 1992 (Figure 1). However, juvenile production from the 1990–1997 year-classes had reached record levels (Figure 3), so that shad population size from 1993 to 2002 was expected to rise dramatically over the pre1993 levels. This apparent discrepancy between observed and expected run size after 1992 is more clearly demon-

* *Corresponding author*: Tom.Savoy@po.state.ct.us

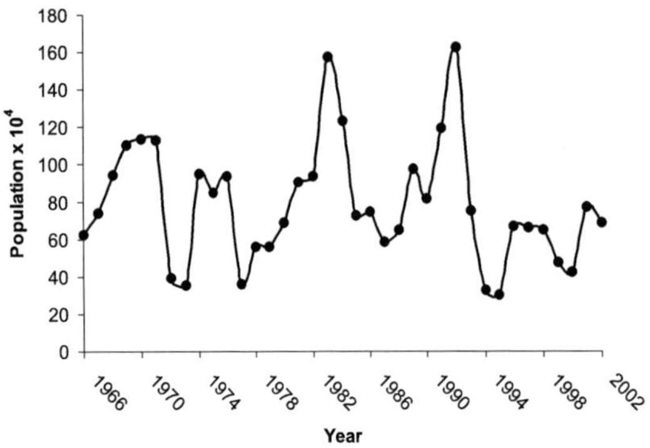

FIGURE 1. Population of adult American shad from 1966 to 2002.

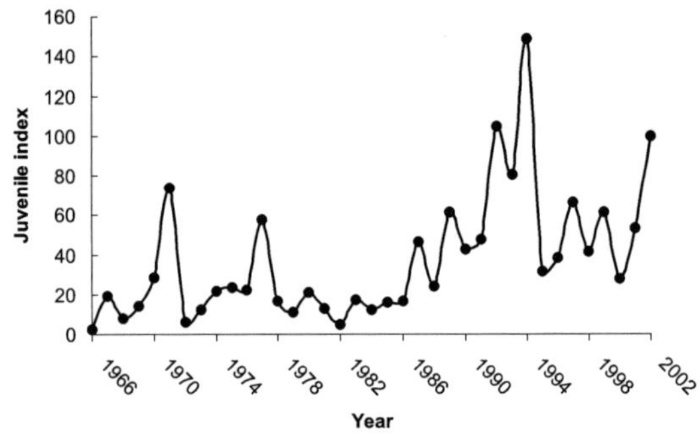

FIGURE 3. Juvenile abundance indices of American shad from 1966 to 2002.

strated by the historic (1971–1982) positive relationship (r = 0.83, P < 0.0005) reported (Crecco and Savoy 1984) between juvenile production in years t-4, t-5, and t-6 and subsequent shad run size in year t. To predict expected shad run sizes ($PredS_t$) from 1993 to 2002, juvenile indices ($JI_{t-4} + JI_{t-5} + JI_{t-6}$) from the 1987–1998 year-classes were substituted into the multiple linear regression equation:

$$\mathrm{PredS}_t = 120{,}187 + 9{,}755 * \mathrm{JL}_{t-4} + 19{,}908 * \mathrm{JL}_{t-5} + 15{,}290 * \mathrm{JL}_{t-6} \quad (1)$$

used to forecast future run size ($PredS_t$) from 1971 to 1992. The resulting predicted run sizes ($PredS_t$) from 1993 to 2002 using equation (1) consistently exceeded the observed shad run sizes by an average (geometric mean) of 1.35 million fish (Table 1). These persistent overpredictions of shad run size in recent years (1993–2002) by the regression model (equation (1) indicates a recent (1993–2002) disconnect between the previously established relationship between juvenile recruitment and future run size. This disconnect also suggests a systematic rise in prerecruit mortality has occurred after 1992 sometime between the juvenile stage and adult recruitment to the spawning population.

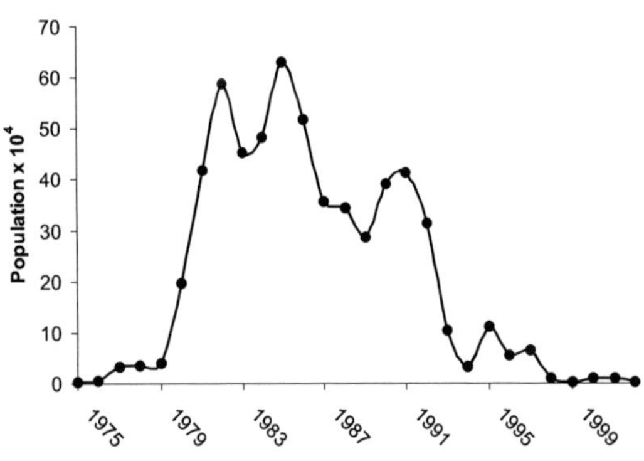

FIGURE 2. Population of adult blueback herring from 1975 to 2002.

Several competing hypotheses may explain the root cause of the recent decline in shad and blueback herring abundance in the Connecticut River. They include (1) overharvest by natal (in-river) and ocean intercept commercial fisheries, (2) stock displacement or enhanced mortality rates of alosine stocks due to colder than normal ocean water temperatures (winter and/or spring), and (3) increased predation mortality on either adult and/or juvenile alosines. The Connecticut Yankee nuclear power station (CY) operated in the lower Connecticut River from 1966 through 1995, but power plant entrainment and thermal effects were not considered here as a realistic working hypothesis. The Connecticut Yankee nuclear power station ceased operation in 1996, and studies conducted in the vicinity of CY (Leggett 1977 and 2004 [1976], this volume; Marcy 2004 [1976], this volume) found no evidence of adverse impact by the plant on American shad and blueback herring stock size and migratory distribution in the Connecticut River.

The objective of this paper is to determine whether or not one of the aforementioned hypotheses could explain the recent drop in the Connecticut River alosine stocks. To examine the overfishing hypothesis, fishing mortality (F) rates on Connecticut River shad were derived from the

TABLE 1. Differences (Resid) between shad population size (VPop), and predicted adult shad population (PredS) size in the Connecticut River from 1993 to 2002 based on the multiple linear regression between juvenile production (SJI) in years t-4, t-5, and t-6 and estimated adult run size (Vpop) from 1971 to 1992 (see Figure 4). The SJI is the arithmeic mean juvenile shad index, GMSJI is the geometric mean juvenile shad index, BJI is the arithmetic mean juvenile blueback index and GMBJI is the geometric mean juvenile blueback herring index. The Bpop is the total number of adult blueback herring lifted annually over the Holyoke Dam.

Year	SJI	GMSJI	BJI	GMBJI	Vpop	Bpop	PredS	Resid
1966	2.7	0.71	-	1.74	621,300	0	–	–
1967	19.4	1.22	52.0	3.06	742,300	356	–	–
1968	8.2	1.93	102.8	3.85	945,800	0	–	–
1969	14.5	2.33	55.6	6.92	1,108,080	10,000	–	–
1970	28.6	2.60	168.7	6.18	1,140,480	1,900	–	–
1971	73.7	4.04	15.6	9.43	1,128,600	302	1,105,674	–
1972	6.4	1.22	28.1	2.82	390,960	188	447,251	–
1973	12.7	–	3.5	–	353,700	302	562,290	–
1974	21.4	–	–	–	952,560	504	717,903	–
1975	23.7	–	–	–	847,530	1,625	919,196	–
1976	22.4	–	–	–	936,900	4,745	804,652	–
1977	57.5	–	–	–	361,890	32,932	556,731	–
1978	16.5	2.45	18.6	5.89	560,700	34,953	669,254	–
1979	11.4	2.63	106.9	18.72	557,010	39,461	971,616	–
1980	21.0	3.85	222.2	29.18	685,080	198,000	1,137,745	–
1981	12.9	2.69	148.9	18.14	909,270	419,733	1,276,019	–
1982	4.8	1.04	74.2	7.99	939,330	586,808	1,381,818	–
1983	17.5	2.27	413.8	21.39	1,574,460	454,242	1,398,521	–
1984	12.6	1.93	576.0	31.51	1,231,110	482,954	743,726	–
1985	15.9	2.99	263.5	32.01	727,560	630,000	783,476	–
1986	17.0	2.68	175.3	17.70	748,440	517,521	703,276	–
1987	46.8	4.43	256.5	22.67	587,520	356,846	541,755	–
1988	24.0	2.51	245.7	17.27	647,640	343,363	591,207	–
1989	61.6	2.20	156.8	10.33	979,440	286,259	713,020	–
1990	43.0	4.18	279.3	17.69	816,480	392,157	729,755	–
1991	47.5	1.85	178.1	15.62	1,195,920	412,380	1,094,781	–
1992	104.6	3.33	154.3	8.26	1,628,039	312,863	1,462,772	–
1993	80.6	3.84	137.0	16.39	749,227	103,447	1,490,481	+741,254
1994	148.9	5.13	123.2	16.95	325,558	31,766	1,328,739	+1,003,181
1995	31.7	0.86	87.2	6.22	303,973	112,070	537,916	+1,233,943
1996	38.6	2.75	58.2	8.34	667,137	55,040	1,997,520	+1,330,382
1997	66.1	3.16	286.2	12.74	659,478	63,945	2,843,091	+2,183,613
1998	41.3	1.71	63.0	5.89	651,358	11,146	3,598,320	+2,946,962
1999	61.5	2.42	110.1	7.56	475,095	2,699	3,042,094	+2,566,998
2000	27.7	2.00	37.3	5.50	427,381	9,537	1,960,797	+1,523,416
2001	53.5	1.38	61.9	3.67	773,114	10,603	1,385,851	+612,737
2002	100.1	2.36	75.7	5.28	686,662	1,939	1,731,590	+1,044,928

ratio of in-river and coastal commercial landings from 1982 to 2001 to the estimated population sizes (Crecco and Savoy 1987; Savoy 2001; Leggett 2004 [1976]). These fishing mortality (F) rates were correlated with trends in shad population size from 1982 to 2001. To test the displacement hypothesis (Jesien et al. 1992), a time series (1982–2001) of adult and juvenile alosine abundance and relative condition (Fulton's) factors were related to changes in monthly and seasonal ocean water temperatures. The predation hypothesis was tested using relative natural mortality estimates for prerecruit (ages 0–6) and subadult (age 0–1) shad and blueback herring from 1982 to 2001. These mortality rates were subsequently correlated to abundance indices for several candidate finfish predators, including striped bass *Morone saxatilis* and bluefish *Pomatomus saltatrix* in Long Island Sound. The potential effects of other potential finfish predators, such as weakfish *Cynoscion regalis*, spiny dogfish *Squalus acanthias*, and smooth dogfish *Mustelus canis*, are also discussed.

Methods

Juvenile and Adult Shad and Blueback Herring Abundance

A long (1966–2002) time series of shad run size in numbers, as well as juvenile relative abundance estimates (catch per unit effort), has been established in the Connecticut River (Tables 1 and 2). Annual shad run size has been estimated by tag recapture studies and expressed by age, sex, and year-class (Crecco and Savoy 1987; Savoy and Shake 1993; Savoy 2002; Leggett 2004 [1976]). The average juvenile abundance index from 1966 to 2002 was expressed by both the arithmetic and geometric mean catch per seine haul from stations located between Essex, Connecticut (river km 16) and Holyoke, Massachusetts (river km 170) (Savoy 2002; Marcy 2004 [1976]). Geometric mean indices for juvenile blueback herring were also developed during juvenile shad surveys from 1979 to 2001. The contribution of adult shad recruitment from each year-class (t) was defined as the sum of all virgin (first time spawner) age-4 and 5 spawners from each year-class in the adult shad populations (Savoy 2001). The time series of juvenile shad indices were positively correlated with adult recruitment (Figure 4; $r = 0.82$, $P < 0.01$) from these year-classes between 1966 and 1988.

Trawl surveys conducted annually in Long Island Sound from April through November (Simpson et al. 1994; Connecticut Department of

TABLE 2. Estimates of relative abundance (mean catch/tow) of age-1 shad (GMCS) and blueback herring (GMCB) in Long Island Sound (LIS), striped bass relative abundance (Stripers) in Long Island Sound and VPA estimate of coastal abundance (mt), and estimates of bluefish relative abundance (Bluefish) in Long Island Sound from 1982 to 2002.

Year	GMCS[a]	GMCB	Stripers LIS[a]	Stripers VPA[b]	Bluefish LIS[a]
1982	–	–	–	3,885	–
1983	–	–	–	2,648	–
1984	–	0.37	0.02	3,926	1.61
1985	1.34	0.16	0.00	4,535	4.16
1986	0.57	0.07	0.00	5,026	3.77
1987	0.92	0.13	0.05	10,438	3.11
1988	0.44	0.53	0.04	10,205	2.20
1989	0.90	0.34	0.06	16,096	1.92
1990	0.34	0.10	0.16	17,771	6.14
1991	0.48	0.04	0.15	21,711	5.59
1992	0.72	0.08	0.22	26,326	8.44
1993	0.32	0.11	0.27	30,635	3.34
1994	0.68	0.93	0.30	34,032	3.07
1995	0.49	0.27	0.59	37,575	4.07
1996	0.48	0.05	0.63	47,795	2.34
1997	1.08	0.75	0.85	49,226	2.35
1998	0.86	0.16	0.97	52,197	1.65
1999	0.80	0.06	1.10	54,708	0.86
2000	0.38	0.06	0.84	58,220	2.18
2001	0.08	0.20	0.61	61,837	2.62
2002	0.61	0.60	1.30	64,142	2.44

(–) Data missing or unavailable
[a] Geometric mean catch taken in Long Island Sound Trawl Survey (Gottschall and Pacileo 2002).
[b] Estimates of spawning stock biomass in weight (metric tons).

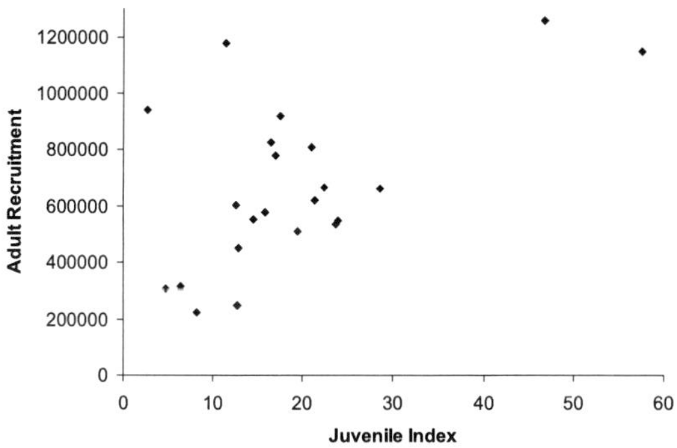

FIGURE 4. Relationship between total adult recruitment and juvenile abundance indices of American shad from 1966 to 1988.

Environmental Protection, unpublished data) have documented the presence of age-1 American shad and blueback herring. Juvenile shad emigrate from the Connecticut River in late summer and fall, but some portion of the juvenile alosine stock remains in Long Island Sound for 1 or 2 years before joining the coastal migratory stocks in the Gulf of Maine during their second or third year of life (Savoy 1993). This migratory pattern noted for Connecticut River shad is consistent with studies conducted off New Jersey (Milstein 1981) and Nova Scotia (Stone and Daborn 1987). Relative abundance of age-1 shad and blueback herring from 1985 to 2001 was expressed as the geometric mean catch per tow in Long Island Sound from the annual Connecticut multispecies trawl survey (Table 2; Gottschall and Pacileo 2002). Although it is possible for juvenile alosines from the Hudson River stock to enter Long Island Sound through the East River, it was assumed that the overall trend in age-1 catch per unit effort (CPUE) from the Long Island Sound Trawl Survey accurately reflects relative abundance of Connecticut River shad and blueback herring in Long Island Sound. Moreover, the time series of juvenile shad indices in the Connecticut and Hudson rivers (New York Department of Environmental Conservation, unpublished data) were strongly positively correlated.

Direct population estimates of adult blueback herring have not been made in the Connecticut River. A relative abundance index for bluebacks (Bpop) was derived based on the total number of adult bluebacks passed annually at the Holyoke Dam during each spring from 1966 to 2001 (Table 1; Figure 2). The fraction of the total adult blueback run passed annually over the Holyoke Dam may have varied annually, but it was assumed that the number of blueback herring lifted was directly proportional to changes in the river-wide blueback population size. Daily lift rates (fish/d) are affected by short-term fluctuations in water temperature and flow conditions, although long-term hydrographic effects would likely average out over the 60-d period in which the lift has been operating, leaving Bpop as a reliable abundance index of adult blueback herring. Moreover, lift rates (mean fish/d) of American shad over the Holyoke Dam from 1966 through 1982 were highly correlated ($r = 0.87$, $P < 0.001$) with river-wide shad population estimates made during the same period from tag-recapture studies (Crecco and Savoy 1987; Leggett 2004 [1976]).

Shad and Blueback Herring Mortality Rates

If finfish predation was directly linked to the current alosine decline in the Connecticut River, a rise in alosine prerecruit natural mortality since 1990 should occur coincident with an increase in the abundance of certain finfish predators. The relative mortality rates between ages 0 and 1 for shad (Seamort) and blueback herring (Herrmort) from Long Island Sound (LIS) were derived for the 1984–2001 year-classes as a log ratio of survival between age-1 shad ($LISAcatch_{st}$) and blueback herring ($LISAcatch_{bt}$) to juvenile shad (SJI_{t1}) and blueback herring (BJI_{t-1}) abundance of that same year-class:

$$\text{Seamort} = -\log\left(\frac{\text{LISAcatch}_{st}}{\text{SJI}_{t-1}}\right) \tag{2}$$

and

$$\text{Hermort} = -\log\left(\frac{\text{LISAcatch}_{ht}}{\text{BJI}_{t-1}}\right) \tag{3}$$

Relative mortality rates of blueback herring (Bmort) between juvenile and adult recruitment (age 4 and 5) were derived as a log ratio of survival between the relative population estimate ($Bpop_t$) in year t (between 1980 and 2001) from the Holyoke Dam and the sum of juvenile blueback herring production in years t-4 and t-5 ($SBJI_t$):

$$\text{Bmort}_t = -\log\left(\frac{\text{Bpop}_t}{\text{SBJI}_t}\right) \tag{4}$$

where $SBJI_t$ = sum of the scaled juvenile blueback herring indices in years *t*-4, *t*-5. Note that the magnitude of $Bpop_t$ values in the numerator of equation (4) (Table 1) is always much higher than that for the juvenile indices ($SBJI_t$) in the denominator. Since Bmort values derived from equation (4) would all be negative values, the juvenile indices ($SBJI_t$) in the denominator were scaled upwards by a constant (100,000) to make resulting mortality values positive integers.

A similar prerecruit mortality index for shad (Tmort) was derived from 1980 through 2001 as a log ratio between the virgin shad population estimate ($VPop_t$) in year *t* and the sum of juvenile shad production in years *t*-4 and *t*-5 ($SSJI_t$):

$$\text{Tmort} = -\log\left(\frac{\text{Vpop}_t}{\text{SSJI}_t}\right) \tag{5}$$

where $SSJI_t$ = sum of the juvenile shad indices in years *t*-4, *t*-5. As was the case for blueback herring mortality estimates (Bmort) from equation (4), positive values of Tmort were produced in equation (5) by multiplying each juvenile index ($SSJI_t$) value by a constant (100,000).

In-River and Coastal Fishing Mortality Rates (*F*)

Overharvest of American shad and blueback herring could have caused the recent stock declines in the Connecticut River if fishing mortality rates (FT) had risen after 1990 coincident with the observed adult stock decline. Shad and blueback herring make extensive coastal migrations, so these stocks are susceptible to several fisheries, including in-river sport and commercial fisheries, coastal intercept fisheries, and ocean bycatch fisheries.

A commercial gill-net fishery and a recreational hook and line fishery have harvested American shad in the Connecticut River since the late 19th century. Commercial shad fishermen are required by law to report their annual gillnet landings to the state by September. Recreational shad landings have been estimated annually from 1980 to 1996 and periodically since then by a roving creel census (Savoy 1998). Prior to 1993, there was a thriving recreational fishery for American shad in the Connecticut River from Enfield, Connecticut (river km 99) to the Holyoke Dam, Massachusetts (river km 140). These sport landings used to comprise as much as 60% of the total in-river landings (Table 3). Recreational shad landings began to fall dramatically after 1995 to a point where harvest estimates from creel surveys were unreliable and imprecise. Recreational and commercial fisheries combined have harvested between 20,000–125,000 fish annually from 1990 to 2002.

There is no commercial fishery for blueback herring in the Connecticut River, and the magnitude of the recreational harvest for blueback herring has never been estimated. Anecdotal evidence suggests that adult blueback herring have been harvested in tributaries and from the main stem to be used for bait to catch striped bass in the river during April and May. Although this recreational bait fishery for bluebacks may have risen since 1992, these landings are thought to be inconsequential relative to overall abundance.

In-river fishing mortality rates (FR) have been estimated from 1970 to 2001 as a log ratio (seasonal fishery; Ricker 1975) of the sum of commercial landings and discards (Table 3) in numbers (CL) plus the river-wide recreational harvest in numbers (RL) divided by total run size (*N*) (Table 1):

$$\text{FR} = -\log\{1 - [(\text{CL} + \text{RL})/N]\} \tag{6}$$

Since male shad are not typically landed due to their low market value, annual commercial landings (CL) in equation (6) were adjusted upward by an average of 50% to reflect under-reporting and male discard mortality. Both Leggett (2004 [1976]) and Crecco et al. (1986) reported that in-river commercial fishermen underreported by 35–67% annually from 1966 to 1983 based on the ratio of tag returns to reported commercial landings.

A coast-wide intercept fishery for American shad expanded from 1975 to 1990, but coast-wide intercept landings have fallen steadily thereafter (Table 3). The coastal intercept fishery harvests a mixed stock of American shad using drift gill nets during late winter and early spring. This commercial intercept fishery is located mainly between South Carolina and New Jersey and has harvested mostly adult shad (size range: 45–60 cm, TL, weighing an average between 1.5 and 2.5 kg; Krantz et al. 1992). The contribution of Connecticut River shad to the coastal intercept fishery between 1980 and 1996 (Table 3) was based on the annual coastal landings from Virginia to Maine

TABLE 3. American shad population estimates (numbers), landings data (numbers) from the coastal intercept fishery (CT Ocean), and Connecticut River commercial (CT Comm) and recreational landings (CT Sport) in numbers. Instantaneous fishing mortality estimates on shad from the coastal intercept fishery (FC), the in-river sport and commercial fisheries (FR) and the combined fishing mortality (FT = FC + FR) were derived from 1980 to 2002. Note that coastal intercept landings on shad are not yet available for 2002.

Year	Total population	CT Ocean[a]	CT Comm	CT Sport	FC	FR	FT
1980	68,5080	49,000	88,329	12,189	0.069	0.159	0.228
1981	90,9270	69,000	97,684	6,877	0.073	0.202	0.275
1982	93,9330	93,000	81,132	44,058	0.094	0.143	0.237
1983	1,574,460	58,000	99,328	99,372	0.036	0.135	0.171
1984	1,231,110	71,000	88,579	71,305	0.056	0.139	0.195
1985	727,560	86,000	89,303	41,160	0.112	0.198	0.309
1986	748,440	82,000	117,770	105,155	0.104	0.354	0.458
1987	587,520	87,000	64,732	93,340	0.138	0.313	0.452
1988	647,640	107,000	77,179	52,906	0.153	0.224	0.377
1989	979,440	87,000	72,996	60,059	0.085	0.146	0.231
1990	816,480	92,000	57,642	37,831	0.107	0.124	0.231
1991	1,195,920	86,000	70,479	85,494	0.069	0.140	0.209
1992	1,628,039	57,000	50,039	120,146	0.034	0.110	0.145
1993	749,227	62,000	32,358	64,855	0.080	0.139	0.218
1994	325,558	44,000	21,212	45,014	0.127	0.227	0.354
1995	303,973	50,000	26,045	14,425	0.152	0.143	0.295
1996	667,137	43,000	27,191	11,000	0.062	0.059	0.121
1997	659,478	50,000	36,496	5,500	0.073	0.066	0.139
1998	651,358	53,000	36,448	6,513	0.078	0.044	0.122
1999	475,095	39,000	19,933	4,571	0.079	0.032	0.111
2000	427,381	36,000	25,000	4,274	0.081	0.067	0.148
2001	773,114	37,000	20,000	7,731	0.047	0.029	0.076
2002	686,662	–	15,000	6,867	–	0.052	–

(–) Data missing/unavailable.
[a] Data reported in pounds, assumed average weight = 4 lbs. for conversion. Estimates taken from Crecco (1997).

and the stock identification data based on tagging and mtDNA results (see Hattala et al. 1997). Specifically, the coastal landings attributed to the Connecticut River shad stock was the sum of the VA–MD coastal harvest (times the predicted Connecticut River contribution of 0.064 and 0.03), the DE–NJ coastal landings (times 0.188), and the NY–NE coastal landings (times 0.50). Since a 50% reporting rate has been used for in-river commercial fisheries, the coastal intercept landings in number (CCL) (assumed average weight = 2.3 kg) from the Connecticut River stock (see Table 3) were also adjusted up to reflect a 50% reporting rate and discard rate of male shad. Coast-wide intercept landings for shad from 1997 to 2001 have not yet been partitioned by river of origin, so we estimated the annual fraction of the Connecticut River stock to the ocean harvest from 1997 to 2001 by multiplying the annual total intercept landings by the average fraction (0.13) of Connecticut River shad in the intercept fishery from 1993 to 1996. Coast-wide intercept landings for American shad are not yet available for 2002.

Given that the intercept fishery removes Connecticut River shad before they enter the river, the coastal fishing mortality rates (FC) on Connecticut River shad were estimated as a log ratio of landings (CLL) to stock size (N) with the coastal landings (CCL) being added to the denominator of equation (7):

$$FC = -\log\left[1 - \left(\frac{CCL}{CCL + N}\right)\right] \qquad (7)$$

The total fishing mortality rate (FT) on Connecticut River shad was estimated between 1980 and 2001 as the sum of the in-river fishing mortality (FR) and the coastal fishing mortality (FC) rates.

If overharvest were an underlying cause for the recent shad stock decline, the total fishing mortality rates (FT) generated by in-river and coastal commercial fisheries should have risen recently (1993–2001) and exceeded the overfishing definitions (F_{30} = 0.43) reported (Crecco 1997) for Connecticut River shad. The Thompson-Bell yield-per-recruit (YPR) model was used to derive F_{30} biological reference points for several Atlantic coastal shad populations as part of a coastwide stock assessment (Crecco 1997). The F_{30} level refers to the fishing mortality rate (F) that generates 30% of the unfished spawning stock (F = 0) biomass per recruit.

Adult blueback herring and alewives are reported as bycatch in the offshore mackerel fishery (Shepherd 1991). Total river herring landings from this fishery have varied without trend from 1984 through 1990 (range: 18–104 metric tons). Although blueback herring bycatch estimates are not available after 1990, total bycatch landing estimates from 1984 to 1990 only account for between 12,000 and 760,000 fish (using 0.1 kg per adult blueback herring) from all coastal stocks. Thus, annual bycatch landings from the mackerel fishery represent a very small fraction of the blueback herring run sizes in the Connecticut River. Crecco and Gibson (1990) reported that between 8 and 10 million adult blueback spawners had entered the river annually between 1980 and 1990. Even if the entire annual blueback herring bycatch in the mackerel fishery was attributed to the Connecticut River stock, this bycatch fishery would have had to expand 10-fold since 1990 to have caused the recent blueback herring decline in the Connecticut River.

Finfish Predator Abundance

Bluefish and striped bass were considered to be the major candidate finfish predators of Connecticut River alosines. Bluefish and striped bass grow rapidly to a size (>65 cm) that can easily prey on adult shad and blueback herring, are highly piscivorous on herring-like prey (Hartman 1993), and consume prey effectively day and night. Moreover, unlike other marine finfish predators, adult (>70 cm) striped bass have been sampled in great numbers well above the salt wedge in the Connecticut River since 1993 (Savoy 1995). Although older (age 2+) weakfish and dogfish may prey on alosines (Hartman and Brandt 1995), weakfish and dogfish were not considered as major predators of alosines for several reasons. Large (>70 cm) weakfish were rarely taken in Long Island Sound since 1993 (Gottschall and Pacileo 2002), and weakfish abundance along the Atlantic Coast has remained relatively low from 1992 to 2001 (Gibson 1994; Kahn 2002). Similarly, dogfish are not likely to be a major predator of alosines because smooth dogfish abundance has declined in Long Island Sound from 1984 through 2001 based on the Connecticut trawl survey, and spiny dogfish abundance has remained persistently low after 1993.

A time series of adult bluefish (ages 1+) relative abundance in Connecticut waters was taken from the Connecticut multispecies trawl survey (Gottschall and Pacileo 2002). The annual bluefish index was expressed as the geometric mean number per tow of adult (age 1+) bluefish from 1984 through 2002 (Figure 5). A time series of striped bass abundance (Figure 6) was based on the geometric mean catch per tow of striped bass (age 3+) in Long Island Sound based on the Connecticut trawl survey from 1984 through 2002 (Gottschall and Pacileo 2002). In addition, coastwide striped bass abundance between 1982 and 2002 were based on stock biomass estimates (SSB expressed in metric tons) of mature female striped bass (ages 6+) from the most recent Virtual Population Analysis (VPA; Carmichael 2002). Since striped bass relative abundance in Long Island Sound between 1984 and 2001 was highly corre-

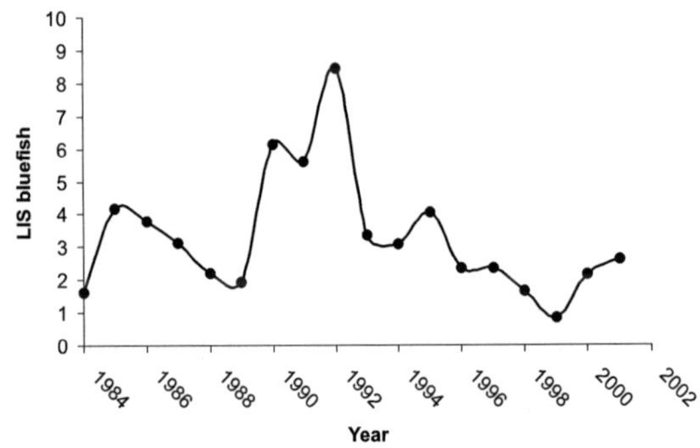

FIGURE 5. Bluefish index of abundance. The geometric mean catch per tow from the LISTS (Long Island Sound Trawl Survey) for fish age 1 and older.

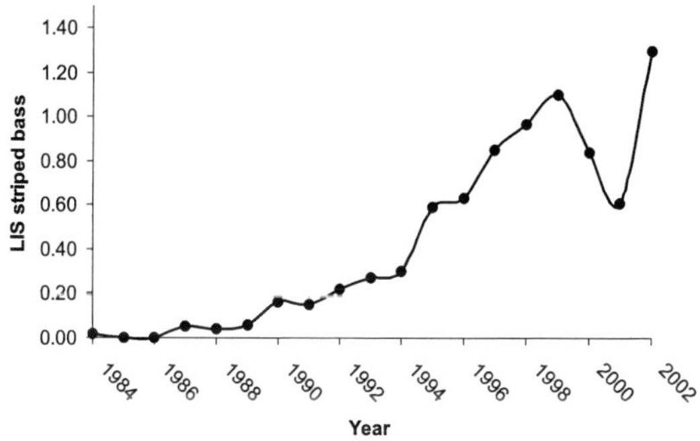

FIGURE 6. Striped bass index of abundance. The geometric mean catch per tow from the LISTS for fish age 3 and older.

lated (r = 0.92, P < 0.0001) to coast-wide striper SSB, only the coastwide SSB estimates from the VPA were used in all regression analyses to test the predation hypothesis.

Striped Bass Abundance

To estimate striped bass population size during May and June in the Connecticut River, Savoy (1995) conducted a mark–recapture study in 1994. Striped bass were captured mainly with electrofishing gear from April through June 1994 throughout the lower Connecticut River (Old Saybrook to Windsor, Connecticut). A total of 346 striped bass from the Connecticut River were captured, measured to total length (cm), and tagged with internal anchor tags (Table 4). Since no commercial fishing is permitted for striped bass in Connecticut waters, most tag recoveries were from recreational fishermen. Striped bass population size (N) in the Connecticut River was estimated by the Petersen equation. To estimate striped bass population size in the Connecticut River from 1995 to 2001, the 1994 stock size (407,300 fish) was adjusted by the ratio of the mean striped bass catch/tow in Long Island Sound in 1994 (Gottschall and Pacileo 2002) to mean catch/tow in the following years (1995–2002).

Because the vast majority of recaptures (93%) and striped bass catches (C) in the river occurred from April through June, the population estimate discussed herein mainly reflected the 1994 spring population abundance of striped bass. In this study, the tag reporting rate by the recreational fishery was assumed to be 60% based on tagging studies of striped bass in Chesapeake Bay (Rugolo

TABLE 4. Population estimate and 95% confidence limits (CI) of striped bass in the Connecticut River from April through June 1994 based on the Petersen mark–recapture method (Savoy 1995).

Statistic	Estimate or number
Number tagged (M)	346 [a]
Number recaptured (R)	26 [b]
Recreational catch in river (C)	30,610 [c]
Population estimate (P)	407,300
95% confidence limits	269,400 to 604,100 [d]

[a] Marked fish were reduced by 50% (from 692 tagged fish to 346 fish) to reflect a combination of factors including tag loss, tag-induced mortality and migration of tagged fish from the river to Long Island Sound.
[b] Number of recaptures increased to reflect a 60% (from 16 recaptures to 26 recaptures) reporting rate.
[c] Striped bass recreational catch in the Connecticut River was taken from the Marine Recreational Fishery Statistical Survey (Connecticut Department of Environmental Protection, unpublished data) and information in the 1994 Volunteer Angler Survey.
[d] 95% confidence limits were based on a Poisson distribution of recaptures (Appendix 2, Ricker 1975).

et al. 1994). The combined effects of tag loss, tag-induced mortality, and migration of tagged stripers from the river was assumed to be 50%. As a result, the number of tagged stripers (Mk) in this study was reduced by 50% (Savoy 1995).

Electrofishing surveys were conducted annually in the Connecticut River in the Windsor Locks area (river km 100–115) from May and June 1994 through 2002 to examine striped bass size structure and to estimate abundance of striped bass by direct proportion with the 1994 population estimate.

Statistical Analyses

To test the overfishing hypothesis, shad relative mortality rates were regressed against the in-river (FR) and coastal (FC) fishing mortality rates separately and together (multiple linear regression). A positive and statistically significant (P < 0.05) slope for fishing effects (FR and FC) would suggest that overfishing was directly related to the shad stock decline. Since there are no fishing mortality (F) estimates on blueback herring, the overfishing hypothesis could not be tested with the blueback mortality rate (Bmort).

The predation hypothesis was tested by linear

regression between the 1980–2001 shad and blueback relative mortality rates and relative bluefish abundance and striped bass SSB. These linear regressions were performed separately for striped bass and bluefish and together (multiple linear regression). A positive and statistically significant ($P < 0.05$) slope for predation effects would suggest that a rise in predation was directly related to stock declines of shad and blueback herring.

Results and Discussion

The River and Ocean Fisheries

Estimates of in-river fishing mortality rates (FR) on Connecticut River shad have remained low but highly variable (FR = 0.03–0.35) between 1980 and 2002 (Table 3). Recent (1996–2002) FR estimates have averaged 0.05 (SE = 0.006) and have dropped by an average of 50% since 1990. This recent drop in FR levels is consistent with a steady decline in gill-net commercial fishing effort (annual gill-net days fished) within the river from 1990 to 2002.

Coastal fishing mortality rates (FC) on Connecticut River shad from the intercept fishery have varied without trend from a low of 0.05 to a high of 0.15 between 1980 and 2001. Total fishing mortality rates (FT), consisting of the sum of in-river (FR) and coastal (FC) fishing mortality, peaked in 1986 at 0.46, but declined steadily since 1994 to the lowest level in the time series (FT = 0.08) in 2001. Since the current (1996–2001) average total fishing mortality rate (FT = 0.12, SE = 0.01) on Connecticut River shad is well below the overfishing definition (F_{30} level of 0.43), there is no evidence that overfishing, either from the in-river or coastal fisheries, was the primary cause for the recent stock decline.

The analysis of coastal shad landings (Hattala et al. 1997) indicated that the annual contribution of Connecticut River coastal landings had fallen steadily from a high of 107,000 fish in 1988 to a low of 36,000 fish in 2000 (Table 3). Coastal landings of this magnitude only comprised 3% to 8% of the estimated average losses (1.35 million shad) to the Connecticut River stock from 1993 to 2001 (Table 2). Even under the worst case scenario, where all coastal intercept landings from 1996 through 2001 were from the Connecticut River stock, the entire coastal intercept fishery could account, on average, 20% to 30% of the estimated losses (1.35 million fish) to the Connecticut River stock between 1993 and 2001 (Table 2).

The coastal intercept fishery could not have caused the recent alosine decline in the Connecticut River, unless catch underreporting from the intercept fishery has risen steadily over time and has exceeded 10 times the reported landings since 1993. The possibility that under-reporting has risen since the mid-1980s to astronomical levels, and that 4–5 thousand metric tons of coastal shad landings had remained hidden from enforcement and management agencies, seems unlikely.

Finally, the condition of the blueback herring adult stock in the Connecticut River after 1990 (Figure 2) is far worse than that for American shad. Relative population size, as indicated by Holyoke lift data, has declined more than 12-fold between 1981 and 2002 (Figure 2). This 95% decline in adult blueback abundance has occurred despite relatively stable juvenile production between 1980 and 1990 (Table 1). Blueback juvenile production had fallen to low levels after 1995, presumably due to extremely low egg production resulting from very low adult stock sizes from 1995 to 2002. This stock collapse cannot be attributed to any directed fisheries, since no known directed ocean fishery currently exists on blueback herring. Moreover, since blueback bycatch landings in the mackerel fishery are presumably from a mixed coastal stock, total blueback losses of between 12,000 and 760,000 fish in the mackerel fishery would not be high enough to account for a stock collapse in the Connecticut River blueback population (Figure 2).

Ocean Temperature Anomalies

It has been hypothesized (Jesien et al. 1992) that a decline in ocean temperatures along the Atlantic coast during winter and spring months since 1990 could have caused (1) a disruption of normal spring migration, resulting in a permanent loss of shad and blueback herring to the Connecticut River and other Atlantic Coast rivers; and/or (2) a displacement of alosine stocks in the ocean to areas of high predation and/or poor food availability. Oceanic temperature changes have been postulated as a cause for New England Atlantic salmon declines, yet these disruptions are reportedly a result of warmer than normal water temperatures. Disruption of the normal migratory patterns due to ocean temperature anomalies requires that American shad and blueback herring follow specific temperature isotherms during their ocean migration route to spawn and during their summer–fall residence time in the Gulf of Maine (Neves and Dupres 1979; Neves 1981). Moreover, for this hypothesis to be plausible, a discernible and persistent shift in ocean temperatures would have had to occur during the early 1990s in the mid and north Atlantic.

The hypothesis that temperature anomalies would disrupt shad migration patterns seems untenable for two reasons. First, although ocean temperatures were reported to have declined since 1990, shad runs were very strong in the Connecticut River in 1990, 1991 and reached record levels in 1992 (Table 1; Figure 1). The decline in Connecticut River shad did not occur in 1990, but instead was evident only after 1993. Second, Dadswell et al. (1987) noted that shad migration routes followed consistent patterns and directions independent of isotherm gradients. Dadswell et al. (1987) also reported that American shad regularly cross thermal barriers and can remain for extended periods in temperatures outside their preferred range. Based on Dadswell's findings, it is unlikely that subtle ocean temperature changes would have sufficiently altered shad and blueback herring migration patterns to cause mortality via starvation of more than 1.5 million adult shad and a complete stock collapse for blueback herring in the Connecticut River.

The possibility of temperature-induced dislocation of alosine stocks in the ocean, or declines in secondary productivity within shad feeding areas, is more difficult to interpret. It is generally believed that unusually cold water temperatures adversely affect fish stocks through a reduction in primary and/or secondary production (Cushing and Dickson 1976). It is expected that a drop in feeding success would result in lower overall fitness, as measured by Fultons condition factor (k) (Ricker 1975), or would limit energy reserves invested in gonadal production, resulting in poor recruitment. A reduction in overall fitness of Connecticut River shad has not occurred since the condition factor (k) of adult shad sampled at the Holyoke fishlift remained relatively stable from 1990 through 2002 (Table 5). Moreover, the production of dominant year-classes in 1993, 1994, and 1997 (Table 1; Figure 3) is inconsistent with the hypothesis that gonadal production from recent adult shad runs has been limited by ocean temperature anomalies.

Predation

Juanes et al. (1993) reported that snapper bluefish (age 0) could cause severe mortality on juve-

TABLE 5. Mean fork length (FL), weight (WT) and sample size (N) of male and female American shad monitored at the Holyoke fishlift.[a] These data were used to calculate Fultons condition factor (k) from 1990 through 2002.

Year	Males N	FL (cm)	WT(g)	k factor[b]	Females N	FL (cm)	WT (g)	k factor[b]
1990	650	42.4	1121.2	0.0147	411	50.5	2134.7	0.0166
1991	889	41.6	1091.4	0.0152	469	47.9	1823.9	0.0166
1992	537	41.8	1057.6	0.0145	624	47.2	1662.7	0.0158
1993	426	42.3	1103.4	0.0146	921	46.9	1639.1	0.0159
1994	229	41.0	1089.4	0.0158	441	47.6	1781.8	0.0165
1995	369	40.1	1018.1	0.0157	179	46.1	1664.7	0.0168
1996	468	40.1	998.7	0.0153	255	45.5	1563.5	0.0165
1997	526	40.7	1059.5	0.0160	433	45.5	1587.7	0.0169
1998	357	40.6	1001.0	0.0149	365	46.0	1563.0	0.0160
1999	315	41.3	1018.0	0.0144	171	45.2	1425.5	0.0154
2000	339	42.5	1189.0	0.0202	307	47.5	1783.0	0.0165
2001	286	41.1	996.0	0.0144	217	46.6	1597.0	0.0161
2002	364	40.7	1045.0	0.0156	416	47.3	1749.0	0.0167

[a] Personal communication from Caleb Slater, Massachusetts Department of Fisheries and Wildlife, Westborough, Massachusetts.

[b] $k = WT/FL^3$

nile alosines in the Hudson River. However, it is unlikely that bluefish predation is the primary cause of the current alosine stock declines in the Connecticut River for two reasons. First, coastwide bluefish abundance within state waters (<6.6 km) has rapidly declined since 1989 (Tercerio 1994; Gibson 2000). Second, there was a significant ($P < 0.05$) inverse relationship between bluefish relative abundance in Long Island Sound and shad ($r = -0.49$, $P < 0.04$) and blueback ($r = -0.51$, $P < 0.03$) mortality rates from 1984 through 2002 (Table 6). However, these statistical relationships between bluefish abundance and shad and blueback mortality were in the opposite direction (negative) from the stated hypothesis (positive direction) linking bluefish abundance to the alosine mortality.

Due to the success of striped bass management, striped bass abundance has risen steadily to record levels in coastal waters from 1992 to 2002 (Crecco 1994; Shepherd 1998; Carmichael 2002). Since recent declines in weakfish and bluefish abundance have occurred, striped bass have become the most abundant nearshore finfish predator. Statistical evidence does support the striped bass predation hypothesis on blueback herring. A significant inverse correlation ($r = -0.91$, $P < 0.001$) exists between the striped bass SSB levels and adult blueback herring abundance from 1966 through 2002 (Table 6). This inverse correlation between potential predator and prey is consistent with a positive and statistically significant relationship ($r = 0.90$, $P < 0.001$) between blueback herring mortality rates (Bmort) occurring between the juvenile and adult stages and striped bass abundance from 1979 to 2002. The relationship between blueback herring mortality from the juvenile to age 1 (Herrmort) and striped bass SSB is also statistically significant but inverse ($r = -0.50$, $P < 0.04$; Table 6). The strong positive correlation between Bmort and striped bass SSB suggests that the rise in blueback mortality and subsequent drop in adult stock size was strongly coupled with a recent increase in striped bass abundance in the Connecticut River.

Statistical evidence in support of the striped bass predation hypothesis is also evident for American shad. Total prerecruit mortality rates (Tmort) on American shad were positively correlated ($r = 0.76$, $P < 0.001$) to striped bass SSB estimates from 1972 to 2002 (Table 6). By contrast, the relationship between mortality rates between ages 0 and age 1 (Seamort) were independent ($r = 0.04$, $P < 0.88$) of striped bass SSB from 1980 to 2002 (Table 6). The significant positive correlation between Tmort and striped bass SSB suggests that the recent rise in adult shad mortality and drop in adult run size in the Connecticut River was strongly coupled with the persistent rise in striped bass abundance since 1993.

The weak statistical relationship between age-1 shad (Seamort) and inverse relationship for blueback herring mortalities (Herrmort) and striped bass abundance (Table 6) may have re-

TABLE 6. Results of linear regression models relating striped bass spawning stock biomass estimates (SSB), and Bluefish (Bluef) relative abundance to shad pre-recruit mortality (Tmort, Seamort) and blueback herring pre-recruit mortality (Bmort, Herrmort). The value *r* is the correlation coefficient, SE = standard error on the slope estimate, and *P* is the probability level that the slope estimate differs from zero.

Species/model	*r*	*P*
Shad:		
Tmort = 0.0001(SSB) + 1.3884	0.76	0.001
Tmort = –0.2527(Bluef) + 3.4134	–0.49	0.04
Seamort = –0.0001(SSB) + 1.4699	0.04	0.88
Seamort = 0.0179(Bluef) + 1.4549	0.04	0.87
Blueback herring:		
BPOP = –8.7777(SSB) + 499631	–0.91	0.001
Bmort = 0.0001(SSB) + 3.8232	0.90	0.001
Herrmort = –0.0001(SSB) + 7.7583	–0.50	0.04
BPOP = 55210(Bluef) + 14526	0.50	0.04
Bmort = –0.4629(Bluef) + 7.8136	–0.51	0.03
Herrmort = 0.0030(Bluef) + 6.8175	0.00	0.99

sulted from several factors. First, unlike American shad, the temporal and spatial distribution of age-1 blueback herring is poorly understood. If the majority of age-0 alosines leave Long Island Sound shortly after emigration from the Connecticut River, the age-1 indices from Long Island Sound (Table 2) may not accurately reflect relative abundance, thereby accounting for the lack of statistical relationship between mortality and striped bass abundance. Second, since age-1 shad and blueback herring from the Hudson River may be present in Long Island Sound, the age-1 indices may reflect the abundance of alosines from both the Connecticut and Hudson rivers, thereby further reducing the statistical fit between age-1 mortality and striped bass abundance. The relatively poor statistical fits (Table 6) may suggest that striper predation is less severe on ages-0 and -1 alosines than on adult alosines. This could occur if the size range (5–10 cm TL) of age-1 shad and blueback herring was below the optimal range of prey selected by most striped bass.

Striped Bass Abundance and Food Habits

Although there is abundant statistical evidence in support of the striped bass predation hypothesis, empirical evidence such as food habit studies of striped bass, striped bass population estimates, and striped bass length-frequency data are also needed to clearly establish causality between striped bass predation and the alosine decline. If striped bass predation is the underlying cause for the recent annual losses of 1.35 million adult shad in the Connecticut River, we would expect (1) that striped bass prey heavily on adult alosines in the river, and (2) that a large (>50,000 fish) population of adult (>71 cm) striped bass has been present since 1993 in the upper Connecticut River during April and May when shad and blueback herring are spawning.

The results of the tagging study revealed a spring (April–June) 1994 population size in the Connecticut River of 407,300 striped bass (95% CI: 269,400–604,100 fish; Table 7). Striped bass in the River ranged in size (TL) from 18 cm to 118 cm. Since striped bass are known to consume finfish prey up to 60% of their own body length (Manooch 1973), striped bass exceeding 43 cm could prey on adult blueback herring (mean length of adult blueback of 25 cm). Using a similar argument for shad, striped bass exceeding 71 cm could consume most male shad (assuming that the mean length of adult male shad is about 43 cm). Based on the 1994 striped bass length-frequency data, about 76% (309,000 fish) of the striper population would be large enough to consume adult bluebacks, and about 48% (197,000 fish) would be large enough to consume most adult male shad. Relative abundance of striped bass in Long Island Sound nearly doubled from 1994 to 2002 (Table 7). Within the Connecticut River, the estimated numbers of 43-cm+ stripers

TABLE 7. Relative abundance (mean catch/tow) of striped bass in Long Island Sound, population etimates of striped bass abundance in the Connecticut River from April through June, % of 71 cm+ stripers in the Connecticut River, and the estimated abundance of 71 cm+ stripers measured in the Connecticut River from 1994 to 2002.

Year	Striped bass CPUE	Striped bass population estimate[a]	Percent > 71 cm	Number > 71 cm
1994	0.30	407,300	45.2	184,100
1995	0.59	801,023	8.7	69,689
1996	0.63	855,330	32.1	274,561
1997	0.85	1,154,017	40.2	463,915
1998	0.97	1,316,937	11.6	152,765
1999	1.10	1,493,433	21.9	327,062
2000	0.84	1,140,440	45.9	523,462
2001	0.61	828,176	33.9	280,752
2002	1.30	1,764,967	25.2	444,772

[a] Estimates of striped bass population size from 1995 to 2002 were derived indirectly by multiplying the Long Island Sound relative abundance index (mean catch/tow) each year between 1994 and 2002 times the magnitude of the 1994 direct population estimate (see Table 4).

tripled (from 309,000 to 1,392,455 fish) between 1994 and 2002 (Table 7), and the abundance of 71-cm+ stripers more than doubled (from 197,000 to 507,800 fish). These abundance data from 1994 to 2002 suggest that a sufficient number of large striped bass were available in the river during springtime to have caused a stock collapse of adult blueback herring after 1995 and a less severe but measurable reduction in shad run size.

While striped bass can, in theory, prey heavily on juvenile shad and blueback herring during the fall, several lines of evidence support the conclusion that striped bass predation occurs mainly on spawning adult alosines. Immigration of striped bass into the Connecticut River during April is coincident with alosine spawning migration. Anecdotal information suggests that not only are striped bass more numerous and larger in the spring than during the fall, they can also be found further upriver during the spring than the fall when stripers concentrate in the estuary and river mouth. The degree of statistical fit (Pearson's *r*) reflecting striped bass predation was much stronger on the adult alosines (Tmort, Bmort) than on the juvenile or age-1 life stages (Seamort, Herrmort; Table 5). Moreover, during the spawning migration, adult shad and blueback herring have an enormous urge to reach their upriver spawning grounds (Leggett 1977; Fay et al. 1983). This strong drive to spawn may hamper the predator avoidance capability of adult alosines, rendering them more susceptible to predation during the migration and spawning phase of their life history. Lower predation risk perhaps occurs on subadult alosines in the ocean, due to their greater capacity to adopt tactics (e.g., schooling, spatial stratification) that may serve to minimize or impede contact with finfish predators. Finally, the fact that most recreational fishermen in the Connecticut River prefer to use live adult blueback herring as bait to catch striped bass during April and May strongly suggests that striped bass prey heavily on adult bluebacks during their upriver migration

Numerous published reports in the Connecticut River and elsewhere document the importance of shad and especially blueback herring as important prey of striped bass (see Raney 1952; Bigelow and Schroeder 1953; Versar, Inc. 1990). Moffit et al. (1982) documented the presence of striped bass in the Connecticut River coincident with adult shad and blueback herring migration. Moreover, Warner and Kynard (1986) reported a high incidence (93%) of alosine parts in the stomachs of striped bass at the base of the Holyoke Dam (river km 140) on the Connecticut River during May and June. Several other studies, including Manooch (1973) and Gardinier and Hoff (1982), noted that striped bass prefer the soft-rayed species such as blueback herring rather than hard-spined species. Stevens (cited in Smith and Wells 1977) reported that adult blueback herring were the main prey item of landlocked striped bass in South Carolina. Both Manooch (1973) and Trent and Hassler (1966) reported striped bass predation on spawning blueback herring.

Most striped bass are present in the Connecticut River for only a few months, but several behavioral and physiological characteristics of striped bass allow them to quickly reduce prey numbers. Striped bass are a highly opportunistic predator (Westin and Rogers 1978). Unlike other marine finfish predators, such as bluefish and weakfish, striped bass are capable of undertaking extensive migrations from saltwater into freshwater in search of food (Warner and Kynard 1986). Raney (1952) found that striped bass can easily consume large numbers of prey at one time, possess rapid digestion rates, exhibit schooling behavior, and tend to ignore other prey items while gorging on a specific body shape (Bigelow and Schroeder 1953). Dovel (1968) suggested that the increase in striped bass abundance in the 1960s was responsible for the decrease in Atlantic croaker *Micropogon undulatus*. The combination of these factors suggests that 300,000–400,000 striped bass (Table 7) could have consumed a large number of adult alosines, leading to the pronounced decline in shad and blueback herring population in the Connecticut River.

Conclusion

We conclude, based on the statistical and empirical evidence, that predation by striped bass provides the best explanation for the recent alosine stock declines in the Connecticut River. There is no evidence that in-river and coastal commercial fisheries have been intense enough to have had a measurable adverse influence on the Connecticut River shad stock. Moreover, fishing mortality rates (*F*) on Connecticut River shad have declined from 1980 to 2001, and recent (1992–1996) *F* estimates have been well below the overfishing definition for American shad (F_{30} = 0.43).

Statistical evidence alone cannot demonstrate causality, but the linkage between the recent rise in alosine mortality and rising striped bass abundance is plausible and convincing when examined with empirical evidence showing the rise in striped bass population size (from 407,300–1,154,000 fish) in the Connecticut River between 1994 and 2001. Moreover, results of striped bass food habits studies are consistent with the hypothesis that adult alosines in the Connecticut River are subject to persistent and widespread predation by striped bass during April and May. Predation effects by striped bass, at high stock size, can dramatically reduce shad and blueback herring abundance independently of in-river and coastal fishing effects. It is important, therefore, that future stock assessments of Atlantic Coast shad and river herring consider time varying natural mortality rates (*M*) brought about by shifts in predation. The impacts of nonfisheries effects need to be integrated into fisheries models to more accurately establish stock status and the effects of fishing. In addition, few recent food habits studies on larger (>71 cm) striped bass have been conducted in the Connecticut River and elsewhere. Future food habits studies should focus more closely on predation by larger (>71 cm) striped bass on alosines. With these data, energetic models can be developed to forecast future striped bass predation effects on alosines in the Connecticut and in other Atlantic Coast rivers.

Acknowledgments

We would like to thank Desmond Kahn and Ken Frank who came through on short notice for reviewing the manuscript. Financial support was provided by the Connecticut Department of Environmental Protection.

References

Bigelow, H. B., and W. C. Schroeder. 1953. Fishes of the Gulf of Maine. U.S. Fish and Wildlife Service. Fisheries Bulletin 53.

Carmichael, J. 2002. Stock assessment of striped bass along the Atlantic Coast through 2001. Report to the Atlantic States Marine Fisheries Commission, Washington, D.C.

Crecco. V. A. 1994. Alternative regulations for the striped bass recreational fishery along the Atlantic Coast consistent with Amendment 5. Connecticut Department of Environmental Protection, Marine Fisheries Division, Old Lyme.

Crecco. V. A. 1997. Stock assessment of American shad from selected Atlantic Coast rivers. Atlantic States Marine Fisheries Commission Special Report, Washington, D.C.

Crecco, V. A., and M. R. Gibson. 1990. Stock assessment of river herring from selected Atlantic Coast rivers. Atlantic States Marine Fisheries Commission Special Report No. 19, Washington, D.C.

Crecco, V. A., and T. F. Savoy. 1984. Effects of fluctuations in hydrographic conditions on year-class strength of American shad (*Alosa sapidissima*) in the Connecticut River. Canadian Journal of Fisheries and Aquatic Sciences 41:1216–1223.

Crecco, V. A., and T. F. Savoy. 1987. Fishery management plan for American shad in the Connecticut River. Connecticut Department of Environmental Protection, Marine Fisheries Division, Old Lyme.

Crecco, V. A., T. F. Savoy, and W. Whitworth. 1986. Effects of density-dependent and climatic factors on American shad (*Alosa sapidissima*) recruitment: a predictive approach. Canadian Journal of Fisheries and Aquatic Sciences 43:457–463.

Cushing, D. H., and R. R. Dickson. 1976. The biological response in the sea to climatic changes. Advances in Marine Biology 14:1–122.

Dadswell, M. J., G. D. Melvin, P. J. Williams, and D. E. Themelis. 1987. Influences of origin, life history, and chance on the Atlantic Coast migration of American shad. Pages 313–330 *in* M. J. Dadswell, R. J. Klauda, C. M. Moffit, R. L. Saunders, R. A. Rulifson, and J. E. Cooper, editors. Common strategies of anadromous and catadromous fishes. American Fisheries Society, Symposium 1, Bethesda, Maryland.

Dovel, W. L. 1968. Predation by striped bass as a possible influence on population size of the Atlantic croaker. Transactions of the American Fishereis Society 97(4):313–319.

Fay, C. W., R. J. Neves and G. B. Pardue. 1983. Species profiles: life histories and environmental requirements of coastal fisheries and invertebrates (mid-Atlantic) striped bass. U.S. Fish and Wildlife Service Biological Services Program 82–11.9, Washington, D.C.

Gardinier, M. N., and T. B. Hoff. 1982. Diet of striped bass in the Hudson River Estuary. New York Fish and Game Journal 29(2):152–165.

Gephard, S., and J. McMenemy. 2004. An overview of the program to restore Atlantic salmon and other diadromous fishes to the Connecticut River with notes on the current status of these species in the river. Pages 287–317 *in* P. M. Jacobson, D. A. Dixon, W. C. Leggett, B. C. Marcy, Jr., and R. R. Massengill, editors. The Connecticut River Ecological Study (1965–1973) revisited: ecology of the lower Connecticut River 1973–2003. American Fisheries Society, Monograph 9, Bethesda, Maryland.

Gibson, M. 1994. Current stock assessment of Atlantic Coast weakfish with the CAGEAN model based on the 1982–1993 catch-at-age data. Report to the Atlantic States Marine Fisheries Commission, Washington, D.C.

Gibson, M. 2000. Stock assessment of Atlantic Coast striped bass through 1999. Report to the Atlantic States Marine Fisheries Commission, Washington, D.C.

Gottschall, K. F., and D. J. Pacileo. 2002. Marine finfish survey: Part 1 Long Island Sound Trawl Survey. Pages 43–139 Job 2 *in* A study of marine recreational fisheries in Connecticut. F-54-R-21 Annual Performance Report. Connecticut Department of Environmental Protection, Hartford.

Hartman, K. J. 1993. Striped bass, bluefish, and weakfish in the Chesapeake Bay: energetics, trophic linkages, and bioenergetics model applications. Doctoral dissertation. University of Maryland, College Park.

Hartman, K. J., and S. B. Brandt. 1995. Predatory demand and impact of striped bass (*Morone saxatilis*), bluefish (*Pomatomus saltatrix*), and weakfish (*Cynoscion regalis*) in the Chesapeake Bay: applications of bioenergetics models. Canadian Journal of Fisheries and Aquatic Sciences 52:1667–1686.

Hattala, K., R. Allen, N. Lazar, and R. O'Reilly. 1997. Stock contributions for American shad landings in mixed stock fisheries along the Atlantic Coast. Atlantic States Marine Fisheries Commission, Washington, D.C.

Jesien, R. V., S. K. Gaichas, J. E. Serafy, and C. H. Hocutt. 1992. Stock composition of coastal Maryland and Virginia American shad fisheries: 1992 tagging study. Maryland Department of Natural Resources, Final Report No. F267–92-008, Annapolis.

Juanes, F., R. E. Marks, K. A. McKown, and D. O. Conover. 1993. Predation by age-0 bluefish on age anadromous fishes in the Hudson River Estuary. Transactions of the American Fisheries Society 122:348–356.

Kahn, D. 2002. Stock assessment of weakfish through 2000, including estimates of stock size on January 1, 2001. Report to the Atlantic States Marine Fisheries Commission, Washington, D.C.

Krantz, C. A., J. P. Mowrer, A. A. Jarzynski, R. V. Jesien and D. R. Weinrich. 1992. Investigation of anadromous alosids in Chesapeake Bay. U.S. Fish and Wildlife Service Report F-37-R, Annapolis, Maryland.

Leggett, W. C. 1977. The ecology of fish migration. Annual Review Ecology and Systematics 8:285–308.

Leggett, W. C. 2004. The American shad, with special reference to its migration and population dynamics in the Connecticut River. Pages 181–238 *in* P. M. Jacobson, D. A. Dixon, W. C. Leggett, B. C. Marcy, Jr., and R. R. Massengill, editors. The Connecticut River Ecological Study (1965–1973) revisited: ecology of the lower Connecticut River 1973–2003. American Fisheries Society, Monograph 9, Bethesda, Maryland. (Originally published in 1976)

Manooch, C. S., III. 1973. Food habits of yearling and adult striped bass, *Morone saxatilis* (Walbaum), from Albemarle Sound, North Carolina. Chesapeake Science 14(2):73–86.

Marcy, B. C., Jr. 2004. Early life history studies of American shad in the lower Connecticut River and the effects of the Connecticut Yankee plant. Pages 155–180 *in* P. M. Jacobson, D. A. Dixon, W. C. Leggett, B. C. Marcy, Jr., and R. R. Massengill, editors. The Connecticut River Ecological Study (1965–1973) revisited: ecology of the lower Connecticut River 1973–2003. American Fisheries Society, Monograph 9, Bethesda, Maryland. (Originally published in 1976)

Milstein, C. B. 1981. Abundance and distribution of juvenile *Alosa* species off southern New Jersey. Transactions of the American Fisheries Society 110:306–309.

Moffit, C. M., B. Kynard, and S. G. Rideout. 1982. Fish passage facilities and anadromous fish restoration in the Connecticut River basin. Fisheries 7(6):2–11.

Neves, R. J. 1981. The offshore distribution of alewife, *Alosa pseudoharengus*, and blueback herring, *Alosa aestivalis*, along the Atlantic Coast. Fisheries Bulletin 79:473–485.

Neves, R. J., and L. Dupres. 1979. The oceanic migration of American shad, *Alosa sapidissima*, along the Atlantic Coast. Fisheries Bulletin 77:199–212.

Raney, E. C. 1952. The life history of the striped bass, *Roccus saxatilis* (Walbaum). Bulletin of the Bingham Oceanographic Collection 14(1):5–97.

Ricker, W. E. 1975. Computation and interpretation of biological statistics of fish populations. Fisheries Research Board of Canada Bulletin 191.

Rugolo, L., V. Crecco, and M. Gibson. 1994. Development of a spawning stock biomass fisheries model for Chesapeake Bay striped bass. Report to the Atlantic States Marine Fisheries Commission, Washington, D.C.

Savoy, T. 1993. Nearshore residence of juvenile clupeids in Long Island Sound. Paper presented at 49th Northeast Fish & Wildlife Conference. Atlantic City, New Jersey.

Savoy, T. 1995. Striped bass investigations in Connecticut waters. A report to the Connecticut Department of Environmental Protection, Fisheries Division, Hartford.

Savoy, T. 1998. Anadromous fish studies in Connecticut waters. Progress Report AFC-25. Connecticut Department of Environmental Protection, Hartford.

Savoy, T. 2001. Connecticut anadromous fish investigations. Progress Report AFC-25. Connecticut Department of Environmental Protection, Hartford.

Savoy, T. 2002. Anadromous fish studies in Connecticut waters. Progress Report AFC-24–1. Connecticut Department of Environmental Protection, Hartford.

Savoy, T., and D. Shake. 1993. Anadromous fish studies in Connecticut waters. Progress Report AFC-21–1. Connecticut Department of Environmental Protection, Hartford.

Simpson, D., M. Johnson, and K. Gottschall. 1994. Marine finfish survey. Job 2 *in* A study of marine recreational fisheries in Connecticut. F54R Annual Progress Report. Connecticut Department of Environmental Protection, Marine Fisheries Divison, Old Lyme.

Shepherd, G. R. 1991. Projection of river herring bycatch in the mackerel fishery. Pages 24–25 *in* Atlantic States Marine Fisheries Commission Report on Shad and River Herring. Washington, D.C. November 1991.

Shepherd, G. R. 1998. Striped bass. In S. H. Clark, editor. Status of fishery resources off the northeastern United States for 1998. NOAA [National Oceanic and Atmospheric Administration] Technical Memorandum NMFS-NE-115, Silver Spring, Maryland.

Smith, W. G., and A. Wells. 1977. Biological and fisheries data on striped bass, *Morone saxatilis* (Walbaum). Technical Series Report 4. National Marine Fisheries Service Northeast Fisheries Center, Woods Hole, Massachusetts.

Stoltc, L. W. 1980. A strategic plan for the restoration of Atlantic salmon to the Connecticut River basin. U.S. Fish and Wildlife Service, Laconia, New Hampshire.

Stone, H. H., and G. R. Daborn. 1987. Diet of alewives, *Alosa pseudoharengus* and blueback herring, *A. aestivalis* (Pices: Clupeidae) in Minas Basin, Nova Scotia, a turbid, macrotidal estuary. Environmental Biology of Fishes 19:55–67.

Tercerio, M. 1994. Stock assessment of Atlantic Coast bluefish. Memorandum to the Pelagic Coastal Subcommittee and Stock Assessment Workshop. Atlantic States Marine Fisheries Commission, Washington, D.C.

Trent, W. L., and W. W. Hassler. 1966. Feeding behavior of adult striped bass, *Roccus saxatilus*, in relation to stages of sexual maturity. Chesapeake Science 7(4):189–192.

Versar, Inc. 1990. Source document for the supplement to the striped bass FMP - Amendment #4, Fisheries Management Report No. 16 of the Atlantic States Marine Fisheries Commission, Washington, D.C. March 1990.

Warner, J., and B. Kynard. 1986. Scavenger feeding by subadult striped bass, *Morone saxatilis*, below a low-head hydroelectric dam. Fisheries Bulletin 84(1):220–222.

Westin, D. T., and B. A. Rogers. 1978. Synopsis of biological data on the striped bass, *Morone saxatilis* (Walbaum) 1792. Marine Technical Report 67, University Rhode Island, Kingston.

American Fisheries Society Monograph 9:379–390, 2004

Stock Assessment of White Perch in the Lower Connecticut River

PENELOPE HOWELL* AND DAVID MOLNAR

Connecticut Department of Environmental Protection, Marine Fisheries Division
Post Office Box 719, Old Lyme, Connecticut 06371-0719, USA

Abstract.—The status of the white perch *Morone americana* population in the lower Connecticut River was assessed in terms of fishing mortality, growth, and its relative contribution to the river's commercial and recreational fisheries. White perch were captured by trap nets set from March to May 1997 and 1998 in three coves. Mean lengths at age were similar among the three sites and between years, as well as to previous studies. This population appears to be relatively fast growing and long-lived compared to other white perch populations. Mark–recapture data indicate movement of white perch throughout the river area studied. White perch overwintering in Hamburg Cove (Haddam) contributes strongly to the quantity of white perch present in spring in other coves to the north and south. The total white perch population in the Connecticut portion of the Connecticut River was estimated at 1.36 million (95% CI 0.96–1.76 million). Average density reported for the Hudson River population was 108,000 fish/km^2 (total area = 295 km^2), four times greater than the average density estimated herein of 27,000 fish/km^2 (total area = 50 km^2). Comparing results herein to previously published studies, it appears that fishing mortality has not increased in the last 30 years. However, the current fishing mortality rate is high enough to be eroding the stock's theoretical compensatory reserve. Currently, the population's age structure is broad, and growth rate comparable or greater than figures recorded in 1975. The stock age structure indicates that mortality in very early ages, probably from natural predation, may be having a greater effect than fishing in depressing total numbers of older white perch.

Introduction

The purpose of this study was to assess the status of the white perch *Morone americana* population in the Connecticut River in 1997–1998, including its relative contribution to the Connecticut River's commercial and recreational fisheries. Population status was measured in terms of age structure and mean length at age, variables that were used to assess fishing mortality and growth throughout the river system. The current exploitation pattern was also measured, including catch per unit effort, total effort, and length frequency of the sport and commercial harvest and discard. The results of this study were compared to work published on this species in 1975–1976 by Maltezos et al. (1979) and in 1965–1969 by Marcy and Richards (1974) and Marcy (2004 [1976], this volume). Changes in these key population characteristics were then examined over the 30 years that the three studies covered.

* *Corresponding author*: penny.howell@po.state.ct.us

Methods

White perch were captured by trap nets set from March to May, 1997 and 1998, in Connecticut River coves (Figure 1): one net was set at each location: Wethersfield (Wethersfield Cove) 70.8 km from the river mouth, Lyme (Hamburg Cove) 14.5 km from the river mouth, and Old Saybrook (North Cove) 3.2 km from the river mouth. Oneida trap nets (trap box 1.8 m high and deep, 1.7 m long; trap wings 6.3 m long; trap lead line 30.7 m long, 1.8 m high) were employed with 19-mm square mesh. This mesh size is effective in retaining white perch 14 cm and larger. Trap nets were checked on average every 3 d. All finfish captured in the nets were identified to species and

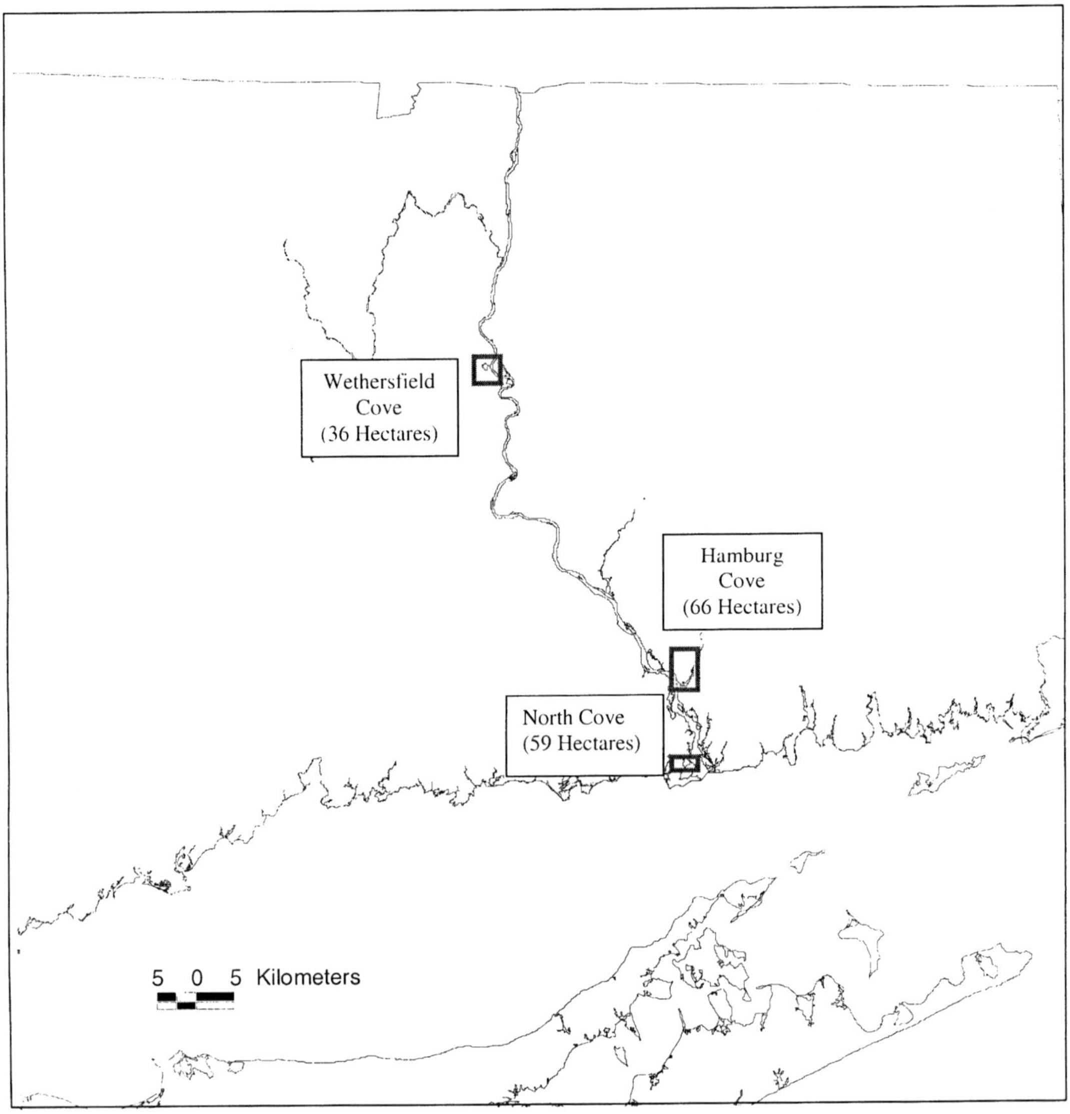

FIGURE 1. White perch study sites on the Connecticut River, 1997–1998. Total area in hectares is listed for each cove.

counted. Total length (mm) was measured for up to 200 white perch per set.

Age Structure

Ten white perch from each centimeter group captured in the nets were used for aging by analysis of scales, with a subsample aged using sagittal otoliths. Scales were prepared following methods described by Mansueti (1961). Otoliths were sectioned by microtome and examined under 40× magnification. Each structure was examined by two readers. All white perch captured were assigned an age based on site-specific age-length keys developed from the aged fish. Instantaneous rates of mortality and survival for all sites were estimated from the overall age frequency from all samples in both years by the catch curve method (Ricker 1975), where the natural log of abundance of each fully recruited age-class (age 3 and greater) was regressed on the corresponding age.

Movement and Population Size

In order to assess whether white perch in the Connecticut River constituted one population or several isolated spawning groups, the degree of movement of mature fish among the study sites was examined. In late March, white perch age 2 and older were fin-clipped. These fish were captured by trap net, using the methods described above, and electrofishing, using methods described in Jacobs and O'Donnell (1996). White perch captured at each site were given a different clip mark and released. Age-1 white perch were not fin-clipped because of handling mortality that was unacceptably high. Fin-clipping mortality was estimated by holding 30 fin-clipped and 30 control white perch in two 1.22-m^3 live cars for 6 weeks. Recaptures were not recorded until 1 month after fin-clipping was completed to allow marked white perch to mix among unmarked white perch. The mark–recapture data were used to estimate the winter and spring populations in each of the three coves. The winter population is defined as the number of white perch present in each cove during March when marking occurred. The spring population is defined as the number of white perch present in each cove during April–May when the marked fish were recaptured. The spring population would correspond to the population of fish available to anglers fishing in each cove in the late spring. Both the winter and spring populations were estimated in two parts, stationary and migratory. The stationary population, remaining in each cove for both the winter and spring period, was estimated using the following equation (Ricker 1975):

$$\text{Stationary population} = (C_s * \text{FR}_s) * (M_s/R_s) \ ,$$

where

C_s = total trap net catch at Site *s*.

FR_s = fraction of total recaptures of fish marked in Site *s* recaptured in Site *s*.

M_s = total fish marked in Site *s* adjusted downward for 6.7% marking mortality.

R_s = fish marked at Site *s* and recaptured at Site *s*.

The migratory population segment, moving between coves after the time of marking but before recapture, was estimated by assuming that the percentage of total recaptures from each site represented the percentage of the total overwintering population present in each cove in spring. In other words, if the recaptures of fish marked at a given site were equally divided between that site and another, then we assumed that half the total overwintering population moved to the second site and was equally divided between two sites in the spring. Having calculated the size of the stationary population segment above, the percentage of the total population that moved to other sites was then calculated by subtraction:

$$\frac{\text{Stationary population at site } s}{\text{\% total recaptures captured at site } s} = \text{Total winter population at site } s$$

For each site,

$$\text{Total population} - \text{stationary} = \text{migratory population}$$

Finally, stationary population estimates for the three coves sampled were expanded to obtain an estimate for the entire river within Connecticut State boundaries. Population estimates for the three coves sampled were converted to a mean density (white perch per hectare) by dividing the total number of white perch in each cove by its total area, as listed in the 1995 Connecticut Department of Environmental Protection (CTDEP) Geographic Information Systems (GIS) database. This mean was then applied to all known white perch spawning coves in the Connecticut River to estimate the spawning population of white perch in the entire river in early spring. White perch moving from the river's main stem into these coves were considered the entire spawning population. This assumption is based on the work of Mansueti (1961) who found that successful spawning required slow-moving warm water. Even though Mansueti reported collecting white perch eggs in swift current waters, none were considered viable based on his laboratory trials. To the extent that this assumption is violated, the population estimate would be smaller than the true population size.

Current Exploitation Pattern

Documentation of the commercial white perch fishery in the Connecticut River was problematic. Since 1981, only 2–12 commercial license holders have reported landing white perch annually from the Connecticut River. All four fishers reporting white perch landings in recent years declined to have their catch measured, and none sold their catch to a licensed dealer. However, in 1997, a commercial gill net confiscated from Hamburg Cove, an area where no commercial fishing is allowed, provided a length frequency of white perch taken by this gear. This mesh size (8.9 cm) is the minimum legal size allowed on the river and the only size commercially used. State regulation requires that reported commercial landings be kept confidential when there are three or fewer fishers involved in a fishery. Therefore, white perch commercial landings are given here as 6-year means, rather than annually, from 1980 to 1998. The recreational white perch fishery in the Connecticut River was surveyed from March to October 1997 and 1998 (see Jacobs et al. 2004, this volume).

The ratio of fishing mortality between the two fisheries was based on estimates of total harvest from the recreational fishery, estimated in the creel census (Jacobs et al. 2004), and commercial landings reported in CTDEP Marine Fisheries Information System. Commercial discard mortality for partially recruited fish was assumed to be 80%, based on personal observation of high mortality with experimental gill nets. Discard mortality for recreational anglers was assumed to be 20%, based on prior experimentation (Maltezos and Howell 1982) with scup *Stenotomus chrysops*, a species of similar size and hardiness. Instantaneous natural mortality was estimated to be 0.25 using Hoenig's (1983) relationship between natural mortality and maximum age. Maximum age was estimated at 17 years based on the oldest documented age of 16 years (AuClair 1960).

Results and Discussion

White perch comprised 42% of the total trap net catch in 1997 and 22% in 1998. A total of 32 species were captured, yielding a total catch of 57,289 fish, including 13,561 white perch during both years (Table 1). White perch were numerically dominant in the catches from Hamburg and North coves, but not Wethersfield Cove.

TABLE 1. Trap net catch by site for the 10 most abundant species captured, 1997–1998. The number of times trap net was checked and processed (N) is given in parentheses.

Wethersfield Cove	Hamburg Cove	North Cove
(N = 31) Total catch = 33,507 fish Total species = 26	(N = 43) Total catch = 15,865 fish Total species = 25	(N = 36) Total catch = 7,917 fish Total species = 20
1. Blueback herring *Alosa aestivalis* = 21,830	1. White perch = 6,797	1. White perch = 6,248
2. Spottail shiner *Notropis hudsonius* = 4,020	2. Yellow perch *Perca flavescens* = 3,930	2. White catfish *Ameiurus catus* = 635
3. Alewife *Alosa pseudoharengus* = 3,824	3. Pumpkinseed *Lepomis gibbosus* = 1,518	3. Alewife = 593
4. Gizzard shad *Dorosoma cepedianum* = 2,131	4. Channel catfish = 965 *Ictalurus punctatus* = 965	4. Blueback herring = 219
5. Channel catfish = 1,589	5. Alewife = 754	5. American shad *Alosa sapidissima* = 46
6. White perch = 517	6. Brown bullhead *Ameiurus nebulosus* = 722	6. Channel catfish = 39
7. White sucker *Catostomus commersoni* = 491	7. Bluegill *Lepomis macrochirus* = 235	7. Striped bass *Morone saxatilis* = 31
8. Common carp *Cyprinus carpio* = 449	8. Golden shiner *Notemigonus crysoleucas* = 210	8. Pumpkinseed = 23
9. Bluegill = 422	9. White sucker = 201	9. American eel *Anguilla rostrata* = 15
10. Pumpkinseed = 275	10. White catfish = 112	10. Golden shiner = 13

Age Structure

Scale versus Otolith Aging

A total of 469 white perch from three sites, ranging in size from 11 to 35 cm total length, were aged using scales and otoliths. The oldest white perch scale-aged was 10 years old, whereas otolith aging resulted in fish as old as 14. Maximum difference between structures was 4 years, with 67% of the structural differences only 1 year. Agreement between readers was higher for otolith sections than for scales: all otolith samples aged were in agreement between the two readers, while scale agreement between readers ranged from a low of 70% (for age 7–10) to a high of 95% (ages 3–6). High rates of agreement between two readers for otolith sections was due to clearly defined annuli for fish of all ages, a condition not consistently seen with scales especially for fish older than age 5. Therefore, all white perch 23 cm in length or greater were aged using age-length keys based on otolith length at age. Difficulty in interpreting scales from white perch older than age 5 calls into question previous work on riverine white perch along the Atlantic Coast where fish were aged using scales only.

Current Age Structure

White perch aged 2–3 years old were most numerous at all three trap net sites in 1997 and 1998 (Table 2). Few white perch older than age 8 were collected at any site either year. The proportion of total catch by age varied across years (Table 2). For example, the proportion of age 3, the dominant age-class, ranged from 55% in 1997 to only 29% in 1998. However, the proportion of age 4 increased in 1998 (21%) compared to 8% in 1997. The age structure for the recreational harvest (Table 3) followed a similar pattern with age as the trap-net age frequency; the dominant age-class was age 3, and the proportion of age-3 fish varied between years. A similar increase in proportion of age-4 fish was evident in 1998 when compared to 1997 for the recreational harvest. These results suggest that, on average, anglers do not select white perch by size above the size recruited to their gear (approximately 14 cm) and, without a legal minimum length limit, that the recreational harvest generally reflects the available population in terms of length.

Compared to other riverine white perch populations in the mid-Atlantic, Connecticut River white perch are relatively long-lived, with a maximum otolith age of 14. Maximum ages reported from other rivers, using scales, were ages 7–9 for the Hudson River (Bath and O'Connor 1982; Klauda et al. 1988), Delaware River (Wallace 1971), and Patuxent River (Mansueti 1961), as well as the Connecticut River 30 years ago (Marcy and Richards 1974; Maltezos et al. 1975). Age data for the Connecticut River population were consistent with Mansueti's (1961) observation of

TABLE 2. White perch age structure for trap net samples, 1997–1998. The total number of white perch captured (*N*) and percent by age is listed. Fish greater than 23 cm were assigned ages based on otolith age analyses, while fish 23 cm and smaller were assigned ages based on scale analyses.

Age	1997 all sites *N*	% by age	1998 all sites *N*	% by age	1997 and 1998 all sites *N*	% by age
1	31	0.7	1,650	18.0	1,681	12.4
2	1,053	23.9	2,161	23.6	3,214	23.7
3	2,401	54.6	2,614	28.5	5,015	36.9
4	350	8.0	1,936	21	2,286	16.8
5	242	5.5	248	2.7	490	3.6
6	195	4.4	221	2.4	416	3.1
7	60	1.4	206	2.2	266	2.0
8	52	1.2	76	0.8	128	0.9
9	16	0.4	4	<0.1	20	0.2
10			39	1.0	39	0.3
11			5	<0.1	5	<0.1
14			2	<0.1	2	<0.1
SUM	4,400		9,162		13,562	

TABLE 3. White perch age structure for the recreational harvest 1997–1998. The total number of white perch (*N*) were apportioned to age-classes based on the percent-at-length of aged white perch. Numbers at age are also given as a percentage (%) of the total catch. Fish greater than 23 cm were assigned ages based on otolith age analyses, while fish 23 cm and smaller were assigned ages based on scale analyses.

Age	1997 catch *N*	% by age	1998 catch *N*	% by age	1997–1998 catch *N*	% by age
1	8	1.0	11	2.7	19	1.7
2	239	32.8	91	22.1	330	29.0
3	330	45.3	128	31.1	458	40.0
4	55	7.6	104	25.3	159	13.9
5	43	5.9	26	6.3	69	6.1
6	37	5.0	15	3.6	52	4.6
7	8	1.0	15	3.6	23	2.0
8	7	1.0	16	3.9	23	2.0
9	1	0.2	3	0.7	4	0.4
10			1.5	0.4	1.5	0.1
11			0.5	0.1	0.5	<0.1
SUM	728		411		1,139	

a latitudinal gradient in longevity of white perch, with longevity increasing from south to north along the Atlantic Coast of North America. Only lacustrine populations of white perch are longer-lived, having been aged to 16 years in Maine (AuClair 1960).

Mean Length-at-Age

Mean lengths at age in this study (Table 4; Figure 2) are similar to data reported by Jacobs and O'Donnell (1996) for the Connecticut River, but greater than Maltezos et al. (1975) reported for these same sites in the 1970s. Mean lengths in this study were also similar to Marcy and Richards (1974) Connecticut River data up to age 5. However, older-aged white perch had larger mean lengths at age. Connecticut River white perch had relatively large mean lengths at age when compared to white perch populations from rivers in the mid-Atlantic (Richkus and Stroup 1987; Klauda et al. 1988). These larger sizes at age suggest that Connecticut River white perch may experience accelerated growth common when total population size is low or that white perch in the other river systems are stunted due to overcrowding. Stunted white perch populations, perhaps influenced by density-dependant mechanisms, have been observed in the Potomac and Susque-hanna rivers in Maryland and in several lakes (Richkus and Stroup 1987).

TABLE 4. White perch mean length-at-age, 1997–1998. Mean length (cm) of white perch captured by trap net each year, and both years combined, from all sites. In 1997, no age 10 were captured (Nc).

Age	1997	1998	1997–1998
1	10.1	10.6	10.3
2	18.7	18.1	18.4
3	21.7	22.4	22.0
4	24.2	24.0	24.1
5	26.5	26.0	26.3
6	26.5	27.3	27.0
7	28.1	27.9	28.0
8	29.4	28.8	29.4
9	33.4	31.2	32.6
10	Nc	35.5	35.5

Mean length at age was similar among the three sites and between years. Because no consistent differences were evident, lengths at age for all sites were pooled. Additionally, fin-clipping showed that white perch move among sampling sites (see below), indicating that any difference in size does not reflect site-specific differences in growth.

Movement and Population Size

In 1997, a total of 2,708 white perch were fin-clipped during the entire sampling period, and 50 were recaptured. These recaptures were used to assess movement within the river. The majority (56%)

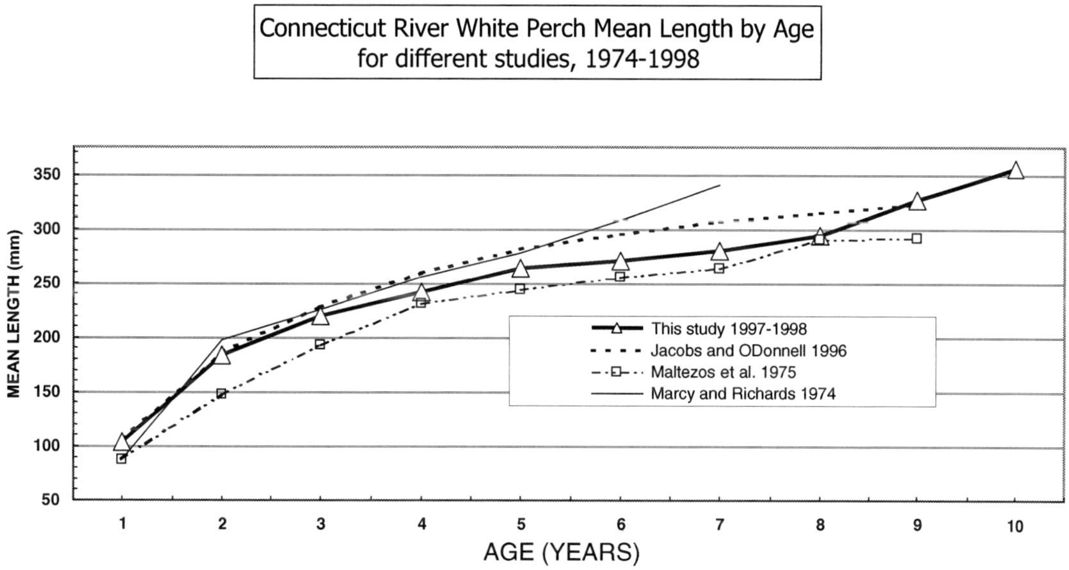

FIGURE 2. Mean length-at-age (mm) for Connecticut River white perch. Lengths given are all total lengths in millimeters for white perch age 1 to 10 years.

of the 50 recaptured fish were taken at the same site where they were fin-clipped and released. This was especially true for the Hamburg Cove white perch where all 22 recaptures originated from that same site. The remaining 44% (22 fish) were taken at a different site, but all originated from Hamburg Cove. Since marked fish were released during the entire sampling period and therefore not given time to mix thoroughly with unmarked fish, population estimates were not calculated with these data. However, the recaptures did indicate that Hamburg Cove fish moved throughout the entire river, a conclusion also drawn from marking at the same site 20 years earlier (Maltezos et al. 1979).

In 1998, a total of 3,280 white perch were fin-clipped during March only, and 55 were subsequently recaptured. These recaptures were used to assess movement within the river as well as calculate a total population number by site and for the entire river (Table 5). Only 50% of marked fish recaptured in Wethersfield Cove were marked there. Similarly, only 68% of recaptures in North Cove were marked there, whereas 90% of Hamburg Cove recaptures were marked there. These percentages were applied to the total fish captured in each cove to estimate the stationary overwintering population in the cove (Table 6). Estimates of each cove's migratory population were based on the distribution of recaptures from each site (column percentages, Table 5): for North Cove and Wethersfield Cove, 91–100% of recaptured fish were marked at the same site. However, only 60% of recaptured fish marked in Hamburg Cove were taken in that cove. These percentages were used to expand the stationary population segment to the total population.

The migratory segment was computed by subtraction. In the case of Wethersfield Cove, no recaptures were taken elsewhere, so the migratory segment was zero. In the case of North Cove, the migratory segment was assigned to Hamburg Cove's spring population based on the two recaptures taken there that were marked in North Cove. In the case of Hamburg Cove, the migratory segment was assigned to both North Cove and Wethersfield Cove in a 5 to 1 ratio, corresponding to the 10 fish and 2 fish marked in Hamburg Cove recaptured in North Cove and Wethersfield Cove, respectively. Although numbers of recaptures were minimum, the relative distribution should reflect the general movement of each cove's fish to the other coves. Movement of white perch from sampled coves to areas not sampled would not change these estimates; however, the estimated total spring population would increase if white perch migrated into a sampled cove from areas not sampled.

When the migratory segment of each popula-

TABLE 5. White perch population estimates for three coves in the Connecticut River, 1998. The number of marked white perch recaptured at three sites in the Connecticut River, 1998, is listed with the percentage recaptured at the same site for row and column totals. Row percentages were then applied to the total number captured to compute the stationary population. Column percentages were applied to the stationary population to compute the total (stationary + migrant) spring population.

Marked in:		North Cove	Hamburg Cove	Wethersfield Cove	Total
Recaptured in:	North Cove	21	10	0	21 (68%)
	Hamburg Cove	2	18	0	20 (90%)
	Wethersfield Cove	0	2	2	4 (50%)
	Total	23 (91%)	30 (60%)	2 (100%)	

	Total catch (*C*)	% same site recaptures (FR)	Number marked (*M*)	Number same site recaptures (*R*)	Stationary population (*C* * FR) * (*M*/*R*)
North Cove	2,339	68%	937	21	70,967
Hamburg Cove	2,651	90%	2,118	18	280,741
Wethersfield Cove	123	50%	29	2	892

	Spring population estimates		
	Stationary population	Migrants moving into cove	Total population
North Cove	70,967	155,360	226,327
Hamburg Cove	280,741	7,043	287,784
Wethersfield Cove	892	31,821	32,713

	Percent total spring population by cove of origin		
	North Cove	Hamburg Cove	Wethersfield Cove
North Cove	31%	69%	0
Hamburg Cove	2%	98%	0
Wethersfield Cove	0	97%	3%

TABLE 6. White perch population estimate for the lower Connecticut River within Connecticut State boundaries. The population estimate is based on mean and standard deviation of densities in sampled coves (spring population from Table 5 divided by cove area in hectares) expanded to all known spawning sites in the river listed below. Area estimates for study sites and known spawning sites* were taken from the CTDEP GIS database.

Total Connecticut River area = 50 km^2
White perch spawning area* = 451 ha (4.5 km^2)
Mean (SD) white perch per hectare = 3,035 (1,860) fish
Population in spawning coves = 1,359,519 fish
95% confidence range of population size = 960,269 – 1,758,768

* Known spawning sites on the Connecticut River (town name given where multiple cove names exist): Scantic River, Podunk River, Farmington River, Hocknum River, Wethersfield Cove, Keeney Cove, Roaring Brook (Glastonbury), Pecausett Pond, Salmon River Cove, Chapman Pond, North Cove (Essex), Lord Cove, North Cove (Old Saybrook).

tion is added to the stationary segment of the cove they moved to (were recaptured in), the degree to which each cove's overwintering population contributes to the spring sport fishery of the other two coves can be examined (Table 5). The resulting spring population percentages indicate that white perch overwintering in Hamburg Cove contribute strongly to the quantity of white perch available to spring anglers fishing other coves. A third of the North Cove spring population wintered in Hamburg Cove. More importantly, almost all of the white perch present in Wethersfield Cove in spring overwintered in Hamburg Cove, 56 km to the south. White perch wintering in Wethersfield did not contribute to spring populations in either southern cove; however, it is possible they could have contributed to spring populations in other areas not included in this study.

The white perch population in the Connecticut portion of the Connecticut River was estimated at 1.36 million, with 95% confidence that the population was between 0.96 and 1.76 million (Table 6). This estimate is small compared to the adjacent Hudson River estimate of 32 million white perch (Klauda et al. 1988). Average density reported for this Hudson population was 108,000 white perch/km^2 (total area = 295 km^2), four times greater than the density estimated here of 27,000 white perch/km^2 (total area = 50 km^2). Possible biases of this method include insufficient time for marked fish to become randomly mixed into the population and lack of intersite movement until the end of the sampling period.

Current Exploitation Pattern

White perch taken in the confiscated gill net averaged 25.5 cm in length (N = 257). Three striped bass (40–51 cm length), a black crappie (30 cm length) and a carp (25 cm length) were also captured. The length frequency of white perch taken in this net (Figure 3) were similar to the frequencies reported by Maltezos et al. (1975) from commercial nets of the same mesh size.

Reported annual commercial landings have varied from 136 to 12,020 kg without trend from 1968 to 1997, except for a brief increase in 1972–1974 to 21,727–30,617 kg. Since 1981, 6-year means have increased slightly from 3,517 kg for 1981–1986 to 3,895 kg for 1987–1992 and 4,448 kg for 1993–1998. At an average weight of 0.23 kg, these landings represent approximately 19,500 fish annually. This small fishery appears stable in terms of total kilograms landed over the last 18 years.

Harvest (catches that are kept) by the Connecticut River recreational fishery (see Jacobs et al. 2004) ranged from 57,871 white perch in 1997 to 39,892 in 1998, from total catches ranging from 72,339 and 63,321 for the 2 years, respectively. White perch catch ranked second and third behind catfish species and striped bass. A 5-year creel census conducted in the Haddam area of the Connecticut River in 1965–1969 (Marcy 2004 [1976]) also ranked white perch second only to brown bullhead, a catfish species. Catch per hour for the shore-based fishery was consistent between 1997 and 1998 (0.52 and 0.57 fish/angler-hour) and very similar to the shore-based catch rate (0.47 fish/angler-hour) reported by Marcy for 1965–1969. Catch per hour for the boat-based fishery varied (1.23–3.03 for 1997–1998) and was generally centered south of Haddam. Marcy (2004 [1976]) also reported that no boat fishery of any magnitude was present in the Haddam area in the 1960s. The shore-based fishery is present throughout the river, accounting for 70% of the total catch in 1997–1998. The highest shore-based catch rate for white perch was consistently south of Haddam (0.91 fish/angler-hour versus 0.26–0.29 for the northern sections), and not surprisingly, 70% of the catch was taken by boat and shore anglers in this southern region. The catch was fairly evenly distributed among seasons surveyed from March through October (see Jacobs et al. 2004: 30% March–April, 30% May–June, 26% July–August, 14% September–October). The winter ice fishery was not included in the recreational survey.

Estimates of Fishing Mortality

Because only 2 years' data were available, no conclusions could be drawn concerning annual changes in mortality or year-class strength. Therefore, abundance at age was pooled for the 2 years to maximize sample size. Regression of natural-log abundance on age resulted in an estimate of instantaneous total mortality (Z) of 0.77 for white perch age 3 and older taken by trap net in 1997–1998, equivalent to an annual mortality estimate (A) of 54%. This average annual mortality rate for 1997–1998 is the same or lower than estimated mortality using age frequencies from the Connecticut River in 1975 and 1965–1966 (Table 7).

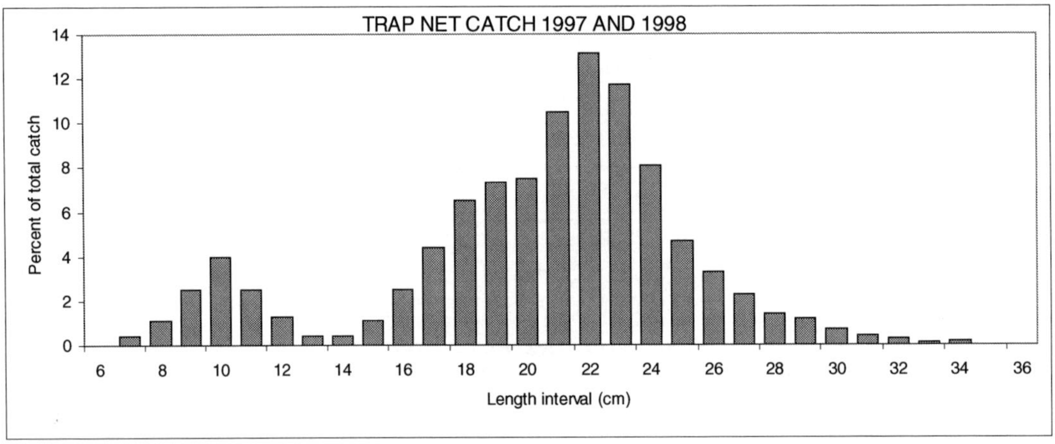

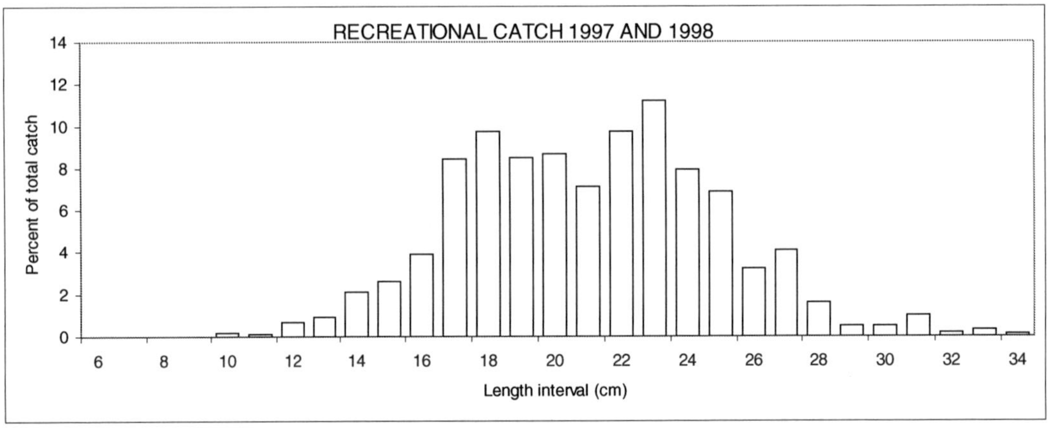

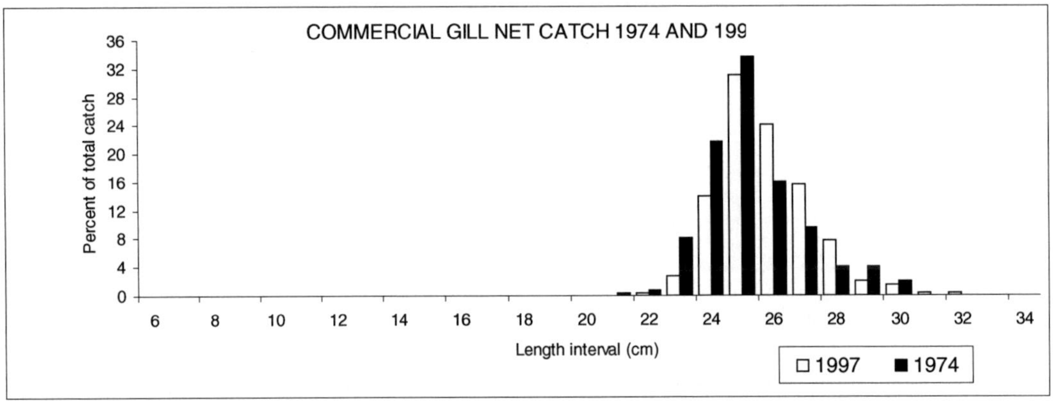

FIGURE 3. Length frequency as a percent for the trap net catch, recreational catch, and commercial catch. Trap net catch length data are from the three sites sampled in this study. Recreational catch length data was taken from Jacobs et al. 2004. The commercial length data are taken from Maltezos et al. (1976) and this study (1997).

TABLE 7. Comparison of white perch age frequencies from Connecticut River trap net catches in 1965–1966, 1975, and 1997–1998. Age frequencies reflect aged catch taken at random (S. Richard, Little Harbor Laboratory, personal communiation) in 1965–1966 as reported in Marcy and Richards (1974), all catches reported in Maltezos et al. (1975) and during this study. Instantaneous total mortality (*Z*) estimates are given for ages 2–7 in 1965–1966, 2–8 in 1975, and 3–11 in 1997–1998.

Year	Age: 2	3	4	5	6	7	8	9	10	11	Z
1965–1966	80	71	60	27	7	1					0.85
1975	142	60	59	41	27	3	1				0.77
1997–1998	3,214	5015	2286	490	416	265	128	20	39	5	0.77

This current estimate is also slightly lower than rates reported by Klauda et al. (1988) for an adjacent population in the Hudson River (*A* = 57–68%). Subtracting natural losses (*M* = 0.25) from total mortality, the remaining fishing mortality (*F*) estimate becomes 0.52, or 37% annually. Division of fishing mortality between the recreational and commercial fisheries, based on current landings, is 71% and 29%, respectively.

Conclusions

The white perch population in the Connecticut River is subject to natural mortality and fishing mortality from both sport and commercial fisheries. It appears that the component of mortality due to fishing has not increased in the last 30 years and may have decreased since the early 1970s. The 1997–1998 estimate of total mortality was the same as that estimated from 1975 age data and less than the 1969 estimate. Additionally, the 1997–1998 recreational catch per hour was similar to that recorded in 1965–1969, and commercial landings and nominal effort remain unchanged over the last 18 years.

A useful reference point by which fishing rates can be put in the context of sustainable yield is the theoretical maximum spawning potential, computed for this population by Howell and Molnar (1999). A fishing rate leaving 20% of this theoretical maximum (F20) is considered a threshold minimum spawning biomass needed to achieve stock replacement and prevent recruitment overfishing for species such as white perch having a relatively long life span (Gabriel 1986; Gabriel et al. 1989). For white perch under conditions of no minimum legal length limit, or a small (17.8 cm) limit as was just recently enacted for the Connecticut River, F_{20} occurs at fishing rates within 10% of the current annual mortality (*F* = 0.5–0.6). It is concluded that this rate of fishing is high enough to be eroding this stock's ability to withstand changes in the environment or increased predation pressures, its so-called "compensatory reserve."

Currently, the population's age structure is broad and growth rate comparable or greater than figures recorded for this population in 1975 as well as for other white perch populations in the mid-Atlantic area. The stock age structure indicates that mortality in very early ages, probably from natural predation, may be having a greater effect than fishing in depressing total numbers of white perch. However, this question cannot be resolved without a more complete time-series of white perch production and survival, in terms of total population numbers and age-specific abundance, in relation to their dominant predators in the Connecticut River.

Acknowledgements

Funding for this research was provided by the state of Connecticut and the Federal Aid to Sport Fish Restoration Program, F54R. We thank D. Vernon Osteen, John Clark, and Robert Sadzinski for their helpful reviews, comments, and suggestions. We are also grateful to Barton C. Marcy, Jr., who served as co-editor of this manuscript.

References

AuClair, R. 1960. White perch in Maine. Maine Department of Inland Fisheries and Game, Departmental Project Report, Augusta.

Bath, D., and J. O'Connor. 1982. The biology of the white perch, *Morone americana*, in the Hudson River Estuary. Fishery Bulletin 80(3):599–610.

Gabriel, W. 1986. Spawning stock biomass per recruit analysis for seven Northwest Atlantic demersal finfish species. National Oceanic and Atmospheric Administration National Marine Fisheries Service,

Woods Hole Reference Document 85-04, Woods Hole, Massachusetts.

Gabriel, W., W. Overholz and M. Sissenwine. 1989. Analysis of spawning stock biomass per recruit: an example for Georges Bank haddock. North American Journal of Fisheries Management 9:383–391.

Hoenig, J. 1983. Empirical use of longevity data to estimate mortality rates. Fisheries Bulletin 81(4):898–905.

Howell, P., and D. Molnar. 1999. Stock assessment of white perch in Connecticut's major rivers. Pages 160–190 *in* CT Department of Environmental Protection. A Study of Marine Recreational Fisheries in Connecticut. Connecticut Department of Environmental Protection, Federal Aid in Sport Fish Restoration Final Report (F54R), Hartford.

Jacobs, R., and E. O'Donnell. 1996. An electrofishing survey of selected Connecticut lakes. Connecticut Department of Environmental Protection, Federal Aid to Sport Fish Restoration Final Report (F57R), Hartford.

Jacobs, R. P., W. A. Hyatt, N. T. Hagstrom, E. B. O'Donnell, E. C. Schluntz, P. Howell, and D. R. Molnar. 2004. Trends in abundance, distribution, and growth of freshwater fishes from the Connecticut River in Connecticut (1988–2002). Pages 319–343 *in* P. M. Jacobson, D. A. Dixon, W. C. Leggett, B. C. Marcy, Jr., and R. R. Massengill, editors. The Connecticut River Ecological Study (1965–1973) revisited: ecology of the lower Connecticut River 1973–2003. American Fisheries Society, Monograph 9, Bethesda, Maryland.

Klauda, R., J. McLaren, R. Schmidt, and W. Dey. 1988. Life history of white perch in the Hudson River Estuary. Pages 69–88 *in* L.W. Barnthouse, R. J. Klauda, D. S. Vaughn and R. L. Kendall, editors. Science, law, and Hudson River power plants. American Fisheries Society, Monograph 4, Bethesda, Maryland.

Maltezos, G., and P. Howell, 1982. Population dynamics and stock assessment of the scup (*Stenotomus chrysops*) from Long Island Sound. Pages 29–56 *in* A Study of Marine Recreational Fisheries in Connecticut. Connecticut Department of Environmental Protection Federal Aid to Sport Fish Restoration Annual Performance Report (F54R), Hartford.

Maltezos, G., R. McMahon, and M. Leslie. 1979. Movements of white perch in the lower Connecticut River. Connecticut Department of Environmental Protection, Fisheries Division Program Final Report (88-309), Hartford.

Maltezos, G., E. Smith, and F. Huntley. 1975. Size and age distribution of the white perch population in the lower Connecticut River. Connecticut Department of Environmental Protection, Fisheries Division Progress Report, Hartford.

Mansueti, R. 1961. Movements, reproduction and mortality of the white perch, *Roccus americanus*, in the Patuxent Estuary, Maryland. Chesapeake Science 2:142–205.

Marcy, B. C., Jr., and F. P. Richards. 1974. Age and growth of the white perch *Morone Americana* in the lower Connecticut River. Transactions of the American Fisheries Society 103:117–120.

Marcy, B. C., Jr. 2004. Fishes of the lower Connecticut River and the effects of the Connecticut Yankee plant. Pages 65–124 *in* P. M. Jacobson, D. A. Dixon, W. C. Leggett, B. C. Marcy, Jr., and R. R. Massengill, editors. The Connecticut River Ecological Study (1965–1973) revisited: ecology of the lower Connecticut River 1973–2003. American Fisheries Society, Monograph 9, Bethesda, Maryland. (Originally published in 1976)

Marcy, B. C. Jr. and F. P. Richards. 1974. Age and growth of the white perch *Morone americana* in the lower Connecticut River. Transactions of the American Fisheries Society 103:117–120.

Richkus, W., and F. Stroup. 1987. Characterization of the biology of and fisheries for Maryland stocks of white perch. Maryland Department of Natural Resources, Tidewater Administration, Annapolis.

Ricker, W. 1975. Computation and interpretation of biological statistics of fish populations. Fisheries Research Board of Canada Bulletin 191.

Wallace, D. 1971. Age, growth, year class strength, and survival rates of the white perch *Morone americana* (Gmelin) in the Delaware River in the vicinity of artificial island. Chesapeake Science 12(4):205–218.

American Fisheries Society Monograph 9:391–405, 2004

The Impact of Enhancement Initiatives on the Structure and Dynamics of the Connecticut River Population of American Shad

WILLIAM C. LEGGETT*,[1]
Department of Biology, Queen's University, Kingston, Ontario K7L 3N6, Canada

THOMAS F. SAVOY[1]
State of Connecticut, Department of Environmental Protection, Marine Fisheries Division Post Office Box 719, Old Lyme, Connecticut 06371, USA.

CHRISTINE A. TOMICHEK[1]
Kleinschmidt Associates, Post Office Box 1050, Deep River, Connecticut 06417, USA

Abstract.—During the interval 1975 to 1981, improvements to existing fishways and new fishway construction at high-head mainstream dams located at river km (rkm) 139, 198, and 228 on the Connecticut River, combined with the breaching of a low head dam at rkm 110, created access by American shad to their historical spawning areas from which they had been excluded since the late 18th century. This resulted in a significant upriver relocation of the main spawning activity by shad, but, to date, has failed to produce the anticipated increase in total population size from the increase in total spawning habitat available. This upriver displacement of spawning resulted in an increase in the length of the spawning migration and a corresponding increase in the energy expended to reach the spawning grounds. The result was higher adult mortality and dramatic reduction in the repeat spawning component of the population and in the mean size and age of adult fish. An increase in the variance in annual population abundance followed this reduction in population age structure. The loss of larger repeat spawning females is estimated to have resulted in a 20% reduction in mean population fecundity and could account for a 14% reduction in annual recruitment to the population. These outcomes are consistent with predictions founded on knowledge of the population dynamics and migration energetics of the species available at the time the enhancement initiatives were begun. Unrelated, but very successful restoration efforts directed to Atlantic Coast populations of striped bass also appear to have negatively impacted the shad population of the Connecticut River through increased predation. The implications of these findings for future management are discussed.

Introduction

The population dynamics and migratory behavior of the Connecticut River population of American shad *Alosa sapidissima* were extensively studied during the interval 1965–1973 as part of a larger investigation of the impact of the heated water effluent from the Connecticut Yankee atomic power plant located at river kilometer (rkm) 31 on the Connecticut River (Merriman and Thorpe [1976] and Part 1 of this volume). The Connecticut River Ecological Study concluded that the operation of the Connecticut Yankee facility had no impact on the behavior or population dynamics of the species (Leggett 2004 [1976], this volume).

Beginning in 1955, and continuing in the 1970s and 1980s, a series of initiatives designed to eliminate artificial barriers to the upstream migration of American shad were undertaken, as were

* *Corresponding author*: wleggett@post.queensu.ca
[1] Authorship determined alphabetically.

efforts to restore the population of one of its major predators, the striped bass *Morone saxatilis*. In this paper, we use the results of studies of the life history dynamics, migration energetics and population dynamics of the American shad populations of the Connecticut and other major Atlantic Coast rivers, conducted as part of, coincident with, and subsequent to the Connecticut River Ecological Study, to assess the impact of the restoration efforts directed toward American shad and striped bass.

Prior to the industrialization of the Connecticut River valley in the late 18th and early 19th century, American shad migrated as far north in the Connecticut River as Bellows Falls (Figure 1) located at rkm 280. A high-head dam constructed at Turners Falls (rkm 198) in 1798 was the first to limit the upstream migration of shad. A subsequent high-head dam constructed at Holyoke (rkm 139) in 1849 further restricted their upstream movements. No shad spawned north of Holyoke between 1849 and 1955 when the first effective fish lift facility was installed at Holyoke Dam (Moffit et al. 1982). A third low-head dam was constructed in 1880 at Enfield (rkm 110). This structure was typically submerged during spring runoff when shad normally migrate upriver, so that while it could delay upstream migration, it did not obstruct passage except during periods of low discharge.

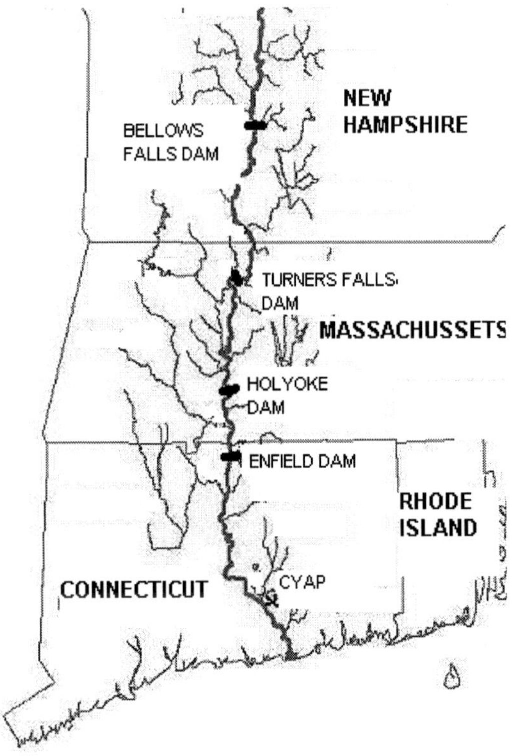

FIGURE 1. Map of Connecticut River illustrating key locations mentioned in text.

There is no direct knowledge of the immediate postconstruction impact of the Turners Falls and Holyoke dams on the Connecticut River shad population beyond the obvious effect of the exclusion of the species from the upper 151 km of its historic range. The species is generally believed to have experienced a decline in abundance following the construction of the Turners Falls and Holyoke dams. As early as 1893, an ineffective fishway was constructed at the Holyoke Dam in an effort to restore access by shad to the upper river. Stocking of hatchery-reared shad was also undertaken in the late 1880s in an effort to increase population size without success. (Gephard and McMenemy 2004, this volume). Details of the magnitude, timing and cause(s) of this 19th century decline in abundance are anecdotal.

Studies of the life history characteristics of American shad, conducted coincident with and as part of the Connecticut River Ecological Study, do, however, provide a basis for assessing the scale of the impact of the construction of these facilities on the age structure and repeat spawning characteristics of the population. American shad populations occur in most major rivers of the Atlantic Coast of North America from the St. Johns River, Florida to the St. Lawrence River, Canada (Scott and Scott 1988). The species is anadromous and exhibits strong home stream fidelity.

The highly developed, river-specific, homing behavior of American shad (Walburg and Nichols 1967; Scott and Scott 1988; Leggett 2004 [1976]), coupled with their extensive range, has resulted in the evolution of genetically based, adaptive, population-specific life histories that differ with the latitude of the home river (Carscadden and Leggett 1975a, 1975b; Leggett and Carscadden 1978; Shoubridge 1978; Bentzen et al. 1989). The main features of this adaptation are a positive relationship between the frequency of repeat reproduction and the latitude of the home river (Figure 2) and an inverse relationship between the proportion of repeat spawning fish in a given river and the mean virgin and lifetime fe-

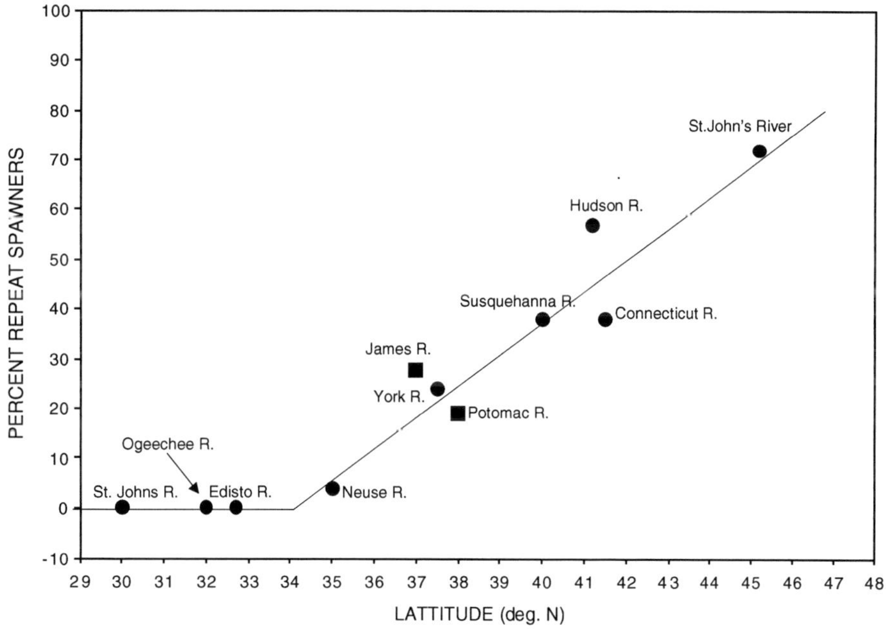

FIGURE 2. Proportion of repeat spawning American shad in populations located on the Atlantic Coast of North America (■ value based on single estimate; ● value based on multiple year estimate).

cundity of individual fish in that population (Leggett and Carscadden 1978). These adaptations respond to latitudinal differences in the predictability of the reproductive environment in the home river and serve to dampen the effect of recruitment variability on the abundance and life history characteristics of individual populations (Leggett 1985).

Analyses of the proportion of repeat spawning shad in the Connecticut River, undertaken in the 1960s and early 1970s, revealed that the proportion of repeat spawners in the Connecticut River population during that interval averaged 38% (±10.2%, n = 12) and exhibited no temporal trend (Leggett and Carscadden 1978). This is consistent with expectation, based on the positive latitudinal trend in the proportion of repeat spawning fish in local populations along the Atlantic Coast in which no significant barriers to migration exist (Figure 2). The pattern of repeat spawning in populations in rivers at similar latitudes that have not been impacted by the effects of dam construction suggests that exclusion of shad from the upper 150 km of its historical range in the Connecticut River did not permanently alter, in a major way, the age structure and repeat spawning characteristics of the population. One reason for this may be that prior to dam construction the majority of the shad spawning in the Connecticut River utilized spawning areas that, subsequent to dam construction, became the only spawning habitat available. There are no data, nor is there anecdotal evidence, with which to assess the historical distribution of spawning activity in the river.

However, given the importance of repeat spawning to the maintenance of population stability (Leggett and Carscadden 1978; Leggett 1985) and the strong latitudinal trend in this feature of the life history in the species (Figure 2), it is reasonable to conclude that the distribution of spawning sites in the Connecticut River would have evolved to produce a repeat spawning component that did not differ dramatically from the overall latitudinal trend in this adaptive life history characteristic.

The impact of the construction of the low head dam at Enfield is more fully understood. Prior to its progressive breaching by natural forces during the mid-1980s, the upstream migration of shad was delayed, but not blocked (except during periods of very low discharge), at Enfield, Connecticut. The aggregations of shad below the dam favored the development of a large and very successful sport fishery (Leggett 2004 [1976]). Spawning surveys conducted prior to the breaching of the dam indicated that the majority of Connecticut River shad spawned below Enfield Dam (Marcy 2004 [1976], this volume). The delay in upstream migration and the associated fishery have virtually disappeared since the dam was fully breached.

Recent Developments of Relevance to the Biology of American Shad

In 1955, under orders from the U.S. Fish and Wildlife Service, the Holyoke Power Company constructed a fish lift facility at the Holyoke Dam, Massachusetts. The increase in the numbers of shad passed at this facility (Figure 3) and in the proportion of the total population migrating above Holyoke (Figure 4) was modest over the period 1955–1974. An expansion of the lift capacity, and improvements to the fish attraction systems and upstream handling facilities implemented in 1975 and 1976 (Henry 1976), resulted in a rapid increase in the number of shad passed and in the proportion of the total population migrating beyond Holyoke Dam. After 1985, the proportion of the total population migrating beyond Holyoke Dam routinely exceeded 40%. To date, there has been no corresponding increase in the size of annual populations of shad entering the river to spawn as a consequence of the expansion of spawning area due to fishway construction (Savoy and Crecco 2004, this volume). Construction designed to double the current upstream fish passage capacity at Holyoke Dam is currently underway.

In 1980 and 1981, new fish passage facilities were opened at Turner's Falls and Vernon dams, respectively. This gave American shad access to their historical spawning areas. (See Gephard and McMenemy 2004, this volume, for a full treatment of the factors leading to fishway construction and the success of these facilities in passing shad). The increase in the numbers of shad migrating beyond Holyoke, and in the proportion of the total shad population beyond Holyoke, was also facilitated by the progressive natural breaching of the Enfield Dam. Downstream fish passage facilities have been constructed Vernon, Turners Falls and Holyoke in Massachusetts. The numbers of adult fish using these facilities has remained below expectation, but it is unclear if this is a problem of design, operation or simply a reflection of the increased postspawning mortality of fish spawning above Holyoke Dam. Telemetry studies of adult shad migrating above Holyoke indicate that the further an individual shad migrates upstream beyond Holyoke

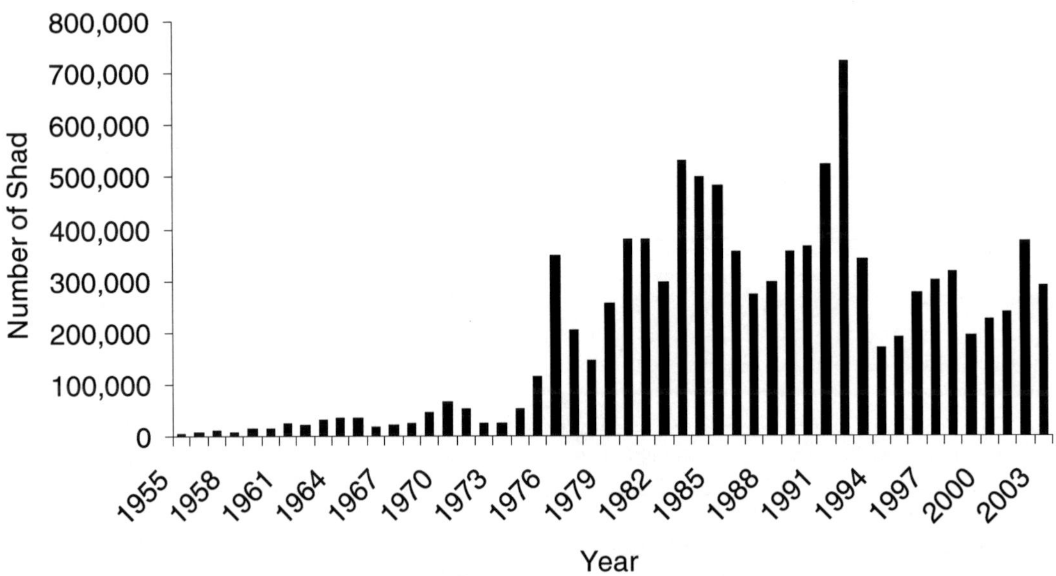

FIGURE 3. Numbers of Connecticut River American shad migrating beyond Holyoke Dam, 1955–2003.

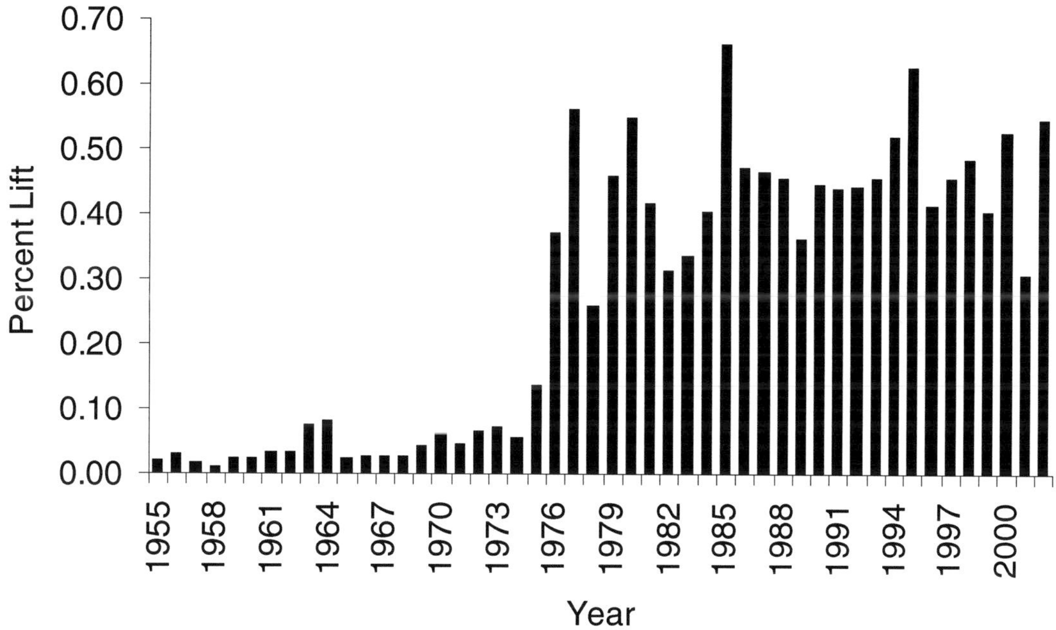

FIGURE 4. Proportion of Connecticut River American shad population migrating beyond Holyoke Dam, 1955–2003.

the less likely it is to pass downstream through Holyoke (R. J. Stira, Northeast Generation Services, personal communication).

Beginning in the late 1980s, another important change resulting from fishery restoration efforts of a completely different kind occurred in the environment encountered by Connecticut River shad during their freshwater migration. A substantial increase in the numbers of adult striped bass in the river, extending as far north as Holyoke, took place (Savoy and Crecco 2004). Adult striped bass were virtually absent from the river during the period of the Connecticut River Ecological Study (1965–1973; Marcy 2004 [1976]). In 1994, the population of adult striped bass in the river was estimated at approximately 400,000. By 1998, the number of adult striped bass in the river was estimated to have increased to as many as 1.4 million (Savoy and Crecco 2004). This increase resulted from the great success of the Atlantic Coast Striped Bass Restoration Program (Essig and Beal 2000). Savoy and Crecco (2004) determined that at least 50% of the adult striped bass in the Connecticut River in 1994 was composed of adults larger than 71 cm total length (TL). Striped bass of this size are capable of feeding on adult male shad and on small first spawning females. Moreover, they are known to prefer soft-rayed species such as herring and shad over hard-spined species (Manooch 1973; Gardinier and Hoff 1982), and the importance of shad and herring as prey of striped bass is well documented (Bigelow and Schroeder 1953; Versar, Inc. 1990). After 1997, an average of more than 300,000 striped bass in the greater than 71-cm-TL size-class were entering the river annually. Savoy and Crecco (2004) concluded that predation by striped bass on adult Connecticut River shad was a major contributor to the post-1992 decline in the population of shad in the river. By way of contrast, they found little statistical support of the hypothesis that predation by striped bass on juvenile shad was linked to this decline (Savoy and Crecco 2004).

Implications of Restoration Activities for the Structure and Size of the Connecticut River Shad Population

The most obvious and dramatic effect of the restoration/conservation initiatives described above was the increase in the distance of the freshwater migration undertaken by individual members of the Connecticut River shad population and the proportion of the total population engaged in this extended migration. Because adult shad do not feed in fresh water (Leggett 1972; Scott and Scott 1988), the entire migration in and out of the river must be fueled by energy accumulated during their marine migrations.

Glebe and Leggett (1981a) studied the total energy expenditure by individual shad during their spawning migration in the Connecticut River. Shad expended 40–60% of their total energy reserves to migrate from the sea to the spawning areas, to spawn (which at the time of their study occurred principally below Enfield Dam), and to complete the downstream migration to the sea.

Glebe and Leggett (1981b) also studied the pattern of energy utilization during the freshwater migrations by shad homing to Atlantic Coast rivers from the St. John's River (Florida) to the Connecticut River. Between rivers, differences in the magnitude of these energy expenditures were significant. Energy expenditures (kJ/kg/km and % of total energy expended) were greatest in the south and became progressively lower toward the north. They concluded that these latitudinal differences in energy allocation to migration were adaptive, and, like the latitudinal differences in repeat spawning and fecundity, formed part of the overall population specific response of individual shad populations to the particular characteristics of the river in which they spawn. This conclusion is consistent with the findings of Shoubridge and Leggett (1978).

The increases in the distance and duration of the spawning migration undertaken by shad in the Connecticut River resulting from fishway construction, and in the levels of predation experienced by adult shad during their freshwater residence, when combined with the insight that comes from research on their biology and ecology, lead to the following predictions with respect to the probable impact of restoration activities on the dynamics of shad in the Connecticut River:

1. The greater migration distances traveled by a significant proportion of the total American shad population of the Connecticut River as a consequence of the breaching of the Enfield Dam and the creation/expansion of lifting capacity at Holyoke, Turners Falls, and Vernon dams will result in higher average energy expenditures by shad during their freshwater migration.
2. These increased energy expenditures will, in turn, lead to higher postspawning mortality in the population at large. This mortality will be reflected in a reduction in the proportion of repeat spawning individuals in the population. The magnitude of this decline is expected to be related to the proportion of the total population undertaking this extended migration.
3. This loss of repeat spawners will, in turn, be reflected in a reduction in the mean size and age of shad in the population and, given that individual fecundity is positively related to size (Leggett and Carscadden 1978), to a reduction in mean population fecundity and total reproductive output.
4. The increased predation imposed on American shad by striped bass will initially accelerate these trends, but subsequently, as size selective predation on smaller males and females intensifies, a reversal in the trends in mean size and mean age may occur.
5. The reduction in the number of year-classes contributing to annual reproductive effort and in the mean fecundity of the population due to elimination of larger repeat spawners will diminish the capacity of the population to compensate for interannual variability in conditions governing recruitment (Leggett and Carscadden 1978; Lorda and Crecco 1987). An overall increase in year-to-year variability in recruitment and spawner abundance, and a gradual decline in the size of the spawning population, is expected.

Each of these predictions is examined and discussed below.

Migration Energetics

Glebe and Leggett (1981a, 1981b) documented the energy expenditure of American shad during their freshwater spawning migration in the Connecticut River in 1971 and 1972 when only a small fraction (order of 5%) of the total population of shad in the Connecticut River migrated beyond Holyoke. The rates of energy expenditure exhibited by shad sampled during the peak of the migration were higher in males (29.3 kJ/kg/km) than in females (20.9 kJ/kg/km). This difference was attributed to size-related differences in swimming efficiency, males being smaller than females. Size-related differences in metabolic rates and swimming efficiency are common in fishes and appear to relate to both physiological and hydrodynamic factors (Winberg 1956; Fry 1971; Brett and Glass 1973; Webb 1975; Trump and Leggett 1980). Energy expenditure per kg/km was consistent throughout the migration. Energy expenditure during the 265 km migration from the river mouth to the spawning grounds and back to saltwater averaged 40% to 60% of the combined somatic and visceral reserves available at the initiation of the freshwater migration.

If one assumes a similar rate of energy expenditure per kg/km for shad migrating as far as Turners Falls in the post-1975 period, the energy expenditure required to sustain the complete migration from river mouth to Turners Falls Dam and back to sea would increase on the order of 56% to 84%, or approximately 40% greater than that calculated during 1971 and 1972 (Glebe and Leggett 1981a, 1981b). Energy expenditures in the range of 60%–70% appear to be lethal in anadromous fishes (Glebe and Leggett 1981b; Jonsson et al. 1997).

Leonard and McCormick (1999) studied the energy use associated with the migration of American shad to spawning sites above Holyoke in 1993 and 1994, by which time 40% to 50% of the population had been spawning above Holyoke for several years. They estimated that the rate of energy expenditure was on the order of 25 kJ/km/fish. The shad sampled by Leonard and McCormick (1999) weighed, on average, 1,633 grams at river entry (assuming equal numbers of males and females sampled in each of the 2 years of the study (See Leonard and McCormick [1999]: Table 1). We estimate that the energy expenditure by shad to complete their migration to and beyond Holyoke in 1993 and 1994 would, when converted to kJ/km/kg as reported by Glebe and Leggett (1981b), be in the range of 40 kJ/km/kg or approximately 35–50% higher than the rates reported by Glebe and Leggett (1981b).

Interannual variability in energy expenditures can be large and can easily explain the differences in rates of energy expenditure reported by Glebe and Leggett (1981a) and Leonard and McCormick (1999). Leonard and McCormick (1999) found that, in 1993, male and female shad expended only 37% to 39% of their total energy reserves to reach their spawning areas. If one assumes the energy required to effect the return migration to the sea approximates 60% of that required for the upstream migration, total energy expenditures would be in the order of 59% to 62%, close to lethal levels. This appears to be a reasonable assumption because (a) while downstream migration is with the current and is therefore likely to be less costly for this reason, it also occurs at higher temperatures, thereby offsetting the benefit of migrating with the current; and (b) Glebe and Leggett (1981a) found that the energy expenditure during the downstream migration averaged 45% to 65% of the upriver expenditure.

In 1994, energy expenditures during the upriver spawning migration were much higher in females, averaging 60% or their total energy reserves. Applying a similar calculation to estimate total expenditure for the complete freshwater migration yields a total energy expenditure of 96%. Leonard and McCormick (1999) attributed this higher energy expenditure in 1994 to 25% higher discharge rates in 1994 relative to 1993. Higher discharge rates or river flow velocities equate to greater energy expenditures per km of upstream migration. Interannual differences in energy expenditure are also likely to be strongly linked to average water temperatures during the migration. Glebe and Leggett (1981a) found that late migrating shad, which experienced higher average water temperatures during migration, exhibited significantly higher rates of energy expenditure (24 kJ/kg/km for early migrants versus 42 kJ/kg/km for late migrants). Shoubridge and Leggett (1978) reported that year-to-year variation in the proportion of repeat spawners in the Connecticut River population was strongly and negatively correlated with average water temperatures during the migration 1 year earlier. They concluded that this effect was related to mortality resulting from the increased energy demands of migrations undertaken at higher temperature. It is likely, therefore, that in years when environmental conditions cause high levels of energy expenditure during the freshwater migration (high river discharge or high temperatures), the levels of mortality experienced by shad migrating to Holyoke and beyond would be elevated. It should be noted that discharge and temperature in the Connecticut River tend to be inversely correlated during the time of the adult shad migration in freshwater. of year (Crecco and Savoy 1985). Thus, in any given year, either discharge or temperature could be the dominant determinant of energy expenditure per kilometer.

In contrast with the findings of Glebe and Leggett (1981a), Leonard and McCormick (1999) found that females expended a greater proportion of their total energy reserves than did males. This is inconsistent with theory (see above) but could result from differences in migration temperatures experienced by males and females. Females typically migrate later than males, and because water temperatures increase steadily during the migration, females experience higher average water temperatures. The major increase in the rates of energy expenditure exhibited by females migrating above Holyoke, and especially above Turners Falls reported by Leonard and McCormick (1999), are consistent with this hypothesis because water temperatures are typically approaching their seasonal maximum during this upstream phase of the migration. Notwithstanding the interannual differences in the pattern and rates energy expenditure as reported by Glebe and Leggett (1981a) and Leonard and McCormick (1999), both studies indicate that the level of energy use experienced by fish migrating above Holyoke may represent the threshold for the maintenance of repeat reproduction in the population.

Proportion of Repeat Spawners

The proportion of repeat spawning males and females in the Connecticut River shad population has been monitored continuously since 1965. Interannual variation is significant and principally a result of interannual variation in recruitment to the population (Leggett 2004 [1976]). Although slightly greater year-to-year variability occurred for males (Figure 5), the proportion of repeat spawners has declined markedly for both males and females. The greatest decline in the proportion of repeat spawning females occurred coincident with the expansion of the fish passage facilities at Holyoke in 1975. From 1965 to 1974 and 1977 on, the proportion of repeat females varied without trend, but exhibited high variance (Figure 5). In contrast, males exhibited a sustained reduction in the proportion of repeat spawners from 1965 through 1974, followed by a brief plateau from 1975 to 1980, which was then followed by a further reduction in the mean value from 1981 through 2003. The latter period, in particular, was characterized by high variance (Figure 5). In both males and females, the pattern of change in the proportion of repeat spawners was significantly and inversely correlated with the proportion of the total population migrating beyond Holyoke (Figures 4 and 5; males: $Y = 33.89 - 25.31X$, df = 70, $r^2 = 0.12$, $P < 0.003$; females: $Y = 28.42 - 23.57X$, df = 70, $r^2 = 0.22$, $P < 0.001$).

The frequency of repeat spawners demonstrated a modest reversal in 1997 in the case of females, and in 1999 in the case of males. These

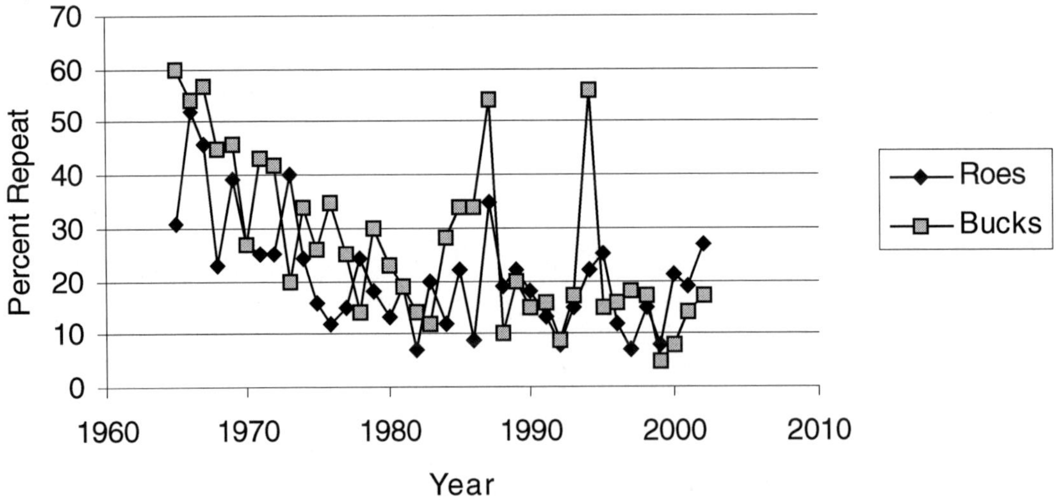

FIGURE 5. Proportion of repeat spawners, by sex, in annual runs of American shad in the Connecticut River, 1965–2003.

changes correspond to a reduction in the numbers of shad utilizing the Turners Falls fishway (R. J. Stira, Northeast Generation Services, personal communication). The reversal is also consistent with the prediction that predation by striped bass could, by selectively increasing the mortality experienced by smaller virgin adults, alter the relative abundance of repeat spawners in the population in a positive manner. The data available are inadequate to determine the extent to which either or both of these factors influenced the increase in repeat spawners noted post-1997.

Individual Sizes at Migration

Mean fork length data for adult shad entering the Connecticut River are available for the periods 1967–1973 and 1985–2002 (Figure 6). From 1985 through 1989–1990, the mean size of female American shad in the Connecticut River were consistent with those observed during the Connecticut River Ecological Study (1967–1973: males, range 47.2–48.4 cm; female, range 50.1–51.0 cm (Leggett 2004 [1976]). By 1985, the mean size of male shad had declined significantly. Beginning in 1990, the mean size of both males and females declined further to a minimum in 1996 for males and 1999 for females (Figure 6). At the nadir, mean sizes of males and females had declined by approximately 15% and 11%, respectively, from the 1965–1973 average lengths. Subsequently, mean individual size began to exhibit a modest annual increase, perhaps due to the size selective predation by striped bass on smaller virgin adults. In 2002, the mean sizes of males and female were approximately 11% and 6%, respectively, below the 1965–1973 average. This result appears to be related to the loss of older repeat spawners from the population. However, this relationship may not be exclusive. Leonard and McCormick (1999) reported that energy expenditures by shad migrating north of Holyoke were proportionally greater in larger fish and were proportionally highest in females, which are larger. Size selective mortality, resulting from this differential energy expenditure could account for some of the reduction in the mean size through its effect on the size of individual shad returning to repeat spawn in subsequent years.

Ages of Migrants

The proportion of male and female shad in annual populations is available for the period 1967–2002. Changes in these proportions have been pronounced and reflect the reduction in the proportion of repeat spawning fish in the population (Figures 7 and 8). Among females, the oldest (ages 7 and 8) age-classes exhibited marked declines following the expansion of the migratory range that resulted from fish lift construction (Figure 7). In males, the proportion of the total population composed of animals aged 6 and older (age-groups

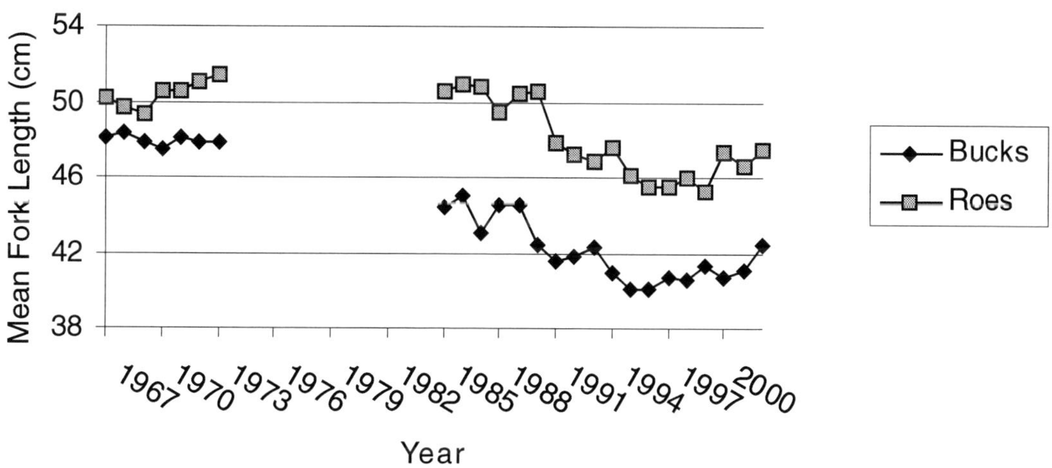

FIGURE 6. Mean fork length of Connecticut River American shad, 1967–2003.

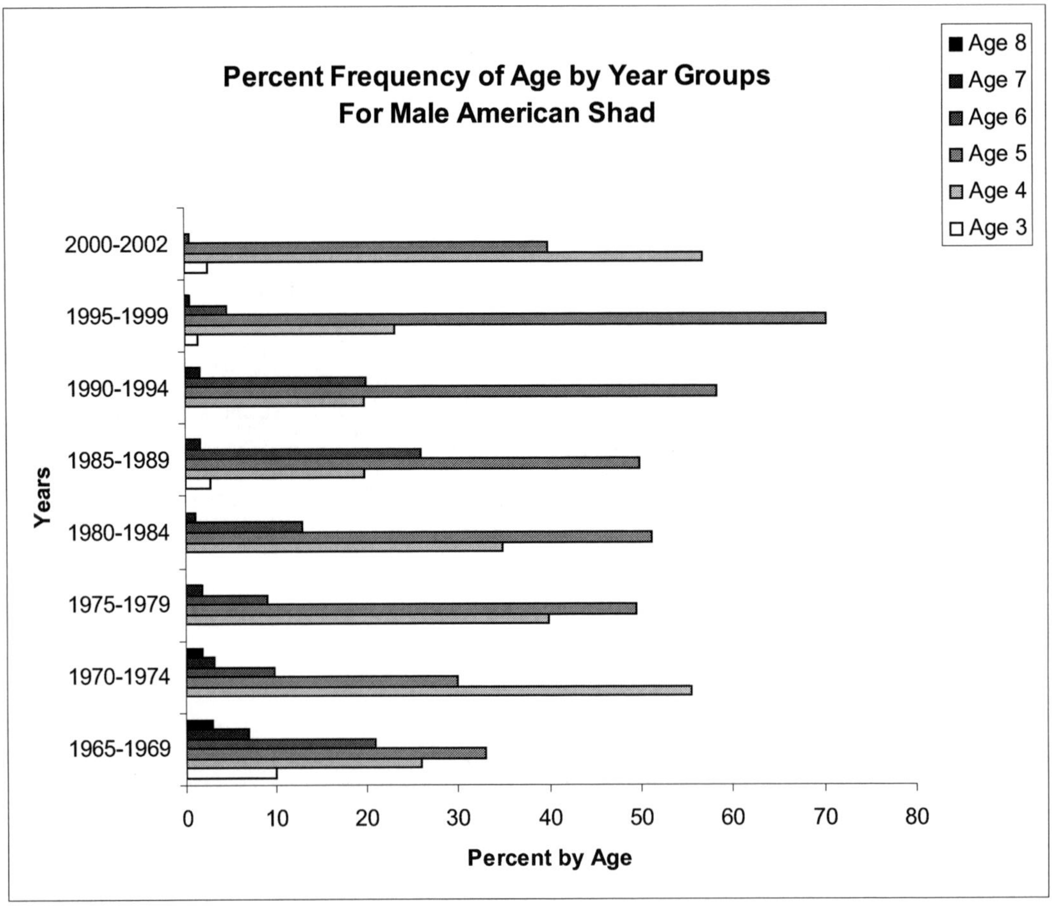

FIGURE 7. Percent frequency by age of male American shad in the Connecticut River, 1965–2002.

dominated by repeat spawning individuals) virtually disappeared (Figure 8).

In males and in females, the proportion of small virgin fish (males age 3 and 4; females age 4 and younger) also declined throughout the 1980s and 1990s. The temporal trend in the proportion of virgin fish (predominantly age-3 and -4 males and age-4 and -5 females) exhibited considerable year-to-year variation. This variability is attributable primarily to year-to-year variability in recruitment, which is superimposed on the overall downward trend in age.

Population Trends

Leggett (2004 [1976]) estimated the size of annual shad populations in the Connecticut River from 1935 to 1964 using data on commercial catches and units of effort in the commercial fishery. Crecco and Savoy (1987) estimated the Connecticut River shad population size for the interval 1966–2002 using lift data from the Holyoke facility. Both the 1935–1964 and 1974–2002 data series were calibrated using mark–recapture studies conducted over several years (Crecco and Savoy 1987; Leggett 2004 [1976]). The mean population size, during the interval 1935–1964, was estimated as 622,000 fish. The mean population size over the interval 1966 to 2002 was also estimated as 622,000 fish. Because the approach used by Crecco and Savoy (1987) and Leggett (2004 [1976]) in their reconstruction of annual population sizes differed in terms of the input data, analyses of similarities or differences in absolute population sizes and analyses based on combining of the two data sets should be approached with caution. However, given the consistency of the method within periods, comparison

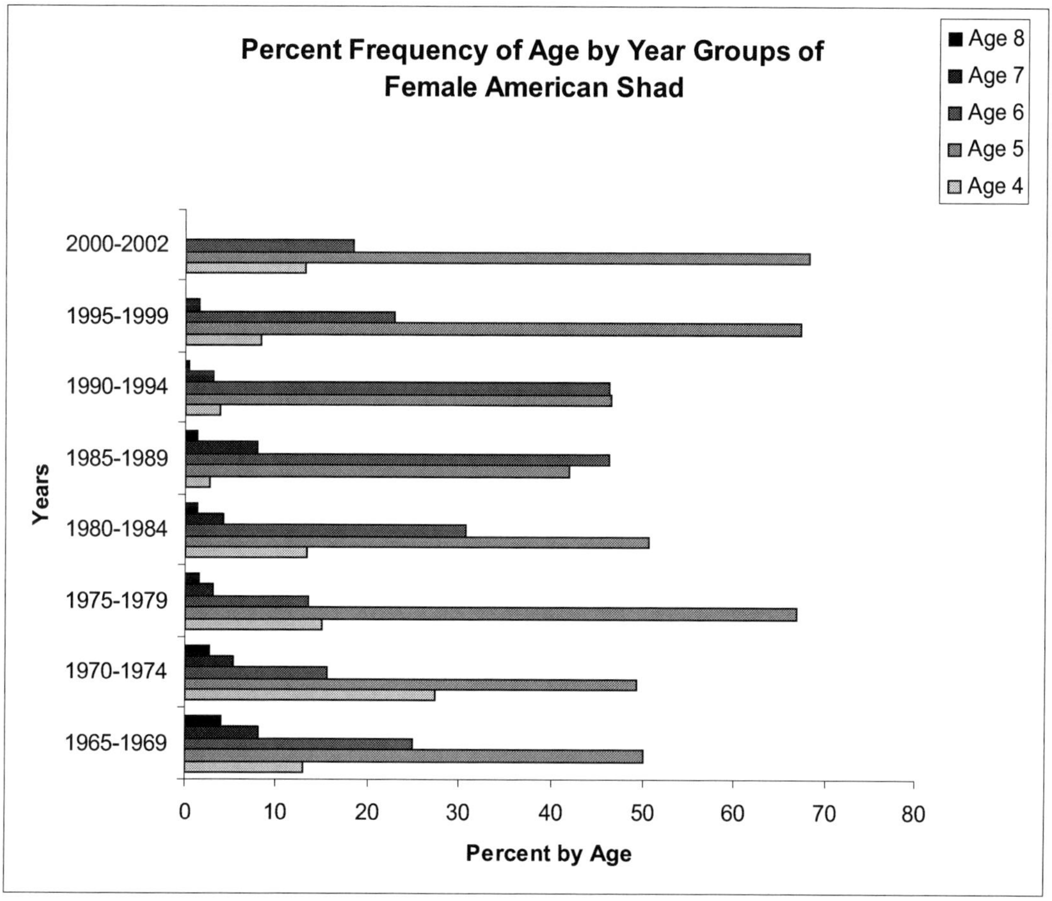

FIGURE 8. Percent frequency by age of female Connecticut River American shad, 1965–2002.

of the variance about the mean population estimates are considered to be both reasonable and reliable. For the interval 1935–1964, the standard deviation (SD) about the annual population size was estimated as 220,000. For the period subsequent to 1975, the year of the doubling of fish handling capacity at Holyoke, the SD about the mean population size increased dramatically to an estimated 343,000. To assess the maximum potential variation in the pre-1975 data series (the control set) over a period equivalent to the 28-year post-1975 interval, we estimated the standard deviation for each of five-hundred 28-year series constructed by drawing 28 individual records at random from within the 1935–1964 time window. The mean SD of that 500 element series was an estimated 223,000 and the range of these 500 SDs was an estimated 206,000–258,000. The estimated SD for the post-1978 period fell well outside the combined ranges for the full 1935–1964 interval and for the 500 simulated 28-year intervals. This finding is consistent with the prediction that a reduction in the frequency of repeat reproduction, and an erosion in the age-structured character of the population would cause it to be more vulnerable to environmental variation and, therefore, to exhibit higher year-to-year variability in population size.

Discussion

Each of the predictions regarding the consequences for the population dynamics of the Connecticut River American shad population of the expansion of the migratory range of shad in the Connecticut River as a result of the construction of fishways, and of the restoration of striped bass on the Atlantic Coast of the United States, appears to have been upheld.

1. The energy expenditure experienced by shad in reaching the spawning grounds increased in proportion to the greater migration distances that resulted from displacing a large portion of the spawning from below Enfield to areas above of rkm 135 (Holyoke).
2. The proportion of repeat spawners in the population declined to well below pre-1975 levels and below the predicted historical levels for populations spawning at this latitude.
3. The proportion of individuals of both sexes in the older age-classes of the population declined post-1975.
4. Interannual variation in the size of the shad population entering the Connecticut River increased significantly post-1975.
5. The proportion of repeat spawners and the mean size of individual shad post-1999 both increased, consistent with predictions relating to the effect of predation by striped bass on smaller members of the adult shad population.

It is probable that the reduction in the proportion of repeat spawners in the population, and the progressive reduction in mean size of individual spawning fish, contributed to the post-1992 reduction in mean population size reported by Savoy and Crecco (2004) and attributed by them to predation by striped bass. Leggett and Carscadden (1978) documented a strong positive relationship between female size and egg production. Using their data, we estimated that the reduction in mean female size post-1975 would account for a 20% reduction in mean population fecundity. This is in addition to, and independent of, the decline in egg production that would result from the overall decline in population numbers that occurred post-1990. The overall impact of this change in per-female egg production on recruitment to subsequent populations is difficult to quantify. In general, the relationship between spawner biomass (a surrogate for population fecundity) and recruitment in fishes is weak, recruitment being modified significantly by environmentally induced interannual variability in the survival of spawned eggs and of larvae (Cushing 1982). This is known to be true for American shad (Crecco and Savoy 1983; Leggett 2004 [1976]). However, a significant stock–recruitment relationship has been shown to exist in the case of Connecticut River shad (Lorda and Crecco 1987; Leggett 2004 [1976]), and therefore, the potential impact cannot be discounted. If one controls for the effect of interannual variability in egg and larval survival, and assumes an average population of 600,000 individuals (the mean population size pre- and post-1975) and a sex ratio of 50/50 (Leggett 2004 [1976]), the reduction in recruitment resulting from the decline in per-spawner fecundity (a product of the elimination of a large part of the larger repeat spawning component of the population) could account for a reduction in recruitment of approximately 14%. This may have contributed, in part, to the failure, to date, of the population to respond numerically to the increased access to upriver spawning habitat created by the construction/expansion of fishways in the 1970s and 1980s.

The data we present are consistent with the hypothesis that the creation of enhanced fish passage capacity at Holyoke, and the construction of new passage facilities at Turners Falls and Vernon, have not yet resulted in the anticipated increases in the overall population size of Connecticut River American shad, which formed part of the justification for the creation of these new and enhanced fish passage facilities. Instead, a progressive upriver displacement of the main geographic location of spawning activity from below to above Holyoke has occurred. This, in turn, has resulted in an overall increase in the per-spawner energetic cost of the migration; a reduction in the repeat spawning component of the population; a reduction in mean age, mean size, and mean population fecundity; and an increase in year to year variability in population size. Each of these effects and their overall impact on population stability are consistent with predictions that are based on the knowledge of the life history characteristics and migration energetics of the population at the time enhancement activities were initiated and, for that matter, from the broader body of ecological literature on the adaptive significance of age-structured populations. (Cole 1954; Holgate 1967; Schaffer and Elson 1975; Leggett and Carscadden 1978).

The impact of higher levels of predation by striped bass, and the probable size selective nature of that predation, is difficult to assess quantitatively given that the primary impact of this predation mortality occurs subsequent to the entry of shad into the river. It is unlikely, therefore, that it would be reflected in changes to the structure of the population at river entry (i.e., reduced numbers or proportions of fish in smaller age-classes). This mortality is concentrated principally on smaller virgin members of the population. Given the data available, the most reliably quantifiable impact would be on population size in subsequent years. Savoy and Crecco (2004) make a convincing case for a negative impact of enhanced predation by striped bass on the mean population size post-1992. The increase in mean fish size and in the proportion of repeat spawners in the population beginning in 1999 may reflect, in part, the effect of high predation rates on smaller members of the male and female population. However, it may also result entirely, or in part, from enhanced recruitment of virgin individuals in those years.

These findings lead us to conclude that a conservative go forward management strategy would call for (a) annual or at least semi-annual determination of the size of the population entering the river to spawn and its size, age, and repeat spawning structure; and (b) adjustments to the numbers of shad passed above Holyoke to ensure first the restoration, and then the maintenance, of an age and repeat spawning structure that is more consistent with pre-1975 levels. This may require that the numbers of shad passed above Holyoke be reduced to levels necessary to achieve this balance, at least until the anticipated positive effects of increased spawning area created by fishway construction is reflected in the size of the population entering the river to spawn. Failure to do so could further erode the proportion of repeat spawners and the number of individuals of older age-classes, thereby placing the population at risk of recruitment failure in the event of a period of several years of successive poor recruitment.

In this context, it should be noted that, while the probability of survival of adult shad that spawn above Holyoke is significantly reduced because of the energetic demands involved in the migration, some survival does occur. For this reason, efforts to ensure the efficacy of downstream passage facilities at Holyoke and beyond must be considered vital to the overall restoration effort.

The experience of the American shad population in the Connecticut River indicates that the enhancement/restoration efforts undertaken within the river had important short-term, and potentially serious long-term effects on the structure and stability of the shad population. There is also good evidence that the highly successful restoration efforts applied to striped bass on the Atlantic Coast of the United States have had important and compounding negative effects on size and structure of the Connecticut River shad population. This is further evidence of the potential difficulties inherent in single species management approaches (Eiriksson 1994; Rothschild et al. 1997; Fujita et al. 1998; Wacker 1999).

Acknowledgements

The authors wish to thank Drs. James E. Carscadden, Marcel Reichert, Victor Crecco, Francis Juanes, and Robert Stira for providing comprehensive reviews; Dr. Caleb Slater and Donald Pugh for proving Holyoke lift data used; and Tim Sisk and Thomas Herra for assistance with graphics.

References

Bentzen P., J. Brown, and W. C. Leggett. 1989. Mitochondrial DNA polymorphism, population structure, and life history variation in American shad (*Alosa sapidissima*). Canadian Journal of Fisheries and Aquatic Science 46:1446–1454.

Bigelow, H. B., and W. C. Schroeder. 1953. Fishes of the Gulf of Maine. U.S. Fish and Wildlife Service. Fisheries Bulletin 53.

Brett, J. R., and N. R. Glass. 1973. Metabolic rates and critical swimming speeds in sockeye salmon, *Oncorynchus nerka*, in relation to size and temperature. Journal of the Fisheries Research Board of Canada 30:379–387.

Carscadden, J. E., and W. C. Leggett. 1975a. Life history variations in populations of American shad, (*Alosa sapidissima*) (Wilson), spawning in tributaries in the St. John River, New Brunswick. Journal of Fish Biology 7:595–609.

Carscadden, J. E., and W. C. Leggett 1975b. Meristic differences in spawning populations of American shad, (*Alosa sapidissima*): evidence for homing to tributaries in the St. John River, New Brunswick. Journal of the Fisheries Research Board of Canada 32:653–660.

Cole, L. C. 1954. The population consequences of life history phenomena. Quarterly Review Biology 29:103–137

Crecco, V. A., and T. F. Savoy. 1983. Daily mortality rates of larval and juvenile American shad (*Alosa sapidissima*) in the Connecticut River and changes in year-class strength. Canadian Journal of Fisheries and Aquatic Science 40:1719–1728.

Crecco, V. A., and T. F. Savoy. 1985. Effects of biotic and abiotic factors on growth and relative survival of young American shad (*Alosa sapidissima*) in the Connecticut River. Canadian Journal of Fisheries and Aquatic Science 42:1640–1648.

Crecco, V. A., and T. F. Savoy 1987. Review of recruitment mechanisms of the American shad: the critical period and match-mismatch hypotheses revisited. Pages 455–468 *in* M. J. Dadswell, R. J. Klauda, C. M. Moffitt, R. L. Saunders, R. A. Rulifson, and J. E. Cooper, editors. Common strategies of anadromous and catadromous fishes. American Fisheries Society, Symposium 1, Bethesda, Maryland.

Cushing, D. H. 1982. Fisheries biology: a study in population dynamics. University of Wisconsin Press, Madison.

Eiriksson, G. 1994. Future challenges for international management of ocean resources sustainable-yield of fish stocks-future international management. Marine Pollution Bulletin 29(6–12):552–555.

Essig, R. J., and R. E. Beal. 2000. Atlantic Coast striped bass recovery: federal aid in sport fish restoration in support of interstate fishery management. Fisheries 25(7):17–18.

Fry, F. E. J. 1971. The effect of environmental factors on the physiology of fish. Pages 1– 98 *in* W. S. Hoar and D. J. Randall, editors. Fish physiology. Academic Press, New York.

Fujita, R. M., T. Foran, and I. Zevos. 1998. Innovative approaches for fostering conservation in marine fisheries. Ecological Applications 8(1)(Supplement S):S130–S150.

Gardinier, M. N., and T. B. Hoff. 1982. Diet of striped bass in the Hudson River Estuary. New York Fish Game Journal 29(2):152–165.

Gephard, S., and J. McMenemy. 2004. An overview of the program to restore Atlantic salmon and other diadromous fishes to the Connecticut River with notes on the current status of these species in the river. Pages 287–317 *in* P. M. Jacobson, D. A. Dixon, W. C. Leggett, B. C. Marcy, Jr., and R. R. Massengill, editors. The Connecticut River Ecological Study (1965–1973) revisited: ecology of the lower Connecticut River 1973–2003. American Fisheries Society, Monograph 9, Bethesda, Maryland.

Glebe, B., and W. C. Leggett. 1981a. Temporal, intrapopulation differences in energy allocation and use by American shad (*Alosa sapidissima*) during the spawning migration. Canadian Journal of Fisheries and Aquatic Science 8:795–805.

Glebe, B., and W. C. Leggett. 1981b. Latitudinal differences in energy allocation and use during the freshwater migrations of American shad (*Alosa sapidissima*) and their life history consequences. Canadian Journal of Fisheries and Aquatic Science 38:806–820.

Henry, S. M. 1976. Development of fish passage facilities for American shad at the Holyoke Dam on the Connecticut River. Pages 289–304 *in* Proceeding of a Workshop on American shad. University of Massachusetts, Amherst.

Holgate, P. 1967. Population survival and life history phenomena. Journal of Theoretical Biology 14:1–10.

Jonsson, N., B. Jonsson, and L. P. Hansen. 1997. Changes in proximate composition and estimates of energetic costs during upstream migration and spawning in Atlantic salmon (*Salmo salar*). Journal of Animal Ecology 66(3):425–436.

Leggett, W. C. 1972. Weight loss in American shad during the freshwater migration. Transactions of the American Fisheries Society 101(3):549–552.

Leggett, W. C. 1985. The role of migrations in the life history evolution of fish. Pages 277–295 *in* M. A. Rankin, editor. Migration: mechanisms and adaptive significance. Contributions to Marine Science, Supplement 7.

Leggett, W. C. 2004. The American shad, with specific reference to its migration and population dynamics in the Connecticut River. Pages 181–238 *in* P. M. Jacobson, D. A. Dixon, W. C. Leggett, B. C. Marcy, Jr., and R. R. Massengill, editors. The Connecticut River Ecological Study (1965–1973) revisited: ecology of the lower Connecticut River 1973–2003. American Fisheries Society, Monograph 9, Bethesda, Maryland. (Originally published in 1976)

Leggett, W. C., and J. E. Carscadden.1978. Latitudinal variation in reproductive characteristics of American shad (*Alosa sapidissima*): Evidence for population specific life history strategies in fish. Journal of the Fisheries Research Board of Canada 35:1469–1478.

Leonard, J. B. K., and S. D. McCormick. 1999. Effects of migration distance on whole-body and tissue-specific energy use in American shad (*Alosa sapidissima*). Canadian Journal Fisheries and Aquatic Sciences 56:1159–1171.

Lorda, E., and V. A. Crecco. 1987. Stock-recruitment relationship and compensatory mortality of American shad in the Connecticut River. Pages 469–482 *in* M. J. Dadswell, R. J. Klauda, C. M. Moffitt, R. L. Saunders, R. A. Rulifson, and J. E. Cooper, editors. Common strategies of anadromous and catadromous fishes. American Fisheries Society, Symposium 1, Bethesda, Maryland.

Manooch, C. S., III. 1973 Food habits of yearling and

adult striped bass, *Morone saxatilis* (Walbaum), from Albemarle Sound, North Carolina. Chesapeake Science 14(2):73–86.

Marcy, B. C. 2004. Early life history studies of American shad in the lower Connecticut River and the effects of the Connecticut Yankee plant. Pages 155–180 *in* P. M. Jacobson, D. A. Dixon, W. C. Leggett, B. C. Marcy, Jr., and R. R. Massengill, editors. The Connecticut River Ecological Study (1965–1973) revisited: ecology of the lower Connecticut River 1973–2003. American Fisheries Society, Monograph 9, Bethesda, Maryland. (Originally published in 1976)

Massengill, R. R. 2004. Benthic fauna: 1965–1967 versus 1968–1972. Pages 43–58 *in* P. M. Jacobson, D. A. Dixon, W. C. Leggett, B. C. Marcy, Jr., and R. R. Massengill, editors. The Connecticut River Ecological Study (1965–1973) revisited: ecology of the lower Connecticut River 1973–2003. American Fisheries Society, Monograph 9, Bethesda, Maryland. (Originally published in 1976)

Massengill, R. R., J. M. Vozarik, D, E. Morgan, and M. Keser. 2004. Comparison of benthic community structure in the lower Connecticut River before, during, and after operation of the Connecticut Yankee power plant. Pages 441–454 *in* P. M. Jacobson, D. A. Dixon, W. C. Leggett, B. C. Marcy, Jr., and R. R. Massengill, editors. The Connecticut River Ecological Study (1965–1973) revisited: ecology of the lower Connecticut River 1973–2003. American Fisheries Society, Monograph 9, Bethesda, Maryland.

Merriman, D., and L. M. Thorpe. 1976. The Connecticut River Ecological Study: the impact of a nuclear power plant. American Fisheries Society, Monograph 1, Bethesda, Maryland.

Moffit, C. M., B. Kynard, and S. G. Rideout. 1982. Fish passage facilities and anadromous fish restoration in the Connecticut River basin. Fisheries 7(6):2–11.

Rothschild, B. J., B. J. Sharov, F. Alexi, and M. Lambert. 1997. Single-species and multispecies management. Pages 141–152 *in* J. Boreman, B. S. Nakashima, J. A. Wilson, and R. L. Kendall, editors. Northwest Atlantic groundfish: perspectives on a fishery collapse. American Fisheries Society, Bethesda, Maryland.

Savoy, T. F., and V. A. Crecco. 2004. Factors affecting the recent decline of blueback herring and American shad in the Connecticut River. Pages 361–377 *in* P. M. Jacobson, D. A. Dixon, W. C. Leggett, B. C. Marcy, Jr., and R. R. Massengill, editors. The Connecticut River Ecological Study (1965–1973) revisited: ecology of the lower Connecticut River 1973–2003. American Fisheries Society, Monograph 9, Bethesda, Maryland.

Schaffer, W. M., and P. F. Elson. 1975. The adaptive significance of variations in life history among local populations of Atlantic salmon in North America. Ecology 56:577–590.

Scott W. B., and M. G. Scott. 1988. Atlantic Fishes of Canada. Canadian Bulletin of Fisheries and Aquatic Sciences No. 219.

Shoubridge, E. A. 1978. Genetic and reproductive variation in American shad. Master's thesis. McGill University, Québec City, Québec.

Shoubridge, E. A., and W. C. Leggett. 1978. Occurrence and adaptive significance of distinct reproductive strategies in local populations of American shad. Pages 25–73 *in* Genetic and reproductive variation in American shad. Final report for Project AFC-10 (Connecticut). Connecticut Department of Environmental Protection, Hartford.

Trump, C., and W. C. Leggett. 1980. Optimum swimming speeds in fish: the problem of currents. Journal of the Fisheries Research Board Canada 37:1086–1092.

Versar, Inc. 1990. Source document for the supplement to the striped bass FMP – Amendment #4, Fisheries Management Report No. 16 of the Atlantic States Marine Fisheries Commission, Washington, D.C. March 1990.

Wacker, H. 1999. Optimal harvesting of mutualistic ecological systems. Resource and Energy Economics 21(1):89–102.

Walburg, C. H., and P. R. Nichols. 1967. Biology and management of the American shad and status of the fisheries. Atlantic Coast of the United States 1960. United States Fish and Wildlife Service, Special Scientific Report for Fisheries, No. 550, Washington, D.C.

Webb, P. W. 1975. Hydrodynamics and energetics of fish propulsion. Fisheries Research Board of Canada, Bulletin 190.

Winberg, G. G. 1956. Rate of metabolism and food requirements of fishes. Nauchn. Tr. Belorunk. Gos. Univ. Im. Vopr. Issled. Lenina, Minsk. (Translated from Russian by Fish. Mar. Serv. Transl. Serv. No 194, 1960).

American Fisheries Society Monograph 9:407–417, 2004

American Shad Early Life History and Recruitment in the Connecticut River: A 40-Year Summary

THOMAS F. SAVOY* AND VICTOR A. CRECCO
Connecticut Department of Environmental Protection, Marine Fisheries Division Post Office Box 719, Old Lyme, Connecticut 06371-0719, USA

BARTON C. MARCY, JR.
Westinghouse Savannah River Company, Environmental Services Building 742-a, Room 125, Post Office Box 616, Aiken, South Carolina 29808, USA

Abstract.—The original studies from 1966 to 1973 (this volume) and from 1974 through 1987 by Connecticut Department of Environmental Protection (CTDEP) provide detailed descriptions of spawning and early life history stages and influences on the recruitment process for the American shad *Alosa sapidissima* in the Connecticut River. Year-class strength of Connecticut River American shad appears to be established by the end of the larval period by a complex interrelationship between biotic and abiotic factors mediated by annual changes in egg production. High June river flows reduce June river temperatures, the growth rates of shad eggs, and early larvae. June river flow, water temperature, and zooplankton availability have been statistically linked to changes in shad larval mortality and year-class variability from 1966 through 1978. Dominant year-classes of shad are most likely to occur when lower than normal June flows are coupled with moderate spawning stocks. The significance of determination of year-class strength during the early larval period is that all subsequent losses during the larval and juvenile stages are depensatory to subsequent adult recruits returning to natal rivers to spawn.

Introduction

This paper reviews the early life history and the influences on the recruitment process of American shad *Alosa sapidissima* along with associated changes that have taken place over the past 40 years in the Connecticut River system. The Connecticut River Ecological Study began in 1965 to assess the potential impact of the Connecticut Yankee Atomic power plant at Haddam, Connecticut, which began commercial operation in 1968 (Jacobson et al. 2004; Merriman and Thorpe 2004 [1976], both this volume). The Connecticut Department of Environmental Protection (CTDEP) continued research efforts on American shad in the Connecticut River from 1979 to 1986, utilizing most of the same sampling stations as were used in the original study. Since 1987, efforts were reduced in scope but continued to monitor annual stock size, juvenile production, and recruitment success.

Many changes have taken place in the Connecticut River basin since the original studies in the late 1960s and early 1970s. These include the operation of the Connecticut Yankee atomic plant (1968–1996, Jacobson et al. 2004) with intake mitigation measures implemented. Passage of the Federal Water Pollution Control in 1972 (now the Clean Water Act) helped make significant improvements in water quality in the river. Additionally, changes made for upstream fish passage at the Holyoke Dam in 1975 resulted in numbers of American shad lifted upstream increasing 10-fold from 36,000 (average from 1965 to 1974) to 327,000 (average from 1975 to 2000). Another change included deterioration of the Enfield Dam at Enfield, Connecticut. Concentrations of shad below the Enfield Dam created one of the premier recreational fisheries for American shad on the East Coast. Breaches in this dam took place as

* *Corresponding author*: Tom.Savoy@po.state.ct.us

early as 1977 (CTDEP, unpublished data), the same year that boat traffic through the canal ceased. No repairs have been made since the mid-1970s, and at least five major breaches are known to have occurred since the late 1980s. Deterioration of the Enfield Dam reduced the pooling of shad below the dam, which supported a significant recreational fishery in the early 1970s (Leggett 2004 [1976], this volume), but breaches have had a greater effect on the fishery than on the distribution of spawning adults.

Accurate and long-term catch and effort statistics have been collected on the commercial shad fishery since 1940, on which several stock assessment studies were based (Fredin 1954; Jones et al. 1976; Crecco 1978; Minta 1980; Leggett 2004 [1976]; Marcy 2004a [1976], this volume). These studies reached the consensus that the annual spawning population size was highly variable, that no serious stock declines were evident since the late 1940s through the early 1990s, and that year-class strength is established before the juvenile stage. These studies and those conducted elsewhere (Chittenden 1976; Klauda et al. 1976) concluded that a greater research commitment was needed on the early life history stages, especially on the root causes of larval mortality, before fishery managers could separate natural from man-made induced fluctuations in recruitment.

Pioneering research efforts in the original Connecticut River Ecological Study reported year-class strength of juvenile shad in the Connecticut River to be significantly correlated with June river flows and water temperatures from 1966 through 1973 ($r = 0.93$; $P < 0.05$; Marcy 2004a [1976]). Marcy (2004a [1976]) also stated that during years of lower discharge flow rates, higher water temperatures, and larger adult shad spawning populations, the production of juvenile shad was higher. High flows and low temperatures during the spawning season in late May and June prolong the development of the eggs and reduce survival. The delay in hatching and high flow rates could exacerbate mortality from enhanced exposure to predation, direct physical causes, or transport to unfavorable areas or salinities. Crecco et al. (1983) utilized information from 1979 to 1982 and reported that the negative exponents for June river flow, the most significant hydrographic variable, support the hypothesis that high and turbulent June flow rates promote unfavorable feeding conditions for first-feeding larvae reducing larval survival and, ultimately, year-class strength.

In this paper, we review and summarize information on the early life history stages (spawning, eggs, larvae, and juveniles) of American shad collected in the Connecticut River from the original 1965–1973 studies (Marcy 2004a [1976]) and subsequent CTDEP studies conducted from 1978 to 2002 from Essex, Connecticut to Turners Falls, Massachusetts, including hydrologic impacts, feeding ecology, changes in the fish assemblage and distribution throughout the river basin, and the subsequent effects of year-class strength on adult recruitment and the annual spawning run. Year-class strength is established well before the end of the larval period, and larval survival appears to mainly controlled by abiotic conditions of river flow and temperature. This paper is meant to serve as a companion paper to the population dynamics of adult shad paper (Leggett et al. 2004, this volume).

Spawning and Larval Studies

In the original 1965–1973 studies, Marcy (2004a [1976]) located the areas where shad spawned in the lower river below Enfield Dam. Collections of shad eggs were generally made twice weekly from April to July in 1967 and 1968 at 61 locations from Enfield Dam (river kilometer [rkm] 85.3) to Essex (rkm 11.3). Marcy (2004a [1976]) found that shad eggs hatch in 71–86 h and travel only 1.6–6.4 km from the point where they were spawned. Shad spawned throughout this area, but primarily above Rocky Hill (rkm 51.5). Major spawning areas (from night observations by boat and egg collections) were located at Windsor Locks (rkm 78.8), Wilson (rkm 74), and Rocky Hill; 11 times more eggs were collected in these areas than in areas downstream. Shad larvae were collected in all months from May through August, but more than 80% were collected in June. Overall, shad eggs ranked eighth and larvae twelfth in abundance in the 1965–1969 ichthyoplankton surveys (690 biweekly plankton tows at 16 stations from Essex to Enfield Dam; Marcy 2004b [1976], this volume). Few comprehensive egg and larval studies have been conducted in the Connecticut River since then, and none can match the earlier efforts in length or breadth of investigation.

Since the original studies, changes in the

Holyoke lift (Henry 1976) and passage of high numbers of adult shad over the Holyoke Dam prompted investigation of American shad population dynamics in the upper river. Layzer (1974), Stira (1976), Kuzmeskus (1977), Cave (1978), and others from the University of Massachusetts located shad spawning sites and sampled eggs between Holyoke Dam and Turner's Falls, Massachusetts. Because of the change in distribution of adult prespawned fish, CTDEP conducted larval sampling from 1978 to 1982 to assist in the evaluation of the recruitment process. Sampling efforts were shifted north of the Holyoke Dam to enhance the number of shad collected given the high number of blueback herring *A. aestivalis* larvae in larval fish collections made below the Holyoke Dam in a 1978 pilot study (CTDEP, unpublished data). Towed bongo nets and an epibenthic sled were also utilized to collect eggs from 1978 to 1980, but no results of these efforts are currently available. Larval collections were made with a 6.1-m plankton bag seine with a funnel to which a plankton bucket (0.25 mm mesh) was attached to help remove larvae from the net. Some shad eggs were collected (fish eggs generally comprised less than 10% of the catch), but collections were focused on larvae.

Aspects of Spawning

Water temperature is considered to be the major hydrographic variable controlling oocyte maturation and the timing of American shad spawning (Leggett 1969; Watson 1970). Spawning occurs in the Connecticut River mainly from mid-May through mid-June at water temperatures between 14°C and 23°C (Crecco and Savoy 1987b; Marcy 2004a [1976]), mainly at night or on overcast days (Whitney 1961; Layzer 1974; Chittenden 1976). Group spawning involving several males and one female shad has been observed during late evening in the Connecticut River (Whitworth and Bennett 1970; Marcy 1972). Those studies reported that spawning fish swim vigorously at the surface forming a closely packed circle, whereby eggs are released into the water and fertilized by several males (Marcy 1972). Female shad release semibuoyant eggs in riffles, shoals near river channels and downstream from the confluence of two rivers (Layzer 1974; Gilmore 1975), but seldom release eggs over silt, mud, or bedrock (Scherer 1974). Fertilized shad eggs roll along the bottom for 1.6–6.4 km (Marcy 1972; Stira 1976), which presumably facilitates oxygen uptake and successful development (Blair 1976).

Female American shad are prolific spawners (150,000–500,000 ova/female), with the absolute fecundity increasing with age, length, and weight (Leggett 1969). The relatively high fecundity of American shad is crucial for perpetuating the stock, since many eggs fail to fertilize (Reed and Russo 1976), and only about 1 out of every 100,000 eggs survives to become a spawning adult (Leggett 1977b). Hatchery studies (Mansueti and Kolb 1953; Meade 1976) have shown that shad eggs ripen gradually in the ovary, allowing spawning to occur over several days. This batch spawning strategy of American shad and many other fishes may be an evolutionary adaptation to enhance the probability that at least some eggs will be released under river flow conditions shown to be favorable to egg and larval survival (Crecco and Savoy 1985a). Maternal age and previous spawning history have been shown to be important in egg and larval survival in other fish species (Knutsen and Tilseth 1985; Rosenberg and Doyle 1986), but has not yet been investigated for American shad. The platoon type recruitment of American shad with several age-classes making up the annual spawning run may serve to mitigate the effects of younger virgin spawners.

After spawning, the surviving adult shad leave the Connecticut River by mid-August (Crecco et al. 1984; O'Leary and Booke 1986) and migrate with subadult shad to their summer feeding grounds in the Gulf of Maine (Walburg and Nichols 1967; Dadswell et al. 1983) or within an area south of Nantucket shoals (Neves and Dupres 1979). When ocean temperatures in the Gulf of Maine drop below 12°C in late November, American shad move south to Florida by following the 12–13°C isotherm (Neves and Dupres 1979). Beginning in February, American shad of various stocks commence their spawning migration up the Atlantic Coast and return to their respective river of origin to spawn (Leggett and Whitney 1972).

Egg Stage

Newly spawned shad eggs are transparent and about 1.8 mm in diameter, but expand to 2.2–3.5 mm after fertilization due to the absorption of water (Mansueti and Hardy 1967; Chittenden

1969). The rate at which shad eggs develop is inversely related to water temperature during May and June (Marcy 2004a [1976]). Shad eggs hatch in 8–12 d at 13–15°C, 6–8 d at 17°C, and 3 d at 24°C (Watson 1970). This inverse relationship between egg incubation period and water temperature is considered to be an evolutionary adaptation, allowing larvae to hatch in synchrony with rising water temperatures and declining river flows shown to be favorable to larval growth and survival (Crecco and Savoy 1985b, 1987a; Marcy 2004a [1976]).

American shad eggs in the Connecticut River experience catastrophic mortality, ranging annually from 24% to 44%/d between 1979 and 1987 (Savoy and Crecco 1988). As a result, only between 5% and 19% of the fertilized eggs survive to hatching. High mortality of fertilized eggs has been attributed to suffocation, fungal infection, predation, and physical damage due to high river flows (Leach 1925; Mansueti and Kolb 1953; Walburg and Nichols 1967). High river flows in the Connecticut River in May and June are inversely related to water temperature, thereby prolonging the duration of the egg stage and indirectly enhancing egg mortality rates.

Larval Stage

Newly hatched shad larvae (7–10 mm total length [TL]) occur during May and June and remain near the river bottom until the yolk sac is nearly absorbed about 3–4 d later (Maxfield 1953; Cave 1978). Following yolk absorption, shad larvae (10–12 mm TL) are transported by river currents into eddies and backwater areas where current velocities are greatly reduced (Cave 1978; Crecco et al. 1983). Within these areas, first-feeding larvae must consume sufficient numbers of crustacean zooplankton and insects (Crecco and Blake 1983), or death from malnutrition usually ensues within 5 d (Wiggins et al. 1984). Moreover, since first-feeding herring larvae deprived of food grow and swim more slowly than well-fed larvae (Hunter 1976), malnourished larvae are more susceptible to predator-induced mortality. There is also direct evidence that hydrographic events during and shortly after spawning modulate the rates of larval predation and malnutrition (Sinclair and Tremblay 1984). Intra-annual fluctuations in rainfall and river flow regulate the May–June temperature gradient, thereby affecting the spatial and temporal match or mismatch between larval production, their food supply, and their predators (Crecco and Savoy 1987a, 1987b).

American shad larval mortality rates were highest (17–26%/d) among first-feeding larvae, and then declining steadily throughout larval development (Crecco et al. 1983; Savoy and Crecco 1988) because older shad larvae (>20 d old) grow rapidly and are better able to locate food and avoid predators. Based on the 1979–1984 survivorship data, 60–80% of newly hatched larvae die within 3–7 d after feeding begins. The larval stage for American shad lasts between 4 and 6 weeks, during which the larvae grow fairly rapidly (0.4 mm/d) to about 22–26 mm TL (Savoy and Crecco 1988). Surviving larvae metamorphose into facultative filter-feeding juveniles by August and resemble the adults in shape and coloration.

Crecco and Savoy (1984) found that the relative abundance of shad larvae and their feeding incidence varied temporally with changes in zooplankton density, precipitation, river flows, and water temperatures from 1979 to 1982. This analysis was designed to determine differences in zooplankton densities and environmental conditions between strong and weak year-classes. They examined the gut content of 1,260 shad larvae captured between 1979 and 1982. Their data suggest that high June river flows and low river temperatures reduce larval feeding success, survival, and ultimately shad year-class strength, providing a reasonable explanation for the lack of a parent–progeny relationship.

Compensatory mortality among first-feeding shad larvae could result from malnutrition (Lasker 1975) only if grazing rates of larvae were capable of depleting local zooplankton abundance. However, spatial changes in zooplankton abundance appear to be more related to hydrographic events than to larval grazing rates (Crecco and Savoy 1984), so malnutrition among first-feeding larvae would probably be a density-independent factor. Given the numerous ways in which hydrographic events can mediate the timing and magnitude of predation and starvation effects, they concluded that it is unlikely that a single mechanism is responsible for the relatively high compensatory mortality of American shad.

Crecco and Savoy (1984) found that cohort-specific growth rates among first-feeding shad larvae (4–14 d) were strongly related to oscillations in river flow, temperature, and zooplankton densities, whereas the same relationships among older larvae (21–28 d) and juveniles were weak. They reported that one explanation of the poor relationship with older larvae was that older larvae are present from July through October when river flows and water temperatures show minor fluctuations relative to May and June. Significant rainfall/increased river flows after the majority of spawning event would result in elevated mortality rates through direct or indirect causes.

Crecco and Blake (1983) looked at the feeding ecology of coexisting larvae of American shad and blueback herring between Enfield Dam in Connecticut and Sunderland, Massachusetts. The results of this study indicate that American shad and blueback herring larvae coexist within the same habitat by utilizing different prey taxa or by selecting different sizes of the same prey taxon. First-feeding (10–12 mm) American shad larvae ate mostly cyclopoid copepodites and chironomid larvae, whereas first-feeding (5–12 mm) blueback herring consumed mostly rotifers. The ability of blueback herring to utilize the abundant rotifer community may have been instrumental in their numerical dominance over American shad in the Connecticut River. Alewife *Alosa pseudoharengus*/blueback herring comprised 87% of the ichthyoplankton and 41% of the juvenile catch compared to 0.04% and 2% for shad, respectively, in the original studies (1965–1972; Marcy 2004b [1976]). Numerical dominance of river herring has also been shown in other East Coast rivers (Loesch and Kriete 1980). However, river herring dominance in the juvenile catch has declined markedly in recent years with the decline in abundance of adult blueback herring (Savoy and Crecco 2004, this volume).

Juvenile Studies

Annual juvenile American shad collections have been made almost continuously from 1966 through 2002. Marcy (2004a [1976]) collected juvenile shad (40–90 mm TL) weekly at 12 stations from 1966 to 1973 from above Holyoke Dam, Massachusetts to Essex, Connecticut. The CTDEP reduced sampling effort somewhat, sampling five to seven of the same stations between Holyoke, Massachusetts (rkm 107) and Essex, Connecticut (rkm 5), from mid-July through mid-October 1978–2002.

Juvenile Stage

Juvenile American shad (40–150 mm TL) are distributed throughout the river from July through November, grazing extensively on aquatic and terrestrial insects, as well as crustacean zooplankton (Leim 1924; Mitchell et al. 1925; Maxfield 1953; Levesque and Reed 1972; Domermuth 1976; Crecco and Blake 1983; Marcy 2004a [1976]). The wide diversity of prey found in the stomachs of juvenile shad suggests that they are opportunistic feeders whose feeding constraints depend mainly on prey size (Maxfield 1953).

Juvenile shad form dense schools and gradually move downriver during late fall (October–November) when river temperatures drop below 16°C (Watson 1968; O'Leary and Kynard 1981; Marcy 2004a [1976]). Their movements throughout the nursery area are influenced by river currents and by their growth rates (Watson 1970). Juvenile shad grow rapidly (0.7–1.0 mm/d) in both fresh and saltwater up to about 150 mm in their first summer (Crecco and Savoy 1985b). Juvenile growth rates from 1966 through 1986 were inversely related to their abundance (Rosen 1982; Crecco et al. 1984), suggesting that juvenile growth is limited by inter- and intra-specific competition for food and space. The fastest growing juvenile shad emigrate to the lower river first (Crecco et al. 1984; Marcy 2004a [1976]). Marcy (2004a [1976]) reported that juvenile shad emigrated from the lower Connecticut River (rkm 1–20) during November and December. O'Leary and Kynard (1981) stated that timing of emigration of juvenile shad from the Holyoke impoundment was dependent upon declining water temperatures and generally took place from late September through November. In recent years, CTDEP sampling (1990–2002, CTDEP, unpublished data) suggests that emigration from the Holyoke Pool takes place as early as August in some years.

Juvenile mortality rates from 1979 through 1984 (Savoy and Crecco 1988) were much lower (1.5–2.5%/d) than mortality rates for eggs and larvae. Although the mortality rates among subadult shad in the ocean has not been directly estimated, Savoy and Crecco (1988) theorized that mortality rates continue to decline with age and, using techniques of Pauly (1980), that oceanic mortality ranges between 30% and 40%/year.

Since juvenile shad can tolerate food deprivation for long periods (Maxfield 1953), the main causes of juvenile mortality are believed to be predation and anthropogenic effects. Taylor and Kynard (1985) estimated that mortality at hydroelectric dams was as high as 90% from turbine-induced mortality, yet other efforts yield estimates of only 1–10% turbine-induced mortality. The lower estimates are more likely given that the Connecticut River stock has persisted despite passing 40–60% of the prespawning adults over the Holyoke Dam since 1976.

Determination of Year-Class Strength and Recruitment Impacts

Marcy (2004a [1976]) noted that low temperatures in June 1968 might have caused poor survival of eggs and larvae. He collected 65% of the eggs in the lower river below Enfield Dam at temperatures below 16°C in 1968. Bradford et al. (1964), Leim (1924), Leach (1925), and Mansueti and Kolb (1953) reported that temperatures below 16°C prolonged the period of egg development and reduced survival. Watson (1970) reported that 20% of the eggs collected above Holyoke Dam in 1968 were dead, which he attributed to widely fluctuating water temperatures. He also noted that many eggs were probably lost from the area due to high river flow.

Crecco and Savoy (1984) concluded that that high larval mortality and year-class failure is linked to high rainfall and high river flows during June. These findings are in accord with the earlier 1966–1973 study (Marcy 2004a [1976]) in which June river flows and water temperatures were significantly correlated with shad year-class strength from 1966 through 1973. Leggett (1977a) also demonstrated that June temperatures and June river flows were related to deviations from the stock–recruitment relationship of American shad.

Lorda and Crecco (1987) chose the June river flow as the major environmental variable because it affects other environmental factors associated with survival of American shad eggs and larvae. High June river flows may transport eggs and larvae to areas of low food density and high predator abundance (Crecco and Savoy 1987b). High river flows in June caused by storm events reduce water temperatures (Crecco and Savoy 1984), thereby delaying the development of eggs and larvae. In years of high May rainfall and river flow, May water temperatures are depressed, postponing the onset of shad spawning (Leggett and Whitney 1972) and larval hatching (Cave 1978) until mid-June. Low flows produce optimal conditions for the high feeding success and survival of American shad larvae (Crecco and Savoy 1984).

Crecco and Savoy (1987a) developed survivorship curves for shad larvae and juveniles from 1979 to 1984. Curves were constructed from seine-caught larvae (42,000–180,000) and juveniles (800–2,300) taken each year and separated into 1-and 3-mm length-groups. They used otolith increments to age between 100 and 400 shad larvae and 80 and 200 juveniles annually from 1979 through 1984. Savoy and Crecco (1987) used the relative distance between the otolith increments to determine age-specific growth rates (mm/d) of both larvae and juveniles. The survivorship curves for shad larvae showed that 65–80% of the larvae died during and shortly after larvae began feeding and that year-class strength is established at that time. Moreover, Savoy and Crecco (1988) found that the relative abundance of 18- to 30-d-old larvae from 1979 through 1984 was closely correlated (r = 0.88; P < 0.05) with the juvenile indices for those years, suggesting that compensatory mortality is mainly confined to the egg and early larval stages.

In addition, juvenile indices of year-class strength from 1966 to 1973 were positively correlated (r = 0.92, P < 0.001) with recruitment levels of adult females 4–6 years later, suggesting that year-class strength of shad is established prior to the juvenile stage. The juvenile abundance indices from 1967 to 1979 were positively correlated (r = 0.90; P < 0.001) with the total number of virgin American shad from these year-classes in the spawning populations (Crecco and Savoy 1985a). Year-class strength of Connecticut River shad appears to be established by the end of the larval period (Crecco et al. 1983).

Other Connecticut River studies since 1973 have demonstrated that larval and juvenile mortality rates decline with age (Crecco et al. 1983, 1986) and that growth and survival rates among discrete larval cohorts are significantly affected by short-term hydrographic and meteorological events (Crecco and Savoy 1985b, 1987b). Declining river flow rates and increase in water temperature coinciding with peak larval abundance of shad lead to high survival and growth of larvae resulting in strong year-class production. By contrast, significant rainfall in May and June are associated with high river flows, reduced water temperatures, and transparency, causing advection of larvae from their preferred habitat and dissipate patches of river zooplankton, thus reducing larval shad survival and leading to year-class failure.

Summary

In summary, numerous research studies from 1966 through 2002 provide detailed descriptions of spawning and early life history stages and influences on the recruitment process for the American shad in the Connecticut River. They show that high June river flows reduce June river temperatures (Marcy 2004a [1976]; Crecco and Savoy 1984), the growth rates of shad eggs (Watson 1968), and early larvae (Crecco and Savoy 1987b). Because slower growing larvae may be susceptible to predation for a longer period of time (Hunter 1976), periods of high flow may indirectly enhance egg and larval predation. Additionally, since high flows reduce the spatial patchiness (Crecco and Savoy 1987a), abundance, and availability of river zooplankton (Whitton 1980; Threlkeld 1986), high flows may result in increased levels of competition among American shad larvae for available prey, especially if shad larvae are capable of depleting local aggregations of edible zooplankton.

Year-class strength of Connecticut River American shad appears to be established by the end of the larval period by a complex interrelationship between biotic and abiotic factors mediated by annual changes in egg production (Leggett 1977a; Crecco et al. 1981, 1983). June river flow, water temperature, and zooplankton availability have been statistically linked to changes in shad larval mortality and year-class variability from 1966 through 1978 (Crecco and Savoy 1984). Dominant year-classes of shad are most likely to occur when lower than normal June flows are coupled with moderate spawning stocks. The significance of determination of year-class strength during the early larval period is that all subsequent losses during the larval and juvenile stages are depensatory to subsequent adult recruits returning to natal rivers to spawn.

A unique data set of 40 years of studies has greatly furthered our understanding of the early life history and influences on the recruitment processes of American shad in the Connecticut River. Long-term monitoring documents the dynamic nature of fish recruitment within a large river system with changes over time. The assemblage and abundance of fish continues to change in the river with natural and anthropogenic effects, and periodic if not annual monitoring is required to ensure that this important American shad stock persists.

Acknowledgments

We would like to thank Martha Mather, Benjamin Letcher, and Robert Stira for reviewing the manuscript. Funding for V. Crecco and T. Savoy provided by the Connecticut Department of Environmental Protection. Funding support for B. Marcy was provided by the Electric Power Research Institute.

References

Blair, A. B. 1976. American shad culture and distribution studies at Harrison Lake National Fish Hatchery. Pages 201–210 *in* Proceedings Workshop on American shad. U.S. National Marine Fisheries Service, Silver Spring, Maryland.

Bradford, A. D., J. G. Miller, and K. Buss. 1964. Progress report summary on phase B-2 to determine by bio-assay techniques the inherent tolerance of the shad during its egg and larval stages to specific environmental factors of the Susquehanna River and its tributaries. Pennsylvania Fisheries Commission, Division of Research and Fish Management, Benner Springs Fish Research Station (Manuscript), Harrisburg.

Cave, J. R. 1978. American shad, *Alosa sapidissima*, larval distribution, relative abundance and movement in the Holyoke Pool, Connecticut River, Massachusetts. Master's thesis. University of Massachusetts, Amherst.

Chittenden, M. E. 1969. Life history and ecology of the American shad, *Alosa sapidissima*, in the Delaware River. Doctoral thesis. Rutgers University, New Brunswick, New Jersey.

Chittenden, M. E. 1976. Weight loss, mortality, feeding, and duration of residence of adult American shad, (*Alosa sapidissima*) in fresh water. Fisheries Bulletin 74(1):151–157.

Crecco, V. 1978. Population dynamics of the American shad (*Alosa sapidissima*) in the Connecticut River, 1940–1977. Connecticut Department of Environmental Protection, Marine Fisheries Division, Old Lyme.

Crecco, V. A., and M. M. Blake. 1983. Feeding ecology of coexisting larvae of American shad and blueback herring in the Connecticut River. Transactions of the American Fisheries Society 112:498–507.

Crecco, V. A., L. Gunn, and T. Savoy. 1981. The Connecticut River shad study, 1981. Progress Report. AFC 13. Connecticut Department of Environmental Protection, Marine Fisheries, Hartford.

Crecco, V. A., and T. Savoy. 1984. Effects of fluctuations in hydrographic conditions on year-class strength of American shad (*Alosa sapidissima*) in the Connecticut River. Canadian Journal Fisheries and Aquatic Sciences 41(8):1216–1223.

Crecco, V. A., and T. Savoy. 1985a. Density-dependent catchability and its potential causes and consequences on Connecticut River shad, *Alosa sapidissima*. Canadian Journal Fisheries and Aquatic Sciences 42(10):1649–1657.

Crecco, V. A., and T. Savoy. 1985b. Effects of biotic and abiotic factors on growth and relative survival of young American shad in the Connecticut River. Canadian Journal Fisheries and Aquatic Sciences 42(10):1640–1648.

Crecco, V. A., T. Savoy, and L. Gunn. 1983. Daily mortality rates of larval and juvenile American shad (*Alosa sapidissima*) in the Connecticut River with changes in year-class strength. Canadian Journal Fisheries and Aquatic Sciences 40:1719–1728.

Crecco, V. A., T. Savoy and L. Gunn. 1984. Population dynamics studies of American shad in the Connecticut River. Connecticut Department of Environmental Protection, Final Report AFC 13, Hartford.

Crecco, V., T. Savoy, and W. Whitworth. 1986. Effects of density-dependent and climatic factors on American shad, *Alosa sapidissima*, recruitment: a predictive approach. Canadian Journal Fisheries and Aquatic Sciences 43(2):457–463.

Crecco, V. A., and T. F. Savoy. 1987a. Recruitment mechanisms of the American shad, *Alosa sapidissima*: test of the critical period and match-mismatch hypothesis. Pages 455–468 *in* M. J. Dadswell, R. J. Klouda, C. M. Moffitt, R. L. Saunders, and R. A. Rulifson, editors. Common strategies of anadromous and catadromous fishes. American Fisheries Society, Symposium 1, Bethesda, Maryland.

Crecco, V. A., and T. F. Savoy. 1987b. Effects of climatic and density-dependent factors on intra-annual mortality of larval American shad. Pages 69–81 *in* R. D. Hoyt, editor. 10th Annual Larval Fish Conference. American Fisheries Society, Symposium 2, Bethesda, Maryland.

Dadswell, M. J., G. D. Melvin, and R. S. Gregory. 1983. Effect of turbidity on the temporal and spatial utilization of the inner Bay of Fundy by American shad (*Alosa sapidissima*) (Pisces: clupeidae) and its relationship to local fisheries. Canadian Journal Fisheries and Aquatic Sciences 40(1):322–330.

Domermuth, R. B. 1976. Summer foods of larval and juvenile American shad, *Alosa sapidissima*, juvenile blueback herring, *Alosa aestivalis*, and pumpkinseed, *Lepomis gibbosus*, in the Connecticut River, between Holyoke and Enfield dams. 1972. Master's thesis. University of Massachusetts, Amhurst.

Fredin, R. A. 1954. Causes of fluctuations in abundance of Connecticut River shad. U.S. Fish Wildlife Service. Fisheries Bulletin 54:247–259.

Gilmore, K. S. 1975. Spawning areas and egg distribution of American shad, *Alosa sapidissima*, (Pisces: Clupeidae) in the Connecticut River between Holyoke and Turners Falls, Massachusetts. Master's thesis. University of Massachusetts, Amherst.

Henry, S. M. 1976. Development of fish passage facilities for American shad at the Holyoke Dam on the Connecticut River. Pages 289–304 *in* Proceeding of a workshop on American shad. University of Massachusetts, Amherst.

Hunter, J. R., editor. 1976. Report on la Colloquium on larval fish mortality studies and their relation to fishery research, Jan. 1975. National Oceanic and Atmospheric Administration Technical Report, National Marine Fisheries Service Circular-395, Silver Spring, Maryland.

Jacobson, P. M., C. Tomichek, and D. Danila. 2004. Twenty years of impingement history: Connecticut Yankee Haddam Neck nuclear power plant. Pages 455–473 *in* P. M. Jacobson, D. A. Dixon, W. C. Leggett, B. C. Marcy, Jr., and R. R. Massengill, editors. The Connecticut River Ecological Study (1965–1973) revisited: ecology of the lower Connecticut River 1973–2003. American Fisheries Society, Monograph 9, Bethesda, Maryland.

Jones, R. A., P. Minta and V. A. Crecco. 1976. A review of American shad studies in the Connecticut River. Pages 135–164 *in* Proceedings of a workshop on American shad. University of Massachusetts, Amherst.

Klauda, R. J., M. Nittel and K. P. Campbell. 1976. Commercial fishery for American shad in the Hudson River: fish abundance and stock trends. Pages 107–134 *in* Proceedings of a workshop on American shad. U.S. National Marine Fisheries Service, Silver Spring, Maryland.

Knutsen, G. M., and S. Tilseth. 1985. Growth, development, and feeding success of Atlantic cod larvae

Gadus morhua related to egg size. Transactions of the American Fisheries Society 114:507–511.

Kuzmeskus, D. M. 1977. Egg production and spawning site distribution of the American shad, *Alosa sapidissima*, (Wilson), in the Holyoke pool, Connecticut River, Massachusetts. Master's thesis. University of Massachusetts, Amherst.

Lasker, R. 1975. Field criteria for survival of anchovy larvae: the relation between inshore chlorophyll maximum layers and successful first feeding. U.S. National Marine Fisheries Bulletin 73:453–462.

Layzer, J. B. 1974. Spawning sites and behavior of American shad, *Alosa sapidissima*, (Wilson), in the Connecticut River between Holyoke and Turners Falls, Massachusetts, 1972. Master's thesis. University of Massachusetts, Amherst.

Leach, G. C. 1925. Artificial propagation of shad. Pages 459–486 *in* U.S. Bureau. Fisheries Document 981, Appendix. VIII, Report, U.S. Commercial Fisheries, Washington, D.C.

Leim, A. H. 1924. The life history of the shad, Alosa sapidissima (Wilson), with special reference to the factors limiting abundance. Contributions to Canadian Biology Volume 2, Part 1, No 11:163–284.

Leggett, W. C. 1969. Studies on the reproductive biology of the American shad (*Alosa sapidissima*). A comparison of populations from four rivers of the Atlantic seaboard. Doctoral thesis. McGill University, Montreal.

Leggett, W. C. 1977a. Density-dependence, density-independence, and recruitment in the American shad (*Alosa sapidissima*) population of the Connecticut River. Pages 3–17 *in* W. Van Winkle, editor. Proceedings of the conference on assessing the effects of power-plant-induced mortality on fish populations. Pergamon Press, New York.

Leggett, W. C. 1977b. The ecology of fish migrations. Annual Review. Ecology and Systematics 8:285–308.

Leggett, W. C. 2004. The American shad, with special reference to its migration and population dynamics in the Connecticut River. Pages 181–238 *in* P. M. Jacobson, D. A. Dixon, W. C. Leggett, B. C. Marcy, Jr., and R. R. Massengill, editors. The Connecticut River Ecological Study (1965–1973) revisited: ecology of the lower Connecticut River 1973–2003. American Fisheries Society, Monograph 9, Bethesda, Maryland. (Originally published in 1976)

Leggett, W. C., T. F. Savoy, and C. A. Tomichek. 2004. The impact of enhancement initiatives on the structure and dynamics of the Connecticut River population of American shad. Pages 391–405 *in* P. M. Jacobson, D. A. Dixon, W. C. Leggett, B. C. Marcy, Jr., and R. R. Massengill, editors. The Connecticut River Ecological Study (1965–1973) revisited: ecology of the lower Connecticut River 1973–2003. American Fisheries Society, Monograph 9, Bethesda, Maryland.

Leggett, W. C., and R. R. Whitney. 1972. Water temperature and the migrations of American shad. U.S. Fish Wildlife Service Fisheries Bulletin 70(3):659–670.

Levesque, R. C., and R. J. Reed. 1972. Food availability and consumption by young Connecticut River shad, *Alosa sapidissima*. Journal of the Fisheries Research Board of Canada 29:1495–1499.

Lorda, E., and V. A. Crecco. 1987. Stock-recruitment relationship and compensatory mortality of American shad in the Connecticut River. Pages 469–482 *in* M. J. Dadswell, R. J. Klauda, C. M. Moffitt, R. L. Saunders, R. A. Rulifson, and J. E. Cooper, editors. Common strategies of anadromous and catadromous fishes. American Fisheries Society, Symposium 1, Bethesda, Maryland.

Loesch, J., and W. H. Kriete. 1980. Anadromous fisheries research program, Virginia. Virginia Institute of Marine Science, Annual Report AFC-10, Gloucester Point.

Marcy, B. C., Jr. 1972. Spawning of the American shad, *Alosa sapidissima* in the lower Connecticut River. Chesapeake Science 13(2):116–119.

Marcy, B. C., Jr. 2004a. Early life history studies of American shad in the lower Connecticut River and the effects of the Connecticut Yankee plant. Pages 155–180 *in* P. M. Jacobson, D. A. Dixon, W. C. Leggett, B. C. Marcy, Jr., and R. R. Massengill, editors. The Connecticut River Ecological Study (1965–1973) revisited: ecology of the lower Connecticut River 1973–2003. American Fisheries Society, Monograph 9, Bethesda, Maryland. (Originally published in 1976)

Marcy, B. C., Jr. 2004b. Planktonic fish eggs and larvae of the lower Connecticut River and the effects of the Connecticut Yankee plant, including entrainment. Pages 125–153 *in* P. M. Jacobson, D. A. Dixon, W. C. Leggett, B. C. Marcy, Jr., and R. R. Massengill, editors. The Connecticut River Ecological Study (1965–1973) revisited: ecology of the lower Connecticut River 1973–2003. American Fisheries Society, Monograph 9, Bethesda, Maryland. (Originally published in 1976)

Mansueti, R. J. and J. D. Hardy, Jr. 1967. Development of fishes of the Chesapeake Bay Region, Part I. An atlas of egg, larval and juvenile stages. In E. E. Deubler, Jr., editor. Natural Resources Institute, University of Maryland, College Park.

Mansueti, R. J. and H. Kolb. 1953. A historical review of the shad fisheries of North America. Chesapeake Biological Laboratory Publication 97, University of Maryland, Solomons.

Maxfield, G. H. 1953. The food habits of hatchery-produced pond-cultured shad *Alosa sapidissima*, reared to a length of two inches. Chesapeake Biological Laboratory Publication No. 98, University of Maryland, Solomons.

Meade, J. W. 1976. Experimental intensive rearing of shad at Van Dyke, Pennsylvania. Pages 211–234 *in* Proceedings of a workshop on American shad. University of Massachusetts, Amherst.

Merriman, D., and L. M. Thorpe. 2004. Introduction. Pages 3–25 *in* P. M. Jacobson, D. A. Dixon, W. C. Leggett, B. C. Marcy, Jr., and R. R. Massengill, editors. The Connecticut River Ecological Study (1965–1973) revisited: ecology of the lower Connecticut River 1973–2003. American Fisheries Society, Monograph 9, Bethesda, Maryland. (Originally published in 1976)

Minta, P. 1980. Connecticut River shad study, 1977–1979. Connecticut Department of Environmental Protection, Federal Aid Project, Final report for AFC-11, Hartford.

Mitchell, P. H., and staff. 1925. A report of investigations concerning shad in the rivers of Connecticut, part 1:7–44. Connecticut State Board of Fish and Game, Hartford.

Neves, R. J. and L. Dupres. 1979. The oceanic migration of American shad, *Alosa sapidissima*, along the Atlantic Coast. Fisheries Bulletin 77:199–212.

O'Leary, J., and H. E. Booke. 1986. Connecticut River anadromous fish investigations. Project Performance Report F-45-R-2. Massachusetts Cooperative Fisheries Research Unit, Amherst.

O'Leary, J., and B. Kynard. 1981. Holyoke Dam fishery studies 1980; Prototype by-pass system for downstream shad passage. Progress Report to Northeast Utilities Service Company, Berlin, Connecticut.

Pauly, D. 1980. On the interrelationship between natural mortality, growth, and mean environmental temperature in 175 fish stocks. Journal du Conseil International pour l'Exploration de la Mer 39(3):175–192.

Reed, R. J. and A. Russo. 1976. American shad research Connecticut River, Massachusetts, 1976. I. Fecundity, egg retention, sex ratio, and age class composition. Massachusetts Cooperative Fisheries Research Unit, University of Massachusetts, Amherst.

Rosen, R. A. 1982. Seasonal cycles, distribution, and biomass of young American shad (*Alosa sapidissima*) in the Holyoke Pool, Connecticut River. Doctoral thesis. University of Massachusetts, Amherst.

Rosenberg, A. A., R. W. Doyle. 1986. Analyzing the effect of age structure on stock-recruitment relationships in herring (*Clupea harengus*). Canadian Journal Fisheries and Aquatic Sciences 43:674–679.

Savoy, T. F., and V. A. Crecco. 1987. Daily increments on the otoliths of larval American shad and their potential use in population dynamics studies. Pages 413–431 *in* R. C. Summerfelt, editor. Age and growth of fish. Iowa State University Press, Ames.

Savoy, T. F., and V. A. Crecco 1988. The timing and significance of density-dependent and density-independent mortality of American shad, *Alosa sapidissima*. Fisheries Bulletin 86(3):467–481.

Savoy, T. F., and V. A. Crecco. 2004. Factors affecting the recent decline of blueback herring and American shad in the Connecticut River. Pages 361–377 *in* P. M. Jacobson, D. A. Dixon, W. C. Leggett, B. C. Marcy, Jr., and R. R. Massengill, editors. The Connecticut River Ecological Study (1965–1973) revisited: ecology of the lower Connecticut River 1973–2003. American Fisheries Society, Monograph 9, Bethesda, Maryland.

Scherer, M. D. 1974. Analysis of factors affecting passage of American shad (*Alosa sapidissima*; Wilson) at Holyoke Dam, Massachusetts, and assessment of juvenile growth and distribution above the dam. Doctoral thesis. University of Massachusetts, Amherst.

Sinclair, M., and M. J. Tremblay. 1984. Timing of spawning of Atlantic herring (*Clupea harengus harengus*) populations and the match-mismatch theory. Canadian Journal Fisheries and Aquatic Sciences 41:1055–1065.

Stira, R. J. 1976. Sinking rate and drift distance of eggs of American shad (*Alosa sapidissima*) in the Connecticut River, Massachusetts. Master's thesis. University of Massachusetts, Amherst.

Taylor, R., and B. Kynard. 1985. Mortality of juvenile American shad and blueback herring passed through a low-head Kaplan hydroelectric turbine. Transactions of the American Fisheries Society 114:430–435.

Threlkeld, S. T. 1986. Lifetable responses and population dynamics of four cladoceran zooplankton during a reservoir flood. Journal Plankton Research 8(4):639–647.

Walburg, C. H., and P. R. Nichols. 1967. Biology and management of the American shad and status of the fisheries. Atlantic Coast of the United States, 1960. U.S. Fish and Wildlife Service Special Scientific Report 550, Washington, D.C.

Watson, J. F. 1968. The early life history of the American shad, *Alosa sapidissima* (Wilson), in the Connecticut River above Holyoke, Massachusetts. Master's thesis. University of Massachusetts, Amherst.

Watson, J. F. 1970. Distribution and population dynamics of American shad, *Alosa sapidissima* (Wilson), in the Connecticut River above Holyoke Dam, Massachusetts. Doctoral thesis. University of Massachusetts, Amherst.

Whitney, R. R. 1961. The Susquehanna Fishery Study, 1957–1960. A report of a study on the desirability and feasibility of passing fish at Conowingo Dam. Department of Research and Education, Contribution 169, Solomons, Maryland.

Whitton, B. A. 1980. River ecology. University of California Press, Berkley.

Whitworth, W. R., and D. H. Bennett. 1970. A limnological study of the lower Farmington River with special reference to the ability of the river to support American shad. Institute of Water Resources, University of Connecticut Report 9, Storrs, Connecticut.

Wiggins, T. A., T. R. Bender, V. A. Mudrak, and J. A. Coll. 1984. The development, feeding, growth and survival of cultured American shad larvae through the transition from endogenous to exogenous nutrition. Pennsylvania Fish Commission, Benner Spring Fish Research Station, Special Publication, Harrisburg.

American Fisheries Society Monograph 9:419–439, 2004

Effect of Connecticut Yankee Power Plant on Population Dynamics of Asiatic Clams and Their Interactions with Native Bivalves

DOUGLAS E. MORGAN*, MILAN KESER, JOHN T. SWENARTON, AND JAMES F. FOERTCH

Millstone Environmental Laboratory, Dominion Nuclear Connecticut, Inc.
Waterford, Connecticut 06385, USA

Abstract.—The Asiatic clam (also known as Asian clam) *Corbicula fluminea*, first reported in the Connecticut River in 1990, was the dominant bivalve around Connecticut Yankee nuclear power station (CY) during the 10-year period (1991–2000) of this study. This population extended its range upriver at least 22 km from 1991 to 1992. *Corbicula* winter survival was positively correlated with winter water temperatures and negatively correlated with spring freshet river flows. For example, no *Corbicula* were found following the two coldest winters of the study (1993–1994 and 1995–1996). Regardless of the level observed for winter population survival, study sites were repopulated each year by August, attributed to passive downstream movement from populations in refugia from winter stresses (e.g., upstream thermal discharges at Middletown and South Meadow power stations, or groundwater seeps). Small clams (2-mm size-class) dominated the CY thermal discharge canal during power plant operation because, by June of each year, discharge temperatures (>35°C) exceeded the upper thermal tolerance for this clam. The CY discharge, therefore, was not an important winter refuge. In years following CY closure (1997–2000), a year-round *Corbicula* population was established in the discharge canal, but abundance at river sites around CY remained similar to levels observed during power plant operation. Our findings suggest that *Corbicula* has established a permanent population in the Connecticut River with little impact on native bivalves up to the last year of this study in 2000.

Introduction

The Asiatic clam (also known as Asian clam) *Corbicula fluminea* is a freshwater clam, native to Southeast Asia, now common in Europe, Africa, the Pacific Islands, and North and South America (McMahon 1982; Counts 1986). In its native range, *C. fluminea* (termed *Corbicula* throughout this report) is a commercial species collected and cultured for human consumption (Chen 1990, Phelps 1993). Once introduced to North America, this clam spread rapidly because of its high reproductive capacity, adult brooding of young to a free-living juvenile stage, fast growth, early maturity, effectiveness of all stages of development to readily disperse by water currents, and tolerance to a wide range of environmental stresses (Gardner et al. 1976; Eng 1979; Sickel 1979, 1986; McMahon 1983). Its ability to clog industrial water systems makes *Corbicula* a very costly problem for water-dependent industries (Isom 1986; McMahon and Lutey 1988). By fouling water systems at electric generating stations, particularly nuclear power stations, *Corbicula* causes significant safety problems as well (Ingram 1959; Goss and Cain 1975; McMahon 1977; Goss et al. 1979; Mattice 1979; USNRC 1981; Johnson et al. 1986; Page et al. 1986).

Corbicula were initially identified at Connecticut Yankee nuclear power station (CY), which is located on the lower Connecticut River (Figure 1), during a cleaning of the service water system in April 1990 (Morgan et al. 1992, 2003). The discovery of *Corbicula* at CY necessitated the development of an effective control methodology. *Corbicula* is highly resistant to short-term application of conventional biocides because of its ability to tightly close its valves during treat-

* *Corresponding author*: douglas_e_morgan@dom.com

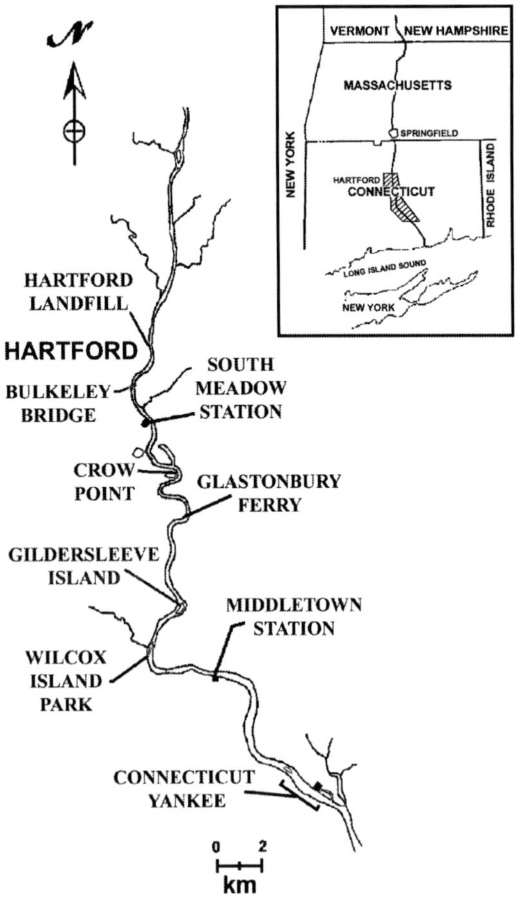

FIGURE 1. *Corbicula fluminea* sampling sites used in surveys conducted in November 1991 and 1992 to determine the extent of up-river clam abundance.

ment (Mattice et al. 1982; Neitzel et al. 1984; Page et al. 1986). However, continuous chlorination is effective in controlling *Corbicula* (Isom 1971; Goss and Cain 1975; Cherry et al. 1986). The continuous low level (<0.1 ppm, Total Residual Chlorine) chlorination of the service water system at CY was approved by the Connecticut Department of Environmental Protection (CTDEP) and began in April 1991. As a condition of the permit, CTDEP required CY to initiate a monitoring study of the *Corbicula* population in the lower Connecticut River. This paper summarizes the results of studies from 1991 to 2000, including data for *Corbicula* distribution, growth, fecundity, and abundance. In addition, native clam and mussel abundance data are presented and examined for possible interactions with *Corbicula*.

Methods

Surveys of the distribution of Connecticut River *Corbicula* population were conducted in November of 1991 and 1992. In 1991, sampling began at CY and continued upriver at intervals of 3–10 km (Middletown station, Wilcox Island Park, Gildersleeve Island, Glastonbury Ferry, Crow Point, and South Meadow station; Figure 1). In 1992, sampling was extended upriver to the Bulkeley Bridge and Hartford Landfill area. In these surveys, sediments were screened using a 3.4 mm, U.S.A. Standard Testing Sieve, which retained clams larger than approximately 6 mm in shell length. All sampling sites were located in the tidal freshwater areas of the Connecticut River.

Corbicula abundance, growth, reproduction and recruitment were studied at five sites in the vicinity of CY from August 1991 to November 2000. Four sampling sites were located in the mainstem river (Figure 2): Haddam Island (A); vicinity of CY intakes (B); down-river of CY intakes (C); and in the vicinity of the CY discharge canal outfall (D). A fifth site was located at the head of the CY discharge canal (E). Water depth at sampling sites ranged from 1 to 6 m at mean low water.

Sampling began in August 1991 and continued annually in May, August, and November through November 2000. Sampling at site D did not start until August 1993. Five 0.1-m^2 bottom samples were obtained at each site using a

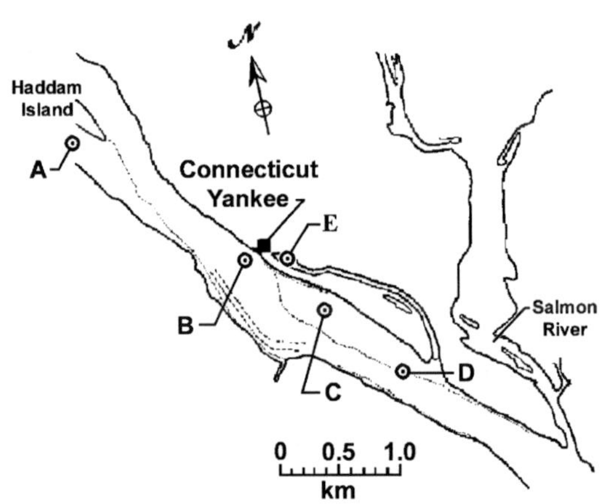

FIGURE 2. Location of the sampling sites (A–E) in the vicinity of CY.

weighted Peterson Grab. Sample processing techniques were similar to those of Gardner et al. (1976). Grab samples were sieved in the field by passing the sample through a series of three screens (6.3-, 2.0-, and 1.0-mm mesh sizes). Clams and sediment retained on the 1-mm screen were subsampled in the field by placing a well-mixed 1-L sample in an elutriator (Magdych 1981) for 3 min at a water flow of 20–30 L/min. The overflow from the elutriator was collected on a 1.0-mm mesh sieve and sorted in the laboratory under a dissecting microscope (10×). Sediment and clams retained on the 6.3- and 2.0-mm screens were taken to the laboratory and washed through a series of five U.S.A. Standard Testing Sieves (19.0-, 12.5-, 6.3-, 3.4-, and 2.0-mm mesh sizes). Size-classes were based on the sieve size in which clams were retained. Clams retained on the 1.0-mm sieve averaged 2.0 mm in shell length; on the 2.0-mm sieve, 4.1 mm; on the 3.4-mm sieve, 6.7 mm; on the 6.3-mm sieve, 14.1 mm; on the 12.5-mm sieve, 19.3 mm; and on the 19-mm sieve, 31.1 mm. The calculated densities (clams/m^2) are used in this study as a measure of abundance. Native bivalve species collected during the processing of these sediment samples were reported only at the total abundance level. Winter survival of clams was defined as the ratio of May clam abundance to November abundance from the previous year, expressed as a percentage.

Several hundred adult clams (>8.0 mm in shell length) were collected from river sites in June 1991 and in May 1992–1994 and held in two lantern-nets, one in the river and the other in the CY discharge canal. Clam growth was monitored monthly from 1991 to 1994 by measuring shell lengths of 12 randomly selected clams from these nets. Each year, monthly measurements were made until total winter mortality occurred in the river (December–February) or summer mortality occurred in the discharge canal (usually June).

Clam fecundity was monitored monthly using techniques of Aldridge and McMahon (1978). Fecundity was assessed on the 12 clams selected for growth estimates in each month that living clams could be found. Each clam was brought to the laboratory and held under static conditions at 20°C for 24 h in a 100-mL beaker filled with filtered Connecticut River water. The number of juveniles released during this period was recorded and used as an index of spawning activity. Additional fecundity assessments were made by dissecting these clams and noting the presence of brood progeny. Maturity of gametes was assessed by removing egg and sperm cells from the gonadal tissues and examining the cells under a compound microscope (400×).

To estimate *Corbicula* recruitment to bottom sediments in the Connecticut River and CY discharge canal, plastic trays (34.5 × 47.0 × 12.5 cm, inside dimensions) were filled with defaunated marine beach sand (<1 mm in grain size). Two sets of four sediment trays were enclosed within wooden frames (84 cm × 112 cm, outside dimensions) partitioned into four compartments (37.0 cm wide × 48.0 cm long × 14.0 cm high) and anchored to the bottom near the mouth of the discharge canal on 30 May 1991 and in the Connecticut River just upriver of the CY intakes (site B) on 6 June 1991. At each site, four trays were placed in each frame; two trays were covered with 6.4 × 6.4 mm plastic mesh screening, and two were not covered. The screen-covered trays were stocked with 50 adult clams (>8 mm), and the uncovered trays were left barren. One replicate tray for each treatment (screen-covered and uncovered) was collected and screened for *Corbicula* during river surveys in August and November 1991. Nearly all clams were dead in May 1992 and sampling was terminated.

Connecticut River flow rates were obtained from the USGS Thompsonville CT monitoring station (USGS 2002). Water temperature was recorded every 15 min at the CY intake and discharge using the Northeast Utilities Environmental Data Acquisition Network. Statistical analyses were performed using SAS version 8 software (SAS Inc., Cary, North Carolina). Abundance data in tables are presented using arithmetic means. Statistical testing (SAS General Linear Models Procedure) of abundance data were performed after log transformation. *T*-tests were performed on log-transformed abundance data, and depending on the results for the equality of variances test, either a pooled or Satterthwaite method was used to determine degrees of freedom. The relationships between winter clam survival and temperature or flow indices were tested using the rank-order Spearman correlation statistic. Growth and reproduction data were not transformed prior to analysis and are presented as arithmetic means.

Results

River Water Flow and Temperature

Annual Connecticut River water flow averaged 509 m^3/s over the 10-year period from 1991 to 2000. Mean monthly flow rates peaked during April in 8 of 10 years with the average April flow rate ranging from a low of 422 m^3/s in 1995 to a high of 2,099 m^3/s in 1993 (Figure 3). Similarly, maximal flow rates for daily averages usually occurred in April and often exceeded 1,700 m^3/s. These values, which were collected at the Thompsonville Connecticut gaging station, represent flows beyond any tidal effects. Semidiurnal tides affect river flow rates and bring an average 425 m^3/s of additional flow to the lower Connecticut River in the vicinity of CY (Merriman and Thorpe 2004 [1976], this volume), causing periodic fluctuations in river height of ~1 m (NSI 1995; Rozsa 2001). Tidal influences are large in relation to natural flow during periods of low river discharge and are absent or nearly absent during freshet conditions (Rozsa 2001; Boyd 2004 [1976], this volume). Daily average ambient water temperatures ranged between −1.7°C and 30.6°C during the 10-year study period (Figure 4). Discharge water temperatures at CY during plant operation were 8–12°C above ambient river temperatures. The maximum discharge flow of 25 m^3/s was directed through a man-made canal 1 km long before mixing with the ambient Connecticut River waters.

Distribution

In November 1991, *Corbicula* was observed as far north in the Connecticut River as Gildersleeve Island, where density averaged 168/m^2 (Figure 1; Table 1). Gildersleeve Island is approximately 16 km upstream of CY, where density averaged 531/m^2. Between these sites, clam densities were 150/m^2 at Middletown station and 12/m^2 at Wilcox Island Park. In November 1992, clams were documented 22 km further north at South Meadow station (30/m^2), which is just south of Hartford,

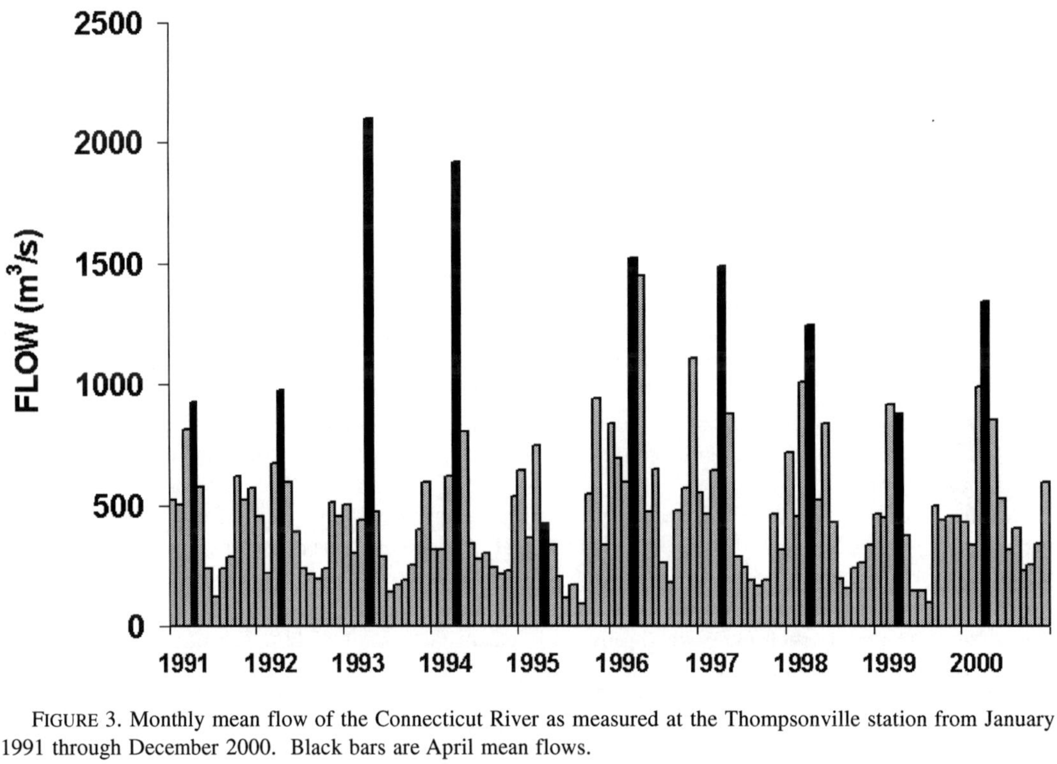

FIGURE 3. Monthly mean flow of the Connecticut River as measured at the Thompsonville station from January 1991 through December 2000. Black bars are April mean flows.

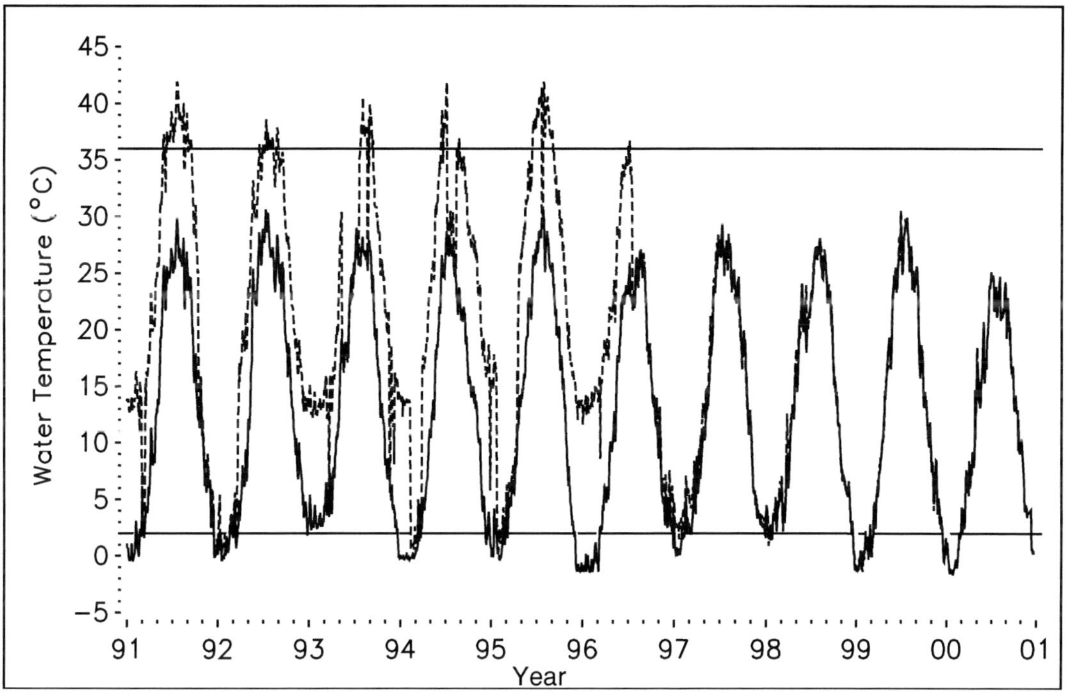

FIGURE 4. Intake (——) and discharge (- - - -) water temperatures at CY from 1 January 1991 to 1 January 2001. Horizontal reference lines represent upper and lower lethal temperature limits for *Corbicula fluminea.* CY shut down in July 1996, as revealed by discharge temperatures tracking intake temperatures.

Connecticut. Clam densities were higher in 1992 than in 1991 at all sites, from 28/m² at Crow Point, approximately 3 km downstream of South Meadow station, to 9,169/m² at Middletown station.

TABLE 1. Mean abundance of *Corbicula fluminea* during the November 1991 and 1992 Connecticut River surveys.

Location	Abundance[a] 1991 clams/m²	1992 clams/m²
Connecticut Yankee (CY)	531	2,182
Middletown station	150	9,169
Wilcox Island Park	12	–[b]
Gildersleeve Island	168	3,300
Glastonbury Ferry	0	–
Crow Point	0	28
South Meadow station	0	30
Bulkeley Bridge	–	0
Hartford landfill	–	0

[a] Abundance estimates for clams > 5.6 mm in shell length (3.4-mm screen).
[b] Site was not sampled.

Abundance

Mean annual total abundance of *Corbicula* at CY sites from 1991 to 2000 was highly variable with significantly ($P < 0.05$) higher abundance in 1992, 1995, and 1999 than any other year of the study, ranging from 1,158/m² in 1995 to 2,610/m² in 1992 (Table 2). Annual abundance in 1998 and 2000 was 326 and 286/m², respectively. These abundances were significantly lower ($P < 0.05$) than those in 1992, 1995 and 1999, but significantly higher ($P < 0.05$) than those in 1993, 1994, 1996, and 1997, which ranged from 45/m² in 1996 to 138/m² in 1997.

Seasonal abundance (i.e., abundance in May, August, and November collections) was also highly variable (Table 2). May abundance was consistently the lowest value observed during the year; 10-year averages for May abundance ranged from 24/m² at site B to 86/m² at site A. Averages for all sites combined show November abundance was highest in 8 of 10 years. November abundance was lowest at site D (365/m²) and highest at site A (2,299/m²).

TABLE 2. Mean total abundance (clams/m^2 ± 95% CI) of *Corbicula fluminea* at each sampling site near the Connecticut Yankee power plant in the Connecticut River from August 1991 to November 2000.

Site	Year										All
Month	1991	1992	1993	1994	1995	1996	1997	1998	1999	2000	years
A											
May	–	124 ± 115	12 ± 33	0	202 ± 161	0	10 ± 21	104 ± 147	146 ± 44	174 ± 284	86 ± 36
Aug.	640 ± 290	5,552 ± 3,980	8 ± 10	54 ± 59	3,052 ± 2,369	44 ± 52	118 ± 120	630 ± 440	4,076 ± 1,500	650 ± 477	1,482 ± 633
Nov.	1,642 ± 724	11,158 ± 4,315	332 ± 347	576 ± 341	2,452 ± 1,768	60 ± 50	648 ± 373	1,498 ± 168	3,552 ± 2079	1,036 ± 238	2,299 ± 959
Average[a]	1,141	5,611	117	210	1,902	35	271	744	2,591	620	1,331
B											
May	–	20 ± 26	0	0	30 ± 37	0	10 ± 28	14 ± 19	86 ± 60	54 ± 41	24 ± 11
Aug.	288 ± 99	982 ± 346	18 ± 31	40 ± 49	1,574 ± 759	32 ± 28	38 ± 27	196 ± 54	1,626 ± 346	196 ± 46	499 ± 186
Nov.	272 ± 215	2,688 ± 659	138 ± 102	110 ± 97	1,874 ± 1,829	80 ± 147	204 ± 73	326 ± 129	1,604 ± 595	278 ± 295	757 ± 289
Average	280	1,230	52	50	1,159	37	84	179	1,105	176	441
C											
May	–	20 ± 26	0	0	164 ± 200	0	8 ± 22	26 ± 46	136 ± 114	26 ± 45	42 ± 25
Aug.	270 ± 208	1,170 ± 645	40 ± 37	84 ± 52	1,934 ± 1,216	100 ± 64	42 ± 34	260 ± 199	1,590 ± 742	376 ± 381	587 ± 220
Nov.	508 ± 180	1,780 ± 1,078	140 ± 64	84 ± 71	1,184 ± 576	34 ± 27	220 ± 350	330 ± 121	956 ± 213	144 ± 152	538 ± 181
Average	389	990	60	56	1,094	45	90	205	894	182	401
D											
May	–	–	–	0	98 ± 73	0	4 ± 111	8 ± 22	8 ± 16	58 ± 71	25 ± 16
Aug.	–	–	74 ± 110	94 ± 61	754 ± 283	48 ± 55	30 ± 23	78 ± 54	1,300 ± 447	242 ± 161	328 ± 147
Nov.	–	–	214 ± 306	130 ± 143	580 ± 339	146 ± 105	294 ± 252	442 ± 56	922 ± 166	192 ± 116	365 ± 96
Average	–	–	144	75	477	65	109	176	743	164	249
All sites[b]											
May	–	55 ± 40	4 ± 9	0	124 ± 56	0	8 ± 8	38 ± 33	94 ± 35	78 ± 58	45 ± 13
Aug.	399 ± 132	2,568 ± 1,548	35 ± 24	68 ± 22	1,828 ± 624	56 ± 22	57 ± 28	291 ± 130	2,148 ± 619	366 ± 139	745 ± 192
Nov.	807 ± 390	5,209 ± 2,647	206 ± 91	225 ± 119	1,522 ± 567	80 ± 38	350 ± 137	649 ± 273	1,758 ± 637	412 ± 189	1,023 ± 284
Average	603	2,610	89	98	1,158	45	138	326	1,334	286	625

[a] Annual averages (clams/m^2) pooling the May, August, and November samples.

[b] Average abundance of clams, pooling the four sites.

Examination of *Corbicula* population structure from 1991 to 2000 shows juvenile clams (2- and 4-mm size-classes) significantly more abundant ($P < 0.05$) than first-year adults (7- and 14-mm size-classes) or clams older than 1 year of age (19- and 31-mm size-classes, Tables 3, 4, and 5). The overall annual average for older size-classes was only 6/m^2 as compared to 282 and 336/m^2 for first-year adult and juvenile size-classes, respectively.

Abundance was also highly variable from site to site (Table 2). Mean abundance at site A (1,331/m^2) was significantly higher than at other river sites. Average annual abundances (clams/m^2) at sites B (441), C (401), and D (249) were not significantly ($P < 0.05$) different from one another. Spatially, mean annual abundance decreased from site A (the northernmost site) to site D, which was the closest to the CY discharge canal. Mean annual abundance of *Corbicula* within the CY discharge canal (Site E) was substantially less than that at river sites and ranged from 25/m^2 in 1993–4,091/m^2 in 1992 (Table 6). Seasonally within the discharge canal, May abundance (0–92/m^2) was significantly ($P < 0.05$) lower than August (0–12,174/m^2) and November (24–880/m^2); November abundance was usually higher than in August.

Mean *Corbicula* abundance for all river sites combined was not significantly different ($P > 0.05$) between CY operational (1991–1996) and postoperational (1997–2000) periods. Similarly, the difference between total *Corbicula* abundance in the discharge canal during operational and postoperational years was not significant ($P > 0.05$). However, abundance in the discarge canal for ≥7 mm size-class clams was significantly higher ($P < 0.05$) in August and November samples during the postoperational period versus the operational period.

Recruitment

Uncovered trays placed in the river without *Corbicula* in June 1991 were colonized by 72 clams (444/m^2) in August and by 133 clams (820/m^2) in November. Size of recruited clams in August ranged from 1.5 to 17.2 mm with a mean shell length of 6.8 mm, and 1.5–22.3 mm in November with a mean shell length of 9.4 mm. Screen-covered trays seeded with 50 adult-size clams/tray (14.1-mm average size) collected upstream of CY sites had 587 clams (3,620/m^2) in August and 1,511 (9,318/m^2) in November. In these trays, clams in August ranged from 1.5 to 19.3 mm with a mean shell length of 7.8, and in November, the size range was 1.5–22.6 mm with a mean shell length of 11.4 mm. In May 1992, 2 live and 145 dead (shell only) clams were recovered from both uncovered trays. In the screened trays, 1,877 clams (all dead) were recovered.

Survival

Average survival of *Corbicula* in the river from November to May for 1992–2000 was 9.9% at site A, 9.3% at site B, 10.4% at site C, and 12.7% at site D (Table 7). Overall winter survival of *Corbicula* (all river sites combined) was positively correlated with average winter water temperature (December–April; $r_s = 0.87$, $P = 0.004$) and negatively correlated with the April flow index ($r_s = -0.91$, $P = 0.002$; Table 8). In addition, winter survival was also negatively correlated to the number of days that water temperatures averaged less than or equal to 1°C ($r_s = -0.73$, $P = 0.04$). *Corbicula* winter survival was not significantly correlated with the number of days that water temperature was less than or equal to 2°C ($r_s = -0.65$, $P = 0.08$).

However, spatial differences among sites were evident between environmental stress indices and winter survival. At site A, the site furthest from the CY, all four indices were significantly correlated with *Corbicula* winter survival: average winter water temperatures ($r_s = 0.92$, $P = 0.001$), peak flows in April ($r_s = -0.74$, $P = 0.036$), the number of days that were less than or equal to 1°C ($r_s = -0.87$, $P = 0.004$) and less than or equal to 2°C ($r_s = -0.82$, $P = 0.013$). At the B site near the plant intakes, winter survival was significantly correlated with average winter water temperature ($r_s = 0.73$, $P = 0.04$) and peak flows in April ($r_s = -0.77$, $P = 0.02$). At C, only average winter water temperature ($r_s = 0.85$, $P = 0.02$) was significantly correlated with survival. At D, the closest site in the river to the CY discharge, winter survival was not significantly correlated with any of the four indices. *T*-tests comparing percentage winter survival during operational and postoperational years showed no differences at any of the four river sites.

Reproduction

Clams within lantern-nets in the Connecticut River showed the presence of eggs and sperm in

TABLE 3. Mean total abundance (clams/m^2 ± 95% CI) of *Corbicula fluminea* in the 2- to 4-mm shell-length size-classes at each sampling site near the Connecticut Yankee power plant in the Connecticut River from August 1991 to November 2000.

Site	Year										All
Month	1991	1992	1993	1994	1995	1996	1997	1998	1999	2000	years
A											
May	–	10 ± 18	12 ± 33	0	176 ± 150	0	8 ± 22	104 ± 147	102 ± 51	6 ± 11	46 ± 24
Aug.	302 ± 236	4,246 ± 3,060	0 ± 0	18 ± 25	1,883 ± 1,937	44 ± 52	72 ± 88	488 ± 449	1,998 ± 547	352 ± 312	940 ± 447
Nov.	520 ± 313	5,924 ± 2,463	232 ± 265	456 ± 328	758 ±1,040	38 ± 37	506 ± 305	710 ± 467	1,228 ± 1,241	292 ± 174	1,066 ± 513
Average[a]	411	3,393	81	158	939	27	195	434	1,109	217	706
B											
May	–	4 ± 11	0	0	28 ± 33	0	10 ± 28	12 ± 20	80 ± 59	26 ± 35	18 ± 10
Aug.	170 ± 52	768 ± 302	0	16 ± 44	866 ± 733	14 ± 14	16 ± 19	156 ± 76	1,104 ± 322	24 ± 31	323 ± 127
Nov.	186 ± 123	2,162 ± 710	100 ± 90	82 ± 67	854 ± 1,149	64 ± 150	152 ± 82	222 ± 118	680 ± 419	74 ± 64	457 ± 202
Average	178	978	33	33	583	26	59	130	621	75	275
C											
May	–	4 ± 11	0	0	156 ± 197	0	6 ± 17	26 ± 46	114 ± 105	14 ± 32	36 ± 24
Aug.	116 ± 123	914 ± 599	18 ± 38	22 ± 20	996 ± 851	88 ± 59	12 ± 27	184 ± 196	814 ± 488	284 ± 306	345 ± 135
Nov.	122 ± 130	994 ± 1,052	74 ± 73	26 ± 24	224 ± 242	14 ± 19	192 ± 348	210 ± 149	222 ± 150	74 ± 88	215 ± 109
Average	119	637	31	16	459	34	70	140	383	124	204
D											
May	–	–	–	0	84 ± 67	0	0	8 ± 22	8 ± 16	38 ± 53	20 ± 13
Aug.	–	–	50 ± 108	40 ± 34	108 ± 73	36 ± 44	10 ± 22	30 ± 28	686 ± 314	190 ± 149	144 ± 76
Nov.	–	–	154 ± 290	92 ± 97	272 ± 87	88 ± 83	252 ± 239	292 ± 82	104 ± 73	88 ± 68	168 ± 44
Average	–	–	102	44	155	41	187	110	266	105	114
All sites[b]											
May	–	6 ± 6	4 ± 9	0 ± 0	111 ± 53	0 ± 0	6 ± 7	38 ± 33	76 ± 30	21 ± 14	30 ± 10
Aug.	196 ± 79	1,976 ± 1,185	17 ± 22	24 ± 12	963 ± 491	46 ± 20	28 ± 21	214 ± 118	1,150 ± 287	238 ± 90	454 ± 132
Nov.	276 ± 132	3,027 ± 1,375	140 ± 76	164 ± 102	527 ± 304	51 ± 33	276 ± 113	358 ± 132	558 ± 312	132 ± 58	493 ± 152
Average	236	1,670	58	63	534	32	103	204	595	130	336

[a] Annual averages (clams/m^2) pooling the May, August, and November samples.
[b] Average abundance of clams, pooling the four sites.

TABLE 4. Mean abundance (clams/m^2 ± 95% CI) of *Corbicula fluminea* in the 7 to 14 mm shell length size-classes at each sampling site near the Connecticut Yankee power plant in the Connecticut River from August 1991 to November 2000.

Site	Years										All
Month	1991	1992	1993	1994	1995	1996	1997	1998	1999	2000	years
A											
May	–	112 ± 98	0	0	26 ± 30	0	0	0	44 ± 31	168 ± 286	39 ± 28
Aug.	338 ± 151	1,284 ± 926	8 ± 10	36 ± 38	1,154 ± 708	0	46 ± 38	142 ± 66	2,068 ± 1,185	290 ± 309	537 ± 224
Nov.	1,092 ± 584	5,172 ± 2,417	100 ± 97	120 ± 65	1,552 ± 786	22 ± 34	176 ± 104	786 ± 458	2,290 ± 1,484	736 ± 219	1,204 ± 472
Average[a]	715	2,139	36	52	911	7	74	309	1,467	398	612
B											
May	–	16 ± 29	0	0	2 ± 6	0	0	2 ± 6	6 ± 11	28 ± 24	6 ± 4
Aug.	118 ± 75	198 ± 96	18 ± 31	24 ± 7	702 ± 51	18 ± 16	22 ± 14	40 ± 25	520 ± 156	72 ± 34	173 ± 67
Nov.	72 ± 91	518 ± 68	36 ± 17	24 ± 41	926 ± 962	16 ± 17	52 ± 10	104 ± 61	918 ± 419	204 ± 240	287 ± 122
Average	95	244	18	16	543	11	25	49	481	101	161
C											
May	–	16 ± 29	0	0	8 ± 6	0	2 ± 6	0	22 ± 14	12 ± 16	6 ± 4
Aug.	154 ± 87	226 ± 60	22 ± 10	62 ± 39	928 ± 497	12 ± 14	30 ± 9	76 ± 42	770 ± 315	92 ± 76	237 ± 98
Nov.	368 ± 60	724 ± 145	66 ± 19	58 ± 58	926 ± 467	20 ± 15	28 ± 22	118 ± 68	730 ± 334	70 ± 77	310 ± 104
Average	261	322	29	40	621	11	20	65	507	58	191
D											
May	–	–	–	0	14 ± 27	0	4 ± 11	0	0	20 ± 25	5 ± 5
Aug.	–	–	24 ± 11	54 ± 35	646 ± 258	12 ± 16	18 ± 16	48 ± 38	608 ± 159	52 ± 24	183 ± 87
Nov.	–	–	48 ± 32	36 ± 47	292 ± 362	58 ± 32	42 ± 27	150 ± 42	804 ± 155	96 ± 73	190 ± 87
Average	–	–	36	30	317	23	21	66	471	56	132
All sites[b]											
May	–	48 ± 36	0	0	12 ± 8	0	2 ± 2	1 ± 1	18 ± 10	57 ± 59	15 ± 8
Aug.	203 ± 71	569 ± 365	18 ± 7	44 ± 13	858 ± 184	10 ± 6	29 ± 9	76 ± 25	991 ± 371	126 ± 72	288 ± 70
Nov.	510 ± 284	2,138 ± 1,360	62 ± 22	60 ± 26	924 ± 320	29 ± 12	74 ± 34	290 ± 160	1,186 ± 412	276 ± 142	514 ± 141
Average	357	918	29	34	598	13	35	122	732	153	282

[a] Annual averages (clams/m^2) pooling the May, August, and November samples.
[b] Abundance of clams, pooling the four sites.

TABLE 5. Mean abundance (clams/m^2 ± 95% CI) of *Corbicula fluminea* in the 19 to 31 mm shell length size-classes at each sampling site near the Connecticut Yankee power plant in the Connecticut River from August 1991 to November 2000.

Site	Years										All
Month	1991	1992	1993	1994	1995	1996	1997	1998	1999	2000	years
A											
May	–	2 ± 6	0	0	0	0	2 ± 6	0	0	0	0.4 ± 0.6
Aug.	0	22 ± 27	0	0	16 ± 26	0	0	0	10 ± 12	8 ± 10	6 ± 3
Nov.	30 ± 51	62 ± 58	0	0	142 ± 184	0	2 ± 6	2 ± 6	34 ± 68	8 ± 10	28 ± 18
Average[a]	15	29	0	0	53	0	1	1	15	5	12
B											
May	–	0	0	0	0 ± 0	0	0	0	0	0	0
Aug.	0	16 ± 21	0	0	6 ± 8	0	0	0	2 ± 6	0	2 ± 2
Nov.	14 ± 27	8 ± 16	2 ± 6	4 ± 7	94 ± 22	0	0	0	6 ± 7	0	13 ± 8
Average	7	8	1	1	33	0	0	0	3	0	5
C											
May	–	0	0	0	0	0	0	0	0	0	0
Aug.	0	30 ± 23	0	0	10 ± 9	0	0	0	6 ± 11	0	5 ± 3
Nov.	18 ± 16	62 ± 16	0	0	34 ± 37	0	0	2 ± 6	4 ± 7	0	12 ± 6
Average	9	31	0	0	15	0	0	1	3	0	6
D											
May	–	–	–	0	0	0	0	0	0	0	0
Aug.	–	–	0	0	0	0	2 ± 6	0	6 ± 11	0	1 ± 1
Nov.	–	–	12 ± 10	2 ± 6	16 ± 19	0	0	0	14 ± 7	8 ± 6	7 ± 3
Average	–	–	6	1	5	0	1	0	7	3	3
All sites[b]											
May	–	0.7 ± 1.4	0	0	0	0	0.5 ± 1	0	0	0	0.1 ± 0.2
Aug.	0	23 ± 10	0	0	8 ± 6	0	0.5 ± 1	0	6 ± 4	2 ± 2	4 ± 1
Nov.	21 ± 15	44 ± 21	4 ± 3	2 ± 2	72 ± 41	0	0.5 ± 1	1 ± 1	15 ± 13	4 ± 13	15 ± 6
Average	10	22	1	1	26	0	1	<1	7	2	6

[a] Annual averages (clams/m^2) pooling the May, August, and November samples.

[a] Abundance of clams, pooling the four sites.

TABLE 6. Mean abundance (clams/m^2 ± 95% CI) of *Corbicula fluminea* in the CY discharge canal (Site E) by size class from August 1991 to November 2000.

		Abundance by size-class			
Year	Month	2–4 mm	7–14 mm	19–31 mm	Total abundance
1991					
	May	–	–	–	–
	Aug.	34 ± 63	0	0	34 ± 63
	Nov.	172 ± 181	6 ± 17	0	178 ± 195
	Year's average	103	3	0	106
1992					
	May	2 ± 6	0	0	2 ± 6
	Aug.	12,174 ± 32,688	0	0	12,174 ± 32,688
	Nov.	96 ± 131	0	0	96 ± 131
	Year's average	4,091	0	0	4,091
1993					
	May	8 ± 10	24 ± 29	0	32 ± 24
	Aug.	0	0	0	0
	Nov.	32 ± 33	10 ± 18	0	42 ± 28
	Year's average	13	11	0	25
1994					
	May	0	0	0	0
	Aug.	36 ± 36	24 ± 29	0	60 ± 59
	Nov.	816 ± 1,303	58 ± 57	6±17	880 ± 1,354
	Year's average	284	27	2	313
1995					
	May	24 ± 31	0	0	24 ± 31
	Aug.	38 ± 54	0	0	38 ± 54
	Nov.	90 ± 202	0	0	90 ± 202
	Year's average	51	0	0	51
1996					
	May	0	8 ± 6	0	8 ± 6
	Aug.	48 ± 81	0	0	48 ± 81
	Nov.	10 ± 12	34 ± 81	0	44 ± 75
	Year's average	19	14	0	33
1997					
	May	60 ± 79	32 ± 31	0	92 ± 66
	Aug.	0	6 ± 17	0	6 ± 17
	Nov.	4 ± 7	18 ± 10	2 ± 6	24 ± 7
	Year's average	21	19	1	41
1998					
	May	18 ± 20	2 ± 6	0	20 ± 23
	Aug.	40 ± 81	16 ± 32	6±11	62 ± 89
	Nov.	106 ± 213	96 ± 77	10±12	212 ± 245
	Year's average	55	38	5	98
1999					
	May	74 ± 35	2 ± 6	0	76 ± 40
	Aug.	4 ± 7	216 ± 196	14 ± 17	234 ± 186
	Nov.	0	124 ± 148	86 ± 92	210 ± 79
	Year's average	26	114	33	173

TABLE 6. Continued

Year	Month	Abundance by size-class 2–4 mm	7–14 mm	19–31 mm	Total abundance
2000					
	May	0	4 ± 11	0	4 ± 11
	Aug.	20 ± 34	8 ± 16	2 ± 6	30 ± 38
	Nov.	162 ± 125	96 ± 100	10 ± 22	268 ± 195
	Year's average	61	36	4	101
10-year average					
	May	21 ± 11	8 ± 5	0	29 ± 12
	Aug.	1,239 ± 2,380	27 ± 22	2 ± 2	1,269 ± 2,378
	Nov.	149 ± 111	44 ± 19	11 ± 10	204 ± 116
	Annual average	485	27	5	517

TABLE 7. Average percentage winter survival of Asiatic clams (*Corbicula*), native clams (Sphaeridae), and native mussels (Unionidae) at four CY sampling sites.

Site	*Corbicula*	Sphaeridae	Unionidae
A	9.9%	47.0%	17.5%
B	9.3%	65.6%	51.3%
C	10.4%	28.5%	33.3%
D	12.7%	28.4%	–
Average	10.6%	42.4%	34.0%

gonadal tissues from May through January (Figure 5). The frequency of clams with eggs ranged from 75% to 100% and with sperm ranged from 52% to 100%. The presence of young in the demibranchs (brooding) and the active releasing of juveniles occurred from June to October. The frequency of clams brooding during this period ranged from 4% in October to 62% in August (Figure 5). In the 4 years of the reproduction study, only one brooding clam was observed in October and three in January. The periods of brooding and release of juvenile clams (spawning) occurred at about the same time, from June to September. Spawning peaked in August with an average release of 2,861 juveniles/clam/d.

During power plant operation, the clams were able to survive in the CY discharge canal from November through May; eggs and sperm were continually present in gonadal tissues (Figure 6). Eggs were present in 100% of clams surveyed and sperm frequency ranged from 8% in Decem-

TABLE 8. Spearman correlation coefficients of percentage winter survival of *Corbicula fluminea* at CY sites versus four indices of winter severity for the Connecticut River from 1992 through 2000[a].

Winter severity index	(Avg. ± SE)	Spearman correlation coefficients by site A	B	C	D	All sites combined
Average winter temp.[b]	(2.93°C ±0.37)	+0.92**	+0.73*	+0.84*	+0.63 ns	+0.87**
No. days ≤1 °C	(54.9 d ±8.75)	−0.87**	−0.63 ns	−0.74 ns	−0.56 ns	−0.73*
No. days ≤2 °C	(70.6 d ±8.12)	−0.82*	−0.52 ns	−0.62 ns	−0.67 ns	−0.65 ns
April flow ≥1,700 m^3/s[c]	(6.4 d ±1.54)	−0.74*	−0.78*	−0.73 ns	−0.69 ns	−0.91**

[a] 1993 data were omitted because of the mortality caused by the March storm Joshua (see text).
[b] Average winter temperature = the annual Dec.–Apr. mean daily Connecticut River temperature at CY.
[c] Number of days in April when the Connecticut River flow equaled or exceeded a daily average of 1,700 m^3/s.
Significance levels: ** = $P < 0.01$; * = $P < 0.05$; ns = $P \geq 0.05$.

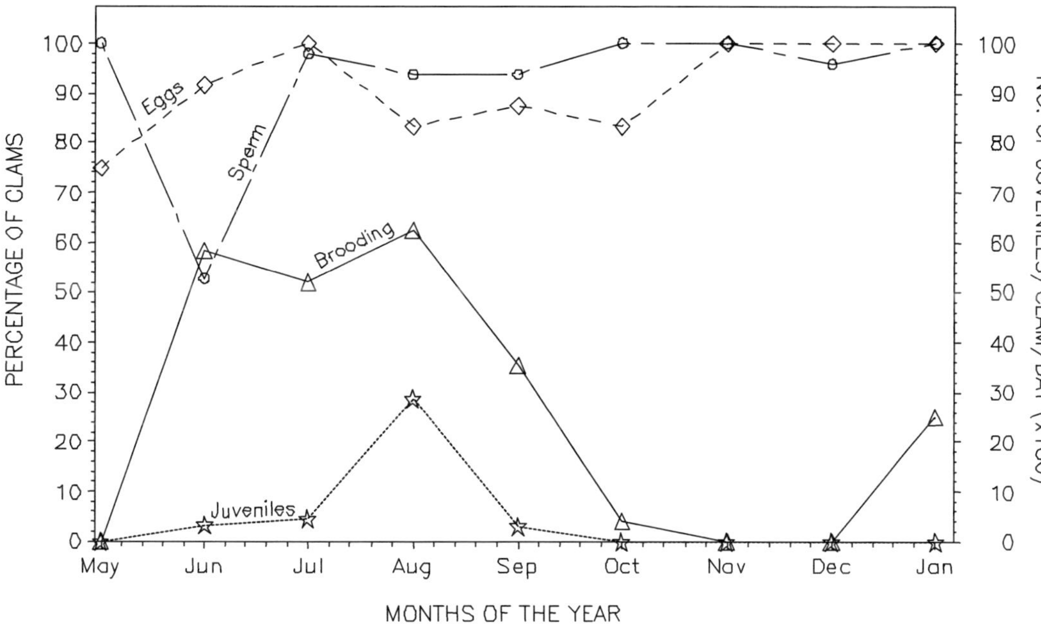

FIGURE 5. Average seasonal reproductive status of *Corbicula fluminea* (1991-1994) at ambient river temperatures. The percentage of adult clams (≥8 mm) with mature gametes or brooding young is plotted using the left axis. Active spawning is plotted using the right axis and is defined as the average number juveniles released over a 24-h period by 12 randomly selected adult clams.

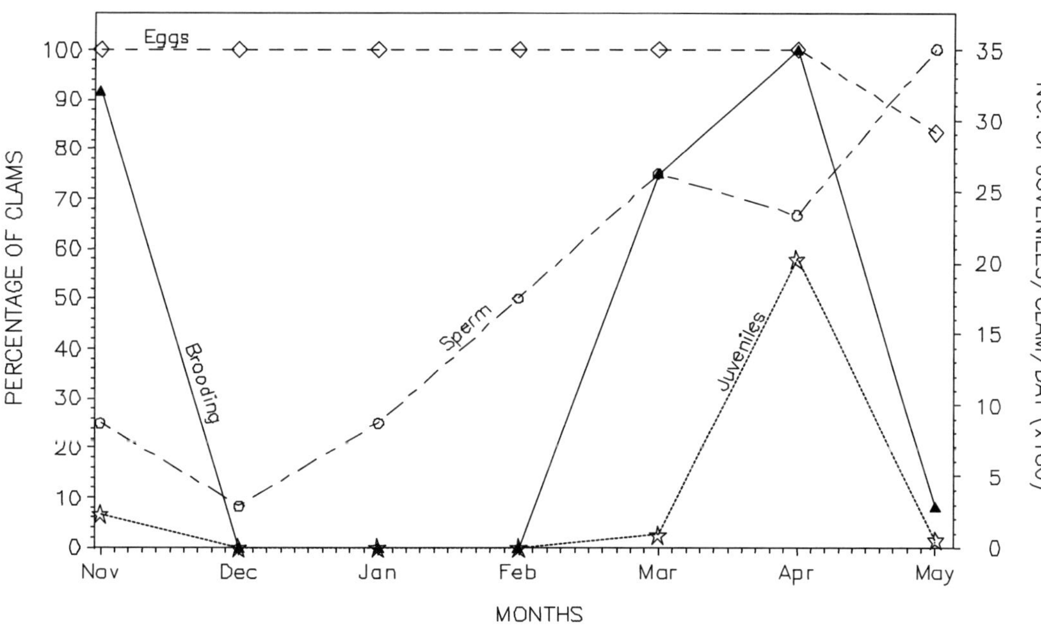

FIGURE 6. Reproductive status of *Corbicula fluminea* held in the CY discharge canal from November 1992 to May 1993, when water temperatures were 8°C to 12°C above ambient river temperatures. The percentage of adult clams (≥8 mm) with mature gametes or brooding young is plotted using the left axis. Active spawning is plotted using the right axis and is defined as the number juveniles released over a 24-h period by 12 randomly selected adult clams.

ber to 100% in May. Continuous brooding and spawning occurred from March through May with brooding adults ranging from 8% to 100%, when water temperature averaged ≥17°C. Brooding and spawning were also observed in November (92% brooding), when temperatures averaged 18.3°C. The sharp decrease in spawning activity in May was the result of power plant shut down beginning on 15 May 1993, which dropped cooling water temperatures from 30°C to 18°C in a single day. Brooding and spawning peaked in June and July as the power outage continued and reproduction of discharge clams followed the pattern observed at ambient river temperatures (17–21°C). On 21 July 1993 the power plant restarted and temperatures increased to >35°C in 4 d. All clams held in the CY discharge canal were dead by 18 August 1993.

Growth

Quadratic regression models fit to captive clam growth data had slope parameters (b1 and b2) that were significantly different from zero ($P < 0.05$; Figure 7). The 1991 growth data showed the best fit to the model ($R^2 = 0.96$) and had the highest peak growth rate of 1.08 mm/week. Peak growth rates in 1992 and 1993 were 0.68 mm/week and 0.44 mm/week, respectively. *Corbicula* growth under ambient conditions during 1991–1993 exhibited seasonal cycles. For example, growth in 1991 between June and July was 1.08 mm/week, between July and August was 0.94 mm/week, and between August and September was 0.60 mm/week. Growth was minimal from September to October (0.10 mm/week) and October to November (0.06 mm/week).

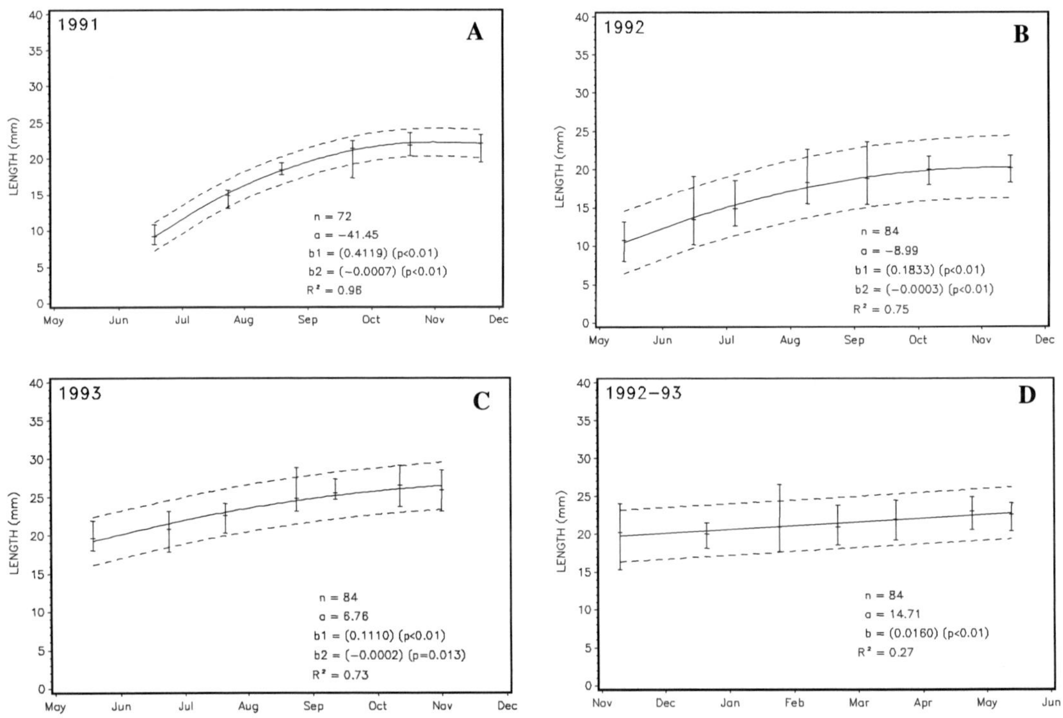

FIGURE 7. Growth of *Corbicula fluminea* at ambient river conditions in 1991, 1992 and 1993 (Panels A–C), as compared to growth at CY discharge (Panel D) conditions from November 1992 to August 1993. The quadratic regression models (1991, 1992, and 1993) are of the form, Length = a + b1(Day) + b2(Day)^2, where "a" is the intercept, "b1" and "b2" are the slopes, and "Day" references the Julian day of the year for 1991, 1992, and 1993. The linear regression model (1992–1993) is of the form, Length = a + b(Day), where "Day" references a cumulative Julian day over the 2-year period 1992–1993. Regression lines are bracketed by 95% confidence limits, and the vertical bars indicate the minimum, maximum, and average shell length by horizontal bars (n = 12).

Clam growth in the CY discharge canal was 0.05 mm/week from November 1992 to February 1993, when water temperatures ranged from 12–21°C (10–12°C above ambient river temperatures; Figure 7). Increased growth rates were observed from February to May (0.14 mm/week), when water temperatures ranged from 12–30°C. Maximum growth rates at this site occurred from May to July (0.36 mm/week), when canal temperatures were similar to those at ambient river sites because of a power plant shut down. All clams died after the power plant restarted in July 1993 and discharge water temperatures exceeded 37°C.

Native Bivalve Abundance

Clams of the family Sphaeridae comprised more than 90% of the native bivalves collected and were not identified below this taxonomic level, while adult mussels (Unionidae) were identified to species. Four mussel species were documented at the Connecticut River sites: eastern elliptio *Elliptio complanata*, alewife floater *Anodonta implicata*, tidewater mucket *Ligumia ochracea*, and eastern pondmussel *L. nasuta*. Of these, *E. complanata* comprised more than 90% of all mussels collected. *Ligumia ochracea* and *L. nasuta*, the rarest mussels, were not collected until 1997 and 1999, respectively.

Mean annual abundances (individuals/m^2) of native clams and mussels were highly variable with clams ranging from 16 to 242 and mussels ranging from 1 to 5 (Figure 8). Abundances of both mussels and clams were highest in 1991. The lowest abundances for clams occurred in 1993, 1996, and 2000, while those for mussels occurred in 1995 and 2000. Mean annual abundances of native clams (Sphaeridae) averaged six times lower than those for *Corbicula*.

Winter survival was considerably greater for native clams and mussels than for *Corbicula* (Table 7). Winter survival of Sphaeridae ranged

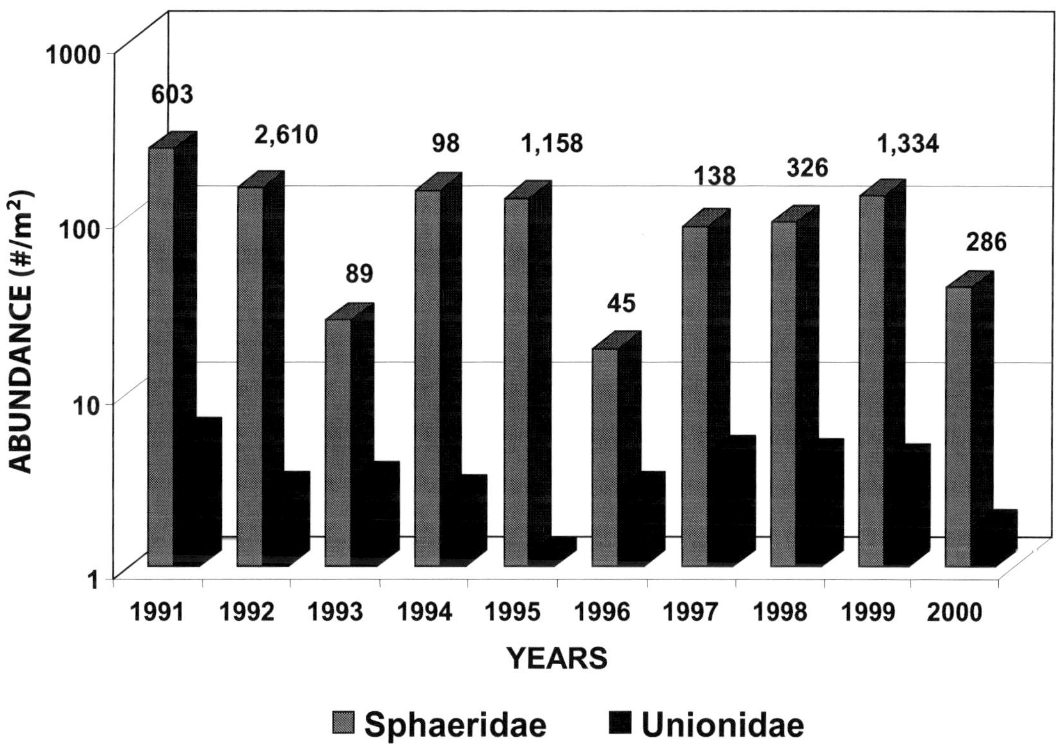

FIGURE 8. Annual abundance ((#/m^2)+1) of native clams (Sphaeridae) and mussels (Unionidae) from 1991 through 2000 at Connecticut River sites in the vicinity of CY. Annual abundance estimates for *Corbicula* are given in numeric values (#/m^2) above each year.

from 28.4% at D to 65.6% at B, averaging 42% for all sites. Unionidae survival ranged from 17.5% at A to 51.3% at B, averaging 34% for sites A–C. Mussels were not collected at D. In comparison, *Corbicula* survival ranged from 9.3% at B to 12.7% at D, averaging 11% overall.

Statistical trend analyses on the log transformed abundance data using a Mann-Kendall statistic showed no significant ($P < 0.05$) trends in abundance for either native clam or mussels from 1991 to 2000. *T*-tests of native clam and mussel seasonal abundance during the CY operational period (1991–1996) versus the pos-operational period (1997–2000) showed no significant differences, except for mussels in November (Table 9).

Discussion

Corbicula has been a major component of macrobenthic communities in the lower Connecticut River following its initial discovery in 1990. *Corbicula* distribution in 1991 extended northward from CY to Gildersleeve Island, a 16 km section of the river, and it spread 22 km further upriver in 1992. Savoy and Benway (2004, this volume) collected sediment grab samples during 2000–2002 approximately 5–10 km upriver from the northern extent reported for *Corbicula* in 1992 (South Meadow station) and did not report the presence *Corbicula*. Upstream movements of *Corbicula* are not unusual and can happen without anthropogenic intervention (Voelz et al. 1998). Rodgers et al. (1977) reported movement of 222 km in 15 years. Once a *Corbicula* population is established upriver, downriver dispersal is more rapid; the clams are able to free themselves from the bottom and are aided by water currents in their movements (Sinclair and Isom 1963; Britton and Morton 1979; Eng 1979; Prezant and Chalermwat 1984; French and Schloesser 1996).

Corbicula winter survival in the Connecticut River was correlated positively with water temperature from December through April, and negatively with high-water flows in spring. No clams survived at any of the sites during the coldest winters of 1993–1994 and 1995–1996, when temperatures were below 2°C for 12–15 weeks. Similarly low survival due to coldwater temperatures was reported by a number of other researchers (Horning and Keup 1964; Bickel 1966; Dreier 1977; Rodgers et al. 1979; Cherry et al. 1980; Graney et al. 1980; Cairns and Cherry 1983). *Corbicula* survival at individual stations was also influenced by distance from the thermal plume discharge. Winter survival at site D, the river site nearest to where the CY discharge mixes with the river, was not significantly correlated with winter water temperature indices. The sites between the CY intakes and discharge canal (sites B and C) were significantly correlated with average winter water temperatures, but not correlated with the number of days water temperatures below 1°C or 2°C. The reason for the poor correlation of these temperature indices (≤2°C and ≤1°C) at sites B, C, and D might be related to the thermal plume mixing with ambient temperature river water during tidal incursions into our sampling sites or the

TABLE 9. Total abundance of *Corbicula fluminea*, native clams (Sphaeridae), and native mussels (Unionidae) in relation to the CY power plant's operational status.

		C. fluminea		Sphaeridae		Unionidae	
Month	Time	Mean[a]	SE	Mean	SE	Mean	SE
May	Operational[b]	37.22	8.48	39.11	13.00	1.22	0.47
	Post-operational[c]	54.50	9.70	45.75	11.49	2.12	0.58
August	Operational	733.33	142.29	147.54	21.90	3.18	0.83
	Post-operational	715.50	120.76	119.00	23.50	1.75	0.50
November	Operational	1,190.09	235.71	110.46	25.77	**0.91	0.42
	Post-operational	792.62	106.96	94.25	16.56	**4.00	1.15

[a] Mean = #/m^2 is calculated by averaging the total number of individuals per grab at the A, B, C, and D for each sample month from 1991 to 2000 and converting that value to #/m^2 ((#/grab) × 10).

[b] CY operational years (1991–1996).

[c] Post-operational years (1997–2000).

** Difference in abundance between time periods is highly significant; Satterthwaite *t*-test, $P < 0.01$.

recruitment of some CY discharge canal clams to these sites closest to the discharge outfall. However, *t*-tests of winter survival data at each site between operation and postoperational years indicated no differences. Although twice each day during low flow conditions the river reverses flow and thermal plume water moves upriver where our sites are located (Rozsa 2001; Boyd 2004 [1976]), differences in site correlations to winter severity indices could be caused by site locations in relation to river conditions other than CY discharge effects. The significant correlations observed at site A between river temperature indices and winter survival of clams indicates no thermal plume effects at this station. During periods of high river flow, usually in the spring and other times of flooding, the flow at CY is unidirectional and not influenced by tidal flows. The *Corbicula* population at CY sites was able to recover after high winter mortalities by August and usually peaked in November. This recovery was driven by passive downriver recolonization from populations that were able to survive the cold winter conditions in refugia (e.g., those known to exist near thermal plumes at Middletown and possibly South Meadow power stations or natural refugia, such as groundwater seeps) (Morgan et al. 2003).

The importance of CY thermal discharge as a refuge for *Corbicula* survival in the Connecticut River during cold winters appears minimal. The clams were not able to maintain year-round populations in the discharge canal because of total mortality due to high discharge temperatures from June through October. Most *Corbicula* that could colonize the discharge canal in winter and spring, therefore, were small and not reproductively mature. If movement of clams to the river sites occurred during March through May, it would occur when conditions were poor for their survival (e.g., low spring water temperatures [<12°C] and high, unidirectional spring flows).

Additional evidence that the CY discharge was not necessary for survival of *Corbicula* populations in the Connecticut River is apparent when *Corbicula* abundance during CY operation (1991–1996) was compared to abundance following the plant closure (1997–2000). Following closure of the CY power plant in 1996, the abundance of *Corbicula* at all sites was not significantly different than during the operational period. In addition, following CY closure, clams were able to survive in the discharge canal throughout the year and significant increases in the abundance of larger size clams were observed.

The growth patterns of *Corbicula* in the Connecticut River are similar to those reported by other researchers. The highest growth rate occurred in the summer at temperatures approximating 25°C, and minimal growth occurred at temperatures below 15°C. Others have reported 10–15°C to be the lower threshold for *Corbicula* growth (Joy 1985; Mattice and Wright 1986; McMahon and Williams 1986a; Stites et al. 1995).

The high *Corbicula* growth rate in 1991 was due to small-sized individuals used in the study. Other researchers have also reported that smaller clams have higher growth rates (O'Kane 1976; Britton et al. 1979; Mattice 1979; Dreier and Tranquilli 1981; Dauble et al. 1985; Joy 1985; Mattice and Wright 1986; McMahon and Williams 1986a, 1986b; Foe and Knight 1986; Morgan et al. 2003). The total growth of *Corbicula* in the Connecticut River was in the range of growth reported by others (Britton et al. 1979; McMahon 1983; Welch and Joy 1984; Mattice and Wright 1986; McMahon and Williams 1986a; French and Schloesser 1991). The growth of *Corbicula* at the CY discharge started in the fall and continued modestly until discharge water temperatures exceeded 37°C in June or July, when all clams died. Others reported that the upper temperature tolerance threshold for this clam between 34–37°C (Mattice and Dye 1976; Dreier 1977; Mattice 1979; Cairns and Cherry 1983; McMahon 1983; McMahon and Williams 1986b).

Brooding and spawning of *Corbicula* in the Connecticut River and CY discharge canal occurred at water temperatures between 15°C and 31°C. Ambient river temperatures remained within this range from June to September, and CY discharge conditions remained within this range in November and April to mid-May. Peak reproduction, as described by brooding and spawning, occurred in August at river temperatures (22–28°C) and in April at CY discharge temperatures (13–24°C). The survival of juveniles released in the discharge canal is questionable, particularly during times of high water flow. Potential recruitment to the river from the discharge canal comes when temperature in the river is low for successful growth and river flow rates are at yearly optimal levels for downriver transport by the swift, unidirectional currents of spring freshets. As a result, they would be exposed to estuarine waters of Long Island Sound and perish due to high salinity.

While *Corbicula* quickly established itself as the dominant bivalve in the Connecticut River, there was little change in native bivalve abundance found in the same sediments. We believe this is due to relatively high winter survival of native clams and mussels when compared to that for *Corbicula*. Similarly, Fuller and Imlay (1976) questioned whether there was any displacement of native Unionidae species following *Corbicula* colonization. Diaz (1994) concluded that *Corbicula* invasions of Northern American tidal freshwater systems in the late 1960s did not displace tidal freshwater species, but rather, these nonnative clams took advantage of underutilized benthic resources. Others suggest filtration activities by high densities of *Corbicula* can lead to alteration of nutrient dynamics in aquatic systems, resulting in negative impacts on native bivalves (Kraemer 1979; Lauritsen 1986; Leff et al. 1990; Phelps 1994). Karatayev et al. (2003) found no correlations between the densities of *Corbicula* and other benthic invertebrates in an East Texas lake.

Low densities of native clams and mussels in the Connecticut River may suggest availability of an unoccupied niche for *Corbicula* to colonize. In addition, high water flow of the Connecticut River may provide a steady supply of nutrients and food for filter feeders, further limiting the effects of *Corbicula* on native species.

Conclusion

Within 2 years of their discovery in the Connecticut River, *Corbicula* colonized 38 km of river upstream of the CY plant. The *Corbicula* population in the vicinity of the CY plant was highly variable within and between years. However, annual densities during plant operation in 1991–1996 were not significantly different from those following the plant closure in 1997–2000. This suggests that the CY thermal discharge did not serve as an important refuge area for *Corbicula* overwintering in the vicinity of the plant. Continuous chlorination (at <0.10 ppm) was effective in preventing *Corbicula* from fouling service water systems at CY. The lack of correlation between presence of *Corbicula* and abundance of native clams and mussels suggest no detrimental effect of *Corbicula* on native species in the Connecticut River during these studies. Finally, the success of *Corbicula* in more northerly ranges of New England may not always be associated with the thermal plumes from industrial facilities, but could also result from other thermal refugia such as groundwater seeps that are above 2°C.

Acknowledgments

This research was conducted by Northeast Utilities Service Company (NU) as a Connecticut Department of Environmental Protection permit requirement for the operation of the Connecticut Yankee nuclear power plant. The authors' gratitude is extended to John Tietjen, Robert Wilce, Guy Hoffman, Ernesto Lorda, Douglas G. Smith, and Edward Roseman for their critical reviews of this paper. Special appreciation is given to Bob Parker, Bill McDermott, and all the NU Production Operation Services Laboratory personnel for their many hours of boat work and sampling. We also extend this appreciation to our colleagues at the Millstone Environmental Laboratory for their support during sample collection and processing.

References

Aldridge, D. W., and R. F. McMahon. 1978. Growth, fecundity, and bioenergetics in a natural population of freshwater clam, *Corbicula manilensis* Philippi, from north central Texas. Journal of Molluscan Studies 44:49–70.

Bickel, D. 1966. Ecology of *Corbicula manilensis* in the Ohio River at Louisville, Kentucky. Sterkiana 23:19–24.

Boyd, W. A. 2004. Hydrology. Pages 27–36 *in* P. M. Jacobson, D. A. Dixon, W. C. Leggett, B. C. Marcy, Jr., and R. R. Massengill, editors. The Connecticut River Ecological Study (1965–1973) revisited: ecology of the lower Connecticut River 1973–2003. American Fisheries Society, Monograph 9, Bethesda, Maryland. (Originally published in 1976)

Britton, J. C., D. R. Coldiron, L. P. Evans, Jr., C. Golightly, K. D. O'Kane, and J. R. TenEyck. 1979. Reevaluation of the growth pattern in *Corbicula fluminea* (Müller). Pages 177–192 *in* J. C. Britton, editor. Proceedings, First International *Corbicula* Symposium. October 13–15, 1977. Texas Christian University, Fort Worth.

Britton, J. C., and B. Morton. 1979. *Corbicula* in North America: the evidence reviewed and evaluated. Pages 249–287 *in* J. C. Britton, editor. Proceedings, First International *Corbicula* Symposium.

October 13–15, 1977. Texas Christian University, Fort Worth.

Cairns, J., Jr., and D. S. Cherry. 1983. A site-specific field and laboratory evaluation of fish and Asiatic clam population responses to coal-fired power plant discharges. Water Science and Technology 15:10–37.

Chen, L. 1990. Aquaculture in Taiwan. Fishing News Books, Cambridge, Massachusetts.

Cherry, D. S., J. H. Rodgers, Jr., R. L. Graney, and J. Cairns, Jr. 1980. Dynamics and control of the Asiatic clam in the New River. Virginia. Bulletin of Virginia Water Resources Research Center 123:1–72.

Cherry, D. S., R. L. Roy, R. A. Lechleitner, P. A. Dunhardt, G. T. Peters, and J. Cairns, Jr. 1986. *Corbicula* fouling and control measures at the CELCO plant, Virginia. Pages 69–81 *in* J. C. Britton and R. S. Prezant, editors. Second International *Corbicula* Symposium. June 21–24, 1983, Hattiesburg, Mississippi. American Malacological Bulletin Special Edition 2.

Counts, C. L., III. 1986. The zoogeography and history of the invasion of the United States by *Corbicula fluminea*. Pages 7–39 *in* J. C. Britton and R. S. Prezant, editors. Second International *Corbicula* Symposium. June 21–24, 1983, Hattiesburg, Mississippi. American Malacological Bulletin Special Edition 2.

Dauble D. D., D. S. Daly, and C. S. Abernethy. 1985. Factors affecting growth and survival of the Asiatic clam, *Corbicula* sp., under controlled laboratory conditions. Pages 134–144 *in* R. D. Cardwell, R. Purdy, and R. C. Bahner, editors. Aquatic Toxicology and Hazard Assessment: Seventh Symposium, ASTM, STP854. American Society for Testing and Materials, Philadelphia.

Diaz, R. J. 1994. Response of tidal freshwater macrobenthos to sediment disturbance. Hydrobiologia 278:201–212.

Dreier, H. 1977. Study of *Corbicula* in Lake Sangchris. Pages 1–52 *in* The Annual Report for Fiscal Year 1976, Lake Sangchris Project. Section 7. Illinois Natural History Survey, Urbana.

Dreier, H., and J. A. Tranquilli. 1981. Reproduction, growth, distribution and abundance of *Corbicula* in an Illinois cooling lake. Pages 378–393 *in* R. W. Lawrimore and J. A. Tranquilli, editors. The Lake Sangchris study: case history of an Illinois cooling lake. Bulletin Illinois Natural History Survey 32.

Eng, L. L. 1979. Population dynamics of the Asiatic clam, *Corbicula fluminea* (Müller), in the concrete-lined Delta-Mendota Canal of Central California. Pages 39–68 *in* J. C. Britton, editor. Proceedings, First International *Corbicula* Symposium. October 13–15, 1977. Texas Christian University, Fort Worth.

Foe, C., and A. Knight. 1986. A thermal energy budget for juvenile *Corbicula fluminea*. American Malacological Bulletin 2:143–150.

French, J. R. P., III, and D. W. Schloesser. 1991. Growth and over-winter survival of the Asiatic clam, Corbicula fluminea, in the St Clair River, Michigan. Hydrobiologia 219:165–170.

French, J. R. P., III, and D. W. Schloesser. 1996. Distribution and winter survival health of Asian clams, *Corbicula fluminea*, in the St Clair River, Michigan. Journal of Freshwater Ecology 11:183–192.

Fuller, S. L. H., and M. J. Imlay. 1976. Spatial competition between *Corbicula manilensis* (Philippi), the Chinese clam (Corbiculidae), and fresh-water mussels (Unionidae) in the Waccamaw River Basin of the Carolinas (Mollusca: Bivalvia). Association of Southeastern Biologists Bulletin 23:60.

Gardner, J. A., Jr., W. R. Woodall, Jr., A. A. Staats, Jr., and J. F. Napoli. 1976. The invasion of the Asiatic clam (*Corbicula manilensis* Philippi) in the Altamaha River, Georgia. Nautilus 90:117–125.

Goss, L. B., and C. Cain, Jr. 1975. Power plant condenser and service water system fouling by *Corbicula*, the Asiatic clam. Presented at the Biofouling Workshop held at Johns Hopkins University, Baltimore, Maryland, June 16–19, 1975.

Goss, L. B., J. M. Jackson, H. B. Flora, B. G. Isom, C. Gooch, S. A. Murray, C. G. Burton, and W. S. Bain. 1979. Control studies on *Corbicula* for steam-electric generating plants. Pages 139–151 *in* J. C. Britton, editor. Proceedings, First International *Corbicula* Symposium. October 13–15, 1977. Texas Christian University, Fort Worth.

Graney, R. L., D. S. Cherry, J. H. Rodgers, Jr., and J. Cairns, Jr. 1980. The influence of thermal discharges and substrate composition on the population structure and distribution of the Asiatic clam, *Corbicula fluminea*, in the New River, Virginia. Nautilus 94:130–135.

Horning, W. B., and L. Keup. 1964. Decline of the Asiatic clam in the Ohio River. Nautilus 78:29–30.

Ingram, W. M. 1959. Asiatic clams as potential pests in California water supplies. Journal of the American Water Works Association 51:363–370.

Isom, B. G. 1971. Evaluation and control of macroinvertebrate nuisance organisms in freshwater industrial supply system. Presented at the 24th Annual Meeting of the Midwest (N. A.) Benthological Society.

Isom, B. G. 1986. Historical review of the Asiatic clam (*Corbicula*) invasion and biofouling of waters and industries in the Americas. Pages 1–5 *in* J. C. Britton and R. S. Prezant, editors. Second International *Corbicula* Symposium. June 21–24, 1983, Hattiesburg, Mississippi. American Malacological Bulletin Special Edition No. 2.

Johnson, K. I., C. H. Henager, T. L. Page, and P. F. Hayes. 1986. Engineering factors influencing *Corbicula* fouling in nuclear service water systems. Pages 47–52 *in* J. C. Britton and R. S. Prezant, editors. Second International *Corbicula* Symposium. June 21–24, 1983, Hattiesburg, Mississippi. American Malacological Bulletin Special Edition 2.

Joy, J. E. 1985. A 40-week study on growth of the Asian clam, *Corbicula fluminea* (Müller), in the Kanawha River, West Virginia. Nautilus 99:110–116.

Karatayev, A. Y., L. E. Burlakova, T. Kesterson, and D. K. Padilla. 2003. Dominance of the Asiatic clam, *Corbicula fluminea* (Müller), in the benthic community of a reservoir. Journal of Shellfish Research 22:487–493.

Kraemer, L. S. 1979. Juvenile *Corbicula*: their distribution in the Arkansas River benthos. Pages 89–97 *in* J. C. Britton, editor. Proceedings, First International *Corbicula* Symposium. October 13–15, 1977. Texas Christian University, Fort Worth.

Lauritsen, D. D. 1986. Assimilation of radiolabeled algae by *Corbicula.* Pages 219–222 *in* J. C. Britton and R. S. Prezant, editors. Second International *Corbicula* Symposium. June 21–24, 1983, Hattiesburg, Mississippi. American Malacological Bulletin Special Edition No. 2.

Leff, L. G., J. L. Burch, and J. V. McArthur. 1990. Spatial distribution, seston removal and potential competitive interactions of the bivalves *Corbicula fluminea* and *Elliptio complanata*, in a coastal plain stream. Freshwater Biology 24:409–416.

Magdych, W. P. 1981. An efficient, inexpensive elutriator design for separating benthos from sediment samples. Hydrobiologia 85:157–159.

Mattice, J. S. 1979. Interactions of *Corbicula* sp. with power plants. Pages 119–138 *in* J. C. Britton, editor. Proceedings, First International *Corbicula* Symposium. October 13–15, 1977. Texas Christian University, Fort Worth.

Mattice, J. S., and L. L. Dye. 1976. Thermal tolerance of the adult Asiatic clam. Pages 130–135 *in* G. W. Esch and R. W. McFarlane, editors. Thermal ecology II. Conference 750425. National Technical Information Service, Springfield, Virginia.

Mattice J. S., R. B. McLean, and M. B. Burch. 1982. Evaluation of short-term exposure to heated water and chlorine for control of the Asiatic clam (*Corbicula fluminea*). Oak Ridge National Laboratory, Environmental Services Division Publication No. 1748, Oak Ridge, Tennessee, operated by Union Carbide Corporation for the Department of Energy.

Mattice, J. S., and L. L. Wright. 1986. Aspects of growth of *Corbicula fluminea.* Pages 167–178 *in* J. C. Britton and R. S. Prezant, editors. Second International *Corbicula* Symposium. June 21–24, 1983, Hattiesburg, Mississippi. American Malacological Bulletin Special Edition 2.

McMahon, R. F. 1977. Shell size-frequency distributions of *Corbicula manilensis* from a clam-fouled steam condenser. Nautilus 91:54–59.

McMahon, R. F. 1982. The occurrence and spread of the introduced Asiatic freshwater clam, Corbicula fluminea, in North America: 1924–1982. Nautilus 96:134–141.

McMahon, R. F. 1983. Ecology of an invasive pest bivalve, *Corbicula.* Pages 505–561 *in* W. D. Russell-Hunter, editor. The mollusca. Volume 6, Ecology. Academic Press, Orlando, Florida.

McMahon, R. F., and R. W. Lutey. 1988. Field and laboratory studies of the efficacy of Poly[oxyethylene(dimethyliminio)ethylene(dimethyliminio)ethylene dichloride] as a biocide against the Asian clam, *Corbicula fluminea.* Presented at Service Water System Reliability Improvement Seminar on October 17–19, 1988, Charlotte, North Carolina. Sponsored by Electric Power Research Institute, Palo Alto, California.

McMahon, R. F., and C. J. Williams. 1986a. A reassessment of growth rate, life span, life cycles and population dynamics in a natural population and field caged individuals of *Corbicula fluminea* (Müller) (Bivalvia: Corbiculacea). Pages 151–166 *in* J. C. Britton and R. S. Prezant, editors. Second International *Corbicula* Symposium. June 21–24, 1983, Hattiesburg, Mississippi. American Malacological Bulletin Special Edition 2.

McMahon, R. F., and C. J. Williams. 1986b. Growth, life cycle, upper thermal limit and downstream colonization rates in a natural population of the freshwater bivalve mollusc, *Corbicula fluminea* (Müller) receiving thermal effluent. Pages 231–239 *in* J. C. Britton and R. S. Prezant, editors. Second International *Corbicula* Symposium. June 21–24, 1983, Hattiesburg, Mississippi. American Malacological Bulletin Special Edition 2.

Merriman, D., and L. M. Thorpe. 2004. Introduction. Pages 3–25 *in* P. M. Jacobson, D. A. Dixon, W. C. Leggett, B. C. Marcy, Jr., and R. R. Massengill, editors. The Connecticut River Ecological Study (1965–1973) revisited: ecology of the lower Connecticut River 1973–2003. American Fisheries Society, Monograph 9, Bethesda, Maryland. (Originally published in 1976)

Morgan, D. E., M. Keser, J. F. Foertch, and J. M. Vozarik. 1992. The Asiatic clam, *Corbicula fluminea*, extends its North American distribution to the Connecticut River (abstract only). Page 523 *in* W. J. Blogoslawski, editor. Aquaculture and environmental stewardship: Milford Shellfish Biology Seminar - 1991. Environmental Management 16:521–529.

Morgan, D. E., M. Keser, J. T. Swenarton, and J. F. Foertch. 2003. Population dynamics of the Asiatic clam, *Corbicula fluminea* (Müller) in the lower Connecticut River: establishing a foothold in New England. Journal of Shellfish Research 22:193–203.

Neitzel, D. A., K. I. Johnson, T. L. Page, J. S. Young, and P. M. Daling. 1984. Bivalve fouling of nuclear power plant service water systems In Volume 1, Correlation of bivalve biological characteristics and raw water system design. U.S. Nuclear Regulatory Commission, NUREG/CR-4070, PNL-5300, Washington, D. C.

NSI (Nautical Software, Inc.). 1995. Tides and currents for Windows, version 2.0. Beaverton, Oregon. Available: nauticalsw@aol.com.

O'Kane, K. D. 1976. A population study of the exotic bivalve *Corbicula manilensis* (Philippi, 1841) in selected Texas reservoirs. Master's thesis. Texas Christian University, Fort Worth.

Page, T. L., D. A. Neitzel, M. A. Simmons, and P. F. Hayes. 1986. Biofouling of power plant service systems by *Corbicula.* Pages 41–45 *in* J. C. Britton and R. S. Prezant, editors. Second International *Corbicula* Symposium. June 21–24, 1983, Hattiesburg, Mississippi. American Malacological Bulletin Special Edition 2.

Phelps, H. L. 1993. *Corbicula* production and marketing techniques in Taiwan. Report to the University of the District of Columbia Research Council, Washington, D.C.

Phelps, H. L. 1994. The Asiatic clam (*Corbicula fluminea*) invasion and system-level ecological change in the Potomac River Estuary near Washington, D.C. Estuaries 17:614–621.

Prezant, R. S., and K. Chalermwat. 1984. Flotation of the bivalve *Corbicula fluminea* as a means of dispersal. Science 225:1491–1493.

Rodgers, J. H., Jr., D. S. Cherry, K. L. Dickson, and J. Cairns, Jr. 1977. Invasion of Asiatic clam, *Corbicula manilensis* in the New River, Virginia. Nautilus 9:43–46.

Rodgers, J. H., Jr., D. S. Cherry, K. L. Dickson, and J. Cairns, Jr. 1979. Invasion population dynamics and elemental accumulation of *Corbicula fluminea* in the New River at Glen Lyn, Virginia. Pages 99–110 *in* J. C. Britton, editor. Proceedings, First International *Corbicula* Symposium. October 13–15, 1977. Texas Christian University, Fort Worth.

Rozsa, R. 2001. Hydrology of the lower Connecticut River. Pages 18–22 *in* G. D. Dreyer and M. Caplis, editors. Living resources and habitats of the lower Connecticut River. December 2001. The Connecticut College Arboretum Bulletin 37, New London, Connecticut.

Savoy, T. F., and J. Benway. 2004. Food habits of shortnose sturgeon collected in the lower Connecticut River from 2000 through 2002. Pages 353–360 *in* P. M. Jacobson, D. A. Dixon, W. C. Leggett, B. C. Marcy, Jr., and R. R. Massengill, editors. The Connecticut River Ecological Study (1965–1973) revisited: ecology of the lower Connecticut River 1973–2003. American Fisheries Society, Monograph 9, Bethesda, Maryland.

Sickel, J. B. 1979. Population dynamics of *Corbicula* in the Altamaha River, Georgia. Pages 69–80 *in* J. C. Britton, editor. Proceedings, First International *Corbicula* Symposium. October 13–15, 1977. Texas Christian University, Fort Worth.

Sickel, J. B. 1986. *Corbicula* population mortalities: factors influencing population control. Pages 89–94 *in* J. C. Britton and R. S. Prezant, editors. Second International *Corbicula* Symposium. June 21–24, 1983, Hattiesburg, Mississippi. American Malacological Bulletin Special Edition 2.

Sinclair, R. M., and B. G. Isom. 1963. Further studies on the introduced Asiatic clam *Corbicula* in Tennessee. Tennessee Stream Pollution Control Board.

Stites, D. L., A. C. Benke, and D. M. Gillespie. 1995. Population dynamics, growth, and production of the Asiatic clam, Corbicula fluminea, in a blackwater river Canadian Journal of Fisheries and Aquatic Sciences 52:425–437.

USGS (United States Geological Survey). 2002. Water resources of the United States, surface-water data for Connecticut, August 15, 2002. Available: http://waterdata.usgs.gov/ct/nwis/sw.

USNRC (United States Nuclear Regulatory Commission). 1981. Flow blockage of cooling water to safety system components by *Corbicula* sp (Asiatic clam) and *Mytilus* sp. (mussel). IE Bulletin no. 81–03, Office of Inspection and Enforcement, United States Nuclear Regulatory Commission, Washington, D.C.

Voelz, N. J., J. V. McArthur, and R. B. Rader. 1998. Upstream mobility of the Asiatic clam *Corbicula fluminea*: identifying potential dispersal agents. Journal of Freshwater Ecology 13:39–45.

Welch, K. J., and J. E. Joy. 1984. Growth rates of the Asiatic Clam, *Corbicula fluminea* (Müller), in the Kanawha River, West Virginia. Freshwater Invertebrate Biology 3:139–142.

American Fisheries Society Monograph 9:441–454, 2004

Comparison of Benthic Community Structure in the Lower Connecticut River before, during, and after Operation of the Connecticut Yankee Power Plant

RONALD R. MASSENGILL*

2434 Trenton Street, Cambria, California 93428, USA

JOSEPH M. VOZARIK, DOUGLAS E. MORGAN, AND MILAN KESER

Millstone Environmental Laboratory, Dominion Nuclear Connecticut, Inc. Waterford, Connecticut 06385, USA

Abstract.—This study was conducted in 2002, 6 years after the Connecticut Yankee Nuclear power plant (CY) was permanently retired, to characterize the postoperational benthic community around CY and compare it to the preoperational (1965–1967) and early operational (1968–1972) communities described in the original Connecticut River Ecological Study. Benthic communities in the lower Connecticut River were dominated by annelids in the preoperational and early operational periods. Mollusks were the numerically dominant phylum in the 2002 postoperational sampling year, primarily due to high abundance of Asiatic clams (also known as Asian clams) *Corbicula fluminea* introduced to the Connecticut River in the 1990s. Trend analyses of total annual abundance and numbers of species showed no significant trend during the 1965–1972 period at stations with silt or sand sediment types. Total annual abundance and number of taxa were highest in 2002 when compared to data from 1965 to 1972. Communities were also more diverse in 2002, with 71 taxa identified compared to 29 and 34 in the preoperational and operational periods, respectively. More insect taxa were identified in both preoperational (11) and operational (16) periods than any other taxonomic group. Insects (38 taxa) were also the dominant taxonomic group in 2002. Analysis of variance (ANOVA) revealed no significant differences between preoperational and operational periods in total annual abundance and numbers of taxa at sand stations. There was no significant difference between preoperational and operational data for the number of taxa at silt stations, but there was a significant increase for total annual abundance. The shift in community structure toward higher numbers of taxa, and total annual abundance during 2002, when compared to data from the preoperational and operational periods, was attributed to improvement of water quality in the Connecticut River during the last 30 years, recovery of the submerged aquatic plant genus *Vallisneria*, and the introduction of *Corbicula fluminea* in the early 1990s.

Introduction

The Connecticut Yankee power plant (CY) was permanently shut down on 22 July 1996. This study examines the Connecticut River benthic infaunal community sampled near CY in July and October 2002 and compares it to observations during the same months in years before (1965–1967) and during (1968–1972) CY operation as reported (Massengill 2004 [1976], this volume) in the original Connecticut River Ecological Study. Benthic communities are useful indicators of environmental conditions in freshwater systems. Community and population-level responses to environmental changes make benthic systems ideal for long-term monitoring studies (Sheehan 1984; Townsend et al. 1987; Weatherley and Ormerod 1990; Strayer and Smith 2001). The tidal freshwater benthic community of the Connecticut River near CY was studied as part of a comprehensive ecological monitoring program

* *Corresponding author*: ab6fm@arrl.net

designed to assess impacts associated with CY operation (Massengill 2004 [1976]). Tidal freshwater areas have ecological value because they serve as nursery grounds for commercially important species such as shad, herring, striped bass *Morone saxatilis*, and eels (Massmann 1954). This study began in 1965, prior to passage of Connecticut Clean Water Act of 1967 and the start-up of CY in 1968, and continued through initial years of CY operation until 1972 when the Federal Water Pollution Control Act Amendment became law. This study was conducted in 2002, 6 years after CY was permanently retired, to characterize the postoperational benthic community around CY and compare it to preoperational and early operational communities described by Massengill (2004 [1976]).

Methods

Site Description

The Connecticut Yankee power plant is located on the eastern shore of the Connecticut River in Haddam, Connecticut, 26 km upstream from Long Island Sound (Figure 1). The river width in the vicinity of CY ranges from 360 to 640 m. The navigation channel is 5–10 m in depth and follows along the western shore of the river opposite the CY intake structure. River bottom near the intake was dredged to a minimum water depth of ~6 m (Marcy 2004 [1976], this volume). An extensive sand flat extends ~1.6 km downstream of the intake until it intersects the discharge canal. A shallow channel (<1 m deep) runs adjacent to the sand flat on the eastern side until it meets with the discharge canal. Sediments downstream from the canal are composed predominantly of silt (Massengill 2004 [1976]). The Connecticut River at CY is tidal, with an amplitude of ~1 m. The saltwater wedge from Long Island Sound extends into the river to about 3 km downstream of CY (Merriman and Thorpe 2004). The annual spring freshet produces high unidirectional river flow; during periods of low precipitation, river flow can reverse due to the influence of tides (Merriman and Thorpe 2004).

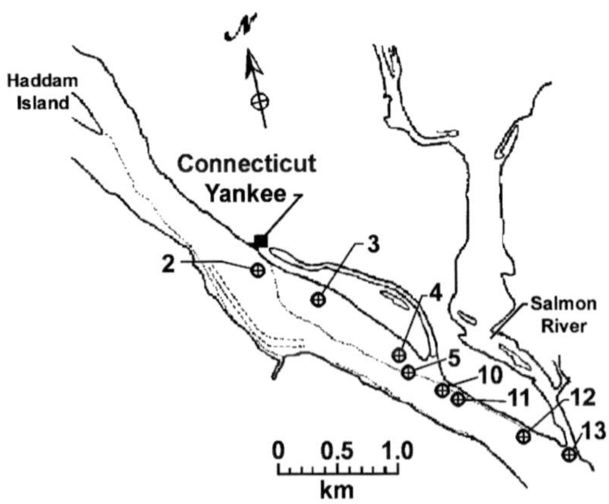

FIGURE 1. Location of Connecticut River benthic sampling stations in the vicinity of CY: stations 2–5 (sand) and stations 10–13 (silt)

Sample Collection

Benthic habitats at eight sites in the vicinity of CY were sampled in July and October 2002 (Figure 1). These sites were previously sampled by Massengill (2004 [1976]) during 1965–1972, and include four stations (2–5) where sediments are sand and four stations (10–13) where sediments are silt. Samples were collected with a weighted Peterson Grab that sampled an area of 0.1 m^2. Three grab samples were collected at each site and each grab was subsampled using three 10-cm-diameter by 5-cm-deep cores, which were pooled together and screened using a 1-mm sieve. Samples were floated using an elutriator (Magdych 1981) to separate organisms from sediment prior to fixing the sample for 48 h in 10% formalin, staining it with rose bengal and preserving it in 70% denatured ethanol. Both the floated portion and remaining sediments were examined using a dissecting microscope (10×) to remove benthic invertebrates, with all organisms identified to the lowest practicable taxon (Merritt and Cummins 1978; Peckarsky et al. 1990; Thorp and Covich 1991; Smith 1995; McCreary Waters and San Giovanni 2002) and counted.

Community Analyses

Annual collections (June and October combined) at each station were compared by calculating the Bray-Curtis similarity index between each pair of years, using the formula (Clifford and Stephenson 1975):

$$S_{jk} = \frac{\sum_{i=1}^{n} 2 \min\left(X_{ij}, X_{ik}\right)}{\sum_{i=1}^{n}\left(X_{ij} + X_{ik}\right)} \quad (1)$$

where S_{jk} is the similarity index between year j and year k; X_{ij} is the log transformed (ln + 1) average abundance of taxon i in year j; X_{ik} is the abundance in year k; and n is the number of taxa in common. A group-average-sorting, clustering algorithm was applied to the resulting similarity matrix (Lance and Williams 1967).

NonMetric Multi-Dimensional Scaling (MDS)

The NonMetric Multi-Dimensional Scaling (MDS) classification techniques used herein were introduced by Shepard (1962) and Kruskal (1964). Multi-Dimensional Scaling was applied to the same similarity matrix used in the classification and cluster analyses. This analysis was performed with the PRIMER suite of programs developed by the Plymouth Marine Laboratory (Warwick and Clarke 1991; Clarke 1993). The purpose of MDS is to construct a map or configuration of samples in a two-dimensional plot. There are no absolute similarity values associated with MDS. Interpretation of sample placement on a configuration/map is relative in nature (i.e., more similar samples are closer together and termed "island"). The iterative process of MDS seeks to position the elements in the configuration/map with the least amount of distortion (stress) derived from the original data matrix. Stress values less than 0.2 indicate a potentially useful two-dimensional picture. Results of the cluster analyses are compared with the ordination plot as a check of the adequacy and mutual consistency of both representations.

Trend Analysis

The nonparametric (i.e., distribution-free) Mann-Kendall test (Hollander and Wolfe 1973) was used to determine whether time-series data on community abundance and numbers of taxa exhibited significant trends over time. Annual means of total density and total numbers of taxa were used as variables in the trend analysis. This approach to trend analysis was suggested by Gilbert (1989) as being particularly well suited for analysis of environmental monitoring data because no distributional assumptions are required and small sample sizes are acceptable. The 2002 annual data were not used in this analysis because the gap between sampling periods was too long to fit the analysis assumptions.

Results

Community Composition and Abundance

Sand Stations

Benthic communities in the vicinity of CY were composed primarily of annelids, insects and mollusks. Annelids were generally the most abundant taxon at sand stations in the preoperational period (Table 1). Annelid densities ranged from 757 to 2,250/m^2 in July and from 365 to 1,367/m^2 in October, and comprised 70% of the individuals collected in July and 55% in October (Figure 2). Mollusks typically ranked second in abundance, ranging from 376 to 1,005/m^2 in July and 314–865/m^2 in October. Mollusks were more abundant than annelids only in the October samples at station 3 (865 versus 365/m^2). Insects were least abundant, ranging from 15 to 102/m^2 in July and 23–190/m^2 in October, and accounted for 3% and 7% of the individuals collected during the July and October sampling periods, respectively (Figure 2).

Annelids continued to be the numerically dominant taxon at all sand stations during the operational period (Table 1) except at station 2, where infauna were absent due to dredging activities. Annelids were found at densities ranging from 1,664 to 2,671/m^2 in July and 1,063–2,617/m^2 in October and comprised 63% and 65% of the total individuals in July and October samples, respectively, during the operational period (Figure 2). Insects ranked second in abundance during the operational period, ranging from 371 to 1,205/m^2 in July and from 416 to 886/m^2 in October and accounting for 22% and 28% of the total individuals collected in July and October, respectively. Mollusks were least abundant, ranging from 195 to 512/m^2 in July and 416–886/m^2 in October and accounting for 11% (July) and 7% (October) of the individuals collected.

The benthic communities at sand stations in 2002 differed from those of the preoperational

TABLE 1. Mean density (#/m^2) of individuals in major taxonomic groups at Connecticut River sand stations in July and October during the pre-operational and operational periods, and 2002.

	Pre-operational		Operational		2002	
	July	October	July	October	July	October
Station 2						
Annelids	2,250	1,367	0	0	257	56
Insects	15	93	0	0	269	84
Mollusks	1,005	598	0	0	539	2,552
Station 3						
Annelids	757	365	1,664	1,121	41	0
Insects	92	23	371	725	29	48
Mollusks	437	865	512	461	314	2,316
Station 4						
Annelids	1,246	1,198	2,149	1,063	64	237
Insects	44	190	579	416	64	290
Mollusks	560	713	322	70	518	2,080
Station 5						
Annelids	1,771	659	2,671	2,617	2,267	2,098
Insects	102	156	1,205	886	1,260	1,399
Mollusks	376	314	195	19	1,159	1,499

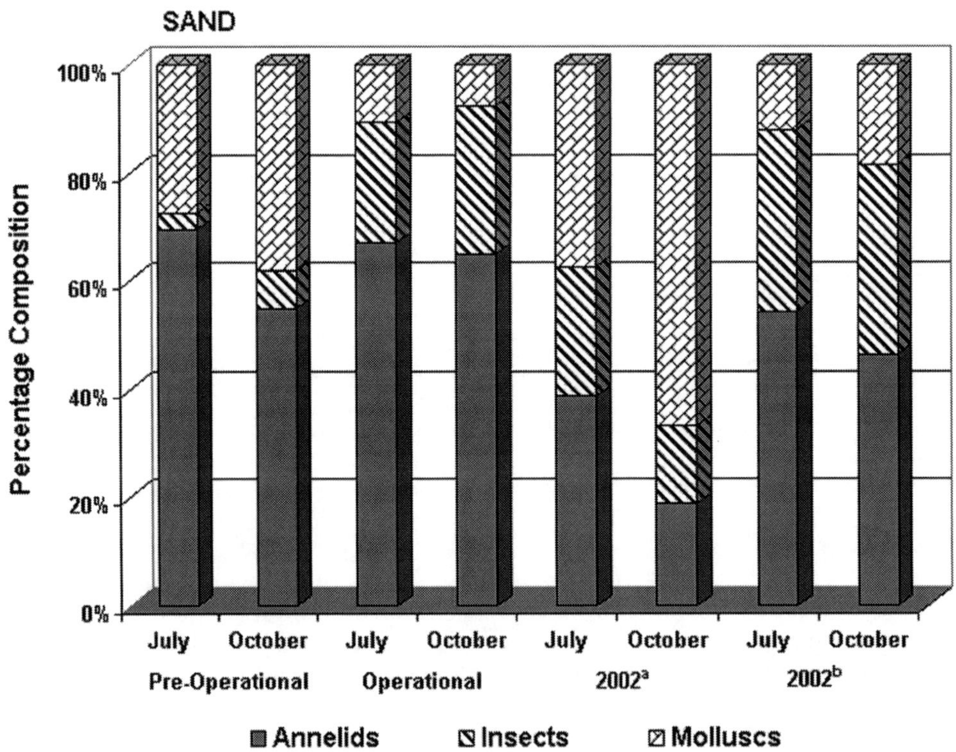

FIGURE 2. Percent composition of annelids, insects and mollusks at sand sites in the vicinity of CY during the pre-operational (1965–1967), operational (1968–1972) and the post-operational (2002) periods. The post-operational year is presented with (2002[a]) and without (2002[b]) *Corbicula fluminea* included in the percentages.

and operational periods, with mollusks being most abundant at stations 2, 3 and 4. Mollusk abundance in 2002 ranged from 314 to 1,159/m^2 in July and 1,499–2,552/m^2 in October (Table 1), and accounted for 37% and 67% of the individuals collected in July and October 2002, respectively (Figure 2). The range in annelid densities at stations 2, 3, and 4 was 41–257/m^2 in July and 0–237/m^2 in October. Annelid density at station 5 was substantially higher (2,098 and 2,267/m^2). Annelids comprised 39% and 19% of the organisms in the July and October sampling periods, respectively, for all stations. Insect abundance ranged from 29 to 1,260/m^2 in July and 48–1,399/m^2 in October and accounted for 24% and 14% of the total individuals collected in the July and October periods, respectively.

Silt Stations

Annelids were the dominant taxon at silt stations in the preoperational period (Table 2). Annelids occurred at densities ranging from 1,517–3,480/m^2 in July and 1,270–2,654/m^2 in October, and accounted for 91% (July) and 79% (October) of the individuals collected in the preoperational period (Figure 3). Insects generally ranked second in abundance (except at station 10 in October), ranging from 98 to 371/m^2 in July and 161–542/m^2 in October and comprising 7% and 16% of the individuals collected in July and October, respectively. Mollusks were generally least abundant, with densities ranging from 33 to 99/m^2 in July and 28–353/m^2 in October and comprising 2% (July) and 5% (October) of the individuals collected at all stations.

Annelids were also the most abundant taxon during the operational period at silt stations (Table 2). Annelid densities were higher than during the preoperational period, ranging from 5,053–10,508/m^2 in July and 3,922–7,775/m^2 in October, accounting for 85% and 86% of the individuals collected during July and October of the operational period, respectively (Figure 3). Insects ranked second and also exhibited high abundance, ranging from 726 to 1,949/m^2 in July and from 529 to 1,798/m^2 in October, and accounted for 15% in July and 14% in October of the individuals collected in the operational period. Mollusks were least abundant and showed a substantial decrease from the preoperational period, ranging from 0 to 13/m^2 in July and 0–36/m^2 in October, accounting for less than 1% of the organisms found at all stations.

Benthic communities at silt stations in 2002 were substantially different from those observed

TABLE 2. Mean density (#/m^2) of individuals in major taxonomic groups at Connecticut River silt stations in July and October during the pre-operational and operational periods, and 2002.

	Pre-operational		Operational		2002	
	July	October	July	October	July	October
Station 10						
Annelids	3,480	2,654	6,697	7,775	985	1,285
Insects	371	161	1,949	529	745	1,242
Mollusks	99	353	13	0	932	1,671
Station 11						
Annelids	2,798	1,907	5,053	6,725	170	406
Insects	103	542	1,308	609	219	446
Mollusks	49	171	0	17	373	3,205
Station 12						
Annelids	2,528	2,520	10,508	6,451	353	1,929
Insects	98	443	1,219	1,798	229	2,411
Mollusks	43	42	0	36	229	1,860
Station 13						
Annelids	1,517	1,270	6,525	3,922	577	692
Insects	230	534	726	1,199	623	553
Mollusks	33	28	0	19	738	2,906

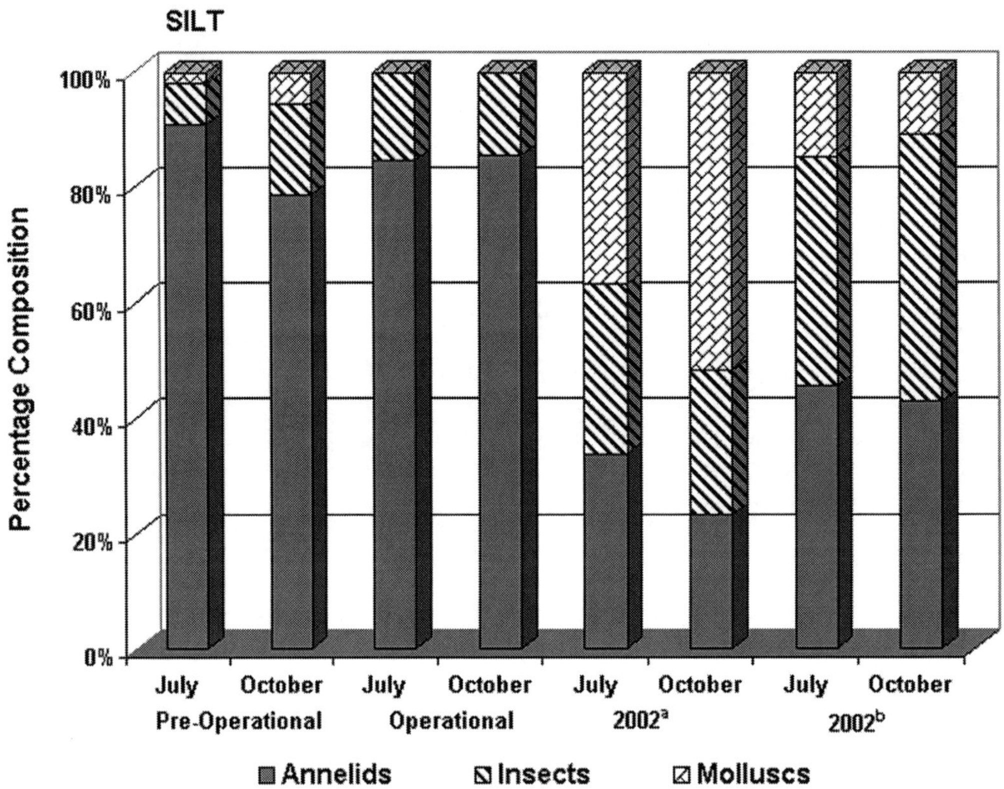

FIGURE 3. Percent composition of annelids, insects and mollusks at silt sites in the vicinity of CY during the preoperational (1965–1967), operational (1968–1972) and the post-operational (2002) periods. The post-operational year is presented with (2002[a]) and without (2002[b]) *Corbicula fluminea* included in the percentages.

during either the preoperational or operational periods. Mollusks were more abundant in 2002 than in any previous period, ranging from 229 to 932/m^2 in July and 1,671–3,205/m^2 in October and comprising 37% and 52% of the total individuals collected in the July and October samples, respectively (Figure 3). Annelids were less abundant than in previous periods, ranging from 170 to 985/m^2 in July and 406–1,929/m^2 in October and accounting for 34% and 23% of the individuals collected in July and October, respectively. Insect densities ranged from 219 to 745/m^2 in July and 446–2,411/m^2 in October, with relative abundance of 29% and 25% in July and October, respectively.

Total Density

Sand Stations

Annual mean densities at CY sand and silt stations were both spatially and temporally variable (Figure 4). The ranges of faunal densities at sand stations (Figure 4A) were 175–1,205/m^2 in the preoperational years, 0–1,298/m^2 in the operational years, 488–3,921/m^2 in 2002 with the Asiatic clam (also known as Asian clam) *Corbicula fluminea* abundance excluded, and 1,413–5,015/m^2 with *C. fluminea*. Faunal densities during 2002 were higher than previous periods both with and without *C. fluminea*. Analysis of variance and trend analyses indicated no significant trends in abundance at any sand station during the period 1965–1972.

Silt Stations

Densities at silt stations (Figure 4B) were generally higher when compared to those at sand stations. The ranges of densities were 248–1,835/m^2 during the preoperational years, 943–2,688/m^2 during the operational years, and 705–3,003/m^2 with *C. fluminea* excluded and 2,434–3,886

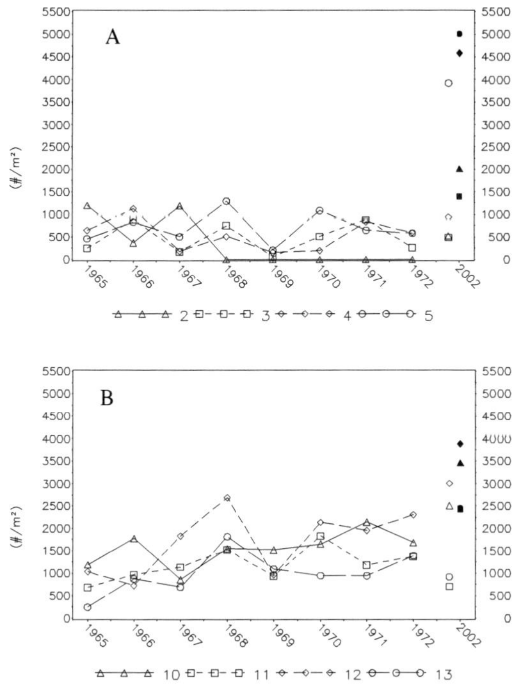

FIGURE 4. Mean number of individuals per m^2 at Connecticut River (A) sand stations (2–5) and (B) silt stations (10–13) in the vicinity of Connecticut Yankee power station in each year from 1965 to 1972 and in 2002. Open symbols in 2002 represent densities excluding *Corbicula fluminea*; filled symbols include *Corbicula fluminea.*

with *C. fluminea* included during 2002. Trend analysis indicated that there were no long-term trends of abundance during the period 1965–1972. However, ANOVA indicated significant differences ($P < 0.002$) in abundance at the silt stations between the preoperational and operational periods. Total faunal abundance was significantly greater ($P < 0.0001$) at silt than at sand stations.

Numbers of Taxa

Sand Stations

The annual number of taxa at CY benthic stations was also variable over time and location (Figure 5). Total number of taxa at sand stations (Figure 5A) was 4–8 in the preoperational years and 0–10 in the operational years. The numbers of taxa in 2002 were higher (8–31). There were no long-term (1965–1972) trends apparent for numbers of taxa at any sand station.

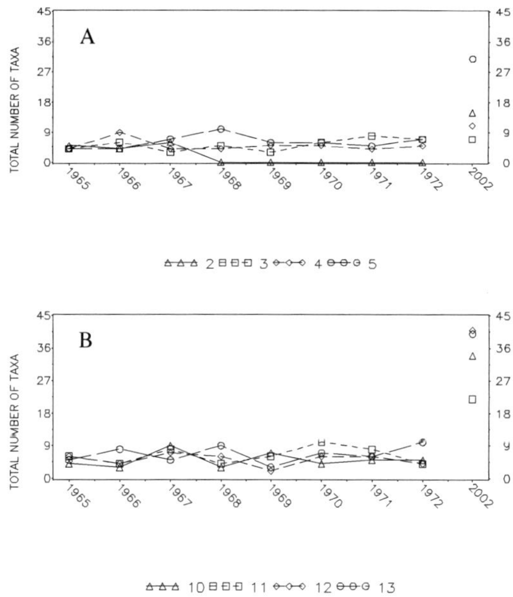

FIGURE 5. Total number of taxa identified at Connecticut River (A) sand stations (2–5) and (B) silt stations (10–13) in the vicinity of Connecticut Yankee power station each year from 1965 to 1972 and 2002.

Silt Stations

The number of taxa at silt stations (Figure 5B) in the preoperational period ranged from 3 to 9 and from 3 to 10 during the operational period. Taxa numbers in 2002 ranged from 19 to 41. No trends in number of taxa were observed at any silt station from 1965 to 1972, and ANOVA showed no significant difference between sand or silt stations.

Presence/Absence of Taxa

A list of taxa collected at all stations in the preoperational, operational, and 2002 surveys is presented in Table 3. A total of 89 taxa were identified in the three survey periods: 29 in the preoperational years (1965–1967), 34 in the operational years, and 71 in the 2002 survey. Insect taxa were the most numerous during the preoperational (11) and operational (16) periods, followed by mollusks (9 taxa for each period). There were 25 taxa in common between the preoperational and operational periods. More insect taxa were observed in 2002 (40) than in previous periods, largely due to substantially higher chironomid diversity. More mollusk taxa (15) were also found in 2002,

TABLE 3. Organisms collected from benthic sand and silt stations during the pre-operational (1965–1967), operational (1968–1972), and post-operational (2002) periods.

Taxon	Pre-operation	Operation	2002
Cnidaria			
Hydrozoa	X	X	
Ectoprocta	X		
Platyhelminthes			
Turbellaria	X	X	X
Nematoda	X	X	X
Annelida			
Polychaeta			
Spionidae			
Marenzelleria viridis		X	
Sabellidae			
Myxicola spp.			X
Manayunkia spp.			X
Oligochaeta			
Tubificidae			
Limnodrilus hoffmeisteri	X	X	X
Branchiura sowerbyi			X
Potamothrix vejdovski			X
Naididae			
Pristine spp.			X
Nais spp.			X
Lumbriculidae			X
Hirudinea			
Glassiphoniidae			
Placobdella spp.			X
Placobdella ornata		X	
Helobdella stagnalis	X	X	
Mollusca			
Gastropoda			
Ferrissia spp.	X	X	
Lymnaea spp.	X	X	
Lyogyrus pupoidea			X
Physa heterstropha	X	X	X
Gyraulus spp.			X
Amnicola limosa			X
Helisoma trivolvis			X
Valvata tricarinata			X
Pelecypoda			
Corbicula fluminea			X
Pisidium compressum	X	X	X
Pisidium nitidum	X	X	X
Sphaerium spp.	X		X
Musculium securis		X	X
Elliptio complanata	X	X	X
Anodonta implicata			*X*
Ligumia ochracea			X
Lampsilis radiata	*X*	*X*	
Lampsilis cariosa	X	X	
Arthropoda			
Crustacea			
Diplostraca			
Illyocryptus acutifrons	*X*	*X*	
Amphipoda			

TABLE 3. Continued

Taxon	Pre-operation	Operation	2002
Gammarus faciatus	X	X	X
Jassa falcata			X
Caprellidae			
Caprella c.f. penantis (= gometrica)			X
Isopoda			
Cyathura polita	X		
Caecidotea spp.			X
Insecta			
Ephemeroptera			X
Baetidae			X
Odonata			X
Macromia spp.		X	X
Gomphus spp.		X	X
Ishnura spp.		X	X
Plecoptera	X	X	X
Hemiptera	X	X	X
Saldidae			X
Megaloptera	X	X	
Trichoptera	X	X	X
Oecetis spp.			X
Mystacides spp.			X
Hydroptilidae	X	X	X
Coleoptera			
Berosus spp.	X	X	
Phanocerus spp.			X
Pronoterus spp.			X
Haliplus spp.			X
Diptera			
Ceratopogonidae		X	X
Heleidae		X	X
Chironomidae			
Cryptochironomus spp.	X	X	X
Chironomus spp.	X	X	X
Microtendipes spp.	X		X
Cryptotendipes spp.	X	X	X
Glyptotendipes spp.			X
Monopelopia spp.			X
Sublettea spp.			X
Ablabesmyia spp.			X
Endochironomus spp.			X
Protanypus spp.			X
Robackia spp.			X
Cricotopus spp.			X
Pseudochironomus spp.			X
Parachironomus spp.			X
Srtictochironomus spp.			X
Orthocladius spp.			X
Procladius spp.	X	X	X
Culicidae		X	X
Tabanidae			X
Chrysops spp.			X
Pelecorhynchidae			X
Glutops spp.			X

compared to previous periods. In addition to the representative freshwater taxa, there were three marine taxa found in the 2002 study that were not found in the previous study. These taxa were the Sabellid polychaete *Myxicola* sp., the Gammaridean amphipod *Jassa falcata*, and the Caprellid amphipod *Caprella c.f. penantis*.

Multivariate Analyses

Cluster Analysis

The clustering dendrogram based on similarities of Connecticut River annual benthic faunal collections from 1965 to 1972 and 2002 shows groupings based on habitat (i.e., sand versus silt) and preoperational (1965–1967), operational (1968–1972), and postoperational (2002) periods (Figure 6A). Three groups, comprising almost all the preoperational and operational period collections, were formed at the 57% similarity level. Group I consisted mostly of sand collections from the preoperational period (of 13 samples, all but 2 were sand, and all but 3 were collected during the preoperational period). This community was characterized by high abundance of the bivalve *Pisidium* sp. and low abundance of the insect *Cryptochironomus* sp.

Group II collections were mostly from Operational years (of 31 samples, all but 12 were in the operational period). This cluster was further divided into group IIa (all silt, mostly collections from 1968, 1972, and half of the 1969 samples) and group IIb (mostly sand samples throughout the operational period). The former subgroup (IIa), was characterized by high abundance of *Cryptochironomus* sp., low abundance of *Procladius* sp. (another chironomid insect), and moderate abundance of the oligochaete *Limnodrilus hoffmeisteri*. The latter subgroup (IIb) shared many of the same species, but with generally lower abundance, especially for *Cryptochironomus* sp.

Group III collections were mostly (9 of 11) silt samples, encompassing the operational years not included in IIa (i.e., 1970, 1971, and the other half of the 1969 samples). This cluster was characterized by relatively high abundance of chironomids, especially *Procladius* sp., and of *Limnodrilus hoffmeisteri*.

Preoperational period silt collections did not form a discrete cluster, but rather were distributed among groups I, IIa, and IIb, as well as a pair of outliers to groups II and III. This pair (silt stations 10 and 11 in 1967) was distinguished by having high abundance of the gastropod *Ferrissia* sp. Another pair of outliers in the cluster dendrogram (stations 3 and 13 in 1972) were dissimilar to other preoperational and operational period collections by having high abundance of the amphipod *Gammarus fasciatus*.

Group IV consisted entirely of collections made in 2002, representing the only postoperational data. They exhibited relatively low similarity among themselves (~40–70%) and very low similarity (~20%) to the other groupings. This dissimilarity was attributed to the high number of species collected in 2002 and to the high abundance of some species, particularly the recently introduced bivalve *Corbicula fluminea*.

Multi-Dimensional Scaling (MDS)

The MDS plot of annual benthic collections from the Connecticut River (the same data set used for the cluster analysis) also illustrates within- and among-group similarities, based on spatial and temporal variables (Figure 6B). Similarity between samples is inversely related to plotted distance between them, projected onto a two-dimensional surface (i.e., similar collections are grouped close together, dissimilar ones farther apart. Generally, this analysis supports conclusions drawn from the above cluster; the preoperational (1965–1967) and operational (1968–1972) period collections form a relatively tight island grouping. In fact, similarity within the group is high enough to overlay collections and obscure labels. There is some separation of sand stations (2–5) to the "south" end of the island, and silt stations (10–13) to the "north." The most dissimilar collections from 1965 to 1972 (stations 3 and 13 in 1972; noted as outliers in the cluster above) are slightly separated to the northeast of the island. Again, the postoperational samples, collected in 2002, exhibit little within-group similarity and very little similarity to the oe-operational or operational years.

Discussion

The contributions of three major taxonomic groupings (annelids, insects, and mollusks) to the benthic community in 2002 were different from those observed during the preoperational and operational periods at CY. Annelids, the numeri-

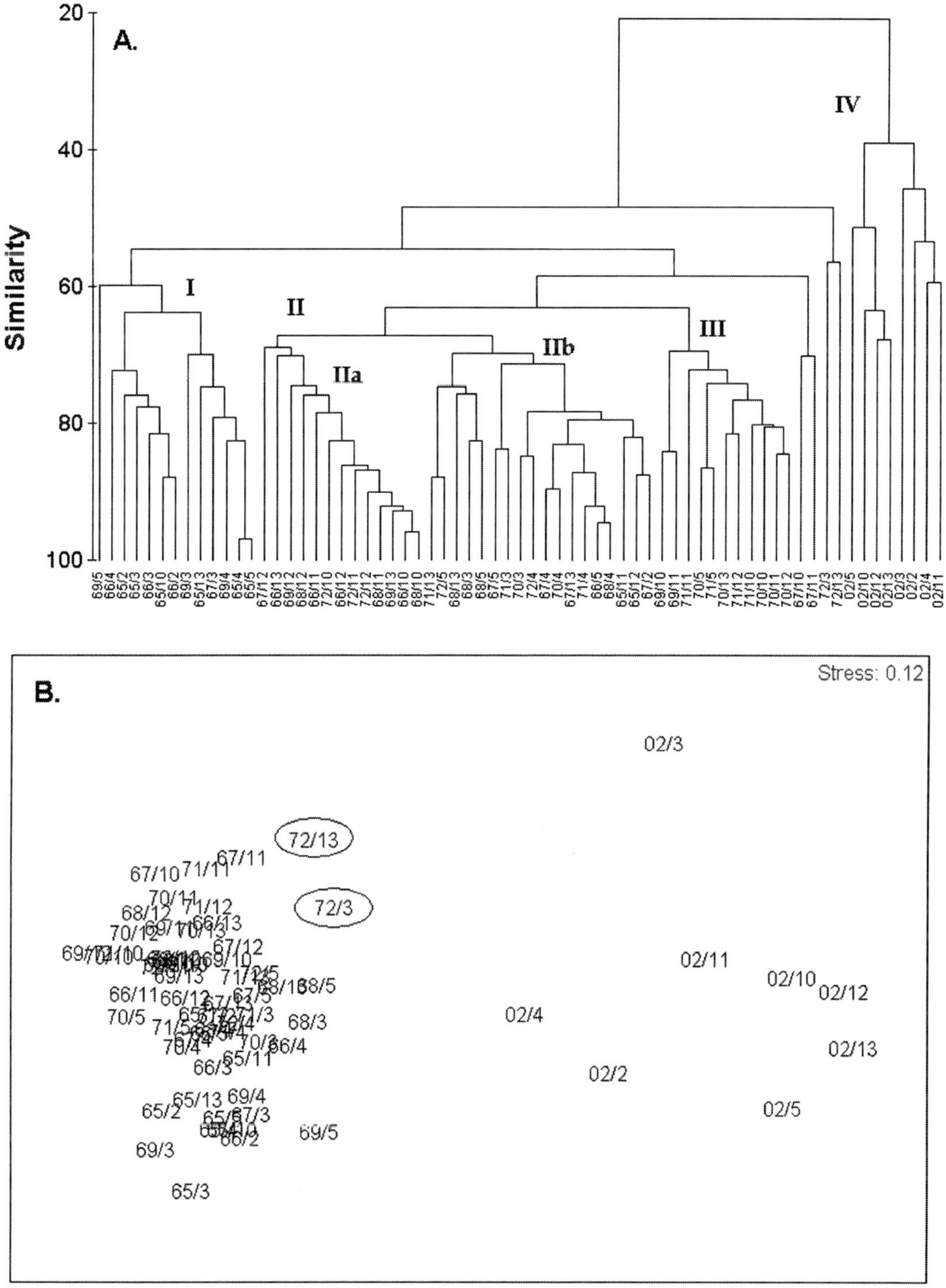

FIGURE 6. Clustering dendrogram (A) year/station entities are indicated along the horizontal axis and Multi-Dimensional Scaling two-dimensional plot (B) based on annual benthic community data during 1965–1972 and 2002 in the vicinity of Connecticut Yankee power station. Station 2 1968–1972 not included, as no organisms were found following intake dredging.

cally dominant phylum in the preoperational and operational periods, were replaced by mollusks in 2002. Insects were also more abundant in 2002 than in previous periods.

Numbers of taxa and total density increased in 2002 relative to the preoperational and operational periods. There was no statistical difference in these parameters between the preoperational and operational periods when station 2 (impacted by intake dredging) was not included in the analyses. This dredging, coupled with scouring resulting from increased water flow in the intake area during CY operation, removed fine sediments, leaving only coarse sand and gravel and no observable infauna for at least 5 years (Massengill 2004 [1976]). Jordan et al. (1976) found that the operation of Surry Nuclear power station in Virginia had only very localized effects on the James River benthic community.

Major changes in the benthic community in the vicinity of CY observed in 2002, 30 years after the preoperational and early operational period studies are likely not related to CY operation. For example, water quality of the Connecticut River has improved dramatically following the passage of the Connecticut Clean Water Act of 1967 (State of Connecticut 1967) and the Federal Water Pollution Control Act Amendment of 1972 (Mullaney 2004, this volume). Subsequent sewage treatment plant upgrades resulted in significant decreases in total phosphorus, organic carbon, and turbidity and increased dissolved oxygen (Trench and Vecchia 2002). No trend in total nitrogen was evident in data collected during 1973–1995 (Trench 2000); however, nitrogen concentrations decreased during the 1990s (Trench and Vecchia 2002).

Disappearance of the submerged macrophyte *Vallisneria americana* in the vicinity of CY in 1968 was attributed to exposure to the thermal effluent (Massengill 2004 [1976]). However, Barrett et al. (1997) reported that *V. americana* was well established in the river near CY in 1994 and 1995 while the power plant still operated. This suggests that the thermal plume may not have been the major cause of the decline of *V. americana* near CY. *V. americana* growth has been reported for temperatures of 19–31.5°C (Meyer et al. 1943; Hunt 1963), and maximum growth in the laboratory occurred between 33°C and 36°C (Lind and Cottam 1969). Elimination of *V. americana* near CY observed during the early operational years (1968–1972) may be related to extensive dredging around the CY intake and other construction activities that resulted in siltation and lower light penetration. Loss of *V. americana* in the Chesapeake Bay was attributed, in part, to elevated water turbidity caused by Tropical Storm Agnes in 1972 (Kerwin et al. 1976). Regardless of the causes, subsequent recovery of *V. americana*, such as that observed in the Connecticut River, can stabilize sediments, provide additional substrate for benthic communities and a food source for many aquatic invertebrates (Gerking 1957).

The introduction of the Asiatic clam, an exotic species first documented in the river at CY in 1990 (Morgan et al. 1992 and 2004, this volume), was a major change to the Connecticut River biota. *Corbicula fluminea* has spread across North America since its introduction on the West Coast around 1930 and is now reported in 37 states in the United States (McMahon 1983; Foehrenbach and Raeihle 1984; Morgan et al. 1992). Soon after its initial colonization of the Connecticut River, *C. fluminea* became the dominant mollusk (Morgan et al. 2003, 2004). It is difficult to ascertain how this species introduction has contributed to overall changes in the Connecticut River benthos since *C. fluminea* appeared 18 years after Massengill (2004 [1976]) completed his work, when other factors mentioned previously (e.g., water quality improvements, recovery of *V. americana*) were already affecting benthic habitats. Morgan et al. (2003) did report that *C. fluminea* had no apparent effect on native bivalve abundance over their 10-year study period (1991–2000). Karatayev et al. (2003) similarly found no correlations between the densities of *C. fluminea* and other benthic invertebrates in an East Texas lake. Diaz (1989) concluded that the *C. fluminea* did not displace other species following invasion of North American rivers, but successfully utilized benthic resources not used by native species. Lower annelid abundance in 2002 at most stations, relative to the preoperation and operation periods may be associated with the introduction of *C. fluminea* or physical environmental factors such as changes in river water quality.

Conclusions

Construction and operation of CY appears to have had minimal impacts on the nearby Connecticut River benthic community. Localized effects near the intake were attributed to dredging activities and, near the discharge canal, to the thermal discharge. The disappearance of *V. americana* in 1968 after CY began operation may be due to increased sedimentation and turbidity related to dredging and plant construction activities. Changes in community structure and composition between 1965 and 1972 studies and this study in 2002 were primarily associated with improvement in water quality over the last 30 years, recovery of *V. americana*, and colonization of the river by *C. fluminea* in the 1990s.

Acknowledgments

We would like to thank those who helped with field collections, laboratory sample processing and with editorial review. Dave Dodge, who collected benthic samples in the 1960s, provided able assistance again in 2002. Jim Foertch provided help on graphics and John Swenarton provided critical editing. For peer review and helpful comments on an earlier version of this paper, we thank John Tietjen, Robert Wilce, and Douglas Smith. In addition, Doug Smith confirmed the freshwater mollusk identifications. Paul Jacobson was the logistical link that led to sampling in 2002. This work was supported by the Electric Power Research Institute.

References

Barrett, J. M., M. Prisloe, L. Giannotti, and N. Barrett. 1997. Distribution and abundance of submerged aquatic vegetation in the lower tidal Connecticut River. Connecticut Department of Environmental Protection, Hartford.

Clarke, K. R. 1993. Non-parametric multivariate analyses of changes in community structure. Australian Journal of Ecology 18:117–143.

Clifford, H. T., and W. Stephenson. 1975. An introduction to numerical classification. Academic Press, New York.

Diaz, R. J. 1989. Pollution and tidal benthic communities of the James River Estuary, Virginia. Hydrobiologia 180:195–211.

Foehrenbach, J., and D. Racihle. 1984. A further range extension of the Asiatic clam. New York Fish and Game Journal 31:224–225.

Gerking, S. D. 1957. A method of sampling the littoral macrofauna and its application. Ecology 38:219–226.

Gilbert, R. O. 1989. Statistical methods for environmental pollution monitoring. Van Nostrand Reinhold Company, New York.

Hollander, M., and D. A. Wolfe. 1973. Non-parametric statistical methods. Wiley, New York.

Hunt, G. S. 1963. Wild celery in the lower Detroit River. Ecology 44:360–370.

Jordan, R. A., R. K. Carpenter, P. A. Goodwin, C. G. Becker, M. S. Ho, G. C. Grant, B. B. Bryan, J. V. Merriner, and A. D. Estes. 1976. Ecological study of the tidal segment of the James River encompassing Hog Point. Special Science Report 78. Virginia Institute of Marine Science, Gloucester Point.

Karatayev, A. Y., L. E. Burlakova, T. Kesterson, and D. K. Padilla. 2003. Dominance of the Asiatic clam, *Corbicula fluminea* (Müller), in the benthic community of a reservoir Journal of Shellfish Research 22:487–493.

Kerwin, J. A., R. E. Munro, and W. W. A. Peterson. 1976. Distribution and abundance of aquatic vegetation in the Upper Chesapeake Bay, 1971–1974. Pages 393–400 *in* J. Davis, editor. The effects of Tropical Storm Agnes on the Chesapeake Bay estuarine system. Chesapeake Research Consortium, Inc., Baltimore, Maryland.

Kruskal, J. B. 1964. Multidimensional scaling by optimizing goodness of fit to a non-metric hypothesis. Psychometrika 29:1–27.

Lance, G. N., and W. R. Williams. 1967. A general theory of classificatory sorting strategies, I. Hierarchical systems. Computation Journal 9:373–380.

Lind, C. T., and G. Cottam. 1969. The submerged aquatics of University Bay: a study in eutrophication. American Midland Naturalist 81:353–369.

Magdych, W. P. 1981. An efficient, inexpensive elutriator design for separating benthos from sediment samples. Hydrobiologia 85:157–159.

Marcy, B. C., Jr. 2004. Early life history studies of American shad in the lower Connecticut River and the effects of the Connecticut Yankee plant. Pages 155–180 *in* P. M. Jacobson, D. A. Dixon, W. C. Leggett, B. C. Marcy, Jr., and R. R. Massengill, editors. The Connecticut River Ecological Study (1965–1973) revisited: ecology of the lower Connecticut River 1973–2003. American Fisheries Society, Monograph 9, Bethesda, Maryland. (Originally published in 1976)

Massengill, R. R. 2004. Benthic fauna: 1965–1967 versus 1968–1972. Pages 43–58 *in* P. M. Jacobson, D. A. Dixon, W. C. Leggett, B. C. Marcy, Jr., and R. R. Massengill, editors. The Connecticut River Ecological Study (1965–1973) revisited: ecology of the lower Connecticut River 1973–2003. American Fisheries Society, Monograph 9, Bethesda, Maryland. (Originally published in 1976)

Massmann, W. H. 1954. Marine fishes in fresh and brackish waters of Virginia rivers. Ecology 35:75–78.

McCreary Waters, N., and C. R. San Giovanni. 2002. Distribution and diversity of benthic macroinvertebrates associated with aquatic macrophytes. Journal of Freshwater Ecology 17:223–232.

McMahon, R. F. 1983. Ecology of an invasive pest bivalve, *Corbicula*. Pages 505–561 *in* W. D. Russell-Hunter, editor. The mollusca. Volume 6, Ecology. Academic Press, Orlando, Florida.

Merriman, D., and L. M. Thorpe. 2004. Introduction. Pages 3–25 *in* P. M. Jacobson, D. A. Dixon, W. C. Leggett, B. C. Marcy, Jr., and R. R. Massengill, editors. The Connecticut River Ecological Study (1965–1973) revisited: ecology of the lower Connecticut River 1973–2003. American Fisheries Society, Monograph 9, Bethesda, Maryland. (Originally published in 1976)

Merritt, W., and K. W. Cummins. 1978. An introduction to the aquatic insects of North America. R. Kendall/Hunt Publishing Co., Dubuque, Iowa.

Meyer, B. S., F. H. Bell, L. C. Thompson, and E. I. Clay. 1943. Effect of depth of immersion on apparent photosynthesis in submersed vascular aquatics. Ecology 24:393–399.

Morgan, D. E., M. Keser, J. F. Foertch, and J. M. Vozarik. 1992. The Asiatic clam, *Corbicula fluminea*, extends its North American distribution to the Connecticut River abstract only). Page 523 *in* W. J. Blogoslawski, editor. Aquaculture and environmental stewardship: Milford Shellfish Biology Seminar – 1991. Environmental Management 16:521–529.

Morgan, D. E., M. Keser, J. T. Swenarton, and J. F. Foertch. 2003. Population dynamics of the Asiatic clam, *Corbicula fluminea* (Müller) in the lower Connecticut River: establishing a foothold in New England. Journal of Shellfish Research 22:193–203.

Morgan, D. E., M. Keser, J. T. Swenarton, and J. F. Foertch. 2004. Effect of Connecticut Yankee power plant on population dynamics of Asiatic clams and their interactions with native bivalves. Pages 419–439 *in* P. M. Jacobson, D. A. Dixon, W. C. Leggett, B. C. Marcy, Jr., and R. R. Massengill, editors. The Connecticut River Ecological Study (1965–1973) revisited: ecology of the lower Connecticut River 1973–2003. American Fisheries Society, Monograph 9, Bethesda, Maryland.

Mullaney, J. R. 2004. Summary of water quality trends in the Conecticut River, 1968–1998. Pages 273–286 *in* P. M. Jacobson, D. A. Dixon, W. C. Leggett, B. C. Marcy, Jr., and R. R. Massengill, editors. The Connecticut River Ecological Study (1965–1973) revisited: ecology of the lower Connecticut River 1973–2003. American Fisheries Society, Monograph 9, Bethesda, Maryland.

Peckarsky, B. L., P. R. Fraissinet, M. A. Penton, and D. J. Conklin, Jr. 1990. Freshwater macroinvertebrates on northeastern North America. Comstock Publishing Associates, Cornell University Press, New York.

Sheehan, P. J. 1984. Effects on community and ecosystem structure and dynamics. Pages 51–99 *in* P. J. Sheehan, D. R. Miller, G. C. Butler and P. Bourdeau, editors. Effects of pollutants at the ecosystem level. Wiley Ltd., New York.

Shepard, R. N. 1962. The analysis of proximities: multidimensional scaling with an unknown distance function. Psychometrika 27:125–140.

Smith, D. G. 1995. Keys to the freshwater macroinvertebrates of Massachusetts. Sunderland, Massachusetts.

State of Connecticut. 1967. Connecticut's Clean Water Act of 1967, an analysis of Public Act 57. Clean Water Task Force, Hartford, Connecticut.

Strayer, D. L., and L. C. Smith. 2001. The zoobenthos of the freshwater tidal Hudson River and its response to the zebra mussel (*Dreissena polymorpha*) invasion. Archives of Hydrobiologia Supplement 139/1 Monograph Studies 1-52.

Thorp, J. H., and A. P. Covich. 1991. Ecology and classification of North American freshwater invertebrates. Academic Press, San Diego, California.

Townsend, C. R., A. G. Hildrew, and K. Schofield. 1987. Persistence of stream invertebrate communities in relation to environmental variability. Journal of Animal Ecology 56:597–613.

Trench, E. C. T. 2000. Nutrient sources and loads in the Connecticut, Housatonic, and Thames river basins. U.S. Department of Interior. U.S. Geological Survey. National Water-Quality Assessment Program. Water-Resources Investigations Report 99-4236. Reston, Virginia.

Trench, E. C. T., and A. V. Vecchia. 2002. Water quality trend analysis and sampling design for streams in Connecticut, 1968–98. United States Geological Survey Water-Resources Investigations Report 02–4011, Reston, Virginia.

Warwick, R. M., and K. R. Clarke. 1991. A comparison of some methods for analysing changes in benthic community structure. Journal of the Marine Biological Association, UK 71:225–244.

Weatherley, N. S., and S. J. Ormerod. 1990. The constancy of invertebrate assemblages in soft-water streams: implications for the prediction and detection of environmental change. Journal of Applied Ecology 27:952–964.

American Fisheries Society Monograph 9:455–473, 2004

Twenty Years of Impingement History: Connecticut Yankee Haddam Neck Nuclear Power Plant

PAUL M. JACOBSON*
Ecological Assessments, LLC, Post Office Box 376, Old Saybrook, Connecticut 06475, USA

CHRISTINE TOMICHEK[1] AND DONALD J. DANILA
Millstone Environmental Laboratory, Dominion Nuclear Connecticut
Rope Ferry Road, Waterford, Connecticut 06385, USA

Abstract.—Estimated number of fish impinged on the Connecticut Yankee (CY) cooling water intake screens was summarized for the period 1968 through 1982. Species were generally impinged in relation to their abundance in the river. White perch *Morone americana*, catfishes, and yellow perch *Perca flavescens* were common in impingement samples and numerically dominant in trawl, seine, and eletrofishing collections. Certain species, such as American shad *Alosa sapidissima*, common carp *Cyprinus carpio*, spottail shiner *Notropis hudsonius*, white sucker *Catostomus commersonii*, sunfishes, and game fish, were impinged in lower numbers than expected based on relative abundance in the Connecticut River. Impingement rates were influenced predominantly by location of the CY intake in relation to species life history and behavior, which determined the relative vulnerability of fish to the cooling water intake. Intake capacity was less important. Fish deterrent devices tested and implemented significantly reduced the number of fish impinged. A fish return installed during 1983 operated for the remaining life of the power plant through July 1996, further minimizing impingement related fish mortality. The information on impingement, life history strategies, relative abundance, and population dynamics of key species, when viewed retrospectively over four decades, suggests that it is possible to make reasoned judgments regarding ecological impact based on site-specific studies and assessments.

Introduction

Merriman and Thorpe (2004 [1976], this volume), in their introduction to this monograph (Part I, page 19), remind us that in the design and preoperational phases of the Connecticut River study, the numbers of fish that would be impinged on the Connecticut Yankee (CY) intake traveling screens was not anticipated. Soon after plant operations began in the fall of 1967, screen washings were removing large numbers of fish from the traveling screens. As a result, systematic efforts to determine the numbers and kinds of species impinged were begun in April 1968. With these counts of the fish washed into the screenwash collection basket, it was possible to make some initial conclusions as to numbers of fish impinged, seasonality, species composition, and the relationship between the numbers of a given species impinged and its local abundance in the river based on trawl and seine samples.

The discussion of impingement and its mitigation provided in Merriman and Thorpe (Part 1 of this volume) was largely limited to the Connecticut River study period. This paper provides additional detail encompassing the period between 1968 and 1987, when estimates of the numbers of fish impinged were made. Comparison between the number of fish impinged and their relative abundance in the river is made possible given additional data collected subsequent to the Connecticut River study. The designs and relative benefit of the devices tested and utilized to minimize impingement impacts are also examined.

* *Corresponding author*: jacobsonpaul@msn.com
[1] Present address: Kleinschmidt Associates, 35 Pratt Street, Suite 201, Essex Connecticut 06426, USA

Methods

Intake Design, Location, and Capacity

The CY cooling water intake is located on the eastern shore of the Connecticut River in Haddam, Connecticut, about 26 km upstream of Long Island Sound. Although the river is tidal (0.75-m amplitude) in this reach, saline waters are limited to locations several kilometers downstream of the intake location. The navigational channel follows along the western shore of the river opposite the intake structure. Immediately downstream of the intake is an expansive sand flat that extends about 1.6 km to its intersection with the discharge canal. The area immediately in front of the intake was dredged to allow for construction and adequate flows. Flows in the river are dominated by the annual spring freshet when water flow is unidirectional and by tidal fluctuations in summer and fall (Merriman and Thorpe 2004 [1976]).

The shoreline intake structure consists of four intake bays, each about 3.6 m in width and 5.5 m deep below mean low water (Figure 1). Each bay is equipped with trash racks and 9.5-mm-mesh traveling screens protecting the condenser and service water pumps located within the structure. Each of the four condenser cooling water pumps had a rated capacity of about 5,867 L/s (208 cfs), and each of the four service water pumps was rated at about 378 L/s (13.4 cfs). Total flow during normal operations with four circulating water pumps and two service water pumps operating was about 24,224 L/s (856 cfs). Approach velocities averaged 36.6 cm/s across the trash racks and 61 cm/s at the traveling screens (USAEC 1973; Merriman and Thorpe 2004 [1976]). Percent of river flow entrained was estimated at 5.5% at minimum average daily flow. Using the Environmental Protection Agency (EPA 2001) calculation method given in their Clean Water Act §316(b) Phase I new facility final rule, about 7.7% and 6.1% of the flood and ebb tidal excursions, respectively, were estimated to be entrained at low river flows.

Impingement Estimates

The frequency of counting impinged fish and the methods used to extrapolate total estimates varied over the period of 1968 through 1987 (Table 1). Generally, during 1968 and 1969, absolute counts were obtained. The traveling screens were washed three times each day, and the number of fish washed into the screenwash basket were collected, counted, measured, and later grouped into three approximate size categories: 0–75 mm, 76–150 mm, and larger than 150 mm. Because the screenwash basket contained 25-mm drain holes, some smaller fish passed through the basket and were not captured in the counts. An experiment using a fine mesh 6-mm net suspended under the basket was conducted in November 1968 to calculate correction factors (or % underestimates) for fish smaller than 75 mm. The total number of fish collected in the net and screenwash basket was divided by the number in the basket to calculate the factor. In general, about 50% of fish 0–75 mm passed through the basket and were not counted (CYAPCO 1971). For any given species and time of year, the percentage of young fish retained in the screen wash basket (or passing through uncounted) was known to be influenced by debris loading and most likely varied widely. As a result, impingement counts reported here for the period of 1970 through 1982 were not adjusted for losses of the smaller size specimens through the basket drain holes. Estimates are based on actual counts of fish taken from the screenwash basket.

From 1975 through 1982, counts of fish in the screenwash basket were made once each week (CYAPCO 1976–1983). Contents of the basket were accumulated over a 24-h period. Fish were identified, counted, and separated into the same three length categories described above. The numbers impinged per month were estimated by averaging the four weekly counts in any one month and multiplying the average by the number of days in the month. During the intervening years, 1970 through 1974, and in 1986 and 1987, counts of fish impinged were made at varying intervals in support of specific experiments designed to evaluate the efficacy of deterrent devices. No impingement sampling was performed from 1983 through 1985 following installation of a fish return system in 1983.

Fish Exclusion Devices

Experimentation with devices to exclude fish from entering the intake structure began when one of four intake forebays was equipped with an array

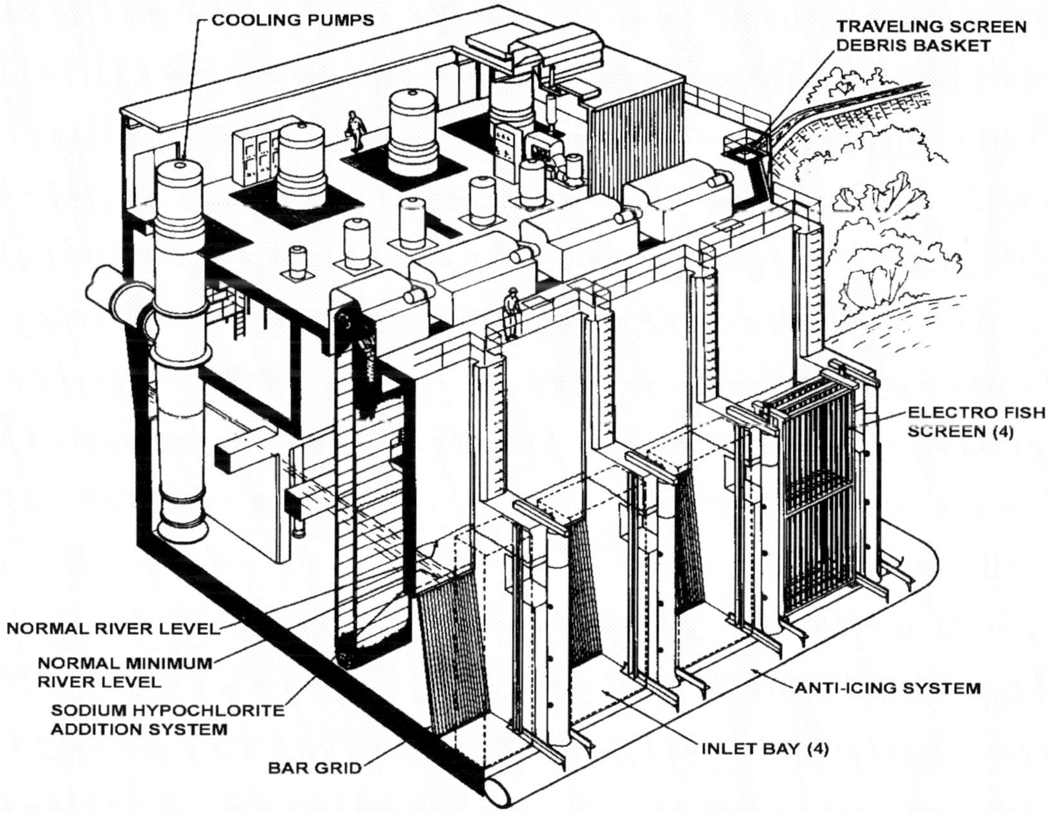

FIGURE 1. The Connecticut Yankee atomic power plant intake structure.

of aluminum pipes as depicted in Figure 2 (CYAPCO 1971, 1974a; Hyman and Pikanowski 1974). The aluminum pipes were arranged in three rows supported by oak timbers. The pipes were 25 mm interior diameter, 5.5 m long, set about 190 mm apart. This deterrent or exclusion device was designed such that the added benefit of electrification of the array could be tested in the future.

The efficacy of this array, without electrification, was tested over the period July through mid-October 1970 (Table 1). Screen washings from Bay A, which contained the triple-row aluminum pipe array, were kept separate from the other three bays. Once again, fish washed into the screen wash basket were identified, counted, and grouped into the three length categories described above. Fish were counted in the morning and evening to ascertain diel differences in impingement rates.

Installation of barriers in each of the four intake forebays occurred in May 1971. Impingement counts were made on a daily basis from 8 March to 16 April and from 27 May to 15 October 1971. Counts were made twice daily as in 1970 and kept separate for each bay. Fish were identified and grouped by length as in previous years.

Electrification of the fish deterrent device was tested in 1972 in an effort to further improve its performance (Hyman et al. 1975). The front row of aluminum pipes in each of the four bays was connected to a fish fence charger. The charger was set at 144 pulses per minute, and 700 V and field strengths were measured with electrodes attached to an oscilloscope. Based on field strengths, it was estimated that a typical 20-cm fish total length would have been subjected to a 0.60–1.20-V differential over its length, whereas a 1-cm fish would experience a 0.26–0.31-V differential. The traveling screens were washed at 6-h intervals, and the sizes and species of all fish collected in the screen wash basket were deter-

TABLE 1. Connecticut Yankee impingement sampling chronology.

Sampling period	Sample type
1968 (April)–1969 (December)	Counts from screenwash basket were taken each shift; lengths recorded as 0–75 mm, 76–150 mm, or >150 mm.
1970 (June)	Exclusion device consisting of three row array of aluminum conduits installed in Bay A for testing.
1970 (July–October)	Counts taken twice daily at 0800 and 2000 hours. Bay A counts kept separate from other three bays; lengths grouped as above.
1971 (May)	All four bays equipped with exclusion device.
1971 (8 March–16 April 27 May–15 October)	Counts made twice daily, kept separate by bay, and grouped into the three length categories above.
1972 (May–December)	Routine counts continued as above; Hyman et al. (1975) undertake test of electricity with aluminum conduit barriers. Counts made four times daily with and without electricity on.
1973	No routine impingement counts made.
1974	No routine sampling; test new electrical barrier installed in front of intake in triangle configuration; samples in June, September, October and November; counts taken four times daily; electrical power on 2 d, off 2 d.
1975–1982	Counts made year round once each week over 24-h period; lengths grouped as above. Counts extrapolated to monthly totals by averaging the weekly counts and multiplying by the number of days in a month.
1986 (May)–1987 (May)	Two bays converted to single-row arrays to assess effectiveness versus bays with three-row arrays. Counts taken over a 24-h period every second week, resulting in 26 collections; fish were grouped into three length categories as above.

mined. The basket was outfitted with a 12-mm-mesh liner to retain smaller fish.

Testing of an electrical device located in front of the intake was performed during 1974 (Hyman et al. 1975). The design (Figure 3) included electrodes positioned on the river surface and bottom in a triangular arrangement such that an electrical field was created over the total depth of the intake. Study methods were previously described by Hyman et al. (1975).

Triple Array versus Single Array

Despite installation of a boom in front of the intake structure in 1980, heavy debris loading continued to require almost continuous maintenance of the barrier consisting of the triple-row array of aluminum pipes. As a result, in 1985, an alternative design of a single-row array of pipes for the fish barrier was proposed to the Connecticut Department of Environmental Protection (CTDEP) coupled with a study to determine the relative performance of the alternative (NUSCo 1985).

In this experiment, the arrays located in forebays A and C were equipped with a single row of aluminum pipes and the other two remained unchanged with three rows each. The screens were washed during the night about 8–10 h before sample collection. Screens in bays B and D were washed first at the time of sample collection, and fish collected were kept separate from the other two bays for comparison. Fish collected were identified and measured. The fish return system sluiceway was replaced with the original collection basket for sampling purposes.

Results

Species Composition and Abundance

During the years impingement was monitored or exclusion devices were tested at CY (1968–1982 and 1986–1987), more than 40 fish species were identified in the impingement collections taken from the cooling water intake screenwash basket

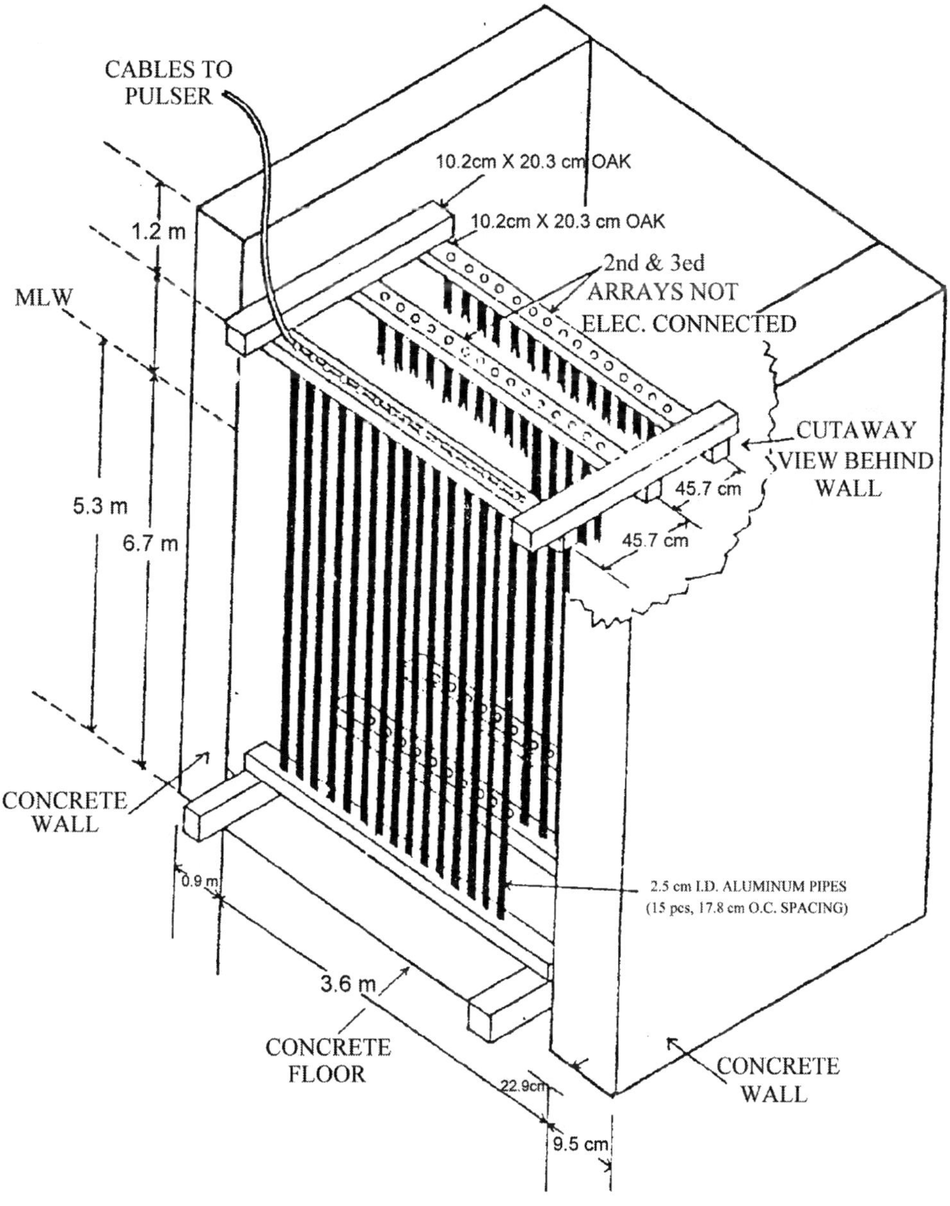

FIGURE 2. Fish exclusion device installed in the Connecticut Yankee forebay in 1970 and 1971 (from Hyman et al. 1975, converted to metrics).

(Table 2) (CYAPCO 1971, 1974a, 1976–1983). An estimated 1.3 million fish were estimated to have been impinged during 1968–1982 (Table 3). As much as half of this total represented the impingement of juvenile Atlantic menhaden *Brevoortia tyrannus* in 1971, when 671,756 individuals were estimated impinged in September and October of that year. In certain years, particularly 1970–1972, impingement counting did not encompass all months. As a result, the annual totals for those years are considered as underestimates. During 1975 through 1982, the number of fish estimated impinged (excluding the 1971 Atlantic menhaden mass impingement events)

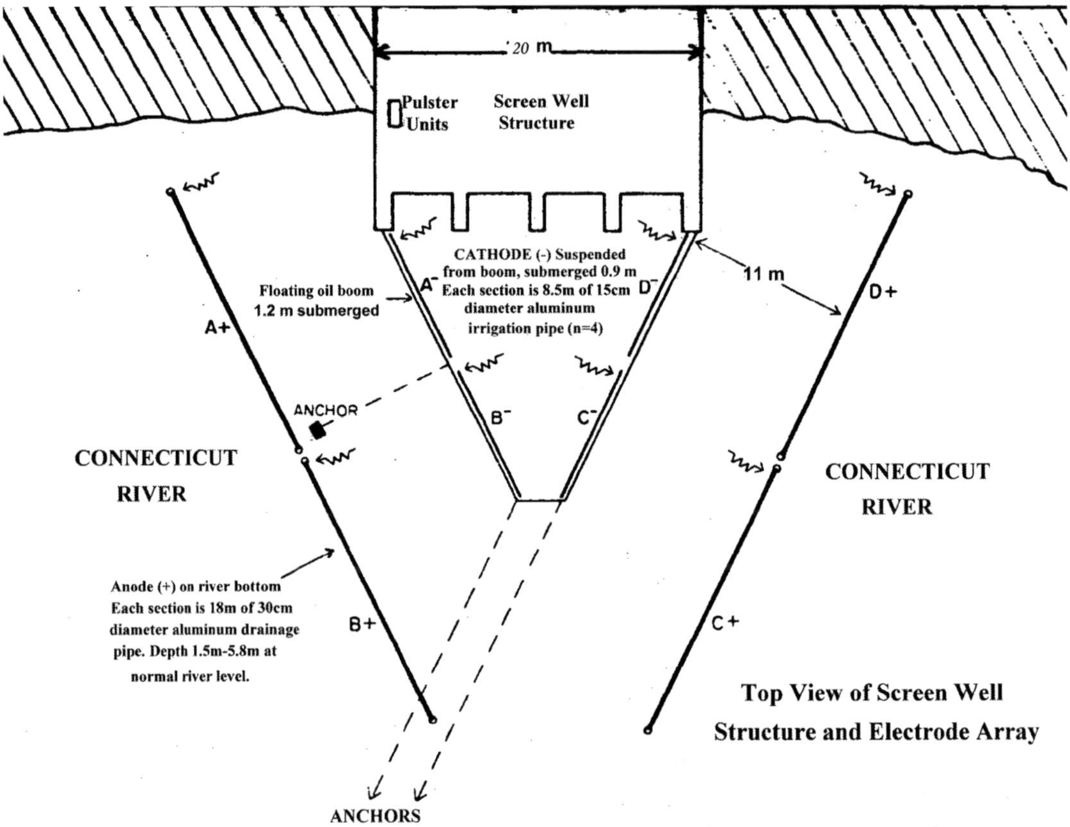

FIGURE 3. Overhead view of screenwell structure and electrode array illustrating electrode configuration and anchoring system tested at Connecticut Yankee in 1974 (from Hyman et al. 1975, converted to metrics).

ranged from 4,353 fish in 1982 to 49,843 fish in 1980. These estimates are significantly lower (CYAPCO 1971; Hyman and Pikanowski 1974) than observed during the period 1968 through 1970 prior to installation of fish deterrent devices within each of the four intake bays (Figure 4).

Six species represented more than 80% of the total fish impinged, excluding the one-time impingement of Atlantic menhaden (Table 3). These included white perch *Morone americana*, spottail shiner *Notropis hudsonius*, white catfish *Ameiurus catus*, blueback herring *Alosa aestivalis*, brown bullhead *Ameiurus nebulosus*, and yellow perch *Perca flavescens*. A comparison with the numbers of fish collected in trawl and seine sampling during the Connecticut River study (Marcy 2004 [1976], this volume) indicated that, in general, fish were impinged in relation to their abundance and seasonal occurrence in the river. The number of white perch impinged annually ranged from 99,412 in 1968 to 205 individuals in 1982; juveniles (<75 mm) were numerically the dominant size-class impinged in the early years of counting (Table 4). The number of white catfish ranged from 55,698 specimens in 1968 to 194 in 1978 and, except for 1968, 1970, and 1971, when large numbers of small specimens were impinged, most were larger than 150 mm. The estimated number of American shad *Alosa sapidissima* ranged from 2,645 specimens in 1976 to 5 in 1972, with the latter year likely reflecting the lower counting effort. Brown bullhead were impinged in fewer numbers than white catfish and ranged from 30,126 specimens in 1971 to 22 in 1981. Sport fishes, including largemouth bass *Micropterus sal-moides* and smallmouth bass *M. dolomieu*, chain pickerel *Esox niger*, northern pike *E. lucius*, and striped bass *Morone saxatilis*, were impinged in relatively low numbers.

TABLE 2. Scientific and common names (from Robins et al. 1991) of fishes impinged at Connecticut Yankee from 1968 to 1982.

Family	Scientific name	Common name
Petromyzontidae (lampreys)	*Petromyzon marinus*	sea lamprey
Acipenseridae (sturgeons)	*Acipenser brevirostrum*	shortnose sturgeon
Anguillidae (freshwater eels)	*Anguilla rostrata*	American eel
Clupeidae (herrings)	*Alosa aestivalis*	blueback herring
	A. pseudoharengus	alewife
	A. sapidissima	American shad
	Brevoortia tyrannus	Atlantic menhaden
Engraulidae (anchovies)	*Anchoa* sp.	anchovy
Cyprinidae (carps and minnows)	*Cyprinus carpio*	common carp
	Luxilus cornutus	common shiner
	Notemigonus crysoleucas	golden shiner
	Notropis hudsonius	spottail shiner
	Pimephales promelas	fathead minnow
Catostomidae (suckers)	*Catostomus commersonii*	white sucker
Ictaluridae (North American catfishes)	*Ameiurus catus*	white catfish
	A. nebulosus	brown bullhead
	Ictalurus punctatus	channel catfish
Esocidae (pikes)	*Esox lucius*	northern pike
	E. niger	chain pickerel
Osmeridae (smelts)	*Osmerus mordax*	rainbow smelt
Salmonidae (trouts and salmons)	*Salmo salar*	Atlantic salmon
Gadidae (cods)	*Microgadus tomcod*	Atlantic tomcod
Fundulidae (topminnows)	*Fundulus diaphanus*	banded killifish
	F. heteroclitus	mummichog
	F. majalis	striped killifish
Atherinopsidae (New World silversides)	*Menidia* spp.	silversides
Gasterosteidae (sticklebacks)	*Gasterosteus aculeatus*	threespine
Percichthyidae (temperate basses)	*Morone americana*	white perch
	M. saxatilus	striped bass
Centrarchidae (sunfishes)	*Lepomis auritus*	redbreast sunfish
	L. gibbosus	pumpkinseed
	L. macrochirus	bluegill
	Micropterus dolomieu	smallmouth bass
	M. salmoides	largemouth bass
	Pomoxis annularis	white crappie
	P. nigromaculatus	black crappie
Percidae (perches)	*Etheostoma olmstedi*	tessellated darter
	Perca flavescens	yellow perch
Pomatomidae (bluefishes)	*Pomatomus saltatrix*	bluefish
Achiridae (American soles)	*Trinectes maculatus*	hogchoker

The size-frequency distribution of the key species impinged reflected their seasonal movements and life histories. Anadromous river herring (blueback herring and alewife) were captured in largest numbers during their downstream migration in late summer and fall. When abundant, American shad juveniles were also found almost exclusively during their fall emigration. Adult shad were collected primarily when they returned downriver in an emaciated postspawn state during late June (CYAPCO 1976–1983). Few adults were impinged during their April and early May upstream migration as they follow the channel along the shore opposite the CY intake structure (Leggett 2004 [1976], this volume).

TABLE 3. Total number of fish impinged at Connecticut Yankee by species and the percent species composition from 1968 to 1982 (excludes 1973 and 1974).

Species	1968	1969	1970	1971	1972	1975	1976	1977	1978	1979	1980	1981	1982	Total	% species composition
Atlantic menhaden	35,971	0	4	671,756	165	0	0	37	118	0	0	0	0	708,051	54.371
White perch	99,412	22,884	40,463	6,870	1,889	744	898	2,675	927	584	1,166	496	205	179,213	13.762
Spottail shiner	5,215	12,137	0	1,128	689	2,726	3,497	8,593	3,681	9,382	38,326	8,359	546	94,279	7.240
White catfish	55,698	3,711	1,398	9,783	1,703	837	434	277	194	285	495	1,321	325	76,461	5.871
Blueback herring	36,237	1,550	23	9,376	1,122	250	2,401	129	3,458	5,239	111	124	472	60,492	4.645
Brown bullhead	6,500	2,746	1,746	30,126	804	701	434	680	1,194	128	29	22	48	45,158	3.468
Yellow perch	8,246	8,769	763	1,817	640	1,436	2,092	2,721	2,297	2,852	2,836	1,425	779	36,673	2.816
Anchovy	0	0	24,185	9	1	0	0	0	22	16	0	0	16	24,249	1.862
Alewife	2,987	2,521	382	3,353	200	171	176	1,604	1,754	76	150	135	524	14,033	1.078
Common shiner	0	5	316	170	1	26	8	273	11,076	47	0	503	20	12,445	0.956
Rainbow smelt	4,038	5,867	0	1,148	2	0	0	0	30	0	23	0	0	11,108	0.853
American eel	5,814	869	337	272	422	172	79	91	45	145	73	142	106	8,567	0.658
American shad	262	693	1,560	811	5	49	2,645	86	60	81	52	681	192	7,177	0.551
Pumpkinseed	1,883	979	465	388	369	234	779	458	204	59	351	201	432	6,802	0.522
Sea lamprey	0	0	0	0	0	14	23	66	54	47	4,873	0	0	5,077	0.390
Golden shiner	683	617	606	255	120	71	119	114	304	151	173	115	15	3,343	0.257
Hogchoker	174	3	3	42	228	27	102	195	1,129	195	248	217	583	3,146	0.242
Banded killifish	2	0	1	10	8	0	16	0	0	16	76	0	0	129	0.010
Redbreast sunfish	0	523	0	0	0	0	38	130	8	147	38	163	45	1,092	0.084
Bluegill	6	48	7	2	9	14	16	159	224	41	509	26	8	1,069	0.082
White sucker	60	72	24	9	2	19	0	32	390	23	56	101	0	788	0.061

TABLE 3. Continued

Species	1968	1969	1970	1971	1972	1975	1976	1977	1978	1979	1980	1981	1982	Total	% species composition
Black crappie	18	33	305	4	31	16	64	31	16	16	80	36	22	672	0.052
Northern pike	156	207	16	48	18	0	16	14	45	16	6	0	0	542	0.042
Tesselated darter	3	0	1	1	8	12	16	8	51	29	67	182	0	378	0.029
Channel catfish	0	0	0	0	0	0	16	254	0	0	0	0	0	270	0.021
Chain pickerel	21	24	6	14	2	7	8	8	20	53	22	15	0	200	0.015
Common carp	12	27	0	2	6	23	40	14	0	0	24	15	0	163	0.013
Striped killifish	3	10	0	21	0	0	0	8	0	0	20	32	6	100	0.008
Atlantic tomcod	50	9	3	34	0	0	0	0	0	0	0	0	0	96	0.007
Largemouth bass	14	0	0	0	1	7	23	0	6	14	24	0	0	89	0.007
Crayfish	0	0	0	0	0	0	0	83	0	0	0	0	0	83	0.006
White crappie	0	0	0	0	0	0	0	53	0	0	0	0	0	53	0.004
Threespine stickleback	1	0	0	39	0	0	0	0	0	0	7	0	0	47	0.004
Fathead minnow	0	0	0	0	0	0	0	38	0	0	0	0	0	38	0.003
Striped bass	1	7	0	5	23	0	0	0	0	0	0	0	0	36	0.003
Bluefish	0	0	0	0	8	0	0	0	23	0	0	0	0	31	0.002
Silversides	0	0	0	0	2	0	0	0	0	0	8	0	0	10	0.001
Atlantic salmon	0	0	0	0	0	0	0	0	0	0	0	0	8	8	0.001
Smallmouth bass	0	0	0	0	0	0	0	0	0	0	0	7	0	7	0.001
Shortnose sturgeon	0	0	0	0	0	0	0	0	1	0	0	0	1	2	0.000
Mummichog	0	0	0	0	0	0	0	0	0	0	0	0	0	0	0.000
Not identified	20	25	0	29	14	0	0	0	0	0	0	0	0	88	0.007
Total	263,487	64,336	72,614	737,522	8,492	7,556	13,940	18,831	27,331	19,642	49,843	14,318	4,353	1,302,265	100

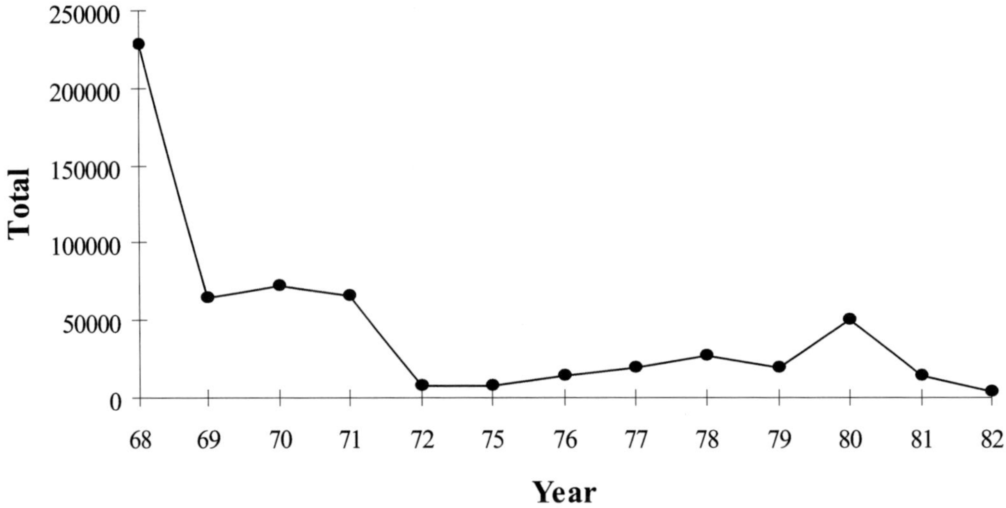

FIGURE 4. Annual Connecticut Yankee impingement, excluding Atlantic menhaden, from 1968 through 1982 (excludes 1973 to 1974).

Fish Exclusion Devices

The first barrier to be examined starting in 1970 consisted of the triple array of aluminum conduits. A simple comparison of the number of fish washed from the traveling screen in Bay A equipped with the barrier to the average number from the other three bays without the barrier indicated that in excess of 70% fewer fish were collected from Bay A (CYAPCO 1971). Whether this was attributed to the location of Bay A at the upstream end of the intake is not known. However, results were sufficiently encouraging that all four bays were equipped with the aluminum arrays in May 1971.

The relative benefit of the deterrent devices was assessed independently by the Essex Marine Laboratory (CYAPCO 1971) and researchers at the University of Rhode Island under the direction of Dr. Saul Saila (Hyman and Pikanowski 1974). In CYAPCO (1971), monthly catch per unit effort (CPUE) was calculated for fish less than 75 mm and fish larger than 75 mm taken by trawl or seine during the Connecticut River Ecological Study in 1968, 1969, and 1971. The species and sizes that did not differ significantly were used to compare the daily intake impingement counts made from 27 May to 15 October 1971 to those counts made during the same periods in 1968 and 1969. Counts of impinged fish less than 75 mm were adjusted upward in this analysis to account for estimated loss of some smaller fish through the basket drain holes. The average reduction in the number of fishes over 75 mm was about 73% when impingement counts in 1971 were compared to the earlier years. Results for fishes less than 75 mm were inconclusive.

For their analysis, Hyman and Pikanowski (1974) selected those species that collectively represented more than 60% of the fish biomass collected during the Connecticut River study. This included white perch, brown bullhead, white catfish, American eel, and yellow perch. Daily impingement counts for these species were summed to calculate weekly totals over the period 12 June through 15 October in years 1968, 1969, and 1971. Only fishes larger than 150 mm were used in the analysis due to the large natural variability of the 0-year size-classes, particularly among the clupeids. The percent reduction in the number of fish impinged in 1971 compared to the average of 1968 and 1969 ranged from 68% for yellow perch to 82% for American eel. While reduction in numbers of American eel and white catfish were not significant statistically, the actual reductions were substantial. Reductions in numbers of brown bullhead, white perch, and white catfish of 81%, 78%, and 79% were achieved, respectively. It was not possible to determine the precise mechanism by which the aluminum arrays deterred fish from entering the intake, but it is likely that the device acted partly as a behavioral barrier. Observation

TABLE 4. Total annual catch for selected species impinged at Connecticut Yankee by size-classes.

Year	American shad <75 mm	American shad 76–150 mm	American shad >150 mm	American shad Total	White perch <75 mm	White perch 76–150 mm	White perch >150 mm	White perch Total	Yellow perch <75 mm	Yellow perch 76–150 mm	Yellow perch >150 mm	Yellow perch Total
1968	120	68	74	262	86,792	3,384	9,236	99,412	2,766	2,959	2,521	8,246
1969	676	2	15	693	4,229	2,035	16,620	22,884	448	6621	1700	8769
1970	1,428	103	29	1,560	27,975	1,741	10,747	40,463	20	61	682	763
1971	803	2	6	811	4,581	216	2,073	6,870	840	482	495	1,817
1972	0	0	5	5	312	36	1,541	1,889	96	111	433	640
1975	22	6	21	49	29	93	622	744	51	645	740	1,436
1976	638	1,952	55	2,645	131	202	565	898	22	866	1,204	2,092
1977	0	8	78	86	121	569	1,985	2,675	23	350	2,348	2,721
1978	0	0	60	60	92	200	635	927	60	877	1,360	2,297
1979	0	0	81	81	84	269	231	584	108	1,254	1,490	2,852
1980	0	15	37	52	148	420	598	1,166	46	1,063	1,727	2,836
1981	16	0	665	681	31	187	278	496	38	371	1,016	1,425
1982	24	0	168	192	7	14	184	205	29	180	570	779
Total	3,727	2,156	1,294	7,177	124,532	9,366	45,315	179,213	4,547	15,840	16,286	36,673

Year	White catfish <75 mm	White catfish 76–150 mm	White catfish >150 mm	White catfish Total	Brown bullhead <75 mm	Brown bullhead 76–150 mm	Brown bullhead >150 mm	Brown bullhead Total
1968	49,641	2,539	3,518	55,698	5,746	197	557	6,500
1969	438	293	2,980	3,711	103	568	2,075	2,746
1970	864	183	351	1,398	1,326	155	265	1,746
1971	9,324	171	288	9,783	29,781	134	211	30,126
1972	400	193	1,110	1,703	314	112	378	804
1973	231	177	429	837	266	167	268	701
1974	46	93	295	434	31	94	309	434
1975	8	27	242	277	126	109	450	680
1976	0	20	174	194	48	308	838	1,194
1977	53	91	141	285	16	6	106	128
1978	62	185	248	495	0	5	24	29
1979	491	405	425	1,321	0	0	22	22
1980	23	89	213	325	6	0	42	48
Total	61,481	4,466	10,414	76,461	37,763	1,855	5,545	45,158

of the array during full plant flow indicated a rapid and audible vibration of the aluminum pipes.

Experiments to test the efficacy of electrifying the aluminum conduits were performed in 1972. Hyman et al. (1975) reported collecting 24 species of fish over the experimental period 18 August to 15 September 1972. Only those fish larger than 10 cm were used in their analysis, as these size fish were more likely to experience a higher body voltage. When catfishes and eels alone were considered, the reduction in the number of fish impinged with the energized deterrent device ranged from 16% to 31% (avg. 20%) across the four bays. Two bays showed statistically significant reductions, the others did not. No apparent up- or downstream pattern was evident across the bays to explain this difference. When all fish species were considered, there was no significant difference in the number of fish impinged, whether the deterrent device was energized or not. Hyman et al. (1975) postulated that an interaction between the energized pipes and sediments may have accounted for the observed response of white catfish and eels compared to the other species.

A second electrical barrier of an entirely different design was tested during June and September through November 1974 (Hyman et al. 1975). The new design considered the likelihood that fish entering the aluminum arrays would be within the bay at the moment they experience the electrical field. Thus, it would be desirable to move the effective field away from the intake and expand its effective range. The resulting design is shown in Figure 3. Cathodes, 8.2 m long, were suspended from an oil slick boom in the form of a triangle. Anodes, 18.3 m long, were located on the river bottom parallel, but outboard of the cathodes. Initial attempts to "tune" the electrical barrier in June led to some improvements, including increasing the electrical pulse rate from 2 to 6 s and hanging chains from the cathode to improve the field coverage. Further changes in field strength were made as the experiments progressed and as fish assemblages shifted due to changing seasonal occurrence and abundance. During these experiments, the existing arrays in the forebays were left in place de-energized.

Hyman et al. (1975) provide details regarding the field strength and other variables tested during the 4 months of study, but in general, as they decreased the electrical field strength to levels below which fish were stunned, performance improved (i.e., fewer fish were impinged). Ultimately, the system, when energized compared to when it was not, was able to achieve a 54% reduction in impingement of all fish larger than 15 cm.

The proposed modification of the triple-row aluminum conduit array was evaluated from May 1986 through May 1987. A total of 26 comparisons were made encompassing 798 specimens representing at least 28 taxa. Catch was dominated by river herring (26%), bluegill (26%), spottail shiner (25%), and yellow perch (9%). In 25 of 26 comparisons, the number of fish impinged was greater for the two bays with the single-row array, representing 62% of the catch. In addition, significantly smaller fish were taken from bays with the single-row array. Consequently, CYAPCO retained the triple-row array in each of the bays and so notified the CTDEP (NUSCo 1987). The fish return system and the triple-row array fish deterrent devices remained in place for the remaining operational life of the power plant through 1996.

Discussion

While the Connecticut River study was largely completed in 1972, 5 years after initial plant operation, impingement sampling continued on a regular schedule from 1975 through 1982. Experimentation with fish exclusion devices occurred during 1970 through 1974. Starting in late 1974, impingement sampling and reporting requirements were incorporated into two new regulatory programs. The Environmental Technical Specifications (CYAPCO 1974b) fashioned after the plant operations Technical Specifications incorporated several environmental requirements, including impingement monitoring, stemming from issuance of the CY Final Environmental Impact Statement by the U.S. Atomic Energy Commission (USAEC 1973). Simultaneously, the CTDEP (CTDEP 1974) issued its first National Pollutant Discharge Elimination System (NPDES) permit for CY in December 1974. The permit set forth conditions "to conduct studies to determine if the cooling water intake structures reflect the best technology available for minimizing adverse environmental impact which may result from entrapment [impingement] and entrainment." The permit also required that "such studies shall contain an evaluation of the location, design, construction and capacity of the intake structures." The Connecticut River Ecological Study, supplemented with ongoing impingement monitoring, provided the basis for CY's response to these requirements (CYAPCO 1974a).

In 1979, the EPA and the CTDEP performed an inspection of several Northeast Utilities (NU) electric generating stations in Connecticut and concluded that impingement could be mitigated with only minor changes to the intake screenwash systems (CTDEP 1979). The agencies proposed that traveling screen wash water be returned directly to the river. As a result, NU performed a feasibility study and concluded that, with minimal cost, a fish return system could be installed at CY by simply modifying the bottom of the screen wash collection basket into a fish return sluiceway (CYAPCO 1980). This system became operational in early 1983, at which time impingement sampling was discontinued. The screenwash basket was modified allowing screen wash water to return directly to the river. Provision was made to replace the original basket when heavy seasonal debris loading resulted in threats to traveling screen operation. Placement of a boom in front of the intake likely minimized screen loading. No experiments were performed to determined survival of fish impinged and subsequently returned to the river in the fish return. However, studies elsewhere have indicated that survival would be high, especially for larger specimens and hardier species, such as white perch, yellow perch, and catfishes (EPRI 1999a, 2003; Taft et al. 2001).

The fish return, along with the aluminum conduit arrays discussed above, operated for the remaining life of the plant through 1996 and provided the basis for the CTDEP determination in 1986 that the CY cooling water intake represented Best Technology Available under §316(b) of the Clean Water Act (CTDEP 1986).

During the years impingement sampling was performed, it was found that species generally were impinged at CY in relation to their abundance and dominance in the River. White perch, catfishes, and yellow perch were common in impingement counts and numerically dominant in resident fish samples collected during the Connecticut River study and in more recent studies (Jacobs et al. 2004, this volume). Exceptions included the common carp, spottail shiner, white sucker, golden shiner, sunfishes, and game fishes such as the smallmouth bass and largemouth bass, which were apparently impinged in fewer numbers than expected based on their relative abundance in trawl and seine samples (Marcy 2004 [1976]) or found more recently in electrofishing collections (Jacobs and O'Donnell 1996; Jacobs et al. 2004). Adult and juvenile American shad were not particularly vulnerable, whereas juvenile Atlantic menhaden and river herring species (undifferentiated alewife and blueback herring) were often impinged in large numbers during peak years of abundance (CYAPCO 1974a, 1973–1983; Merriman and Thorpe 2004). For example, the occurrence of Atlantic menhaden in 1971 was highly atypical and was also seen at the Millstone power station along the Connecticut coast, when an estimated 50 million juvenile menhaden were entrained or impinged in the summer and fall of 1971 (NUSCo 1983).

Perspective on the numbers of fish impinged was drawn from information on sport and commercial catches as well as from direct population estimates. White perch exhibit considerable movement throughout the lower river (Howell and Molnar 1999) as they move between coves and, as a result, may become vulnerable to impingement. The total number of white perch estimated impinged annually ranged from about 99,000 fish in 1968 to 200 individuals in 1982. By contrast, Howell and Molnar (1999 and 2004, this volume) estimated the total population of white perch in the Connecticut portion of the Connecticut River to be on the order of 1.6 million fish, based on tag and recapture estimates made in 1998. Harvest from the recreational fishery ranged from 57,871 white perch in 1997 to 39,892 in 1998. They found that catch rates in their study period were comparable to those reported by Marcy (2004 [1976]). Over the period 1965–1969, he estimated the recreational catch of white perch in the Haddam area of the Connecticut River at 21,000. Howell and Molnar (1999) reported that commercial catches of white perch varied from 136 to 12,020 kg annually from 1968 to 1997, except for peaks of 21,727–30,617 kg during 1972–1974, with biomass remaining stable up to the present. These landings were estimated to represent about 19,600 fish annually.

From routine sampling of fish in the CY heated discharge canal, Marcy (2004 [1976]) was also able to estimate relative number of fish residing in the canal. During the period 1968–1972, a total of 25,646 brown bullhead were captured in trawls, while a total of 7,304 white catfish were caught. Population estimates of these two species based on tag and recapture of fish in the canal indicated that as many as 31,000 brown bullhead and 39,000 white catfish may have overwintered there annually. Recreational catch of catfish species in the lower Connecticut River, south of Hartford, was also reported by Howell and Molnar (1999) for 1997 and 1998. They reported total catches for white and channel catfish and brown bullhead of 78,465 and 66,908 fish, respectively, for the 2 years. Annual estimates of impinged bullhead ranged from about 6,500 fish in 1968 to 22 fish in 1981.

The apparent lack of juvenile American shad smaller than 150 mm in the impingement counts during the period 1977–1982 may be partially due to a change in personnel performing the counts and the difficulty distinguishing juvenile shad from blueback herring. The difference in counts cannot be accounted for by any large-scale decrease in abundance as measured by the annual juvenile shad seine survey (Savoy 1999; Savoy et al. 2004, this volume). In general, however, the impingement of young shad appears low compared to the numbers that must be available to sustain an annual Connecticut River run of between 300,000 and 1,500,000 adult spawners (Savoy et al. 2004).

The trend toward an increased number of adult American shad impinged after 1977 cannot be entirely accounted for by population changes. The annual population estimates during 1977–1979 were at relatively lower levels (490,000 adults average for the 3 years) compared to subsequent years when the adult population size trended upwards toward a record high of 1,574,460 migrating adults in 1983 (Savoy 1999; Savoy et al. 2004). However, during 1981 and 1982, when the number of adult shad impinged was the largest observed over the counting period, there was a doubling of the population size compared to the 1977–1979 period and a corresponding increase in lifting at the Holyoke Dam, the first dam on the main-stem Connecticut River. Extended upriver migration may weaken emigrating adult shad making them more susceptible to impingement. This extended migration, coupled with a large increase in adults, may have contributed to the higher impingement rates the last 2 years that counts were taken.

A species not studied during the Connecticut River study was the endangered shortnose sturgeon. Shortnose sturgeon are known to move throughout the lower river at various times of the year. For example, they move past the CY power plant to the lower estuarine portion of the river in late spring to forage and then northward beyond CY to concentration and spawning areas during summer and fall (Savoy 1999 and 2004, this volume; Halavik and Jacobson 2002). However, it is unlikely that CY has had a measurable impact on this organism. During 1968–1982, when routine impingement counts were made, only two specimens were observed in the CY screenwash basket. One fish, 665 mm in length, was collected in September 1978 and another, 560 mm in length, was collected in April 1982. Whether these two specimens were alive or dead at the time of impingement is not known. No shortnose sturgeon were collected by trawl or seine sampling during the Connecticut River study. The number of shortnose sturgeon in the lower Connecticut River may have increased since the 1970s and is estimated at 750+ fish (USEPA 1995; Gephard 2001; Savoy 2004).

Other species of interest in the Connecticut River include Atlantic salmon, striped bass, and, increasingly, largemouth bass. An effort to reintroduce Atlantic salmon has been underway since the mid-1960s (Gephard and McMenemy 2004, this volume). The program depends on the stocking of hatchery-reared fry and smolts, which emigrate as smolts down the Connecticut River past CY. During the period when routine impingement counts were made, a total of eight smolts were captured, all in May 1982. An additional three smolts were captured during the special behavioral barrier studies performed in 1986 and 1987. In 1982, the estimated run of smolts emigrating down the Connecticut River was 220,000, and in 1986, it was 316,000 (S. Gephard, Connecticut Marine Fisheries, personal communication). Stocking continues as part of the restoration program, and impinged smolts may have escaped notice on days when no counts were made or since 1982 after counting was discontinued. However, the overall number impinged is likely to have remained low. No returning adults were reported impinged.

Similarly, neither striped bass nor smallmouth or largemouth bass appear vulnerable to impingement, and few have been found in impingement samples. However, striped bass are a striking example of how population changes can alter conclusions reached from earlier studies. Since implementation of the striped bass management plan and subsequent amendments and the ban on commercial fishing in the mid-Atlantic states, the numbers observed in the Connecticut River have risen sharply (Gephard 2001; Gephard and McMenemy 2004). In excess of 1,200 specimens were counted in the Holyoke fish lift in 1993 and again in 2001, and a productive fishery now exists in the lower river. Striped bass was the numerically dominant sport fish taken in the lower Connecticut River in 1998 based on creel census data (Howell and Molnar 1999). Whether, given the large increase in striped bass abundance, impingement rates and mortality of this species would have remained low is an important question. The relatively large size of these fish and their swimming ability (critical swimming speed ranged between 32 and 86 cm/s, as listed in EPRI 2000) would suggest that impingement of striped bass at CY, if operating today, would remain low (EPRI 2003) and survival in the fish return system high, but confirmation of these assumptions would be necessary. Only 30 striped bass were impinged at Millstone power station Unit 2 from 1976 through 1987 (NUSCo 1988). This nuclear power plant is larger than CY and is located on the coast of Long Island Sound about 14 km east of the mouth of the Connecticut River.

Why largemouth bass or other smaller centrarchids are impinged in relatively low numbers is not immediately apparent. The CY intake structure is a typical shoreline installation, yet with the exception of spottail shiner, does not appear to disproportionately affect species that are common to nearshore habitats. Electrofishing performed by Jacobs et al. (2004) suggests that redbreast sunfish and pumpkinseed are abundant in the near shore areas of the lower river yet few were impinged. Their movements may be sufficiently limited that exposure to the intake zone of influence is negligible.

Largemouth bass are most abundant in upper reaches of the lower Connecticut River above Hartford, but are also commonly collected in the shallow nearshore waters of the Haddam area. A comparison of the intake approach velocity of the CY plant (36.6 cm/s) to the average critical swimming velocity of largemouth bass (range of 20–49.7 cm/s) listed in EPRI (2000) suggests that largemouth bass likely have the ability to avoid the CY intake should they encounter it. Critical velocity is a measure of the ability of a fish to sustain its maximum swimming velocity.

Fish Exclusion Devices

Experimentation with deterrent devices to reduce the number of fish impinged was initiated voluntarily by CY in 1970, soon after the magnitude of impingement was understood. The barriers tested included an array of aluminum conduits in the intake forbays outboard of the trash racks (CYAPCO 1971; Hyman and Pikanowski 1974). These arrays significantly reduced (>70%) the number of fish larger than about 75 mm impinged and were utilized to reduce impingement for the remaining operational life of the plant through 1996. The array performed similarly with or without electrification.

While additional experiments demonstrated the feasibility of using electrical fields as a deterrent to fish as they enter the cooling water intake zone of influence (Richardson and Dixon 2004, this volume), an operational system as described by Hyman et al. (1975) was never permanently installed at CY due to icing and heavy seasonal debris loading, which precluded the placement of such a system without extensive maintenance. Furthermore, significant overall reductions in impingement had been achieved with the existing aluminum piping arrays. A permanent boom located upstream and in front of the CY intake was put in place in 1980, with extensive support structures to deflect heavy debris and thereby protect the existing deterrent device (aluminum pipe array) within the forebays and ensure operation of the station's service water system.

Successful deployment of the triple array deterrent device appears to be unique to CY. In-depth reviews of fish protection devices by EPRI (1999a) and Taft (2000) indicated that this type of barrier had not been used elsewhere and that, in general, behavioral barriers have demonstrated limited success in preventing impingement at typical cooling water intake structures utilized at electric generating stations. Whether the CY fish deterrent device acted as a physical or behavioral barrier is not clear. The location of the aluminum arrays in the forebay likely acted as a behavioral barrier limiting entry of fish into the intake structure through the trash racks. The aluminum conduits also exhibited a discernable vibration and sound audible at least to the human ear and may have elicited avoidance cues to fish as they entered the cooling water intake flow field.

Impingement Revisited

Determining the factors that influence rates of impingement and impingement survival is critical to selection of sites for electric generating stations and for assessing impacts of existing cooling water systems. A growing body of literature exists describing these factors and their use in ecological impact assessments (Bailey et al. 1999; EPRI 1999a, 1999b, 2000, 2003; Taft 1999; Wisniewski 2000; Taft et al. 2001). Important attributes include location of the intake relative to fish concentrations, species' temporal and spatial occurrence, swimming ability and physiological condition of susceptible organisms, intake design and capacity; water quality, and hydraulic influences. None acts independently. The experience at CY suggests that, among the factors known to influence impingement rates, location, in relation to species life history and behavior, is probably the most important. Capacity (CY entrained about 6% to 7% of ebb and flood tidal excursions at mean low river flow) and design (shoreline intake with traditional 9.5-mm-mesh traveling screens) appeared less important as certain key species' life stages were more or less susceptible depending on their seasonal movements and abundance in relation to the intake location. Adult American shad are present in the river during their spring spawning migration and postspawn emigration. Migrating upstream to spawning areas, they use the navigational channel across the river from the CY cooling water intake and are impinged in relatively low numbers during this period. During the early summer postspawn emigration, however, adult shad were impinged in somewhat larger numbers as they move downstream in a weakened condition.

American shad and river herring spawn well upstream of CY. Yet the juveniles move past the intake during their fall downstream emigration, when they become susceptible to impingement. Despite this migration pattern, overall impingement rates for juvenile shad, though likely underestimated, were low nevertheless. White perch and white catfish adults, the dominant resident fish, move extensively throughout the river during various times of the year and were impinged in larger numbers due to their relative abundance. Seasonal migrants, such as striped bass, enter the river primarily as larger specimens and move upstream well past the CY intake as far north as the Holyoke Dam in Massachusetts. These fish were historically impinged in very low numbers and, most likely, can avoid the intake flows due to their strong swimming ability. However, the number of striped bass in the river has increased dramatically, and additional information on impingement rates post 1982 would be needed to test this assumption if CY were still operational. The occurrence of 0-year size striped bass in impingement samples was rare (only two < 150 mm were collected, and only one of these was < 75 mm), given that no direct evidence of striped bass spawning in the Connecticut River has been found to date (Gephard and McMenemy 2004).

The endangered shortnose sturgeon migrate past CY as they move well downriver toward the estuarine portion of the lower river during spring freshet and return upriver to concentration and spawning areas well upstream of CY during late summer and fall. As a result, they do not appear to concentrate near the power station (Savoy 2004). In addition, their limited numbers and benthic orientation likely make them less susceptible to impingement.

Following completion of the Connecticut River study in 1972 and, in particular, following the cessation of impingement counting in 1982, there have been observable changes in the population dynamics of the migratory fishes and in management strategies with respect to restoration of key species such as Atlantic salmon (Gephard 2001; Gephard and McMenemy 2004; Jacobs et al. 2004; Savoy and Crecco 2004, this volume). It has been found that natural or anthropogenic large scale changes in fishery dynamics overwhelm impacts attributed to cooling water intakes (EPRI 1999b; Barnthouse 2000; LaJeone and Monzingo 2000; Michaud 2000). With respect to the CY experience, information collected during the early studies (1965–1972), from impingement counting (1968–1987), and from installation of deterrent devices provided a basis upon which to retrospectively evaluate the importance of potential plant impacts as ecological changes have occurred. The information on life history strategies, abundance, and dominance of fishes of the lower Connecticut River obtained from the initial Connecticut River Studies and information collected in the intervening years (this volume), in conjunction with direct estimates of plant impact from impingement, suggest that it is possible not only to initially quantify plant impacts, but also make reasoned judgments about the ecological consequences of long-term power plant operations. The importance of site-specific studies and evaluations is particularly evident in the case of CY, where location and species-specific life history strategies are the dominant factor affecting overall plant impacts. As suggested by Merriman and Thorpe (2004 [1976]), these studies demonstrate the fallacy of categorical a priori decision-making on intake design criteria.

Acknowledgments

Collection of the impingement data summarized in this report was funded by the Connecticut Yankee Atomic Power Company and Northeast Utilities Service Company (NUSCo). Impingement sampling during 1968 and 1969 was performed by the Essex Marine Laboratory staff and later by NUSCo staff, including the co-authors of this paper, and by NUSCo contractors, notably Mervin Roberts of Old Lyme, Connecticut. Experimental testing of the behavioral barriers was performed by Saul Saila and his students of the University of Rhode Island. We are also grateful for the constructive reviews of this paper provided by William Dey, Douglas Dixon, Ned Taft, Stephen Summer, John Tietjen, and Robert Wilce. Special thanks are extended to Barton C. Marcy, Jr., lead editor.

References

Bailey, D. E., J. J. Loos, and E. S. Perry. 1999. Studies of cooling water intake structure effects at Potomac Electric Power Company generating station. Pages 8-1–8-29 *in* Proceedings: 1998 EPRI Clean Water Act Section 316(b) Technical Workshop. Coolfont Conference Center. Electric Power Research Institute, Report TR-112613, Palo Alto, California

Barnthouse, L. 2000. Impacts of power-plant cooling water systems on estuarine fish populations: the Hudson River after 25 years. Pages 342–348 *in* J. Wisniewski, editor. Power plants & aquatic resources: issues and assessment. Environmental Science & Policy 3(Supplement 1). Elsevier, New York.

CTDEP (Connecticut Department of Environmental Protection). 1974. NPDES Permit, Connecticut Yankee Atomic Power Company. Connecticut Department of Environmental Protection, Hartford.

CTDEP (Connecticut Department of Environmental Protection). 1979. Letter from R. L. Smith Asst. Director to W. G. Council, Vice President, Northeast Utilities Service Company dated December 10, 1979. Connecticut Department of Environmental Protection, Hartford.

CTDEP (Connecticut Department of Environmental Protection). 1986. NPDES Permit, Connecticut Yankee Atomic Power Company. Connecticut Department of Environmental Protection, Hartford.

CYAPCO (Connecticut Yankee Atomic Power Company). 1971. Aquatic cost benefit, Appendix E to the Environmental Report, Operating License Stage, Docket 50–213. Connecticut Yankee Atomic Power Company, Haddam Neck plant, Haddam Neck, Connecticut.

CYAPCO (Connecticut Yankee Atomic Power Company). 1974a. Summary of the environmental assessment of the once-through condenser cooling system. Submitted to the State of Connecticut Department of Environmental Protection by Northeast Utilities Service Company, Hartford.

CYAPCO (Connecticut Yankee Atomic Power Company). 1974b. Environmental Technical Specifications. Appendix B to the Facility Operating License DPR-61, Docket 50–213. Haddam Neck, Connecticut.

CYAPCO (Connecticut Yankee Atomic Power Company). 1976–1983. Haddam Neck plant, Annual Environmental Operating Reports, Part A: Non-radiological Report. Operating License No. DPR-61, Docket No. 50–213. Haddam Neck, Connecticut.

CYAPCO (Connecticut Yankee Atomic Power Company). 1980. Feasibility of bypass facilities to return impinged fish to the source water, Haddam Neck station. Submitted to the Connecticut Department of Environmental Protection by Northeast Utilities Services Company, Haddam Neck.

EPA (Environmental Protection Agency). 1995. Biological assessment for the shortnose sturgeon, *Acipenser brevirostrum*, for the Connecticut River in Massachusetts. U.S. EPA Region 1, Water Quality Branch, Boston.

EPA (Environmental Protection Agency). 2001. National pollutant discharge elimination system: regulations addressing cooling water intake structures for new facilities. Federal Register 66(243).

EPRI (Electric Power Research Institute). 1999a. Fish protection at cooling water intakes: Status report. Electric Power Research Institute, Report TR-114013, Palo Alto, California.

EPRI (Electric Power Research Institute). 1999b. Procedural guidelines for evaluating alternative fish protection technologies to meet Section 316(b) requirements of the Clean Water Act. Electric Power Research Institute, Report No. 1000551, Palo Alto, California.

EPRI (Electric Power Research Institute). 2000. Technical evaluation of the utility of intake approach velocity as an indicator of potential adverse environmental impact under Clean Water Act Section 316(b). Electric Power Research Institute, Report No. 1000731, Palo Alto, California.

EPRI (Electric Power Research Institute). 2003. Evaluating the effects of power plant operations on aquatic communities. Summary report of impingement survival studies. Electric Power Research Institute, Report No. 1007821, Palo Alto, California.

Gephard, S. 2001. Status of diadromous fish populations in the Connecticut River, thirty years later. In Proceedings, Electric Power Research Institute (EPRI) Workshop, Connecticut River Ecological Study – revisiting the impact of a power plant, November 15–16, 2001, University of Connecticut. Electric Power Research Institute, Palo Alto, California. CD Order #1006900.

Gephard, S., and J. McMenemy. 2004. An overview of the program to restore Atlantic salmon and other diadromous fishes to the Connecticut River with notes on the current status of these species in the river. Pages 287–317 *in* P. M. Jacobson, D. A. Dixon, W. C. Leggett, B. C. Marcy, Jr., and R. R. Massengill, editors. The Connecticut River Ecological Study (1965–1973) revisited: ecology of the lower Connecticut River 1973–2003. American Fisheries Society, Monograph 9, Bethesda, Maryland.

Halavik, T., and R. Jacobson. 2002. Fisheries of the Connecticut River Estuary and tidal wetlands. Pages 56–61 *in* G. D. Dreyer and M. Caplis, editors. Living resources and habitats of the lower Connecticut River. Bulletin 37. The Connecticut College Arboretum, New London.

Howell, P., and D. Molnar. 1999. Stock assessment of white perch in the Connecticut River. Part I. Pages 167–185, and An angler survey of the Connecticut portion of the Connecticut River Part II. Pages 190–205 *in* A study of marine recreational fisheries in Connecticut. Job 7. Federal Aid to Sport Fish Restoration. Connecticut Department of Environmental Protection, Hartford.

Howell, P., and D. Molnar. 2004. Stock assessment of white perch in the lower Connecticut River. Pages 379–390 *in* P. M. Jacobson, D. A. Dixon, W. C. Leggett, B. C. Marcy, Jr., and R. R. Massengill, editors. The Connecticut River Ecological Study (1965–1973) revisited: ecology of the lower Connecticut River 1973–2003. American Fisheries Society, Monograph 9, Bethesda, Maryland.

Hyman, M. A. M., W. H. Mowbray, and S. B. Saila. 1975. The effects of two electrical barriers on the entrainment of fish at a freshwater nuclear power plant. Pages 205–237 *in* S. B. Saila, editor. Fisheries and energy production: a symposium. Lexington Books, D. C. Heath and Co., Lexington, Massachusetts.

Hyman, M. A. M., and R. A. Pikanowski. 1974. Statistical analysis of fish counts at Connecticut Yankee Atomic Power Company Haddam Neck, Connecticut. University of Rhode Island, Marine Experiment Station, Kingston.

Jacobs, R. P., N. T. Hagstrom, W. Hyatt, E. B. O'Donnell, and E. Schluntz. 2004. Trends in abundance, distribution, and growth of freshwater fishes from the Connecticut River in Connecticut (1988–2002). Pages 319–343 *in* P. M. Jacobson, D. A. Dixon, W. C. Leggett, B. C. Marcy, Jr., and R. R. Massengill, editors. The Connecticut River Ecological Study (1965–1973) revisited: ecology of the lower Connecticut River 1973–2003. American Fisheries Society, Monograph 9, Bethesda, Maryland.

Jacobs, R. P., and E. B. O'Donnell. 1996. An electrofishing survey of selected Connecticut Lakes. Federal Aid to Sport Fish Restoration, F-61-T-13. Connecticut Department of Environmental Protection, Hartford.

LaJeone, L. J. and R. G. Monzingo. 2000. 316(b) and Quad Cities Station, Commonwealth Edison Company. Pages 313–322 *in* J. Wisniewski, editor. Power plants & aquatic resources: issues and assessment. Environmental Science & Policy 3(Supplement 1). Elsevier, New York.

Leggett, W. C. 2004. The American shad , with special reference to its migration and population dynamics in the Connecticut River. Pages 181–238 *in* P. M. Jacobson, D. A. Dixon, W. C. Leggett, B. C. Marcy, Jr., and R. R. Massengill, editors. The Connecticut River Ecological Study (1965–1973) revisited: ecology of the lower Connecticut River 1973–2003. American Fisheries Society, Monograph 9, Bethesda, Maryland. (Originally published in 1976)

Marcy, B. C., Jr. 2004. Fishes of the lower Connecticut River and the effects of the Connecticut Yankee plant. Pages 65–124 *in* P. M. Jacobson, D. A. Dixon, W. C. Leggett, B. C. Marcy, Jr., and R. R. Massengill, editors. The Connecticut River Ecological Study (1965–1973) revisited: ecology of the lower Connecticut River 1973–2003. American Fisheries Society, Monograph 9, Bethesda, Maryland. (Originally published in 1976)

Merriman, D., and L. M. Thorpe. 2004. Introduction. Pages 3–25 *in* P. M. Jacobson, D. A. Dixon, W. C. Leggett, B. C. Marcy, Jr., and R. R. Massengill, editors. The Connecticut River Ecological Study (1965–1973) revisited: ecology of the lower Connecticut River 1973–2003. American Fisheries Society, Monograph 9, Bethesda, Maryland. (Originally published in 1976)

Michaud, D. T. 2000. Wisconsin Electric's experience with fish impingement and entrainment studies. Pages 333–340 *in* J. Wisniewski, editor. Power plants & aquatic resources: issues and assessment. Environmental Science & Policy 3(Supplement 1). Elsevier, New York.

NUSCo (Northeast Utilities Service Company). 1983. Section 5.5.3.3.4.3 Atlantic menhaden. Pages 5.1-36–5.1-47 *in* Millstone nuclear power station Unit 3 Environmental Report. Operating License Stage volume 2. Submitted to the United States Nuclear Regulatory Commission, Hartford, Connecticut.

NUSCo (Northeast Utilities Service Company). 1985. Haddam Neck plant, request for modification of intake fish protection device. Letter from R. A. Reckert (Northeast Utilities Service Company) to R. Moore (Connecticut Department of Environmental Protection) dated December 10, 1985. Hartford, Connecticut.

NUSCo (Northeast Utilities Service Company). 1987. Haddam Neck plant, request for modification of intake fish protection device. Letter from R. A. Reckert (NUSCo) to R. Moore (CTDEP) dated June 30, 1987. Hartford, Connecticut.

NUSCo (Northeast Utilities Service Company). 1988. Fish ecology studies. Pages 255–307 *in* Monitoring the Marine Environment of Long Island Sound at Millstone nuclear power station, Waterford, Connecticut. Three-unit operational studies, 1986–1987. Hartford, Connecticut.

Richardson, J., and D. A. Dixon. 2004. Modeling the hydraulic zone of influence of Connecticut Yankee nuclear power plant's cooling water intake structure. Pages 513–524 *in* P. M. Jacobson, D. A. Dixon, W. C. Leggett, B. C. Marcy, Jr., and R. R. Massengill, editors. The Connecticut River Ecological Study (1965–1973) revisited: ecology of the lower Connecticut River 1973–2003. American Fisheries Society, Monograph 9, Bethesda, Maryland.

Robins, C. R., R. M. Bailey, C. E. Bond, J. R. Brooker, E. A. Lachner, R. N. Lea, and W. B. Scott. 1991. Common and scientific names of fishes from the United States and Canada. Fifth edition. American Fisheries Society, Special Publication 20, Bethesda, Maryland.

Savoy, T. 1999. Anadromous fish studies in Connecticut waters. Progress report AFC-25. Connecticut Department of Environmental Protection, Hartford.

Savoy, T. F. 2004. Population estimate and utilization of the lower Connecticut River by shortnose sturgeon. Pages 345–352 *in* P. M. Jacobson, D. A. Dixon, W. C. Leggett, B. C. Marcy, Jr., and R. R. Massengill, editors. The Connecticut River Ecological Study (1965–1973) revisited: ecology of the lower Connecticut River 1973–2003. American Fisheries Society, Monograph 9, Bethesda, Maryland.

Savoy, T. F., and V. A. Crecco. 2004. Factors affecting the recent decline of blueback herring and American shad in the Connecticut River. Pages 361–377 *in* P. M. Jacobson, D. A. Dixon, W. C. Leggett, B. C. Marcy, Jr., and R. R. Massengill, editors. The Connecticut River Ecological Study (1965–1973) revisited: ecology of the lower Connecticut River 1973–2003. American Fisheries Society, Monograph 9, Bethesda, Maryland.

Savoy, T. F., V. A. Crecco, and B. C. Marcy, Jr. 2004. American shad early life history and recruitment in the Connecticut River: a 40-year summary. Pages 407–417 *in* P. M. Jacobson, D. A. Dixon, W. C. Leggett, B. C. Marcy, Jr., and R. R. Massengill, editors. The Connecticut River Ecological Study (1965–1973) revisited: ecology of the lower Connecticut River 1973–2003. American Fisheries Society, Monograph 9, Bethesda, Maryland.

Taft, E. P. 2000. Fish protection technologies: a status report. Pages 349–359 *in* J. Wisniewski, editor. Power plants & aquatic resources: issues and assessment. Environmental Science & Policy 3(Supplement 1). Elsevier, New York.

Taft, E. P., W. P. Dey and L. W. Barnthouse. 2001. Factors affecting the potential for adverse environmental impact from cooling water withdrawals. Report to the Utility Water Act Group Hunton & Williams Inc., Washington, D.C.

Taft, N. 1999. Thresholds: can the potential for environmental impacts be determined on the basis of plant design or operational variables. Pages 16-1–16-11 *in* Proceedings: 1998 EPRI Clean Water Act Section 316(b) Technical Workshop. Coolfont Conference Center. Electric Power Research Institute, Report TR-112613, Palo Alto, California.

USAEC (United States Atomic Energy Commission). 1973. Final environmental impact statement related to the Haddam Neck (Connecticut Yankee) nuclear power plant. Docket No. 50–213, Washington, D.C.

Wisniewski, J., editor. 2000. Power plants & aquatic resources: issues and assessment. Environmental Science & Policy 3(Supplement 1). Elsevier, New York.

American Fisheries Society Monograph 9:475–500, 2004

Developments in Fishery Science with Special Reference to Power Plant Impact Assessment Methods and the Connecticut River Ecological Study: 1973–Present

SAUL B. SAILA

University of Rhode Island, Narragansett Bay Campus, Narragansett, Rhode Island 02882, USA

ERNESTO LORDA*

Enviro-Stat Consulting Services, Old Lyme, Connecticut 06371, USA

Abstract.—The purpose of this contributed paper is to provide a brief review of new developments in fishery science and ecology that are thought to be relevant to any future study similar to the original Connecticut River Ecological Study. The data and results from each section of the original study have been examined with the goal of making suggestions for possible improvements in the design, analysis, and interpretation of both observational data as well as planned experiments. These suggestions are based on the application of new developments in statistics, data acquisition, and analysis, as well as advances in fishery science. From the above, it was determined that predictive techniques related to methods primarily developed for fisheries management may, in combination with computer simulation models in some cases, be preferable to conventional monitoring and before–after/control–impact (BACI) experimental designs for assessing impacts at the population level. Assessing impacts at the community level still requires monitoring studies, the analysis of which can be now greatly facilitated by improved nonparametric multivariate statistical methods and readily available computer programs. Artificial intelligence based methods also show great promise as flexible tools that will help solve complex impact assessment and fishery management problems that involve a high degree of uncertainty. Finally, a brief overview of the new discipline of fisheries oceanography is provided to illustrate the integration of new technologies such as satellite imagery, modern mapping (GIS), and massive global weather databases into fishery science.

Introduction

The major purpose of this paper is to provide a review of the newer methodologies in fishery science and ecology that have been developed since the publication of the original Connecticut River Ecological Study in 1976. Some of the new developments are thought to be relevant to any future study of a similar nature to the original study as well as to assessing adverse environmental impacts of cooling water intake structures under Clean Water Act §316(b) regulations in general.

This paper is divided into three units: a methodology section that briefly describes some of the newer developments broadly grouped as statistical methods, computer simulation techniques, and artificial intelligence, including some of their applications; a second section that introduces the new discipline of "fisheries oceanography"; and a third section that provides a brief review and suggestions related to each of the chapters in the original Connecticut River Ecological Study. This is followed by a closing brief discussion and a list of references cited.

Our suggestions and comments are not inclusive with respect to each aspect of the original Connecticut River Ecological Study, nor do they include all new developments since that time. Instead, they reflect some of the interests of the authors who believe they are relevant to efforts in obtaining more rigorous results from any study of

* *Corresponding author*: lorder@snet.net

this type that may be made in the near future. It is also thought that this document may complement available Electric Power Research Institute (EPRI) reports, such as EPRI's (1999) catalog of assessment methods for evaluating effects of power plants on aquatic communities.

Newer Methodologies

Statistical Methods

The last 20 years have seen a revolution in statistical methodology fostered by the increasing availability of inexpensive computer power and better statistical software. Additionally, new methods that rely on raw computer power have made it possible to approach analytical problems previously intractable. This material is intended to briefly introduce fisheries scientists and managers to some of the new statistical methods that may be of utility in their work. Emphasis is placed on those methods developed since the publication of the original American Fisheries Society monograph. We believe that an excellent and readable book on applications of statistics to environmental science (monitoring and impact assessment) is provided by Manly (2000). This book covers several important topics, including the choice of sample sizes and various sample design strategies. Our section on statistics is constrained to only a few topics, which we hope will expand the excellent material covered by Manly. Among these topics are a further brief discussion of models for analysis of variance (ANOVA), including mixed-effects models and before–after/control–impact (BACI) designs for analyzing monitoring data; multivariate methods such as clustering and multidimensional scaling (MDS); power analysis and sample size determination; and computer intensive methods (resampling). An introduction to meta-analysis is also provided to include ecology and fishery related examples in addition to the material provided in Manly's book.

Analysis of Variance and Experimental Design

When monitoring data conform to a complete, balanced design with equal numbers of observations in each treatment, it is relatively easy to conduct ANOVA. A number of statistical computing procedures will aid in estimation and interpretation of balanced data. Regrettably, it is rare when monitoring projects generate data that are completely balanced. An excellent review of ANOVA applications and experimental designs in marine biology is provided by Underwood (1981), the robustness of ANOVA for environmental monitoring applications was investigated by Grieb (1985), and the use of ANOVA for unbalanced data are reviewed by Shaw and Mitchell-Olds (1993). Another useful approach for data obtained from field studies collected over time is the repeated measures split plot ANOVA design. This has been effectively described by Maceima et al. (1994).

Before–after/control–impact sampling designs proposed by Green (1979) have been widely used in investigations of environmental impacts upon populations of organisms. The concept involved is that an anthropogenic disturbance on the impact location will cause a different pattern of change from before to after it starts compared with natural change in a control area. The associated statistical tests to detect that change (i.e., the time–space interaction) are based on the temporal replication of some measure of difference between paired control and impacted sites. However, a number of researchers (Underwood 1981, 1994; Hurlbert 1984; Stewart-Oaten et al. 1986, 1992; Smith et al. 1993; Heffner et al. 1996) have criticized the lack of proper replication in most applications and have demonstrated other problems and limitations of the conventional BACI designs. Underwood (1994) has suggested mixed ANOVA models with fixed and random effects to reduce bias in the interaction test and asymmetrical experimental designs, which tend to provide more effective comparisons. An example of an application of mixed effects ANOVA models in impact assessment work is described in NAESCO (1994). In this study, a BACI design with replicated control and impacted sites was analyzed with sampling times and sites as random effects and only the "before–after" effect as fixed. In this ANOVA, the relevant *F* ratios were calculated from the expected mean square values (i.e., variance components) as described in Underwood (1981), and the evidence for impact was provided by the interaction between locations and the "before–after" effect. More recently and in a different type of application in marine ecology, Bolam et al. (2002) studied the relationship between macroinvertebrate species richness and ecosystem functions in a soft-bottom intertidal system using a two-way mixed ANOVA model.

In brief, much of what is required for the efficient detection of environmental impacts in a monitoring context is nicely summarized by Osenberg et al. (1994). This work includes consideration of natural variability, effect size and power analysis, and statistical assumptions.

Multivariate Methods for Community Analysis

Scaling, or the ordering of multivariate data, represents a fundamental descriptive approach to ecological data analysis (Pielou 1984). Spatial and temporal changes in invertebrate communities or fish assemblages can be addressed by hierarchical clustering, nonmetric MDS, principal components analysis (PCA), and correspondence analysis (CA), among other statistical techniques. The intent is to summarize patterns in species composition and related environmental variables, in addition to a host of other important applications.

Cluster analysis (Everitt 1980) seeks to organize information about variables so that relatively homogeneous groups or clusters can be formed. The main outcome of a cluster analysis is a grouping of data into self-similar groups. Both PCA and CA are ordination techniques that seek to reduce dimensionality of multivariate data and map composite variables or associated samples to describe their relationships (Greenacre 1984). Nonmetric multidimensional scaling is an exploratory numerical technique concerned with ranked data sets, so that it makes few assumptions about the form of the data or the interrelationship of the samples. The starting point in MDS is the similarity or dissimilarity matrix among the samples. This can be whatever similarity matrix is biologically relevant to the questions being asked of the data. The purpose of MDS can be simply stated: it constructs a "map" or configuration of the samples in a specified number of dimensions, which attempts to satisfy all the conditions imposed by the rank similarity matrix. For instance, if sample A has higher similarity to sample B than it does to sample C, then sample A will be placed closer on the map to sample B than it is to sample C. The merits of MDS techniques for ordination are now well established in ecological applications, and comparative studies have consistently demonstrated their reliability (Kenkel and Orloci 1986). In comparison with earlier multivariate techniques, such as PCA and CA, MDS has a number of practical advantages stemming from its flexibility and lack of assumptions. None of the above techniques is new; the newest MDS was introduced almost 40 years ago by Kruskal (1964) and demonstrated by Everitt (1978). A more recent review of these techniques can be found in Legendre and Legendre (1998). What is new now is the availability of excellent computer software to apply them to community analyses and impact assessment work.

We believe that, at present, the software of choice for statistical analysis of community data is the package of Plymouth Routines in Multivariate Ecological Research (PRIMER) developed by Clarke and Warwick (1994) at the Plymouth Marine Laboratory, Plymouth, UK. This computer program is available now for the Windows environment (Clarke and Warwick 2001), including routines for the following multivariate analyses: clustering (CLUSTER), ordination and nonmetric MDS, PCA, hypothesis testing on similarity matrices (ANOSIM) and other permutation tests (RELATE), linking biotic patterns to environmental variables (BIOENV), identifying species responsible for observed community pattern (SIMPER, BVSTEP), comparison of ordinations (2nd stage MDS), and dominance curves and biodiversity indices, including new measures based on taxonomic relatedness of species (TAXDTEST). Applications of these techniques have been reported by Clarke and Green (1988); Smith and Iles (1988), Clarke (1990, 1993), Clarke and Ainsworth (1993), Clarke et al. (1993), Clarke and Warwick (1994), Dye (1998), Underwood and Chapman (1998), and DNC (2003b).

The last reference above illustrates the use of the PRIMER routine MDS applied to the analysis of the abundance of more than 100 macroalgal and invertebrate taxa that occur in a rocky shore community monitored since 1982 at four sites in the vicinity of Millstone power station, Waterford, Connecticut. The analysis provides a comparison of community states at a potentially impacted site to those observed at the other three unimpacted sites. The resulting MDS map shown in Figure 1 has a low stress level (0.09), indicating a good representation of the ordination in two dimensions (Clarke 1993), and reveals two distinct

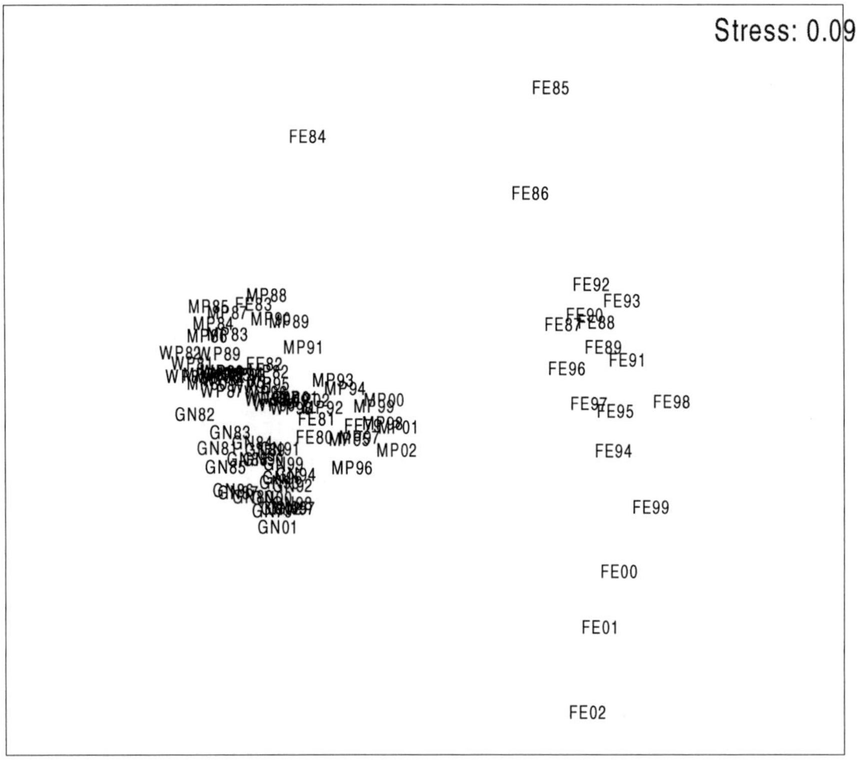

FIGURE 1. Two-dimensional MDS plot for annual percent cover data from rocky intertidal transects at stations FE, GN, and WP (1979–2002), and MP (1982–2002) in the vicinity of Millstone power station, Waterford, Connecticut.

groupings with several outliers. Most collections fall into a relatively tight grouping that comprises all the annual collections at the unimpacted sites, along with a few annual collections from the impacted site (labeled as FE) prior to the widening of the cooling water discharge structure to accommodate increased thermal effluent flow (1979–1983). The compactness of this cluster, with considerable overlap among collections, suggests a distinct community state relative to the impacted areas (FE). The changing community states at FE are illustrated by the relationships among the remaining collections in the MDS map. Annual collections at FE from 1984 to 1986 are spatially distinct due to dramatic community changes at FE following the increase of thermal flow. The other grouping in the MDS map includes FE annual collections from 1988 to 2002, years of development and maintenance of the thermal community established since the increase of thermal flow in 1983.

Although other statistical software packages such as PC-ORD developed at Oregon State University include many of the routines from the Primer package, the statistical methods of the Primer software are explained in nonmathematical terms in the methods manual and by the specific examples that are provided. Additional information on PRIMER current version 5 and free trial software can be downloaded at http://www.primer-e.com. A complete description of the PC-ORD software is provided by McCune and Mefford (1997), and the recent study of Scarsbrook (2002) uses PC-ORD to apply nonmetric MDS to relate lotic invertebrate community changes to variability in environmental conditions. Additional information from the developers of PC-ORD can be obtained at http://home.centurytel.net/~mjm.

Power Analysis and Sample Size Determination

Statistical power analysis has been a well-developed concept in theoretical statistics for many years (Winer 1971; Cohen 1988; Murphy and Myors 1998), but it was seldom mentioned in relation to ecological studies, environmental monitoring, and impact assessment work until the 1980s (Toft and Shea 1983; Rottenberry and Wiens 1985; Gerrodette 1987; Green 1989; Osenberg et al. 1994; Schlese and Nelson 1996; Schmitt and Osenberg 1996; Steidl and Thomas 2001). Statistical power is defined as 1-β, where β is the probability of failing to reject the null hypothesis H_0, when in fact H_0 is false. Therefore, in the context of impact assessment methodology, statistical power is the probability of correctly rejecting H_0 when there is an "effect" or impact. The purpose of this section is to emphasize that statistical power concepts should be applied regularly by fisheries scientists and impact assessment workers, as suggested by Peterman (1990) and Fairweather (1991), respectively.

There are several types of power analysis that clearly can be applied with great advantage to fishery science in general and to monitoring studies and the assessment of environmental impacts in particular (Peterman and M'Gonigle 1992). First, statistical power analysis can be applied before the start of data collection (a priori analysis) or after completion of the fieldwork (a posteriori analysis). Second, power analysis can be used to estimate relevant parameters of a study other than the power of a given test. For instance, the sample size required to achieve a desired power can be estimated a priori in many cases (Mapstone 1995; Ortiz 2002). Similarly, power analysis can be applied to estimate the sample size and/or experimental design required to detect some specified magnitude of an effect or change (Vaughan and Van Winkle 1982; Green 1989). When no analytical solutions exist, power analysis can still be applied using Monte Carlo simulation methods. A good example of this is the application of power analysis to assessing trends in fish populations by Peterman and Bradford (1987) and the application of power analysis on the monitoring of bull trout stocks by Maxell (1999). More recently, Trenkel and Rochet (2003) present methods for testing the null hypothesis of nonsignificant impact of fishing on a fish community and for calculating the statistical power of their tests; Lindley et al. (2000) describe the application of power analysis to monitoring the recovery of endangered winter chinook salmon; and Underwood and Chapman (2003) incorporate the concept of statistical power in their discussion of sampling designs for assessment of environmental impacts.

In general, statistical power is some function of the chosen α value for a test, the effect size being tested, the number of observations available, and their sample variance. Power increases with increasing sample size, effect size, and higher α value and declines with increasing sampling variance. Specific forms of this power function are available depending on the particular statistical model or test being used (i.e., *t*-test, *F*-test in two-way ANOVA, or more complex statistical models). In general, except for a simple *t*-test, the computations required can be rather lengthy. Also, distribution tables for the required noncentral parameter values are not available in most cases and need to be calculated or approximated (Cohen 1988). Computer programs to carry out the computations needed for statistical power analyses were only available at academic mainframe computers until recently. However, with the advent of fast and inexpensive desktop computers during the 1990s, many specialized programs for power analysis are now widely available for the Windows environment. Thomas and Krebs (1997) provide an excellent review of 29 computer programs, either stand-alone programs for power and sample size calculations or general purpose statistical software with built-in power analysis capabilities. They evaluated three aspects of each program: scope, ease of use, and ease of learning. The review concluded that PASS (Heintze 2000), among others, included all the power analyses likely to be needed in actual applications, while being ranked as the top software in terms of both ease of use and ease of learning. This software in its latest version can be downloaded for a free trial from http://www.ncss.com/pass.html together with information on program capabilities and tests descriptions. The PASS user's manual also provides examples for each test available. Nevertheless, we suggest that fisheries scientists interested in the subject also review the other software discussed by Thomas and Krebs (1997).

Computer Intensive Methods —Resampling

Diaconis and Efron (1983) coined the term "computer intensive methods" in their very informative and readable article in *Scientific American* on resampling methods and their applications. The resampling approach addresses a key problem in statistics: how to infer the "truth" from a sample of data that may be incomplete or drawn from an ill-defined population. For example, what is the best estimate of the true value of a variable from which we have 10 different measurements but no other information? The average of the 10 measurements may be a good answer, but it is not always the best answer and one may want to know how good an estimate it is. A resampling approach provides a way of using the data contained in a single sample to estimate the accuracy of the average or some other statistical measure. The idea is to extract as much information as possible from the data on hand by resampling with replacement to produce many new samples of simulated data. The payoff from such intensive computation is freedom from two limiting factors that dominate traditional statistical theory: the assumption that the data conform to a bell-shaped curve, called the normal distribution, and the need to focus on statistical measures whose theoretical properties can be analyzed mathematically. Resampling with replacement from a single sample to determine sample statistical variability was first suggested independently by Simon et al. (1976) and Efron (1979). Resampling methods including the bootstrap, the jackknife, and many randomization tests for statistical hypotheses have been developed in the last 20 years. They now form a large body of new statistical techniques known as computer intensive methods and also "new statistics" (Efron 1981, 1982; Efron and Tibshirani 1991, 1993; Peterson 1991; Simon and Bruce 1991; Hjorth 1994; Davison and Hinkley 1997; Good 2000; Manly 1999, 2000; and Dixon 2001).

The bootstrap, like the jackknife and cross-validation, is a technique that uses Monte Carlo simulation to estimate standard errors (Efron 1979). It has been applied to many problems for which the variability of a statistic cannot be assessed analytically. In less than two decades, the bootstrap has grown from an obscure object of theoretical study to a widely used resampling method with broad applications in ecology (Kohavi 1995) and numerous real-world situations. It involves resampling the original observations in a suitable way, to construct "pseudo-data" on which the estimator of interest is exercised. More specifically, the theoretical distribution of an unobservable disturbance or error term is approximated by the distribution of a set of observable residuals. Measures of variability, confidence intervals, and even estimates of bias may then be calculated. It is tempting to set confidence limits for a parameter estimate, mean fish abundance for instance, by moving 1.96 standard errors in either direction from the estimate. This relies on asymptotic normality and often performs well when a large number of independent observations are available. In many cases, however, the sample size is not large enough or the underlying distribution is clearly non-normal and the above method gives a confidence level that is far short of the nominal level. The bootstrap method is particularly well suited for these situations, and it has been successfully applied to similar problems in fisheries and wildlife work (Buckland 1980 and 1983). Also, Norris and Pollock (1996) described bootstrap-based methods to incorporate model uncertainty in estimating variances in mark and recapture studies. They illustrated the use of these methods with real fishery data on whitefish from Canada's Shakespeare Island Lake. More recently, Chen and Irvine (2001) applied the bootstrap to evaluate uncertainties in model parameters of stock–recruitment relationships incorporating environmental data.

The jackknife requires fewer calculations than the bootstrap, but it also seems less flexible and at times less dependable (Efron 1982; Diaconis and Efron 1983). However, Meyer et al. (1986) reported almost identical results from both resampling methods in estimating the uncertainty in population growth rates of *Daphnia pulex* in a laboratory study. Although brief, the review of bootstrap and jackknife applications in ecology provided by Dixon (2001) is very informative, including practical details and suggestions on where and when not to use them.

Randomization or permutation tests are a computer-intensive statistical technique that predates computers. R. A. Fisher introduced the idea in the 1930s as an argument supporting Student's *t*-test. Modern computer power makes permutation tests practical to apply to many statistical problems today. Permutation tests are free of mathematical assumptions and form an important class of modern statistical methods ideally suited for testing hypotheses in situations where parametric theory would not apply. A comprehensive overview of permutation tests is given by Edgington (1987). However, Noreen (1989) provides a more readable introduction with many numerical examples and simple computer algorithms to implement the tests. Other recent references with applications are Adams et al. (1997), Manly (1999), and Good (2000, 2001). Permutation tests are used in some computer programs to test for differences between similarity matrices and clusters in community analysis; see ANOSIM in the PRIMER programs (Clarke and Warwick 1994). Permutation tests can also be used to test hypotheses in linear statistical models such as multivariate ANOVA and other complex designs in applications where the normal statistical assumptions cannot be met (Anderson 2001).

Resampling methods can be easily implemented with statistical packages that allow the user to create macros or enter code (e.g., SAS, MatLab, S-Plus, Minitab, Excel, etc.), but macros and functions to simplify the process are increasingly available. Some of the best implementations include the JACKBOOT macro from SAS (SAS Institute 1995), the boot library for S-Plus (Davison and Hinkley 1997), and the bootstrap functions for S-Plus (Efron and Tibshirani 1993). However, we found the computer program "Resampling Stats" (Simon 1999) a most convenient tool for applying resampling methods. This program runs in the Windows operating system; it has a very readable user guide and includes many actual numerical examples of various resampling applications of bootstrap and permutation tests. There is also a plug-in module that adds all the "Resampling Stats" capabilities to the Microsoft spreadsheet program EXCEL (Blank et al. 1999). This software is available from Resampling Stats, Inc., Arlington, Virginia; free trial versions of the programs can be obtained at http://www.resample.com.

Meta-Analysis

We define meta-analysis as a collection of statistical methods for providing information from many independent studies in order to reach general conclusions. It is believed that these methods were first introduced to ecologists by Gurevitch and Hedges (1994). However, these methods had already influenced research synthesis in medicine and in the social sciences at earlier dates. A book edited by Cooper and Hedges (1994) provides a good general reference on the statistical and methodological aspects of the subject. More recently, Rosenberg et al. (1999) describe the statistical software MetaWin 2.0 for conducting meta-analyses using fixed and mixed ANOVA models and resampling tests.

The basic idea of meta-analysis is the concept that progress in any discipline, such as fishery science or environmental monitoring, depends upon the ability to reach general conclusions from a relatively large body of disparate research. The way that meta-analysis reaches general conclusions from a broad base of research is different from the approach typically used by ecologists and fishery scientists. In the conventional approach, results of different studies are compared and evaluated in a so-called research review. In meta-analysis, one begins by representing the outcomes of each experiment or study by a quantitative index of effect size. The effect size is chosen to reflect differences between experiment and control groups, or the degree of relationship between the independent and dependent variables in a way that is independent of sample size and the scale of measurement.

In fishery science, the approach of examining estimates of various population parameters from different studies involving fishes with similar life history is now often referred to as meta-analysis (NRC 1998). In such studies, the parameter estimates are then used to derive predictive relationships or statistical distributions for population or assessment models of the same or even different fishes as long as they share some traits critical for the intended application. A clear limitation of this approach is that the life history parameters used in developing the predictive relationships may come from studies using different estimation methods and/or data in different ranges of life history parameters. In those cases, the bias associated with the predictive relationships developed is generally unknown. Quinn and Deriso (1999), in their brief review of this subject, suggest that the predictive relationships should never be used to extrapolate beyond the range of the parameter estimates used in the meta-analysis.

Many applications of this type of meta-analysis have already been made in fishery science. The frequently cited work of Pauly (1980) was not labeled as a meta-analysis by the author, but his statistical relationship between natural mortality, growth parameters, and temperature derived from 175 fish stocks was an early classical application of the meta-analysis concept. Similarly,

Peterson and Wroblewski (1984) derived a power function relating dry weight at age and mortality of pelagic fishes, and Hoenig (1983) estimated natural mortality in fishes using longevity data. For a review of methods in other studies that have synthesized life history parameters to estimate fish mortality, see Vetter (1988). Some of the more interesting recent meta-analysis applications have been made by R. A. Myers of Dalhousie University (Myers et al. 1995, 2001; Myers and Barrowman 1996; Myers 1997). His application to recruitment variability of fishes is reported in Myers (1997), and his meta-analysis of the carrying capacity of north Atlantic cod stocks using empirical Bayes techniques is described in Myers et al. (2001). Liermann and Hillborn (1997) also used a Bayesian meta-analysis in their study of depensation in fish stocks, and Adams et al. (1997) applied resampling methods for testing hypotheses in meta-analysis of ecological data. Another interesting application is due to Cote et al. (2001) where the effect of marine reserves on protecting fish stocks was assessed. Some even more recent meta-analysis applications include Miller and Methot (2002) who modeled the ratio between swept area biomass and actual biomass of rockfishes using a Bayesian age-structured meta-analysis of several stocks. Another meta-analysis was made by Dorn (2002) to jointly estimate the stock–recruit relation of many stocks of rockfish. Denney et al. (2002) used combined life history and population data for many stocks of commercially exploited fishes to test relationships between life history parameters and the rate of population increase at low abundance, and Mueter et al. (2002, 2003) estimated effects of ocean temperature on survival rates of three species of Pacific salmon across 120 different stocks. In summary, it appears that meta-analysis will play an important role in future management of fisheries, as well as in monitoring and assessment of ecological impacts.

Computer Simulation Techniques and Population models

The widespread use of computers and modern statistical methods is providing direct links between deterministic models that describe a process and the noisy and limited data, often indirect, that are available to parameterize them and evaluate hypotheses. In all modeling applications of this type, computer simulation techniques rooted in the Monte Carlo method are common place nowadays. The most common applications deal with testing new estimation methods and modeling approaches with simulated data (Rykiel 1996); simulation of population dynamics that incorporate parameter uncertainty (Perry et al. 2003); probabilistic estimation of risk (Ferson et al. 1991); and sensitivity analysis of parameters in complex dynamic models. A very recent good example of the latter is Ellner and Fieberg (2003) who survey general simulation-based methods for assessing the uncertainty in model-based forecasts of population viability analysis (PVAs) of Pacific salmon.

Also based on computer simulation, a system level modeling approach that has received increasing recent attention has been implemented in a package of computer programs known as ECOPATH with ECOSIM (EwE). Complete information on this software and downloads are available at http://www.ecopath.org. The "Sea Around Us" Project of the Fisheries Centre, University of British Columbia, utilizes ecosystem modeling with above mentioned software to characterize the status of north Atlantic Ocean ecosystems. In general, the capabilities of this modeling system relate to evaluating how fisheries and the environment impact ecosystems.

Computer Population Models

Computer population models specifically developed for assessing power plant impacts on fish populations have been available in mainframe computer versions since the early 1970s. One of the earliest models was that of Hess et al. (1975), including both a hydrodynamic simulation model and a population model of the Niantic River, Connecticut, winter flounder *Pseudopleuronectes americanus* stock. Many different approaches have been used since then to simulate the dynamics of fish populations, with models gaining in detail and analytical complexity as more computer power became available at the desktop (Suter et al. 1993). One commonly used type of model known as "age-structured" is based on the Leslie matrix (Tuljapurkar and Caswell 1997; Caswell 2001). These models incorporate age-specific mortality and reproductive rates and, often, explicit density-dependent regulatory mechanisms. They assume that age is correlated with vital rates and that these rates are invariant in the case of deterministic models. Some models are stochastic and incorporate parameter uncertainty by means of Monte Carlo simulation techniques (Lorda et al. 2000; Perry et al. 2003). Still others add a probabilistic risk analysis component to assess losses resulting from entrainment and impingement (Lorda et al. 2000). Since we could provide here only a very limited overview of the many models developed for fisheries management and for impact assessment purposes during the last 25 years, we direct the interested readers to the excellent review provided in EPRI (1999). Another recent review of fish population dynamics and related simulation models is provided by Quinn and Deriso (1999). Early applications of deterministic population matrix models to fisheries and power plant impact evaluation include Vaughan and Saila (1976), Horst (1977), Saila and Lorda (1977), Rago (1980), and Vaughan (1981). Applications of stochastic matrix models to fish populations and impact assessment have been described most recently by Goodyear and Christensen (1984), Barnthouse et al. (1990), Ferson et al. (1991), Lorda et al. (2000), and DNC (2003a). Some computer programs of stochastic age-structured population models have become available. The DOS-based RAMAS-Age software developed by Applied Biomathematics of Setauket, New York was described in Ferson et al. (1991) and Saila et al. (1991). This program is now available in a new Windows version (Akcakaya 1998). More information about RAMAS-Age can be found at http://www.ramas.com.

Other special applications of population models and computer simulation techniques include sensitivity analysis of population parameters and calibration of parameters via comparison of model predictions to actual field data. Both applications involve running many population simulations with different parameter values, in order to measure model response to each parameter change in a sensitivity analysis, or to establish the set of parameters that result in model output most consistent with the data in a calibration study. Quinlan and Crowder (1999) used a modified Leslie matrix model to explore the sensitivity of long-term population growth rates of Atlantic menhaden *Brevoortia tyrannus* to changes in vital rates. In DNC (2003a), the mortality rate of juvenile winter flounder was calibrated by comparing model population projections to a time series of flounder abundance data. In an application to power plant impact assessment, the effectiveness of several larval entrainment mitigation alternatives was evaluated by running many simulations at different rates of larval entrainment to determine what reduction level was more cost-effective (DNC 2001).

A recently introduced and totally different class of models is the "individual-based" population models discussed in Judson (1994) and DeAngelis et al. 1994). These models seek to describe the dynamics of a population by tracking the attributes of individual population members (e.g., a single fish). Individual-based models (IBMs) are being increasingly utilized in fisheries science and ecology. The attraction of this approach seems to be related to the premise that IBMs may capture emergent effects on a macroscopic level while only implementing simple interactions between elements on a microscopic level. However, IBMs are not without problems, as noted by Grimm (1999). Our opinion is that IBMs are very complex with the ensuing problem that it is difficult to obtain reliable parameter estimates and also salient general results from their application. In summary, we believe that the modeling of complex systems defeats its purpose if the complexity cannot be reduced.

Because of the above reasons and because IBMs require extensive databases for their implementation, actual application of this class of models to power plant impact assessments has been very limited. One reported application (Rose et al. 1996a) consisted of an individual-based young-of-the-year model of winter flounder coupled with an age-structured matrix model for the adult population.

Finally, there have been attempts to develop multispecies models of fish populations in the past (Riffenburgh 1969; Anderson and Ursin 1977; Pope 1989), but again, actual applications to either fisheries management or impact assessment work are problematic because of the intense data requirements and the lack of reliable population parameter estimates for most fishes. A very good discussion of this type of model and its future in fisheries science was recently provided by Rose et al. (1996b).

Artificial Intelligence Methods

Fuzzy logic, neural networks, genetic algorithms, and simulated annealing are artificial intelligence methodologies that are included in the concept of soft computing. These methodologies have already generated considerable interest among fisheries scientists, as will be indicated by some examples. This interest appears to have risen because the above-mentioned methods are able to address a new set of fishery- and ecology-related problems that are more loosely defined and often involve a high degree of operational uncertainty. An early review and discussion of computerized artificial intelligence methods for application in fisheries research was provided by Saila (1996).

Soft computing includes these methodologies operating alone or in various combinations. They can be applied to problems where the complexity, nonlinearity, uncertainty, and intrinsic properties of real processes were previously considered as limitations for the further understanding of complex fishery related and ecological systems. Soft computing techniques are the starting point for new inferential mechanisms and nonconventional algorithms for the development of more tolerant, robust, and effective management strategies and operational methodologies in fisheries and environmental impact studies. Soft computing is quite different from traditional computing. The latter involves serial architecture, the strong point of which is its ability to multiply, add, and perform other mathematical operations on numbers rapidly and precisely. For example, traditional computers can quickly find the answer to X times Y if you tell it the value of X and Y. Traditional computer programs can perform difficult computations very quickly, and they may use complex algorithms to process numerical data. However, a numerical program cannot tell you that X plus $X = 2X$. Soft computing programs differ from traditional programs in that they understand not just symbols or information, but the relationships between symbols. Soft computing provides algorithms that have the capability to reason, value, and discriminate, rather than just calculate. These new structures of calculation are based on logic with a varying degree of uncertainty and are inspired by natural processes such as selection, aggregation, and cooperation. The basic principle of soft computing is its combined use of new computational methods that allow it to achieve a higher tolerance toward imprecision and approximation. This results in software/hardware products that are robust and better integrated. The systems derived from these new computing technologies are considered to be a new frontier of artificial intelligence with high potential for application in environmental impact studies.

Some selected fishery science-related applications involving soft computing will be briefly described. However, it should be clearly understood that these do not consist of a comprehensive bibliography of soft computing applications to fisheries. An excellent general introduction to soft computing is provided by Fortuna et al. (2001). This book provides several applications of soft computing in detail, although not in a fisheries context.

Fuzzy Logic

A recent fisheries related application of fuzzy logic is provided by Mackinson (2001) with an example of how both local and scientific knowledge can be integrated in a heuristic model. This model is developed in the framework of a fuzzy logic expert system and uses linguistic statements written in natural language to capture and combine knowledge sources in the form of "if ... then" rules. The model uses these rules to provide qualitative and quantitative information on the structure, dynamics, and distribution of schools of migrating adult herring during different stages of their life cycle. In an earlier paper, Mackinson et al. (1999) described the application of fuzzy logic to the analysis of stock–recruitment relationships in fish populations. Another application of fuzzy logic is described by Chen (2001) who uses it to detect environmental regimes in fish stock–recruit relationships. He indicates that the fuzzy logic model consistently outperformed more traditional approaches to stock–recruit relationships because the fuzzy logic model addresses uncertainty better than the traditional approaches. In a previous report (Chen et al. 2000), a fuzzy logic model with a genetic algorithm was developed, which also overcomes some of the inherent uncertainties in the fish stock–recruitment process. Comparisons of a traditional Ricker stock–recruit model and a neural network model demonstrated the superior capability of the fuzzy logic model for addressing problems of uncertainty and vagueness in both the data and the stock–recruit relationship. This example also illustrates how artificial intelligence techniques can be combined for increased effectiveness. Another interesting application of fuzzy logic to fish behavior is illustrated by Dagorn et al. (1999). In this case, a fuzzy logic model of tuna behavior near fish aggregation devices was developed to reproduce individual differences in horizontal movements observed from ultrasonic telemetry experiments. A further example of the application of fuzzy logic is indicated by Saila et al. (1997) to represent parametric uncertainty in data. In this instance, fuzzy arithmetic was applied to fuzzy numbers constructed from empirical data to provide upper and lower bounds to estimated equivalent adults of selected species from entrainment losses. A final example is provided by Saila (1992) in the application of fuzzy graph theory to successional analysis of a multispecies trawl fishery. It seems evident from these selected examples that fuzzy logic has already shown a high potential for further interesting applications in fishery science and future monitoring activities.

Neural Networks

Neural networks have also received a considerable amount of attention for some time by fisheries scientists and ecologists, as is evident from the many applications of this tool. Scardi (2001) illustrates neural networks as a powerful tool for phytoplankton primary production modeling and indicates the value of this approach in other ecological applications. A more fisheries-related application of neural networks is by Reyjol et al. (2001) in which neural network models were utilized to compare microhabitat use by three fish species. Brosse et al. (2001a) describe a nonsupervised neural network model (Kohonen self-organizing map) that displays high dimensional data sets in lower dimensional space. Self-organizing maps are compared with principal components analysis to assess spatial occupancy by fish assemblages in order to detect trends. The advantages of the self-organizing maps were clearly demonstrated. The work of Brosse et al. (2001b) includes a neural network predictive model for the distribution and abundance of several fish species to macrohabitat scales in a French lake. Lek et al. (1996) illustrate the use of neural networks in assessing the role of environmental variables in trout abundance models. Nunnari et al. (1998) applied neural techniques for modeling time-series of atmospheric pollution data and introduced the concept of neuro-fuzzy networks. Brosse et al. (1999) describe the use of neural networks to assess fish abundance and distribution in a mesotrophic lake. More recently, Zhou (2003) applied a neural networks model to forecast salmon escapement and concluded that the neural networks model generally outperformed the more conventional moving average and ARIMA models. Huse and Ottersen (2003) also applied a neural network to predict recruitment of northeast Arctic cod *Boreogadus saida* and discussed the benefits of using neural network models as a forecasting tool in fisheries assessment.

The above include only a small sample of the numerous applications available regarding the use of neural networks in fishery science. It is evident that neural networks are the most frequently used soft computing tool to date and their value as a predictive tool for fisheries applications and monitoring activities was recognized fairly quickly. See, for example, Saila (1996) for a brief introduction to various artificial intelligence tools, including neural networks; see also the more recent review of neural networks as a tool in ecological modeling by Lek and Guegan (1999).

Genetic Algorithms

In a mathematical model of fish schooling a genetic algorithm was utilized by Hiramatsu et al. (2000) to develop an intrinsic model of structure from experimental data on the schooling behavior of fish species. A fuzzy logic model and a genetic algorithm were utilized for the purpose of analyzing fish stock–recruitment relationships by Chen et al. (2000), as indicated previously. This model overcomes some of the uncertainties inherent in the stock–recruit process. Huse and Gjoesaeter (1999) also used a neural networks model that was trained with a genetic algorithm using ecological data of importance to capelin. The developed model was used to estimate abundance with high predictive ability. Huse et al. (1999) described a method for implementing behavior in individual-based models by using neural networks and genetic algorithms. This combination, termed individual-based neural network-genetic algorithm (ING), is described and some studies are presented to demonstrate solutions based on optimization.

Simulated Annealing

Walters and Deriso (2000) analyzed catch per unit effort (CPUE) for bigeye tuna with regression tree methods. A simulated annealing algorithm was used to summarize the regression tree based results by partitioning the fishing grounds into regions where the trends in CPUE were similar. It was concluded that simulated annealing can be useful for designing strata in sampling programs such as the one described. A second example of the application of simulated annealing is described in Matear (1995). In this study, a simulated annealing optimization algorithm is formulated to optimize parameters of ecosystem models. Among other advantages, the simulated annealing optimization analysis provides a means for assessing the degree of model complexity necessary to model the available data.

Fisheries Oceanography: a System View of Fishery Science

Physical properties of water, whether from the ocean, estuaries, rivers or lakes, clearly influence biological events at many scales. These range from microns, at which scale viscosity has a significant effect on the ability of phytoplankton to take up nutrients, to large ocean basin scale events, such as the North Atlantic Oscillation (NAO), which relates to the winter values of the barometric pressure differences between Iceland and the Azores (Ottersen et al. 2001; Attrill and Power 2002; Greene and Pershing 2003). In the past, there has been a tendency for biological events to be studied in isolation, and physical and chemical events were also studied separately. More recently, fisheries scientists have begun to search for evidence of the influence of the environment or hydrodynamic phenomena on the process of biological production. This type of study has been termed by some as fisheries oceanography and by others as a system level modeling approach. Regardless of the terminology used, it is clear that a system level approach is being increasingly utilized for complex problems involving interactions of physical, chemical, and biological events.

There is strong circumstantial evidence that large-scale oceanographic processes have a major influence on the rise and fall of fish stocks worldwide (Cushing 1982; Shepherd et al. 1984; Laevastu 1993). The implications for management of the stocks appear to be profound, calling for little less than a revolution in the approach to the problem (Harrison and Parsons 2000). At present, the evidence is mainly of a correlational nature, with all the uncertainty that this entails. One approach to reducing the uncertainty is to search for mechanisms linking the physical processes to the biological consequences and test for their occurrence in the ecosystems supporting the fish of interest. The work of Harris et al. (1992) in Tasmania is a good example of the application of the most up-to-date techniques to understanding the problem. Taking full advantage of satellite imagery and of the most recent atmosphere–ocean models, they have been able to understand the changes in Tasmanian waters in the light of large-scale events. They have been able to devise sampling programs to test the linkages between changes in physical oceanography and events in the food chains leading to commercial fish stocks. They are now in a strong position to forecast changes in the fishery on the basis of atmospheric events (Mann 1993).

A few examples follow to further indicate some applications of this approach in a fisheries context. However, these are only indicative of this subject area, which is rapidly expanding. Bakun (1990) explored the probable consequences of global climate changes on coastal upwelling systems, including biota. Cury and Roy (1989) suggest that the recruitment of larval fish in upwelling areas is maximized when winds are neither so weak that there is insufficient upwelling of nutrients, but not so strong that turbulence prohibits the formation of algal aggregations, which are required by fish larvae for adequate growth and survival. Rose and Summers (1992) investigated relationships among long-term fish abundance, hydrographic variables, and pollution indicators in estuaries. More recently, Rose (2000) extensively discusses analytical problems encountered when attempting to relate environmental variables to fish population dynamics; Mueter et al. (2002) investigated the effects of regional-scale sea surface temperatures and large-scale climate anomalies associated with the Pacific Decadal Oscillation (PDO) on the survival rates of Pacific salmon species; and the review paper of Stenseth et al. (2003) discuss the application of various climate indices and the effect of regional climate changes on ecological processes.

Suggestions Related to Specific Sections of the Original Connecticut River Ecological Study

Hydrography

The major new development in the collection and analysis of hydrographic data are the ready availability of water quality instrumentation, which is self-contained and which records data digitally. This contrasts markedly with the analog recorders used in the original study. These were able to record far fewer parameters and only in an analog fashion. The participants in the original hydrographic study are to be congratulated for obtaining considerable success with equipment of limited capability. Today's digital instrumentation is engineered for portability, dependable performance, and versatility. Water quality instruments available at present may contain sensors that not only include temperature, but also dissolved oxygen, pH, specific conductance, turbidity, chlorophyll, ambient light, ammonium/ammonia, turbidity, nitrate, and chlorides. Data from digital data loggers can be gathered for periods ranging from minutes to months at various sampling intervals and can be quickly and easily downloaded to computers, which makes the gathering and analysis of virtually all hydrographic data much easier than it was in the past. In addition, there are recently developed sophisticated computer algorithms available for statistical estimation and interpretation of trends in water quality and other hydrographic time series data. These lead to much better quality and diversity in this form of data. A potentially useful approach to the analysis of water quality time series data involves the use of wavelet analysis. This form of analysis seems particularly well suited for detecting transient effects in time series data. Walker (1999) provides a very readable introduction to wavelets and their scientific applications. This book is suggested to readers unfamiliar with this method. Wavelets are an interesting extension of Fourier analysis, which is the mathematical analysis of the frequency content of signals. Fourier analysis has already been applied in several fishery-related studies.

Oxygen Balance in the Condenser-Cooling Water System of the Connecticut Yankee Plant

Much of what has been suggested for the previous section is also believed to apply to this one. First, it should be recalled that currently available digital data loggers record not only dissolved oxygen, but also the degree of oxygen saturation. The degree of saturation is an important consideration in biological studies. In this study, the acquisition of samples from selected locations at a considerably higher sampling frequency could have further extended the value of these oxygen balance studies. Comparisons of data time series by Fourier analysis and/or wavelet analysis would have been helpful. As indicated previously, a number of software programs are currently available for the analysis of water quality time series data. In conclusion, it is suggested that currently available instrumentation and data analysis procedures would have markedly improved the interpretation and value of these oxygen balance studies.

Benthic Fauna: 1965–1967 versus 1968–1972

Substantial improvements in the quantitative analysis of invertebrate assemblage data have occurred since the original Ecological Study. Of special relevance is a software package entitled Primer, Version 5. This package is currently available from PRIMER-E Ltd, Plymouth, UK (http://www.primer-e.com). Although earlier versions are also useful, this latter version is more user-friendly and operates in a Windows environment. The Primer package consists of a range of various graphical and multivariate features for analyzing matrices of species by sample abundance (or biomass, percent cover, presence/absence) that arise in the biological monitoring of environments as well as more fundamental studies in ecology, together with physical and environmental data. Although several statistical software packages include most of the routines from the Primer package, the statistical methods of the Primer software are explained in nonmathematical terms in the methods manual and by the specific examples that are provided (Clarke and Warwick 2001). Spatial and temporal changes in invertebrate communities can be addressed by hierarchical clustering, nonmetric multidimensional scaling, and principal components analysis to summarize patterns in species composition and environmental variables, in addition to a host of other important applications.

Studies found in the Dominion Millstone Environmental Laboratory's 2001 Annual Report entitled "Monitoring the Marine Environment of Long Island Sound at Millstone Power Station, Waterford, CT" (DNC 2002) contain excellent examples of the application of multivariate methods to the study of benthic assemblages. Of additional interest is a report of Wright (1995) describing the developments and use of a system for predicting macro invertebrate fauna in flowing waters. This is a good example of the utility of biological techniques for assessing aquatic environmental perturbations. A more recent paper by Ravera (2001) provides good insights into the use of diversity, similarity, and biotic indices for environmental monitoring purposes.

Entrainment of Zooplankton at the Connecticut Yankee Plant

This section of the Connecticut River Ecological Study described some mechanical and thermal factors that may affect survival of zooplankton (exclusive of fish eggs and larvae) during passage through the cooling water condensers of the power plant. Estimates were also made of the proportion of the river zooplankton entrained by the power plant. Emphasis was placed on the larger forms of zooplankton. It is generally recognized that large rivers, such as the Connecticut River near the power plant, differ from lakes with respect to a number of physical and chemical attributes. Rivers lack thermal stratification and contain high concentrations of inorganic particulates and nutrients, and their plankton populations experience continuous downstream displacement. Two factors of relevance to this section were (a) zooplankton biomass in large rivers have a sustainable biomass less than would be expected based on algal abundance (Pace et al. 1992); and (b) rivers also differ from lakes in having a zooplankton mixture where rotifers are numerically dominant and crustaceans relatively scarce (Hynes 1970). It is believed that the scarcity of zooplankton relative to phytoplankton in large rivers is probably due to the comparatively slow population growth rates of the zooplankton. The observed scarcity of crustaceans relative to rotifers in the zooplankton of large rivers is thought to have a similar explanation because rotifers reproduce at a considerably faster rate than the crustacean zooplankton.

This section of the original ecological study is unique in the sense that it emphasizes zooplankton mainly as crustaceans and places no emphasis on fish eggs and larvae, which are of considerably greater significance from the point of view of ecological impact assessment. The relatively small proportion of zooplankton (as defined above) that passes through the power plant was low, on the order of 4% of that which passes in the river. This type of work has been largely replaced nowadays by increased emphasis on fish eggs and larvae and the effect of their mortality on the adult population. An excellent review of various models, including equivalent adult, equivalent yield to a fishery and biomass lost, is provided in EPRI (2002a) and Dey (2003). These reviews cover the various approaches in sufficient detail to provide the current "state of the art" with reference to the estimation of entrainment losses. An earlier reference by Ulanowicz (1975) still seems to cover the mechanical effects of water flow on fish and larvae effectively, and it is believed to apply equally well to zooplankton in general. We believe that further research to quantify damage to zooplankton by entrainment in riverine environments does not seem to be justified at present.

Fishes of the Lower Connecticut River and the Effects of the Connecticut Yankee Plant

This section described the results of an intensive survey for four years prior to and during about 4 years of operation of the Connecticut Yankee facility. It clearly described the sport fishery in the general area of the power plant, as well as some effects of the power plant. It was a well-planned and executed study that provided excellent baseline material. If this study were to be repeated today, major suggestions would involve the nature of the data analysis for which some recommendations follow.

We believe that utilization of recent multivariate analysis software designed for community analyses would be helpful in examining the dynamics of fish assemblages. An example is the application of MDS analysis to the fish assemblage in the Niantic River, Connecticut described by Tomichek and Roseman (2003). Neural network models and principal component analysis could also be used to describe fish assemblages, as shown in Brosse et al. (2001a). Also, planning of the survey design and analysis prior to the execution would help avoid redundancy and provide more explicit tests of hypothesis than those tested in the original study. In general, we feel that *t*-tests and ANOVAs can be overused in marine biological and fisheries studies where an abundance of collected data often result in too many statistical tests that are not fully relevant to the problem being investigated. In many cases, it may be sufficient to provide tables and simple descriptive statistics to summarize and interpret the data followed by a discussion closely related to the study goals.

Several other developments have taken place that are believed to be of considerable value to that portion of the original ecological study where only scales were used for aging fish. For example, a recent review of age determination and validation methods by Campana (2001) should be very helpful. This review clearly highlights the best methods for aging accuracy and quantifying aging precision. In addition, the reduction of aging errors is demonstrated through the use of quality control and monitoring. This type of information is essential for effective and comprehensive analysis of growth studies.

Weisberg (1986) first described a linear model for growth comparisons based on scale increments. He also discussed and illustrated a method for inferring changes in length from available data. In a later paper, Weisberg (1993) presented a family of linear models that compare fish growth as reflected in increments on scales or other hard parts, and the results can be divided into size effects as well as external environmental factors. Obviously, this approach would have been helpful for studies related to the power plants effects and the influence of the warmwater canal on the growth of selected species such as the brown bullhead and channel catfish.

In addition to the above suggestions, the nonlinear fitting of conventional growth curves has advanced considerably. Von Bertalanffy and Richards growth functions as well as others are now relatively easy to fit with commonly available software. Using hard part increment data to estimate age and environmental effects is comparatively easy at present. Weisberg (1993) has also provided a special purpose program that can fit all the models indicated in his paper. This program is available for a nominal fee from the Minnesota Sea Grant Institute, University of Minnesota (item RR30 in the Sea Grant publications list found at www.seagrant.umn.edu). It should also be recognized that detailed information can be extracted from the analysis of otoliths of early life history stages of fish. These studies are thought to be very relevant to entrainment analyses at power plants because both growth data and mortality data can be extracted from otoliths (Geffen 1987). It is evident that these and other recent developments involving fish growth models and environmental effects available currently could have substantially improved the quantitative analysis of the power plant effect on various fish species.

Planktonic Fish Eggs and Larvae of the Lower Connecticut River and the Effects of the Connecticut Yankee Plant, Including Entrainment

This section of the original ecological study provided an excellent description of early life history stages of the more important fish species in the vicinity of the plant. This description included estimates of fish size, occurrence, relative abundance, and water temperature ranges during the major part of the growing season of sampled fishes. It also examined immediate mortality resulting from passage through the plant cooling system for selected species, but without projecting effects of these losses at the adult population level or their impact on future recruitment.

Although no detailed information on natural mortality of early life history stages for most sampled fishes was available at the time, further studies would now focus on the effects of entrainment on adult population dynamics applying various methods previously suggested for estimating equivalent adult losses (EPRI 2002a; Dey 2003). Also, larval otolith studies could provide information leading to estimates of growth and mortality (Geffen 1987).

Early Life History Studies of American Shad in the Lower Connecticut River and the Effects of the Connecticut Yankee Plant

This section was one of the earliest contributions to our knowledge on the early life history stages of the American shad *Alosa sapidissima* and provided information not available at that time. Among the new work relevant to the initial study are the methods developed by Crecco et al. (1983) to estimate daily mortality rates of larval and juvenile shad and an investigation of climatic effects on intra-annual mortality of larval shad (Crecco and Savoy 1987). These two studies used up-to-date statistical methods and detailed analyses to support their findings. Stock–recruitment models and resampling techniques to estimate prediction error were also applied by Lorda and Crecco (1987) to assess the effects of power plant-related juvenile mortality on the Connecticut River American shad run. A current assessment of juvenile shad abundance and year-class strength is provided herein by Savoy et al. (2004, this volume).

The American Shad, with Special Reference to Its Migration and Population Dynamics in the Connecticut River

There has been a great deal of work done on the American shad stock spawning in the Connecticut River since the studies reported in the initial Ecological Study. However, those studies represented at that time the most extensive investigation conducted of the dynamics of the Connecticut River shad population, its migration rates, timing, and distribution, and its stock–recruitment relationship. The latter was based on the Ricker model and featured stock and recruits data adjusted for repeat spawners. Although the study acknowledged the influence of river temperatures and flow conditions on juvenile production and recruitment, the Ricker stock–recruit model did not incorporate any of those factors explicitly.

Among the new work relevant to the initial study are the methods developed by Crecco et al. (1983) to estimate daily mortality rates of larval and juvenile shad; an investigation of climatic effects on intra-annual mortality of larval shad (Crecco and Savoy 1987); and the development of a stock–recruitment model for the Connecticut River shad incorporating river flow effects on year-class strength (Lorda and Crecco 1987). A new approach for integrating environmental time series into stock assessment models has been suggested very recently by Maunder and Watters (2003). Their approach incorporates process error explicitly in the model to reduce bias and it could be easily adapted for the shad recruitment model with river flow effects similar to the one described by Lorda and Crecco (1987).

The analysis of shad movements derived from conventional and ultrasonic tagging described in the original study could be enhanced by careful study of new approaches described in a recent book by Turchin (1998). Also recent availability of relatively inexpensive global positioning systems (GPS) for fieldwork would facilitate the tracking of tagged fish by reducing effort and increasing precision.

Other developments relevant to this section of the original ecological study include the great progress made in model development and software to analyze mark–recapture data. In addition to the now classic Jolly–Seber model (Jolly 1965) available in various programs versions for mainframe computers and DOS microcomputers, two new major programs for Windows desktop computers are now available to fisheries scientists. These computer programs, POPAN-5 and MARK, include several new routines to analyze mark–recapture data in addition to various versions of the Jolly–Seber model. POPAN-5 is described by Arnason et al. (1998) and is available as a "freeware" at the University of Manitoba site http://www.cs.umanitoba.ca/~popan. The massive program MARK is briefly described by White and Burnham (1999). A complete description of the program is available from the primary author at http://www.cnr.colostate.edu/~gwhite/mark/mark.htm. A secondary web site for information on MARK is http://canuck.dnr.cornell.edu/mark/. Free versions of MARK can be downloaded at either site.

Recurrent Mass Mortalities of the Blueback Herring in the Lower Connecticut River

The material in this section of the Connecticut River Ecological Study provided a comprehensive record of observations related to excessive mortality of fish in the lower Connecticut River.

This excessive mortality appeared to be associated almost exclusively with a single species, namely adult blueback herring *Alosa aestivalis*. It was suggested that the specificity of the mortality minimized the probability of a toxic pollutant as a cause for the mortality. It was observed by a visual examination of a scatter plot of the relation between dissolved oxygen and water temperature that early morning combinations of high temperature and low dissolved oxygen values coincided with high mortality events of the blueback herring. A straight line was drawn through this scatter diagram as a separator between adverse and tolerable conditions. The slope and intercept of this line was then approximated, and the model parameters were used for predicting excessive mortality occurrences. It is suggested that a more rigorous and precise estimator could be derived for this prediction based on multivariate classification procedures or a neural network model. These approaches are suggested for future studies of this nature. Although slight mortality was noted for some other species, these deaths were believed to be unrelated to the excessive blueback mortality observations.

Discussion

Recent developments in fishery science, ecology, statistics and computer science have substantially contributed to more sophisticated and effective power plant impact assessment methods. We have attempted to describe some of these developments, which were considered to be particularly relevant to environmental impact issues. Although the original Connecticut River Ecological Study made many contributions to basic science and to our understanding of the effects of a power plant on the Connecticut River ecosystem, the progress in scientific knowledge and methodologies since this initial study has dramatically improved the opportunity for more effective planning and execution of studies related to assessing the extent of adverse environmental impacts (AEI) caused by power plant cooling water systems.

Section 316(b) of the Clean Water Act (enacted in 1972) requires an assessment of the potential for AEI from cooling water withdrawal at steam-electric power plants. A wide variety of methods have been used for such assessment over the last 30 years, and until recently, the emerging consensus was that the assessment should focus on direct effects to populations and communities in the source water bodies. More recently, however, the emphasis has been changing to include ecological risk estimation either by means of probabilistic risk analysis or by alternative approaches such as assessment/determinations, based on multiples lines of evidence as in Barnthouse et al. (2003) or as proposed by Van Winkle et al. (2003). A variation of the treatment of uncertainty based on multiples lines of evidence is the strength-of-evidence analysis described in USEPA (2000). There has been a great deal of work carried out in recent years attempting to summarize, review, classify, and help select §316(b) assessment methods for each particular situation, both in freshwater and marine environments. The impetus for all this work has been the U.S. Environmental Protection Agency (USEPA) recent Phase I §316(b) rulemaking for new cooling water intake structures (CWIS) and its counterpart Phase II rulemaking for existing CWIS expected in early 2004. Although there is extensive literature on this subject, we will mention here only three recent publications of Electric Power Institute, Inc. that very much summarize the current state of research on methods and applications for §316(b) determinations. They are a catalog of assessment methods for power plant effects (EPRI 1999); a review of guidelines for selecting assessment methods (EPRI 2002a); and an excellent review of current risk assessment methodology for §316(b) determinations (EPRI 2002b). Additional relevant research is reported in a recently published book (Dixon et al. 2003) covering materials presented at a symposium on assessment of AEI from power plants during the 2001 Annual Meeting of the American Fisheries Society held at Phoenix, Arizona.

We have noted in previous sections that some new statistical methods and software can now greatly facilitate studies to quantify entrainment and impingement losses, minimization of which still are the major requirements of the new §316(b) rules. Our review of assessment methods actually applied in a wide range of situations indicates that quantifying the effects of fish entrainment and impingement at the population level should involve both empirical observations and model-derived predictions, including quantification of uncertainty. A simple and flexible class of impact assessment models for fishes at the population level is the "equivalent adult" or "equivalent loss" model, which has been suggested and applied for many years to quantify AEI (Horst 1975; Boreman et al. 1981; Rago 1984; Jensen et al. 1988; Saila et al. 1997). We believe that these models are still a valid approach to quantify AEI in §316(b) determinations. The various forms of equivalent loss models, their data input requirements, and their assumptions and limitations have been recently reviewed in detail by Dey (2003). Although assessing impacts at the community level still requires long-term monitoring studies and the application of sound principles of experimental ecology (Underwood 1996), these studies are now greatly facilitated by improved nonparametric multivariate statistical methods and readily available computer programs (Clarke 1993; Clarke and Warwick 1994).

We believe that science applied to the solution of AEI issues can lead to the development of new tools and concepts, which then accelerate the overall progress of ecology and other environmental sciences. A few illustrations of such developments have been briefly mentioned in the previous sections of this paper. Although both fishery science as well as ecology have benefited from scientific studies related to impact assessment in the past, it is becoming increasingly evident that a shift in emphasis toward a dynamical systems approach to problem solving has taken place since the original Connecticut River Ecological Study. Clark and Mangel (2000) illustrate this change in approach with their book entitled "Dynamic State Variable Models in Ecology." This book describes system models that are analyzed by a dynamic programming algorithm as well as other computational algorithms, such as Monte Carlo simulations. Computers are a natural tool for mathematical modeling, and it would be impossible to apply such models as indicated in Clark and Mangel's book without them.

Walters (1996) has outlined many of the new opportunities that computers make possible with respect to fishery science. Of particular importance in our context is his listing of things that can be done today with computers, but which would have been virtually impossible only a few decades ago. One of these is the ability to run many population simulations with different parameter values in order to establish the set of parameter combinations that are more consistent with the data. This has been noted in our discussions of new developments and computer simulation models. Another concept, namely that the development of ecosystem simulations in a format which can be easily used and understood in the form of simulation games can be of considerable value to future managers and decision makers. It was also suggested that one of the fundamental issues in fisheries currently involves a study of the relative importance of climate and environmental factors as they affect fisheries. This subject area is embraced by Fisheries Oceanography, and its importance is expected to grow in the future.

Saila (1996) has summarized some future developments in artificial intelligence as related to fisheries. These future directions tend to parallel the descriptions provided herein with respect to neural networks, genetic algorithms, and other artificial intelligence tools. We conclude with the optimistic statement that the future will permit us to more precisely and effectively predict consequences of environmental perturbations as well as their causes.

Acknowledgments

We are indebted to Webster Van Winkle and William A. Richkus for their helpful review comments and suggestions on an early version of the manuscript. We also thank our editors William C. Leggett and Douglas A. Dixon for their help and guidance during the review process.

References

Adams, D. C., J. Gurevitch, and M. S. Rosemberg. 1997. Resampling tests for meta-analysis of ecological data. Ecology 78:1277–1283.

Akcakaya, H. R. 1998. RAMAS metapop: viability analysis for stage-structured metapopulations (version 3.0). Applied Biomathematics, Setauket, New York.

Anderson, K. P., and E. Ursin. 1977. A multispecies extension of the Beverton and Holt theory of fishing, with accounts of phosphorus circulation and primary production. Meddelelser Fra Danmarks Fiskeriog Havundersøgelser, N. S 7:319–435.

Anderson, M. J. 2001. Permutation tests for univariate or multivariate ANOVA and regression. Canadian Journal of Fisheries and Aquatic Sciences 58:626–639.

Arnason, A. N., C. J. Schwarz, and G. Boyer. 1998. POPAN-5: A data maintenance analysis system for mark-recapture data. Scientific Report, Department of Computer Science, University of Manitoba, Winnipeg, Canada.

Attrill, M. J., and M. Power. 2002. Climatic influence on a marine fish assemblage. Nature (London) 417:275–278.

Bakun, A. 1990. Global climate change and intensification of coastal upwelling. Science 247:198–201.

Barnthouse, L. W., D. G. Heimbuch, V. C. Anthony, R. W. Hilborn, and R. A. Myers. 2003. Indicators of AEI applied to the Delaware Estuary. Pages 165–184 *in* D. A. Dixon, J. A. Veil, and J. Wisniewski, editors. Defining and assessing adverse environmental impacts from power plant impingement and entrainment of aquatic organisms. A. A. Balkema Publishers, Lisse, The Netherlands.

Barnthouse, L. W., G. W. Sutter, I. I., and A. E. Rosen. 1990. Risks of toxic contaminants to exploited fish populations: influence of life history, data uncertainty, and exploitation intensity. Environmental Toxicology and Chemistry 9:297–311.

Blank, S., C. Seiter, and P. Bruce. 1999. Resampling stats in Excel. Resampling Stats, Inc., Arlington, Virginia.

Bolam, S. G., T. F. Fernandes, M. Huxman. 2002. Diversity, biomass, and ecosystem processes in the marine benthos. Ecological Monographs 72:559–615.

Boreman, J., C. P. Goodyear, and S. W. Christensen. 1981. An empirical methodology for estimating entrainment losses at power plants sited on estuaries. Transactions of the American Fisheries Society 110:253–260.

Brosse, S., J. F. Guegan, J. N. Tourenq, and S. Lek. 1999. The use of neural networks to assess fish abundance and spatial occupancy in the littoral zone of a mesotrophic lake. Ecological Modelling 120:299–311.

Brosse, S., J. L. Giraudel, and S. Lek. 2001a. Utilization of non-supervised neural networks and principal component analysis to study fish assemblages. Ecological Modelling 146(1–3):159–166.

Brosse, S., S. Lek, and C. C. Townsend. 2001b. Abundance, diversity, and structure of freshwater invertebrates and fish communities: an artificial network approach. New Zealand Journal of Marine and Freshwater Research 35(1):135–145.

Buckland, S. T. 1980. A modified analysis of the Jolly-Seber capture-recapture model. Biometrics 36:419–435.

Buckland, S. T. 1983. Monte Carlo methods for confidence interval estimation using the bootstrap technique. BIAS 10:194–212.

Campana, S. E. 2001. Accuracy, precision, and quality control in age determination, including a review of the use and abuse of age validation methods. Journal of Fish Biology 59(2):197–242.

Caswell, H. 2001. Matrix population models: construction, analysis and interpretation. 2nd edition. Sinauer Associates, Inc. Publishers, Sunderland, Massachusetts.

Chen, D. G. 2001. Detecting environmental regimes in fish stock recruitment relationships by fuzzy logic. Canadian Journal of Fisheries and Aquatic Sciences 58(11):2139–2148.

Chen, D. G., N. B. Hargreaves, D. M. Water, and Y. Liu. 2000. A fuzzy logic model with genetic algorithm for analyzing fish stock-recruitment relationships. Canadian Journal of Fisheries and Aquatic Sciences 57(9):1878–1887.

Chen, D. G., and J. R. Irvine. 2001. A semiparametric model to examine stock-recruitment relationships incorporating environmental data. Canadian Journal of Fisheries and Aquatic Sciences 58:1178–1186.

Clark, C. W., and M. Mangel. 2000. Dynamic state variable models in ecology. Oxford University Press, Inc., New York.

Clarke, K. R. 1990. Comparisons of dominance curves. Journal of Experimental Marine Biology and Ecology 138:143–157.

Clarke, K. R. 1993. Non-parametric multivariate analyses of changes in community structure. Australian Journal of Ecology 18:117–143.

Clarke, K. R., and M. Ainsworth. 1993. A method of linking multivariate community structure to environmental variables. Marine Ecology Progress Series 92:205–219.

Clarke, K. R., and H. R. Green. 1988. Statistical design and analysis for a 'biological effects' study. Marine Ecology Progress Series 46:213–226.

Clarke, K. R., and R. M. Warwick. 1994. Change in marine communities: An approach to statistical analysis and interpretation. Plymouth Marine Laboratory, Plymouth, UK.

Clarke, K. R., and R. M. Warwick. 2001. Change in marine communities: An approach to statistical analysis and interpretation (2nd edition). PRIMER-E, Plymouth, UK.

Clarke, K. R., R. M. Warwick, and B. E. Brown. 1993. An index showing breakdown of seriation, related to disturbance, in a coral-reef assemblage. Marine Ecology Progress Series 102:153–160.

Cohen, J. 1988. Statistical power analysis for the behavioral sciences (2nd edition). L. Erlbaum Associates, Hillsdale, New York.

Cooper, H. M., and L. V. Hedges, editors. 1994. Handbook of research synthesis. The Russell Sage Foundation, New York.

Cote, I. M., I. Mesqueira, and J. D. Reynolds. 2001. Effects of marine reserve characteristics on the protection of fish populations: a meta-analysis. Journal of Fish Biology 59:178–189.

Crecco, V., and T. Savoy. 1987. Effects of climatic and density-dependent factors on intra-annual mortality of larval American shad. Pages 69–81 *in* R. D. Hoyt, editor. 10th Annual Larval Fish Conference. American Fisheries Society, Symposium 2, Bethesda, Maryland.

Crecco, V., T. Savoy, and L. Gunn. 1983. Daily mortality rates of larval and juvenile American shad (*Alosa sapidissima*) in the Connecticut River with changes in year-class strength. Canadian Journal of Fisheries and Aquatic Sciences 40:1719–1728.

Cury, P., and C. Roy. 1989. Optimal environmental window and pelagic fish recruitment success in upwelling areas. Canadian Journal of Fisheries and Aquatic Sciences 46:670–680.

Cushing, D. H. 1982. Climate and fisheries. Academic Press, New York.

Dagorn, L., E. Jasse, P. Bach, and A. Bertrand. 1999. Modeling tuna behaviour near floating objects: from individuals to aggregations. Pages 203–211 *in* Mechanisms and effects of the aggregation of tuna by Fish Aggregating Devices (FADs). Volume 13(4) Aquatic Living Resources, Elsevier, Paris.

Davison, A. C., and D. V. Hinkley. 1997. Bootstrap methods and their application. Cambridge Series in Statistical and Probabilistic Mathematics No. 1, Cambridge University Press, Cambridge, UK.

DeAngelis, D. L., K. A. Rose, and M. A. Huston. 1994. Which individual-based approach is most appropriate for a given problem? Pages 390–410 *in* S. A. Levin, editor. Frontiers in mathematical biology. Lecture notes in biomathematics. Volume 100. Springer-Verlag, New York.

Denney, N. H., S. Jennings, and J. D. Reynolds. 2002. Life-history correlates of maximum population growth rates in marine fishes. Proceedings of the Royal Society, London Series B 269:2229–2237.

Dey, W. P. 2003. Use of equivalent loss models under Section 316(b) of the Clean Water Act. Pages 247–263 *in* D. A. Dixon, J. A. Veil, and J. Wisniewski, editors. Defining and assessing adverse environmental impacts from power plant impingement and entrainment of aquatic organisms. A. A. Balkema Publishers, Lisse, The Netherlands.

Diaconis, P., and B. Efron. 1983. Computer-intensive methods in statistics. Scientific American 248:116–130.

Dixon, D. A., J. A. Veil, and J. Wisniewski, editors. 2003. Defining and assessing adverse environmental impacts from power plant impingement and entrainment of aquatic organisms. A. A. Balkema Publishers, Lisse, The Netherlands.

Dixon, P. M. 2001. The bootstrap and the jackknife: describing the precision of ecological studies. Pages 267–288 *in* S. Scheiner and J. Gurevitch, editors. Design and analysis of ecological experiments. 2nd edition. Oxford University Press, Inc., New York.

DNC (Dominion Nuclear Connecticut, Inc.) 2001. The effectiveness of reducing entrainment mortality of fish eggs and larvae at MNPS. Pages 1–82, Section II, Chapter 4 *in* An evaluation of cooling water system alternatives. Report submitted to the Connecticut Department of Environmental Protection, Dominion Nuclear Connecticut, Inc., Waterford, Connecticut. August 2001. Waterford, Connecticut.

DNC (Dominion Nuclear Connecticut, Inc.) 2002. Monitoring the marine environment of Long Island Sound at Millstone Power Station, Waterford, Connecticut: annual report 2002. Waterford, Connecticut.

DNC (Dominion Nuclear Connecticut, Inc.) 2003a. Stochastic simulation of the Niantic River winter flounder stock. Pages 95–112 *in* Annual Report 2002. Monitoring the marine environment of Long Island Sound at Millstone Power Station, Waterford, Connecticut. April 2003. Waterford, Connecticut

DNC (Dominion Nuclear Connecticut, Inc.) 2003b. Rocky intertidal studies. Pages 203–235 *in* Annual Report 2002. Monitoring the marine environment of Long Island Sound at Millstone Power Station, Waterford, Connecticut. April 2003. Waterford, Connecticut.

Dorn, M. W. 2002. Advice on West Coast rockfish harvest rates from Bayesian meta-analysis of stock–recruitment relationships. North American Journal of Fisheries Management 22(1):280–300.

Dye, A. H. 1998. Community-level analyses of long-term changes in rocky littoral fauna from South Africa. Marine Ecology Progress Series 164:47–57.

Edgington, E. S. 1987. Randomization tests, 2nd edition. Dekker, New York.

Efron, B. 1979. Bootstrap methods: another look at the jackknife. The Annals of Statistics 7:1–26.

Efron, B. 1981. Nonparametric standard errors and confidence intervals (with discussion). Canadian Journal of Statistics 9:139–172.

Efron, B. 1982. The jackknife, the bootstrap, and other resampling plans. CBMS-NSF Regional Conference Series in Applied Mathematics, Monograph 38. Society for Industrial and Applied Mathematics, Philadelphia.

Efron, B., and R. Tibshirani. 1991. Statistical data analysis in the computer age. Science 253:390–395.

Efron, B., and R. J. Tibshirani. 1993. An introduction to the bootstrap. Monographs on Statistics and Applied Probability 57. Chapman and Hall, Inc., New York.

Ellner, S. P., and J. Fieberg. 2003. Using PVA for management despite uncertainty: effects of habitat, hatcheries, and harvest on salmon. Ecology 86(6):1359–1369.

EPRI (Electric Power Research Institute). 1999. Catalog of assessment methods for evaluating the effects of power plant operations on aquatic communities. Electric Power Research Institute, Technical Report 112013, Palo Alto, California.

EPRI (Electric Power Research Institute). 2002a. Evaluating the effects of power plants on aquatic communities: Guidelines for selection of assessment methods. Electric Power Research Institute, Technical Report 1005176, Palo Alto, California.

EPRI (Electric Power Research Institute). 2002b. Evaluating the effects of power plant operations on aquatic communities: An Ecological Risk Assessment framework for Clean Water Act 316(b) determinations. Electric Power Research Institute, Technical Report 1005337, Palo Alto, California.

Everitt, B. 1978. Graphical techniques for multivariate data. Heinemann, London.

Everitt, B. 1980. Cluster analysis (2nd edition). Heinemann, London.

Fairweather, P. G. 1991. Statistical power and design requirements for environmental monitoring. Australian Journal of Marine and Freshwater Research 42:555–567.

Ferson, S., R. Akcakaya, L. Ginzburg, and M. Krause. 1991. Use of RAMAS to estimate ecological risk: two fish species case studies. Prepared by Applied Biomathematics, Inc. Setauket, New York, for Electric Power Research Institute, Palo Alto, California. EPRI Report EN-7176.

Fortuna, L., G. Rizzotto, M. Lavorgna, G. Nunnari, M. G. Xibilia, and R. Caponetto. 2001. Soft computing. Springer-Verlag, London.

Geffen, A. G. 1987. Methods of validating daily increment deposition in otoliths of larval fish. Pages 223–240 *in* R. C. Summerfelt and G. E. Hall, editors. Age and growth of fish. Iowa State University Press, Ames.

Gerrodette, T. 1987. A power analysis for detecting trends. Ecology 68(5):1364–1372.

Good, P. 2001. Resampling methods: a practical guide to data analysis (2nd edition). Birkhauser, Boston.

Good, P. I. 2000. Permutation tests: a practical guide to resampling methods for testing hypotheses (2nd edition). Springer-Verlag New York, Inc., New York.

Goodyear, C. P., and S. W. Christensen. 1984. Bias-elimination in fish population models with stochastic variation in survival of the young. Transactions of the American Fisheries Society 113:627–632.

Green, R. 1989. Power analysis and practical strategies for environmental monitoring. Environmental Research 50:195–205.

Green, R. H. 1979. Sampling design and statistical methods for environmental biologists. Wiley & Sons, New York.

Greenacre, M. J. 1984. Theory and applications of correspondence analysis. Academic Press, London.

Greene, C. H., and A. J. Pershing. 2003. The flip-side of the North Atlantic Oscillation and modal shifts in slope-water circulation patterns. Limnology and Oceanography 48(1):319–322.

Grieb, T. M. 1985. Robustness of the ANOVA model in environmental monitoring applications. EPRI Report EA-4015 prepared by Tetra Tech, Inc., Lafayette, California. Electric Power Research Institute, Inc., Palo Alto, California.

Grimm, B. 1999. Ten years of individual-based modeling in ecology: what have we learn and what can we learn in the future? Ecological Modelling 115:129–148.

Gurevitch, J., and L. V. Hedges. 1994. Meta-analysis: Combining the results of independent experiments. Pages 347–369 *in* S. M. Scheiner and J. Gurevitch, editors. Design and analysis of ecological experiments. Chapman and Hall, New York.

Harris, G. P., F. B. Griffiths, and L. A. Clementson. 1992. Climate and fisheries off Tasmania: interactions of physics, food chains and fish. Pages 107–121 *in* A. I. L. Payne, K. H. Brink, K. H. Mann, and R. Hilborn, editors. Benguela trophic functioning. South African Journal of Marine Science 12.

Harrison, J. P., and T. R. Parsons, editors. 2000. Fisheries oceanography: an integrative approach to fisheries ecology and management. Fish and Aquatic Resources Series 4, Blackwell Scientific Publications, Ltd. Osney Mead, Oxford, UK.

Heffner, R. A., M. J. Butler, I. V., and C. Keelan Reilly. 1996. Pseudoreplication revisited. Ecology 77(8):2558–2562.

Hess, K. W., M. P. Sissenwine, and S. B. Saila. 1975. Pages 1–30 *in* S. B. Saila, editor. Fisheries and energy production: a symposium. Heath, Lexington, Massachusetts.

Heintze, J. L. 2000. PASS-2000: number cruncher statistical systems. Kaysville, Utah.

Hiramatsu, K., S. Shikasho, and K. Mori. 2000. Mathematical modeling of fish schooling of Japanese Medaka using basic behavioral patterns. Journal of the Faculty of Agriculture, Kyusu University 45(1):237–253.

Hjorth, J. S. U. 1994. Computer intensive statistical methods, validation model selection and bootstrap. Chapman and Hall, London.

Hoenig, J. M. 1983. Empirical use of longevity data to

estimate mortality rates. U.S. Fishery Bulletin 82:898–903.

Horst, T. J. 1975. The assessment of impact due to entrainment of ichthyoplankton. Pages 107–118 *in* S. B. Saila, editor. Fisheries and energy production: a symposium. Heath, Lexington, Massachusetts.

Horst, T. J. 1977. Effects of power station mortality on fish population stability in relationship to life history strategy. Pages 297–310 *in* W. Van Winkle, editor. Assessing the effects of power-plant induced mortality on fish populations. Proceedings of a conference in Gatlinburg, Tennessee, May 1977. Pergamon, New York.

Hurlbert, S. H. 1984. Pseudoreplication and the design of ecological field experiments. Ecological Monographs 54:187–211.

Huse, G., and H. Gjoesaeter. 1999. A neural network approach for predicting stock abundance of the Barents Sea Capelin. Sarsia 84(5–6):457–464.

Huse, G. E., and G. Ottersen. 2003. Forecasting recruitment and stock biomass of Northeast Arctic cod using neural networks. Scientia Marina 67(Supplement 1):325–335.

Huse, G., E. Strand, and J. Giske. 1999. Implementing behaviour in individual-based models using neural networks and genetic algorithms. Evolutionary Ecology 13(5):469–483.

Hynes, H. B. N. 1970. The ecology of running waters. The University of Toronto Press, Toronto.

Jensen, A. L., R. H. Reider, and W. P. Kovalak. 1988. Estimation of production forgone. North American Journal of Fisheries Management 8:191–198.

Jolly, G. M. 1965. Explicit estimates from capture-recapture data with death and immigration stochastic model. Biometrika 52:225–247.

Judson, O. P. 1994. The rise of the individual-based model in ecology. Trends in Ecology and Evolution 9:9–14.

Kenkel, N. C., and L. Orloci. 1986. Applying metric and nonmetric multidimensional scaling to some ecological studies: some new results. Ecology 67:919–928.

Kohavi, R. 1995. A study of cross validation and bootstrap for estimation and model selection. Pages 1137–1143 *in* Proceedings of the 14th International Joint Conference on Artificial Intelligence. Morgan Kauffman Publishers, Inc., San Francisco.

Kruskal, J. B. 1964. Multidimensional scaling by optimizing goodness of fit to a nonmetric hypothesis. Psychometrika 29:1–27.

Laevastu, T. 1993. Marine climate, weather and fisheries. Halsted Press, New York.

Legendre, P., and L. Legendre. 1998. Numerical ecology. Elsevier, Amsterdam.

Lek, S., and J. F. Guegan. 1999. Artificial neural networks as a tool in ecological modeling, an introduction. Ecological Modelling 120:65–73.

Lek, S., A. Roland, I. Dimopoulos, and M. Dalecoste. 1996. Role of environmental variables in trawl abundance models using neural networks. Aquatic Living Resources 9:23–29.

Liermann, M., and R. Hillborn. 1997. Depensation in fish stocks: a Bayesian meta-analysis. Canadian Journal of Fisheries and Aquatic Sciences 54:1976–1984.

Lindley, S. T., M. S. Mohr, and M. H. Prager. 2000. Monitoring protocol for Sacramento River winter chinook salmon, *Oncorhynchus tshawytscha*: application of statistical power analysis to recovery of an endangered species. Fishery Bulletin 98:759–766.

Lorda, E., and V. A. Crecco. 1987. Stock-recruitment relationship and compensatory mortality of American shad in the Connecticut River. Pages 469–482 *in* M. J. Dadswell, R. J. Klauda, C. M. Moffitt, R. L. Saunders, R. A. Rulifson, and J. E. Cooper. Common strategies of anadromous and catadromous fishes. American Fisheries Society, Symposium 1, Bethesda, Maryland.

Lorda, E., D. J. Danila, and J. D. Miller. 2000. Application of a population dynamics model to the probabilistic assessment of cooling water intake effects of Millstone Nuclear Power Station (Waterford, CT) on a nearby winter flounder spawning stock. Environmental Science & Policy 3:S471–S482.

Maceima, M. J., P. W. Bettoli, and D. R. DeVries. 1994. Use of a split-plot ANOVA design for repeated measures fishery data. Fisheries 19(3):14–20.

Mackinson, S. 2001. Integrating local and scientific knowledge: an example in fisheries science. Environmental Management 27(4):533–545.

Mackinson, S., M. Vasconcellos, and N. Newlands. 1999. A new approach to the analysis of stock-recruitment relationships: "model-free estimation" using fuzzy logic. Canadian Journal of Fisheries and Aquatic Sciences 56:686–699.

Manly, B. F. J. 1999. Randomization, bootstrap and Monte Carlo methods in biology (2nd edition). Chapman and Hall/CRC, Boca Raton, Florida.

Manly, B. F. J. 2000. Statistics for environmental science and management. Chapman and Hall/CRC, Boca Raton, Florida.

Mann, K. H. 1993. Physical oceanography, food chains, and fish stocks: a review. ICES Journal of Marine Science 50:105–119.

Mapstone, B. 1995. Scalable decision rules for environmental impact studies: effect size, Type I and Type II errors. Ecological Applications 5(2):401–410.

Matear, R. J. 1995. Parameter optimization and analysis

of ecosystems models using simulated annealing: a case study at station P. Journal of Marine Research 53(4):571–607.

Maunder, M. N., and G. M. Watters. 2003. A general framework for integrating environmental time series into stock assessment models: model description, simulation testing, and example. Fishery Bulletin 101:89–99.

Maxell, B. A. 1999. A power analysis on the monitoring of bull trout stocks using redd counts. North American Journal of Fisheries Management 19:860–866.

McCune, B., and M. J. Mefford. 1997. PC-ORD. Multivariate analyses of ecological data. MjM Software, Gleneden Beach, Oregon.

Meyer, J. S., C. G. Ingersol, L. L. McDonald, and M. S. Boyce. 1986. Estimating uncertainty in population growth rates: jackknife vs. bootstrap techniques. Ecology 67(5):1156–1166.

Miller, R. B., and R. D. Methot. 2002. Age-structured meta-analysis of U.S. West Coast rockfish (*Scorpaenidae*) populations and hierarchical modeling of trawl survey catchabilities Canadian Journal of Fisheries and Aquatic Sciences 59(2):383–392.

Mueter, F. J., R. M. Peterman, and B. J. Pyper. 2002. Opposite effects of ocean temperature on survival rates of 120 stocks of Pacific salmon (*Oncorhynchus* spp.) in northern and southern areas. Canadian Journal of Fisheries and Aquatic Sciences 59:456–463.

Mueter, F. J., R. M. Peterman, and B. J. Pyper. 2003. Corrigendum: Opposite effects of ocean temperature on survival rates of 120 stocks of Pacific salmon (*Oncorhynchus* spp.) in northern and southern areas. Canadian Journal of Fisheries and Aquatic Sciences 60:757.

Murphy, K. R., and B. Myors. 1998. Statistical power analysis. Lawrence Erlbaum Associates, Inc., Mahwah, New Jersey.

Myers, R. A. 1997. Recruitment variation in fish population assessment using meta-analysis. Pages 431–468 *in* R. E. Chambers and E. A. Trippel, editors. Early life history and recruitment in fish populations. Chapman and Hall, London.

Myers, R. A., and N. J. Barrowman. 1996. Is fish recruitment related to spawner abundance? Fishery Bulletin 94:707–724.

Myers, R. A., J. Bridson, and N. J. Barrowman. 1995. Summary of worldwide stock and recruitment data. Canadian Technical Report of Fisheries and Aquatic Sciences 2024. National Research Council of Canada, Ottawa.

Myers, R. A., B. R. MacKenzie, K. G. Bowen, and N. J. Barrowman. 2001. What is the carrying capacity for fish in the ocean? A meta-analysis of population dynamics of north Atlantic cod. Canadian Journal of Fisheries and Aquatic Sciences 58:1464–1476.

NAESCO (North Atlantic Energy Service Co.). 1994. Seabrook environmental studies, 1993. A characterization of environmental conditions in the Hampton-Seabrook area during the operation of Seabrook Station, Seabrook, New Hampshire.

Noreen, E. W. 1989. Computer intensive methods for testing hypotheses: an introduction. Wiley, New York.

Norris, III, J. L., and K. H. Pollock. 1996. Including model uncertainty in estimating variances in multiple capture studies. Environmental and Ecological Statistics 3:235–244.

NRC (National Research Council). 1998. Improving methods for fish stock assessment. National Academy Press, Washington, D.C.

Nunnari, G., A. F. M. Nucifora, and C. Randieri. 1998. The application of neural techniques to the modelling of time-series of atmospheric pollution data. Ecological Modelling 111:187–205.

Ortiz, M. 2002. Optimum sample size to detect perturbation effects: the importance of statistical power analysis – a critique. Marine Ecology 23(1):1–9.

Osenberg, C. W., R. J. Schmitt, S. J. Holbrook, K. E. Abu-Saba, and R. Flegal. 1994. Detection of environmental impacts: natural variability, effect size, and power analysis. Ecological Applications 4(1):16–30.

Ottersen, G., B. Planque, A. Belgrano, E. Post, P. C. Reid, and N. C. Stenseth. 2001. Ecological effects of the North Atlantic Oscillation. Oecologia 128:1–14.

Pace, M. L., S. F. Findlay, and D. L. Lints. 1992. Zooplankton in advective environments: the Hudson River community and comparative analysis. Canadian Journal of Fisheries and Aquatic Sciences 49:1060–1069.

Pauly, D. 1980. On the interrelationships between natural mortality, growth parameters, and mean environmental temperature in 175 fish stocks. Journal du Conseil Internationelle pour l'Exploration du Mer 39(2):175–192.

Perry, E., G. Seegert, J. Vondrusca, T. Lohner, and R. Lewis. 2003. Modeling possible cooling water intake system impacts on Ohio River fish populations. Pages 56–78 *in* D. A. Dixon, J. A. Veil, and J. Wisniewski, editors. Defining and assessing adverse environmental impacts from power plant impingement and entrainment of aquatic organisms. A. A. Balkema Publishers, Lisse, The Netherlands.

Pielou, E. C. 1984. The interpretation of ecological data. A primer on classification and ordination. Wiley & Sons, New York.

Peterman, R. M. 1990. Statistical power analysis can improve fisheries research and management. Canadian Journal Fisheries and Aquatic Sciences 47:1–15.

Peterman, R. M., and M. J. Bradford. 1987. Statistical power of trends in fish abundance. Canadian Journal of Fisheries and Aquatic Sciences 44:1879–1889.

Peterman, R., and M. M'Gonigle. 1992. Statistical power analysis and the precautionary principle. Marine Pollution Bulletin 24(5):231–234.

Peterson, I. 1991. Pick a sample. Science News 140:56–58.

Peterson, I., and J. S. Wroblewski. 1984. Mortality rates of fishes in the pelagic ecosystem. Canadian Journal of Fisheries and Aquatic Sciences 41:1117–1120.

Pope, J. G. 1989. Multispecies extensions to age-structured assessment models. Pages 102–111 *in* E. F. Edwards and B. A. Megrey. Mathematical analysis of fish stock analysis. American Fisheries Society, Symposium 6, Bethesda, Maryland.

Quinlan, J. A. and L. B. Crowder. 1999. Searching for sensitivity in the life history of Atlantic menhaden: inferences from a matrix model. Fisheries Oceanography 8(Supplement 2):124–133.

Quinn, T. J. II, and R. B. Deriso. 1999. Quantitative fish dynamics. Oxford University Press, New York.

Rago, P. J. 1980. A temporally distributed matrix model for inshore fish populations in Lake Michigan. M.S. thesis. Colorado State University, Ft. Collins.

Rago, P. J. 1984. Production foregone: an alternative method for assessing the consequences of fish entrainment and impingement at power plants and water intakes. Ecological Modelling 24:79–111.

Ravera, O. 2001. A comparison between diversity, similarity, and biotic indices applied to the macroinvertebrate community of a small stream: the Ravella River (Como Province, Northern Italy). Aquatic Ecology 35:97–107.

Reyjol, Y., P. Lim, A. Balaud, and S. Lek. 2001. Modeling of microhabitat used by fish in natural and regulated flows in the River Garonne (France). Ecological Modelling 146(1–3):131–146.

Riffenburgh, R. H. 1969. A stochastic model of inter population dynamics in marine ecology. Journal of the Fisheries Research Board of Canada 26:2843–2880.

Rose, K. 2000. Why are quantitative relationships between environmental quality and fish populations so elusive? Ecological Applications (10) 2:367–385.

Rose, K., and J. K. Summers. 1992. Relationships among long-term fish abundances, hydrographic variables, and gross pollution indicators in northeastern US estuaries. Fisheries Oceanography 1:281–293.

Rose, K. A., J. A. Tyler, R. C. Chambers, G. MacPhee, and D. J. Danila. 1996a. Simulating winter flounder population dynamics using coupled individual-based young-of-the-year and age-structured adult models. Canadian Journal of Fisheries and Aquatic Sciences 53:1071–1091.

Rose, K. A., J. A. Tyler, D. SinghDermot, and E. Rutherford. 1996b. Multispecies modeling of fish populations. Pages 194–222 *in* B. Megrey and E. Moksness, editors. Computers in fisheries research. Chapman and Hall, New York.

Rosenberg, M. S., D. C. Adams, and J. Gurevitch. 1999. MetaWin 2.0. Statistical software for conducting meta-analysis: fixed effect models, mixed effects models, and resampling tests. Sinauer Associates, Sunderland, Massachusetts.

Rottenberry, J. T., and J. A. Wiens. 1985. Statistical power analysis and community-wide patterns. The American Naturalist 125:164–168.

Rykiel, E. J. 1996. Testing ecological models: the meaning of validation. Ecological Modelling 90:229–244.

Saila, S. B. 1992. Application of fuzzy graph theory to successional analysis of a multispecies trawl fishery. Transactions of the American Fisheries Society 121(2):211–233.

Saila, S. B. 1996. Guide to some computerized artificial intelligence methods. Pages 8–40 *in* B. A. Megrey and E. Moksness, editors. Computers in fisheries research. Chapman and Hall, London.

Saila, S. B., and E. Lorda. 1977. Sensitivity analysis applied to a matrix model of the Hudson River striped bass population. Pages 311–332 *in* W. Van Winkle, editor. Assessing the effects of power-plant induced mortality on fish populations. Proceedings of a conference in Gatlinburg, Tennessee, May 1977. Pergamon, New York.

Saila, S. B., E. Lorda, J. D. Miller, R. A. Sher, and W. H. Howell. 1997. Equivalent adult estimates for losses of fish eggs, larvae, and juveniles at Seabrook Station with use of fuzzy logic to represent parameter uncertainty. North American Journal of Fisheries Management 17:811–825.

Saila, S., B. Martin, S. Ferson, L. Ginzburg, and J. Millstein. 1991. Demographic modeling of selected fish species with RAMAS. Electric Power Research Institute Report EN-7178. Electric Power Research Institute, Palo Alto, California.

SAS Institute. 1995. Jackboot macro. SAS Institute, Inc., Cary, North Carolina.

Savoy, T. F., V. A. Crecco, and B. C. Marcy, Jr. 2004. American shad early life history and recruitment in the Connecticut River: a 40-year summary. Pages 407–417 *in* P. M. Jacobson, D. A. Dixon, W. C. Leggett, B. C. Marcy, Jr., and R. R. Massengill, editors. The Connecticut River Ecological Study (1965–1973) revisited: ecology of the lower Connecticut River 1973–2003. American Fisheries Society, Monograph 9, Bethesda, Maryland.

Scardi, M. 2001. Advances in neural network modeling of phytoplankton primary production. Ecological Modelling 146(1–3):33–45.

Scarsbrook, M. R. 2002. Persistence and stability of lotic invertebrate communities in New Zealand. Freshwater Biology 47:417–431.

Schlese, W., and W. Nelson. 1996. A power analysis of methods for assessment of the change in seagrass cover. Aquatic Botany 53:227–233.

Schmitt, R. J., and C. W. Osenberg editors. 1996. Detecting ecological impacts: concepts and applications in coastal habitats. Academic Press, San Diego, California.

Shaw, A. G., and T. Mitchell-Olds. 1993. ANOVA for unbalanced data: an overview. Ecology 74(6):1638–1645.

Shepherd, J. G., J. G. Pope, and R. D. Cousens. 1984. Variations in fish stocks and hypothesis concerning their links with climate. Rapports et Proces-Verbaux des Reunions, Conseil International pour l'Exploration de la Mer 185:255–267.

Simon, J. L. 1999. Resampling: The new statistics (2nd edition). Resampling Stats, Inc., Arlington, Virginia.

Simon, J. L., D. T. Atkinson, and C. Shevokas. 1976. Probability and statistics: Experimental results of a radically different teaching method. The American Mathematical Monthly 83:733–739.

Simon, J. L., and P. Bruce. 1991. Resampling: a tool for everyday statistical work. Chance 4(1):22–32.

Smith, E. P., D. R. Orvos, and J. Cairns, Jr. 1993. Impact assessment using the before-after-control-impact (BACI) model: concerns and comments. Canadian Journal of Fisheries and Aquatic Sciences 50:627–637.

Smith, N. J., and K. Iles. 1988. A graphical depiction of multivariate similarity among sample plots. Canadian Journal of Forestry Research 18:467–472.

Steidl, R. J., and L. Thomas. 2001. Power analysis and experimental design. Pages 14–36 *in* S. M. Scheiner and J. Gurevitch, editors. Design and analysis of ecological experiments, 2nd edition. Oxford University Press, Inc., New York.

Stenseth, N. Ch., G. Ottersen, J. W. Hurrell, A. Mysterud, M. Lima, K-S, Chan, N. G. Yoccoz, and B. Adlandsvik. 2003. Studying climate effects on ecology through the use of climate indices: the North Atlantic Oscillation, El Nino Southern Oscillation and beyond. Proceedings of the Royal Society, London, series B 270:2087–2096.

Stewart-Oaten, A., J. R. Bence, and C. W. Osenberg. 1992. Assessing effects of unreplicated perturbations: no simple solutions. Ecology 73(4):1396–1404.

Stewart-Oaten, A., W. W. Murdoch, and K. R. Parker. 1986. Environmental impact assessment: "pseudoreplication" in time? Ecology 67(4):929–940.

Suter, G. W., L. W. Barnthouse, S. M. Bartell, T. Mill, D. Mackay, and S. Paterson. 1993. Ecological risk assessment. 1st edition. Lewis Publishers, Boca Raton, Florida.

Thomas, L., and C. J. Krebs. 1997. A review of statistical power analysis software. Bulletin of the Ecological Society of America 78(2):126–139.

Toft, K., and P. Shea. 1983. Detecting community-wide patterns: estimating power strengthens statistical inference. The American Naturalist 122(5):618–625.

Tomichek, C. A., and E. F. Roseman. 2003. Changes in fish and macroinvertebrate assemblages of the Niantic River Estuary, Connecticut 1974–2000. Proceedings of American Fisheries Society Estuarine Symposium in Baltimore, Maryland, August 19–22, 2002. American Fisheries Society, Bethesda, Maryland.

Trenkel, V. M., and M-J. Rochet. 2003. Performance of indicators derived from abundance estimates for detecting the impact of fishing on a fish community. Canadian Journal of Fisheries and Aquatic Sciences 60:67–85.

Tuljapurkar, S., and H. Caswell editors. 1997. Structured population models in marine, terrestrial and freshwater systems. Chapman and Hall, New York.

Turchin, P. 1998. Quantitative analysis of movement: measuring and marking population distribution in animals and plants. Sinauer Associates, Sunderland, Massachusetts.

Ulanowicz, R. E. 1975. The mechanical effects of water flow on fish eggs and larvae. Pages 79–88 *in* S. B. Saila, editor. Fisheries and energy production: a symposium. Lexington Books, Lexington, Massachusetts.

Underwood, A. 1996. Detection, interpretation, prediction and management of environmental disturbances: some roles for experimental marine ecology. Journal of Experimental Marine Biological Ecology 200:1–27.

Underwood, A. J. 1981. Techniques of ANOVA in experimental marine biology and ecology. Oceanography and Marine Biology Annual Reviews 19:513–605.

Underwood, A. J. 1994. On beyond BACI: sampling designs that may reliably detect environmental disturbances. Ecological Applications 4(1):3–15.

Underwood, A. J., and M. G. Chapman. 1998. Spatial

analyses of intertidal assemblages on sheltered rocky shores. Australian Journal of Ecology 23:138–157.

Underwood, A. J., and M. G. Chapman. 2003. Power, precaution, Type II error and sampling design in assessment of environmental impacts. Journal of Experimental Marine Biology and Ecology 296:49–70.

USEPA (U. S. Environmental Protection Agency). 2000. Stressor identification guidance document. Office of Water and Office of Research and Development, EPA/822/B-00/025, Washington, D.C..

Van Winkle, W., W. P. Dey, S. M. Jinks, M. S. Bevelhimer, and C. C. Coutant. 2003. A blueprint for the problem formulation phase of EPA-type ecological risk assessments for 316(b) determinations. Pages 264–290 *in* D. A. Dixon, J. A. Veil, and J. Wisniewski, editors. Defining and assessing adverse environmental impacts from power plant impingement and entrainment of aquatic organisms. A. A. Balkema Publishers, Lisse, The Netherlands.

Vaughan, D. S. 1981. An age structure model of yellow perch in western Lake Erie. Pages 189–216 *in* D. G. Chapman and V. F. Galluci, editors. Quantitative population dynamics. International Co-Operative Publishing House, Fairland, Maryland.

Vaughan, D. S., and S. B. Saila. 1976. A method for determining mortality rates using the Leslie matrix. Transactions of the American Fisheries Society 105:380–383.

Vaughan, D. S., and W. Van Winkle. 1982. Corrected analysis of the ability to detect reductions in year class strength of the Hudson River white perch (*Morone americana*) population. Canadian Journal of Fisheries and Aquatic Sciences 39:782–785.

Vetter, E. F. 1988. Estimation of natural mortality in fish stocks: a review. U.S. Fishery Bulletin 86:25–43.

Walker, J. S. 1999. A primer on wavelets and their scientific applications. Chapman and Hall/CRC, Boca Raton, Florida.

Walters, C. J. 1996. Computers and the future of fisheries. Pages 229–238 *in* B. A. Mergrey and E. Moksness, editors. Computers in fisheries research. Chapman and Hall, London.

Walters, G., and R. Deriso. 2000. Catches per unit effort of bigeye tuna: a new analysis with regression trees and simulated annealing. Bulletin, Inter-American Tropical Tuna Commission 21(8):531–552.

Weisberg, S. 1986. A linear model approach to backcalculation of fish length. Journal of the American Statistical Association 81, No. 396. Applications: 922–929.

Weisberg, S. 1993. Using hard-part increment data to estimate age and environmental effects. Canadian Journal of Fisheries and Aquatic Sciences 50(6):1229–1237.

White, G. C., and K. P Burnham. 1999. Program MARK: survival estimation from populations of marked animals. Bird Study 46(Supplement):120–138.

Winer, B. 1971. Statistical principles in experimental design. McGraw-Hill, New York.

Wright, J. F. 1995. Development and use of a system for predicting the macroinvertebrate fauna in flowing waters. Australian Journal of Ecology 20:181–197.

Zhou, S. 2003. Application of artificial neural networks for forecasting salmon escapement. North American Journal of Fisheries Management 23:48–59.

American Fisheries Society Monograph 9:501–511, 2004

Feasibility of Applying a Watershed Model to the Connecticut River for Total Maximum Daily Load Analysis and Watershed Planning

CARL W. CHEN*

Systech Engineering, Inc.
3180 Crow Canyon Place, Suite 260, San Ramon, California 94583, USA

ROBERT A. GOLDSTEIN

Electric Power Research Institute,
Post Office Box 10412, Palo Alto, California 94303, USA

JOEL HERR

Systech Engineering, Inc.,
3180 Crow Canyon Place, Suite 260, San Ramon, California 94583, USA

Abstract.—Point source discharges, land use changes, agricultural activities, and atmospheric deposition can contribute to water quality impairments within the Connecticut River and Long Island Sound. The Clean Water Act requires determination of total maximum daily loads of pollutants that will meet water quality criteria for protecting a water body's designated uses. A watershed model, the Watershed Analysis Risk Management Framework (WARMF), was applied to the Hockanum watershed, a small tributary to the Connecticut River. The WARMF simulated stream flow, temperature, dissolved oxygen, pH, ammonia, nitrate, and phosphorus comparable to observed data. The application of the WARMF creates a roadmap for stakeholders to analyze watershed behavior, make watershed management decisions, and develop consensus on a management plan. The Hockanum application of the WARMF provides information about data availability and techniques for extending the model domain to the entire Connecticut River, which is 660 km (410 mi) long with a total drainage area of 29,540 km^2 (11,400 mi^2).

Introduction

When the Connecticut River Ecological Study was designed and conducted, the assessment of ecological impacts focused on the Connecticut Yankee nuclear generating station. For the most part, interactions among individual sources of pollution (e.g., other point sources such as sewage treatment plants and industrial facilities) and disturbances (e.g., fishing, nonpoint source runoff) were not considered. Today, we recognize that there are interactions among sources (point and nonpoint) that affect the efficacy and cost efficiency of water resource management strategies. We also recognize that the protection of aquatic organisms must involve watershed protection to provide suitable habitat. These changes in perspective have been driven both by the evolution of ecology as a science and achievements that have been made in improving the quality of our nation's waters through point source controls.

As population and demand for fresh, clean water increase throughout the United States, there is an ever-growing need to create and implement watershed-based, water resource management plans that take into consideration the diverse needs and objectives of all stakeholder groups. Development of such plans will require stakeholder involvement and the shifting of decision-making processes from command and control to

* *Corresponding author:* carl@systechengineering.com

stakeholder consensus building. A prime example in the Clean Water Act is the calculation, allocation, and implementation of total maximum daily loads (TMDLs) of pollutants to meet designated uses for surface waters.

In anticipation of society's need to adopt a watershed-based management approach that incorporates stakeholder participation, the Electric Power Research Institute (EPRI) has sponsored the development and testing of the Watershed Analysis Risk Management Framework (WARMF) as a decision support system for TMDL analysis and watershed planning. It can potentially be applied to the entire Connecticut River watershed, which flows from Canada through Vermont, New Hampshire, Massachusetts, and Connecticut to Long Island Sound.

It has been recognized that point source discharges, land use changes, agriculture activities, and atmospheric deposition can contribute to water quality impairments within the Connecticut River and the Long Island Sound (Gasworks et al. 1997). A watershed analytic framework such as the WARMF can help track pollutants (heat, sediment, phosphate, nitrogen species, pathogens, and mercury) throughout a watershed from their sources to their fates. It can also help evaluate the effectiveness of an integrated strategy (including atmospheric deposition control, point source control, flow diversion, best management practices, and pollution trading) to reduce pollution loads to Long Island Sound.

The objective of this paper is to demonstrate how the WARMF can be used to evaluate alternative water resource management strategies for the Connecticut River. The feasibility study for such application was carried out in cooperation with the Connecticut Department of Environmental Protection (CTDEP), the Northeast Utilities Service Company, and the University of Connecticut.

WARMF Application Methodology

The formulations of the WARMF have been documented (Chen et al. 1998, 2001). The model has been peer reviewed according to EPA guidance (Keller 2000). The WARMF has been applied to the Catawba River basin of North and South Carolina, the Truckee River basin of California and Nevada, the Cheat River basin of West Virginia, the Chartier Creek basin of Pennsylvania, the Oostanuala Creek basin of Tennessee, the St. Louis River basin of Minnesota, the Nemadji River basin of Wisconsin and Minnesota, the Blue River basin of Colorado, the Mica Creek basin of Idaho, and the Santa Clara River basin of California. Following is a brief description of the general methodology.

Application of the WARMF requires a front-end effort that includes site-specific model setup, data compilation, and model calibration. During this effort, the WARMF is used to delineate the entire river basin into a network of land catchments (with land use, vegetation and soil layers), river segments, and lakes (with water layers). The network facilitates the routing of water and its associated chemicals from land to river, river to reservoir, reservoir back to river, and, if necessary, ultimately to estuary. Water balance, heat budget, and mass balance of chemical constituents are performed to simulate snow hydrology, soil hydrology, surface runoff, soil erosion, pollutant accumulation on land, dissolution of pollutants into surface runoff for nonpoint source loading, and hydrology and water quality of the receiving water.

The network of land catchments, river segments, and lakes can be delineated to the appropriate spatial resolution. The level of resolution will depend on the nature of the issues being addressed. In applying the WARMF, it is not necessary to have uniform spatial resolution across the entire river basin. Subregions with septic systems or subregions proposed for new development can be delineated into a fine resolution to mesh with the coarser resolution of the entire river basin.

A feasibility study was conducted to test the applicability of the WARMF to a small watershed tributary to the Connecticut River. The Hockanum River watershed with a drainage area of 198 km^2 (or 77 mi^2) was chosen for the test. The pilot study was designed to obtain information about data availability and technique to extend the model domain to the entire Connecticut River, which is 660 km (410 mi) long with a total drainage area of 29,540 km^2 (11,400 mi^2).

Results

Model Setup

Geographic Information System (GIS) data for the Hockanum River watershed were obtained from the Connecticut Department of Environmental

Protection (CTDEP). The data layers received included hydrography, hydrography subset, geographic names information system, drainage basins, soils, public water supply reservoirs, public water supply watersheds, public water supply wells, sanitary waste water sewage treatment plants, leachate and wastewater discharge inventory, and land use/land cover. The United States Geological Survey (USGS) Digital Elevation Model (DEM) was downloaded from the USGS Web site.

We imported CDEP DEM data into the WARMF and then USGS DEM data. The watershed delineated with the CDEP DEM did not match the one delineated with the USGS DEM. We used the USGS 7.5-min series topographic maps to reconcile the discrepancy. Figure 1 shows the final basin map created by the WARMF.

As shown in the basin map, the Hockanum River watershed has two sections. The upstream section is the small subwatershed at the upper right (northeast) corner that drains into Lake Shinepsit. The outflow of the lake empties into the lower Hockanum River. The black dot shown in East Hartford County on the basin map is the downstream gaging station where flow and water quality data are available for model calibration.

The WARMF delineated the entire watershed into 27 land catchments, 26 stream segments and one lake with 1-m-thick layers of water. The DEM data were processed by the WARMF to calculate area, slope, and aspect of land catchments. Also calculated were length and upstream and downstream riverbed elevations of stream segments. The WARMF processed GIS data layers to determine locations and names of watershed landmarks, locations of wastewater discharges, and land uses in land catchments.

Meteorology, air quality, precipitation quality, point source discharges, and reservoir release data are required to drive the model. Meteorological data for Hartford Brainard Airport, Mansfield Hollow, Storrs, Stafford Springs, and North Somers were obtained from a meteorology CD published by Earth/Info. The meteorological

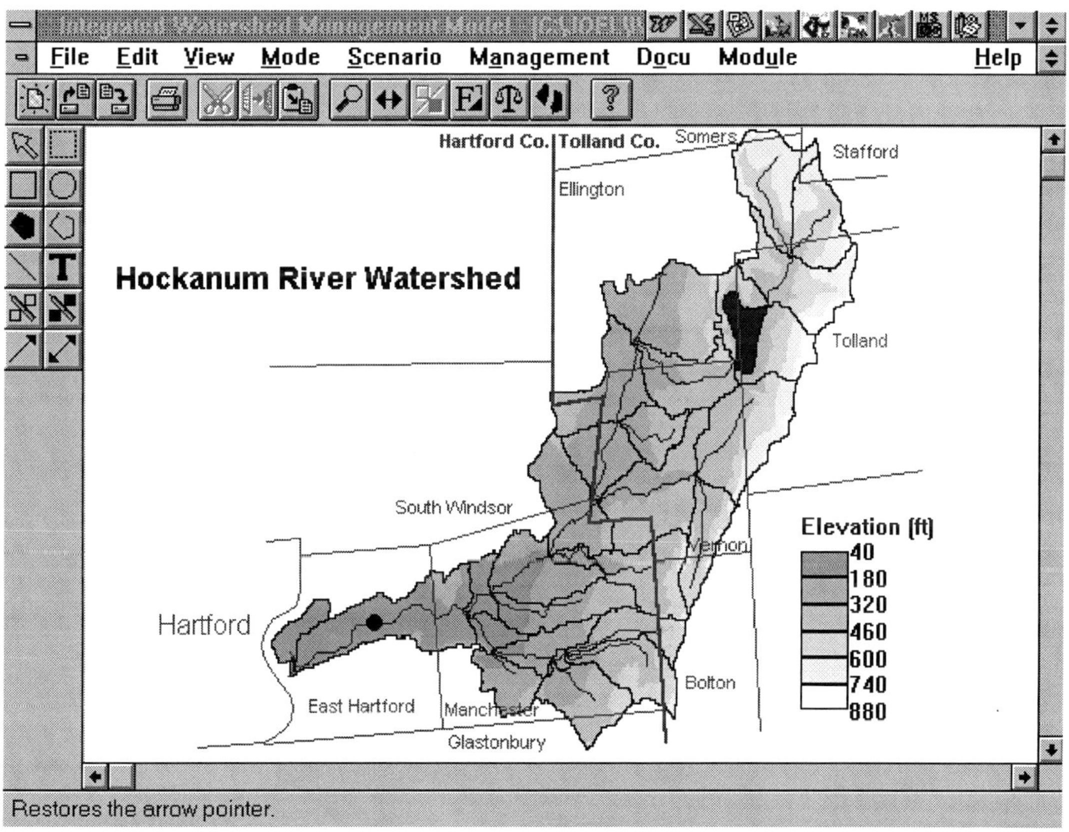

FIGURE 1. Hockanum River watershed map generated by the Watershed Analysis Risk Management Framework.

data for Hartford Bradley Airport were downloaded from the website of the National Climatic Data Center. Data for Rockville station were too scarce to use.

There are data gaps in the meteorological data set for all stations. Meteorological time series for pairs of stations were compared, and the data from one station were used to fill the gaps of another station. By this procedure, complete time series of meteorological data for Hartford Bradley Airport, Hartford Brainard Airport, Storrs, and Stafford Springs were prepared for input to the model.

The National Atmospheric Deposition Program (NADP) commonly monitors air and precipitation quality for the United States. Unfortunately, there is no NADP station in the state of Connecticut. Precipitation chemistry data are available from the West Point station in New York. University of Connecticut researchers have established three monitoring stations in the state. Based on 1991 to 1993 data, spatial and temporal variations of atmospheric deposition of sulfur, nitrogen, and phosphorus have been estimated for the three monitoring stations in Connecticut (Yang et al. 1996). One of the stations is Shenipsit State Forest. The precipitation chemistry data of West Point, New York and the deposition estimates for the Shenipsit station were combined to generate the air and precipitation quality data for input to the WARMF.

Point source discharge data were obtained from the GIS data layer of the CDEP. Reservoir release data were not available. A constant lake outflow was assumed.

The WARMF can simulate flow and water quality parameters for various locations in the watershed. For some locations and times, there are observed data to check against the model results. Figures 2–8 present the comparison of simulated and observed flow, temperature, pH, dissolved oxygen, ammonia, nitrate, and total phosphorus for the Hockanum River at East Hartford. The simulated values match observed values reasonably well. These good matches were achieved mostly with default values for model coefficients, many of which were developed for the watersheds in the Adirondacks of New York.

Model Usage

The WARMF is not just a watershed model. It is a decision support system. The WARMF performs scientific analyses, but it also includes a road map that leads stakeholders through a process to evalu-

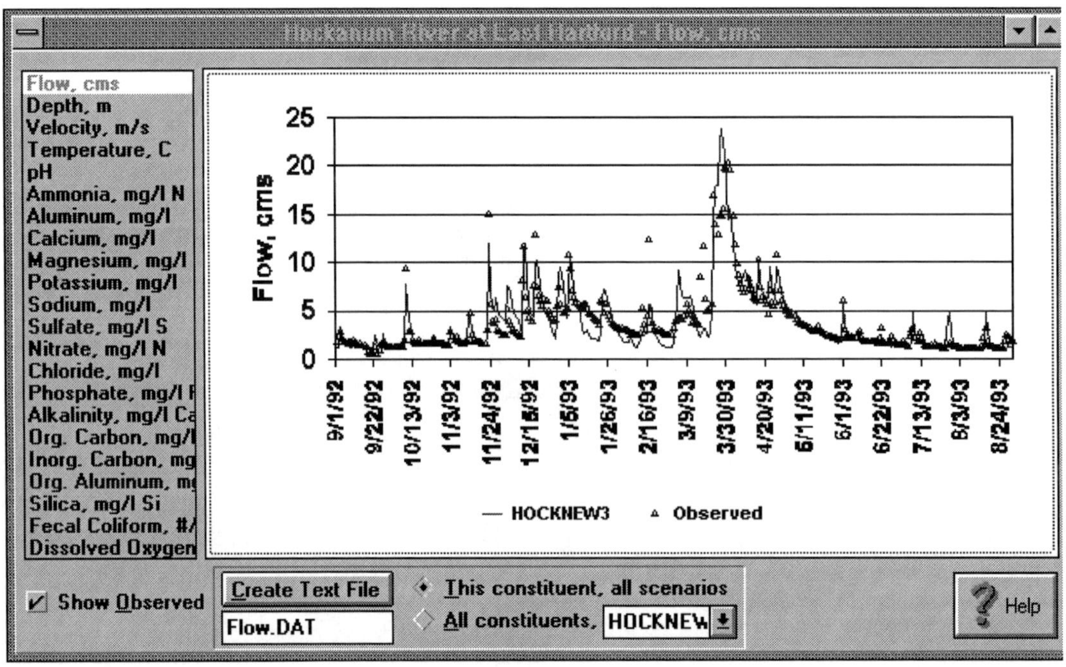

FIGURE 2. Simulated and observed flow of Hockanum River at East Hartford.

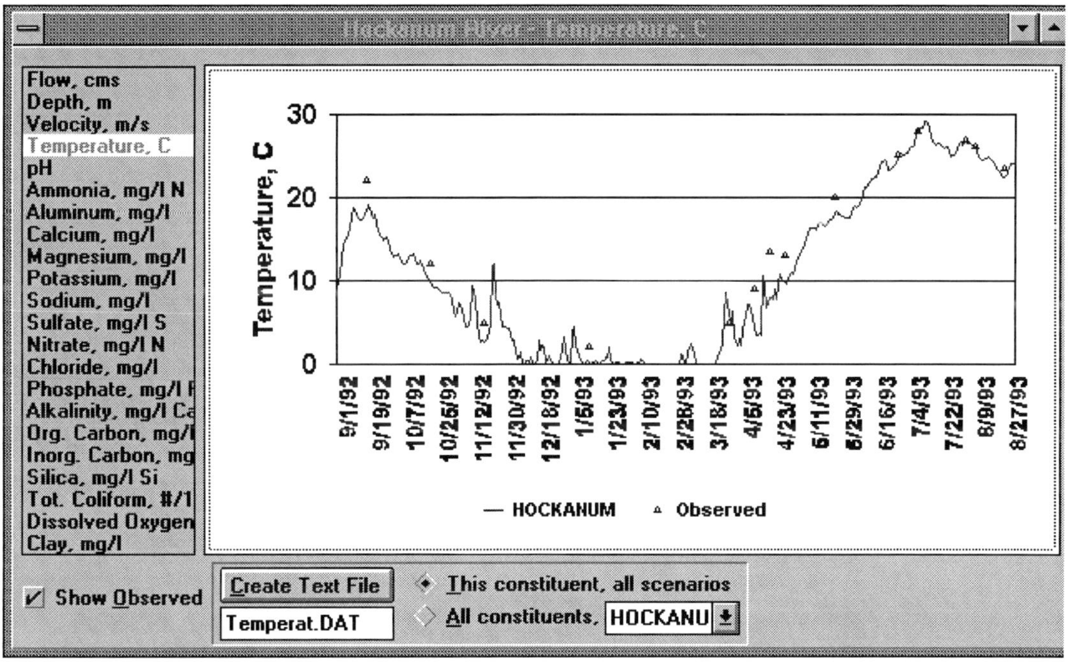

FIGURE 3. Simulated and observed temperature of Hockanum River at East Hartford

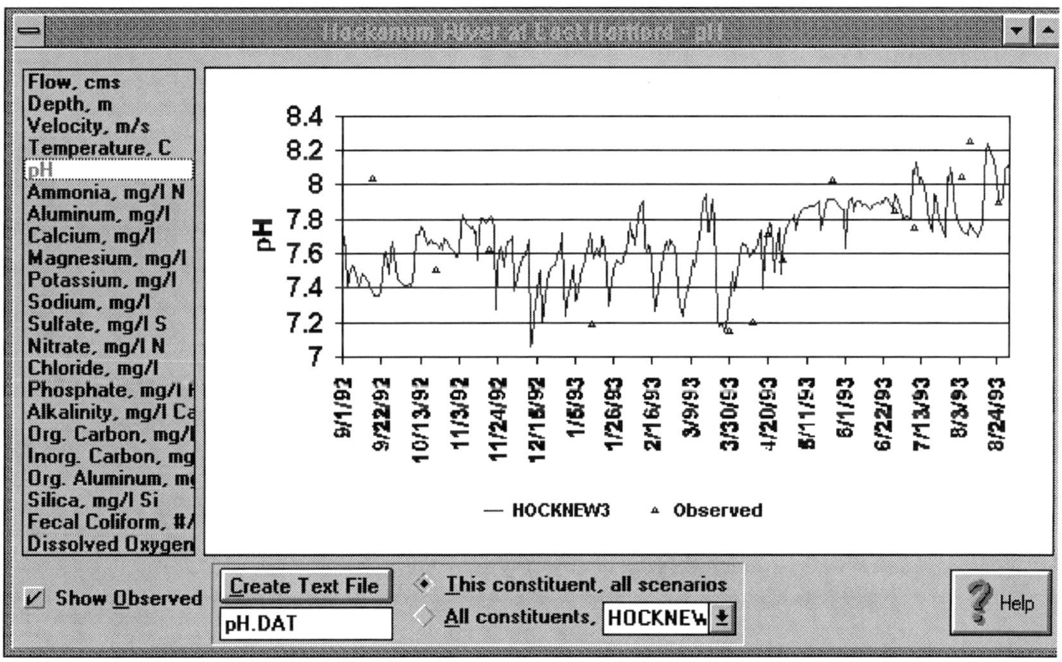

FIGURE 4. Simulated and observed pH of Hockanum River at East Hartford.

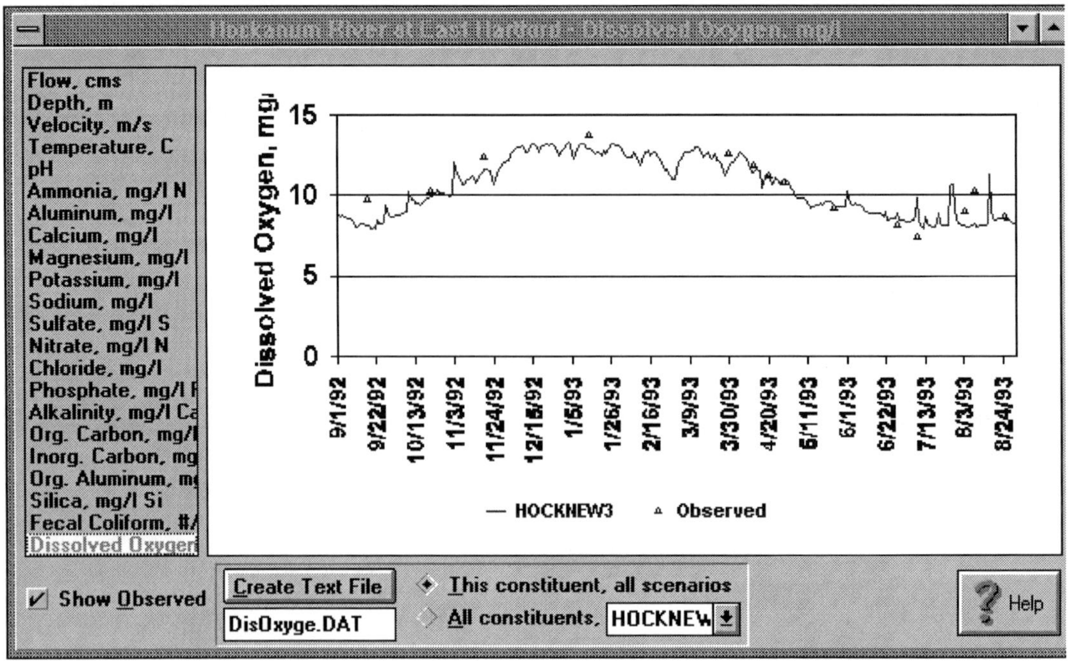

FIGURE 5. Simulated and observed dissolved oxygen of Hockanum River at East Hartford.

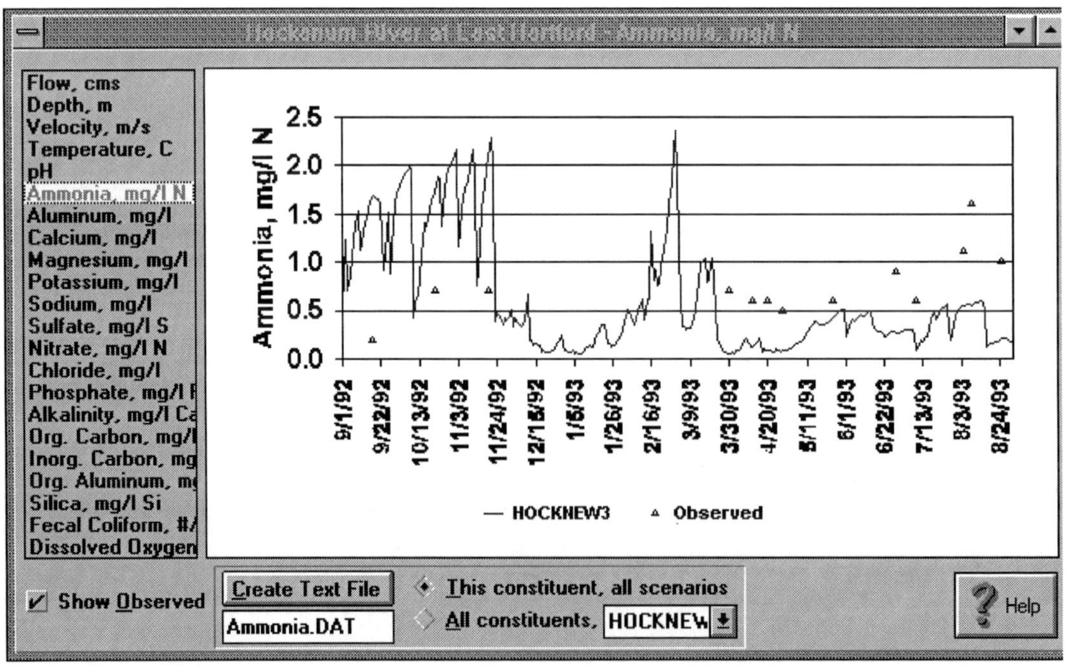

FIGURE 6. Simulated and observed ammonia of Hockanum River at East Hartford.

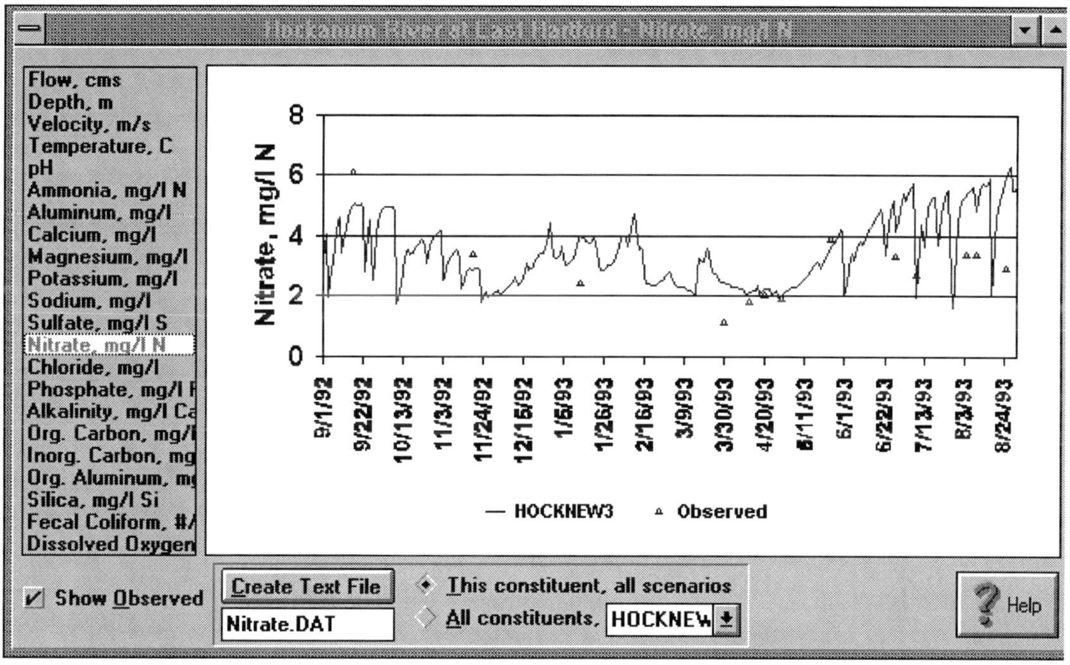

FIGURE 7. Simulated and observed nitrate of Hockanum River at East Hartford.

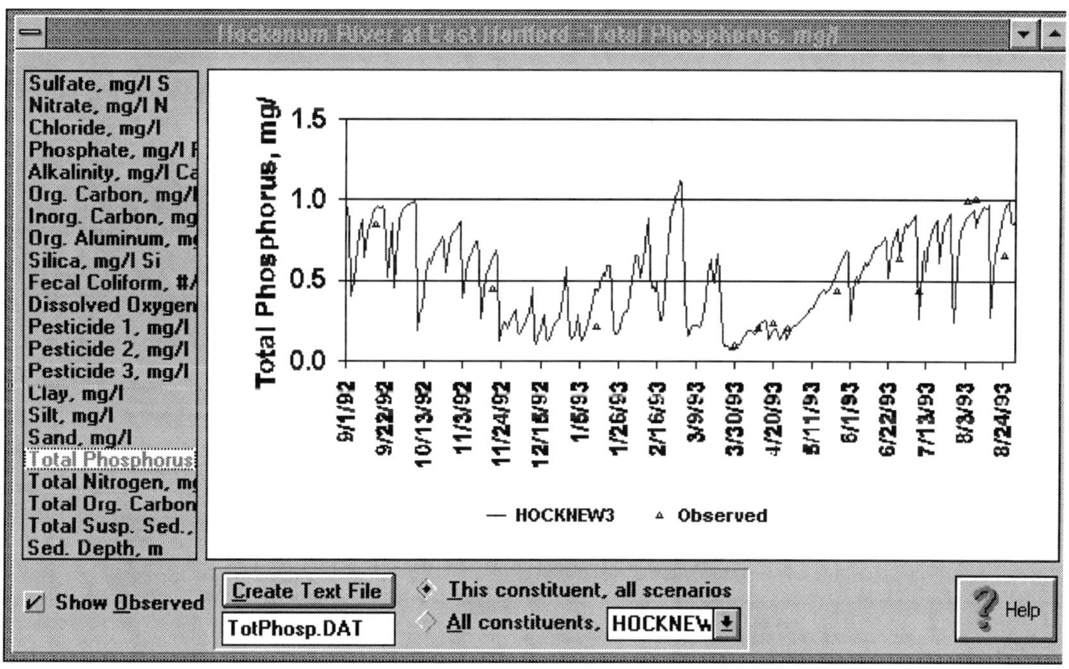

FIGURE 8. Simulated and observed total phosphorus of Hockanum River at East Hartford.

ate management options, make decisions, and reach consensus.

Figure 9 shows the consensus road map. The object of the road map is to guide stakeholders through the seven step processes of organizing themselves, developing a work plan, identifying water quality issues, learning how their watershed works, formulating management alternatives, analyzing cost and effectiveness, and voting and negotiating to reach a consensus on an implementation plan. For each step, there are a number of tasks to perform. Buttons on the computer screen represent these tasks. Stakeholders can click a button on the road map, and the WARMF will provide a dialogue box to follow.

For example, the pull down menu for the "Edit" button in Step 5 (management alternatives) of the road map was selected in Figure 9. The WARMF shows that a management alternative can involve a change in point sources, nonpoint sources, atmospheric deposition, and diversions. By selecting point source, a dialogue box to modify point sources will appear. The user can change management options individually or in combination simply by clicking a button.

After specification of a management plan, the user can go to Step 4 and click the "Simulate" button. The WARMF will perform the simulation for the management alternative. The user can then click the "Loading" button to view the pollution load from various subwatershed regions.

Figure 10 shows calculated regional loadings of nitrate for the Hockanum River watershed. On the computer screen, the four subcatchments that make up the watershed are differentiated by color, as are the point and nonpoint source loadings. In the paper, the figure appears in black and white. There are four bar charts shown in the figure. Each bar is associated with a different subwatershed. The length of a bar is proportional to the total nitrate loading in the subcatchment. The lighter shaded portion of a bar represents the proportion of loading originating from nonpoint sources and the darker shaded portion, the proportion from point sources. Three of the bars are entirely lightly shaded, indicating that most regions in the Hockanum River watershed contribute only nonpoint source loads of nitrate. For the developed subwatershed in East Hartford, the bar has two shades: the lighter shaded portion on top indicates the smaller nonpoint source contribution, and the darker shaded portion on the bottom indicates the larger point source contribution. By clicking on an indi-

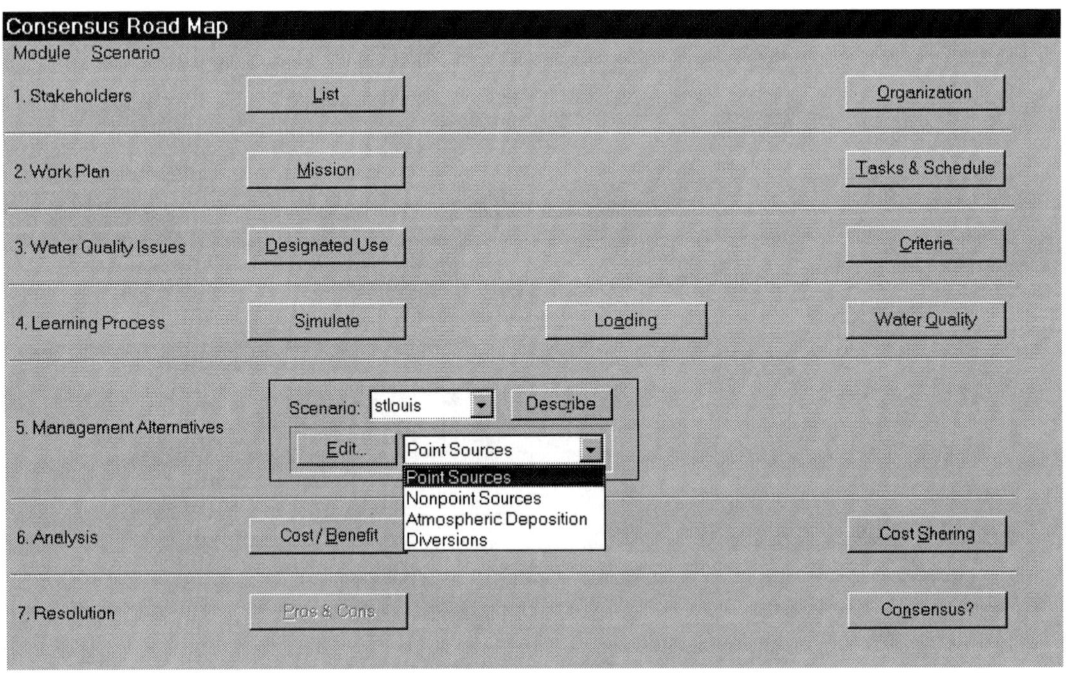

FIGURE 9. Consensus road map for the development of a watershed management plan.

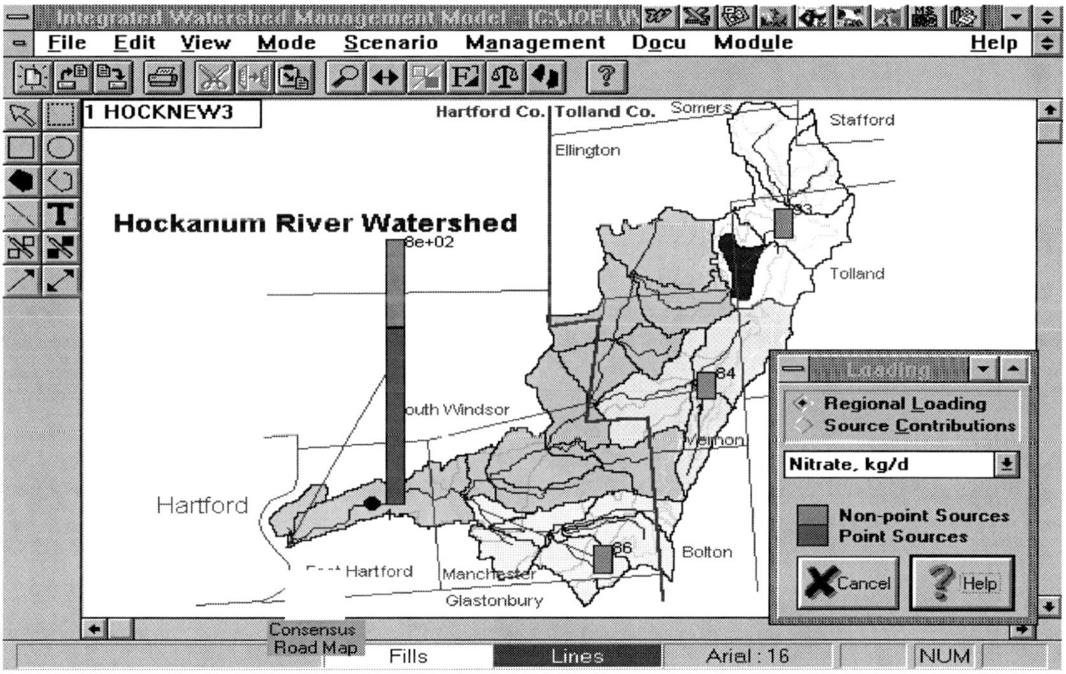

FIGURE 10. Nitrate loadings for four subcatchments of Hockanum River watershed.

vidual bar shown on the computer screen of Figure 10, a stakeholder will create a spreadsheet of point and nonpoint source loads from various land uses within the subcatchment.

After learning where pollution loads come from, stakeholders can click on the "Water Quality" button of Step 4 to see the water quality consequences of these loads. The WARMF will color river stretches that meet criteria for their intended uses in green and those that do not in red. Thus, stakeholders can relate pollution load to water quality in a visual fashion. They can modify the management strategy to change the distribution and amount of river miles that are pictured as green.

Extension to Connecticut River Watershed

The WARMF can be applied to the entire Connecticut River basin. The WARMF divides a large river basin into subwatersheds to achieve good resolution. The WARMF performs simulations for the subwatersheds, one at a time, with a dynamic allocation of memory. This technique eliminates the computer memory problem associated with simulation of large watersheds such as the Connecticut River.

There are two different ways to approach TMDL simulation for the Connecticut River watershed. One approach is to include the entire river basin from Canada to Long Island Sound in an analysis. Another approach is to limit the application to the Connecticut River basin within the state of Connecticut. Figure 11 shows the basin map if the WARMF is applied to the lower Connecticut River within the state of Connecticut. For this approach, it is necessary to have good flow and water quality data for the boundary between Massachusetts and Connecticut.

Conclusion

This feasibility study shows that the WARMF can simulate flow, temperature, dissolved oxygen, pH, ammonia, nitrate, and phosphorus for the Hockanum River, a small tributary to the Connecticut Rivcr. The network of land catchments, river segments, and lakes is scaleable to the entire Connecticut River. The new capability being added to simulate mercury species (Chen 2003) will enhance the WARMF's usefulness in addressing the current issue regarding fish accumulation of mercury.

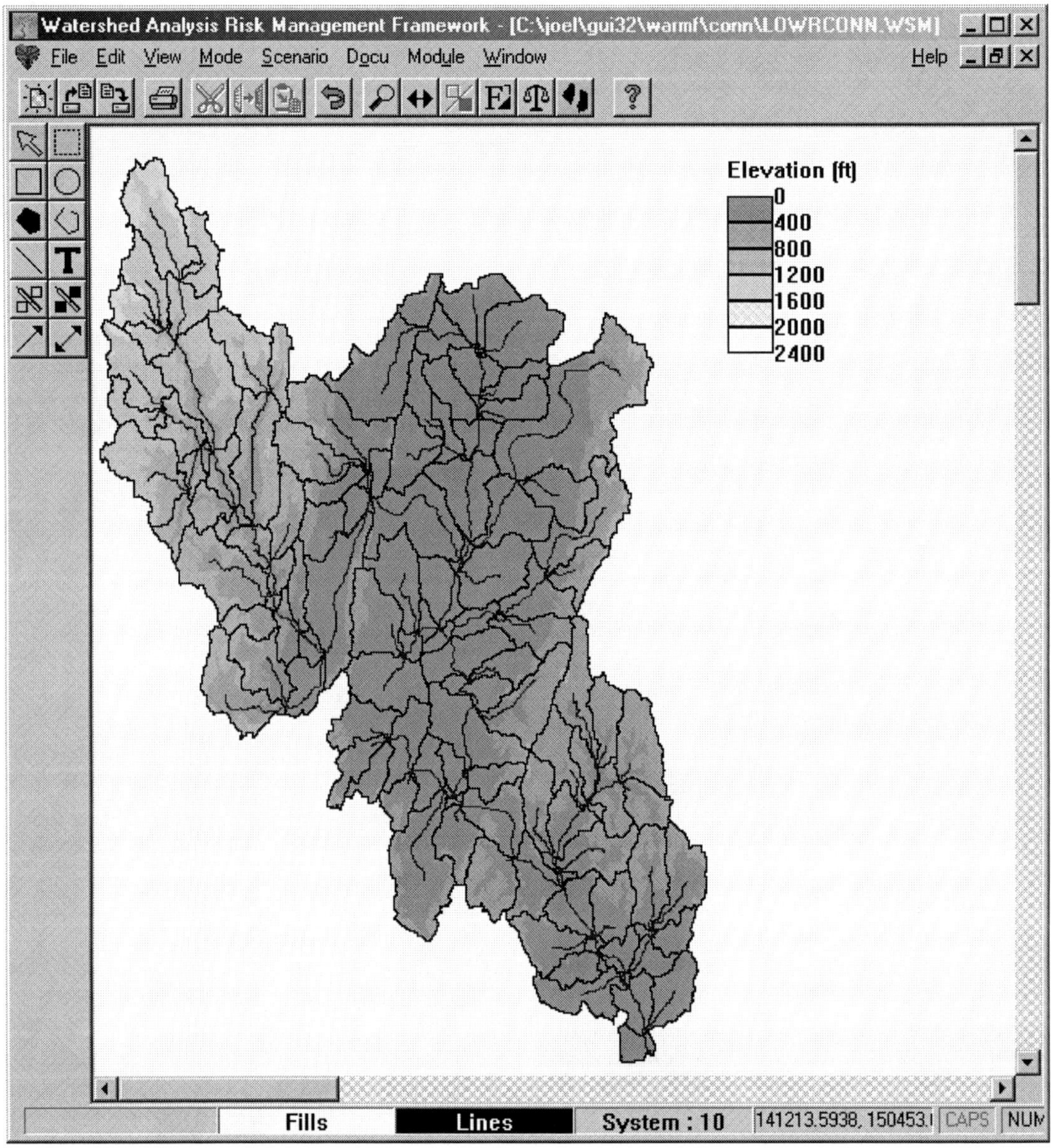

FIGURE 11. Application of the Watershed Analysis Risk Management Framework to the Connecticut River basin within Connecticut.

The future holds many challenges for environmental scientists and water resource managers. They need to make their results accessible to nonprofessional policy makers and stakeholders. They need to consider phenomena and processes operating nonuniformly over large watersheds. They need to integrate effects of multiple point and nonpoint sources, while at the same time differentiate each source's contribution. They need to conduct robust analyses of combinations of multiple control, ecological enhancement, and water quality trading options. The WARMF is a tool that can aid them to meet these challenges.

Acknowledgements

The feasibility study discussed in this paper was funded by the Electric Power Research Institute of Palo Alto, California and the Northeast Utilities Service Company of Connecticut. The Connecticut Department of Environmental Protection (CTDEP) and the University of Connecticut cooperated and contributed their data to the study. Dr. Arturo Keller, University of California - Santa Barbara, Dr. George Ice, National Council for Air and Stream Improvement (NCASI), Inc., Corvalis, Oregon, and an anonymous reviewer provided valuable comments and suggestions on an earlier draft to improve the paper.

References

Chen, C. W. 2003. Mercury TMDL Study for St. Louis River basin, Nemadji River basin, and North Coast River basins of Minnesota. Systech Engineering, Inc., Progress Report No. 4, San Ramon, California.

Chen C. W., J. Herr, and L. Weintraub. 2001. Watershed Analysis Risk Management Framework (WARMF): update one. A decision support system for watershed analysis and total maximum daily load calculation, allocation, and implementation. Electric Power Research Institute, Technical Report 1005181, Palo Alto, California.

Chen, C. W., J. Herr, and L. Ziemelis. 1998. Watershed Analysis Risk Management Framework: a decision support system for watershed approach and total maximum daily load calculation. Electric Power Research Institute, Technical Report 110709, Palo Alto, California.

Gasworks, N. A., R. W. Hearth, and L. J. Halting. 1997. Atmospheric deposition of nitrogen oxides onto the landscape contributes to coastal eutrophication in the northeast United States. Environmental Science and Technology 31:1995–2004.

Keller, A. 2000. Peer review of the Watershed Analysis Risk Management Framework (WARMF). An evaluation of WARMF for TMDL applications by independent experts using USEPA guidelines. Electric Power Research Institute, Technical Report 1000252, Palo Alto, California.

Yang, X., D. Miller, X. Xu, L. H. Yang, H. Chen, and N. P. Nikolaidis. 1996. Spatial and temporal variations of atmospheric deposition in interior and coastal Connecticut. Atmospheric Environment 30(22):3801–3810.

American Fisheries Society Monograph 9:513–524, 2004

Modeling the Hydraulic Zone of Influence of Connecticut Yankee Nuclear Power Plant's Cooling Water Intake Structure

JOHN RICHARDSON*
Alden Research Laboratory, Holden, Massachusetts, USA

DOUGLAS A. DIXON
Electric Power Research Institute, Palo Alto, California, USA

Abstract.—Ecological studies performed in the late 1960s and throughout the 1970s sought to identify the impact of the Connecticut Yankee nuclear power plant (CY) located on the banks of the Connecticut River. These original studies relied on intensive biological sampling, hydrography, and inference based on observations. Numerical modeling was, at that time, in its infancy, and comprehensive dynamic simulations of river, lake, and estuarine systems required computer resources beyond the reach of most individuals and institutions. Today, detailed hydraulic analyses over large scales are practical, and unified approaches can now be used to study complex flow problems involving different near-field and far-field dynamics. For example, where scale models would have been used to study the dispersion of effluents from a heated water discharge in the past, today, this kind of analysis can be done on the computer using computational fluid dynamics (CFD) techniques. This paper demonstrates how the hydraulic zone of influence (HZI) of the CY cooling water intake structure (CWIS) can be determined from the results of CFD investigation. The analysis procedures are general and can be used for HZI or intake area of influence determination on other water body types (e.g., lakes, rivers, salt wedge estuaries, and open coastal locations).

Introduction

Ecological studies performed in the late 1960s and throughout the 1970s sought to identify the impact of the Connecticut Yankee nuclear power plant (CY) located on the banks of the Connecticut River (Merriman 1970; Merriman and Thorpe 2004 [1976], this volume). These original studies relied on intensive biological sampling, hydrography, and inference based on observations. Numerical modeling was, at that time, in its infancy, and comprehensive dynamic simulations of river, lake, and estuarine systems required computer resources beyond the reach of most individuals and institutions.

Recent developments in computer technology are providing new ways to integrate different kinds of data and to promote informed decision making (e.g., this is the emphasis of *Hydroinformatics*—an expression coined in the 1980s by Michael Abbott, former director of the Danish Hydraulic Institute). As a result, initiatives like the EPA's Total Maximum Daily Load (TMDL) Program are being devised and a new generation of analysis tools are being implemented (e.g., Chen et al. 2004, this volume). Hydraulic modeling has developed similarly since the original Connecticut Yankee Ecological Study was performed.

Today, detailed hydraulic analyses over large scales are practical, and unified approaches can now be used to study complex flow problems involving different near-field and far-field dynamics. For example, where scale models would have been used to study the dispersion of effluents from a heated water discharge in the past, today, this kind of analysis can be done on the computer using computational fluid dynamics (CFD) techniques.

Emerging hydraulic modeling techniques,

* *Corresponding author*: jrichardson@bluehillhydraulics.com

like CFD, coupled with biological and hydrologic information, can be used to estimate the entrainment of organisms passing by cooling water intake structure (CWIS) from far-field locations (i.e., to determine the hydraulic zone of influence [HZI] or "area of influence" of a CWIS). State-of-the-art techniques such as those described herein will contribute to future ecological studies by reducing costs (e.g., to optimally design field programs that minimize data collection and clarify data needs), improving the interpretation of data, and promoting insight (e.g., what-if scenarios can be used to establish cause and effect relationships).

The analysis presented in this paper demonstrates the use of *FLOW-3D* (Flow Science 2003) to determine the HZI of the CY CWIS. The simulation was conducted as a case study for an EPRI sponsored project designed to demonstrate the use of commercially available CFD programs applied to the problem of HZI determination (EPRI 2004). In this analysis, *FLOW-3D* was used to estimate the entrainment of passive organisms originating at locations upstream of the CY CWIS. As part of the EPRI study, the HZI was determined for power plant CWIS located at five other water body type locations (i.e., main-stem reservoir, tidal estuary, river, coastal ocean station, and Great Lake station) in addition to CY. Results from additional EPRI case studies (discussed later) illustrate other ways in which the HZI for CWIS can be visualized.

Study Area

The Connecticut Yankee nuclear power plant is located on the eastern bank of the Connecticut River at river kilometer 25.7. The study area (and domain of the numerical model) extends about 1.6 km upstream from the CWIS and about 5.0 km downstream from the CWIS. These are the approximate limits of the thermal plume produced by the power plant (Merriman 1970; Merriman and Thorpe 2004 [1976]). The flow in this portion of the river is affected by the tide; however, saline water from Long Island Sound do not intrude into the study area.

The Connecticut Yankee nuclear power plant withdraws approximately 25.5 m^3/s of water from the river and raises its temperature to a maximum of 12.5°C above ambient before returning it to the river. This diversion amounts to as much as 6% of the river flow during low-flow conditions (Massengill 2004 [1976], this volume).

Cooling water is withdrawn from a shoreline location through a single structure containing four inlet bays (Jacobson et al. 2004, this volume). While not demonstrated in this study, CFD models that include intake details such as the bar grids and fish screens, and extend into the far field can be developed through the use of grid nesting techniques, multiblock gridding procedures, or unstructured grids.

Heated water is discharged from the power plant into a channel that carries the effluent back to the river. The passage of water through the plant and down the discharge canal is described by Merriman and Thorpe (2004 [1976]), Merriman (1970), and Jacobson et al. (2004).

Flow in the Connecticut River changes direction at the location of the CY CWIS during periods of low discharge. For this reason, low river flow conditions were chosen for this study to demonstrate the ability of a fully three-dimensional CFD model to simulate tidally influenced river flows affected by the discharge of heated water from a power plant. Table 1 identifies the simulation conditions considered in the analysis.

Methods

Computational Model

All of the simulation results presented in this paper were developed within the framework of the *FLOW-3D* computer software system (Flow Science 2003; version 8.1.1). *FLOW-3D* is a CFD software package developed by Flow Science, Inc. located in Santa Fe, New Mexico. *FLOW-3D* has been commercially available for about 20 years. Originally used to investigate fuel sloshing by the aerospace industry, today the program is used extensively by engineers and scientists engaged in the study of free-surface flows. In fact,

TABLE 1. Simulation conditions.

Parameter	Value
Mean river flow rate	425 m^3/s
Water temperature	30°C
Tidal range	1.5 m
Intake/discharge flow rate	25.5 m^3/s
Discharge water temperature	41°C

FLOW-3D is the only commercially available CFD program that fully implements Hirt and Nichols' (1981) volume-of-fluid (VOF) method for the computation of free-surface flows.

The roots of *FLOW-3D* may be traced back to original developments at the Los Alamos National Laboratory (LANL) beginning in the early 1960s. Many basic numerical techniques originated there for the solution of compressible and incompressible flow problems. Of particular interest are techniques for describing the behavior of free surface flows, a technique for solving both compressible and incompressible flow problems with a single solution method and with new types of grid and geometry models. Variations of these basic algorithms are, today, resident in the *FLOW-3D* software system.

At the time the original CY ecological study was performed, numerical simulations of the kind discussed in this research could not be carried out due to limitations in computing speed and data storage. Today, the typical personal computer (PC) provides enough resources to solve all but the most demanding problems.

Numerical Approximations

Numerical analysis of the kind used in this study involves the solution of governing equations for fluid motion, mass/volume conservation, turbulence (production, advection, and dissipation), energy conservation, and thermal buoyancy. In this study, the Reynolds-averaged Navier-Stokes equations were used as the governing equations for fluid motion. Therefore, the computed results provide estimates of fluid velocities in three-dimensions. Connecticut River water was also assumed to be incompressible (a standard assumption).

The Renormalized-Group (RNG) method was used for turbulence closure (Yakhot and Orszag 1986; Yakhot and Smith 1992). This approach applies statistical methods for a derivation of the averaged equations for turbulence quantities, such as turbulent kinetic energy and its dissipation rate. RNG-based models have wider applicability than the standard two equation k - e model. In particular, the RNG model is known to describe more accurately low intensity turbulence flows and flows having strong shear regions.

An equation for energy conservation was used to account for the transfer of heat from the discharge water to the receiving water, and the influence of buoyant forces associated with density variations in the flow were approximated with a linearized model for buoyant flow (i.e., one that neglects the thermal expansion of the water). All of these features are part of the commercial *FLOW-3D* program.

Analysis

The procedure used to set up the numerical model and to calculate tidally influenced flows in the Connecticut River consists of four basic operations: (1) model setup, (2) definition of boundary conditions, (3) flow simulation, and (4) visualization and analysis. The data required for the analysis is typically available and of no more detail than the data required to carry out more simplified studies (except for the fact that more topographic/bathymetric information will be used, if available).

In this study, topographic/bathymetric data describing the Connecticut River in the vicinity of the CY CWIS were taken from USGS topographic maps, NOAA navigational charts, and detailed bathymetric data provided by Dominion Nuclear Connecticut (DNC). This digital information was imported directly into the CFD program, which transformed the data into a three-dimensional surface model (Figure 1). In the final preprocessing operation a structured, rectangular, grid was superimposed over the surface model and intersections between the grid and the surface model were calculated. A FAVOR method was then used to assign blockages to cells occupied in full or in part by "solid ground" (Hirt and Sicilian 1985). The completed mesh consisted of about 260,000 control volumes with horizontal dimensions of 30 m and vertical dimensions of 0.6 m (in the main body of the river).

Conditions on the flow entering and leaving the model domain (i.e., the computational mesh) were assigned at open boundary locations. These included the up and downstream limits of the modeled region, the location of the CWIS, and the location of the heated water discharge. Table 2 provides a list of conditions specified at each open boundary.

Parameter values, required for the specification of boundary conditions, are normally provided by plant operating records (CWIS and discharge flow rates and temperature), in situ flow measurements (e.g., USGS gauging station data),

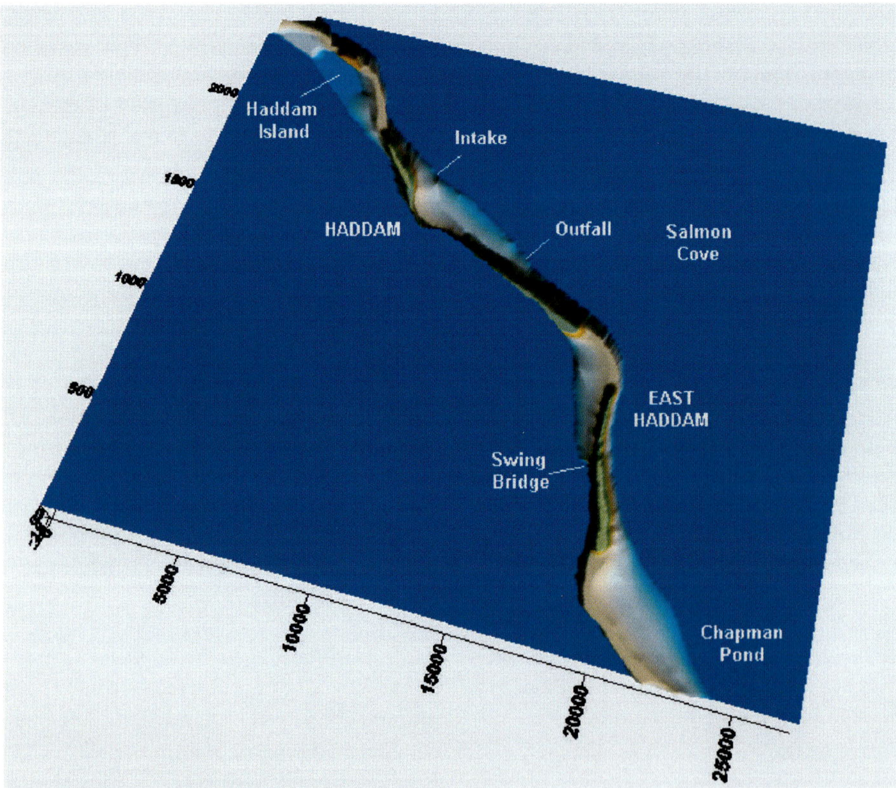

FIGURE 1. Digital terrain model (grid not shown for clarity).

or the results of other larger scale flow calculation (e.g., more detailed simulation grids can be nested within larger more coarsely defined two- or three-dimensional grids—this is referred to as grid nesting). In this study, conditions at the upstream and downstream river cross sections were derived from the results of two-dimensional tidal flow simulations. That is, time-series of water surface elevations and velocities at locations corresponding to the upstream and downstream river cross-sections were determined from the results of the two-dimensional flow calculations and used to "force" the flow through the three-dimensional model.

With the computational mesh constructed and boundary conditions defined, numerical simulations of river flows and the movement of passive organisms through the modeled region were performed. In this study, three-dimensional velocities in a 6.5-km reach that surrounds CY were calculated over a period of seven successive tidal cycles (about 3.5 d). To determine the HZI of the CY CWIS, scalar quantities, representing concentrations of passive organisms, were released into the flow from different locations upstream of the power plant. An advection/dispersion algorithm was used to calculate the movement of the "numerical surrogates" past the CWIS towards the open ocean. The concentration of organisms in the intake flow was monitored during the simulations. At the conclusion of the simulations, the percentage of organisms entrained by the CWIS could be calculated since the total organism population at the beginning of the simulations was known. In some respects, this type of analysis is similar to that used to determine the movement and extent of thermal plumes, only in reverse. That is, scalar quantities are released from

TABLE 2. Boundary conditions.

Upstream/downstream river cross sections	• Velocity • Water surface elevation • Water temperature
CWIS	• Flow rate (sink)
Discharge canal	• Flow rate (source • Water temperature

remote locations far from the CWIS and the amount of material entering the CWIS is calculated. This is in contrast to a thermal plume analysis, where a scalar quantity (heated water) is released from a single discharge point and the evolution of the thermal plume in surrounding waters is estimated.

The results of the studies were postprocessed after the computer calculations were complete and used to promote subsequent analyses. One of CFD's greatest strengths relates to its ability to provide detailed datasets that can later be "visualized." Figure 2 shows a view of the Connecticut River from above. In this figure, the bathymetry of the river is colored by elevation, and red "wire-meshing" is used to delineate the extent of the thermal plume defined as a surface of constant temperature (both, the temperature used to define the plume and the color scheme used to color bed were arbitrarily chosen). Visual renderings, such as the ones shown in Figure 2, can be used to explain complex flow phenomena and to aid in the interpretation of field data. Standard plot types include vector diagrams, streamline plots, and contour plots colored by scalar quantities such as speed, temperature, and concentration.

The quantitative data produced by the simulations was used to determine where the source water that enters the CWIS comes from and to determine the likelihood of entrainment of passive organisms coming from remote locations. Additional what-if scenarios could also be performed

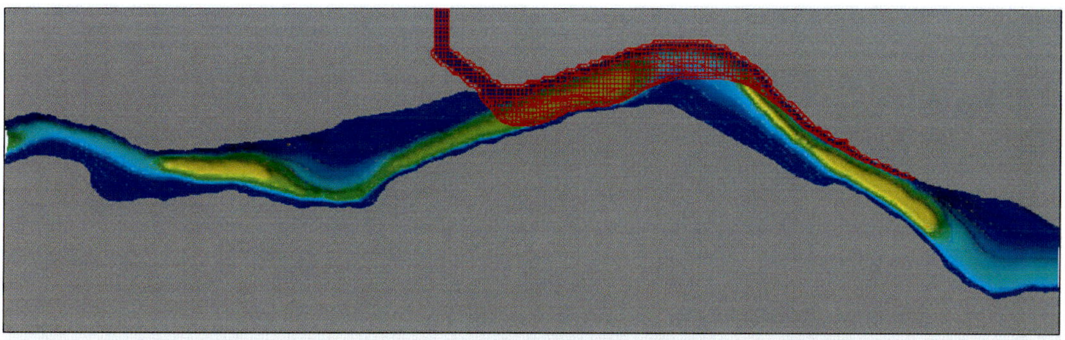

(a)

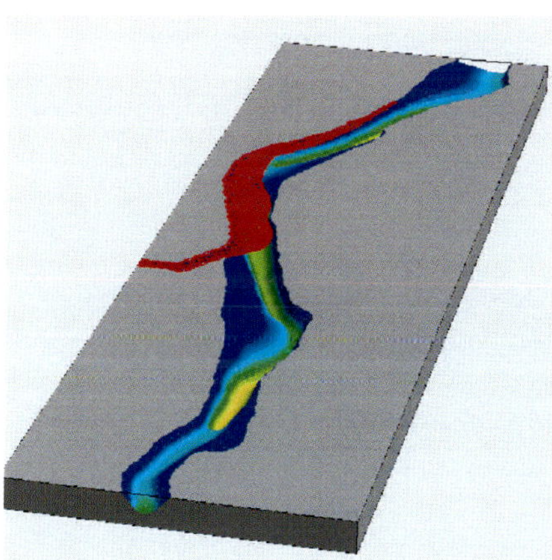

(b)

(c)

FIGURE 2. Thermal plume, low tide slack water: (a) plan-view, (b) oblique-view, and (c) close-up showing area of cold water beneath the thermal plume (river bottom colored by depth, thermal plume is red)

to determine how entrainment responds to changes in CWIS location and design.

Results

Hydraulics

To demonstrate the ability of a commercial CFD program to simulate time-varying flows in the Connecticut River, a low-flow condition was chosen to accentuate the effect of the tide on local flow conditions (see Table 1). At this location, the effects of the thermal plume must be accounted for, since the presence of the thermal plume affects flow patterns in the vicinity of the CY CWIS during flood tides for this low-flow condition. Figure 3 shows the extent of the thermal plume at low slack water.

The size, shape, and location of the thermal plume compares closely with infrared photographs provided by Merriman (1970). The results of CFD analyses can also be compared to other more detailed information. For example, velocity distributions measured with the latest generation of acoustic Doppler current profilers (ADCP) data can be used to develop boundary conditions for CFD analyses.

The results of CFD analyses can be shown in a form similar to that used to present data collected with modern field equipment. Figure 4 shows calculated time histories of organism concentration, water temperature, and local velocities near the CWIS. These time series are similar to those that can be produced by measurements made with velocimeters, thermometers, and fluorometers. It is relatively easy to compare the results of CFD analysis with field data since many measured quantities are computed explicitly in the numerical simulations.

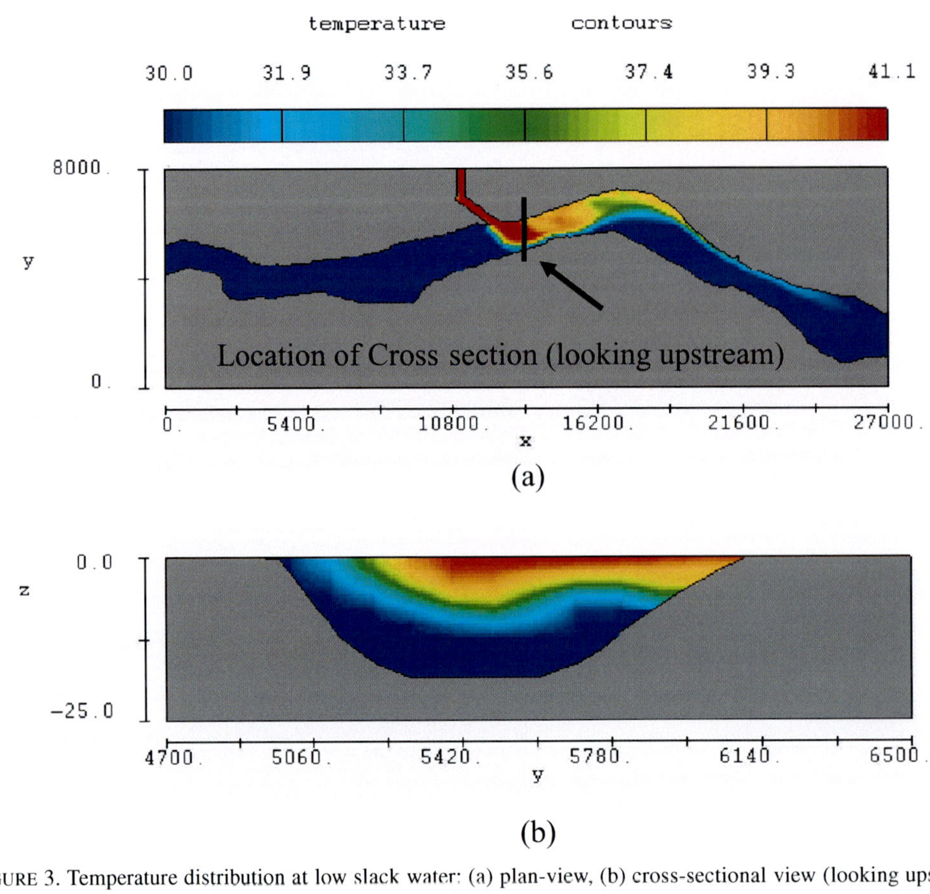

FIGURE 3. Temperature distribution at low slack water: (a) plan-view, (b) cross-sectional view (looking upstream; colored by temperature—degrees Celsius, length units—feet)

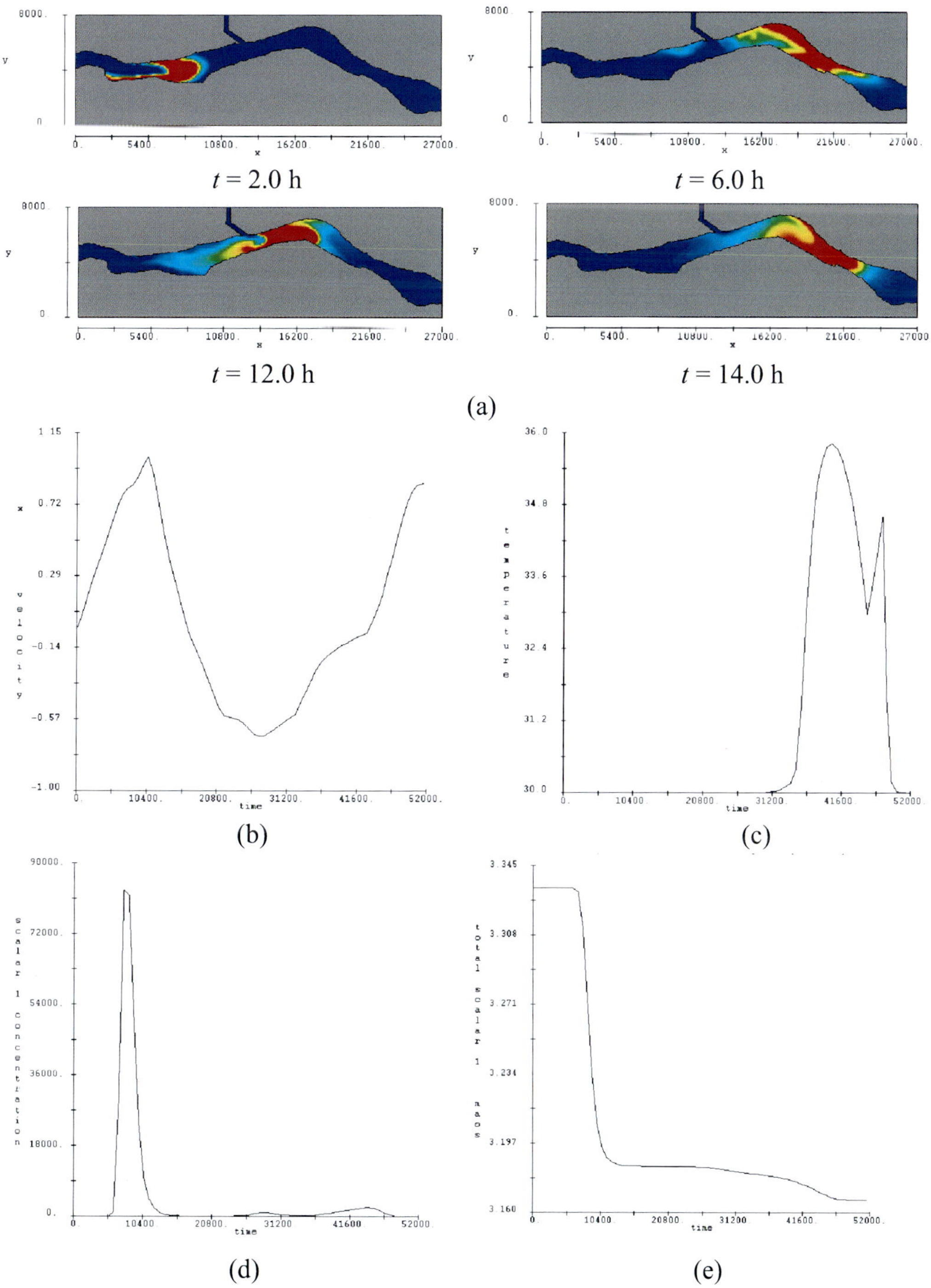

FIGURE 4. Computed movement of organisms past CY – CWIS: (a) concentration of organisms, (b) velocity in direction of flow near intake, t = s, (c) water temperature near intake, degrees Celsius, (d) concentration of organisms near intake, and (e) total mass of organisms in computed region (units of concentration and total mass are arbitrary).

Hydraulic Zone of Influence

The HZI of a CWIS is defined as the region in a water body where the probability of entrainment (i.e., passage of a passive or somewhat passive particle, such as fish eggs and larvae, into and through the power plant cooling water condenser system) is high. For a condition where the flow does not change with time, the HZI is defined by the area from which the CWIS withdraws water directly. For example, if a CWIS is located on a river where the flow is steady, then the HZI can be shown by identifying streamlines of flow that enter the CWIS. Figure 5 shows the HZI for the Tennessee Valley Authority Browns Ferry Nuclear Power Plant Case Study delineated for a steady-flow condition (EPRI 2004). In this figure, streamlines that enter the CWIS are colored red and the volume of water occupied by the streamlines is, in fact, the HZI of the CWIS for this flow condition.

When a CWIS is located on a water body where the direction of flow changes with time, the HZI of the CWIS can no longer be visualized with streamlines. Entrainment percentages for the Dominion Millstone nuclear power plant (Waterford, Connecticut on Long Island Sound) are shown in Figure 6 (EPRI 2004). To identify the HZI of the CWIS at this coastal location, known concentrations of scalar quantities (tracer material) were released into the flow at 25 different locations. The amount of tracer material that entered the CWIS (located on the eastern shore of Niantic Bay near Millstone Point) was calculated for each release. The entrainment percentage was then expressed as a ratio between the amount of material that entered the intakes and the total amount of material released at each location. The results in Figure 6 show that majority of flow entering the CWIS at Millstone originates in Niantic Bay rather than Long Island Sound.

Since the direction of flow in the river changes twice a day, for the modeled conditions, the HZI

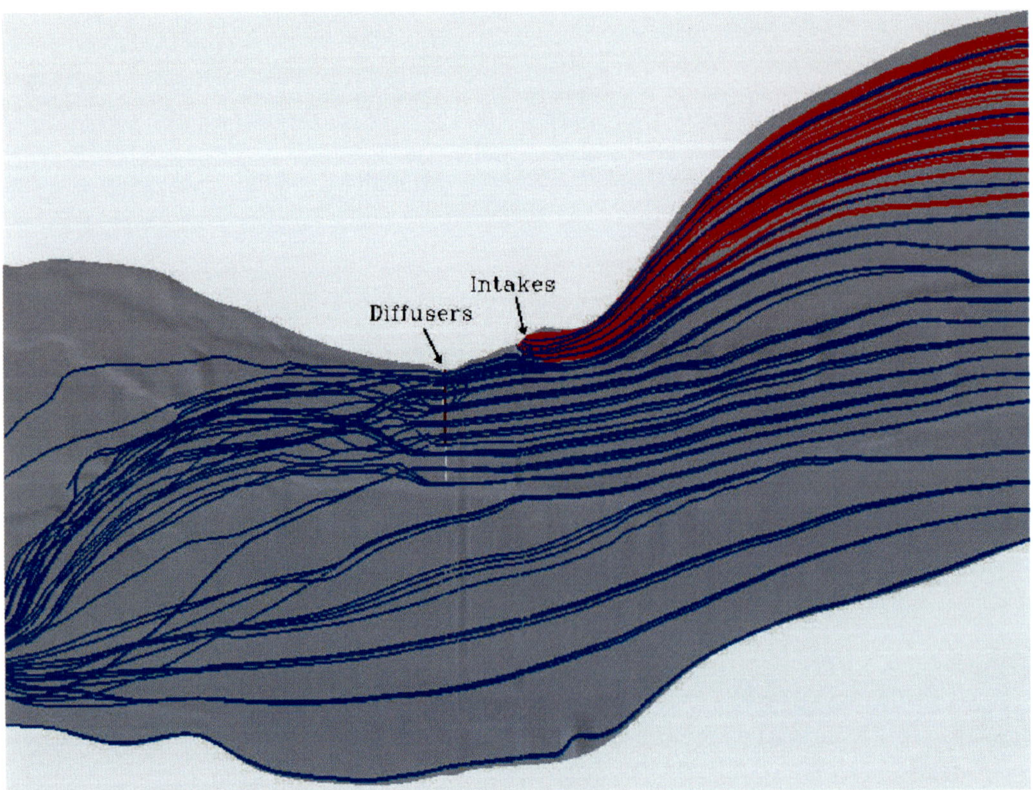

FIGURE 5. Hydraulic zone of influence, Tennessee Valley Authority Browns Ferry nuclear power plant on Lake Wheeler, an impounded portion of the Tennessee River (EPRI 2004)

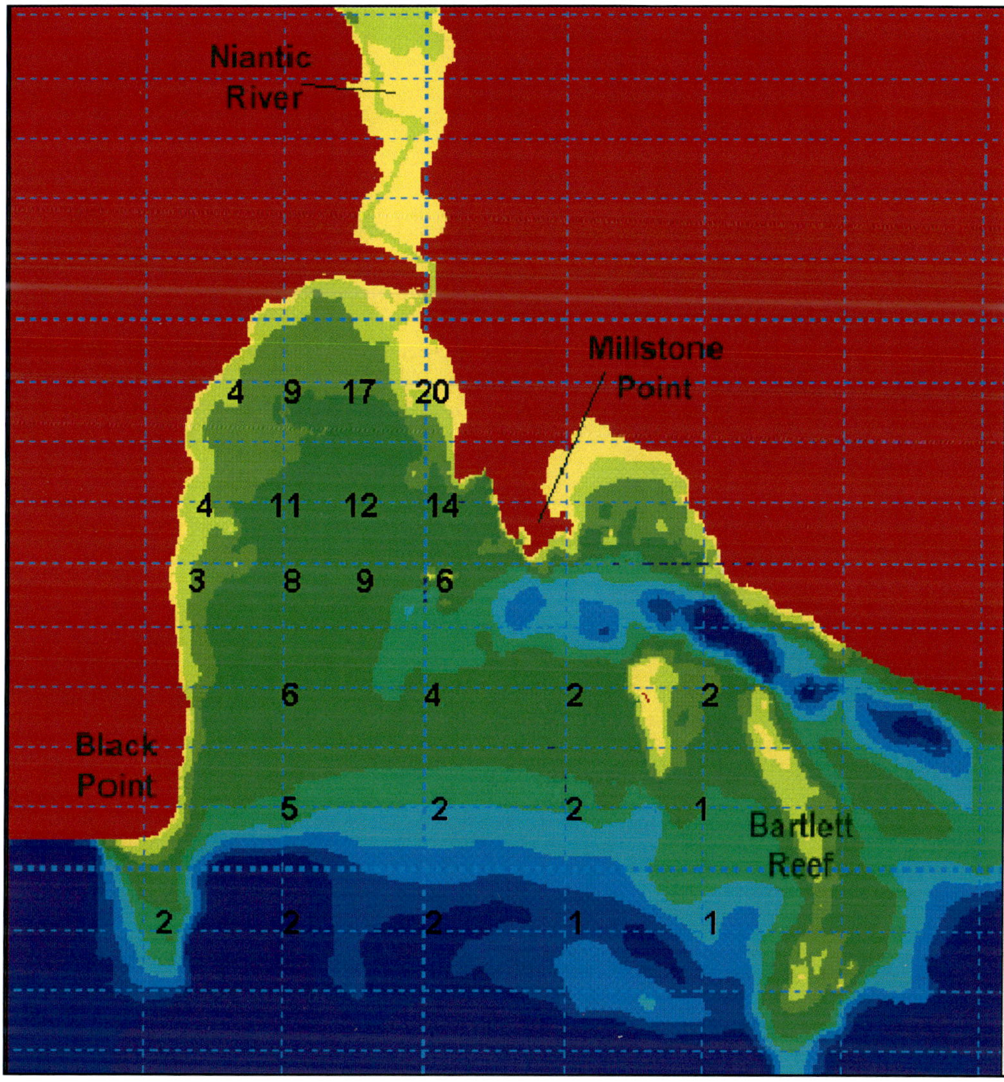

FIGURE 6. Hydraulic zone of influence, Dominion Millstone nuclear power plant, Waterford, Connecticut on Niantic Bay and Long Island Sound (entrainment percentage calculated 6d after release) (EPRI 2004).

for Connecticut Yankee's CWIS could not be effectively visualized with streamlines. For this reason, the HZI for Connecticut Yankee is better illustrated as a mapping of entrainment percentages similar to the one developed for Millstone.

Figure 7 shows the HZI of the CY CWIS. To identify the HZI at these two river cross sections, known concentrations of scalar quantities were released into the river at 19 different locations. The amount of "material" that entered the plant intakes from each location was calculated. Entrainment percentages were then expressed as the ratio between the amount of material that entered the intakes and the total amount of material released at each location.

At the cross section closest to the CWIS, the results in Figure 7 show that more water is withdrawn from the same side of the river as the intakes are located. Somewhat more water is also withdrawn from locations near the surface of the river. At the upstream cross section, the difference between the amount of water withdrawn from the surface and the amount of water withdrawn from the bottom is less (i.e., the results show vertical

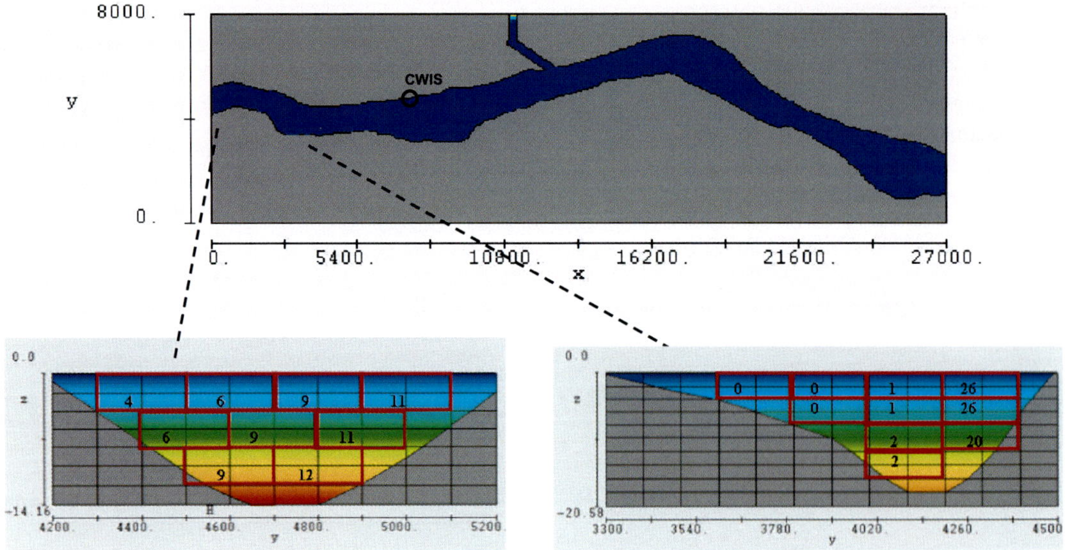

FIGURE 7. Hydraulic zone of influence, Connecticut Yankee nuclear power plant (entrainment percentage).

mixing). However, a difference in results across the channel still exists. If one were to produce similar results from far enough upstream, then even this lateral difference would disappear (i.e., at this new location the distance between the release point an the intakes would be great enough to mix the flow completely.

The techniques used to show the HZI of a CWIS can also be used to draw certain conclusions about the entrainment of organisms. For example, if a population of organisms was uniformly distributed in a body of water (and if they were neutrally buoyant and passive), then the entrainment of those organisms would be the same as the percentage of water removed by the CWIS. However, if a population of organisms was not uniformly distributed, then the entrainment of those organisms could be significantly different from the percentage of water removed by the CWIS. In the case of Connecticut Yankee, if a population of organisms spawned from a river location opposite the CWIS location and not far upstream of the plant, minimal numbers of these organisms would be entrained by the plant intakes (whereas, one might be inclined to be more concerned about the proximity of the spawning ground to the CWIS without the benefits of this type of study). It should also be noted that the results of this study compare well with Marcy (2004 [1976], this volume) and Massengill's (2004 [1976]) ecological sampling and entrainment studies conducted on the Connecticut River. Their results suggest that greater quantities of water are withdrawn from the side of the river where the CWIS is located—just as the numerical solutions do.

Discussion

Owing to continued advances in computer technology, detailed three-dimensional simulations of complex environmental flows can now be performed with desktop computing systems. The results of these analyses provide insights regarding the cause-and-effect relationships that govern the movement of water through natural systems and into CWIS. The programs that enable this type of study have a strong theoretical basis and many were, in fact, under development at the time of the original Connecticut River Ecological Study. Thirty years later, these programs are referred to commercially as CFD software packages.

CFD software packages are designed to be "general purpose" flow solvers and can be applied to the study of a great many different kinds of fluid dynamics problems. As applied to once-through cooling systems, a single CFD analysis can be used to determine the HZI of a CWIS and to study the behavior of discharge structures (i.e., thermal plume analysis) for power plants located on any type of water body.

If the Connecticut Yankee nuclear power plant were designed and sited today, CFD tools could be used to

a. *Perform what-if scenarios:* to site the intake and discharge structures and to assess differences in the performance and potential impact of different designs,
b. *Optimize field data collection programs:* to target important areas for baseline sampling within the HZI,
c. *Improve the interpretation of field-data:* CFD analysis can provide discrete time series information similar to field data and continuous time series information for the entire computed region,
d. *Evaluate the potential impact of plant operations on threatened or endangered species located near an intake*,
e. *Estimate the entrainment of organisms migrating from far-field locations*, and
f. *Augment TMDL studies:* to improve the quality of information required for effective decision-making (see Chen et al. 2004, this volume).

With the HZI of a CWIS defined, the next step in the evolution of this type of analysis will be to include environmental and behavioral realism (e.g., diurnal variations in organism abundance, swimming ability, and organism behavior).

In the next 30 years, the use of CFD to aid in the assessment of environmental impact should increase dramatically due to the still rapid improvements in microchip processing speeds, the ability to use the results of CFD studies to explain complex flow problems, the results of new biological research, and the continued development of the analysis tools by dedicated developers.

Acknowledgments

The contributions of Don Danila, Paul Jacobson, and Chris Tomichek at the Millstone Environmental Laboratory are gratefully acknowledged as well as the review comments provided by C.W. (Tony) Hirt and his colleagues at Flow Science, Michael Chelminski at Woodlot Alternatives, and Ted Chang at Texas A&M. Ned Taft's (Alden Research Laboratory) guidance on this project is also recognized.

References

Chen, C. W., R. A. Goldstein, and J. Herr. 2004. Feasibility of applying a watershed model to the Connecticut River for total maximum daily load analysis and watershed planning. Pages 501–511 *in* P. M. Jacobson, D. A. Dixon, W. C. Leggett, B. C. Marcy, Jr., and R. R. Massengill, editors. The Connecticut River Ecological Study (1965–1973) revisited: ecology of the lower Connecticut River 1973–2003. American Fisheries Society, Monograph 9, Bethesda, Maryland.

EPRI (Electric Power Research Institute). 2004. Using computational fluid dynamics techniques to define the hydraulic zone of influence of cooling water intake structures. Electric Power Research Institute, Report 1005528, Palo Alto, California

Flow Science. 2003. *FLOW-3D*® User's Manual version 8.1.1. Flow Science, Santa Fe, New Mexico.

Hirt, C. W., and B. D. Nichols. 1981. Volume of fluid (VOF) method for the dynamics of free boundaries. Journal of Computational Physics 39(1):201–225.

Hirt, C. W., and J. M. Sicilian. 1985. A porosity technique for the definition of obstacles in rectangular cell meshes. Fourth International Conference on Numerical Ship Hydrodynamics, Washington, D.C, September 1985.

Jacobson, P. M., C. Tomichek, D. J. Danila. 2004. Twenty years of impingement history: Connecticut Yankee Haddam Neck nuclear power plant. Pages 455–473 *in* P. M. Jacobson, D. A. Dixon, W. C. Leggett, B. C. Marcy, Jr., and R. R. Massengill, editors. The Connecticut River Ecological Study (1965–1973) revisited: ecology of the lower Connecticut River 1973–2003. American Fisheries Society, Monograph 9, Bethesda, Maryland.

Marcy, B. C., Jr. 2004. Planktonic fish eggs and larvae of the lower Connecticut River and the effects of the Connecticut Yankee plant, including entrainment. Pages 125–153 *in* P. M. Jacobson, D. A. Dixon, W. C. Leggett, B. C. Marcy, Jr., and R. R. Massengill, editors. The Connecticut River Ecological Study (1965–1973) revisited: ecology of the lower Connecticut River 1973–2003. American Fisheries Society, Monograph 9, Bethesda, Maryland. (Originally published in 1976)

Massengill, R. R. 2004. Entrainment of zooplankton at the Connecticut Yankee plant. Pages 59–64 *in* P. M. Jacobson, D. A. Dixon, W. C. Leggett, B. C. Marcy, Jr., and R. R. Massengill, editors. The Connecticut River Ecological Study (1965–1973) revisited: ecology of the lower Connecticut River 1973–2003. American Fisheries Society, Monograph 9, Bethesda, Maryland. (Originally published in 1976)

Merriman, D. 1970. The calefaction of a river. Scientific American 222(5):42–52.

Merriman, D., and L. M. Thorpe. 2004. Introduction. Pages 3–25 *in* P. M. Jacobson, D. A. Dixon, W. C. Leggett, B. C. Marcy, Jr., and R. R. Massengill, editors. The Connecticut River Ecological Study (1965–1973) revisited: ecology of the lower Connecticut River 1973–2003. American Fisheries Society, Monograph 9, Bethesda, Maryland. (Originally published in 1976)

Yakhot, V., and S. A. Orszag. 1986. Renormalization group analysis of turbulence. I. Basic theory Journal of Scientific Computing 1:1–51.

Yakhot, V., and L. M. Smith. 1992. The renormalization group, the e-expansion and derivation of turbulence models. Journal of Scientific Computing 7:35–61.

American Fisheries Society Monograph 9:525–538, 2004

Summary and Conclusions

PAUL M. JACOBSON*[,1]
Ecological Assessments, LLC, Post Office Box 376, Old Saybrook, Connecticut 06475, USA

WILLIAM C. LEGGETT[1]
Department of Biology, Queen's University, Kingston, Ontario, K7L 4V1, Canada

BARTON C. MARCY, JR.[1]
Westinghouse Savannah River Company, Environmental Services, Building 742-A Post Office Box 616, Aiken, South Carolina 29808, USA

RONALD R. MASSENGILL[1]
2434 Trenton Street, Cambria, California 93428, USA

Summary

Introduction: The Connecticut River Ecological Study Monograph "Revisited"

This "revisited" monograph is divided into two parts. Results of the Connecticut River Ecological Study (the "Study"), published in 1976 as American Fisheries Society Monograph Number One (Merriman and Thorpe 1976), are reprinted herein (Part I) to facilitate comparison with the results of subsequent 30 years of research on the ecology of the river as reported in Part II of this monograph. Papers providing overviews of advances in analytical impact assessment tools previously unavailable during the Connecticut River Ecological Study are also contained in Part II. This summary provides a historical perspective authored by the original investigators.

The genesis of this revisitation of the ecology of the Connecticut River, and the conclusions of the Study, is to be found in a workshop sponsored by the Electric Power Research Institute (EPRI) held at the University of Connecticut, Avery Point Marine Sciences Campus, in November 2001 (EPRI 2001). The purpose of the workshop was to assess the degree to which the conclusions of the original study were supported by information collected subsequent to its completion. The information presented there suggested that the general conclusion of the Study (i.e., that the Connecticut Yankee Haddam Neck nuclear powered electric generating station [CY] operation had a very limited temporal and spatial impact on the ecology of the lower Connecticut River) was correct. At the same time, a review of the U. S. Environmental Protection Agency's (EPA) recently promulgated Clean Water Act (CWA) §316(b) rule (Phase I) regulating cooling water intakes at new facilities indicated that the CY's original design, location, and capacity would not meet the new technology standards (EPA 2001). Consequently, under existing rules, it could not be built and operated in its original configuration.

Subsequently, EPA published CWA §316(b) standards for determining Best Technology Available for minimizing adverse environmental impact at existing (Phase II) cooling water intakes (EPA 2004). These regulations establish performance standards as a percentage reduction in entrainment of organisms in the cooling water system and impingement of larger specimens on the intake screens. Standards vary by water body type and capacity utilization of the power station. However, the final rule allows for site-specific determinations that performance standards set a priori in the rule are more stringent than

* *Corresponding author*: jacobsonpaul@msn.com
[1] Authorship determined alphabetically.

necessary given a comparative assessment of the cost of alternatives and the relative biological or ecological benefit. Such determinations inherently require a detailed understanding of the ecology of the source waters, including information on the population dynamics of key species and the relationship between aquatic assemblages and cooling water intake location.

It was within this context that we concluded that the Study experience provided a unique window through which to review and reassess what began as North America's first large-scale ecological impact assessment experiment. The question posed, and examined further in this monograph, was whether, upon reflection, the ecological studies conducted over the three decades that have elapsed since the termination of the Study in 1973 provide important insights into the degree to which the original study correctly identified and evaluated the potential long-term ecological impacts of the CY facility and the extent to which subsequent conservation efforts have affected the ecology of the Connecticut River.

The Connecticut River Ecological Study

The Study was carried out during the years 1965 to 1973. It examined the potential impact of the operation of CY on the ecology of the lower Connecticut River. The Connecticut Yankee Haddam Neck nuclear powered electric generating station, the largest nuclear plant in the world at that time, began commercial operation in 1968 and was shut down in July 1996. Decommissioning was begun in December 1996. No comprehensive review of the Study conclusions regarding the immediate and probable long-term impacts of the operation of CY has been conducted since its completion in 1973.

There now exists a considerable body of literature on power plant cooling water impact assessments (EPRI 1999a, 1999b, 2002; Wisniewski 2000; Dixon et al. 2003). However, in 1965, at the initiation of the Study, no such information was available. The Study was, therefore, the first to evaluate the impact of large-scale heated water discharge and cooling water withdrawal on the ecology of a tidal river. It was also unique by way of its origin, scientific foundation, and comprehensive scope. The Study's 3-year preoperational and 5-year operational time-frame allowed for development of a baseline description of the ecology of the lower Connecticut River against which data collected during operations were compared.

The Study provided fundamental, new information on the spawning of anadromous clupeids and of many key resident species (Leggett 2004 [1976]; Marcy 2004 [1976]; both this volume). This information included the timing and distribution of eggs and larvae during dispersion following spawning. Biotic and abiotic factors influencing the timing, movement, and energetics of adult American shad *Alosa sapidissima* migration along the Atlantic Coast and in the Connecticut River were investigated for the first time (Leggett 2004 [1976]). The benthic invertebrate communities of the lower river were described for the first time, and their relationship to habitat types was determined (Massengill 2004a [1976], this volume). And, while an inventory of resident fish species throughout Connecticut had been developed (Whitworth et al. 1968), a comprehensive list of species present, their relative abundance, distribution, and movements in the lower Connecticut River, which were previously unknown, were documented for the first time (Marcy 2004 [1976]). Reflective of additional research conducted since the 1965–1973 study, an updated species list of fishes is herein included as Appendix I of this volume.

A key feature of the Study was its continuity and scientific underpinnings. While many special studies, descriptive in nature, such as the determination of fish egg and larval distribution at locations significantly upstream of CY (75 km), were completed prior to the termination of the Study, the basic impact assessment sampling, which conformed to what became known later as the BACI (Before–After, Control–Impacted) design (Green 1979), remained unchanged throughout. This made possible quantitative comparisons between stations over time both before and after the plant became operational.

Sampling design and the analytical approaches used to assess the data generated were overseen by an independent Ecological Advisory Committee (EAC) composed of internationally recognized scientific experts in relevant aquatic and marine disciplines (see page 16, Merriman and Thorpe 2004b [1976], this volume). This committee met biannually throughout the Study, assisted with the development of the scientific program, provided peer review of reports and publications, and evaluated study conclusions. Most investigations of the abundance and ecology of the resident and anadromous fishes of the river also benefited from the active cooperation and involvement of personnel of the Connecticut State Board of Fisheries and Game.

The independence and scientific rigor of the study are reflected in the fact that more than 50 peer-reviewed papers describing the work undertaken were published in scientific journals (Appendix II of this volume). This both established the credibility of the inquiry and provided a significant database for use in subsequent assessments of the river's ecology and the impacts of plant operation.

The sampling design employed in the Study focused on what were, at the time, believed to be the most probable impacts of the three-dimensional temperature characteristic of the thermal plume. Initially, these included the potential for blockage or delay of the upstream migration of shad past CY to spawning areas located farther upstream and their return downriver as juveniles, the possible displacement of resident fish from the plume area and related thermal impacts on growth and reproduction, and changes to the benthic communities occupying the shallow reaches immediately surrounding the CY discharge.

However, because the Study had, as a guiding principle, a commitment to ensuring that participating scientists, aided by the EAC, had the freedom to respond to emerging areas of scientific interest and need and to design and execute studies required to fill gaps in the initial scope of work, numerous special studies were executed in response to our evolving understanding of the impact of plant operation. These included analyses of impacts related to entrainment of zooplankton and fish eggs and larvae and to the impingement of larger fish. Impingement sampling was initiated in 1968 and studies of the entrainment of zooplankton and fish eggs and larvae in 1970.

Early in the Study, it also became evident that thorough evaluation of the effects of CY operation would require that the data being generated by the monitoring programs be evaluated in the light of a more detailed population specific understanding of the ecology of key species. The result was a series of studies that examined the life history of American shad, not only in the Connecticut River, but along its entire Atlantic coastal range, and of white perch *Morone americana* and other key resident fish species. This program flexibility and the scientific freedom and encouragement given to the Study's principal investigators were critical to the comprehensive nature of the Study's findings and conclusions and to its success.

Significance of the Updated Monograph

The goal of this updated monograph is, therefore, to revisit and evaluate the Study's conclusions regarding the potential short and long-term impacts of CY operation on the ecology of the Connecticut River and to serve as a guide for resource managers faced with developing regulatory and conservation strategies.

Despite the considerable shoreline and riverine development that has occurred throughout northeastern United States in the last three decades, the lower Connecticut River valley remains largely undeveloped and retains much of its original character and beauty. This is in large part due to intensive efforts by many individuals and agencies that have worked to preserve its largely rural characteristic and its natural and relatively undisturbed biotic communities. In addition to the very significant water quality improvements (Mullaney 2004, this volume) resulting from measures taken during the last 30 years in response to the Clean Water Act, the river has been the beneficiary of numerous conservation designations, including its status as an American Heritage River, a Nature Conservancy "Last Great Place," and a part of the Silvio O. Conte National Fish and Wildlife Refuge. In 1994, the tidal wetlands of the lower river were also selected as one of the nation's Wetlands of International Importance under the Ramsar Convention (Caplis 2001; Maloney et al. 2001; Wetherell 2002).

These designations and corresponding preservation efforts lend added importance to the development of a comprehensive understanding of the current ecological status of the river and of the efficacy of conservation and regulatory initiatives past and present.

Connecticut Yankee Regulatory History

The origin of the Connecticut River Ecological Study predated the current regulatory framework found within §402, §316(a), and §316(b) of the CWA. The Study owed its existence to a mandate by the Connecticut Water Resources Commission that such investigations be undertaken as part of an approval to dredge and construct the cooling water intake. In 1974, a §316(a) determination that more stringent control measures limiting the size and temperature of the thermal discharge were unwarranted was made on issuance of the first National Pollutant Discharge Elimination System (NPDES) permit. This determination was based on the results of the Connecticut River Ecological Study. Subsequently, the results of the Study together with impingement monitoring data collected at CY through 1982 combined with the installation of fish protection devices, notably an intake behavioral barrier and a fish return system (Jacobson et al. 2004b, this volume), led the Connecticut Department of Environmental Protection (CTDEP) in 1986 to issue a §316(b) determination that the once-through cooling water system at CY represented best technology available for minimizing adverse environmental impact. This occurred upon renewal of the NPDES permit (CTDEP 1986).

The findings of the Study, in combination with mitigating measures taken by CY with respect to impingement, contributed to decisions by the Nuclear Regulatory Commission (then the AEC) and the CTDEP to authorize continued operation of CY's once-through cooling water system. These authorizations proceeded without controversy, in stark contrast to that which surrounded the development of other well-known power projects such as those being developed during the same era on the lower Hudson River (Barnthouse et al. 1988) and in the Delaware River (Barnthouse et al. 2003). There, resolution of differing stakeholder views of the potential adverse environmental impact (AEI) of plant construction and operation were, and remain, difficult and controversial.

The scientific and regulatory successes associated with the Study added to our conviction that the Study, as updated in this monograph, could serve to inform the ongoing dialog regarding the uncertainties associated with assessment of ecological impacts related to cooling water withdrawals and the relative efficacy of alternative technologies or performance measures being advanced with the objective of protecting aquatic resources (Nagle and Morgan 2000; Dixon et al. 2003; Super and Gordon 2003).

The Connecticut Yankee Power Plant

CY was a single-unit 600 MWe pressurized water reactor. When it began operation, it was the largest nuclear power plant in the world. The cooling water intake, located on the eastern shore of the Connecticut River in Haddam Neck, Connecticut, consisted of four intake bays, each about 3.6 m in width and 5.5 m deep below mean low water. The intake provided approximately 24,220 L/s of combined condenser cooling and service water flow. The intake velocity through the 9.5-mm-mesh intake traveling screens was approximately 61 cm/s. The cooling water discharge exited the plant into a discharge canal that rejoined the river 1.8 km downstream of the intake structure. The delta T (i.e., the added thermal load to the ambient water temperature) across the plant was 12.5°C at full power.

The three-dimensional character and distribution of the thermal plume was influenced by the seasonal changes in river flow. The spring average daily flow in the Connecticut River during the 8-year period of the Study was approximately 700 m^3/s (Merriman and Thorpe 2004a [1976], this volume). During these freshet conditions, the plume extended downstream along the eastern shore of the river opposite the navigation channel located along the western shore of the river. When freshwater flow is at seasonal lows in late summer and fall (about $140 m^3/s$), the plume moved both downstream and upstream in response to tidally driven currents. In this reach of the river, variation in river height due to tide averages 0.76 m (Merriman 1970; USAEC 1973; Merriman and Thorpe 2004a [1976]). During low seasonal river flows, about 7.7% and 6.1% of the flood and ebb tidal excursions, respectively, were estimated entrained by CY.

Conclusions

Insights Gained from 37 Years of River Studies

The ecology of the Connecticut River has undergone dramatic changes during the 30 years that have elapsed since the termination of the Study. The available evidence indicates that the river has responded in a highly dynamic manner to species-specific and long-term water quality management actions. The evidence also points to the occurrence of major natural changes such as the occurrence and subsequent dominance of invasive species (Morgan et al. 2004, this volume). Water temperature increases (Jacobson et al. 2004a, this volume) that may or may not be natural appear to be consistent with trends found throughout the North Atlantic (Strong 1989; Barnett et al. 2001; Ottersen et al. 2001; Keser et al., in press). In some instances, observed changes in the ecology of the Connecticut River are so dramatic as to overwhelm perturbations linked to direct anthropogenic effects. Most importantly, when evaluated some 30 years later, in the context of significant improvements in analytical tools, the original Study conclusions regarding the level of impact have been largely confirmed. There is no evidence of any long-term impact of the operation of CY on the ecology of the lower Connecticut River.

General Ecological Trends and Long-Term Changes in the Dynamics of the River

Among the key attributes of the ecology of the lower river that have changed over the past 37 years are

- a measurable improvement in water quality;
- a trend of increasing water temperature;
- changes in relative dominance among benthic invertebrate species due to the invasion of the Asiatic clam (also known as Asian clam) *Corbicula fluminea* contrasted by an overall increase in density and number of benthic taxa;
- a recovery of the submerged aquatic plant (*Vallisneria* sp.) in the plume area before and subsequent to plant shutdown;
- a decrease in abundance of the migratory alosids (e.g., American shad, alewives *A. pseudoharengus*, and blueback herring *A. aestivalis*) believed to be due to predation associated with increased striped bass *Morone saxatilis* abundance— the result of restoration efforts directed toward this species along their Atlantic coastal range;
- the occurrence of alosids such as gizzard shad *Dorosoma cepedianum* and hickory shad *A. mediocris* typically associated with the mid-Atlantic and southern rivers along the eastern U.S. coast;
- a shift in the relative abundance of catfishes, with white catfish *Ameiurus catus* and channel catfish *Ictalurus punctatus* now dominating where bullheads *A. nebulosus* once prevailed;
- a change in the population structure and stability of the population of American shad as a consequence of the upriver displacement of spawning to sites above Holyoke Dam following the expansion of fishway capacity there and beyond; and
- the importance of the lower Connecticut River as essential habitat for the endangered shortnose sturgeon *Acipenser brevirostrum* has been clarified.

Collectively, these observations have important implications for future regulatory and resource management and, as a result, identify additional long-term monitoring and research needs.

Acquiring Baseline Information and Establishing Study Design

This retrospective review of the Study has also provided important insight into the strengths and weaknesses of the original study, insight that may prove useful in the design and execution of future impact assessments within the Connecticut River watershed and elsewhere. Perhaps the most important attribute of the Study is that it described, for the first time, basic ecological information that was previously unknown. If we are to effectively evaluate and understand the potential for long-term ecological changes resulting from anthropogenic impacts, the assessment must be carried out within the framework of information derived from classic and, in many cases, descriptive knowledge of the hydrological and biological characteristics of the environment in question.

The design of the original investigations conducted at CY was determined, in large part, by predicted thermal plume distributions. Sampling stations for benthic infauna and resident fishes were established within the predicted plume distribution and control sites were located upstream and downstream of the affected areas at locations that were, to the maximum degree possible, similar in depth, substrate, and other important habitat characteristics. This was done to facilitate comparison of the ecological differences between control and affected stations both before and during operations. This BACI design (Green 1979; Saila and Lorda 2004, this volume) allowed for temporal and spatial comparison of ecological changes, bound by the geographic extent and time span of the sampling.

Subsequent investigations of the habitat and species dynamics in regions of the river above and below the area investigated in the original Study have revealed a dynamic mosaic of habitats and animal communities. In retrospect, a more expansive study of the river's ecology that encompassed these diverse habitats located well downstream and upstream of CY would have yielded a more comprehensive insight into the ecology of the river and its variety of aquatic flora and fauna and would have enhanced the quality of the assessment of the overall impact of plant operation. In particular, such studies would have allowed an assessment of the degree to which the habitat in the CY reach of the river was otherwise unique.

To provide a specific example, knowledge of the existence and habitat use of the shortnose sturgeon was noticeably absent from the original study—a product of the limited range of the sampling area investigated. As a consequence, the importance of critical sturgeon habitat in the lower estuary during and after freshet was not understood, and the potential impact of plant operation on their migration past CY to upriver areas of concentration was not assessed. Subsequent impingement studies (Jacobson et al. 2004b) coupled with new information on sturgeon habitat developed in later studies (Savoy 2004; Savoy and Benway 2004; both this volume) suggest that, fortuitously, CY had little or no impact on the long-term well being of this species.

We now also have a better appreciation for the importance of the varied wetland habitats located in the tidal reaches of the lower Connecticut River, and while it would have been difficult to infer a relationship between plant operational impacts at CY, these habitats and their contribution as nursery areas and refugia are none-the-less important (Barrett et al. 1997). Clearly, the contribution of such habitats must be understood, particularly in the light of the potential impacts of future large water withdrawal projects requiring mitigation or natural resource offsets such as wetland preservation and restoration (EPRI 2003).

Utilizing Impact Assessment Techniques

Knowledge of the impact of CY entrainment on fish eggs and larvae was aided by studies of egg and larval distributions carried out well upstream (75 km) and in the immediate vicinity of CY (Marcy 2004 [1976]). The general understanding that shad congregated and spawned above Hartford in the vicinity of the Enfield Dam provided the impetus to investigate alosid egg and larval distribution throughout the lower river. This work revealed that spawning of alosids and some of the resident species of concern occurred well upstream of the power plant. As a consequence, an extremely small proportion of the total quantity of fish eggs present in the river were entrained. However, the tidal character of the river in the area of CY, and the resulting reversal of flows during summer and early fall, did cause fish larvae to concentrate in relatively higher numbers in the vicinity of the plant. An important unknown, however, even today, is the extent to which tributaries to the main river provide spawning habitat and contribute to the overall recruitment success of anadromous and resident fish species.

The entrainment studies performed by Marcy (2004 [1976]) and Massengill (2004b [1976]) were vital to the characterization of the mechanisms leading to the CY related entrainment mortality (i.e., thermal, mechanical, or biocide or a combination) and to quantification of the proportion of available fish larvae that became entrained. These original estimates of entrainment losses have since been corroborated by hydraulic zone of influence (HZI) modeling (Richardson and Dixon 2004, this volume). In the future, studies of larval distribution and dispersal coupled with state-of-the-art HZI modeling would facilitate refined a priori estimates of larval entrainment. In addition, increased emphasis on the application of equivalent adult loss, production foregone and age structured population dynamics models (EPRI 1999a, 2002; Saila and Lorda 2004) have greatly improved the assessment of the loss of fish eggs and larvae due to entrainment. These tools were not yet well developed at the time of the Study.

The advent of HZI modeling, coupled with improved in situ temperature monitoring, now allows refined predictive mapping and calibration of thermal plumes. This advancement is best demonstrated by a comparison of information presented by Merriman (1970) and Boyd (2004 [1976], this volume) compared to the advances in fluid dynamics modeling demonstrated by Richardson and Dixon (2004). There is no question that a better a priori assessment of the probable three-dimensional properties of the CY plume would have improved the definition of sampling sites and frequency and the quality of the information that was ultimately available with which to assess the relationship between plume conditions and the ecological changes that were observed. This is particularly true in the shallow reaches along the eastern shore between the CY intake and the exit of the discharge canal where the disappearance of *Vallisneria* was initially believed to have resulted from the thermal plume. However, the resurgence of this rooted submerged aquatic plant in the early 1990s, during plant operation (Barrett et al. 1997), strongly suggests that other ecological changes perhaps related to siltation may have contributed to its decline during the early years of the Study. In retrospect, formal studies of *Vallisneria* abundance and distribution in relation to temperature could have provided a more effective quantitative characterization of the impact of thermal stress on this species.

The review of currently available analytical methods provided by Saila and Lorda (2004) clearly documents the dramatic advances in the sophistication of computer-aided statistical methods available for assessing the impacts of anthropogenic forcing of natural systems. Had these advances been available and applied to the original Study data, it is likely that even greater insights into the dynamics of the ecology of the lower Connecticut River would have been forthcoming. Massengill et al. (2004, this volume) applied several of these new methods to the original benthic data set and to the new data collected in 2002. These included community analyses, multi-dimensional scaling, cluster analyses, and nonparametric trend analysis. Interestingly, while Massengill (2004a [1976]) originally reported moderate decreases in abundance, biomass, and percent species composition at benthic infaunal stations affected by the plume, the statistical analysis subsequently applied (Massengill et al. 2004) to control and thermally affected stations sampled during the 1965–1972 period lead to the conclusion that the differences noted may not have been significant. The analyses also revealed that in 2002, the benthos of the region was characterized by significantly greater numbers of taxa and a greater total abundance of individuals. Mollusks and insects were more abundant in 2002 relative to the earlier (1965–1973) period, which was dominated by annelids.

Studies of the food habits of bullheads (Massengill 1973) and white perch (Marcy 2004 [1976]) performed during the original studies also provided insight into changes in the character of the benthic infauna and, to a limited extent, in the phytoplankton and zooplankton. It is now clear, however, that assessment of the impact of plant operation on the invertebrate communities in the affected area would have benefited from a better correlation between the various media sampled. Trophic relationships were not examined in a comprehensive manner, and this limited the resulting understanding of relative ecological risk and the overall impact of CY operation on system productivity.

Evaluating Population Dynamics

While the essential characteristics of the now classic BACI study design formed the basis for the assessment of CY impact, the original investigators and the EAC also appreciated the need for special studies of population dynamics for key resident fish species such as catfish (including channel, white catfish, and bullhead) and white perch. Classic fisheries information, including population size, food habits, movement, and age and growth, were collected and described. Much of the resulting knowledge of the biology of white catfish and bullhead in the Connecticut River was specific to the effects of overwintering in the CY heated water discharge canal on body condition, maturation, and movement (Marcy 2004 [1976]). The review of new analytical methods in this monograph by Saila and Lorda (2004) suggest that predictive population dynamics models for these key species would, in retrospect, have improved our understanding of the population status of these species and have allowed the development of predictions of long-term impact based on age-specific estimates of mortality.

On the other hand, the more traditional studies of the biology of white perch conducted during and subsequent to the Study (Howell and Molnar 2004, this volume; Marcy 2004 [1976]) clearly indicate that the population size and age structure of this species has not changed meaningfully over the 30 years since the original Study was concluded.

The research into the population dynamics of American shad conducted within the overall context of the Study now stands as a pioneering effort, not only in terms of impact assessment, but also in providing a fundamental understanding of migration timing, rates of migration, energetics, life history strategies, and other factors affecting recruitment in this species (See Appendix I of this volume and Leggett 2004 [1976]). The level of sophistication applied in these studies allowed development of quantitative estimates of year-class strength and recruitment success in relation to environmental variables and adult stock size. Later, age-structured models were developed that provided a basis for further research and management efforts aimed at sustaining and enhancing the Connecticut River shad population. The utility of long-term population dynamics and year-class strength data for key species in the assess- ment of anthropogenic impacts and enhancement activities directed toward long-lived species is dramatically highlighted by these studies of American shad that had their genesis with the Study and continue to this day under the direction of the CTDEP (Leggett et al. 2004; Savoy et al.2004, both this volume).

Establishing the Monitoring Philosophy

If we were to be asked, knowing what we now know after 37 years of involvement with studies of the river and its ecology, to stipulate a general philosophy for ecological impact assessment, our collective response would be to emphasize the following study attributes:

- Studies of the population dynamics of key species of concern
- Baseline (preplant operations) studies of sufficient scope to understand the differing habitat types and species life stages that occupy them, as well as, seasonal and annual variation in species distribution and abundance
- Operational studies using the traditional BACI design and fixed/random effects analysis of variance models
- Supplemental special short-term investigations deemed necessary to understand the trophic relationships and patterns of energy transfer in the ecosystem in question.

Moreover, we believe that a fundamental underpinning of successful efforts at impact assessment is the scientific credibility and openness of the process and the long-term stability of the program design and staffing. The scientific credibility of the Study derived from the freedom and the encouragement given to its participating scientists to publish the results of their work in the peer reviewed literature and from independent oversight provided by the EAC. Continuous long-term monitoring, employing consistent sampling methods, and sampling locations at selected stations representing key habitats and communities, beginning well before the onset of plant operations and continuing well into the operational phase, provided the stability and continuity necessary to allow credible evaluation of the probable causes of changes observed.

Moreover, the design of the studies, and the sampling strategies applied to them, must be founded on a comprehensive baseline of information that characterizes the source water with sufficient scope to bound the possible spatial impacts, locate critical habitats, build linkages among trophic levels, and understand overall system productivity. Finally, given the sophistication of today's analytical methods, the development of a priori hypotheses, based on simulation modeling, provides the basis for a much more rigorous scientific approach to the assessment of probable impacts and their significance than was employed during the Study.

We are also strongly of the view that the flexibility to augment planned studies or to incorporate special studies of key species or unforeseen impacts (as was the case in the original Study) is essential to the effective assessment of potential or realized impacts. It is important to note, also, that these additions should not be undertaken to the detriment of long-term monitoring.

Given the dramatic natural changes that have occurred in the ecology of the lower Connecticut River during the last 37 years as discussed herein, the traditional approach to monitoring (i.e., 3–5 years of study before and during operation) may not be sufficient to effectively characterize ecological impacts in the long-term because the baseline against which such impacts are measured may be constantly changing. We, therefore, suggest that if short-term change or lack thereof is determined, additional studies, albeit of lesser scope and frequency, should be continued to ensure that the consequences of long-term natural changes are fully integrated into the final assessment of impact.

Site Location in Relation to the Impact of Cooling Water Withdrawal

The findings of the Connecticut River Ecological Study are consistent with the commonly accepted principle that the location of a water withdrawal source is the most important determinant of ecological impact (Taft et al. 2001). When viewed in relation to the ecological effects related to operation of CY, it is reasonable to conclude that, at this site at least, a priori criteria for cooling water intake design or capacity such as those contained in EPA (2001) are less relevant than plant location in defining the potential for adverse environmental impact.

Assessing Adverse Environmental Impact (AEI)

The most controversial of the issues arising from the ongoing U.S. dialog and EPA rulemaking effort (Nagle and Morgan 2000; EPA 2001, 2004) relative to the setting of national standards for minimizing ecological impacts associated with cooling water intakes is the notion of adverse environmental impact (AEI) (Dixon et al. 2003). Effective definition of AEI and the ability to establish criteria for site-specific assessment of AEI appear to be intractable. This difficulty has led to the development of technology-based performance standards (EPA 2001, 2004). The standards for existing facilities are expressed as a percentage reduction in entrainment (60–90%) and impingement mortality (80–95%) relative to a theoretical baseline at the facility that assumes a shoreline intake, traveling water screens with 9.5-mm mesh, and no other fish protection technology in place. Meeting these standards will almost certainly require the application of technologies and operational measures that effectively eliminate cooling water flow or prevent organisms from entering the intake and/or the designation of natural resource offsets or enhancements designed to mitigate the impact of entrainment and impingement. The experience of the Study suggests that such extreme and costly measures may not be warranted in all cases. This is particularly so given the advances in sampling and analytical methodologies and the resulting increased ability to predict, detect and correct serious anthropogenic impacts.

Several authors have suggested alternative approaches to addressing the management implications of AEI that are consistent with our conclusions relating to the Study experience. These have generally been expressed either in the context of real or predicted ecological impacts, typically compared to some historical baseline or in the context of sustaining a natural resource or preserving an implied human value. For example, Barnthouse et al. (2003) and Van Winkle et al. (2003) have advocated the use of "weight of evidence" or "multiple benchmarks" that encompass the traditional ecological measures grounded in resource management principles such as balanced communities, long-term trends, and stock stability (i.e., recruitment success). Similarly, Bailey and Bulleit (2003) argue that the science of fisheries management and related quantitative tools, set in a risk management framework, are sufficiently robust to achieve site-specific, case-by-case balancing of ecological risks and benefits protective of our natural resources. In a similar vein, Seegert (2003) has argued that by designating aquatic life uses and expressing impacts in terms of attainment of such uses, it is possible to achieve a workable framework for determining AEI, while Coutant (2000) has proposed determining AEI based on an understanding of what is "normative" for an ecological system and assessing whether an impact drives the ecosystem outside this normative range. Bulleit (2000) advanced a case for the application of sustainable development as a principle or standard for determining AEI.

Super and Gordon (2003) have advanced a contrasting point of view. They argue that defining an ecological baseline is impossible due to "ever-changing interactions in an ecological community," and therefore, the approach is incapable of providing a basis for assessing AEI. Nor, say

Super and Gordon, is it reasonable to assert that the impact of cooling water withdrawals should be allowed to consume a portion of the natural resource resilience or compensatory reserve inherent in natural ecosystems, since this reserve has evolved as a means of protecting and mitigating perturbations related to the dynamics of natural ecological systems.

The experience of the operation of the Connecticut Yankee facility, as reflected in the results of the Study, and of subsequent work conducted to date and discussed herein, provide a useful perspective on AEI. The three decades of studies of the ecology of the Connecticut River and of the impact of the operation of CY on the river and its ecology (summarized in this document) clearly indicate that that it is possible to effectively assess ecological impact using multiple lines of evidence derived from the sound application of basic scientific principles. Furthermore, since the completion of the Study in 1973, great progress has been made in developing ever more sophisticated scientific approaches to the assessment of ecological impacts (EPRI 1999a, 2002; Wisniewski 2000; Dixon et al. 2003; Chen et al. 2004, this volume; Richardson and Dixon 2004; Saila and Lorda 2004). The availability of these tools and approaches strengthen the above conclusion.

We conclude that provided well-developed and clearly stated ecological endpoints and hypotheses form the framework for ecological research designed to assess ecological impacts, it *is* possible to effectively predict, assess, and, where necessary, mitigate unacceptable ecological impacts. In doing so, it is also possible to achieve use attainment, population stability, community balance, fisheries yield, regulatory compliance, and/or other endpoints that may be desired.

Suggested Additional Reading

For lessons learned from other comprehensive studies of the ecological impact of cooling water withdrawals and heated effluents, the reader is referred to Barnthouse et al. (1988), Klauda et al. (1988a, 1988b), EPRI (1999a, 1999b), Barnthouse (2000), Richkus and McLean (2000), Wisniewski (2000), EPRI (2002), Barnthouse et al. (2003), and Dixon et al. (2003). Regarding information relative to the determination of AEI, the reader is referred to Wisniewski (2000) and Dixon et al. (2003).

References

Bailey, D. E. and K. A. N. Bulleit. 2003. Defining "adverse environmental impact" and making 316(b) decisions: a fisheries management approach. Pages 143–164 *in* D. A. Dixon, J. A. Veil, and J. Wisniewski, editors. Defining and assessing adverse environmental impact from power plant impingement and entrainment. Swets & Zeitlinger, B.V. Lisse, The Netherlands.

Barnett, T. P., D. W. Pierce and R. Schnur, R. 2001. Detection of anthropogenic climate change in the world's ocean. Science 292:270–274.

Barnthouse, L. W. 2000. Impacts of power plant cooling systems on estuarine fish populations: the Hudson River after 25 years. Pages 341–348 *in* J. Wisniewski, editor. Power plants & aquatic resources: issues and assessment. Environmental Science & Policy 3(Supplement 1). Elsevier Science, New York.

Barnthouse, L. W., D. G. Heimbuch, V. C. Anthony, R. W. Hilborn and R. A. Myers. 2003. Indicators of AEI applied to the Delaware Estuary. Pages 165–184 *in* D. A. Dixon, J. A. Veil, and J. Wisniewski, editors. Defining and assessing adverse environmental impact from power plant impingement and entrainment. Swets & Zeitlinger, B.V. Lisse, The Netherlands.

Barnthouse, L. W., R. J. Klauda, D. S. Vaughan, R. L. Kendall. 1988. Science, law and Hudson River power plants: a case study in environmental impact assessment. American Fisheries Society, Monograph 4, Bethesda, Maryland.

Barrett, J. M., and M. Prisloe, L. Giannotti and N. Barrett. 1997. Distribution and abundance of submerged aquatic vegetation in the lower Connecticut River. Submitted to the Connecticut Department of Environmental Protection, Hartford.

Boyd, W. A. 2004. Hydrography. Pages 27–36 *in* P. M. Jacobson, D. A. Dixon, W. C. Leggett, B. C. Marcy, Jr., and R. R. Massengill, editors. The Connecticut River Ecological Study (1965–1973) revisited: ecology of the lower Connecticut River 1973–2003. American Fisheries Society, Monograph 9, Bethesda, Maryland. (Originally published in 1976)

Bulleit, K. A. N. 2000. Achieving sustainable development: a review of the environmental balancing act in two clean water act regulatory programs. Pages xv–xx *in* J. Wisniewski, editor. Power plants & aquatic resources: issues and assessment. Environmental

Science & Policy 3(Supplement 1). Elsevier Science, New York.

Caplis, M. 2001. Introduction. Pages 7–10 *in* G. D. Dreyer and M. Caplis, editors. Living resources and habitats of the lower Connecticut River. The Connecticut College Arboretum, Bulletin 37, New London.

Chen, C. W., R. A. Goldstein, and J. Herr. 2004. Feasibility of applying a watershed model to the Connecticut River for total maximum daily load analysis and watershed planning. Pages 501–511 *in* P. M. Jacobson, D. A. Dixon, W. C. Leggett, B. C. Marcy, Jr., and R. R. Massengill, editors. The Connecticut River Ecological Study (1965–1973) revisited: ecology of the lower Connecticut River 1973–2003. American Fisheries Society, Monograph 9, Bethesda, Maryland.

Coutant, C. C. 2000. What is "normative" at cooling water intakes? Defining normalcy before judging adverse. Pages 37–42 *in* J. Wisniewski, editor. Power plants & aquatic resources: issues and assessment. Environmental Science and Policy 3(Supplement 1). Elsevier Science, New York.

CTDEP (Connecticut Department of Environmental Protection). 1986. NPDES Permit, Connecticut Yankee Atomic Power Company, CTDEP, Hartford.

Dixon, D. A., J. A. Veil, and J. Wisniewski, editors. 2003. Defining and assessing adverse environmental impact from power plant impingement and entrainment. Swets & Zeitlinger, B.V. Lisse, The Netherlands.

EPA (Environmental Protection Agency). 2001. National pollutant discharge elimination system: regulations addressing cooling water intake structures for new facilities. Federal Register 66:243(18 December 2001):65255–65345.

EPA (Environmental Protection Agency). 2004. National pollutant discharge elimination system: regulations to establish requirements for cooling water intake structures at phase II existing facilities. Federal Register 69:131(9 July 2004):41576.

EPRI (Electric Power Research Institute). 1999a. Catalogue of assessment methods for evaluating the effects of power plant operations on aquatic communities. EPRI Report No. TR-112013, Palo Alto, California.

EPRI (Electric Power Research Institute). 1999b. Proceedings: 1998 EPRI Clean Water Act Section 316(b) technical workshop. Coolfont Conference Center. EPRI Report No. TR-112613, Palo Alto, California.

EPRI (Electric Power Research Institute). 2001. Connecticut River Ecological Study Workshop: revisiting the impact of a power plant. November 15–16, the University of Connecticut Marine Sciences Institute at Avery Point, Groton. EPRI Report (Compact disc of workshop presentations and notes) 1006900, Palo Alto, California.

EPRI (Electric Power Research Institute). 2002. Evaluating the effect of power plants on aquatic communities: guidelines for selection of assessment methods. EPRI Report No. 1005176, Palo Alto, California.

EPRI (Electric Power Research Institute). 2003. Enhancement strategies for mitigating potential operational impacts of cooling water intake structures: approaches for enhancing environmental resources. EPRI Report No. 1007454, Palo Alto, California.

Green, R. H. 1979. Sampling, design and statistical methods for environmental biologist. John Wiley and Sons, New York.

Howell, P., and D. Molnar. 2004. Stock assessment of white perch in the lower Connecticut River. Pages 379–390 *in* P. M. Jacobson, D. A. Dixon, W. C. Leggett, B. C. Marcy, Jr., and R. R. Massengill, editors. The Connecticut River Ecological Study (1965–1973) revisited: ecology of the lower Connecticut River 1973–2003. American Fisheries Society, Monograph 9, Bethesda, Maryland.

Jacobson, P. M., C. C. Fredette, and N. Barrett. 2004a. Introduction: Connecticut River watershed management—past, present, and future. Pages 263–272 *in* P. M. Jacobson, D. A. Dixon, W. C. Leggett, B. C. Marcy, Jr., and R. R. Massengill, editors. The Connecticut River Ecological Study (1965–1973) revisited: ecology of the lower Connecticut River 1973–2003. American Fisheries Society, Monograph 9, Bethesda, Maryland.

Jacobson, P. M., C. A. Tomichek, and D. Danila. 2004b. Twenty years of impingement history: Connecticut Yankee Haddam Neck nuclear power plant. Pages 455–473 *in* P. M. Jacobson, D. A. Dixon, W. C. Leggett, B. C. Marcy, Jr., and R. R. Massengill, editors. The Connecticut River Ecological Study (1965–1973) revisited: ecology of the lower Connecticut River 1973–2003. American Fisheries Society, Monograph 9, Bethesda, Maryland.

Keser, M. J., J. T. Swenarton, and J. F. Foertch. In press. Effects of thermal input and climate change on growth and mortality of *Ascophyllum nodosum* (*Fucales, Phaeophyceae*) in eastern Long Island Sound (USA). Journal of Phycology.

Klauda, R. J., L. W. Barnthouse, and D. S. Vaughan. 1988b. What we didn't learn about the Hudson River, why, and what it means for environmental assessment. Pages 329–335 *in* L. M. Barnthouse, R. J. Klauda, D. S. Vaughan, and R. L. Kendall. Science, law and Hudson River power plants: a case study in environmental impact assessment. American Fisheries Society, Monograph 4, Bethesda Maryland.

Klauda, R. J., L. W. Barnthouse, and D. S. Vaughan. 1988a. What we learned about the Hudson River: journey toward an elusive destination. Pages 316–328 *in* L. M. Barnthouse, R. J. Klauda, D. S. Vaughan, and R. L. Kendall. Science, law and Hudson River power plants: a case study in environmental impact assessment. American Fisheries Society, Monograph 4, Bethesda Maryland.

Leggett, W. C. 2004. The American shad, with special reference to its migration and population dynamics in the Connecticut River. Pages 181–238 *in* P. M. Jacobson, D. A. Dixon, W. C. Leggett, B. C. Marcy, Jr., and R. R. Massengill, editors. The Connecticut River Ecological Study (1965–1973) revisited: ecology of the lower Connecticut River 1973–2003. American Fisheries Society, Monograph 9, Bethesda, Maryland. (Originally published in 1976)

Leggett, W. C., T. F. Savoy, and C. A Tomichek. 2004. The impact of enhancement initiatives on the structure and dynamics of the Connecticut River population of American shad. Pages 391–405 *in* P. M. Jacobson, D. A. Dixon, W. C. Leggett, B. C. Marcy, Jr., and R. R. Massengill, editors. The Connecticut River Ecological Study (1965–1973) revisited: ecology of the lower Connecticut River 1973–2003. American Fisheries Society, Monograph 9, Bethesda, Maryland.

Maloney, T., and eight coauthors. 2001. Tidewaters of the Connecticut River, an explorers guide to hidden coves and marshes. River's End Press, LLC, Old Saybrook, Connecticut.

Marcy, B. C., Jr. 2004. Fishes of the lower Connecticut River and the effects of the Connecticut Yankee plant. Pages 65–124 *in* P. M. Jacobson, D. A. Dixon, W. C. Leggett, B. C. Marcy, Jr., and R. R. Massengill, editors. The Connecticut River Ecological Study (1965–1973) revisited: ecology of the lower Connecticut River 1973–2003. American Fisheries Society, Monograph 9, Bethesda, Maryland. (Originally published in 1976)

Massengill, R. R. 1973. Change in feeding and body condition of brown bullheads overwintering in the heated effluent of a power plant. Chesapeake Science 14:138-141.

Massengill, R. R. 2004. Benthic fauna: 1965–1967 versus 1968–1972. Pages 43–58 *in* P. M. Jacobson, D. A. Dixon, W. C. Leggett, B. C. Marcy, Jr., and R. R. Massengill, editors. The Connecticut River Ecological Study (1965–1973) revisited: ecology of the lower Connecticut River 1973–2003. American Fisheries Society, Monograph 9, Bethesda, Maryland. (Originally published in 1976)

Massengill, R. R. 2004b. Entrainment of zooplankton at the Connecticut Yankee plant. Pages 59–64 *in* P. M. Jacobson, D. A. Dixon, W. C. Leggett, B. C. Marcy, Jr., and R. R. Massengill, editors. The Connecticut River Ecological Study (1965–1973) revisited: ecology of the lower Connecticut River 1973–2003. American Fisheries Society, Monograph 9, Bethesda, Maryland. (Originally published in 1976)

Massengill, R. R., J. M. Vozarik, D, E. Morgan, and M. Keser. 2004. Comparison of benthic community structure in the lower Connecticut River before, during, and after operation of the Connecticut Yankee power plant. Pages 441–454 *in* P. M. Jacobson, D. A. Dixon, W. C. Leggett, B. C. Marcy, Jr., and R. R. Massengill, editors. The Connecticut River Ecological Study (1965–1973) revisited: ecology of the lower Connecticut River 1973–2003. American Fisheries Society, Monograph 9, Bethesda, Maryland.

Merriman, D. 1970. The calefaction of a river. Scientific American 222(5):42–52.

Merriman, D., and L. M. Thorpe, editors. 1976. The Connecticut River Ecological Study: the impact of a nuclear power plant. The American Fisheries Society, Monograph 1, Bethesda, Maryland.

Merriman, D., and L. M. Thorpe. 2004a. Introduction. Pages 3–25 *in* P. M. Jacobson, D. A. Dixon, W. C. Leggett, B. C. Marcy, Jr., and R. R. Massengill, editors. The Connecticut River Ecological Study (1965–1973) revisited: ecology of the lower Connecticut River 1973–2003. American Fisheries Society, Monograph 9, Bethesda, Maryland. (Originally published in 1976)

Merriman, D., and L. M. Thorpe. 2004b. Original preface. Pages xiii–xv *in* P. M. Jacobson, D. A. Dixon, W. C. Leggett, B. C. Marcy, Jr., and R. R. Massengill, editors. The Connecticut River Ecological Study (1965–1973) revisited: ecology of the lower Connecticut River 1973–2003. American Fisheries Society, Monograph 9, Bethesda, Maryland. (Originally published in 1976)

Morgan D. E., M. Keser, J. T. Swenarton, and J. F. Foertch. 2004. Effect of Connecticut Yankee power plant on population dynamics of Asiatic clams and their interactions with native bivalves. Pages 419–439 *in* P. M. Jacobson, D. A. Dixon, W. C. Leggett, B. C. Marcy, Jr., and R. R. Massengill, editors. The Connecticut River Ecological Study (1965–1973) revisited: ecology of the lower Connecticut River 1973–2003. American Fisheries Society, Monograph 9, Bethesda, Maryland.

Mullaney J. R. 2004. Summary of water quality trends in the Connecticut River, 1968–1998. 2004. Pages 273–286 *in* P. M. Jacobson, D. A. Dixon, W. C. Leggett, B. C. Marcy, Jr., and R. R. Massengill, editors. The Connecticut River Ecological Study (1965–1973) revisited: ecology of the lower Connecticut River 1973–2003. American Fisheries Society, Monograph 9, Bethesda, Maryland.

Nagle, D. G., and J. T. Morgan, Jr. 2000. A draft regulatory framework for analyzing potential adverse environmental impact from cooling water intake structures. Pages ix–xiv *in* J. Wisniewski, editor. Power plants & aquatic resources: issues and assessment. Environmental Science and Policy 3(Supplement 1). Elsevier Science, New York.

Ottersen, G., B. Planque, A. Belgrano, E. Post, P. C. Reid, and N. C. Stenseth. 2001. Ecological effects of the North Atlantic Oscillation. Oecologia 128:1–14.

Richardson, J., and D. A. Dixon. 2004. Modeling the hydraulic zone of influence of Connecticut Yankee nuclear power plant's cooling water intake structure. Pages 513–524 *in* P. M. Jacobson, D. A. Dixon, W. C. Leggett, B. C. Marcy, Jr., and R. R. Massengill, editors. The Connecticut River Ecological Study (1965–1973) revisited: ecology of the lower Connecticut River 1973–2003. American Fisheries Society, Monograph 9, Bethesda, Maryland.

Richkus, W. A., and R. McLean. 2000. Historical overview of the efficacy of two decades of power plant fisheries impact assessment activities in Chesapeake Bay. Pages 283–293 *in* J. Wisniewski, editor. Power plants & aquatic resources: issues and assessment. Environmental Science and Policy 3(Supplement 1). Elsevier Science, New York.

Saila, S. B, and E. Lorda. 2004. Developments in fishery science with special reference to power plant impact assessment methods and the Connecticut River Ecological Study: 1973–present. Pages 475–500 *in* P. M. Jacobson, D. A. Dixon, W. C. Leggett, B. C. Marcy, Jr., and R. R. Massengill, editors. The Connecticut River Ecological Study (1965–1973) revisited: ecology of the lower Connecticut River 1973–2003. American Fisheries Society, Monograph 9, Bethesda, Maryland.

Savoy, T. F. 2004. Population estimate and utilization of the lower Connecticut River by shortnose sturgeon. Pages 345–352 *in* P. M. Jacobson, D. A. Dixon, W. C. Leggett, B. C. Marcy, Jr., and R. R. Massengill, editors. The Connecticut River Ecological Study (1965–1973) revisited: ecology of the lower Connecticut River 1973–2003. American Fisheries Society, Monograph 9, Bethesda, Maryland.

Savoy, T. F., and J. Benway. 2004. Food habits of shortnose sturgeon in the lower Connecticut River from 2000 through 2002. Pages 353–360 *in* P. M. Jacobson, D. A. Dixon, W. C. Leggett, B. C. Marcy, Jr., and R. R. Massengill, editors. The Connecticut River Ecological Study (1965–1973) revisited: ecology of the lower Connecticut River 1973–2003. American Fisheries Society, Monograph 9, Bethesda, Maryland.

Savoy, T. F., V. A. Crecco, and B. C. Marcy, Jr. 2004. American shad early life history and recruitment in the Connecticut River: a 40-year summary. Pages 407–417 *in* P. M. Jacobson, D. A. Dixon, W. C. Leggett, B. C. Marcy, Jr., and R. R. Massengill, editors. The Connecticut River Ecological Study (1965–1973) revisited: ecology of the lower Connecticut River 1973–2003. American Fisheries Society, Monograph 9, Bethesda, Maryland.

Seegert, G. 2003. Using attainment of the designated aquatic life use to determine adverse environmental impact. Pages 136–142 *in* D. A. Dixon, J. A. Veil, and J. Wisniewski, editors. Defining and assessing adverse environmental impact from power plant impingement and entrainment. Swets & Zeitlinger, B.V. Lisse, The Netherlands.

Strong, A. E. 1989. Greater global warming revealed by satellite-derived sea-surface temperature trends. Nature 338:642–645.

Super, R. W., and D. K. Gordon. 2003. Minimizing adverse environmental impact: how murky the waters? Pages 213–230 *in* D. A. Dixon, J. A. Veil, and J. Wisniewski, editors. Defining and assessing adverse environmental impact from power plant impingement and entrainment. Swets & Zeitlinger, B.V. Lisse, The Netherlands.

Taft, E. P., W. P. Dey, and L. W. Barnthouse. 2001. Factors affecting the potential for adverse environmental impact from cooling water withdrawals. Report to the Utility Water Act Group. Hunton & Williams Inc., Washington, D.C.

USAEC (United States Atomic Energy Commission). 1973. Final Environmental Impact Statement related to the Haddam Neck (Connecticut Yankee) nuclear power plant. Docket No. 50-213, Washington, D.C.

Van Winkle, W., W. P. Dey, S. M. Jinks, M. S. Bevelhimer and C. C. Coutant. 2003. A blueprint for the problem formulation phase of EPA type ecological risk assessments for 316(b) determinations. Pages 264–286 *in* D. A. Dixon, J. A. Veil, and J. Wisniewski, editors. Defining and assessing adverse environmental impact from power plant impingement and entrainment. Swets & Zeitlinger, B.V. Lisse, The Netherlands.

Wetherell, W. D., editor. 2002. This American river, five

centuries of writing about the Connecticut. University Press of New England, Lebanon, New Hampshire.

Whitworth, W. R., P. L. Berrien, and W. T. Keller. 1968. Freshwater fishes of Connecticut. State Geological and Natural History Survey of Connecticut. Division of Agriculture and Natural History, Bulletin 101, Hartford, Connecticut

Wisniewski, J., editor. 2000. Power plants & aquatic resources: issues and assessment. Environmental Science and Policy 3(Supplement 1). Elsevier Science, New York.

Appendix I

Lower Connecticut River Fish Species List

COMPILED BY BARTON C. MARCY, JR.

Eighty-three freshwater, anadromous, and marine fish species (of 35 families) were sampled from the lower Connecticut River (1965–2003). References are shown below.

Family name Common name	Scientific name
Petromyzontidae (lampreys)	
Sea lamprey	*Petromyzon marinus*
Rajidae (skates)	
Little skate***	*Raja erinacea*
Acipenseridae (sturgeons)	
Shortnose sturgeon	*Acipenser brevirostrum*
Atlantic sturgeon	*A. oxyrinchus*
Amiidae (bowfins)	
Bowfin	*Amia calva*
Anguillidae (freshwater eels)	
American eel	*Anguilla rostrata*
Clupeidae (herrings)	
American shad	*Alosa sapidissima*
Alewife	*A. pseudoharengus*
Blueback herring	*A. aestivalis*
Atlantic menhaden	*Brevoortia tyrannus*
Atlantic herring	*Clupea harengus*
Hickory shad	*A. mediocris*
Gizzard shad	*Dorosoma cepedianum*
Engraulidae (anchovies)	
Bay anchovy	*Anchoa mitchilli*
Cyprinidae (carps and minnows)	
Fathead minnow	*Pimephales promelas*
Common carp	*Cyprinus carpio*
Golden shiner	*Notemigonus crysoleucas*
Common shiner	*Luxilus cornutus*
Spottail shiner	*Notropis hudsonius*
Creek chub	*Semotilus atromaculatus*
Fallfish	*S. corporalis*
Goldfish	*Carassius auratus*
Eastern blacknose dace*	*Rhinichthys atratulus*
Longnose dace*	*R. cataractae*
Bridle shiner*	*Notropis bifrenatus*
Catostomidae (suckers)	
White sucker	*Catostomus commersonii*
Ictaluridae (North American catfishes)	
Channel catfish	*Ictalurus punctatus*

Family name Common name	Scientific name
Brown bullhead	*Ameiurus nebulosus*
White catfish	*A. catus*
Yellow bullhead**	*A. natalis*
Esocidae (pikes)	
Chain pickerel	*Esox niger*
Northern pike	*E. lucius*
Redfin pickerel	*E. americanus*
Umbridae (mudminnows)	
Central mudminnow	*Umbria limi*
Osmeridae (smelts)	
Rainbow smelt	*Osmerus mordax*
Salmonidae (trouts and salmons)	
Atlantic salmon	*Salmo salar*
Rainbow trout	*Oncorhynchus mykiss*
Brown trout	*S. trutta*
Brook trout	*Salvelinus fontinalis*
Gadidae (cods)	
Atlantic tomcod	*Microgadus tomcod*
Burbot*	*Lota lota*
Fundulidae (topminnows)	
Mummichog	*Fundulus heteroclitus*
Striped killifish	*F. majalis*
Banded killifish	*F. diaphanous*
Cyprinodontidae (pupfishes)	
Sheepshead minnow	*Cyprinodon variegatus*
Atherinopsidae (New World silversides)	
Atlantic silverside	*Menidia menidia*
Inland silverside	*M. beryllina*
Gasterosteidae (sticklebacks)	
Threespine stickleback	*Gasterosteus aculeatus*
Fourspine stickleback	*Apeltes quadracus*
Syngnathidae (pipefishes)	
Northern pipefish	*Syngnathus fuscus*
Triglidae (searobins)	
Northern searobin***	*Prionotus carolinus*
Striped searobin***	*P. evolans*
Cottidae (sculpins)	
Grubby***	*Myoxocephalus aeneus*
Cyclopteridae (lumpfishes)	
Lumpfish***	*Cyclopterus lumpus*
Moronidae (temperate basses)	
Striped bass	*Morone saxatilis*
White perch	*M. americana*
Centrarchideae (sunfishes)	
Smallmouth bass	*Micropterus dolomieu*
Largemouth bass	*M. salmoides*
White crappie	*Pomoxis annularis*
Black crappie	*P. nigromaculatus*
Bluegill	*Lepomis macrochirus*
Redbreast sunfish	*L. auritus*
Pumpkinseed	*L. gibbosus*

Family name Common name	Scientific name
Longear sunfish**	*L. megalotis*
Rock bass	*Ambloplites rupestris*
Banded sunfish****	*Enneacanthus obesus*
Percidae (perches)	
Tessellated darter#	*Etheostoma olmstedi*
Yellow perch	*Perca flavescens*
Walleye	*Sander vitreus*
Pomatomidae (bluefishes)	
Bluefish	*Pomatomus saltatrix*
Carangidae (jacks)	
Crevalle jack	*Caranx hippos*
Lookdown***	*Selene vomer*
Sciaenidae (drums and croakers)	
Weakfish	*Cynoscion regalis*
Spot***	*Leiostomus xanthurus*
Northern kingfish***	*Menticirrhus saxatilis*
Pholidae (gunnels)	
Rock gunnel***	*Pholis gunnellus*
Ammodytidae (sand lances)	
American sand lance***	*Ammodytes americanus*
Scophthalmidae (turbots)	
Windowpane	*Scophthalmus aquosus*
Paralichthyidae (sand flounders)	
Summer flounder	*Paralichthys dentatus*
Smallmouth flounder***	*Etropus microstomus*
Fourspot flounder***	*P. oblongus*
Pleuronectidae (righteye flounders)	
Winter flounder***	*Pseudoplueronectes americanus*
Achiridae (American soles)	
Hogchoker	*Trinectes maculatus*

* Denotes species collected in tributaries of the Connecticut River (Jacob, Connecticut Department of Environmental Protection, unpublished data).
** Denotes species collected from Connecticut River at Holyoke, Massachuysetts. From fish bypassed and lifted at Holyoke Dam facilities (July–November 1995) (Holyoke Water Power 1997).
*** Denotes species collected from juvenile shad and trawl sampling in lower Connecticut River (Savoy, Connecticut Department of Environmental Protection, unpublished data).
**** Denotes collection at one Connecticut River location at "an inexact locality from a previous survey" (Whitworth et al. 1968).
# The Johnny darter *Etheostrum nigrum* was identified as separate species from the tessellated darter *E. olmstedi* by Cole (1967). Lee et al. (1980) concluded that the only species likely to be in Connecticut waters is the tesselated darter. The collections of Johnny darter in the original study by Marcy (1965–1973) are likely to be the tesselated darter and not the Johnny darter.

Sources for Connecticut River Fish Species List

Cole, C. F. 1967. A study of the eastern Johnny darter, *Etheostoma olmstedi storoer* (Teleostei, Percidae). Chesapeake Science. 8(1):28–51.

Holyoke Water Power. 1997. Holyoke Project Water Quality Survey, Appendix B of Exhibit E, Holyoke Water Power Company License Application to the Federal Energy Regulatory Commission, Volume III. Hydroelectric Project No. 2004, Holyoke, Massachusetts.

Howell, P., and D. R. Molnar. 2004. Stock assessment of white perch in the lower Connecticut River. Pages 379–390 *in* P. M. Jacobson, D. A. Dixon, W. C. Leggett, B. C. Marcy, Jr., and R. R. Massengill, editors. The Connecticut River Ecological Study (1965–1973) revisited: ecology of the lower Connecticut River 1973–2003. American Fisheries Society, Monograph 9, Bethesda, Maryland.

Jacobs, R. P., W. A. Hyatt, N. T. Hagstrom, E. B. O'Donnell, E. C. Schluntz, P. Howell, and D. R. Molnar. 2004. Trends in abundance, distribution, and growth of freshwater fishes from the Connecticut River in Connecticut (1988–2002). Pages 319–343 *in* P. M. Jacobson, D. A. Dixon, W. C. Leggett, B. C. Marcy, Jr., and R. R. Massengill, editors. The Connecticut River Ecological Study (1965–1973) revisited: ecology of the lower Connecticut River 1973–2003. American Fisheries Society, Monograph 9, Bethesda, Maryland.

Jacobson, P. M., C. A. Tomichek, and D. Danila. 2004. Twenty years of impingement history: Connecticut Yankee Haddam Neck nuclear power plant. Pages 455–473 *in* P. M. Jacobson, D. A. Dixon, W. C. Leggett, B. C. Marcy, Jr., and R. R. Massengill, editors. The Connecticut River Ecological Study (1965–1973) revisited: ecology of the lower Connecticut River 1973–2003. American Fisheries Society, Monograph 9, Bethesda, Maryland.

Lee, D. S., C. R. Gilbert, C. H. Hocutt, R. E. Jenkins, D. E. McAllister, and J. R. Stauffer, Jr. 1980. Atlas of North American freshwater fishes. North Carolina Biological Survey Publication 1980-12, Raleigh.

Marcy, B. C., Jr. 2004. Fishes of the lower Connecticut River and the effects of the Connecticut Yankee plant. Pages 65–124 *in* P. M. Jacobson, D. A. Dixon, W. C. Leggett, B. C. Marcy, Jr., and R. R. Massengill, editors. The Connecticut River Ecological Study (1965–1973) revisited: ecology of the lower Connecticut River 1973–2003. American Fisheries Society, Monograph 9, Bethesda, Maryland.

Whitworth, W. R., P. L. Berrien, and W. T. Keller. 1968. Freshwater fishes of Connecticut. State Geological and Natural History Survey of Connecticut, Bulletin 101, Department of Agriculture and Natural Resources, University of Connecticut, Storrs.

Appendix II

Connecticut River Ecological Study Bibliography: Related Papers by the Original (1965–1973) Principle Investigators

Boyd, W. A. 1970. Connecticut River Ecological Study: water temperatures 1966-1969. Essex Marine Laboratory Report 1, Essex, Connecticut.

Boyd, W. A. 1972. Connecticut River Ecological Study: dissolved oxygen concentrations 1966-1970. Essex Marine Laboratory Report 3, Essex, Connecticut.

Boyd, W. A. 1972. Connecticut River Ecological Study: the extent and temperature of the effluent plume from Connecticut Yankee atomic power plant at Haddam Neck, CT. Essex Marine Laboratory Report 4, Essex, Connecticut.

Buck, J. D. 1970. Connecticut River microbiology summary report. U.S. Depart- ment of the Interior, Federal Water Pollution Control Administration, Contract No. 14-12-177, Washington, D.C.

Carscadden, J. E., and W. C. Leggett. 1975. Life history variations in populations of American shad, *Alosa sapidissima* (Wilson), spawning in tributaries of the St John River, New Brunswick. Journal Fish Biology 7:595–609.

Carscadden, J. E., and W. C. Leggett. 1975. Meristic differences in spawning populations of American shad, *Alosa sapidissima*: evidence for homing to tributaries in the St. John River, New Brunswick. Journal of the Fisheries Research Board of Canada 32(5):653–660.

Dodson, J. J. 1973. The behavior of American shad (*Alosa sapidissima*, Wilson) during the final saltwater stages of the homing migration to the Connecticut River. Doctoral dissertation. McGill University, Montreal.

Dodson, J. J., and W. C. Leggett 1973. Behavior of adult shad (*Alosa sapidissima*) homing to the Connecticut River from Long Island Sound. Journal of the Fisheries Research Board of Canada 30:1847–1860.

Dodson, J. J., and W. C. Leggett. 1974. Role of olfaction and vision in the behavior of American shad (*Alosa sapidissima*) homing in the Connecticut River from Long Island Sound. Journal of the Fisheries Research Board of Canada 31:1607–1619.

Dodson, J. J., W. C. Leggett, and R. A. Jones. 1972. The behavior of adult American shad (*Alosa sapidissima*) during migration from salt to fresh water as observed by ultra sonic tracking techniques. Journal of the Fisheries Research Board of Canada 29:1445–1449.

Essex Marine Laboratory. 1972. Connecticut River study: hydrology data, 1966-1972. Essex, Connecticut.

Essex Marine Laboratory. 1972. Possible effects of various diversions from the Connecticut River. Essex Marine Laboratory 2, Essex, Connecticut.

Foerster, J. W., F. R. Trainor, and J. D. Buck. 1974. Thermal effects on the Connecticut River: phycology and chemistry. Water Pollution Control Federation 46(9):2138–2152.

Glebe, B. D. 1977. Weight loss and associated energy expenditure of American shad (*Alosa sapidissima*, Wilson) during the freshwater spawning migration. Doctoral dissertation. McGill University, Montreal.

Glebe, B. D., and W. C. Leggett. 1976. Weight loss and associated energy expenditure of American shad during the freshwater migration. U.S. National Marine Fisheries Service, Project AFC-8. McGill University, Essex Marine Laboratory and Department of Biology, Montreal.

Glebe, B. D., and W. C. Leggett. 1981. Latitudinal differences in energy allocation and use during the freshwater migrations of American shad (*Alosa sapidissima*)and their life history consequences. Canadian Journal of Fisheries and Aquatic Sciences 38:806–820.

Glebe, B. D., and W. C. Leggett. 1981. Temporal, intrapopulation differences in energy allocation and use by American shad (*Alosa sapidissima*) during the spawning migration. Canadian Journal Fisheries and Aquatic Sciences 38:795–805.

Leggett, W. C. 1969. Studies on the reproductive biology of the American shad (*Alosa sapidissima*, Wilson). A comparison of populations from four rivers of the Atlantic seaboard. Doctoral dissertation. McGill University, Montreal.

Leggett, W. C. 1969. A study of the reproductive potential of the American shad (*Alosa sapidissima*) in the Connecticut River, and of the possible effects of natural and man induced changes in the population structure of the species on its reproductive success. Essex Marine Laboratory, Essex, Connecticut.

Leggett, W. C. 1972. Weight loss in American shad (*Alosa sapidissima*, Wilson) during the freshwater migration. Transactions of the American Fisheries Society 101(3):549–552.

Leggett, W. C. 1973. The migrations of the shad. Scientific American 228(3):92–98.

Leggett, W. C. 1977. Density dependence, density independence, and recruitment in the American shad (*Alosa sapidissima*) population of the Connecticut River. Pages 3–16 *in* W. Van Winkle, editor. Assessing the effects of power plant induced mortality on fish populations. Pergamon Press, New York.

Leggett, W. C. 1977. The ecology of fish migrations. Annual Review Ecology and Systematics 8:285–308.

Leggett, W. C. 1977. Ocean migration rates of American shad (*Alosa sapidissima*). Journal of the Fisheries Research Board of Canada 34(9):1422–1426.

Leggett, W. C. 1985. The role of migrations in the life history evolution of fish. Pages 277–295 *in* M. A. Rankin, editor. Migration: mechanisms and adaptive significance. Contributions in Marine Science 27(Supplement).

Leggett, W. C., and J. E. Carscadden. 1978. Latitudinal variation in reproductive characteristics of American shad (*Alosa sapidissima*): evidence for population specific life history strategies in fish. Journal of the Fisheries of the Research Board of Canada 35(11):1469–1478.

Leggett, W. C., and R. A. Jones. 1970. Net avoidance behavior in American shad (*Alosa sapidissima*) as observed by ultrasonic tracking techniques. Journal of the Fisheries Research Board Canada 28:1167–1171.

Leggett, W. C., and R. A. Jones. 1973. Connecticut River Ecological Study, a study of the rate and pattern of shad migration in the Connecticut River - utilizing sonic tracking apparatus. Essex Marine Laboratory Report 5, Essex, Connecticut.

Leggett, W. C., and R. N. O'Boyle. 1976. Osmotic stress and mortality in adult American shad during transfer from saltwater to freshwater. Journal Fisheries Biology 8:459–469.

Leggett, W. C., and C. L. Trump. 1978. Energetics of migration in American shad. Pages 370–377 *in* K. Schmidt-Koenig and W.T. Keeton. Animal migration, orientation, and homing. Springer-Verlag, Berlin, Heidelberg, New York.

Leggett, W. C., and R. R. Whitney. 1972. Water temperatures and the migrations of the American shad. Fishery Bulletin 70(3):659–670.

Marcy, B. C., Jr. 1969. Age determination from scales of *Alosa pseudoharengus*, (W.) and *Alosa aestevalis*, (M.) in Connecticut waters. Transactions of the American Fisheries Society 98(4):622–630.

Marcy, B. C., Jr. 1970. Planktonic fish eggs and larvae of the lower Connecticut River and its tributaries. Essex Marine Laboratory, Essex, Connecticut.

Marcy, B. C., Jr. 1971. Survival of young fish in the discharge canal of a nuclear power plant. Journal of the Fisheries Research Board of Canada 27:1057–1060.

Marcy, B. C., Jr. 1972. Spawning of the American shad, *Alosa sapidissima* in the lower Connecticut River. Chesapeake Science 13(2):116–119.

Marcy, B. C., Jr. 1973. Vulnerability and survival of young Connecticut River fish entrained at a nuclear power plant. Journal of the Fisheries Research Board of Canada 30(8):1195–1203.

Marcy, B. C., Jr., and R. C. Galvin. 1973. Winter-spring sport fishery in the heated discharge of a nuclear power plant. Journal of Fisheries Biology 5:1–7.

Marcy, B. C., Jr., P. M. Jacobson, and R. L. Nankee. 1972. Observations on the reactions of young American shad to a heated effluent. Transactions of the American Fisheries Society 101(4):740–743.

Marcy, B. C., Jr., and F. P. Richards. 1974. Age and growth of the white perch, *Morone americana*, in the lower Connecticut River. Transactions of the American Fisheries Society 103(1):117–120.

Massengill, R. R. 1973. Change in feeding and body condition of brown bullheads overwintering in the heated effluent of a power plant. Chesapeake Science 14:138–141

Merriman, D., editor. 1965–1973. The Connecticut River Ecological Study semi-annual progress reports, Numbers 1–16, to the Connecticut Water Resources Commission and the Connecticut Department of Environmental Protection, Hartford, Connecticut.

Merriman, D. 1970. The calefaction of a river. Scientific American 222(5):42–52.

Merriman, D. 1970. Does industrial calefaction jeopardize the ecosystem of a long tidal river? Symposium on Environmental Aspects of Nuclear Power Stations. United Nations, International Atomic Energy Commission, New York.

Merriman, D. 1972. Power stations and aquatic ecology. Pages 48–52 *in* Proceedings of the American Power Conference 34, Chicago, April 1986.

Merriman, D. 1973. The calefaction of the Connecticut River, U.S.A. Transactions of the New York Academy of Sciences 35(1):59–65.

Merriman, D. 1973. Some biological aspects of aquatic thermal discharges. Pages 154–164 *in* The First World Congress on Water Resources, a Symposium on Thermal Pollution. Water Resources Association, Chicago.

Moss, S. A. 1970. The response of young American shad to rapid temperature changes. Transactions of the American Fisheries Society 99(2):381–384.

Rankin, J. S., J. D. Buck, and J. W. Foerster. 1974. Thermal effects on the microbiology and chemistry of the Connecticut River – a summary. Pages 350–355 *in* J. W. Gibbons and R. R. Sharitz, editors. Thermal ecology. U.S. Atomic Energy Commission, Technical Information Center, Oak Ridge, Tennessee.

Richards, F. P., and R. M. Ibara. 1978. The preferred temperatures of the brown bullhead, *Ictalurus nebulosus*, with reference to its orientation to the discharge canal of a nuclear power plant. Transactions of the American Fisheries Society 107(2):288–294.

Schubel, J. R., and B. C. Marcy, Jr. 1978. Power plant entrainment. A biological assessment. Academic Press, Inc., New York.